REALM OF THE UNIVERSE

SECOND EDITION

GEORGE O. ABELL

University of California, Los Angeles

SAUNDERS COLLEGE
Philadelphia

Saunders College
West Washington Square
Philadelphia, PA 19105

Published by Saunders College/Holt, Rhinehart and Winston

Library of Congress Cataloging in Publication Data

Abell, George Ogden, 1927–
Realm of the universe.

Bibliography: p.
Includes index.
1. Astronomy. I. Title.
QB45.A163 1980 520 79-27788
ISBN 0-03-056796-3

Printed in United States of America

0123 032 987654321

ACKNOWLEDGMENTS

COVER PHOTOGRAPHS:
FRONT: Jupiter and two of its satellites, photographed by Voyager 1. *(NASA/JPL)*
BACK: The Orion nebula. *(Hale Observatories)*

HALFTITLE: Jupiter and two of its satellites (Io, left, and Europa), photographed by Voyager 1 on February 13, 1979. *(NASA/JPL)*
FRONTISPIECE: The peculiar elliptical galaxy NGC 5128 in Centaurus. *(© Association of Universities for Research in Astronomy, Inc. The Cerro Tololo Inter-American Observatory)*
Additional credits are included in captions on pertinent pages of the text.
The photographs credited to the Hale Observatories in the color sections of this book are copyrighted by California Institute of Technology and Carnegie Institute of Washington. Reproduced by permission from the Hale Observatories.

PREFACE

These are terribly exciting times in astronomy! We are seeing the universe in the light of X-rays; interplanetary space probes are carrying our eyes to the other planets so that we may explore them at close range, and we are finding countless marvels; indirectly, we believe we are probing the very creation of the universe itself!

Needless to say, we hardly know it all. We do not know, for example, whether the universe will expand forever, or whether it will one day stop its enlarging and give way to a contraction. We do not know for certain whether intelligent life exists anywhere else in this vast universe (or, for that matter, even on earth). We do not yet know which, if any, of the various possible models for the mechanism that drives the quasars is correct. These are all subjects of investigation at the frontier of knowledge. There are many unknowns at the frontier and that's why science is so exciting.

As we explore that elusive frontier, we chip away at the outstanding problems, and eventually they crumble and fall to join the well understood and well trodden gravel in the scientific foreground. As they do so, we learn a bit more, not only about Nature, but about those laws that describe her behavior. Throughout the history of science a very great deal has been learned; we have come an enormous distance from the science of antiquity, and even from the time of Newton. In fact, it is scarcely half a century since Edwin Hubble showed that there are other galaxies in the universe! Since then, we have learned of the quantum nature of the universe, of nuclear energy (which keeps stars shining), and of the synthesis of the atoms that comprise our very bodies in the collapsed cores of giant stars!

After science comes technology, the application of science to the utility of our daily lives, and all of us have been touched very deeply by that! Indeed, to many people, the technological explosion must seem miraculous. We have all become so accustomed to technological miracles that many of us have come to expect that anything—absolutely anything!—is possible. But only those well acquainted with the science relevant to a technology can evaluate the legitimacy of alleged new "breakthroughs." Of course the real scientific excitement is at the frontier, but only those thoroughly familiar with the subject at hand can say what is really frontier science, what has long been well understood, and what is utter nonsense. As science progresses and the frontier pushes ever onward, the techniques become ever more sophisticated; even those in a field tend to develop a jargon that only they understand. It is a regrettable fact that most scientists working at the frontier of knowledge can converse only with those in their narrow area of specialization. Most of us cannot even read, let alone understand, technical journal articles dealing with subjects outside our own fields of expertise.

How, then, can we expect the lay public—even those with very sophisticated university educations—to be able to judge true science from utter rubbish. Of course they cannot, for the most part, any more than any of us scientists can do so in subjects alien to us. But almost everyone is interested in the grand questions of science—those dealing with the origin and evolution of the universe, with the beginning of our solar system, and with ourselves. They (we all) are sitting ducks for the quacks (the scientific quacks have replaced the medicine men of old), who are quick and ready with simple and readily understood answers to all of our great questions. We would not be likely to pay much heed to them, except that they present their products in the name of "science," and we have become so conditioned to scientific miracles that we are about ready to accept almost anything (perhaps *anything*) if it is presented in a sufficiently arcane language.

The extremely successful pitch of the scientific quacks is omnipresent: Television is full of it in the guise of pseudodocumentaries and "specials" (the disclaimer stating that the foregoing is simply "entertainment" goes by far too fast to be noticed by most). Newspaper columns, especially on the front and early pages, describing sensational scientific "breakthroughs" far outnumber the qualifying or disqualifying counterstatements on back pages. Magazines feature pseudoscientific claims, and books on the best-seller lists regularly reap fortunes from the gullible.

Surely no one (or very few, at least) knowingly wants to be conned. But who can blame the ordinary citizen for being misled by claims made in the name of science, especially when he or she has so little access to factual scientific information? Yet, we are living in a time when rational decisions on the part of the citizenry may well be essential to our survival. In the ever increasing complexity of our technological society, it has never been more essential to the survival of democracy that everyone have at least some understanding of basic science, and more important: what, exactly, science is, how it works, and what its limitations are. Astronomy has the great advantage of being an extraordinarily cogent and exciting branch of science—the branch where things are currently "going on." It is also fascinating in its own right, it deals with questions of fundamental interest to us all, and it provides very many examples of the application of science to the acquisition of new knowledge.

That's what this book is all about. It is, in part, a survey of the astronomical universe as we understand it today. But more, it attempts (at least I hope it does!) to show where science gets its clout. It is not only interesting to know that Alpha Centauri is 4 light years away, but also, *how do we know* this astonishing fact? I have tried to explain the working of science, and the nature of scientific law. I have a (possibly futile) hope that if we all can understand the nature of science and scientific investigation, we will be in a better position to evaluate the plethora of information—valid and invalid—besieging us in the name of science.

There is little question that real facts of science put the pseudoscientific claims to shame! Since the last edition of *Realm of the Universe*, we have discovered X-ray binary objects (stars?), binary pulsars that appear to be radiating gravitational waves, volcanoes on Io (the innermost large satellite of Jupiter), an anisotropy in the microwave background radiation (explained near the end of the book) that probably indicates the motion of our Galaxy through the universe, evidence that *probably rules out* life on Mars (but not on planets revolving about other stars), and a host of other things that have necessitated a virtual rewriting of this text.

Originally, *Realm of the Universe* was a brief edition, intended for a one quarter or one semester course, of my fuller text, *Exploration of the Universe*, intended for a longer course, or for a far more in-depth shorter course. This time, however, I have completely revised *Realm* before even touching *Exploration* (whose revision is my next big project!). Many new sections and even completely new chapters are not condensations of a fuller account, but are written especially for the audience of general college and university students (and even laypeople) for whom *Realm* is intended.

Realm, like *Exploration of the Universe*, is still essentially traditional in its approach. Material is grouped by subject matter, starting with the inner part of the universe and working out (with some historical digression to bring to light the human nature of scientific exploration). This way, we start with familiar concepts and gradually introduce the more alien ideas. I have another text, however, *Drama of the Universe*, that is organized around grand themes (gravitation, modern physics, the distance ladder, evolution, search for life, and cosmology) rather than around specific subject matter. Some professors prefer one approach, and some the other. I am not sure of my own views; both approaches have their validity, and much depends on the style and technique of the teacher.

This new edition of *Realm of the Universe*, like *Drama*, is almost completely devoid of mathematical formalism. To be sure, astronomy is a mathematical and exact science (or at least, tries to be), and every practicing astronomer must live and breathe mathematical equations. I emphatically do *not* want to misrepresent astronomy as a descriptive subject that does not require the rigor of mathematics and physics. Every student in astronomy who aspires to be a professional in the field must devote himself to a very high degree of specialization in mathematical techniques and modern physics, the study for which must begin in high school and continue through several years of graduate study. But this is no reason for the nonscience student or layperson to be deprived of an opportunity to learn something of the nature of the universe and of the rules by which the universe seemingly must "play the game." *Realm* is primarily for these people, not science students, and it is my hope that, in reading it, some will feel a bit of the enthusiasm felt by those of us who are privileged to work in what I regard to be the supreme scientific endeavor.

Realm, like my other textbooks, has benefitted greatly from the help of many other people. I particularly wish to thank Tom Gehrels, Harland Epps, Lawrence Aller, Jonathan Katz, Julian Schwinger, Holland Ford, Paul Routly, Roger Ulrich, Bruce Margon, Dave Pierce, Mirek Plavec, Riccardo Giacconi, and many others as well; some have supplied data for me, and others have supplied valuable conversation that has helped me formulate my ideas more clearly. In particular, three people were extremely helpful to me in the very first edition of my first text in astronomy, and their sound advice has been of great value to me in all subsequent printings and editions and offshoots; they are Paul Routly, Dan Popper, and the late Paul Wylie. Their advice must have been sound, considering the extent to which many of my ideas have been adopted by other authors!

For this edition of *Realm of the Universe*, special thanks are due to my very patient editor, Karen Nelson, who put up with many trials thrust upon her by the reorganization of the publishing company, and still managed to be good natured at my constant ill-tempered complaints to her; she has done a Herculean job! And finally, a *very* special thanks to my good wife, Phyllis, who not only put up with me (and my fights with Karen) but was ever faithful and tireless (well, *almost* tireless) in her help with organization, rubber cementing old commentary in with new, with proofreading, and with innumerable other tasks too odious to mention; seriously, she has been *terrific*!

But for all of those best efforts, nobody's perfect, and this book, like all others (even, alas, my own) has errors. I am, and shall continue to be, extremely grateful to those many good people who write to me to give suggestions and to call careless slips to my attention so that they can be corrected in future printings.

I hope that my efforts will enable you to share the excitement that I feel for astronomy!

George O. Abell
Encino, California
January 1980

CONTENTS

CHAPTER 1

Galileo Galilei (1564–1642) symbolized the beginning of modern scientific inquiry, wherein we perform experiments or make observations to ask Nature her ways, rather than deciding how things must be on the basis of preconceived notions.

REALM OF THE UNIVERSE

Religion deals with absolutes—beliefs that must be accepted on faith. Science accepts nothing on faith, but does not deal with absolutes. Of course individual scientists may be religious, and I think a belief common to the religions of most, if not all, scientists is that scientific laws—the rules of the game that Nature plays—are truly universal, the same everywhere. When you think of it, you realize that the universality of scientific laws is a really marvelous concept.

But science itself is only a *method* by which we attempt to understand nature and how it behaves. Specifically, we attempt to understand things in terms of *models* that correctly describe the behavior of nature. In their tentative stages these models are called *hypotheses*. Eventually, when a body of hypotheses that have been at least partially checked out has been pieced together into a self-consistent system, that system becomes a *theory*. Some models and theories—for example, general relativity and quantum mechanics—are quite mathematical and abstract. Others, such as Newtonian gravitation or the explanation of eclipses, are more easy to visualize or even are susceptible to mechanical representation.

No scientific theory, however, should be confused with absolute *truth*, for science does not concern itself with truth; it merely represents nature with descriptive models. The procedure is simple: a phenomenon or an experimental result is observed. One or more hypotheses are advanced that enable us to understand the phenomenon or experimental result in ways that do not violate other observations or experiments. Finally, a hypothesis must be susceptible to testing by further observations or experiments. This last point is crucial; if there is no possible way of testing a hypothesis, it may be interesting speculation in other areas of human thought, but it does not belong in the realm of science.

The real power of science, of course, is its success. If science does not give us absolute truth, it nevertheless can give us considerable insight into the workings of nature. It has also made possible an explosive development of technology. In fact, people have become so used to technological miracles that many have come to believe that science can do anything, and that everything is possible. Most of us are intrigued by dramatic conjectures, but in areas about which we know little we are not equipped to judge what is legitimate and what is worthless. Too often we substitute gullibility for open-mindedness. It is important, therefore, to learn at least some of the limitations of science and to gain some understanding of what is possible and what is not.

We have found that some things are not only possible, but have been done—say, television broadcasting and travel to the moon. There are other things that we have not yet done, but that we presume to be in the realm of possibility, such as manned flights to Mars. A large class of things that may or may not be possible—we lack enough information to say—include such questions as whether any life exists on Mars or whether the universe oscillates between alternate expansions and contractions. Finally, there are many things that we can say are impossible—for example, for an object to travel faster than light, for a comet to stop the earth from turning, or for the sun to be hollow and cool inside. At least such things are impossible in terms of rules of nature that we have discovered, such as the conservation of momentum and the laws of thermodynamics. To be sure, we cannot guarantee that these laws will not suddenly cease to apply, but we can nevertheless base predictions on them with at least as much certainty as the prediction that the sun will rise tomorrow. Much appreciation of nature can be achieved by learning some of its rules and being able to recognize some propositions that clearly violate them.

What a fine reason to study astronomy—the oldest and in many ways the grandest of sciences!

AN OVERVIEW OF THE UNIVERSE

Astronomy, despite its great popular appeal, is particularly alien to most nonscientists. It also has a lot of jargon and involves a lot of ideas. When you first encounter these, there is a danger of your getting bogged down with details. To help prevent this from happening, we begin with a brief overview of what the universe is like. You may want to return to this thumbnail sketch of the scale of the universe from time to time, to keep things in perspective.

Even this overview might require you to imagine distances on a scale you have never thought about before, and numbers larger than any you have encountered. Do not worry for now about precise figures, but strive to keep clear the distinction between such quantities as 10, 1000, 1 million, 1000 million, and 10^{33} (the figure 1 followed by 33 zeros). I am told that there are primitive tribes whose people have not learned to count to more than two. They say, "one, two, many." Most Americans think of any big number as a million, but there is an incredible difference between 1 million and 1 million million (called a *trillion* in the United States and a *billion* in England; to avoid confusion between American and English usage, I shall not use names of numbers whose order of magnitude is larger than a million, but shall use the simpler powers-of-ten notation; it is described in Appendix 3).

Our earth is approximately spherical and has a diameter of nearly 13,000 kilometers (8000 miles). Light, with the speed of 300,000 kilometers per second (km/s) can travel seven times the earth's circumference in one second. A traveler by commercial airplane takes about two days to go once around the earth, and an astronaut in orbit can do the same thing in about 100 minutes.

Our nearest astronomical neighbor is the moon. We can draw the earth and moon to scale on the same diagram (Figure 1.1). The moon's distance from the earth is about 30 times the earth's diameter, or about 384,000 km (239,000 miles). Light takes about one and one-third seconds to make the journey. The reader may recall hearing live radio conversations between astronauts on the moon and the staff at Mission Control in Houston. After a simple question from Houston, there would be a delay of at least three seconds before even a "yes" or a "no" answer. The reason is not that the astronaut was just casually thinking it over, but that it took the radio waves (which travel with the speed of light) that long to make the round trip.

Whereas the moon revolves once about the earth each month, the earth revolves once about the sun each year. The sun is about 150 million km away, or about 400 times as far as the moon. Light requires about eight minutes to come from the sun. Since the sun is nearly 1.5 million km in diameter, it is also about 100 solar diameters distant. We could, with care, draw the sun and earth to scale on one diagram, but then the earth would be only a point and even the orbit of the moon could not be shown to scale unless the diagram was drawn on a very large sheet of paper.

Two planets, Mercury and Venus, revolve about the sun on orbits inside the earth's; hence, at times we are closer to them than we are to the sun. Venus, the planet that can come nearest the earth, at its closest is 100 times as far away as the moon. Six other planets revolve about the sun outside the earth's orbit. The largest, Jupiter, is about five times the earth's distance from the sun. We call the average distance from earth to sun the *astronomical unit* (AU); Jupiter's orbit is thus about 5 AU in radius. Jupiter has ten times the earth's diameter, but only one-tenth the sun's diameter and one-thousandth the volume of the sun. Pluto, the most distant planet, is 40 AU from the sun. Light travels to

Figure 1.1
The earth (left) and moon, drawn to scale.

Figure 1.2
A portion of the planet Mercury, as observed by the spacecraft Mariner 10 on March 29, 1974. *(NASA/JPL)*

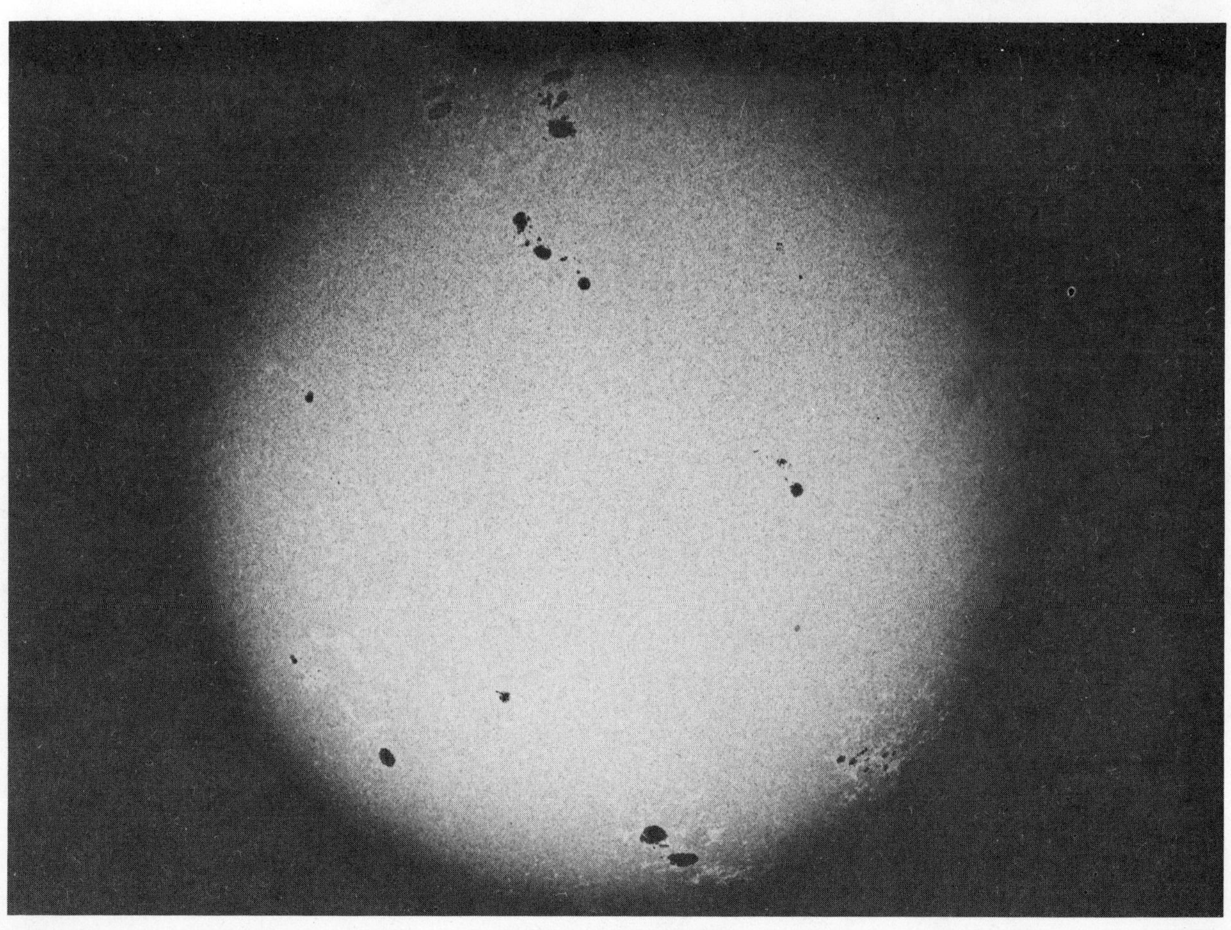

Figure 1.3
The Sun. *(Hale Observatories)*

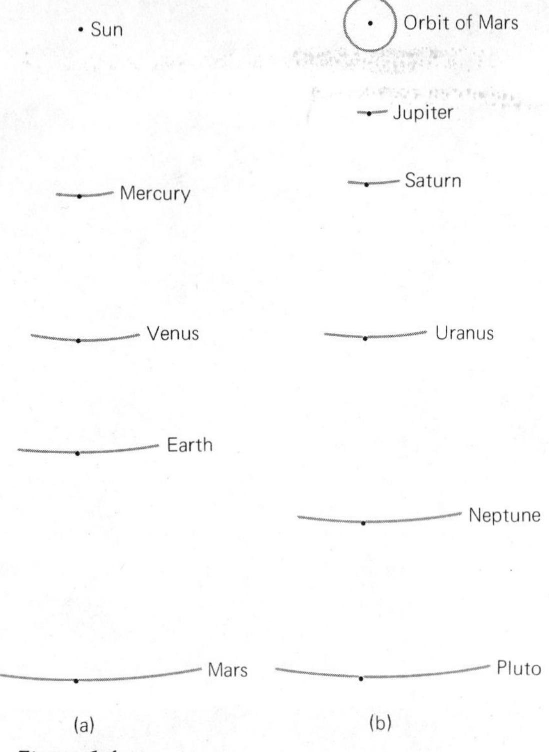

Figure 1.4
The distances of the planets: *(a)* the inner planets. The sun's size is shown to the same scale as the planets' orbits, but the planets themselves could not be seen on this scale; *(b)* the outer planets to the scale of the orbit of Mars.

the earth from the planets in periods ranging from a few minutes to a few hours.

The sun is a typical star. The nearest star beyond the sun is about 300,000 AU away, or about 7000 times as far from the sun as is Pluto, the outermost planet. On a scale map showing the orbits of the planets about the sun (Figure 1.4), the stars cannot be shown. If the earth's orbit were represented as only 2 millimeters (mm), in radius, Pluto's orbit would be about 16 centimeters (cm) in diameter and the nearest star would have to be placed 0.6 km away. On this scale, most of the hundred or so nearest stars would be 1 to 2 km away. Remember that light takes eight minutes to come from the sun. It takes about four years to come from the nearest star. The distances to stars are thus often expressed in ***light years*** (LY), the time light takes to come from them to us.

Suppose we make a rough scale drawing, showing the stars within 10 LY of the sun. In Figure 1.5(a), the circle represents a sphere 10 LY in radius centered on the sun. Roughly ten stars are included. Now we change scale. In Figure 1.5(b), the sphere of 10 LY is the small center circle and the larger circle represents a sphere 100 LY in radius—ten times as large. In that sphere, we would have approximately 10,000 stars. Similarly, the 100-LY-radius sphere is the small circle in our next change of scale, Figure 1.5(c), while the larger circle represents a sphere 1000 LY in radius,

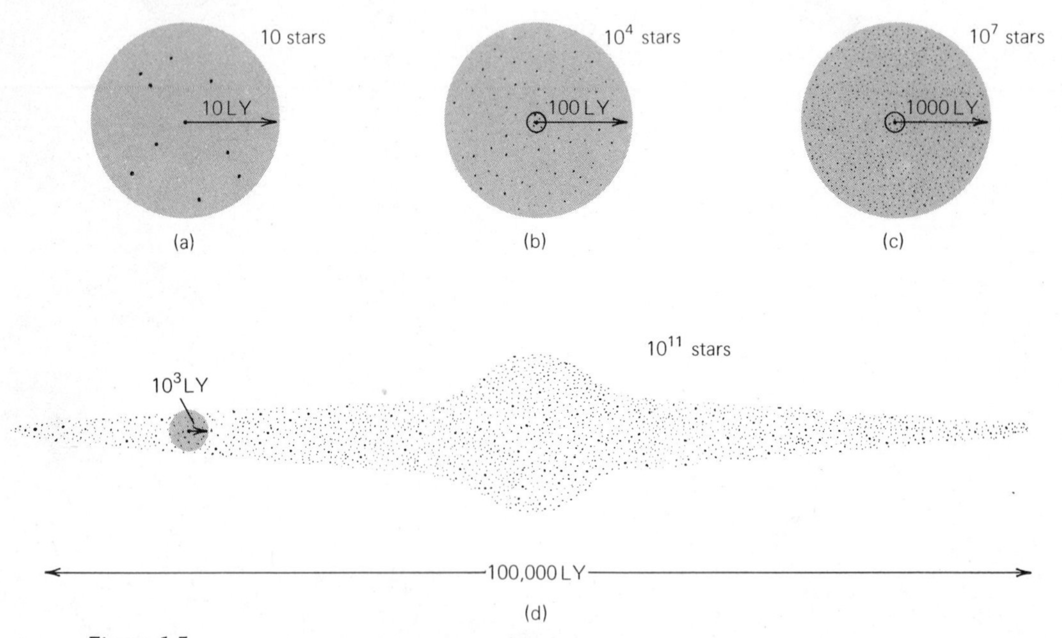

Figure 1.5
The distribution of stars around the sun within *(a)* 10 LY; *(b)* 100 LY; *(c)* 1000 LY; *(d)* the Galaxy.

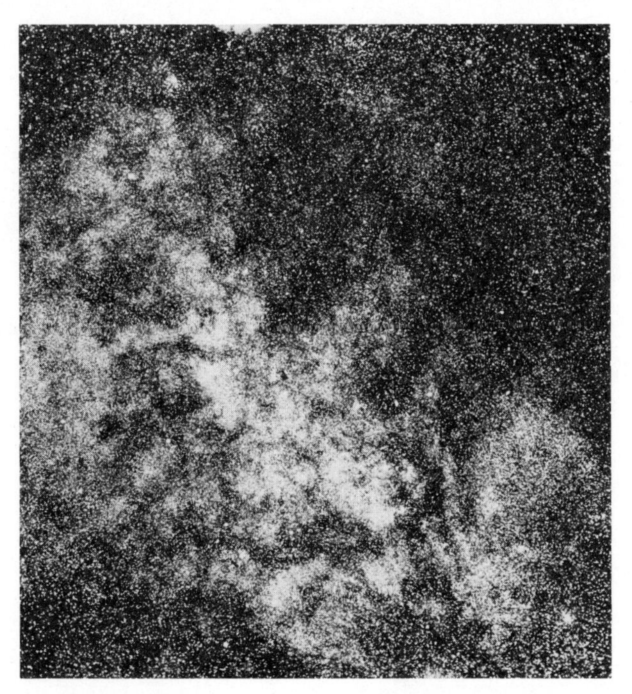

Figure 1.6
The Milky Way in the region of Sagittarius. *(Yerkes Observatory)*

within which there are 10 million (10^7) stars. In our next change of scale, Figure 1.5(d), the stars thin out, but in some directions before others. The sun is part of a wheel-shaped system of stars [shown edge-on in Figure 1.5(d)] about 100,000 LY in diameter, containing 10^{11} stars (100 thousand million), of which the sun is typical. We call the system our *Galaxy*.

Perhaps many, possibly even a majority, of the stars in the Galaxy have planetary systems, as does the sun; we do not know, for we cannot detect planets revolving about other stars — they are too insignificant to see, even with our greatest telescopes. When we look edge-on through our Galaxy we find so many stars in our line of sight that the more remote ones give just a glow of light in a circular band around the sky — the *Milky Way* (Figure 1.6). We do not see directly through the entire Galaxy to its far rim with ordinary light, because the interstellar space is not completely empty. It contains a sparse distribution of gas (mostly hydrogen) intermixed with microscopic particles, which we call interstellar dust. This material is so extremely sparse that interstellar space is a far, far better vacuum than any we can produce in terrestrial laboratories. Yet the dust, extending over thousands of light years, obscures the light from stars that lie far away in the Galaxy.

There are many double stars — pairs of stars revolving about each other — and even triple and multiple star systems. More than a thousand clusters of stars have been cataloged. These star clusters have anywhere from a few dozen to a hundred thousand member stars each, and range in diameter from a few LY to several hundred LY.

A half century ago, most astronomers thought our Galaxy constituted the entire universe. Now, however, we know that it is an insignificant part. Other comparable galaxies stretch as far in space as we can see — at least 1000 million of them within the reach of our present telescopes. These galaxies tend to occur in clusters. Our own Galaxy is part of a small cluster of about 20 members extending over a region of space about 3 million LY in diameter; we call the cluster our Local Group (Figure 1.8). At 10 to 15 million LY from the Local Group are other small clusters, and at about 50 million LY (or more) is the nearest rather important

Figure 1.7
The spiral galaxy M81 in Ursa Major. The small round images along the edges of the photograph (especially in the upper left and lower right) are of foreground stars in our own Galaxy. We look out past and beyond them to see M81. *(Lick Observatory)*

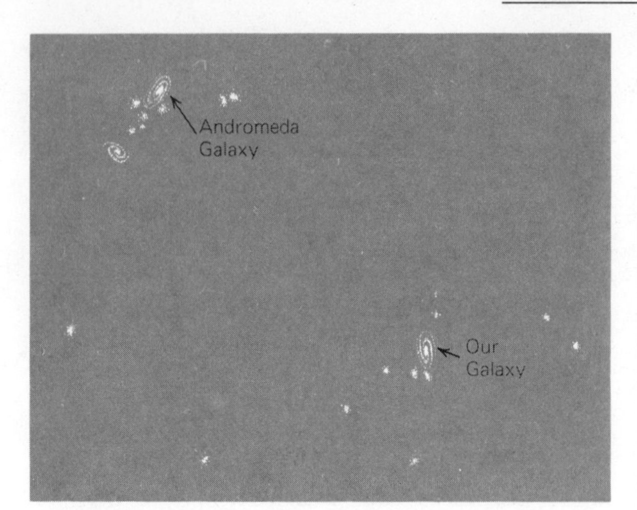

Figure 1.8
Schematic diagram of the Local Group, approximately to scale.

cluster, which contains at least 1000 member galaxies. Some 400 to 500 million LY away we find the first really great cluster, which has at least 10,000 galaxies as members. Within that distance, there must be hundreds or thousands of clusters as tiny as the Local Group. Yet we can identify great clusters of galaxies at distances of thousands of millions of light years. Nearly 3000 of these great clusters are cataloged.

At distances that are probably several times greater still, we find the quasars. We do not know for sure yet what the quasars are, but most astronomers think that they are bright regions in the centers of galaxies otherwise too remote to see. In any event, the quasars emit enough light that we can detect them, as faint points of light, even though most astronomers believe many to be at distances as great as 10^{10} (10 thousand million) LY or more. If we tried to show the most remote quasars on the scale we used for our Local Group in Figure 1.8, they would have to be drawn nearly 1 km away.

Still beyond the most remote quasars lies the farthest part of the universe we can detect. We can see it only by means of feeble radio waves coming from all directions in space. Because of the finite speed of light, radiation arriving from great distances must have originated in the past. We see the remote quasars as they were when light left them thousands of millions of years ago. Still earlier, we have reason to believe, the

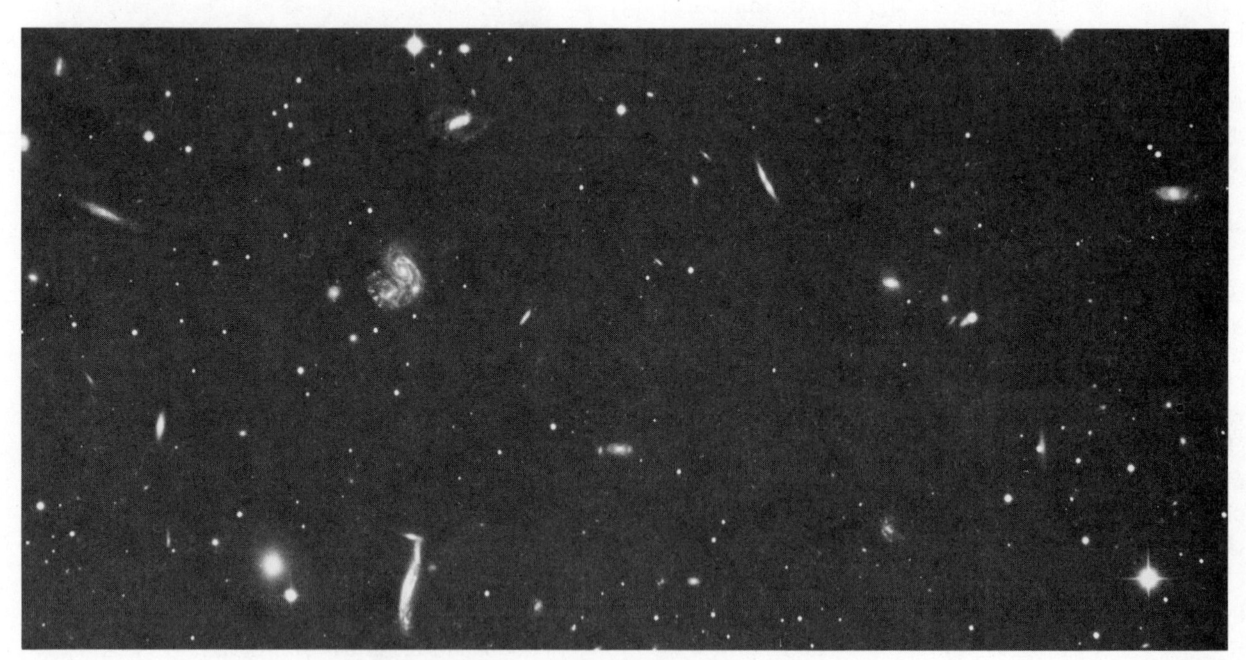

Figure 1.9
A cluster of galaxies in Hercules. *(Hale Observatories)*

universe, exploding in a gigantic fireball, began expanding. That feeble radio radiation from the remote distance and remote past is thought to be all that we can now observe of the primeval fireball that began the universe as we know it today.

AN INNER VIEW OF THE UNIVERSE

The foregoing discussion should impress on you that the universe is extraordinarily empty. Indeed, taken as a whole, the universe, on the average, is ten thousand to a million times as empty as (that is, less dense than) our own Galaxy. Yet, as we have seen, the Galaxy is mostly empty space. Recall the interstellar gas in our Galaxy; typically, there is about one atom of hydrogen in each cubic centimeter of space. In contrast, the air that we breathe has about 10^{19} times as much material per cubic centimeter, and we think of the air as pretty empty stuff. Solid matter, such as our own bodies or this page, is tremendously dense in comparison.

Yet even the familiar solids are mostly space. If we could take such a solid apart, piece by piece, we would eventually reach the molecules of which it is formed. Molecules are the smallest particles that matter can be divided into while still retaining its chemical properties. They are in turn composed of atoms, the building blocks of all matter. Nearly 100 different kinds of atoms exist in nature, but most of them are rare and only a handful account for more than 99 percent of everything with which we come into contact.

All atoms consist of a central, positively charged nucleus surrounded by negatively charged electrons. The bulk of the matter of an atom is in the nucleus, which consists of a certain number of positive protons and a roughly equal number of electrically neutral neutrons, all tightly bound together. In its regular condition, an atom has as many electrons around the nucleus as protons in the nucleus. Protons and electrons have charges that are equal in magnitude but opposite in sign, so they cancel each other and the atom is electrically neutral. Different kinds of atoms, responsible for the different kinds of chemical elements, are distinguished by the number of protons in their nuclei (or electrons outside). Thus the simplest kind of atom, hydrogen, has one proton and one electron. Helium has two of each, oxygen has eight of each, and so on.

In the atoms in a gas the electrons move around the nuclei according to special rules, but in a solid the nuclei and electrons are all interlocked in a relatively rigid structure. In any case, the volume of space occupied by an atom is set by the distance of the electrons from the nucleus. This distance is huge compared to the size of the nucleus itself. Sometimes atoms are compared to miniature solar systems, in which the electrons around the nuclei are described as analogous to the planets around the sun. The analogy is poor, however, first because the laws that govern the motions of the planets are different from those that govern the motions of the parts of atoms. In addition, the scale is much different. Recall that the distance of the earth from the sun is about 100 times the sun's diameter. The distance from an atomic nucleus to the electrons, however, is typically 100,000 times the size of the nucleus! A typical atom is far more empty than the entire solar system out to the distance of Pluto. Our densest solids are almost completely empty space.

Under some conditions, however, atoms can be ionized—that is, they can lose some or all of their electrons. In a solid, the electrons are bound rigidly, but in a gas, if the atoms are completely ionized (that is, if they have lost all their electrons) the atomic nuclei remaining and the freed electrons can be crowded much closer than in a solid. Thus gases can be compressed to enormously high densities compared to solids. In our exploration of the universe, we shall encounter incredible vacuums throughout great expanses of space, but we shall also encounter phenomena that we believe to be associated with matter packed to the densities of atomic nuclei—or perhaps even closer. For example, we believe that pulsars (Chapter 17) are *neutron stars* in which matter is packed 10 to 1000 million million times as densely as in water. Matter may also collapse into *black holes* of still higher densities. Our astronomical laboratory—the universe—allows us to study matter over an astonishing range of conditions.

THE FORCES OF NATURE

Despite the seeming complexity of the totality of natural phenomena, we have so far found only four fundamental forces. These four forces appear to hold the parts of the universe together and to cause all of the events in it that we observe.

The most familiar force is *gravitation*, an attraction of all matter for all other matter. It keeps us on the earth and holds the earth together. It binds the planets in their orbits about the sun. It pulls on all parts of the universe, slowing down, we think, its expansion. The force of gravitation between two objects is greater the greater are their masses, and weaker the farther away they are. Despite its familiarity, gravitation is an incredibly weak force. We are ordinarily aware of the gravitational effect of only a very large amount of matter—such as the earth. The most delicate laboratory measurements are required to detect the force of gravitation between small objects.

The second force is the *electric* or *magnetic force*. Whether an electric or magnetic attraction is felt depends on the relative motion of the objects. Electromagnetic forces attract or repel, but only between charged objects or particles. Most matter is electrically neutral, so that only in rather unusual circumstances is the electromagnetic force noticed in everyday life. The nuclei and electrons of an atom, however, are electrically charged, and the electromagnetic force binds them together. It is also the electromagnetic force that holds together the latticed structure of solid matter. The atoms and molecules of the book you are reading are held together by this force. The same force holds our bones and muscles together and keeps the floor rigid so that we do not fall through it.

The electromagnetic force is enormously strong compared to the gravitational one. To be sure, there is a force of gravitation between the nucleus of an atom and its electrons. In the case of the hydrogen atom, however, the electromagnetic force between the nucleus and single electron exceeds their mutual gravitational attraction by more than a factor of 10^{39} (or 1 followed by 39 zeros).

The other two forces are the *strong* and *weak* nuclear forces. The strong nuclear force is the strongest of all, about 100 times as strong as the electromagnetic force. It binds together the parts of the nucleus of an atom (protons and neutrons, or *nucleons*). Although it produces an extremely powerful bond between nucleons, it is significant over only very small distances within the nucleus itself; the strong nuclear force can be thought of as a sort of "nuclear glue."

The weak nuclear force is roughly a million times weaker than the electromagnetic force, and is now understood to be related to it. The weak nuclear forces come into play in nuclear reactions involving radioactivity and in the decay of neutrons into protons and electrons.

Throughout the following chapters we shall return frequently to the forces of nature and the laws by which they behave. So far as we have determined, those same forces exist, and work by the same rules, and on identical kinds of atoms, everywhere in the universe.

SUMMARY REVIEW

The method of science

The objects in the universe: the solar system; astronomical unit; sun; light year; stars; the Milky Way; the Galaxy; galaxies; Local Group; clusters of galaxies; quasars; the expanding universe

Structure of the atom: atomic nucleus; protons; neutrons; electrons; scale of sizes in the atom

Forces of Nature: gravitation; electromagnetic force; strong nuclear force; weak nuclear force

EXERCISES

1. Give one or more examples that illustrate the application of the scientific method.

2. From the data given in this chapter, calculate how many circles the size of the earth's orbit would be required to reach across the length of a diameter of the Galaxy, if they were laid out barely touching each other, as in a chain. How many such circles would it take to reach a typical quasar?

3. If we were to try to communicate, say by radio waves, with a hypothetical inhabitant of a planet revolving about a star 100 LY away, how long would we have to wait after transmitting a question before we could expect to receive an answer?

4. If radio waves from a typical quasar contained coded messages from some superintelligence, how long ago would the broadcast of those messages have taken place?

5. If a gas is *ionized*—that is, if the electrons are stripped from the nuclei of its atoms, so that each freed electron moves about in the gas the same way an individual molecule would—the gas can be compressed to a far higher density than ordinary solids can. Explain why. (Assume that high temperatures or other factors prevent ions from reuniting with electrons, thereby becoming neutral atoms.)

6. The electromagnetic force causes a repulsion between the positively charged protons in an atom, but the strong nuclear force still binds them together in the atomic nucleus, along with a comparable number of neutrons. But the strong force is effective only between particles in virtual contact, while the hundred-times-weaker electromagnetic force acts over a considerable distance. Does this suggest to you why atomic nuclei with roughly 100 or more protons are not stable, and disintegrate? Explain.

CHAPTER 2

Nicolaus Copernicus (Mikolaj Kopernik) (1473–1543) did not prove that the earth revolves about the sun, but he presented compelling arguments that turned the tide of cosmological thought.

CRYSTALLINE SPHERES

From prehistoric times man must have looked at the sky. In some cultures, he discovered some of the regularities in the celestial motions. Indeed, some very early civilizations developed highly sophisticated astronomical sciences. Of course, ancient astronomy was based on naked-eye observations, but it is fascinating to see how far that astronomy advanced.

THE CELESTIAL SPHERE

If we gaze upward at the sky on a clear night, we cannot avoid the impression that the sky is a great hollow spherical shell with the earth at the center. The early Greeks regarded the sky as just such a *celestial sphere;* some apparently thought of it as an actual sphere of a crystalline material, with the stars embedded in it like tiny jewels. The sphere, they reasoned, must be of very great size, for if its surface were close to the earth, as one moved from place to place he would see an apparent angular displacement in the directions of the stars.

Of course, at any one time we see only a hemisphere overhead, but we can easily envision the remaining hemisphere, that part of the sky that lies below the horizon. If we watch the sky for several hours, we see that the celestial sphere appears to turn around us gradually. It is only an illusion, though, caused by the rotation of the earth, which carries us under successively different portions of the sphere. Following along with us must be our *horizon,* that circle in the distance at which the ground seems to dip out of sight, providing a demarcation between earth and sky. (The horizon may, of course, be hidden from view by mountains, trees, buildings, or in large cities, smog.) As our horizon tips down in the direction that the earth's rotation carries us, stars hitherto hidden beyond it appear to rise. In the opposite direction the horizon tips up, and stars hitherto visible appear to set. Analogously, as we round a curve in a mountain road, new scenery comes into view while old scenery disappears behind us.

The direction around the sky toward which the earth's rotation carries us is *east;* the opposite direction is *west.* The ancients, unaware of the earth's rotation, imagined that the celestial sphere rotated about an axis that passed through the earth. As it turned, it carried the stars up in the east, across the sky, and down in the west.

Celestial Poles and Celestial Equator

Some stars do not rise or set. As seen from the Northern Hemisphere, there is a point in the sky some distance above the northern horizon about which the whole celestial sphere appears to turn. As stars circle about that point, those close enough to it can pass beneath it without dipping below the northern horizon. A star exactly at the point would appear motionless in the sky. Today the star *Polaris* (the North Star) is within 1° of this pivot point of the heavens.

That pivot point is along an extension of the line through the earth's North and South poles. In the opposite direction in the sky is another such pivot point. As the earth rotates about its polar axis, the sky appears to turn in the opposite direction about those *north* and *south celestial poles.* Halfway between them, and separating the sky into its north and south halves, is the *celestial equator*—exactly analogous to the earth's equator.

An observer at the North Pole of the earth would see the north celestial pole directly overhead (at his *zenith*). The stars would all appear to circle about the

Figure 2.1
Time exposure showing trails left by stars as a consequence of the apparent rotation of the celestial sphere. *(Lick Observatory)*

sky parallel to the horizon, none rising or setting. An observer at the earth's equator, on the other hand, would see the celestial poles at the north and south points on his horizon. As the sky apparently turns about these points, all the stars appear to rise straight up in the east and set straight down in the west. For an observer at an arbitrary place in the Northern Hemisphere (for example, in Greece) the north celestial pole would appear above the northern horizon at an angular height, or *altitude*, equal to that observer's distance north of the earth's equator (that is, his *latitude*). The stars that were not always above the horizon would rise at an oblique angle in the east, arc across the sky in a slanting path, and set obliquely in the west.

Rising and Setting of the Sun

The sun always has some position in the sky—that is, on the celestial sphere. When the apparent rotation of the sphere carries the sun above the horizon, the brilliant sunlight scattered about by the molecules of the earth's atmosphere produces the blue sky that hides the stars that are also above the horizon. The early Greeks were aware that the stars were there during the day as well as at night.

The Greeks were also aware, as were the Chinese, Babylonians, and Egyptians before them, that the sun gradually changes its position on the celestial sphere, moving each day about 1° to the east among the stars. Of course, the daily westward rotation of the celestial sphere (or eastward rotation of the earth) carries the sun, like everything else in the heavens, to the west across the sky. Each day, however, the sun rises, on the average, about four minutes later with respect to the stars; the celestial sphere (or earth) must make just a bit more than one complete rotation to bring the sun up again. The sun, in other words, has an independent motion of its own in the sky, quite apart from the daily apparent rotation of the celestial sphere.

In the course of one year the sun completes a circuit of the celestial sphere. The early peoples mapped the sun's eastward journey among the stars. This apparent path of the sun is called the *ecliptic* (because eclipses can occur only when the moon is on or near it—see Chapter 9). The sun's motion on the ecliptic is in fact merely an illusion produced by another motion of the earth—its annual revolution about the sun. As we look at the sun from different places in our orbit, we see it projected against different stars in the background, or we would, at least, if we could see the stars in the daytime; in practice, we must deduce what stars lie behind and beyond the sun by observing the stars visible in the opposite direction at night. After a year, when we have completed one trip around the sun, it has apparently completed one circuit of the sky along the ecliptic. We have an analogous experience if we walk around a campfire at night; we see the flames appear successively in front of each of the people seated about the fire.

Figure 2.2
Time exposure showing star trails in the region of the north celestial pole. The bright trail near the center was made by Polaris (the North Star). *(Yerkes Observatory)*

It was also known by the ancients that the ecliptic does not lie along the celestial equator, but is inclined to it at an angle of about 23½°. This angle is called the *obliquity* of the ecliptic, and was measured with surprising accuracy by several ancient observers. The obliquity of the ecliptic, as we shall see (Chapter 4), is responsible for the seasons, as well as for the familiar tilt in the axes of terrestrial globes.

Fixed and Wandering Stars

The sun is not the only moving object among the stars. The moon and each of the five planets visible to the unaided eye—Mercury, Venus, Mars, Jupiter, and Saturn—also change their positions in the sky from day to day. The moon, being the earth's nearest celestial neighbor, has the fastest apparent motion; it completes a trip around the sky in about 1 month. During a single day, of course, the moon and planets all rise and set, as do the sun and stars. But like the sun, they also have independent motions among the stars, superimposed on the daily rotation of the celestial sphere. The Greeks distinguished between what they called the *fixed stars*, the real stars that appeared to maintain fixed patterns among themselves throughout many generations, and the *wandering stars* or *planets*. The word *planet* means "wanderer." Today, we do not regard the sun and moon as planets, but the ancients applied the term to all seven of the moving objects in the sky. Much of ancient astronomy was devoted to observing and predicting their motions. In fact, they give us the names for the seven days of our week; Sunday is the sun's day, Monday the moon's day, and Saturday is Saturn's day. We have only to look at the names of the other days of the week in the Romance languages to see that they are named for the remaining planets.

The individual paths of the moon and planets in the sky all lie close to the ecliptic, although not exactly on it. The reason is that the paths of the planets about the sun, and of the moon about the earth, are all in nearly the same plane, as if they were marbles rolling about on the top of a table. The planets and moon are always found in the sky within a narrow belt 18° wide centered on the ecliptic called the *zodiac*. The apparent motions of the planets in the sky result from a combination of their actual motions and the motion of the earth about the sun, and consequently they are somewhat complex.

The famous philosopher Pythagoras (who died ca. 497 B.C.), pictured a series of concentric spheres, in

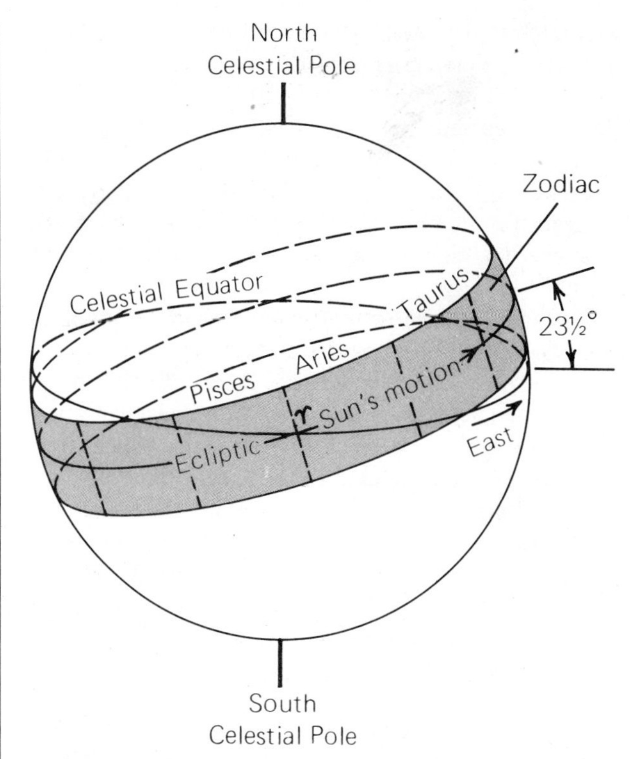

Figure 2.3
The celestial sphere, with the celestial poles, equator, and ecliptic shown. The observer views the celestial sphere from inside, at its center. The sun, moon, and planets are never far from the ecliptic in the sky, and therefore move in a zone centered on the ecliptic—the zodiac. For astrological purposes the zodiac is divided into 30° sectors called *signs*. The first sign, *Aries*, lies immediately to the east of the *vernal equinox*, ♈, one of the two points where the ecliptic crosses the celestial equator. The next sign east of Aries is *Taurus*, and the twelfth sign, just west of the vernal equinox, is *Pisces*.

which each of the seven moving objects—the planets, the sun, and the moon—was carried by a separate sphere from the one that carried the stars, so that the motions of the planets resulted from independent rotations of the different spheres about the earth. The friction between them gave rise to harmonious sounds, the *music of the spheres*, which only the most gifted ear could hear.

Constellations

The backdrop for the motions of the "wanderers" in the sky is the canopy of stars themselves. Like the Chinese and the Egyptians, the Greeks had divided the sky into *constellations*, apparent configurations of stars. Modern astronomers still make use of these constellations to denote approximate locations in the sky, much as geographers use political areas to denote the locations of places on the earth. The modern boundaries between the modern constellations are imaginary lines in the sky running north-south and east-west, so that every point in the sky falls in one constellation or another.

Many of the 88 recognized constellations are of Greek origin and bear names that are Latin translations of those given them by the Greeks. Today, the lay person is often puzzled because the constellations seldom resemble the people or animals for which they were named. In all likelihood the Greeks themselves did not name groupings of stars because they resembled actual people or objects, but rather named sections of the sky in *honor* of the characters in their mythology, and then fitted the configurations of stars to the animals and people as best they could.

GREEK SCIENCE

Contrary to popular opinion, educated people knew the earth is round at least 2000 years before Columbus. Belief in a spherical earth may have stemmed from the time of Pythagoras. Even then it was realized that the moon shines by reflected sunlight, and the curved shape of the *terminator*, the demarcation line between the moon's illuminated and dark portions, showed that the moon must be a sphere. The sphericity of the earth might have seemed to follow by analogy. In any case the belief that the earth is round never disappeared from Greek thought.

The writings of Aristotle (384–322 B.C.) describe how the progression of the moon's *phases*—its changing shape during the month—results from our seeing different portions of the moon's illuminated hemisphere during the month. Aristotle knew that the sun has to be more distant from earth than the moon is because occasionally the moon passes exactly between the earth and sun and temporarily hides the sun from view (*solar eclipse*). Now, of course, only the half of the moon turned toward the sun is lit—having daylight—at any time. The apparent shape of the moon in the sky depends simply on how much of that side is turned our way.

Aristotle also cited two convincing arguments that the earth is round. First is the fact that during a lunar eclipse, as the moon enters or emerges from the earth's shadow, the shape of the shadow seen on the moon is always round (Figure 2.4). Only a spherical object always produces a round shadow. If the earth were a disk, for example, there would be some occasions when the sunlight would be striking the disk edge on, and the shadow on the moon would be a line.

As a second argument, Aristotle explained that northbound travelers observed hitherto invisible stars to appear above the northern horizon and other stars to disappear behind the southern horizon. Southbound travelers observed the opposite effect. The only possible explanation is that the travelers' horizons had tipped to the north or south, respectively, which indicates that they must have moved over a curved surface of the earth.

Greek astronomers of the Alexandrian school in the third century B.C. had measured the distance to the moon, and Aristarchus of Samos (ca. 310–230 B.C.) devised an ingenious (although incorrect) method to find

Figure 2.4
Partially eclipsed moon moving into the earth's shadow. Note the curved shape of the shadow. *(Yerkes Observatory)*

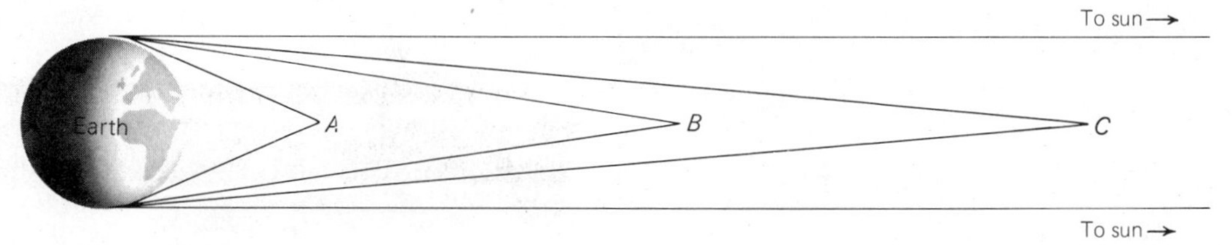

Figure 2.5
The more distant an object, the more nearly parallel are the rays of light from it.

how much more distant the sun is than the moon (he obtained the ratio 20; the correct value is 400). Aristarchus, in fact, even believed that the earth revolves about the sun, a notion that did not gain general acceptance for nearly two millennia.

One reason for rejecting the motion of the earth goes back to Aristotle: if the earth moved about the sun we would be observing the stars from successively different places along our orbit and their apparent directions in the sky would then change continually during the year. The argument was advanced many times since, even by Tycho Brahe in the late sixteenth century.

Any apparent shift in the direction of an object as a result of a motion of the observer is called *parallax*. An annual shifting in the apparent directions of the stars that results from the earth's orbital motion is called *stellar parallax*. For the nearer stars it is observable with modern telescopes (see Chapter 12), but it is impossible to measure with the naked eye because of the great distances of even the nearest stars.

Measurement of the Earth by Eratosthenes

The first fairly accurate determination of the earth's diameter was made by Eratosthenes (276–195 or 196 B.C.), another astronomer of the Alexandrian school. The sun is so distant from the earth compared to its size that the sun's rays intercepted by all parts of the earth approach it along essentially parallel lines. Imagine a light source near the earth, say at position A in Figure 2.5. Its rays strike different parts of the earth along diverging paths. From a light source at B, or at C, still farther away, the angle between rays that strike extreme parts of the earth is smaller. The more distant the source, the smaller is the angle between the rays. For a source *infinitely* distant, the rays travel along parallel lines. The sun is not, of course, infinitely far away, but light rays striking the earth from a point on the sun diverge from each other by at most an angle of less than one-third of a minute of arc (⅓′), far too small to be observed with the unaided eye. As a consequence, if people all over the earth who could see the sun were to point at it, their fingers would all be pointing in the same direction—they would all be parallel to each other. The concept that rays of light from the sun, planets, and stars approach the earth along parallel lines is vital to the art of celestial navigation—the determination of position at sea.

Eratosthenes noticed that at Syene, Egypt (near modern Aswân), on the first day of summer, sunlight struck the bottom of a vertical well at noon, which indicated that Syene was on a direct line from the center of the earth to the sun. At the corresponding time and date in Alexandria, 5000 stadia north of Syene (the *stadium* was a Greek unit of length), he observed that the sun was not directly overhead but slightly south of the zenith, so that its rays made an angle with the vertical equal to 1/50 of a circle (about 7°). Yet the sun's rays striking the two cities are parallel to each other. Therefore (see Figure 2.6), Alexandria must be one-fiftieth of the earth's circumference north of Syene, and the earth's circumference must be 50×5000, or 250,000, stadia.

It is not possible to evaluate precisely the accuracy of Eratosthenes' solution because there is doubt as to which of the various kinds of Greek stadia he used. If it was the common Olympic stadium, his result was about 20 percent too large. According to another interpretation, he used a stadium equal to about ⅙ km, in which case his figure was within 1 percent of the correct value of 40,000 km. The diameter of the earth is found from the circumference, of course, by dividing the latter by π.

Later Greek Astronomy

The greatest astronomer of pre-Christian antiquity was Hipparchus, who was born in Nicaea in Bithynia.

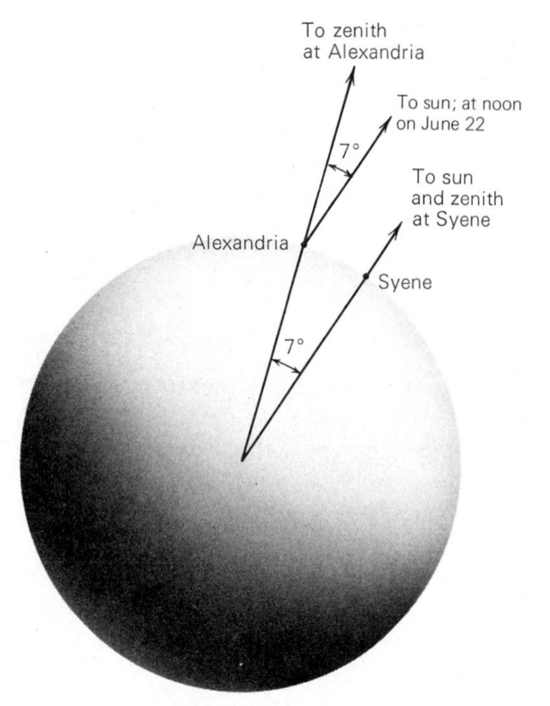

Figure 2.6
Eratosthenes' method of measuring the size of the earth.

He carried out his work at Rhodes, and possibly also at Alexandria, in the period from 160 to 127 B.C.

Hipparchus erected an observatory on the island of Rhodes and built instruments with which he measured as accurately as possible the directions of objects in the sky. He compiled a star catalog of about 850 entries. He designated for each star its celestial coordinates, that is, quantities analogous to latitude and longitude that specify its position (direction) in the sky. He also divided the stars according to their apparent brightness into six categories or *magnitudes*, and specified the magnitude of each star. In the course of his observations of the stars, and in comparing his data with older observations, he made one of his most remarkable discoveries—that the position in the sky of the north celestial pole had altered over the previous century and a half. Hipparchus correctly deduced that the direction of the axis about which the celestial sphere appears to rotate continually changes. The real explanation for the phenomenon is that the direction of the earth's rotational axis changes slowly because of the gravitational influence of the moon and the sun, much as a top's axis describes a conical path as the earth's gravitation tries to tumble the top over. This variation of the earth's axis, called *precession*, requires about 26,000 years for one cycle.

The last great Greek astronomer of antiquity was Claudius Ptolemy (or Ptolemaeus), who flourished about 140 A.D. He compiled a series of 13 volumes on astronomy known as the *Almagest*. All of the *Almagest* does not deal with Ptolemy's own work, for it includes a compilation of the astronomical achievements of the past, principally of Hipparchus. In fact, it is our main source of information about Greek astronomy. The *Almagest* also contains the contributions of Ptolemy himself.

Ptolemy's most important original contribution was a geometrical representation of the solar system that predicted the motions of the planets with considerable accuracy. Hipparchus, having determined by observation that earlier theories of the motions of the planets did not fit their actual behavior, and not having enough data on hand to solve the problem himself, instead massed observational material for posterity to use. Ptolemy supplemented the material with new observations of his own and with it produced a cosmological hypothesis that endured until the time of Copernicus.

The complicating factor in the analysis of the planetary motions is that their apparent wanderings in the sky result from the combination of their own motions and the earth's orbital revolution. Notice, in Figure 2.7(a), the orbit of the earth and the orbit of a hypothetical planet farther from the sun than the earth. The earth travels around the sun in the same direction as the planet and in nearly the same plane, but has a higher orbital speed. Consequently, it periodically overtakes the planet, like a faster race car on the inside track. The apparent directions of the planet, seen from the earth, are shown at successive intervals of time along lines $AA'A''$, $BB'B''$, and so on. In Figure 2.7(b), we see the resulting apparent path of the planet among the stars. From positions B to D, as the earth passes the planet, it appears to drift backward, to the *west* in the sky, even though it is actually moving to the *east*. Similarly, a slowly moving car appears to drift backward with respect to the distant scenery when we pass it in a faster-moving car. As the earth rounds its orbit toward position E, the planet again takes up its usual eastward motion in the sky. The temporary westward motion of a planet as the earth swings between it and the sun is called *retrograde* motion. (During and after its retrograde motion, the planet's apparent path in the sky does not trace exactly over itself because of the slight inclinations be-

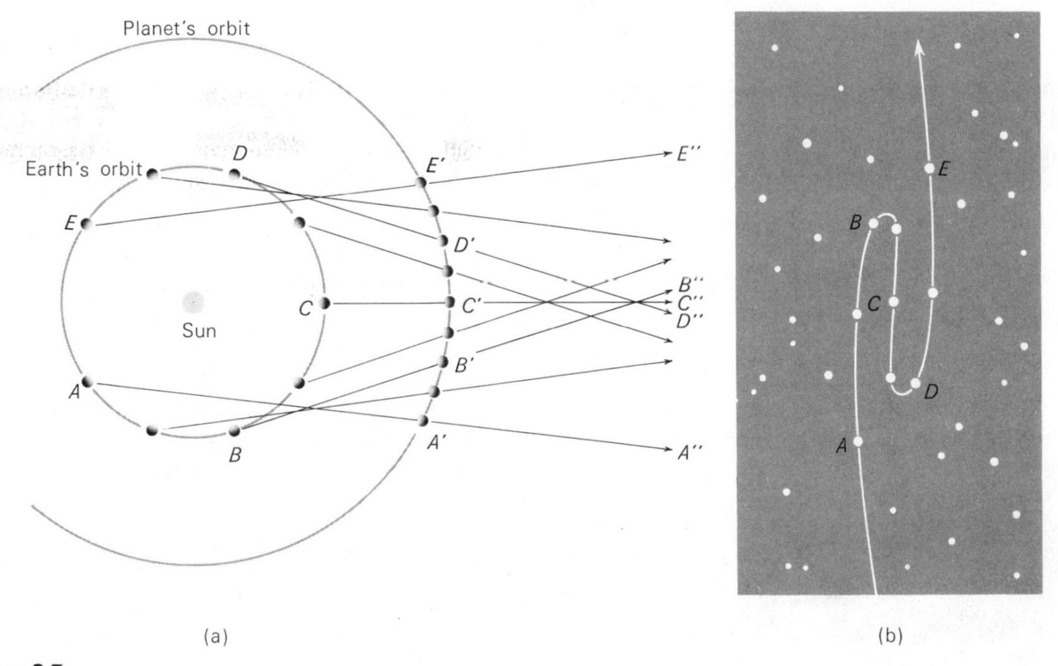

Figure 2.7
Retrograde motion of a superior planet. *(a)* Actual positions of the planet and earth; *(b)* the apparent path of the planet as seen from the earth, against the background of distant stars.

tween the orbits of the earth and other planets. Thus, the retrograde path is shown as an open loop in Figure 2.7.) Obviously, it is difficult to find an explanation for retrograde motion on the hypothesis that the planet is revolving about the earth.

Ptolemy solved the problem by having a planet P (Figure 2.8) revolve in a small orbit, called an *epicycle*, about C. The center of the epicycle C in turn revolved in a path called the *deferent* about the earth. When the planet is at position x, it is moving in its epicycle orbit in the same direction as the movement of point C about the earth, and the planet appears to be moving eastward. When the planet is at y, however, its epicyclic motion is in the opposite direction to the motion of C. By choosing the right combination of speeds and distances, Ptolemy succeeded in having the planet moving westward at the right speed at y and for the correct interval of time. However, we shall see in the next chapter that the planets, like the earth, travel about the sun in orbits that are ellipses, not circles, and their actual behavior cannot be represented accurately by so simple a scheme of uniform circular motions. Consequently, Ptolemy placed the deferent eccentrically, that is, with its center slightly away from the earth at A. Furthermore, he had the center of the epicycle, C, move at a uniform angular rate, not around A, or E, but around a point B, called the *equant*, on the opposite side of A from the earth.

It is a tribute to the genius of Ptolemy as a mathematician that he was able to conceive such a complex system to account successfully for the observations. His hypothesis, with some modifications, was accepted as absolute authority throughout the Middle Ages, un-

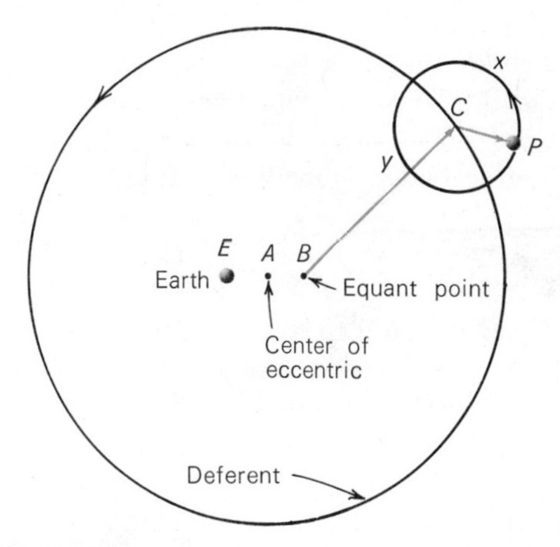

Figure 2.8
Ptolemy's system of deferent, epicycle, eccentric, and equant.

til it finally gave way to the heliocentric theory in the seventeenth century. In the *Almagest,* however, Ptolemy made no claim that his cosmological model described reality. He intended his scheme rather as a mathematical representation to predict the positions of the planets at any time. Modern astronomers do the same thing with algebraic formulas. Our modern mathematical methods were not available to Ptolemy; he had to use geometry.

ASTROLOGY

Modern research has shown that all matter in the universe is composed of atoms—and the same kinds of atoms. Thus our probes that landed on Mars, Vikings 1 and 2, and our telescopic spectra of the light from the most remote quasars indicate that, whatever we do not yet understand about Mars and the quasars, at least they are made of the same stuff that makes up our own bodies. It is a grand discovery, for it suggests a beautiful unity of the universe.

Still, we cannot fault the ancients for assuming that the luminous orbs in the sky, the stars and planets, are made of "heavenly" substances, and not of the "earthly" elements we find at home. In fact, the realization that celestial worlds are actually worlds and not ethereal substance is relatively recent in the history of science. Small wonder, then, that the ancients regarded the planets (including the sun and moon), which alone moved about among the stars on the celestial sphere, as having special significance. Thus the planets came to be associated with the gods of ancient mythologies; in some cases, they were themselves thought of as gods. Even in the comparatively sophisticated Greece of antiquity, the planets had the names of gods, and were even credited with having the same powers and influences as the gods whose names they bore. From such ideas grew the primitive religion of astrology.

Astrology began, we think, in Babylonia a millennium or so before Christ. The Babylonians, believing that the planets and their motions influenced the fortunes of kings and nations, practiced what we call *mundane* astrology. When the Babylonian culture was absorbed by the Greeks, their astrology gradually influenced the entire western world, and eventually spread to the Orient as well. By the third or second century B.C., the Greeks democratized astrology by developing the tradition that the planets bore their influence on every individual. In particular, they believed that the configuration of the planets at the moment of a person's birth affected his personality and fortune. This form of astrology, known as *natal* astrology, reached its acme with Ptolemy in the second century A.D. Ptolemy, as famous for his astrology as for his astronomy, compiled the *Tetrabiblos,* a four-volume treatise on astrology that remains the "bible" of the subject even today.

The Horoscope

The key to natal astrology is the *horoscope,* a chart that shows the positions of the planets in the sky at the moment of an individual's birth. The charting of a horoscope, as of any map, requires the use of coordinates. The celestial coordinates used by astrology, in antiquity as well as today, are analogous to, and share a common origin with, those used by astronomers. First, the planets (including the sun and the moon—classed as planets by the ancients) have to be located in the sky with respect to the fixed stars on the celestial sphere. Second, the constantly turning celestial sphere, with its stars and the planets, must have its orientation specified with respect to the earth at the time and place of the subject's birth.

The positions of the planets on the celestial sphere are determined by locating them in the zodiac—the belt centered on the ecliptic that contains the planets. For the purposes of astrology, the zodiac is divided into twelve sectors called *signs,* each 30° long. The signs have their origin at the place on the ecliptic where the sun, in its annual journey about the sky, crosses from the south half to the north half of the sky at the beginning of Spring (about March 21). The north and south halves of the celestial sphere are separated by the *celestial equator,* halfway between the north and south celestial poles, and the place where the sun crosses it on the first day of Spring is called the *vernal equinox.* The first zodiacal sign is the 30° sector of the zodiac, centered on the ecliptic, immediately to the east of the vernal equinox; that first sign is called *Aries,* so the vernal equinox is also known as the *first point of Aries,* symbolized by ♈, the horn symbol for Aries, the Ram. The subsequent eleven signs are, in order to the east (the direction of the sun's annual motion): Taurus, Gemini, Cancer, Leo, Virgo, Libra, Scorpio, Sagittarius, Capricorn, Aquarius, and Pisces. All but Libra (the Scales) are named for animals or people; *zodiac* means "the zone or circle of the animals." A horoscope shows the position of each planet in the sky by indicating its position in the appropriate sign of the zodiac.

As the celestial sphere turns (due to the rotation of the earth), the vernal equinox, and the entire zodiac with it, moves across the sky to the west, completing a circuit of the heavens each day. Thus, locating the planets in the zodiac does not tell where they are in the sky with respect to the horizon of a particular place. The latter is accomplished through a knowledge of *sidereal time*, a measure of how far the vernal equinox has progressed since it passed from the eastern to the western half of the sky. Once the location in the sky of the vernal equinox (or sidereal time) is known, the positions of all of the signs, and hence of the planets, are known. To specify the latter, the astrologers define a system of twelve *houses*, each fixed with respect to the zenith and horizon of a particular place on earth.

The twelve houses divide the sky into twelve sectors, beginning with the eastern horizon. The first house is that part of the sky immediately below the eastern horizon; objects within the first house are the ones that will rise within the next two hours or so. The second house is the next one below the first; the third through sixth houses are the remaining ones below the horizon, the sixth containing objects that have set within approximately the past two hours. Houses seven through twelve stretch across the upper half of the sky from west to east.

A complete horoscope is usually represented by a circle denoting the center of the zodiac (the ecliptic) with the twelve houses indicated as sectors inside the circle. The signs and their boundaries are also located on the horoscope, as well as the positions of the seven planets. Sometimes the positions of the more conspicuous stars of the zodiac are indicated as well. Figure 2.10 shows my horoscope.

If the axis of the earth maintained an absolutely fixed orientation in space, the vernal equinox would be fixed among the stars on the celestial sphere. As we have seen, however, precession (discovered by Hipparchus) produces a slow change in the direction toward which the earth's axis points, which causes the vernal equinox to slide westward about the ecliptic. Originally the signs of the zodiac had the same names as the constellations that they coincided with. Because of precession, however, the signs no longer line up with the constellations of the same names; thus the sign of Aries is now in the constellation of Pisces.

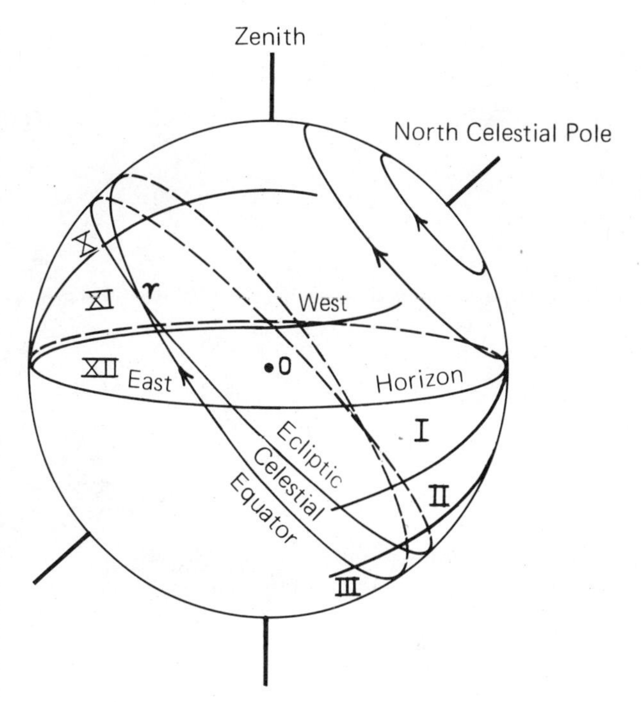

Figure 2.9
To an observer, *O*, on the earth, the zenith and horizon are fixed, and the north celestial pole (for a northern hemisphere observer) is some distance above the northern horizon (at an altitude equal to the observer's northern latitude). The apparent turning of the celestial sphere makes all points in the sky move about in paths parallel to the celestial equator (curved arrows in the figure). The boundaries between the houses are fixed in the sky, and the turning celestial sphere carries stars and planets across these boundaries from house to house. The precise definition of the houses varies from one system of astrology to another; the boundaries between some of the houses shown here follow the definition of Placedius. The house numbers are given in Roman numerals. At the instant for which the figure is appropriate, the sidereal time would be about 20½ hours, and the vernal equinox is in the eleventh house.

Interpretation of the Horoscope

There are many rules for the interpretation of the horoscope, many or most of which (at least in Western schools of astrology) are derived from the *Tetrabiblos* of Ptolemy. Each sign, each house, and each planet, the latter supposedly acting as a center of force, is associated with particular matters.

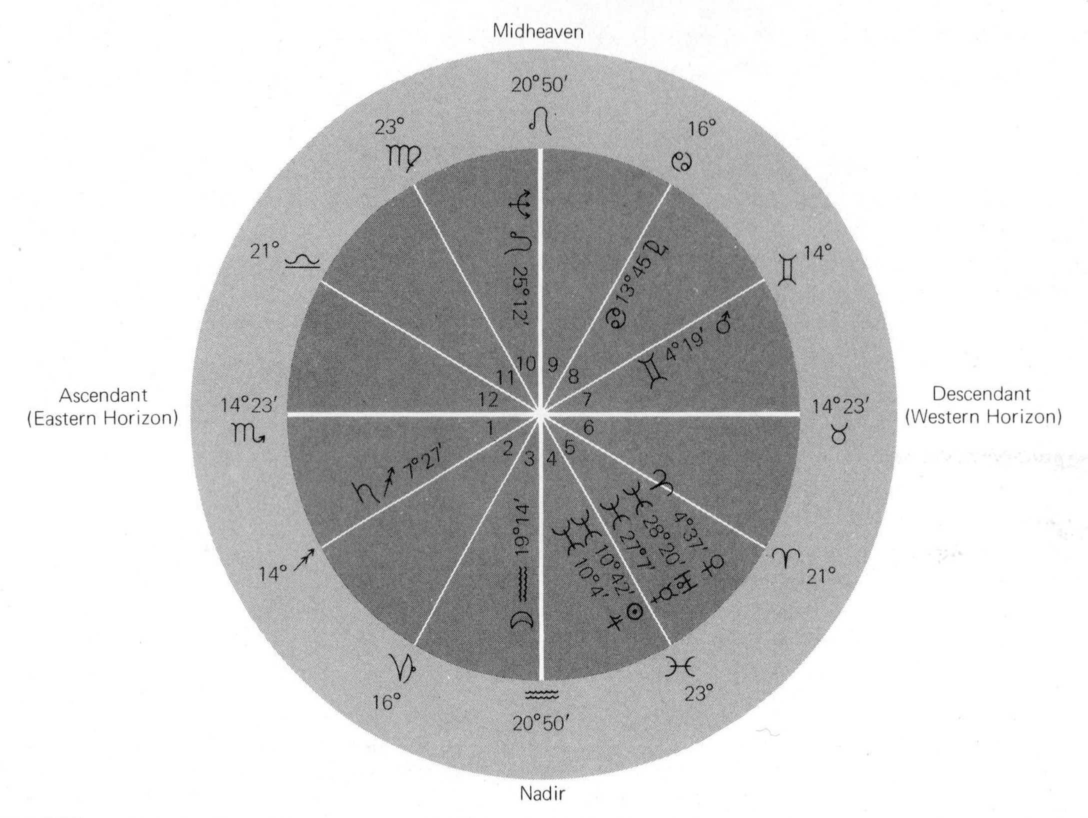

Figure 2.10
Natal horoscope of the author, who was born in Los Angeles, California, on March 1, 1927, at 10:50 p.m., P.S.T. The 12 pie-shaped sectors represent the 12 houses, and the outer circular zone represents the zodiac. The definition of houses used in preparing this horoscope is that of Placedius, in which, as the rotating celestial sphere carries the planets around the sky, each place in the zodiac spends equal times in each of the six houses above the horizon *(diurnal houses)* and also in the six houses below the horizon *(nocturnal houses)*; however, the time required for an object to pass through a diurnal house is not the same as that required for it to pass through a nocturnal house except for objects on the celestial equator. The boundaries between houses *(cusps)* intersect the ecliptic in the zodiacal signs indicated by their symbols in the outer circular zone. The number beside each sign symbol is the angular distance of the cusp from the beginning of that sign. The position of each planet is shown in the house it occupied at the instant of the author's birth. Beside the symbol for the planet is the symbol of the zodiacal sign it was also in at that time, and the angular distance of the planet from the beginning of that sign. The places where the horizon intersects the zodiac are shown, and also the highest point of the ecliptic in the sky *(midheaven)* and its lowest point below the horizon, *nadir* (this astrological definition of the nadir is slightly different from the astronomical one, in which the nadir is directly opposite the zenith).

Thus if you were born between March 21 and April 19, the sun was in the sign of Aries, and you are known as an *Aries* (even though the sun was actually in the constellation of Pisces at the time). You are supposed to be dominant, inventive, and impatient. Just as important, however, is what was in the *ascendancy*—that is, in the first house (about to rise). If this were the moon, for example, you would be moody and tend toward procrastination. Now if at the same time Mars were in the sign of Virgo, you would be shrewd and appear impulsive. But the *aspects* of the planets also matter: if any are, say, 0°, 60°, 90°, 120°, or 180° apart from each other in the zodiac. Thus, if Mercury is in conjunction with Jupiter in your natal horoscope it is a good omen, unless, at the same time, Venus is in evil aspect with Saturn (for example, 90° away from it), in which case things are bad for you. It matters where each planet and sign is with respect to each house, and where each planet is with respect to each other planet, and whether a planet is entering or leaving a house or sign, or is near a boundary thereof (*cusp*).

In other words, the interpretation of a horoscope is a very complicated business, and whereas the rules may be standardized, how each rule is to be weighed and applied is a matter of judgment—and "art." It also means that it is very difficult to tie astrology down to specific predictions.

The interpretation of an individual's horoscope, charted for the time and place of his birth, is natal astrology; his characteristics and fortunes, presumably, depend on his natal horoscope. Another branch of the subject is *horary* astrology, which purports to answer specific questions by casting a horoscope for the time and place at which the question was first posed. Horary astrology might be used, for example, to find whether one's mate were having an affair with another, or whether the coming Monday would be a good time for a particular business deal.

We recall that the signs of the zodiac, because of precession, are slowly slipping westward with respect to the constellations. The traditional astrology, nevertheless, is based on the moving *signs*, not the constellations; it is called *tropical astrology*. In a way it has a rationale, because the seasons themselves depend on the sun's position with respect to the equinoxes and solstices on the ecliptic.

On the other hand, there is a school of astrology that is based on the positions of the planets in the *constellations* rather than in the signs; it is called *sidereal astrology*. Today the first point of Aries (or vernal equinox) is in the constellation Pisces; it has slid westward through nearly an entire width of a sign in the past two millennia since the constellations and signs were named. Soon it will have passed entirely through Pisces and into the constellation Aquarius. At that time we will be said to have entered the *Age of Aquarius*. The precise time that this will occur, however, depends on where the boundaries lie between the constellations of Pisces and Aquarius; old star maps are noncommittal, and the modern constellation boundaries were drawn up arbitrarily by the International Astronomical Union in 1928.

Value of Astrology

Today, with our knowledge of the nature of the planets as physical bodies, composed as they are of rocks and fluids, it is hard to imagine that the directions of these planets in the sky at the moment of one's birth could have anything to do with his personality or future. The gravitational influence of the moon and sun on tides is unquestionable, but tides produced on a person by a book in his hand are millions of times as strong as those produced by all the planets combined. The sun's light and heat are obviously of great importance to us, but even minute variations in the sun's irradiation are millions of times as great as the combined light of the planets. Jupiter (and to a lesser extent, the other planets) has a strong magnetic field and emits radio waves, but their detection requires magnetometers carried on space probes and large radio telescopes. The feeble radio signals from a small 1000-watt transmitter 100 miles away reach us with a strength millions of times as great as the radio waves from Jupiter, and can be picked up by a pocket transistor radio. Even the magnet in the loudspeaker of that radio produces around the listeners a magnetic field enormously stronger than Jupiter's. Moreover, the distances of the planets from the earth vary greatly, and any gravitational and radiation effects would vary as the inverse square of their distances—factors ignored by astrology.

Astrology would have to argue that there are unknown forces exerted by the planets that depend on their configurations with respect to each other and with respect to arbitrary coordinate systems invented by man—forces for which there is not a whit of solid evidence. Are astronauts on the moon similarly affected by the same kind of force exerted by the earth? Or is the earth, alone, subject to these unknown laws of nature?

In the most orthodox astrology, one's entire life (and death) is predetermined by his natal horoscope. If a man dies in an auto accident at the age of 63 because someone else ran a stoplight, are we supposed to assume that all of the complicated chain of events that led to the circumstances of his being in that accident were blueprinted by the planets at the instant of his birth, but that all would have been different if he had been born two hours later? Most of us would find this assumption so incredulous that we would need the most overwhelming evidence of its validity before taking it seriously. In the tens of centuries of astrology, no such evidence has been presented.

One could argue, on the other hand, that astrology only works statistically; that other influences—heredity and environment, for example—are important too, and that astrological influences are only important as tendencies, everything else being equal. In that case the reality of astrological effects could only be tested statistically. From time to time astrologers have presented statistical "proofs" of astrology, but not one survives objective scientific scrutiny. A recent and exhaustive study[1] of the astrological literature by an Australian astrologer failed to turn up a single piece of evidence that he regarded as verification of traditional astrology.

In retrospect, we can understand the belief in astrology on the part of ancient peoples who thought the heavenly bodies to be made of celestial material different from the elements that compose the earth, and to be placed in the sky by their gods for the benefit of mankind. In the light of modern knowledge, the astrological claims seem so farfetched as to be ludicrous. Because we would not expect the supposed influences, even in a statistical sense, we would want solid evidence and demonstrable predictions. Physical scientists and others who have investigated the subject with the hope of finding some grain of validity in it have found negative results. Virtually all scientists reject astrology as an unfounded superstition. Yet it continues to appeal to the popular fancy. The hope of predicting the future by magical or mystical means, and perhaps of transferring one's responsibilities and the blame for one's failures and misfortunes to an omnipotent power, continues to be a strong attraction. Moreover, it may simply be "fun" to speculate about the unknown and unprovable no matter how little basis there may be for it. Many astrologers today acknowledge that astrology cannot be proven by statistics or by experiment, but assert that it must be "known" or "realized" as knowledge or truth. In this context, it is outside the realm of science, and no rational argument based on the rules of science is relevant. To many astrology is a religion, and hence is outside the scope of our consideration here.

One fact remains: The practice of astrology in ancient times required the knowledge of the motions of the planets to construct horoscopes for past or future events. The quest to find a mechanism for charting the planets, joined with a natural curiosity about nature, stimulated centuries of observations and calculations, leading—as we shall see—to our modern technology.

[1] *Recent Advances in Natal Astrology*, by Geoffrey Dean, *et al.*, published under the aegis of The Astrological Association, 1977. Distributed in North America by Para Research, Inc., Whistlestop Mall, Rockport, MA. 01966.

THE BIRTH OF MODERN ASTRONOMY

Astronomy made no major advances in medieval Europe, where the prevailing philosophy was acceptance of dogmatic authority. Medieval cosmology combined the crystalline spheres of Pythagoras (as perpetuated by Aristotle) with the epicycles of Ptolemy. Astrology was widely practiced, however, and an interest in the motions of the planets was thus kept alive. Then came the Renaissance; in science the rebirth was clearly embodied in Nicholas Copernicus.

Nicholas Copernicus (in Polish, Mikolaj Kopernik, 1473–1543) was born in Torun on the Vistula in Poland. His training was in law and medicine, but Copernicus' main interest was astronomy and mathematics. By the time he had reached middle age, he was well known as an authority on astronomy.

Copernicus' great contribution to science was a critical reappraisal of the existing theories of cosmology and the development of a new model of the solar system. His unorthodox idea that the sun, not the earth, is the center of the solar system had become known by 1530, chiefly through an early manuscript circulated by him and his friends.

His ideas were set forth in detail in his *De Revolutionibus*, published in the year of his death. Supervision over the publication of the book fell into the hands of a Lutheran preacher named Osiander, who was probably responsible for the augmented title of the work—*De Revolutionibus Orbium Celestium* (On the Revolutions of the Celestial Spheres). Osiander wrote a preface, which he neglected to sign, implying that the theory set forth in the book was only a convenient calculating scheme. The preface is almost certainly in contradiction to Copernicus' own feelings.

In *De Revolutionibus*, Copernicus set forth certain postulates from which he derived his system of planetary motions. His postulates include the assumptions that the universe is spherical and that the motions of the heavenly bodies must be made up of combinations of uniform circular motions; thus Copernicus was not free of all traditional prejudices. Yet, he evidently found something orderly and pleasing in the heliocentric system, and his defense of it was elegant and persuasive. His ideas, although not widely accepted until more than a century after his death, never disappeared and were ultimately of immense influence.

A person moving uniformly is not necessarily aware of his motion. We have all experienced the phenomenon of seeing an adjacent train, car, or ship

Armillary Sphere of Antonio Santucci delle Pomerance, made for the Grand Duke Ferdinando I Medici in 1593. *(Istituto e Museo di Storia della Scienza di Firenze)*

Telescopes donated by Galileo to the Grand Duke Ferdinando II and to his brother, the Prince Leopoldo. The longest has a wooden tube covered with paper, a focal length of 1.33 m, and an aperture of 26 mm. *(Istituto e Museo di Storia della Scienza di Firenze)*

Newton's birthplace at Woolsthorpe. The apple tree in the foreground grew from the stump of the one standing in Newton's time. *(Photograph by the author)*

First editions of some books of great historical interest. *(Crawford Library; Courtesy, Astronomer Royal for Scotland, Royal Observatory Edinburgh)*

appear to change position, only to discover that it is we who are moving. Copernicus argued that the apparent annual motion of the sun about the earth could be represented equally well by a motion of the earth about the sun, and that the rotation of the celestial sphere could be accounted for by assuming that the earth rotates about a fixed axis while the celestial sphere is stationary. To the objection that if the earth rotated about an axis it would fly into pieces, Copernicus answered that if such motion would tear the earth apart, the even faster motion (because of its greater size) of the celestial sphere required by the alternative hypothesis would be even more devastating to it.

The important assumption Copernicus made in *De Revolutionibus* is that the earth is but one of six (then known) planets that revolve about the sun. Given this, he was able to work out the correct general picture of the solar system. He placed the planets, starting nearest the sun, in the order Mercury, Venus, Earth, Mars, Jupiter, and Saturn. Further, he deduced that the nearer a planet is to the sun, the greater is its orbital speed. Thus the retrograde motions of the planets were easily understood without the necessity for epicycles. Also, Copernicus worked out the correct approximate scale of the solar system.

Copernicus did not prove that the earth revolves about the sun. In fact, with some adjustments the old Ptolemaic system could have accounted as well for the motions of the planets in the sky. But the Ptolemaic cosmology was clumsy and lacked the beauty and symmetry of its successor. Copernicus made the earth an astronomical body, which brought a kind of unity to the universe. It was, to borrow from Neil Armstrong, a "giant step for mankind." It is fitting that the quinquecentennial celebration of the birth of Copernicus was honored by scientists throughout the world in 1973, for the Copernican revolution marked the origin of modern science and of our contemporary understanding of the universe.

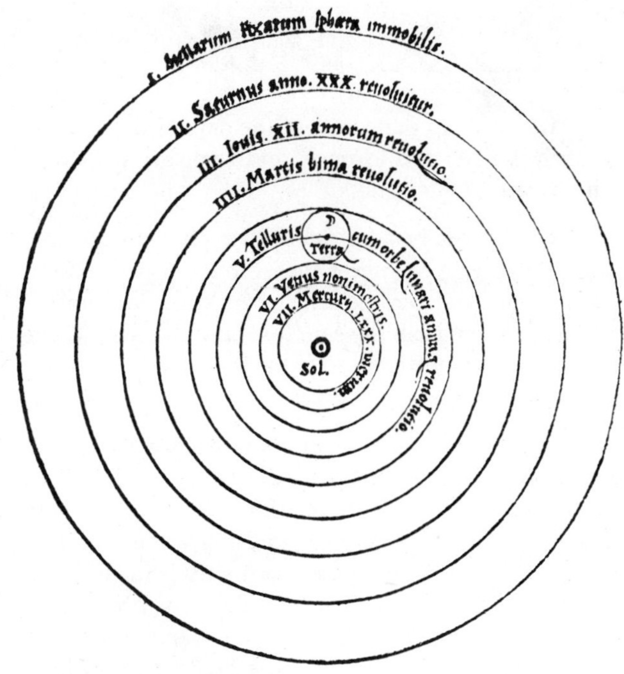

Figure 2.11
Plan of the solar system in the First Edition of Copernicus' *De Revolutionibus. (Crawford Collection, Royal Observatory Edinburgh)*

SUMMARY REVIEW

The celestial sphere: celestial poles; celestial equator; apparent rotation of the sky; rising and setting of the sun and stars; constellations

Motions of the sun, moon, and planets: ecliptic and zodiac

Early Greek science: Aristotle on the shape of the earth; measurement of the earth by Eratosthenes; Hipparchus and the discovery of precession; Aristarchus' heliocentric view

Ptolemaic cosmology: epicycles; deferent; eccentric; equant

Astrology: mundane, natal, and horary astrology; tropical and sidereal astrology; horoscope; signs; houses; aspects of planets in the horoscope; interpretation of the horoscope; Ptolemy's *Tetrabiblos;* value of astrology and lack of evidence for its validity

Copernicus' system: *De Revolutionibus;* the heliocentric hypothesis

EXERCISES

1. Look up the names of the days of the week in French, Italian, and Spanish, and compare them with the names of the planets.

2. As seen by a terrestrial observer, which (if any) of the following can never appear in the opposite direction in the sky from the sun? in the same direction? at an angle of 90° from the sun? (a) Mars; (b) a star; (c) the sun; (d) Earth; (e) Jupiter; (f) the moon; (g) Venus; (h) Mercury.

3. Suppose you are on a strange planet, and observe, at night, that the stars do not rise and set, but circle parallel to the horizon. Now you walk in a constant direction for 8000 miles, and at your new location on the planet you find that all stars rise straight up in the east and set straight down in the west, perpendicularly to the horizon.

(a) How could you determine the circumference of the planet without any further observations?
(b) What evidence is there that the Greeks could have done what you suggest?
(c) What *is* the circumference of the planet, in miles?

4. On March 13, there will be a total eclipse of the moon. About what time will the moon rise that night?

5. Where on earth (a) are all stars at some time visible above the horizon? (b) is only half the sky ever above the horizon?

6. Why can an eclipse of the sun never occur on the day following an eclipse of the moon? (*Hint:* see Chapter 9.)

7. Give some everyday examples of parallax.

8. Suppose Eratosthenes had found that at Alexandria at noon on the first day of summer the line to the sun makes an angle of 45° with the vertical. What then would he have found for the earth's circumference?

9. One aspect of the planets that is regarded favorably by astrology is the *trine*—when two planets are 120° away from each other in the zodiac. Give two examples of trine aspects that can never occur.

10. Many people try to use pseudostatistical arguments to justify their belief in a pseudoscience. Try the experiment of flipping a coin ten times and then recording the number of heads that turn up. Do this experiment 100 or more times (several people can flip coins at the same time and then pool results, thereby saving labor). Prepare a table showing how many times no head was obtained (ten tails in a row), how many times one head was obtained, how many times two heads, and so on. Make a graph showing the same data. What was the most frequent number of heads? What fraction of the time were less than three or more than seven heads obtained? If an event occurring only one percent of the time is enough to arouse your suspicions, how many heads would you have to obtain in a single experiment to question the honesty of the coin? How do these results apply to some of the arguments used by those who believe in astrology?

11. Parallaxes of stars were not observed by ancient astronomers. How can this fact be reconciled with the heliocentric hypothesis of the solar system proposed by Copernicus?

The earth rising over the lunar landscape. *(NASA photograph by Apollo 17 astronauts in lunar orbit)*

CHAPTER 3

Sir Isaac Newton (1642–1727) had the insight to realize that the force that makes planets fall around the sun and the force that makes apples fall to the ground are different manifestations of the same thing: gravitation.

GRAVITATION

Although gravitation is the weakest force known in nature, it is extremely important to us, because gravitation holds together the earth, the solar system, the Galaxy, and perhaps the universe itself. It is also the first of the natural forces whose mathematical properties were discovered. Moreover, under extreme circumstances gravitation can become an extremely strong force; modern investigations of gravitational theory (Chapter 18) lead to the prediction of such esoteric objects as *black holes*.

But our story of gravitation begins where we left off in the last chapter. This greatest achievement of Isaac Newton culminated from the contributions of astronomical investigations of Brahe and Kepler and from the physics of Galileo.

LAWS IN THE CELESTIAL REALM

Three years after the publication of *De Revolutionibus*, Tycho Brahe (1546–1601) was born of a family of Danish nobility. Tycho (as he is generally known) developed an early interest in astronomy and as a young man made significant astronomical observations. The reputation of the young Tycho Brahe as an astronomer gained him the patronage of Frederick II, and in 1576 Tycho was able to establish a fine astronomical observatory on the Danish island of Hveen, off the coast of Elsinor.

Tycho's observations at Hveen included a continuous record of the positions of the sun, moon, and planets. His daily observations of the sun, extending over years and comprising thousands of individual sightings, led to solar tables that were good to within 1′. He reevaluated nearly every astronomical constant and determined the length of the year to within one second. His extensive and precise observations of planetary positions enabled him to note that the positions of the planets varied from those given in published tables, and he even noted regularities in the variations.

But Tycho was an extravagant and cantankerous fellow, and he accumulated enemies, especially among government officials. Thus when his patron, Frederick II, died in 1597, Tycho was forced to leave Denmark. He took up residence near Prague, where he became Court Astronomer to the Emperor Rudolf of Bohemia. He took with him some of his instruments and most of his records. There Tycho Brahe spent the remaining years of his life. In 1600, the year before his death, he secured the assistance of a most able young mathematician, Johannes Kepler.

Kepler (1571–1630) was born in Weil-der-Stadt, Württenberg (southwestern Germany). He attended college at Tübingen and studied for a theological career. There he learned the principles of the Copernican system, and became converted to the heliocentric hypothesis. Kepler, a Protestant refugee from his Catholic homeland, went to Prague to serve as an assistant to Tycho, who set him to work trying to find a satisfactory theory of planetary motion—one that was compatible with the long series of observations made at Hveen.

But Tycho, being jealous of young Kepler's brilliance, was reluctant to provide him with much material at any one time for fear Kepler would discover the secrets of the universal motions by himself, thereby robbing Tycho of some of the glory. It was not until after Tycho's death that Kepler obtained possession of the majority of the priceless records. Their study occupied most of Kepler's time for the next 25 years.

Kepler's most detailed study was of Mars, for which the observational data were the most extensive. He published the first results of his work in 1609 in *The New Astronomy*, or *Commentaries on the Motions of Mars*. It is there that we find his first two laws of planetary motion. Their discovery was a profound step in the development of modern science.

Kepler discovered that the orbit of Mars was in the shape of a curve known as an *ellipse*. Next to the circle, the ellipse is the simplest kind of closed curve, belonging to a family of curves known as *conic sections* (Figure 3.1). A conic section is simply the curve of intersection between a hollow cone (whose base is presumed to extend downward indefinitely) and a plane that cuts through it. If a plane is perpendicular to the axis of the cone (or parallel to its base), the intersection is a circle. If the plane is inclined at an arbitrary angle, but still cuts completely through the surface of the cone, the resulting curve is an ellipse. If the plane is parallel to a line in the surface of the cone, it never quite cuts all the way through the cone, and the curve of intersection is open at one end. Such a curve is called a *parabola*. If the plane is inclined at an even smaller angle to the axis of the cone, an open curve results that is called a *hyperbola*. The ellipse, then, ranges from a circle at one extreme to a parabola at the other. The parabola separates the family of ellipses from the family of hyperbolas.

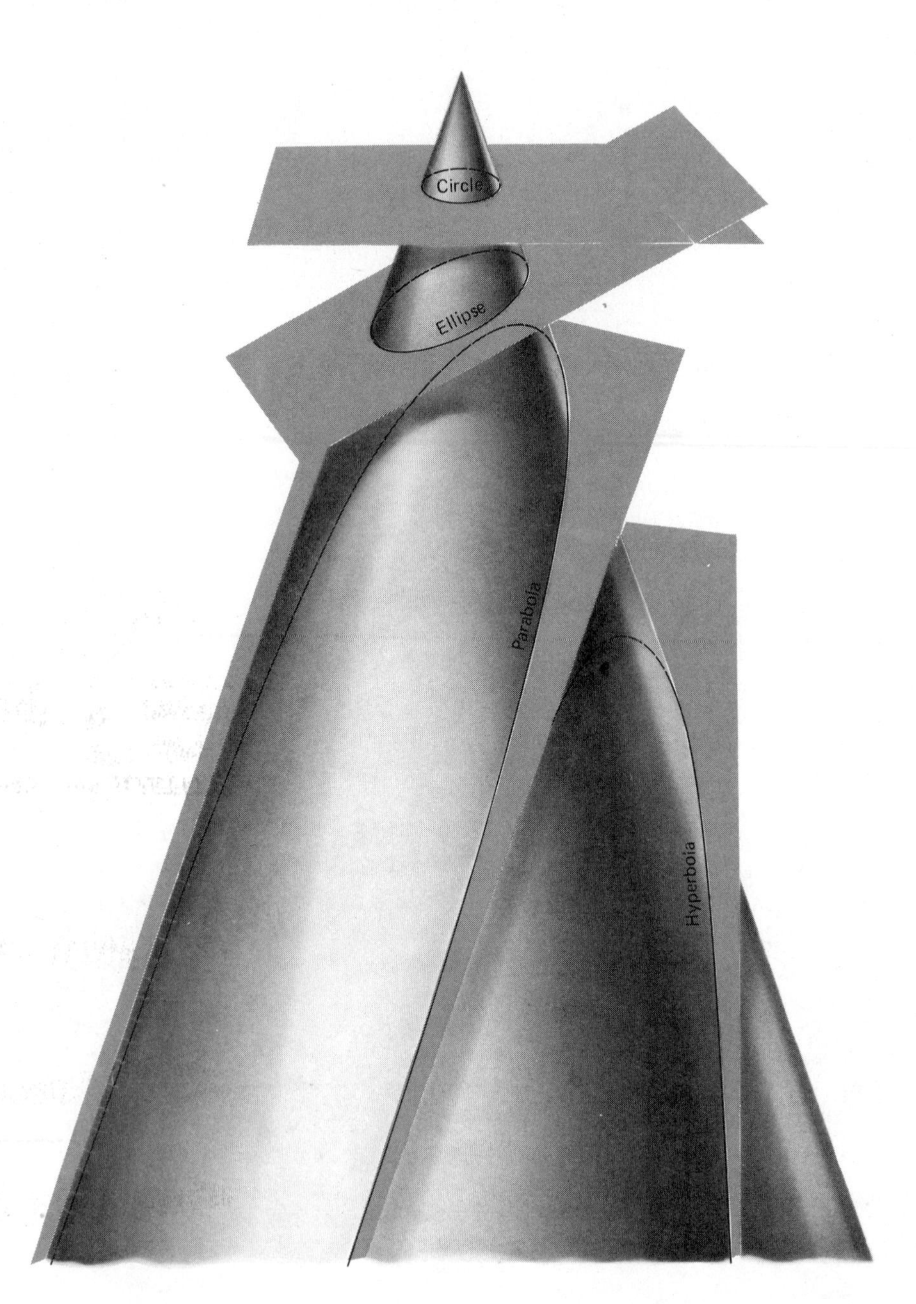

Figure 3.1
Conic sections.

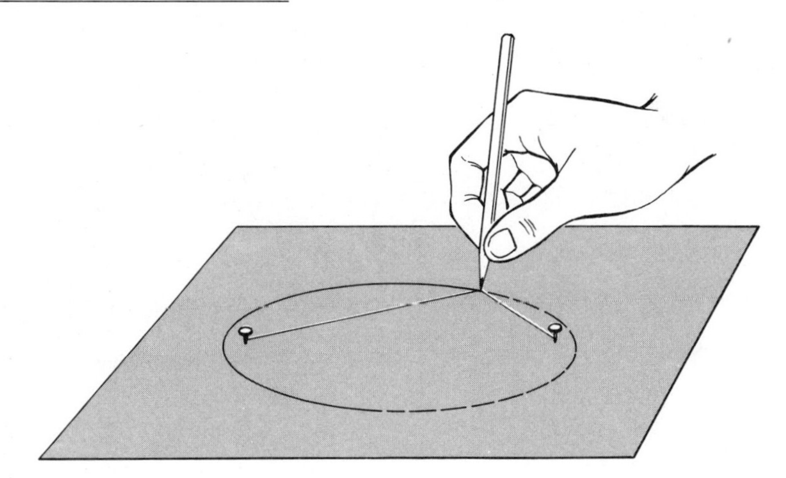

Figure 3.2
Drawing an ellipse.

An interesting and important property of an ellipse is that from *any point* on the curve, the sum of the distances to two points inside the ellipse, called the *foci* of the ellipse, is the same. This property suggests a simple way to draw an ellipse. The ends of a length of string are tied to two tacks pushed through a sheet of paper into a drawing board, so that the string is slack. If a pencil is then pushed against the string, so that the string is held taut, and then slid against the string around the tacks (Figure 3.2), the curve that results is an ellipse; at any point where the pencil may be, the sum of the distances from the pencil to the two tacks is a constant length—the length of the string. The tacks, of course, are at the two foci of the ellipse.

The maximum diameter of the ellipse is called its *major axis*. Half this distance, that is, the distance from the center of the ellipse to one end, is the *semimajor axis*. The *size* of an ellipse depends on the length of the major axis. The *shape* of an ellipse depends on how close together the two foci are compared to the major axis. The ratio of the distance between the foci to the major axis is called the *eccentricity* of the ellipse. If an ellipse is drawn as described above, the length of the major axis is the length of the string, and the eccentricity is the distance between the tacks divided by the length of the string (Figure 3.2). If the foci (or tacks) coincide, the ellipse is a circle; a circle is, then, an ellipse of eccentricity zero. Ellipses of various shapes are obtained by varying the spacing of the tacks (as long as they are not farther apart than the length of the string). If one tack is removed to an infinite distance, and if enough string is available, "our end" of the resulting, infinitely long ellipse is a parabola. A parabola has an eccentricity of unity. An ellipse is completely specified by its major axis and its eccentricity.

Kepler found that Mars has an orbit that is an ellipse and that the sun is at one focus (the other focus is empty). The eccentricity of the orbit of Mars is only about 0.1; the orbit, drawn to scale, would be practically indistinguishable from a circle. It is a tribute to Tycho's observations and to Kepler's perseverance that he was able to determine that the orbit is an ellipse at all.

Kepler also found that Mars speeds up as it comes closer to the sun and slows down as it pulls away from the sun. Kepler expressed this relation by imagining that the sun and Mars are connected by a straight, elastic line. As Mars travels in its elliptical orbit around the sun, in equal intervals of time the areas swept out in space by this imaginary line are always equal (Figure 3.3). This relation is commonly called the *law of areas*.

In 1619 Kepler found a simple algebraic relation between the lengths of the semimajor axes of the planets' orbits and their periods of revolution about the sun (*sidereal periods*). The relation is now known as Kepler's *third, or harmonic, law*. Kepler eventually showed that all the known planets obeyed the simple laws he had discovered, now known as his three laws of planetary motion:

KEPLER'S FIRST LAW: Each planet moves about the sun in an orbit that is an ellipse, with the sun at one focus of the ellipse.

KEPLER'S SECOND LAW (THE LAW OF AREAS): The straight line joining a planet and the sun sweeps out equal areas in space in equal intervals of time.

KEPLER'S THIRD LAW: The squares of the sidereal periods of the planets are in direct proportion to the cubes of the semimajor axes of their orbits.

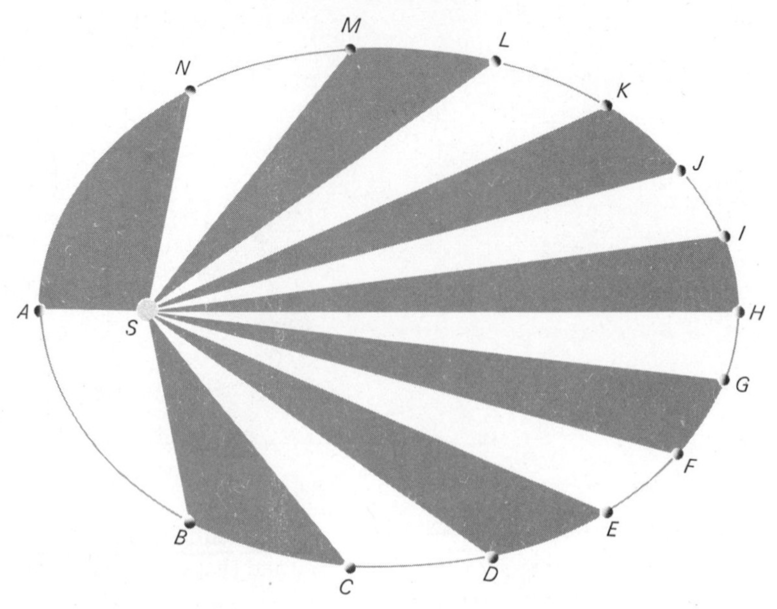

Figure 3.3
Law of equal areas. A planet moves most rapidly on its elliptical orbit when it is at position A, nearest the focus of the ellipse, S, where the sun is. The planet's orbital speed varies in such a way that in equal intervals of time it moves distances AB, BC, CD, and so on, so that regions swept out by the line connecting it and the sun (shaded and clear zones) are always the same in area.

LAWS IN THE TERRESTRIAL REALM

While Kepler was unveiling some of the secrets in the celestial realm, his great Italian contemporary, Galileo Galilei (1564–1642) was making important strides to our understanding of the behavior of things on the earth.

Galileo, as he is usually called, was born in Pisa. Like Copernicus, he began training for a medical career, but he had little interest in the subject and later switched to mathematics. He held a faculty position at the University of Pisa and later at the University of Padua, and eventually became mathematician to the Grand Duke of Tuscany in Florence. While at Padua he became famous throughout Europe as a brilliant lecturer and as the foremost scientific investigator.

Galileo's greatest contributions were in the field of ***mechanics***, the study of motion and the actions of forces on bodies. It was familiar to all persons then as it is to us now that if a body is at rest it tends to remain at rest, and requires some outside influence to start it in motion. Rest was thus generally regarded as the *natural state of matter*. Galileo showed, however, that rest was no more natural than motion. If an object is slid along a rough horizontal floor, it soon comes to rest, because friction between it and the floor acts as a retarding force. However, if the floor and object are both highly polished, the body, given the same initial speed, will slide farther before coming to rest. On a smooth layer of ice, it will slide farther still. Galileo reasoned that if all resisting effects could be removed (for example, the friction between the body and the floor or ground, or the air) it would continue in a steady state of motion indefinitely. In fact, he argued, not only is a force required to start an object moving from rest, but a force is also required to slow down, stop, speed up, or change the direction of a moving object.

Galileo also studied the way bodies accelerate, that is, change their speed, as they fall freely, or roll down inclined planes. He found that such bodies accelerate uniformly; that is, in equal intervals of time they gain equal increments in speed. Galileo formulated these newly found laws in precise mathematical terms that enabled one to predict, in future experiments, how far and how fast bodies would move in various lengths of time.

Sometime in the 1590s Galileo adopted the Copernican hypothesis of the solar system. In Roman Catholic Italy, this was not a popular philosophy, for the Church authorities still upheld the ideas of Aristotle and Ptolemy. It was primarily because of Galileo that in 1616 the Church issued a prohibition decree which stated that the Copernican doctrine was "false and absurd" and was not to be held or defended.

The prevailing notion of the time was that the celestial bodies belonged to the realm of the heavens where all is perfect, unchanging, and incorruptible. Perpetual circular motion, being the "perfect" kind of motion, was regarded as the natural state of affairs for those heavenly bodies. Once Galileo had established the principle of inertia—that on the earth bodies in undisturbed motion remain in motion—it was no longer necessary to ascribe any special status to the fact that the planets remain perpetually in orbit. By the same token, even the earth could continue to move, once started. What *does* need to be explained is why the planets move in curved paths around the sun rather than in straight lines. Evidently, Galileo was sufficiently steeped in Aristotelian concepts that he accepted uniform circular celestial motion without subjecting the planets to the same objective scrutiny that he applied in his terrestrial experiments.

In answer to the common objection that objects could not remain on the earth if it were in motion, Galileo noted that if a stone is dropped from the masthead of a moving ship it does not fall behind the ship and land in the water beyond the stern, but rather lands at the foot of the mast, for the stone already has a forward inertia gained from its common motion with the ship before it is dropped. In an analogous way, objects on the earth would not be swept off and left behind if the earth were moving, for they share the earth's forward motion.

Galileo's Astronomical Observations

It is not certain when the principle was first conceived of combining two or more pieces of glass to produce an instrument that enlarged images of distant objects, making them appear nearer. At any rate, the first telescopes that attracted much notice were made by the Dutch spectacle maker Hans Lippershey in 1608. Galileo heard of the discovery in 1609, and without ever having seen an assembled telescope he constructed one of his own with a three-power magnification, which made distant objects appear three times nearer and larger. He quickly built other instruments, his best with a magnification of about 30 diameters.

It was a fairly obvious step to apply the newly invented telescope to celestial observations. The idea may have occurred to others about the same time as it did to Galileo, or possibly earlier. Galileo, however, realized the importance of careful and persistent study of the objects he viewed. In 1610 he startled the world by publishing a list of his remarkable discoveries in a small book, *Sidereal Messenger* (*Sidereus Nuncius*).

Figure 3.4
Galileo's house, and prison during his last years. The plaque between the windows invites passers by to stop and contemplate, with respect, this home of the great watcher of the skies. *(Courtesy, Professor G. Godoli, Arcetri Observatory, Florence)*

Galileo found that many stars too faint to be seen with the naked eye became visible with his telescope. In particular, he found that some nebulous blurs resolved into many stars (for example, the Praesepe in Cancer) and that the Milky Way was made up of multitudes of individual stars. He found four satellites or moons revolving about Jupiter with periods ranging from just under 2 days to about 17 days (ten other satellites of Jupiter have been found since). This discovery was particularly important because it showed that there could be centers of motion that in turn are in motion. It had been argued that if the earth were in motion the moon would be left behind, because it could hardly keep up with a rapidly moving planet. Yet here were Jupiter's satellites doing exactly that!

Another important telescopic discovery that strongly supported the Copernican view was the fact that Venus goes through phases like the moon, showing that it must revolve about the sun, so that we see

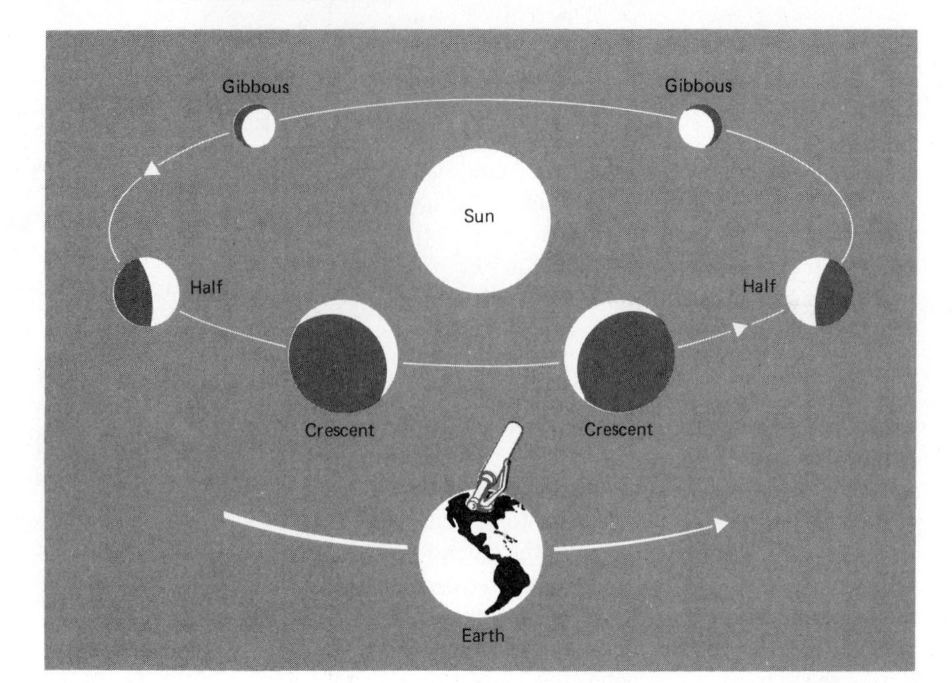

Figure 3.5
Phases of an inferior planet.

different parts of its daylight side at different times (Figure 3.5).

Galileo observed the moon, and saw craters, mountain ranges, valleys, and flat dark areas that he guessed might be water (the dark *maria*, or "seas," on the moon were thought to be water until long after Galileo's time). Not only did these discoveries show that the heavenly bodies, regarded as perfect, smooth, and incorruptible, do indeed have irregularities, as does the earth, but they showed that the moon is not so dissimilar to the earth, which suggested that the earth, too, could belong to the realm of celestial bodies.

Dialogue on the Great World Systems

Despite the decree of 1616 forbidding him to hold or defend the Copernican system, Galileo still hoped to convert his countrymen to the heliocentric view. He finally prevailed upon Pope Urban VIII, an old friend, to allow him to publish a book that explained fully all arguments for and against the Copernican system, not for the purpose of extolling it, but merely to examine it, and to show those of other nationalities that Italians were not ignorant of new theories.

The book appeared in 1632 under the title *Dialogue on the Two Great World Systems* (*Dialogo dei Due Massimi Sistemi*). The *Dialogue* was written in Italian (not Latin) to reach a large audience, and is a magnificent and unanswerable argument for Copernican astronomy. It is in the form of a conversation, lasting 4 days, among three philosophers: Salviati, the most brilliant and the one through whom Galileo generally expresses his own views; Sagredo, who is usually quick to see the truth of Salviati's arguments, and Simplicio, an Aristotelian philosopher who brings up all the usual objections to the Copernican system, which Salviati promptly shows to be absurd.

It is pointed out in the preface to the *Dialogue* that the arguments to follow are merely a mathematical fantasy, and that divine knowledge assures us of the immobility of the earth. This was thinly cloaked irony, however, and Galileo's enemies acted quickly to build a case against him. He was called before the Roman Inquisition on the charge of believing and holding doctrines that are false and contrary to the Divine Scriptures. Urban VIII, apparently won over by Galileo's enemies, failed to come to his aid. Galileo, then nearly seventy, was forced to plead guilty and deny his own doctrines. His sentence was very light for that time, consisting of confinement to his own home in Florence for the last 10 years of his life. During this time he completed his last book, *The Two New Sciences*, which describes much of his early investigation of solids and mechanics.

The *Dialogue*, meanwhile, took its place along with Copernicus' *De Revolutionibus* on the *Index of Prohibited Books*, where it remained until 1835.

NEWTON'S GREAT SYNTHESIS OF CELESTIAL AND TERRESTRIAL SCIENCE

It took the genius of Isaac Newton (1643–1727) to unify natural laws. Newton realized that the force that makes apples and other things fall on the earth and the force that makes the moon fall perpetually around the earth are different manifestations of the *same* force—gravitation.

Newton was born at Woolsthorpe Manor in Lincolnshire, England, almost exactly one year after the death of Galileo. (Newton was born on Christmas Day, 1642, according to the calendar in use at his time, but by the modern Gregorian calendar his birth date was January 4, 1643.)

Newton entered Trinity College at Cambridge in 1661 and eight years later was appointed Lucasian Professor of Mathematics, a post the he held during most of his productive career. As a young man in college, he became interested in natural philosophy, as science was called then. He worked out many of his ideas on mechanics and optics during the plague years of 1665 and 1666.

Eventually Newton's friend, Edmond Halley, prevailed on him to collect and publish the results of his investigations in mechanics and gravitation. The result was *Philosophiae Naturalis Principia Mathematica*. The *Principia*, as the book is generally known, was published at Halley's expense in 1687.

At the very beginning of the *Principia*, Newton states three laws that he presumes to govern the motions of all objects:

I. Every body continues in a state of rest, or of uniform motion in a straight line, unless it is compelled to change that state by forces impressed upon it.

II. The change of motion is proportional to the force impressed; and is made in the direction of the straight line in which that force is impressed.

III. To every action there is always an equal and opposite reaction: or, the mutual actions of two bodies upon each other are always equal, and act in opposite directions.

In the original Latin, those three laws contain only 59 words, but those few words set the stage for modern science. We shall have to examine them carefully.

Newton's first law is a statement of one of the great conservation laws—the *conservation of momentum*—a measure of the motion of a body. The law states simply that in the absence of any outside in-

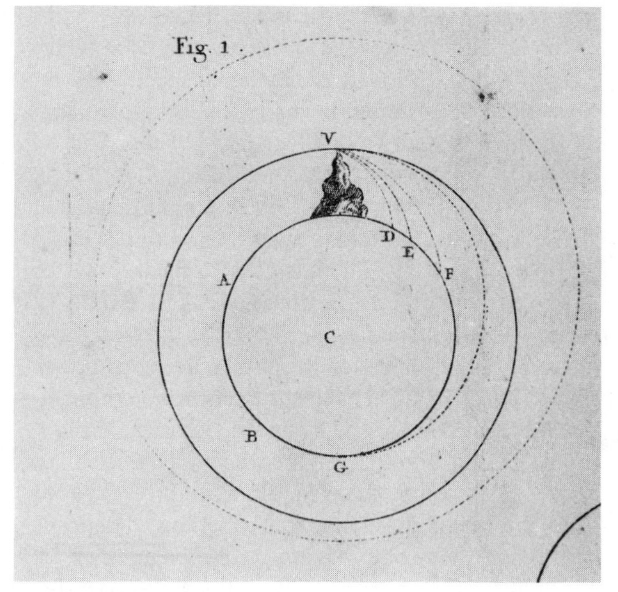

Figure 3.6
A diagram by Newton in his *De mundi systematic*, 1731 Edition. *(Crawford Collection, Royal Observatory Edinburgh)*

fluence, a body's momentum remains unchanged. That is, a stationary object stays put and a moving object keeps moving. Momentum depends on three things: the first is *speed*—how fast a body moves (*zero* if it is stationary). The second is its *direction*. The term *velocity* describes both speed *and* direction. For example, 20 mi/hr due south is velocity, while 20 mi/hr is speed. The third factor in momentum is what Newton called *mass*. Mass is that property of a body that gives it inertia—that is, the ability to resist change in motion. Mass is a measure of the actual amount of matter in a body. Our intuitive notion of *weight* is what we mean by mass. Technically, weight is something different—it is the gravitational attraction of the earth (or another body) on the object. But when we speak of weight, we usually mean mass. Newton defined momentum as the product of mass and velocity. A body's momentum stays the same unless the body is disturbed.

The momentum of a body can change only under the action of an outside influence. Newton's second law defines *force* as that quantity that changes momentum. Force has both magnitude and direction, and the momentum changes in the direction of the applied force. A change of either mass or velocity or both can result from a force, but in most examples a body's mass is not changed, so that a force on it results in a change of velocity. This means that a force is required to change either the speed or the direction of a body or both—that is, to start it moving, to speed it up, to slow it down, to stop it, to change its direction, or to do all these things. Any such change in velocity is called *acceleration*.

Newton's third law is the most profound. Basically, it is a generalization of the first law, but it also gives us a way to define mass. If we consider a system of two or more objects, isolated from outside influences, Newton's first law suggests that the total momentum of the system of objects should remain constant. Therefore, any change of momentum within the system must be balanced by another change that is equal and opposite to it, so that the momentum of the entire system is not changed. Thus, all forces occur as *pairs* of forces that are mutually equal to and opposite each other. If one object exerts a force on another object, it must be exerted by something else, and the object will exert an equal and opposite force on that something. Suppose a boy jumps off a table down to the ground. The force pulling him down is a mutual gravitational force between him and the earth. Both he and the earth suffer the same total change of momentum because of the influence of this mutual force. Of course, the boy does most of the moving; because of the greater mass of the earth, it can experience the same change of momentum by accelerating only a negligible amount.

A more obvious manifestation of the mutual nature of forces between objects is familiar to all who have played baseball. The recoil of the bat shows that the ball exerts a force on the bat during the impact, just as the bat does on the ball. The momentum imparted to the bat by the ball is transmitted through the batter to the earth, so the acceleration produced is far less than that suffered by the ball. Similarly, when a rifle is discharged, the force pushing the bullet out the muzzle is equal to that pushing backward upon the gun and marksman.

Here, in fact, is the principle of jet engines and rockets—the force that discharges the exhaust gases from the rear of the rocket is balanced by a force that shoves the rocket forward. The exhaust gases need not push against air or the earth; a rocket operates best of all in a vacuum. Incidentally, astronauts on a rocket ship must take care how they dispose of waste material. For example, if garbage were discharged through a port, the reaction force would accelerate the rocket slightly off course. Thus we can see how we can measure mass. Initially, some object must be adopted as a standard and said to have *unit mass*—for example, one cubic centimeter of water under specified conditions of temperature and pressure. The latter is defined as one *gram (g)*. Then the mass of any other object can be compared to it by measuring the relative accelerations produced when the same force acts on each of the two. We assure that each is subjected to the same force by isolating them from other objects and producing equal and opposite changes in their momenta with an *internal* force—say, by separating them with a compressed spring placed between them and released by an internal triggering device, or by detonating a small charge between them. Thus, the third law permits an *operational definition* of mass. This is called the *inertial* definition of mass.

Having found a way to measure mass, we can now express the value of a force numerically. The most common unit of force is the *dyne*, which is the force needed to give a mass of 1 g an acceleration of 1 cm/s/s.

Density

It is important not to confuse mass, volume, and *density*. Volume is simply a measure of the physical space occupied by a body, say in cubic centimeters or liters. In short, the volume is the "size" of an object—it has nothing to do with its mass. A lady's wrist watch and an inflated balloon may both have the same mass, but they have very different volumes.

The watch and balloon are also very different in *density*, which is a measure of how much mass is contained within a given volume. Specifically, it is the ratio of mass to volume. The units of density are usually expressed in grams per cubic centimeter (g/cm^3). Sometimes density is given in terms of the density of water (1 g/cm^3), in which case it is called *specific gravity*. Iron has a specific gravity of 7.9 or a density of 7.9 g/cm^3; gold has a specific gravity of 19.3. To sum up, then, *mass* is "how much," *volume* is "how big," and *density* is "how tightly packed."

*Angular Momentum

Another great conservation law is that of *angular momentum*. Whereas the ordinary momentum of a body is simply its mass times its velocity, its angular momentum provides a description of its motion about some fixed point or origin. Specifically, the angular momentum of the body is its distance from that origin, times its mass, times that part of its speed that moves it at right angles to the direction to the origin. If there is no force on the body other than one exactly toward or away from the origin, Newton's laws dictate that the angular momentum is conserved (the simple proof of this statement is given in standard texts in mechanics). Thus a body constrained to move on a circular path will, in the absence of frictional forces, move at a constant speed indefinitely. A planet moving on an elliptical path about the sun, on the other hand, changes its distance from the sun, so to conserve its angular momentum it must speed up as it comes nearer the sun and slow down as it moves farther away—exactly as stated in Kepler's second law. In fact, Kepler's second law is nothing but a geometrical manifestation of the conservation of the angular momentum of planets moving about the sun.

A rigid body rotating about an internal axis can be thought of as a composite of many component parts. Each part of that body is constrained to revolve about a point on that axis in a circular path. The forces causing this constraint are the cohesive forces binding the atoms and molecules of the body together. The unbalanced forces are all radial, provided that there is no external force acting in such a way as to make a part of the body try to move faster or slower about its axis of rotation. Consequently, in the absence of any outside force tending to change its rotation a rotating body will continue to rotate indefinitely with constant angular momentum. Thus planets, such as the earth, keep spinning.

The rotation rate of a body, again assuming that there are no forces, can only change if that body can rearrange its parts in such a way as to move some of them farther away from or closer to the axis of rotation (the origin of angular momentum). This is exactly what is accomplished by figure skaters; suppose a skater is spinning on the tip of her skate with arms outstretched. If she holds her body rigidly until friction with the air and ice produce enough force to slow her down, she rotates at a constant rate. However, if she pulls her arms in to her body, some parts of her mass are closer to their rotation axes, and hence would decrease in angular momentum unless she compensates by spinning faster. Thus, a figure skater can start a spin with her arms out, and then pull them in, thereby spinning faster so that her angular momentum is conserved. She can slow down again by pushing her arms (or a free leg) out from her body.

The conservation of angular momentum is an important concept to an understanding of the formation of the solar system with its planets and their satellites, including the rings of Saturn, Uranus, and Jupiter, and even to understanding the formation of galaxies. We shall refer to it again in coming chapters.

Acceleration in a Circular Orbit

It might be assumed that some force or power is required to keep the planets in motion. However, Galileo argued from the principle of inertia (Newton's first law of motion) that once started, the planets would remain in motion—that the state of motion for planets was as natural as for terrestrial objects. What does require explanation, however, is why the planets move in nearly circular orbits rather than in straight lines (the latter motion would eventually carry them away from the vicinity of the solar system). Galileo had not considered this problem.

By Newton's time, a number of investigators had considered the problem of circular motion. The correct solution to the problem was first published (in 1673) by the Dutch physicist Christian Huygens (1629–1695). However, Newton had found the solution independently in 1666.

For a body to move in a circular path rather than in a straight line, it must continually suffer an acceleration toward the center of the circle. Such an acceleration is called *centripetal acceleration*. The central force that produces the centripetal acceleration (*centripetal force*) is, for a planet, an attraction between the planet and the sun. For a stone whirled about at the end of a string, the centripetal force is the tension in the string. With the help of Newton's laws of motion and some elementary mathematics we can calculate how great that central force has to be.[1]

UNIVERSAL GRAVITATION

Newton's grandest concept is that of universal gravitation. He could not say why there is a gravitational force between all material bodies in the universe (nor can we), but he brilliantly described how that force operates. It is obvious that the earth exerts a force of

---

[1] If a particle of mass m moves with a speed v on the circumference of a circle of radius r, the centripetal force is given by the formula

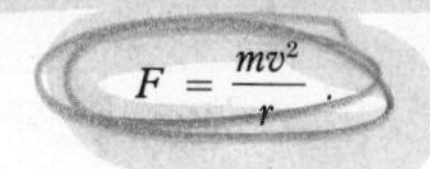

$$F = \frac{mv^2}{r}.$$

attraction upon all objects at its surface. This is a mutual force; a falling apple and the earth are pulling on each other. Newton reasoned that this force of attraction between the earth and objects on or near its surface might extend as far as the moon and produce the centripetal acceleration required to keep the moon in its orbit. He further speculated that there is a general force of attraction between *all* material bodies. If so, the attractive force between the sun and each of the planets could provide the centripetal acceleration necessary to keep each in its respective orbit.

Thus Newton hypothesized that there is a universal attraction between all bodies everywhere in space. Next he had to determine the mathematical nature of the attraction and test the hypothesis by using it to predict *new* phenomena.

By Newton's hypothesis, gravitation is universal; thus the sun and planets should exert a gravitational force on each other. The precise mathematical description of that gravitational force must dictate that the planets move exactly as Kepler had observed them to (and as codified in his three laws). Moreover, at the same time, the law of gravitation must predict the correct behavior of falling bodies on the earth, as observed by Galileo. How then must the gravitational force depend on distance for these conditions to be met? The answer to this question involved mathematics not yet developed. But this did not deter Isaac Newton.

He invented the branch of mathematics he called *fluxions* (we know it as *differential calculus*) and, with the power of this new method of analysis, found that Kepler's first law (that the planets have elliptical orbits with the sun at one focus) requires that the force of attraction between a planet and the sun must vary inversely as the square of the distance separating them (that is, the attraction of a planet toward the sun is less in proportion to the square of the planet's distance). The fact that the attraction between a planet and the sun is along the line between them also predicts that the planets should move according to Kepler's second law—that is, should obey the law of areas. A consideration of his laws of motion also led Newton to the hunch that the force of gravitation between any two bodies should be greater in proportion to the mass of each, and hence to the product of their masses.

Newton's law of gravitation was observed to hold for the planets. To be general, however, the gravitational theory should predict the observed acceleration toward the earth of the moon, falling about the earth at a distance of 60 earth radii, and also of an object dropped near the earth's surface. The gravitational attraction of the earth for an object at (or near) its surface (we have seen) is the object's *weight*. Does Newton's gravitational law predict the correct weights for objects at the earth's surface?

The weight of a man or other object at the surface of the earth is the result of the simultaneous attractions of the many parts of the earth pulling on him from many different directions. Exactly what is the resultant gravitational effect of the many parts of a sphere, each pulling independently upon a mass outside the surface of the sphere? Here was a difficult problem to which Newton had to find a solution before he could test his law of gravitation.

To solve the problem, Newton had to calculate the force between an object on the surface of the earth and each infinitesimal piece of the earth, and then calculate how all of these forces combined. It was necessary for him to invent and use a new method of mathematics which he called *inverse fluxions* (today we call it *integral calculus*). Fortunately, the solution to the problem gives a beautifully simple result. A spherical body acts gravitationally as though all its mass were concentrated at a point at its center (Figure 3.7).[2] This means that we can consider the earth, the moon, the sun, and the planets as geometrical points as far as their gravitational influences are concerned.

Thus we see that Newton's tentative law of gravitation predicts that the acceleration of an object toward the earth should be proportional to the earth's mass and inversely proportional to the square of its distance from the center of the earth. Objects at the surface of the earth (one earth radius from its center) are observed to accelerate downward at 980 cm/s per second (32 ft/s/s). Thus the moon, 60 earth radii from its center, should experience an acceleration toward the earth that is 60^2 or 3600 times less, that is, about 0.272 cm/s/s (0.0089 ft/s/s). This is precisely what Newton's formula for the centripetal acceleration of the moon predicts. Newton's law of gravitation is dramatically confirmed! The success of our modern space program, which is based on Newtonian gravitational theory, is a further testimony to its power.

The actual formula for the gravitational force of attraction (in dynes) between two objects of masses m_1 and m_2, and separation (in cm) d, is

---

[2] Strictly, the statement is correct only if the density distribution within the body is spherically symmetrical.

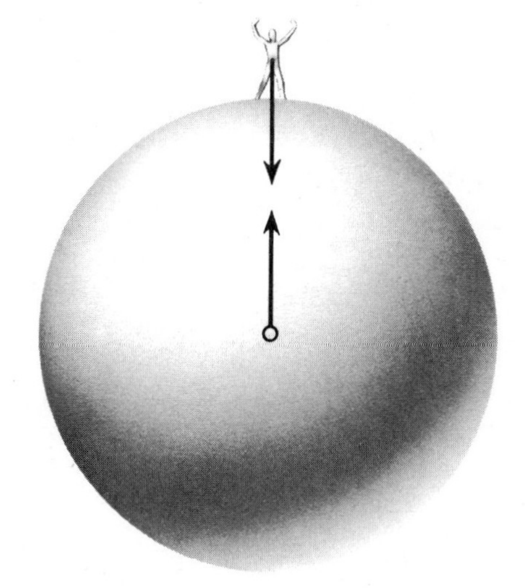

Figure 3.7
Attraction of a sphere is as though all its mass were concentrated at its center.

$$F = G \frac{m_1 m_2}{d^2},$$

where G, the constant of proportionality in the equation, is a number called the *constant of gravitation*. The value of G has to be determined by laboratory measurement of the attractive force between two material bodies. If metric units are used (grams for mass, centimeters for distance, and dynes for force), G has the numerical value 6.67×10^{-8}.[3]

ORBITAL MOTION

If an object is dropped toward the ground, at the end of one second it has accelerated to a speed of 980 cm/s. Its average speed during that second is 490 cm/s. Thus it drops, in one second, through a distance of 490 cm. The moon, which accelerates only 1/3600 as much, drops toward the earth only about 490/3600 cm in one second, or about 0.14 cm. In other words, as the moon moves forward in its orbit for one second and travels about 1 km it falls 0.14 cm toward the earth away from a straight-line path. However, because of the earth's curvature, the ground has fallen away under the moon by that same distance of 0.14 cm, so the moon is still the same distance from the earth. In this way the moon literally "falls around the earth." In the period of about one month it has "fallen" through one complete circuit of the earth and is back to its starting point (Figure 3.8).

We can easily understand orbital motion in terms of Newton's laws of motion and gravitation. At any given instant, the orbital speed of the moon would tend to carry it off in space in a straight line tangent to its orbit. The gravitational attraction between the earth

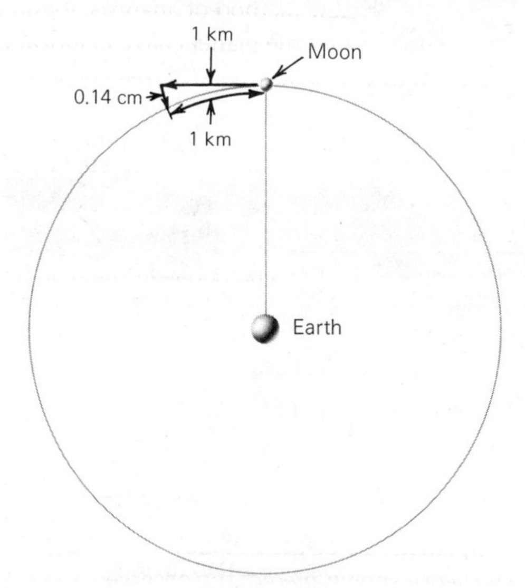

Figure 3.8
Velocity and acceleration of the moon.

[3] The notation 6.67×10^{-8} means 0.0000000667; see Appendix 3.1.

and the moon provides the proper centripetal force to accelerate the moon into its nearly circular path. Consequently, the moon continually falls toward the earth without getting any closer to it. The orbital motions of the planets about the sun are similarly explained.

The Fable of "Centrifugal Force"

Orbital motion is often incorrectly described as being a balance between two forces, the force of gravitation pulling the moon toward the earth, and an outward "centrifugal force" that keeps the moon from falling into the earth. This explanation, however, is misleading.

If one whirls a rock around over his head at the end of a string, he feels a "tug" in the string. This tug is the tension in the string. It is the result of the mutual pull between his hand and the stone. He provides, through the string, the centripetal force needed to accelerate the stone into a circular path. The stone tugs with a mutual force on the person's hand. However, there is no outside force pulling the stone away from him. If there were, and the string were to break, removing the central or centripetal force, the stone would move radially away from the person. Actually, if the string broke, the stone would fly off in a straight line in a direction *tangent* to its former circular path, continuing in the direction it was moving at the instant the string broke, in accord with Newton's first law of motion.

In short, there is no such thing as centrifugal (or outward) force in orbital motion. No force is pulling outward on the moon to balance the attraction between the moon and earth. If there were such a balance of forces, the moon would either not move at all or would move in a straight line, as predicted by Newton's first law. It is, in fact, just the *unbalanced* gravitational force that causes the moon to move in its nearly circular path around the earth. Of course, this discussion of the moon's motion applies equally to the planets' motions about the sun. (A "fictitious" centrifugal force, however, which arises from the introduction of a rotating coordinate system, is sometimes useful in the solution of more advanced problems in mechanics.)

There is, to be sure, a reaction to the earth's pull on the moon, as predicted by Newton's third law. The reaction to the earth's gravitational force on the moon is the moon's equal but opposite gravitational attraction for the earth. In other words, both pull mutually and centrally *toward* each other (there is no *outward* "reaction" force). The earth and moon suffer the same change of momentum as they revolve around each other, but the earth, being much more massive than the moon, accelerates far less.

Center of Mass

According to Newton's third law, the total momentum of an isolated system is conserved; that is, all changes of momentum within it are balanced. We can, therefore, define a point within the system that remains fixed (or moves uniformly) as if the entire mass of the system were concentrated at that point. It is called the *center of mass*. It can be shown that the center of mass of a complex body (which is a collection of point masses joined rigidly together) is that point about which the body balances when placed near a gravitating body; thus it is also often called the *center of gravity*.

The center of mass (or gravity) for two bodies is given a special name: the *barycenter*. It lies on a line connecting the centers of the bodies, and is so located that the distance of each body from that point is in inverse proportion to its mass.

The center of mass about which the earth and moon mutually revolve can be located by making careful observations of the other planets—especially of Mars—and, more recently, by tracking space probes. Small monthly periodic variations in the apparent motion of Mars result from the earth's monthly revolution about the barycenter. From the size of those variations we find that the barycenter is about 4700 km from the center of the earth, or about 1700 km below its surface. The moon is about 81 times as far from the barycenter and so is correspondingly less massive than the earth. Similarly, because of the sun's far greater mass, the center of mass of a system of revolving bodies consisting of the sun and a planet lies very close to the center of the sun—in most cases, within the sun's surface.

Newton's Derivation of Kepler's Laws

Kepler's laws of planetary motion are *empirical* laws; that is, they describe the way the planets are *observed* to behave. Kepler himself did not know of the more fundamental laws or relationships from which his three laws of planetary motion follow. On the other hand, Newton's laws of motion and gravitation were proposed by him as the basis of all mechanics.

Thus Newton was able to derive Kepler's laws from them.

Newton found that under the influence of their mutual gravitational forces, two objects should move about each other in paths that are some kinds of *conic sections*. Whether those orbits are hyperbolas or ellipses depends on the relative speeds of the bodies at a given separation; if the speed of one with respect to the other is great enough, the bodies will have too much energy to revolve in closed orbits, and will pass each other by on hyperbolic paths. Unless one body can be slowed down with respect to the other by an outside force (or by firing a retrorocket, as in a space vehicle), it can never be captured in a closed orbit by the other. On the other hand, if its speed does not have at least a critical value, its path will be a closed ellipse (a circle is theoretically possible). Then the two bodies never escape each other unless an outside force intervenes (or one body is accelerated by firing a rocket). The planets, of course, are permanently trapped in elliptical orbits around the sun; if they had hyperbolic orbits they would long since have been gone.

In any case, since the gravitational force between two bodies is directed centrally along a line connecting them, their angular momentum is conserved, and the law of areas (Kepler's second law) is obeyed—whether their orbits are closed ellipses or open hyperbolas.

Newton found that if two bodies revolve about each other in closed orbits, their period of revolution depends not only on the semimajor axes (sizes) of their orbits, as Kepler had found in discovering his third law, but also on their combined masses. The actual formula[4] Newton obtained is that the sum of the masses of the two bodies is proportional to the cube of the semimajor axis of the orbit of one with respect to the other divided by the square of their period of mutual revolution. The sun's mass is so great compared to that of any planet (even Jupiter has only 0.001 the mass of the sun) that the sum of the masses of the sun and any planet is almost the same, whichever planet is chosen. Thus Kepler could not have been aware that the masses mattered at all, and they did not appear in his formulation. However, Newton's precise version of Kepler's third law provides us with a means of calculating the masses of any pair of objects in mutual revolution—for example, a planet and its satellite (Chapter 8), or two stars in a double-star system (Chapter 14).

Artificial Satellites

An artificial satellite is a manmade object that is in orbit around the earth, but it is an astronomical body in its own right. If some of the artificial satellites that have been launched are temporary astronomical objects (as was Skylab), it is because they dip into the atmosphere of the earth during some portions of their revolutions. The friction of the air causes a satellite to lose energy so that eventually it spirals into the denser part of the atmosphere where friction usually heats it until it burns up completely, although some dense pieces may survive to ground. If an artificial satellite is launched so that its entire orbit is outside the earth's atmosphere, it will remain in orbit indefinitely as an astronomical body.

To illustrate how a satellite is launched, imagine a man on top of a high mountain, firing a rifle in a direction exactly parallel to the surface of the earth (Figure 3.9—adapted from a similar diagram by Newton). Imagine, further, that the friction of the air could be removed, and that all hindering objects, such as other mountains, buildings, and so on, are absent.

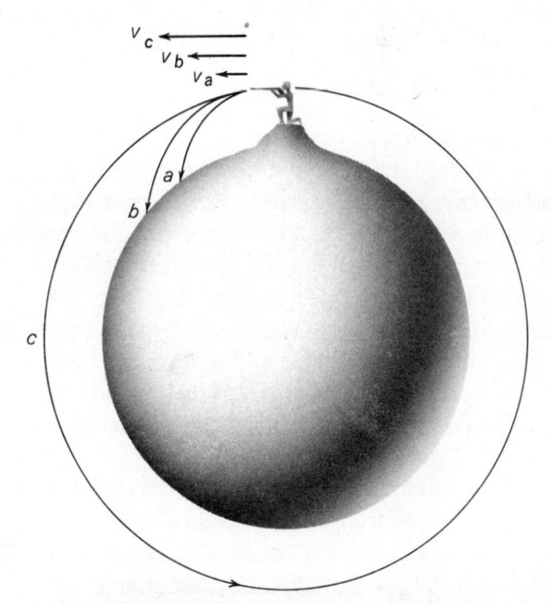

Figure 3.9
Firing a bullet into a satellite orbit.

[4] $m_1 + m_2 = \text{constant} \times a^3/P^2$, where m_1 and m_2 are the masses, a is the semimajor axis, and P the period.

Then the only force that acts on the bullet after it leaves the muzzle of the rifle is the gravitational force between the bullet and earth.

If the bullet is fired with muzzle velocity v_a, it will continue to have that forward speed, but meanwhile the gravitational force acting upon it will accelerate it downward so that it strikes the ground at *a*. However, if it is given a higher muzzle velocity v_b, its higher forward speed will carry it farther before it hits the ground, for, regardless of its forward speed, its downward gravitational acceleration is the same. Thus this faster-moving bullet will strike the ground at *b*. If the bullet is given a high enough muzzle velocity v_c, as it accelerates toward the ground, the curved surface of the earth will cause the ground to tip out from under it so that it remains the same distance above the ground, and "falls around" the earth in a complete circle. This is another way of saying that at a critical speed v_c the gravitational force between the bullet and earth is just sufficient to produce the centripetal acceleration needed for a circular orbit about the earth. The speed v_c, the *circular satellite velocity* at the surface of the earth, is about 8 km/s (5 mi/s).

Novelist Jules Verne anticipated earth satellites long ago. In one of his stories an enemy force was planning to bomb a city with a gigantic cannon ball. However, the cannon ball was propelled with too great a speed—in fact, the circular satellite velocity—so it passed harmlessly over the city and on into a circular orbit around the earth.

Possible Satellite Orbits

Suppose that a missile is shot up to an altitude of a few hundred miles, then turned so that it is moving horizontally, and finally given a forward horizontal thrust. It will proceed in an orbit the size and shape of which depend critically on the exact direction and speed of the missile at the instant of its "burnout," that is, the instant when the thrust supplied by its fuel is shut off. For simplicity, suppose that it is moving exactly horizontally, or parallel to the ground, at burnout. The possible kinds of orbits it can enter are shown in Figure 3.10.

If the missile's burnout speed is less than the circular satellite velocity, its orbit will be an ellipse, with the center of the earth at one focus of the ellipse. The *apogee* point of the orbit, that point that is *farthest* from the center of the earth, will be the point of burnout; the *perigee* point (closest approach to the center of the earth) will be halfway around the orbit from burnout.

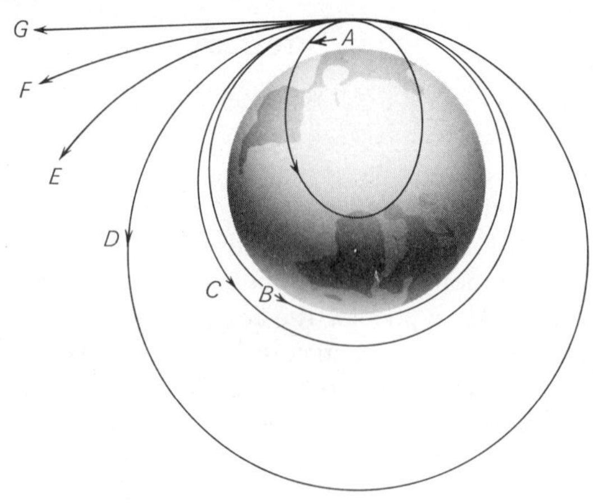

Figure 3.10
Various satellite orbits that result from different burnout velocities but that are all parallel to the earth's surface.

If the burnout speed is substantially below the circular satellite velocity, most of its elliptical orbit will lie beneath the surface of the earth (orbit *A*), where, of course, the satellite cannot travel; consequently, it will traverse only a small section of its orbit before colliding with the surface of the earth (or more likely, burning up in the dense lower atmosphere of the earth). If the burnout speed is just slightly below the circular satellite velocity, the missile may clear the surface of the earth (orbit *B*), although its orbit will probably lie too low in the atmosphere for the satellite to be long-lived.

If the burnout speed were exactly the circular satellite velocity, a circular orbit centered on the center of the earth would result (orbit *C*). It is extraordinarily unlikely that a missile could be given so accurate a direction and speed that a perfectly circular orbit could be achieved. A slightly greater burnout speed will produce an elliptical orbit with *perigee* at burnout point and apogee halfway around the orbit (orbit *D*).

A burnout speed equal to the *velocity of escape* from the earth's surface, that is, the *parabolic velocity* (about 11 km/s, or 7 mi/s), will put the missile into a parabolic orbit that will just enable the vehicle to escape from the earth into space (orbit *E*). A still higher burnout speed will produce a hyperbolic orbit in which the missile escapes the earth with energy to spare (orbit *F*). The higher the burnout speed, the nearer the orbit will be to a straight line (orbit *G*).

Interplanetary Probes

We have now learned the principles of space travel. Rockets, once they have left the earth, are astronomical bodies. They obey the same laws of celestial mechanics as the planets and natural and artificial satellites. In other words, rockets or space probes travel in orbits. If the space vehicles carry auxiliary rocket engines and extra fuel, it may be possible to alter their orbits at will, but the principles remain the same.

We shall illustrate one particular kind of space trajectory by showing one of the many possible ways to reach the planet Mars. The orbit to Mars we show is that which requires the expenditure of the least energy as the rocket leaves the earth and is thus the most economical of fuel. The orbits of the highly successful United States *Mariner*, *Pioneer*, and *Viking* Venus and Mars probes, and of the similar Soviet probes, were nearly of this type.

Suppose, for simplicity, that the orbits of Earth and Mars are circles centered on the sun (when the slight ellipticity of planetary orbits is taken into account, the problem is similar but slightly more complicated). The least-energy orbit that will take us to Mars is an ellipse tangent to the earth's orbit at the space vehicle's *perihelion* (closest approach to the sun) and tangent to the orbit of Mars at the vehicle's *aphelion* (farthest point from the sun) (Figure 3.11).

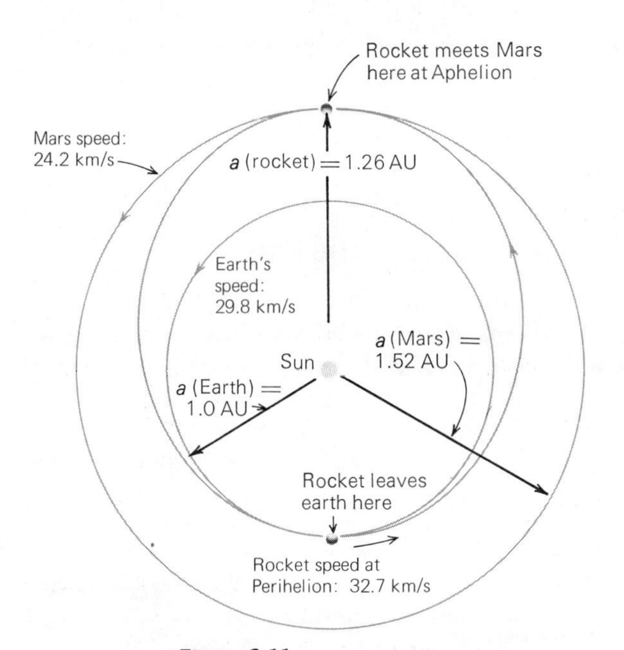

Figure 3.11
Least-energy orbit to Mars.

The earth is traveling around the sun at the right speed for a circular orbit. For us on the earth to enter the elliptical orbit to Mars, we must achieve a speed, in the same direction as the earth is moving, that is slightly greater than the earth's circular velocity (which is about 29.8 km/s). The required speed turns out to be slightly under 33 km/s. Since the earth is already moving 29.8 km/s, we need to leave the earth with the proper speed and direction so that when we are far enough from it that its gravitational influence on us is negligible compared to the sun's, we are still moving in the same direction as the earth with a speed relative to it of about 3 km/s.

We have now entered an orbit that will carry us out to the orbit of Mars. The time required for the trip can be found from Kepler's third law, because our spaceship is a planet. The period required to traverse the entire orbit is 1.41 years. The time required to reach the aphelion point (Mars' orbit) is half of this, or about 8½ months. The trip will have to be planned very carefully so that when we reach the aphelion point of the least-energy orbit, Mars will be there at the same time. Space probes generally carry rocket engines that can be activated by radio command from the earth, so that minor corrections in their trajectories can be made, as necessary, for them to achieve their missions.

From the foregoing discussion it is obvious that the earth and the planet to be visited must be at a critical configuration at the time of launch, in order that the space vehicle meet the planet at the other end of the vehicle's heliocentric orbit. In practice it is not necessary, and is seldom feasible, to launch the rocket at exactly the proper instant to achieve the least-energy orbit. However, there is a short range of time (typically a few weeks) during which a *nearly* least-energy orbit can be achieved. The length of this time period, called a "window" in space jargon, depends on the thrust capabilities of the available rockets (that is, on how much energy, above the least possible needed, can be supplied by the rocket). "Windows" for Mars journeys occur at intervals of about 780 days; those for Venus trips at intervals of about 584 days.

MORE THAN TWO BODIES AND NONSPHERICAL BODIES

Until now we have considered the sun and a planet, or a planet and one of its satellites, as a pair of mutually revolving bodies. Actually, the planets (and different satellites of a planet) exert gravitational forces upon each other as well. These interplanetary attractions cause slight variations from the orbits that would be expected if the gravitational forces between planets were neglected. Unfortunately, the problem of treating the motion of a body that is under the gravitational influence of two or more other bodies, called the *multi-body* or *n-body problem*, is very complicated.

If the exact position of each other body is specified at any given instant, we can calculate the combined gravitational effect of the entire ensemble on any one member of the group. Knowing the force on the body in question, we can find how it will accelerate; a knowledge of its initial velocity, therefore, is enough to calculate how it will move in the next instant of time, and thus to follow its motion. However, the problem is complicated by the fact that the gravitational acceleration of one body depends on the positions of all the other bodies in the system. Since they, in turn, are accelerated by all the members of the system, we must simultaneously calculate the acceleration of each particle produced by the combination of the gravitational attractions of all the others to follow the motions of all of them, and hence of any one. Such extremely complex calculations have been carried out, with electronic computers, to follow the evolution of hypothetical star clusters of up to at least 1000 members.

Although computations of the type just described can be carried out, in principle, to study the motion of any one member of a group or cluster of bodies, it is not possible to write an equation that will describe the trajectory (or orbit) of that body for all time, as it is in the two-body problem (in which the orbits are always conic sections). Consequently, the *n*-body problem is often said to have no solution. Actually, by numerical calculation many problems can be solved to the desired precision, although for some problems of importance, such as the evolution of the solar system, even the biggest electronic computer is not adequate. In principle, however, the *n*-body problem is not solvable only in the sense that a single equation does not describe the motion.

Fortunately, the many-body problem can be solved rather accurately when a given body feels predominantly the gravitational force of one other mass. The motion of a planet around the sun, for example, is determined mainly by the gravitational force between it and the sun, the force between it and any of the other planets being very small in comparison. Thus the influences of the other planets can be regarded as small corrections to be applied to the two-body solution; these corrections are called *perturbations*.

The complex mathematical techniques required to calculate these perturbative corrections posed a challenge to astronomy in the late eighteenth century. Consequently some of the greatest minds in physics, mathematics, and astronomy—Karl Friedrich Gauss, Marquis Pierre Simon de Laplace, and, a century later, Simon Newcomb—turned their talents to this developing field of mathematical astronomy. By the middle of the nineteenth century, many intricate problems could be handled almost routinely. For example, analysis of the perturbations of the planet Uranus, discovered in 1781, led to the prediction of the existence of an eighth planet, Neptune (Chapter 10). The telescopic observation of Neptune in 1846, within a degree of the predicted position, was an outstanding triumph for gravitational theory.

The application of perturbation theory is not restricted to isolated objects in the solar system. Bodies with spherical symmetry act, gravitationally, as point masses, for which the gravitational influences are easily calculated. In nature, however, most bodies are not exactly spherical, and the simple two-body theory does not give precise results. If the shape of a body deviates only slightly from a sphere, we usually approximate its gravitational influence by that produced by a point mass and treat the small effects of its asphericity as perturbations. A common cause of the deformation of a star or planet from a perfect sphere is its rotation. Rapidly rotating planets, such as Jupiter, are noticeably flattened. The rotational flattening of the earth is slight but is important.

The Shape of the Earth

Because of the earth's rotation, the inertia of its constituent parts tends to make them fly off tangentially into space. Therefore, each particle on and in the earth must be undergoing a constant *centripetal* acceleration to keep it in place. Of course, it is the earth's gravitation that provides this acceleration. Only a small part of the gravitational force on an object at the earth's surface is required to provide the centripetal acceleration that keeps it on the ground; the remainder is the object's actual *weight*. In other words, the rotation of the earth slightly reduces the weights of objects on the earth's surface (and the interior parts of the earth itself as well), because some part of the gravi-

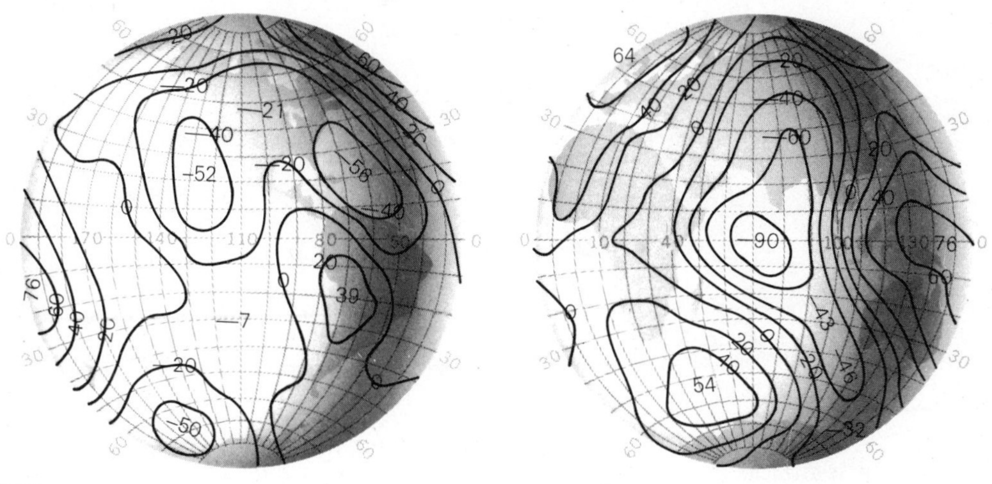

Figure 3.12
Height, in meters, of the mean surface of the earth above (positive numbers) and below (negative numbers) the surface of an oblate spheroid with a flattening of 1 part in 298.25. *(Courtesy of William Kaula, UCLA)*

tational force is used to provide the centripetal acceleration.

On the other hand, a body exactly on the axis of the earth has zero speed and suffers no centripetal acceleration. Therefore, at the poles of the earth (they are the ends of its rotational axis) the full force of gravity goes into the weight of objects. Thus the force of gravity varies slightly over the surface of the earth, with the result that it has the shape of an *oblate spheroid*, a slightly flattened sphere that has an elliptical cross section. That is, the earth is flattened at the poles and bulged out in the equatorial regions. The actual oblateness of the earth is small. Its diameter from pole to pole is only 43 km (27 mi) less than through the equator, about 1 part in 298.

The equatorial bulge of the earth is responsible for a deformation in the earth's gravitational field from that which would be produced by a point mass or spherically symmetrical earth. These deformations are especially important near the surface of the earth and produce conspicuous perturbations in the orbits of low-altitude earth satellites. Since the motions of the earth satellites depend on the precise shape of the earth, careful studies of perturbations on satellite orbits enable us to derive the earth's shape rather accurately.

Tides

Early in history it was realized that tides are related to the moon, because the daily delay in high tide ("high water") is the same as the daily delay in successive transits of the moon across the local meridian. A satisfactory explanation of the tides, however, awaited the theory of gravitation, supplied by Newton.

For the moment, we ignore the flattening of the earth due to its rotation; the tidal effects described below are actually superimposed on the earth's oblateness. Our planet can be regarded as being composed of a large number of particles, each of unit mass, all bound together by their mutual gravitational attraction and cohesive forces. The gravitational forces exerted by the moon at several arbitrarily selected places in the earth are illustrated in Figure 3.13. These forces differ slightly from each other because of the earth's finite size; all parts are not equally distant from the moon, nor are they all in exactly the same direction from the moon. If the earth retained a perfectly spherical shape, the resultant of all these forces would be that of the force on a point mass, equal to the mass of the earth, and located at the earth's center. This is approximately true, because the earth is nearly spherical, and it is this resultant force on the earth that causes it to accelerate each month in an elliptical orbit about the barycenter of the earth-moon system.

The earth, however, is not *perfectly* rigid. Consequently, the differences between the forces of the moon's attraction on different parts of the earth (called *differential* forces) cause the earth to distort slightly. The side of the earth nearest the moon is attracted toward the moon more strongly than is the center of

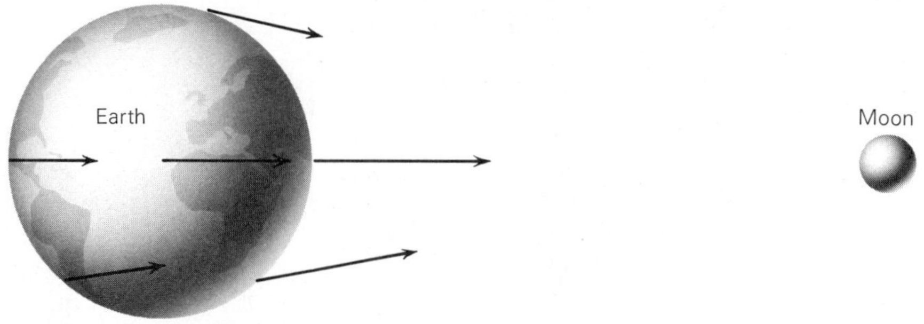

Figure 3.13
The moon's attraction on different parts of the earth.

the earth, which, in turn, is attracted more strongly than is the side of the earth opposite the moon. Thus, the differential forces tend to "stretch" the earth slightly into a *prolate spheroid* (like a football) with its long diameter pointed toward the moon.

If the earth were fluid, like water, it would distort until the moon's differential forces over different parts of its surface came into equilibrium with the earth's own gravitational forces pulling it together. Calculations show that in this case the earth would distort from a sphere by amounts ranging up to nearly one meter (two or three feet). Measurements have been made to investigate the actual deformation of the earth, and it is found that the solid earth does distort, as would a liquid, but only about one-third as much because of the high rigidity of the earth's interior. The tidal distortion of the solid earth amounts at its greatest to only about 20 cm (8 in.).

Thus the earth does not distort enough to achieve an equilibrium shape, and the moon's differential gravitational forces are not completely balanced by the earth's own gravitation. Hence objects at its surface experience tiny horizontal tugs, tending to make them slide about. These so-called *tide-raising* forces are too insignificant to notice or to affect solid objects or crustal rocks. But they do affect the waters in the oceans.

The actual accelerations of the ocean waters caused by the tide-raising forces are very small. These forces, acting over a number of hours, however, produce motions of the water that result in measurable tidal bulges in the oceans. Water on the lunar side of the earth is drawn toward the sublunar point (the point on the earth where the moon appears in the zenith), piling up water to greater depths on that side of the earth, with the greatest depths at the sublunar point. On the opposite side of the earth, water is drawn in the *opposite* direction, producing a tidal bulge on the side of the earth opposite the moon (Figure 3.14).

Note that the tidal bulges in the oceans do not result from the moon compressing or expanding the water, nor from the moon lifting the water "away from the earth." Rather, they result from an actual flow of water over the earth's surface, toward the regions below and opposite the moon, causing the water to pile up to greater depths at those places.

The tidal bulge on the side of the earth *opposite* the moon often seems mysterious to students who picture the tides as being formed by the moon "lifting the water away from the earth." What actually happens, of course, is that the differential gravitational force of the moon on the earth tends to stretch the earth, elongating it slightly toward the moon. The solid earth distorts slightly, but because of its high rigidity, not enough to reach complete equilibrium with the tidal forces. Consequently, the ocean, moving freely over the earth's surface, flows in such a way as

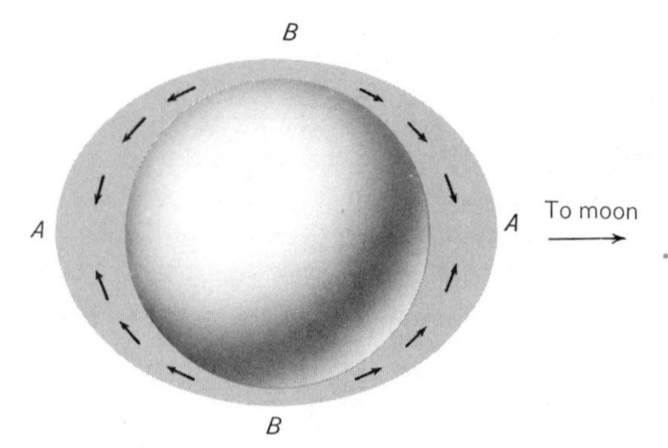

Figure 3.14
Tidal bulges in the "ideal" oceans.

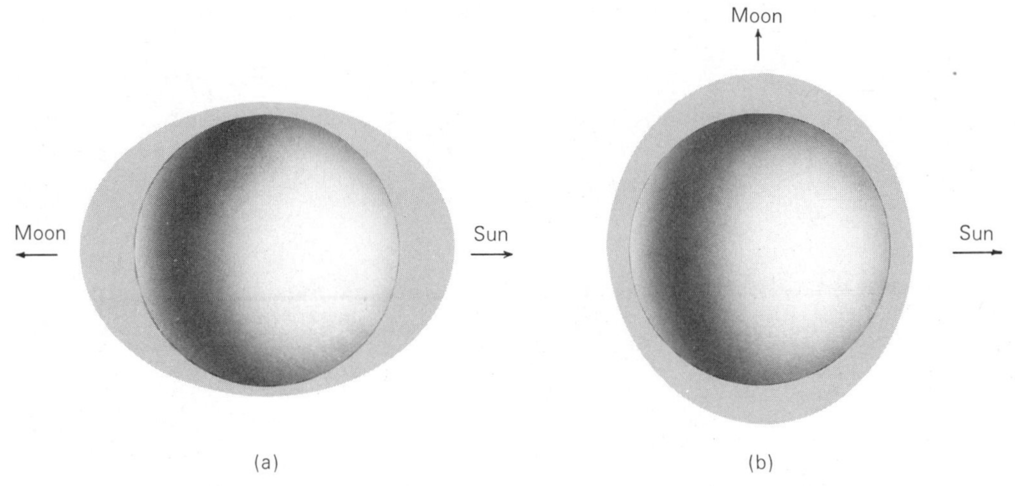

Figure 3.15
(a) Spring tides; *(b)* neap tides.

to increase the elongation and piles up at points under and opposite the moon.

In the idealized picture described above, not actually realized even in the largest oceans, the tides would cause the depths of the ocean to range through only a few feet. The rotation of the earth would carry an observer at any given place alternately into regions of deeper and shallower water. As he was being carried toward the regions under or opposite the moon where the water was deepest, he would say, "the tide is coming in"; when carried away from those regions, he would say, "the tide is going out." During a day, he would be carried through two tidal bulges (one on each side of the earth) and so would experience two "high tides" and two "low tides."

The sun also produces tides on the earth, although the sun is less than half as effective a tide-raising agent as the moon. Actually, the gravitational attraction between the sun and the earth is about 150 times as great as that between the earth and the moon. However, the sun is so distant that it attracts all parts of the earth with almost equal strength. The moon, on the other hand, is close enough for its attraction on the near side of the earth to be substantially greater than its attraction on the far side. In other words, its *differential gravitational pull* on the earth is greater than the sun's, even though its total gravitational attraction is less.

If there were no moon, the tides produced by the sun would be all we would experience, and the tides would be less than half as great as those we now have. The moon's tides, therefore, dominate. On the other hand, when the sun and moon are lined up, that is, at new moon or full moon, the tides produced by the sun and moon reinforce each other and are greater than normal. These are called *spring tides*. Spring tides (which have nothing to do with spring) are approximately the same, whether at new moon or full moon, because tidal bulges occur on both sides of the earth—the side *toward* the moon (or sun) and the side *away* from the moon (or sun).

In contrast, when the moon is at first quarter or last quarter (at right angles to the sun's direction), the tides produced by the sun partially cancel the tides of the moon, and the tides are lower than usual. These are *neap tides*. Spring and neap tides are illustrated in Figure 3.15.

The "simple" theory of tides, described in the preceding paragraphs, would be sufficient if the earth were completely surrounded by very deep oceans, and if it rotated very slowly. However, the presence of land masses stopping the flow of water, the friction in the oceans and between oceans and the ocean floors, the rotation of the earth, the variable depth of the ocean, winds, and so on, all complicate the picture.

The earth's rapid rotation causes the tide-raising forces within a given mass of water to vary too rapidly for the water to adjust completely to them. These forces, however, recurring periodically, set up oscillations, making the oceans slosh back and forth in their basins, so that the water over a large area rises and lowers in step. Consequently, the highest water does not necessarily occur when the moon is highest in the sky (or lowest below the horizon), but rather when the

Figure 3.16
The Bay of Fundy at high and low tides. *(Tourism—New Brunswick)*

oscillations of the ocean, produced by the tidal forces acting upon it, pile up the water to its greatest depth at that location. The latter depends critically upon the shape and depth of the adjacent ocean basin.

Thus, the times and the heights of high tide vary considerably from place to place on the earth. The tidal flow of waters over shallow parts of the ocean, and into irregular coastal regions, causes the tides there to range generally far more than the meter or less that they do in the deep oceans. In particular, where the tidal flow of water is funneled shoreward by a V-shaped inlet that narrows back away from the sea, the difference between high and low water may be especially large. Such a place is the Bay of Fundy between New Brunswick and Nova Scotia in Southeast Canada, where the tidal range sometimes exceeds 15 m (about 50 ft).

Tides also occur in the atmosphere. These atmospheric tides are complicated by weather phenomena, but in principle they are the same as earth and ocean tides. They are, obviously, of importance to meteorologists.

Some people have speculated that lunar tidal forces acting on the fluids in the human body might affect the physical and mental states of men and women. The moon's tidal force on the oceans, to be sure, is appreciable, but in that case the tidal force is acting over the entire 8000-mile diameter of the earth. The differential pull of the moon on different parts of an individual person, on the other hand, is extraordinarily slight; the moon's tides can affect the weights of our bodily fluids by only about one part in 3×10^{13} (or 30 trillion). This book is exerting a tidal force on you that is thousands of times as great as that of the moon on you.

Precession

The plane of the earth's equator, and thus of its equatorial bulge, is inclined at about 23½° to the plane of the ecliptic (Chapter 2), which, in turn, is inclined at 5° to the plane of the moon's orbit. The differential gravitational forces of the sun and moon upon the earth not only cause the tides but also attempt to pull the equatorial bulge of the earth into coincidence with the ecliptic.

The latter pull is illustrated in Figure 3.17. The solid arrows represent the attraction of the moon on representative parts of the earth. The part of the earth's equatorial bulge nearest the moon is pulled more strongly than the part farthest from the moon, and the earth's center is pulled with an intermediate force. The dashed arrows show the differential forces with respect to the earth's center. Note how they tend not only to "stretch" the earth toward the moon, but also to pull the equatorial bulge toward the plane of the ecliptic. The differential force of the sun, although

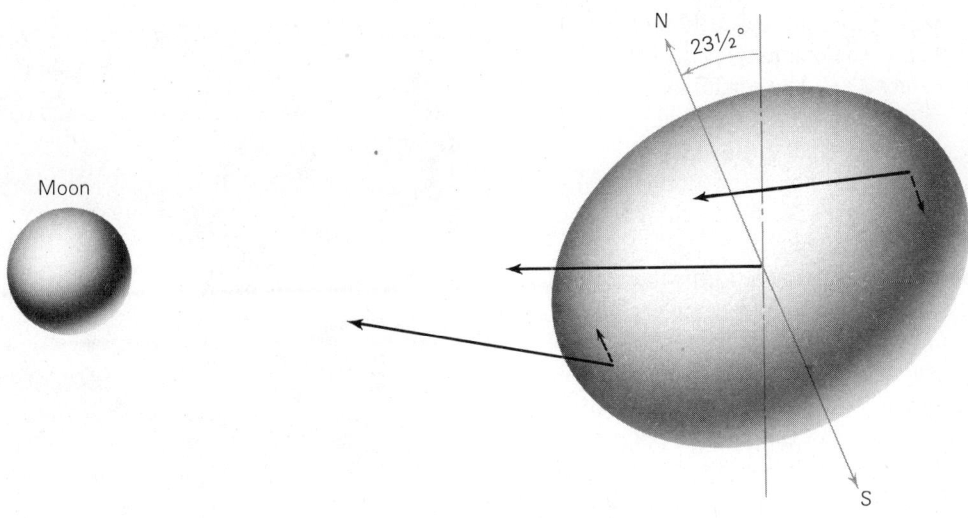

Figure 3.17
Differential force of the moon on the oblate earth tends to "erect" its axis.

less than half as effective, does the same thing. Thus, the gravitational attractions of the sun and the moon on the earth act in such a way as to attempt to *change the direction of the earth's axis of rotation,* so that it would stand perpendicular to the orbital plane of the earth. To understand what actually takes place, we must digress for a moment to consider what happens when a similar force acts upon a top or gyroscope.

The axis of the top (a simple form of gyroscope) pictured in Figure 3.18 is not perfectly vertical. Thus its weight (the force of gravitation between it and the earth) tends to topple it over. We know from watching a top spin, however, that until the spin of the top is slowed down by friction, the axis does not change its angle of inclination to the vertical (or to the floor), but rather describes a conical motion (a cone about the vertical line passing through the pivot point of the top). This conical motion of the top's axis is called *precession*. It may at first seem mysterious, but when a detailed analysis is carried out on the effect of forces on all parts of the spinning rigid top, precession is seen to follow naturally from Newton's simple laws.

The differential gravitational forces of the sun and moon on the earth's equatorial bulge tend to pull the earth's axis into a direction approximately perpendicular to the ecliptic plane. Like a top, however, the

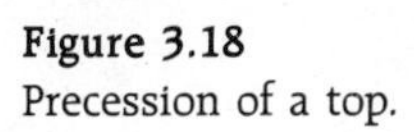
Figure 3.18
Precession of a top.

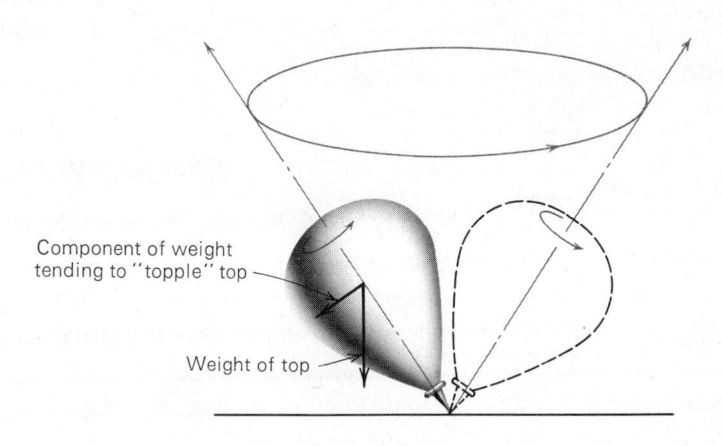

earth's axis does not yield in the direction of these forces, but precesses. The obliquity of the ecliptic remains approximately 23½°. The earth's axis slides along the surface of an imaginary cone, perpendicular to the ecliptic, and with a half-angle at its apex of 23½° (see Figure 3.19). The precessional motion is exceedingly slow; one complete cycle of the axis about the cone requires about 26,000 years.

Precession does not affect the cardinal directions on the earth nor the positions of geographical places that are measured with respect to the earth's rotational axis, but only the orientation of the axis with respect to the celestial sphere. It does, therefore, affect the positions among the stars of the celestial poles, those points where extensions of the earth's axis intersect the celestial sphere. In the twentieth century, for example, the north celestial pole is very near Polaris. This was not always so. In the course of 26,000 years, the north celestial pole will move on the celestial sphere approximately along a circle of about 23½° radius, centered on the pole of the ecliptic (where the perpendicular to the earth's orbit intersects the celestial sphere). This motion of the pole is shown in Figure 3.20. In about 12,000 years, the celestial pole will be fairly close to the bright star Vega. It was by noting the very gradual changes in the positions of stars with respect to the celestial poles that Hipparchus discovered precession in the second century B.C. (Chapter 2).

This chapter, of course, has hardly exhausted the subject of gravitation. In future chapters we shall encounter some of its remarkable effects. We shall also see (Chapter 18) how Einstein's general theory of relativity provides a whole new way of looking at gravitation.

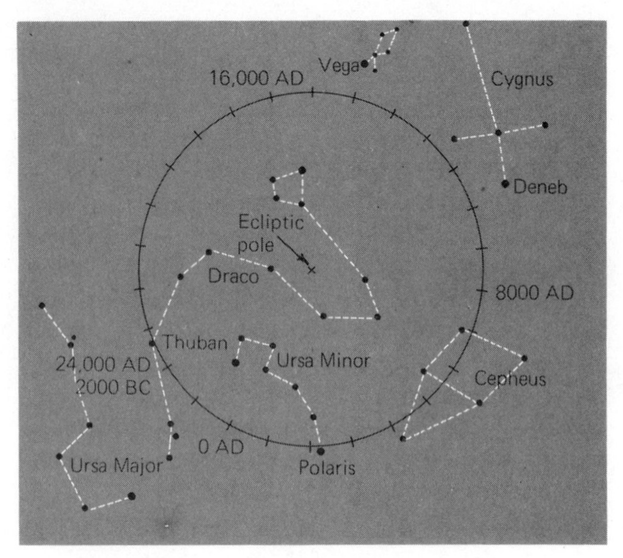

Figure 3.20
Precessional path of the north celestial pole among the northern stars.

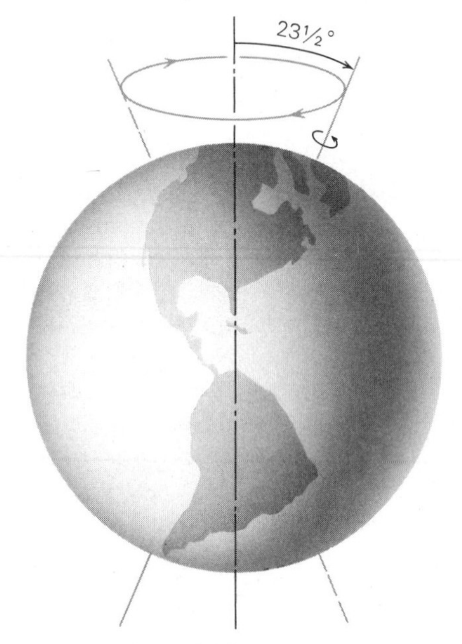

Figure 3.19
Precession of the earth.

SUMMARY REVIEW

Tycho Brahe and his observations of the planets

Kepler, and Kepler's three laws of planetary motion: 1. Orbits are ellipses; 2. Law of areas; 3. Harmonic law

Conic sections: ellipse; hyperbola; parabola; semimajor axis; eccentricity

Galileo: *Dialogue on the Two Great World Systems; Sidereal Messenger;* phases of Venus; satellites of Jupiter

Newton, and Newton's laws of motion and of universal gravitation: 1. Inertia; 2. Force; 3. Action and reaction; mass; density; volume; angular momentum; acceleration in a circular orbit

Orbital motion: centripetal and centrifugal force; center of mass; artificial satellites; perigee and apogee; interplanetary probes; n-body problem

Perturbations; shape of the earth; oblateness; tides; differential force; spring and neap tides; precession

EXERCISES

1. A friend tells you that the earth cannot be rotating, as the scientists claim, because if a point on the equator were being carried eastward at about 1000 mi/hr, as claimed, a baseball pitcher could throw a ball straight up in the air, and the earth and pitcher would move to the east out from under the ball, which would land some distance behind (or to the west) of the pitcher. How might you straighten out this friend of yours?

2. Suppose you weigh 100 pounds at sea level. What would you weigh if a) the earth were suddenly to double in radius, its mass remaining the same? b) the earth were suddenly to double in mass, its radius remaining the same?

3. Suppose you tie a string to a stone and whirl it around over your head at the end of the string. Now you hold up a razor blade directly in front of your face so that you cut the string just as the stone is out at the end of the string directly in front of you. Ignore the pull of the earth's gravity and the friction of the air.

(a) Just after the string is cut, what forces are acting on the stone?

(b) Which way does the stone move after the string is cut? Why?

4. Why is it nonsense to speak of "force of forward motion"?

5. What is the major axis of a circle of radius 13 cm?

6. Suppose you observe a double star system, and measure the size of the orbit of one of the stars about the other and also their period of mutual revolution. Can you suggest a way in which you might determine the masses of the stars?

7. What is the distance of a planet from the sun (in AU) if the planet has a period of eight years?

8. Can a quantity of lead be more massive than an inflated balloon? Can it be less massive? Can it have a greater volume? a smaller volume? a greater density? a smaller density? Give examples.

9. How many accelerators are there on a standard passenger car? Explain?

10. What does a pan balance measure? a spring balance?

11. Two bodies 5 m apart have masses of 10 g and 40 g, respectively. How far is their mutual center of mass from the 10-g body?

12. Would it be feasible to launch an artificial satellite that was intended to revolve about the earth in an orbit of semimajor axis 3 AU? Why?

13. A ballistic missile is a missile that is given an initial thrust and then allowed to coast to its target. Describe how it can be regarded as an earth satellite. What is its orbit like? Where are the foci? What are some complications that must be taken into account during launch?

14. Describe how we would have to launch a rocket in order to make it hit the sun. After it is far from the earth, but still essentially the same distance from the sun that the earth is, what are its speed and direction with respect to the earth? Why should it be difficult to send a very large payload to the sun?

15. Strictly speaking, should it be a 24-hour period during which there are two high tides? If not, what should the period be?

16. Does a bicycle offer another example of precession? Explain. *Hint:* Consider how a rider can steer by leaning to one side.

CHAPTER 4

Simon Newcomb (1835–1909) was the great American astronomer who laid a foundation of precise positional astronomy based on his measurements of the motions of the earth. *(U. S. Naval Observatory)*

"BUT NEVERTHELESS, IT MOVES!"

It has been written that Galileo, following the retraction of his belief in the rotation and revolution of the earth, as ordered by the Roman Inquisition, said under his breath, "But nevertheless, it moves." The story itself is apocryphal, but of course Galileo knew that the earth is in motion. In fact, the earth has many motions, most of them unsuspected by him. In this chapter we consider some of the effects of the earth's motion: the seasons, time, and date. We postpone to Chapter 9 a description of the earth's properties as a planet.

The most important motions of the earth are its *rotation* and its *revolution*. In astronomy "rotation" means a turning about an axis through a body, and "revolution" means a motion about an external body; the earth rotates on its axis and revolves about the sun.

EARTH AND SKY

Locating Places on the Earth

We denote positions of places on the earth by a system of coordinates on the earth's surface. The earth's axis of rotation, which fixes the locations of its North and South Poles, is the basis for such a system.

A *great circle* is any circle on the surface of a sphere whose center is at the center of the sphere. The earth's *equator* is a great circle on the earth's surface halfway between the North and South Poles. We can also imagine a series of great circles that pass *through* the North and South Poles. These circles are called *meridians;* they intersect the equator at right angles.

A meridian can be imagined passing through an arbitrary point on the surface of the earth (see Figure 4.1). This meridian specifies the east-west location of that place. The *longitude* of the place is the number of degrees, minutes, and seconds of arc along the equator between the meridian passing through the place and the one passing through Greenwich, England, the site of the old Royal Observatory. Longitudes are measured either to the east or west of the Greenwich meridian from 0° to 180°. (The convention of referring longitudes to the Greenwich meridian is of course arbitrary.) As an example, the longitude of the benchmark in the clock house of the Naval Observatory in Washington, D.C., is 77°03′56″.7 W. Note in Figure 4.1 that the number of degrees along the equator between the

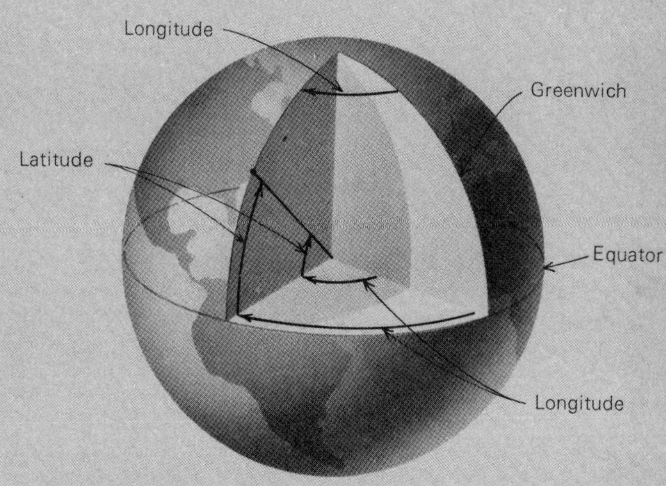

Figure 4.1
Latitude and longitude of Washington, D.C.

meridians of Greenwich and Washington is also the angle at which the planes of those two meridians intersect at the earth's axis.

The *latitude* of a place is the number of degrees, minutes, and seconds of arc measured along its meridian from the equator to the place. Latitudes are measured either to the north or south of the equator from 0 to 90°. As an example, the latitude of the above-mentioned benchmark is 38°55′14″.0 N. Note that the latitude of Washington is also the angular distance between it and the equator as seen from the center of the earth.[1]

Locating Places in the Sky

In denoting positions of objects in the sky, it is often convenient to make use of the fictitious *celestial sphere*, a concept, we recall, that many early peoples accepted literally. We saw in Chapter 2 that the sky appears to rotate about points in line with the North and South Poles of the earth—the *north celestial pole* and the *south celestial pole*. Halfway between the celestial poles, and thus 90° from each, is the *celestial equator*, a great circle on the celestial sphere that is in the same plane as the earth's equator; it would appear to pass directly overhead for a person on the equator of the earth. Great circles passing through the celestial poles and intersecting the celestial equator at right angles (analogous to meridians on the earth) are called *hour circles*.

Now the celestial equator and hour circles are fixed on the celestial sphere and rotate with it. On the other hand, an observer's overhead point, or *zenith* (defined as opposite the direction of a plumb bob) and his *horizon* (90° from his zenith) are fixed with respect to *him*, and stay put in his sky. Also at rest with respect to a stationary observer is his *celestial meridian* (or, simply, *meridian*). It is the great circle passing through the celestial poles and the zenith (and also through the point straight down, opposite the zenith, called the *nadir*), dividing the east half of the sky from the west. It coincides with the projection of his terrestrial meridian, as seen from the earth's center, onto the celestial sphere. The celestial meridian intercepts the horizon at the *north* and *south* points. Halfway between these north and south points on the horizon are the *east* and *west* points.

As the earth turns, carrying the observer, he and his terrestrial meridian move under the celestial sphere; therefore, his celestial meridian sweeps continuously around the sky. To us on the earth, though, it appears as if the sky is doing the turning, carrying the sun and stars around, so that they pass our "stationary" meridian each day.

It helps to visualize these circles in the sky if we imagine that the earth is a hollow transparent spherical shell with the terrestrial coordinates (latitude and longitude) painted on it. Then we imagine ourselves at the center of the earth, looking out through its transparent surface to the sky. The terrestrial poles, equator, and meridians will be superimposed upon the celestial ones.

There are several systems of coordinates, analogous to latitude and longitude, in common use by astronomers to denote positions of objects on the rotating celestial sphere. The interested reader will find them defined in Appendix 7. The most important such system is that of *right ascension* and *declination*, regularly used in star catalogs.

The stars appear fixed on the celestial sphere, and the planets are nearly fixed for short times, but the daily rotation of the celestial sphere makes the positions of objects change rapidly in the sky with respect to one's meridian and horizon. The ancient astrologers defined the houses (Chapter 2) to indicate the approximate temporary locations of planets with respect to the horizon. The most common modern system is that of *altitude* and *azimuth*. Great circles passing through the zenith (*vertical circles*) intersect the horizon at right angles. Imagine a vertical circle through a particular star (Figure 4.2). The *altitude* of that star is the number of degrees along this circle from the horizon up to the star. It is also the angular "height" of the star as seen by the observer.

The *azimuth* is the number of degrees along the horizon to the vertical circle of the star from the north point. It is measured to the east (clockwise to one looking down from the sky) along the horizon from 0° to 360°.

---

[1] Strictly, this is *geocentric* latitude. Because of the earth's oblate shape, there are several ways to define latitude. The *geodetic* (or *geographical*) latitude commonly used is defined as the angle between the equatorial plane and the perpendicular to the mean "sea-level" surface of the earth at the place in question. It may differ by several minutes from geocentric latitude. *Astronomical* latitude and longitude, obtained directly from astronomical observations, may differ from geodetic latitude and longitude by a few seconds of arc because of the deflection of the plumb bob by mountains or other crustal irregularities.

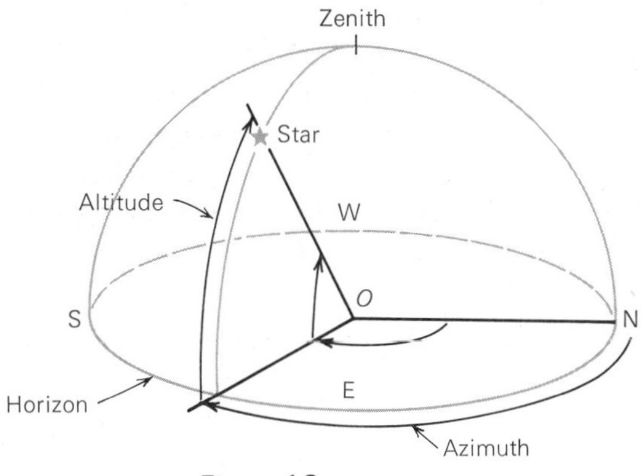

Figure 4.2
Altitude and azimuth.

The Orientation of the Celestial Sphere

It is easy to determine the orientation of the celestial sphere with respect to the zenith and horizon of a particular observer on the earth. At the North (or South) Pole, the problem is very simple indeed. The north celestial pole, directly over the earth's North Pole, appears at the zenith. The celestial equator, 90° from the celestial poles, lies along the horizon. As the earth rotates, the sky turns about the zenith, and the stars neither rise nor set, but circle parallel to the horizon. Only that half of the sky that is north of the celestial equator is ever visible to an observer at the North Pole. Similarly, an observer at the South Pole would see only the southern half of the sky (see Figure 4.3). The daily paths of stars (or other objects) as they move around the sky are called their ***diurnal circles***.

At the equator the problem is almost as simple. The celestial equator, in the same plane as the earth's equator, passes through the zenith and, since it runs east and west, it intersects the horizon at the east and west points. The celestial poles, being 90° from the celestial equator, must be at the north and south points on the horizon. There, as the sky turns, all stars rise and set; they move straight up from the east side of the horizon and set straight down on the west side. During a 24-hour period, all stars are above the horizon exactly half the time (Figure 4.4).

Evidently at points on the earth between the equator and poles, one of the celestial poles must be a certain distance above the horizon. Consider a northern observer (Figure 4.5). His angular distance from the North Pole, the angle z at the center of the earth (in Figure 4.5), is 90° less his latitude. But since the north celestial pole is in a direction parallel to the axis of the earth, the angular distance of the north celestial pole from the observer's zenith must also be z. Inspection of the figure shows, therefore, that the altitude of the north celestial pole is equal to the observer's latitude. The same is true of the south celestial pole for observers in southern latitudes.

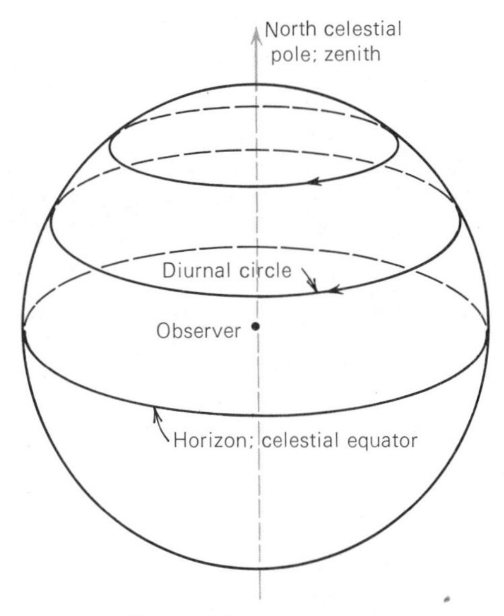

Figure 4.3
Sky from the North Pole.

For an observer between the equator and North Pole, say at 34° north latitude, the situation is as de-

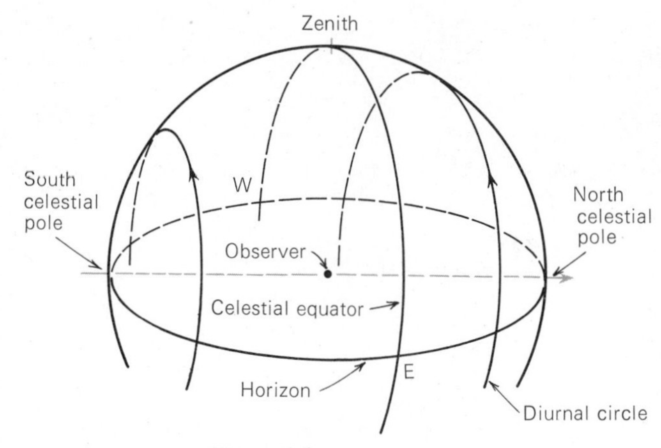

Figure 4.4
Sky from the equator.

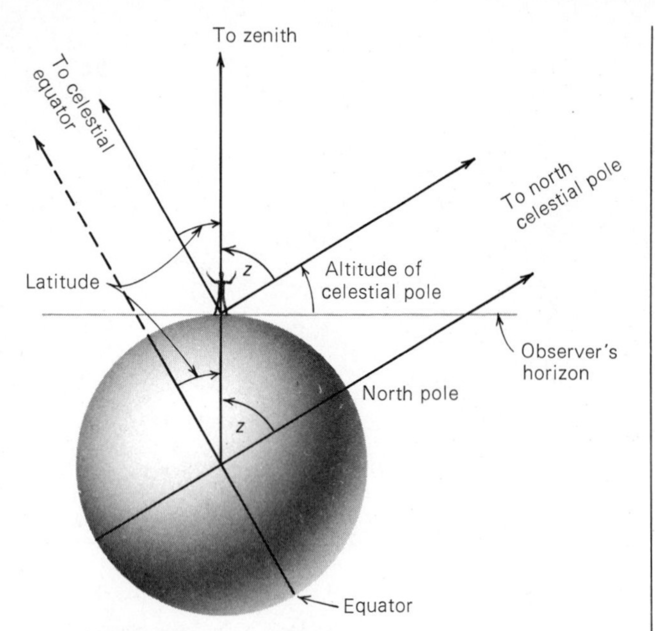

Figure 4.5
The altitude of the celestial pole equals the observer's latitude.

picted in Figure 4.6. Here the north celestial pole is 34° above the observer's northern horizon. The south celestial pole is 34° *below* the southern horizon. As the earth turns, the stars appear to circle parallel to the celestial equator. The whole sky seems to pivot about the north celestial pole. For this observer, stars within 34° of the North Pole can never set. They are always above the horizon, day and night. This part of the sky is called the *north circumpolar zone*. To observers in the United States, the Big and Little Dippers and Cassiopeia are examples of star groups that are in the north circumpolar zone. On the other hand, stars within 34° of the south celestial pole never rise. That part of the sky is the *south circumpolar zone*. To most U.S. observers, the Southern Cross is in that zone. At the North Pole, half the sky is circumpolar above the horizon and half below.

Stars north of the celestial equator, but outside the north circumpolar zone, lie above the horizon in the greater parts of their daily paths, or diurnal circles; hence they are up more than half the time. Stars *on* the celestial equator are up exactly half the time, for their diurnal circle is the celestial equator. Because it is a great circle, exactly half of it must be above the horizon. Stars south of the celestial equator, but outside the south circumpolar zone, are up less than half the time.

THE SEASONS

The earth's orbit around the sun is an ellipse, its distance from the sun varying by about 3 percent. However, the changing distance of the earth from the sun is *not* the cause of the seasons. The seasons result because the plane in which the earth revolves is not the same as the plane of the earth's equator. The planes of the equator and ecliptic are inclined to each other by about 23½°, the *obliquity of the ecliptic* (Chapter 2).

Globes of the earth are usually mounted with the earth's axis tilted from the vertical. This tilt is the same angle of 23½°, for that is the angle the earth's

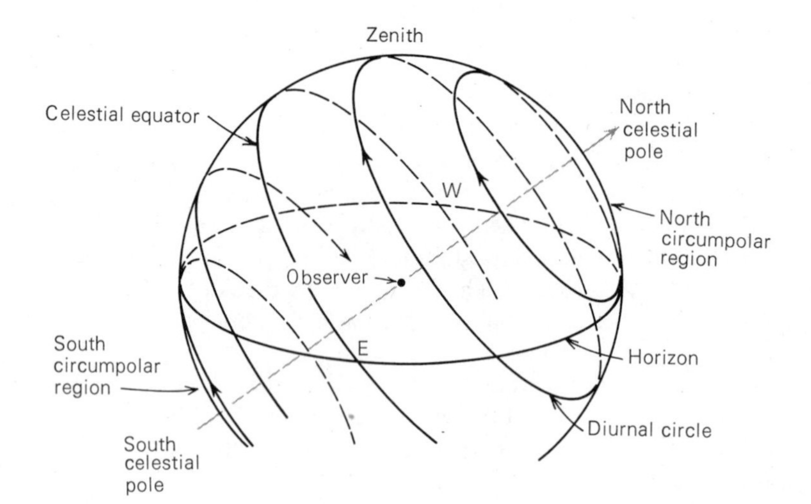

Figure 4.6
Sky from latitude 34° N.

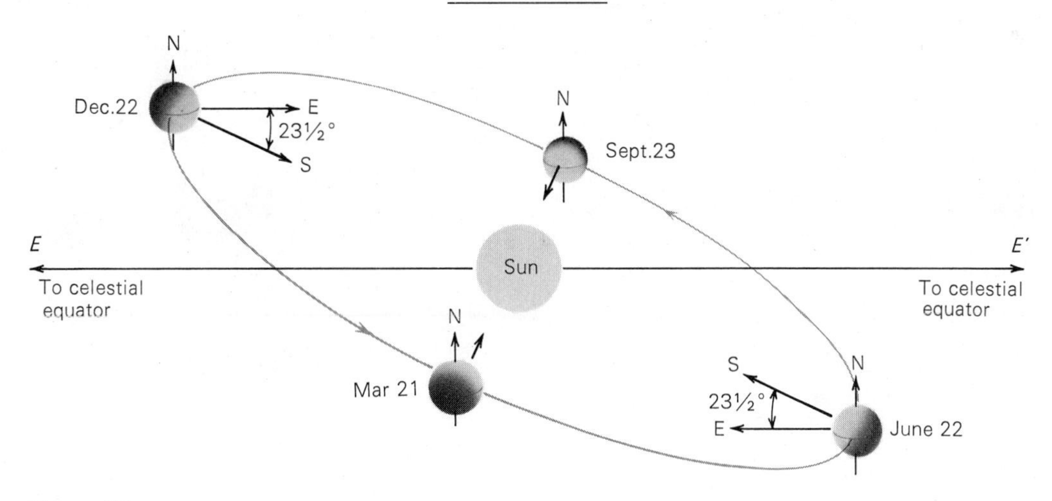

Figure 4.7
The seasons are caused by the inclination of the plane of the earth's orbit to the plane of the equator.

axis must make with the perpendicular to the plane of its orbit around the sun. The result of the obliquity of the ecliptic is that the northern hemisphere is inclined *toward* the sun in June and away from it in December.

The Seasons and Sunshine

Figure 4.7 shows the earth's path around the sun. The line EE' is in the plane of the celestial equator. In the figure the earth appears to pass alternately above and below this plane, but the celestial sphere is so large, and the celestial equator so far away, that a line from the center of the earth through the earth's equator always points to the celestial equator.

We see in the figure that on about June 22 (the date of the *summer solstice*), the sun shines down most directly upon the northern hemisphere of the earth. It appears 23½° *north* of the equator and thus on that date passes through the zenith of places on the earth that are at 23½° north latitude. The situation is shown in detail in Figure 4.8. To an observer on the equator, the sun appears 23½° north of the zenith at noon. To a person at a latitude 23½° N, the sun is overhead at noon. This latitude on the earth, at which the sun can

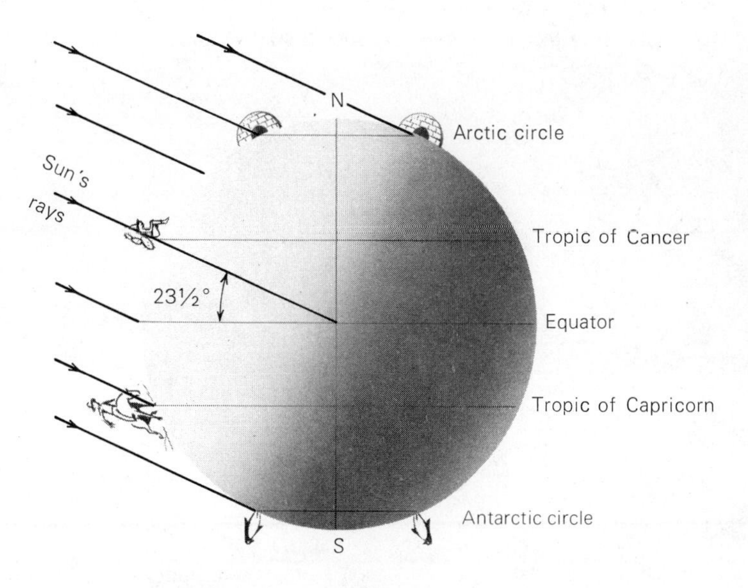

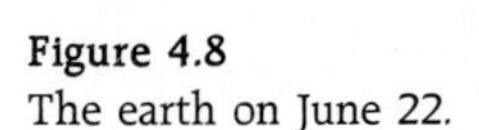
Figure 4.8
The earth on June 22.

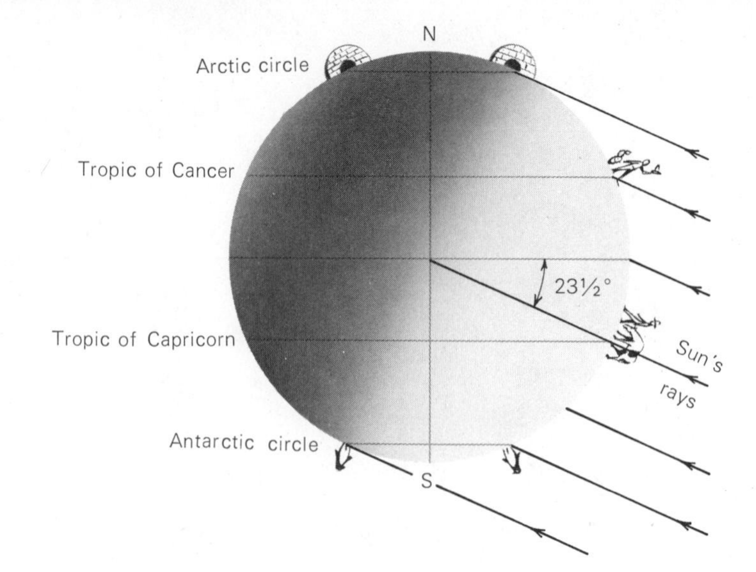

Figure 4.9
The earth on December 22.

appear at the zenith at noon on the first day of summer, is called the *Tropic of Cancer*. We see also in Figure 4.8 that the sun's rays shine down past the North Pole; in fact, all places within 23½° of the pole, that is, at a latitude greater than 66½° N, have sunshine for 24 hours on the first day of summer. The sun is as far north on this date as it can get; thus, 66½° is the southernmost latitude where the sun can ever be seen for a full 24-hour period (the *midnight sun*); that circle of latitude is called the *Arctic Circle*.

During this time, the sun's rays shine very obliquely on the southern hemisphere. In fact, all places within 23½° of the South Pole—that is, south of latitude 66½° S (the *Antarctic Circle*)—have no sight of the sun for the entire 24-hour period.

The situation is reversed six months later, about December 22 (the date of the *winter solstice*), as shown in Figure 4.9. Now it is the Arctic Circle that has a 24-hour night and the Antarctic Circle that has the midnight sun. At latitude 23½° S, the *Tropic of Capricorn*, the sun passes through the zenith at noon. It is winter in the northern hemisphere, summer in the southern.

Finally, we see in Figure 4.7 that on about March 21 and September 23 the sun appears to be in the direction of the celestial equator, and, on these dates, the equator itself is the diurnal circle for the sun. Every place on the earth then receives exactly 12 hours of sunshine and 12 hours of night. These points, where the sun crosses the celestial equator, are called the *vernal* (spring) *equinox* and *autumnal* (fall) *equinox*. *Equinox* means "equal night."

Figure 4.10 is a map in which the sky is shown

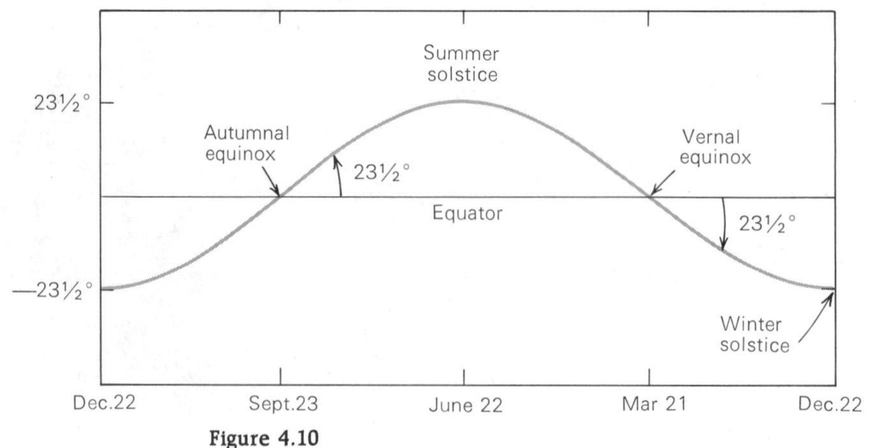

Figure 4.10
Plot of the ecliptic around the celestial equator.

flattened out, as in a Mercator projection of the earth. The equator runs along the middle of the map, and the ecliptic is shown as a wavy line crossing the equator at the two equinoxes. Both equator and ecliptic are, of course, great circles, but they cannot both be shown as straight lines on a flat surface. Notice that the ecliptic intersects the equator at an angle of 23½°, and that its northernmost extent is 23½° north of the equator (the summer solstice) and its southernmost extent is 23½° south of the equator (the winter solstice).

Figure 4.11 shows the aspect of the sky at a typical latitude in the United States. During the spring and summer, the sun is north of the equator and is thus up more than half the time. A typical spot in the United States, on the first day of summer (about June 22), receives about 14 or 15 hours of sunshine. Also, notice that the sun appears *high* in the sky, and so in these seasons the sunlight is more direct, and thus more effective in heating than in the fall and winter when the sun appears at a lower altitude in the sky.

In the fall and winter the sun is south of the equator, where most of its diurnal circle is below the horizon, and so it is up less than half the time. On about December 22, a typical city at, say 30° to 40° north latitude, receives only nine or ten hours of sunshine. Also, the sun is low in the sky; a bundle of its rays is thus spread out over a larger area on the ground (Figure 4.12) than in summer. Because the energy is spread out over a larger area, there is less for each square meter, and so the sun at low altitudes is less effective in heating the ground.

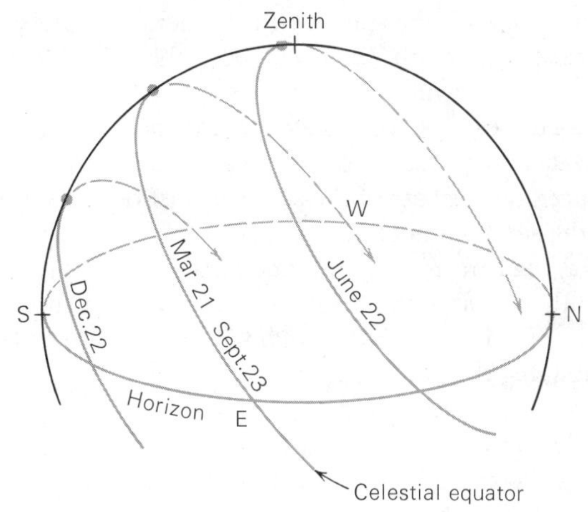

Figure 4.11
Diurnal paths of the sun for various dates at a typical place in the United States.

The Seasons at Different Latitudes

At the equator all seasons are much the same. Every day of the year, the sun is up half the time, so there are always 12 hours of sunshine at the equator.

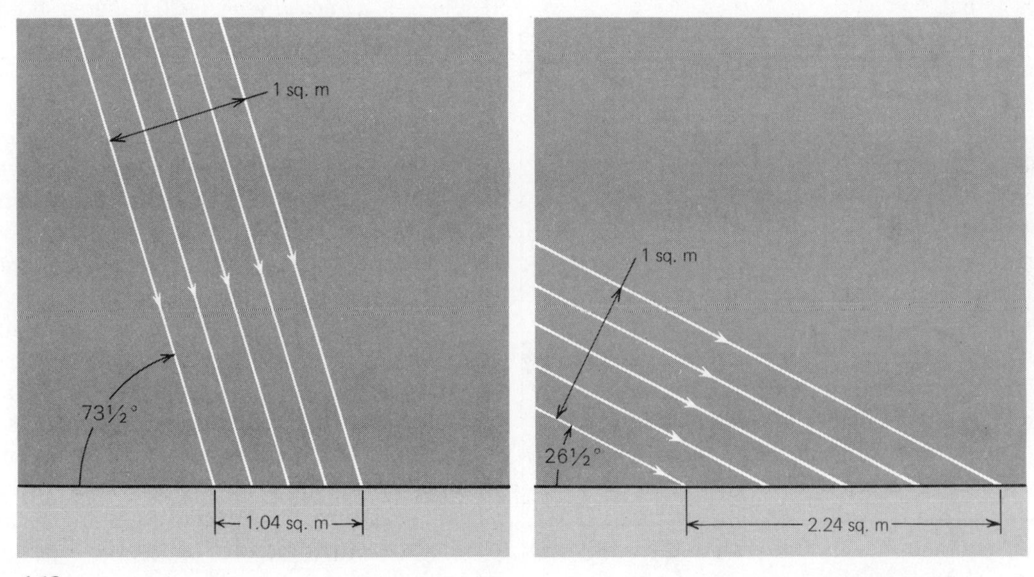

Figure 4.12
Effect of the sun's altitude. When the sun is low in the sky, its rays are more oblique to the ground and are over a larger area than when the sun is high in the sky.

About June 22, the sun crosses the meridian 23½° north of the zenith, and about December 22, 23½° south of the zenith.

The seasons become more pronounced as one travels north or south of the equator. At the Tropic of Cancer, on the date of the summer solstice, the sun is at the zenith at noon. On the date of the winter solstice, the sun crosses the meridian 47° south of the zenith. At the Arctic Circle, on the first day of summer, the sun never sets, but at midnight can be seen just skimming the north point on the horizon. About December 22, the sun does not quite rise at the Arctic Circle, but just gets up to the south point on the horizon at noon. Between the Tropic of Cancer and the Arctic Circle, the number of hours of sunshine and the noon altitude of the sun range between these two extremes.

We recall that at the North Pole, all celestial objects that are north of the celestial equator are always above the horizon and, as the earth turns, circle around parallel to it. The sun is north of the celestial equator from about March 21 to September 23, and so at the North Pole the sun rises when it reaches the vernal equinox and sets when it reaches the autumnal equinox. There are six months of sunshine at the Pole. The sun reaches its maximum altitude of 23½° about June 22; before that date it climbs gradually higher each day, and after that it drops gradually lower. A navigator can easily tell when he is at the North Pole, for there the sun circles around the sky parallel to the horizon, getting no higher or lower (except gradually as the days go by).

In the southern hemisphere, the seasons are reversed from those in the north. While we are having summer in the United States, in Australia it is winter. Furthermore, in the southern hemisphere, the sun crosses the meridian generally to the *north* of the zenith. In Buenos Aires, you would want a house with a good *northern* exposure.

The earth, in its elliptical orbit, reaches its closest approach to the sun about January 4. It is then said to be at ***perihelion***. It is farthest from the sun, at ***aphelion***, about July 5. We see, then, that the earth is closest to the sun when it is winter in the north. However, it is summer in the southern hemisphere when the earth is at perihelion, and the earth is farthest from the sun during the southern hemisphere's winter. Therefore, we might expect the seasons to be somewhat more severe in the southern hemisphere than in the northern. However, there is more ocean area in the southern hemisphere; this and other topographical factors are more important in their influence on climate than is the earth's changing distance from the sun. We shall see that for Mars, whose orbit is considerably more eccentric than the earth's, the same kind of situation does have a pronounced effect upon the seasons.

Precession of the Equinoxes

As the earth's axis precesses in its conical motion (Chapter 3), the equatorial plane retains (approximately) its 23½° inclination to the ecliptic plane; that is, the obliquity of the ecliptic remains constant. However, the intersections of the celestial equator and the ecliptic (the equinoxes) must always be 90° from the celestial poles (because all points on the celestial equator are 90° from the celestial poles). Thus, as the poles move because of precession, the equinoxes slide around the sky, moving westward along the ecliptic. This motion is called the ***precession of the equinoxes***. The angle through which the equinoxes move each year, the ***annual precession***, is 1/26,000 of 360°, or about 50″. Each year as the sun completes its *eastward* revolution about the sky with respect to, say, the vernal equinox, that equinox has moved *westward*, to meet the sun, about 50″. Since it takes the sun about 20 minutes to move 50″ along the ecliptic (or, more accurately, because it takes the earth that long to move through an angle of 50″ in its orbit about the sun), the apparent revolution of the sun with respect to the equinoxes, a ***tropical year***, is 20 minutes shorter than the revolution with respect to the stars, called a ***sidereal year***.

Precession has no important effect on the seasons. The earth's axis retains its inclination to the ecliptic, so the northern hemisphere is still tipped toward the sun during one part of the year and away from it during the other. Our calendar year is based on the beginnings of the seasons (the times when the sun reaches the equinoxes and solstices), so spring in the northern hemisphere still begins in March, summer in June, and so on. The only effect is that as the precessional cycle goes on, a given season will occur when the earth is in gradually different places in its orbit with respect to the stars. In the twentieth century, for example, Orion is a *winter constellation;* we look out at night, away from the sun, and see Orion in the sky during the winter months. In 13,000 years, half a precessional cycle later, it will be summertime when we look out in the same direction, away from the sun, and

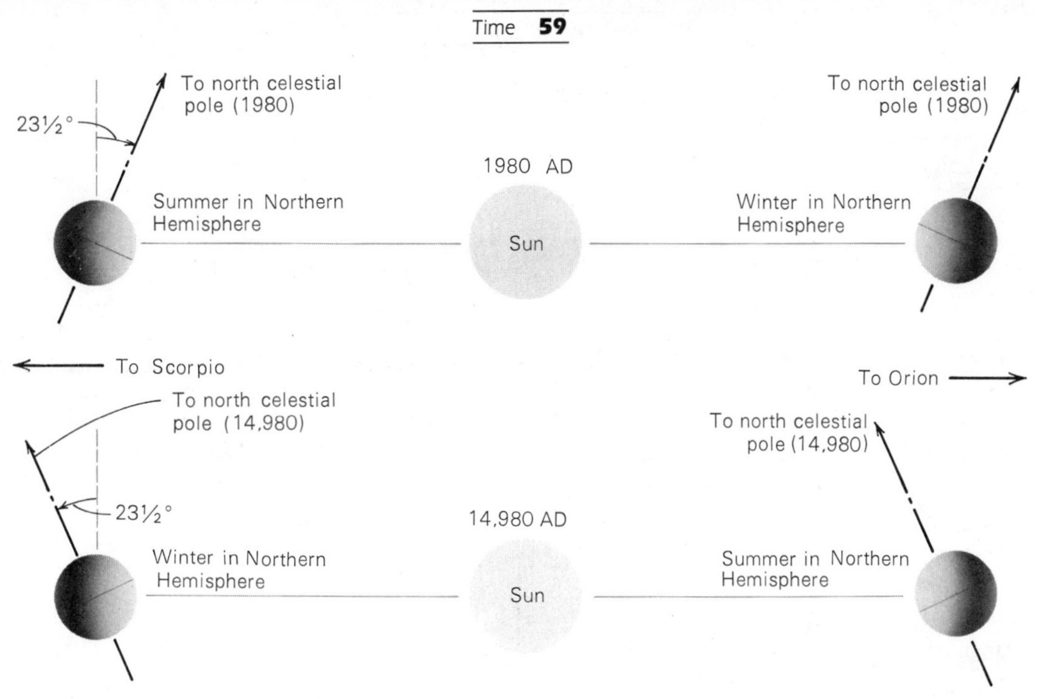

Figure 4.13
Summer and winter constellations change due to precession.

see Orion. Similarly, Scorpio is a summer constellation now, whereas in the year 15,000 it will be a winter constellation (see Figure 4.13).

Twilight

We all know that the sky does not immediately darken when the sun sets. Even after the sun is no longer visible from the ground, the upper atmosphere of the earth can catch some of the rays of the setting sun and scatter them helter-skelter, illuminating the sky. Gases in the earth's atmosphere are dense enough to scatter appreciable sunlight up to altitudes of about 300 km (200 mi). The sun must be at least 18° below the horizon for all traces of this postsunset or presunrise sky light (*twilight*) to be absent. At latitudes near the equator, where the sun rises and sets nearly vertically to the horizon, twilight lasts only a little over one hour. However, at far northern and southern latitudes, the sun rises and sets in a much more oblique direction and takes correspondingly longer to reach a point 18° below the horizon, so twilight may last for two hours or more. In the far northern countries twilight lasts all night in the summertime. At the North Pole there are six weeks of twilight in the late winter before sunrise and again in the early fall after sunset.

TIME

The measurement of time is based on the rotation of the earth. To say "what time it is" is to specify the orientation of the local meridian and horizon with the celestial sphere.

The Passage of Time—Hour Angle

Time is reckoned by the angular distance around the sky that some convenient reference object has moved since it last crossed the meridian. The motion of that point around the sky is like the motion of the hour hand on a 24-hour clock. The angle measured to the west along the celestial equator from the local celestial meridian to the hour circle passing through any object (for example, a star) is that object's *hour angle*. *Time* can be defined as the *hour angle of the reference object*.

Hour angle (or time) is generally measured in hours, minutes, and seconds of time (h, m, and s) rather than in degrees, minutes, and seconds of arc (°, ′, and ″), but the two kinds of angular measure are entirely equivalent—see Table 4.1. Thus, for example, an hour angle of 8^h is the same as 120°.

TABLE 4.1 Conversion Between Units of Time and Arc

Time Units	Arc Units
24^h	360°
1^h	15°
4^m	1°
1^m	15′
4^s	1′
1^s	15″

We have already seen that one kind of time is sidereal time, needed for the preparation of horoscopes (Chapter 2). Sidereal time, defined as the hour angle of the vernal equinox, is widely used today by astronomers and navigators, as well as by the astrologers of antiquity. On the other hand, for most practical purposes, it is far more convenient to define time by the sun. Thus for everyday business we use *solar time*. A complication, however, is that solar time and sidereal time progress at different rates, because a day by the sun is not the same as a day by the vernal equinox.

The Solar and Sidereal Day

The solar day is the period of the earth's rotation with respect to the sun. The sidereal day is, instead, the time required for the earth to make a complete rotation with respect to the vernal equinox. Technically, the term "sidereal day" is a misnomer, because the vernal equinox slowly shifts its position in the sky as a result of precession. This movement is so slow, however, that a sidereal day is within 0.01 s of the true period of rotation of the earth with respect to the stars.

A solar day is slightly longer than a sidereal day, as a study of Figure 4.14 will show. Suppose we start a day when the earth is at A, with the sun on the meridian of an observer at point O on the earth. The direction from the earth to the sun, AS, if extended, points in the direction C among stars on the celestial sphere. After the earth has made one rotation with respect to the stars, the same stars in direction C will again be on the local meridian to the observer at O. However, because the earth has moved from A to B in its orbit about the sun during its rotation, the sun has not yet returned to the meridian of the observer but is still slightly to the east. The vernal equinox is so

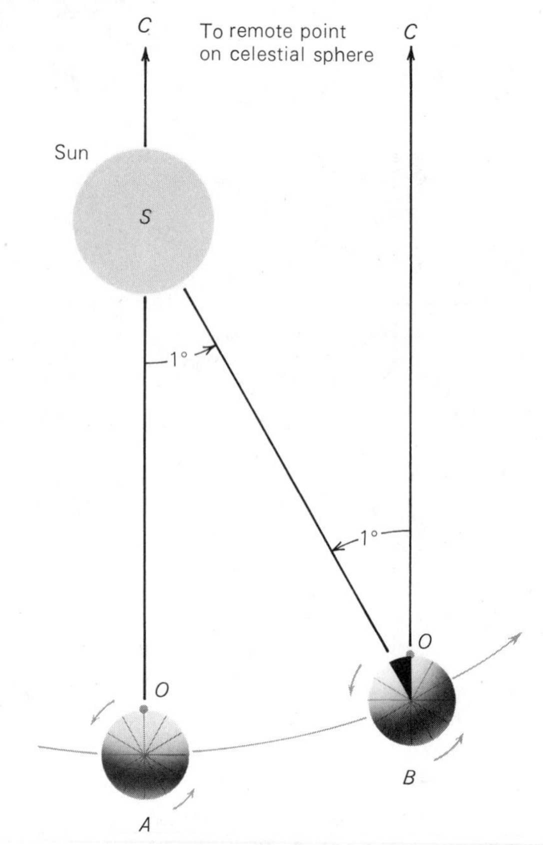

Figure 4.14
Sidereal and solar day.

nearly fixed among the stars that the earth has completed, essentially, one *sidereal* day, but to complete a *solar* day it must turn a little more to bring the sun back to the meridian.

There are about 365 days in a year and 360° in a circle; thus the daily motion of the earth in its orbit is about 1°. This 1° angle, ASB, is nearly the same as the additional angle over and above 360° through which the earth must turn to complete a solar day. It takes the earth about four minutes to turn through 1°. A solar day, therefore, is about four minutes longer than a sidereal day.

Each kind of day is subdivided into hours, minutes, and seconds. A unit of solar time (hour, minute, or second) is longer than the corresponding unit of sidereal time by about 1 part in 365. In units of solar time, one sidereal day is $23^h56^m4^s.091$. The period of the earth's rotation with respect to the stars is $23^h56^m4^s.099$ in solar time units.

Apparent Solar Time

Just as sidereal time is reckoned by the hour angle of the vernal equinox, so *apparent solar time* is determined by the hour angle of the sun. At midday, apparent solar time, the sun is on the meridian. The hour angle of the sun is the time *past midday* (*post meridiem,* or P.M.). It is convenient to start the day not at noon, but at midnight. Therefore the elapsed apparent solar time since the beginning of a day is the hour angle of the sun *plus* 12 hours. During the first half of the day, the sun has not yet reached the meridian. We designate those hours as *before midday* (*ante meridiem,* or A.M.). We customarily start numbering the hours after noon over again, and designate them by P.M. to distinguish them from the morning hours (A.M.). On the other hand, it is often useful to number the hours from 0 to 24, starting from the beginning of the day at midnight. For example, in various conventions, 7:46 P.M. may be written as 19^h46^m, 19:46, or simply 1946.

Apparent solar time, defined as the hour angle of the sun plus 12 hours, is the most obvious and direct kind of solar time. It is the time that is kept by a sundial. In a sundial, a raised marker, or *gnomon,* casts a shadow whose direction indicates the hour angle of the sun. Apparent solar time was the time kept by man through many centuries.

The exact length of an apparent solar day, however, varies slightly during the year. The eastward progress of the sun in its annual journey around the sky is not uniform because of the slightly varying speed of the earth in its orbit and the fact that the earth's axis of rotation is not perpendicular to the plane of its revolution. Thus the amount by which the sun appears to shift to the east during one sidereal day is not exactly the same every day of the year, so that the extra amount through which the earth must rotate after a sidereal day to complete a solar day is not always the same. Apparent solar time, in other words, does not advance at a uniform rate.

The apparent solar day is always *about* four minutes longer than a sidereal day, but because of the sun's variable daily progress to the east, the precise interval varies by up to one-half minute one way or the other. The variation can accumulate after a number of days to several minutes. After the invention of clocks that could run at a uniform rate, it became necessary to abandon the apparent solar day as the fundamental unit of time. Otherwise, all clocks would have to be adjusted to run at a different rate each day.

Mean Solar Time

Mean solar time is based on the *mean solar day,* which has a duration equal to the *average* length of an apparent solar day. Mean solar time is defined as the hour angle of the mean sun plus 12 hours, where the *mean sun* is a fictitious point in the sky that moves uniformly to the east along the *celestial equator,* with the same average eastern rate as the true sun. In other words, mean solar time is just apparent solar time averaged to be uniform.

Although mean solar time has the advantage of progressing at a uniform rate, it is still inconvenient for practical use. Recall that it is defined as the hour angle of the mean sun. But hour angle refers to the local celestial meridian, which is different for every longitude on earth. Thus observers on different north-south lines on the earth have a different hour angle of the mean sun, hence a different mean solar time. If mean solar time were strictly observed, a person traveling east or west would have to reset his watch continually as his longitude changed, if it were always to read the local mean time correctly. For instance, a commuter traveling from Oyster Bay to New York City would have to adjust his watch as he rode through the East River tunnel, because Oyster Bay time is actually about 1.6 min more advanced than that of Manhattan.

Standard Time

Until near the end of the last century, every city and town in the United States kept its own local mean time. With the development of railroads and the telegraph, however, the need for some kind of standardization became evident. In 1883 the nation was divided into four time zones. Within each zone, all places keep the same time, the local mean solar time of a standard meridian running more or less through the middle of each zone. Now a traveler resets his watch only when the time change has amounted to a full hour. For local convenience, the boundaries between the four time zones are chosen to correspond, as much as possible, to divisions between states. Mean solar time, so standardized, is called *standard time.* The standard time zones in the United States (not including Alaska and Hawaii) are Eastern Standard Time (EST), Central Standard Time (CST), Mountain Standard Time (MST), and Pacific Standard Time (PST), which respectively keep the mean times of the meridians of 75°, 90°, 105°, and 120° west longitude. Hawaii and Alaska both keep the time of the meridian 150° west longitude, two hours less advanced than Pacific Standard Time. Other nations, of course, have also adopted standard time systems.

The local mean solar time of the meridian running through Greenwich, England, is called *universal time*. Data are usually given in navigational tables, such as the *Nautical Almanac*, for various intervals of universal time.

To take advantage of the maximum amount of sunlight during waking hours, most states in this country, as well as many foreign nations, keep what is called *daylight saving time* (or *summer time*) during the spring and summer, and sometimes during fall and winter as well. Daylight saving time is simply the local standard time of the place *plus* one hour.

The International Date Line

The fact that time is always more advanced to the east presents a problem. Suppose a man moving eastward travels around the world. He passes into a new time zone, on the average, for about every 15° of longitude he travels and, each time, he dutifully sets his watch ahead an hour. By the time he has completed his trip, he has set his watch ahead through a full 24 hours, and has thus *gained a day* over those who stayed home. Let us look at the problem another way. It is 3 hours later in New York than in Berkeley, California. In London, it is 8 hours later than in Berkeley; in Tokyo, still farther to the east, it is 17 hours later. Because San Francisco is about 100° east of Tokyo, it follows that in San Francisco, a few kilometers to the west across the San Francisco Bay from Berkeley, it is 7 hours later than at Tokyo, or 24 hours—one day—later than at Berkeley! One might suppose that by going around the world to the east or west enough times, we could go into the future or past as far as we desired, creating the equivalent of a time machine.

The solution to the dilemma is the *international date line*, set by international agreement along the 180° meridian of longitude. The date line runs about down the middle of the Pacific Ocean, although it jogs a bit in a few places to avoid cutting through groups of islands and through Alaska. By convention, at the date line, the date of the calendar is changed by one day. If a person crosses the date line from west to east, so that he is advancing his time, he compensates by decreasing his date; if he crosses from east to west, he increases his date by one day.

Summary of Time

We may briefly summarize the story of time as follows:

1. The ordinary day is based on a rotation of the earth with respect to the sun (not the stars).
2. The length of the apparent solar day is slightly variable because of the earth's variable orbital speed and the obliquity of the ecliptic.
3. Therefore, the average length of an apparent solar day is defined as the mean solar day.
4. Time based on the mean solar day is mean solar time. It is defined as the local hour angle of the mean sun plus 12 hours. The mean sun is an imaginary sun that revolves annually on the celestial equator at a perfectly uniform eastward rate.
5. The mean solar time at any one place, called *local mean time*, varies continuously with longitude, so that two places a few kilometers east and west of each other have slightly different times.
6. Therefore time is standardized, so that each place in a certain region or zone keeps the same time—the local mean time of the standard meridian in that zone.
7. In many localities, standard time is advanced by one hour during part or all of the year to take advantage of the maximum number of hours of sunshine during the waking hours. This is daylight saving time.

Time Standards

The precise time is now broadcast by shortwave radio throughout the world by several agencies—in the United States by the U.S. Naval Observatory and the National Bureau of Standards. The time of day is defined by astronomical observations. We keep track of the passage of time, however, with highly accurate atomic clocks that regulate time intervals by the frequency of radiation absorbed or emitted by a particular isotope of the cesium atom (cesium 133). Internationally coordinated time kept by such clocks is called *atomic time*.

With precise modern time standards it is possible to monitor small changes in the rate of the earth's rotation. Some of these irregularities in the earth's spin are due to seasonal shifts of air masses and ice and snow deposits over the earth, but the cause of some others is not yet known. It is known, however, that friction in the tides is causing the earth to slow its rotation very gradually. This effect results in a lengthening of the day by about 0.0016 second per century.

The slow increase in the period of the earth's rotation as well as the short-term irregularities cause universal time (obtained from the earth's rotation) to deviate from and generally fall behind the far more accurate atomic time. Thus once or twice a year (as needed) an extra second (a *leap second*) is inserted in a day to bring the two times into average agreement.

DATE

The natural units of the calendar are the day, based on the period of rotation of the earth; the month, based on the period of revolution of the moon about the earth; and the year, based on the period of revolution of the earth about the sun. Difficulties in the calendar have resulted from the fact that these three periods are not commensurable—that is, one does not divide evenly into any of the others.

The period of revolution of the moon with respect to the stars, the *sidereal month,* is about 27⅓ days. However, the interval between corresponding phases of the moon, the more obvious kind of month, is the moon's period of revolution with respect to the sun, the *synodic month,* which has about 29½ days (Chapter 9).

The period of revolution of the earth with respect to the vernal equinox, that is, with respect to the beginnings of the various seasons, is the *tropical* year. Its length is 365.242199 mean solar days, or $365^d5^h48^m46^s$. Our calendar, to keep in step with the seasons, is based on the tropical year. Because of precession, the tropical year is slightly shorter than the sidereal year.

Early Calendars

Even in the earliest cultures man was concerned with the keeping of time and the calendar. Particularly interesting are monuments left by Bronze Age people in northwestern Europe, especially in the British Isles. The best preserved of the monuments is Stonehenge (Figure 4.15), about 13 km from Salisbury in southwest England. It is a complex array of stones, ditches, and holes arranged in concentric circles. Carbon dating and other studies show that Stonehenge was built during three periods ranging from about 2500 B.C. to about 1700 B.C. Some of the stones are aligned with the directions of the sun and moon during their risings and settings at critical times of the year (such as the beginnings of summer and winter), and it is generally believed that at least one function of the monument was connected with the keeping of a calendar.

Figure 4.15
Stonehenge. *(Courtesy E. C. Krupp, Griffith Observatory)*

The Maya in Central America are also known to have been concerned with the keeping of time. The Mayan calendar was more sophisticated and complicated than calendars in use in Europe. Apparently, the Maya did not attempt to correlate their calendar accurately with the length of the year or lunar month. Rather, their calendar was a system for keeping track of the passage of days and for counting time far into the past or future. Among other purposes, their calendar was useful for predicting astronomical events—for example, the positions of Venus in the sky.

Our calendar derives from the Greek and Roman calendars dating from at least the eighth century B.C. They led, eventually, to the Julian calendar, introduced in 46 B.C., which approximated the tropical year by 365.25 days, fairly close to the actual value of 365.242199 mean solar days. Julius Caesar, acting on the advice of the astronomer Sosigenes, achieved his approximation by declaring years to have 365 days each, with the exception of every fourth year. The leap year, which was to have one extra day, bringing its length to 366 days, thus brought the average length of the calendar year to 365.25 days.

Even the Julian calendar, however, differs from the tropical year by 11^m14^s, an amount which accumulates over the centuries to an appreciable error.

The dates of important religious holidays had been fixed by the Council of Nicaea in 325 A.D. These holidays were based on the beginnings of certain seasons. For example, Easter, according to the rule adopted, falls on the first Sunday after the fourteenth day of the moon (almost full moon) that occurs on or after March 21. At that time March 21 was the date of the vernal equinox. (The Sunday *after* full moon was specified intentionally to avoid the possibility of an occasional coincidence with the Jewish Passover.)

Now, between 45 B.C. and 325 A.D., the date of the vernal equinox had slipped back from March 25 to March 21. This was because the Julian year, with an average length of 365¼ days, is 11^m14^s longer than the tropical year of $365^d5^h48^m46^s$. The slight discrepancy had accumulated to just over three days in those four centuries.

The Gregorian Calendar

By 1582, that 11 minutes and 14 seconds per year had added up to another ten days, so that the first day of spring was occurring on March 11. If the trend were allowed to continue, eventually Easter and the related days of observance would be occurring in early winter. Pope Gregory XIII therefore felt it necessary to institute a further calendar reform.

The Gregorian calendar reform consisted of two steps. First, ten days had to be dropped out of the calendar to bring the vernal equinox back to March 21, its date at the time of the Council of Nicaea. This step was expediently accomplished. By proclamation the day following October 4, 1582, became October 15.

The second feature of the new Gregorian calendar was that the rule for leap year was changed so that the average length of the year would more closely approximate the tropical year. Gregory decreed that three out of every four century years, all leap years under the Julian calendar, would be common years henceforth. The rule was that only century years divisible by 400 should be leap years. Thus, 1700, 1800, and 1900, all divisible by four, and thus leap years in the old Julian calendar, were *not* leap years in the Gregorian calendar. On the other hand, the years 1600, and 2000, both divisible by 400, are leap years under both systems. The average length of this Gregorian year, 365.2425 mean solar days, is correct to about one day in 3300 years.

The Catholic countries immediately put the Gregorian reform into effect, but countries under control of the Eastern Church and most Protestant countries did not adopt it until much later. It was 1752 when England and the American colonies finally made the change. The year 1700 had been a leap year in the Julian calendar but not in the Gregorian; thus the discrepancy between the two systems had become 11 days. By parliamentary decree, September 2, 1752, was followed by September 14. Although special laws were passed to prevent such breaches of justice as landlords collecting a full month's rent for September, there were still riots, and people demanded their 11 days back. To make matters worse, in England it had been customary to follow the ancient practice of starting the year on March 25, the date of the vernal equinox in 45 B.C., when the Julian calendar was introduced. In 1752, however, the start of the year was moved back to January 1, so in England and the colonies 1751 had no months of January and February, and had lost 24 days of March! We mark George Washington's birthday on February 22, 1732, but at the time of his birth, a calendar would have read February 11, 1731. Russia did not abandon the Julian calendar until the time of the Bolshevik revolution. The Russians then had to omit 13 days to come into step with the rest of the world.

THE MANY MOTIONS OF THE EARTH

In this chapter we have discussed in detail two of the earth's motions: rotation and revolution. However, there are many other motions of the earth, some dealt with in other chapters. Here, for completeness, we summarize these motions:

1. The earth *rotates* daily on its axis.
2. The earth periodically shifts slightly with respect to its axis of rotation (*variation in latitude*).
3. The earth *revolves* about the sun.
4. The gravitational pull of the sun and moon on the earth's equatorial bulge causes a very slow change in orientation of the axis of the earth called *precession* (Chapter 3).
5. Because the moon's orbit is not quite in the plane of the ecliptic, and because of a slow change in orientation of the moon's orbit, there is a small periodic motion of the earth's axis superimposed upon precession, called *nutation*.
6. Actually, it is the center of mass of the earth-moon system, or *barycenter*, that revolves about the sun in an elliptical orbit. Each month the center of the earth revolves about the barycenter (Chapter 3).
7. The earth shares the motion of the sun and the entire solar system among its neighboring stars. This *solar motion* is about 20 km/s (Chapter 12).
8. The sun, with its neighboring stars, shares in the general *rotation of the Galaxy*. Our motion about the center of the Galaxy is about 300 km/s (Chapter 15).
9. The universe is expanding, with all of its galaxies moving away from all the others (Chapter 21); thus our entire Galaxy is in motion with respect to other galaxies.
10. In addition to the uniform expansion of the universe, galaxies have smaller independent motions of their own. Recent observations show that our Galaxy has such an independent motion of about 500 km/s (Chapter 21), superimposed on those motions resulting from the uniform expansion of the universe.

In view of these many motions of the earth, we might wonder what the *real* speed of the earth is in *absolute space*. Newton did suppose that there exists an absolute space, but Einstein showed that the concept can have no meaning. There is no center to the universe, and no special coordinate system that is absolute. Even in principle we can only speak of *relative* motion—for example, of our own speed with respect to something else. The nonexistence of an absolute reference frame in space, with respect to which absolute motions of objects could be defined, is one of the fundamental postulates of special relativity (Chapter 7).

SUMMARY REVIEW

Coordinates on the earth: latitude and longitude; meridian; east and west

Coordinates in the sky: celestial poles; celestial equator; hour circles; zenith; horizon; nadir; celestial meridian; right ascension and declination; altitude and azimuth

Orientation of the sky: altitude of the north celestial pole equals the latitude; the sky from the North or South Pole; the sky from the equator; diurnal circles; circumpolar stars

The seasons: obliquity of the ecliptic; vernal and autumnal equinoxes; summer and winter solstices; Tropics of Cancer and Capricorn; Arctic and Antarctic Circles; midnight sun; reversal of the seasons in the southern hemisphere; precession of the equinoxes; tropical year; twilight

Time: hour angle; solar and sidereal days; sidereal time; apparent solar time; A.M. and P.M.; mean solar time; standard time; daylight saving time; universal time; international date line; atomic time

Date and calendar: sidereal and synodic months; tropical and sidereal years; the Roman calendar; the Julian calendar; leap year; the Gregorian calendar; Council of Nicaea

Summary of the motions of the earth

EXERCISES

1. If a star rises in the northeast, in what direction does it set?

2. What is the latitude of (a) the North Pole? (b) the South Pole? (c) a point two-thirds of the way from the equator to the South Pole?

3. Why has longitude no meaning at the South Pole?

4. Tell where you are on the earth from the following descriptions:
(a) The stars rise and set perpendicularly to the horizon.
(b) The stars circle the sky parallel to the horizon.
(c) The celestial equator passes through the zenith.
(d) In the course of a year, all stars are visible.
(e) The sun rises on September 23 and does not set until March 21.

5. Suppose you observe a star 30° from the south celestial pole pass through your zenith. Which of the following will be above your horizon sometime during a 24-hour period? (a) Big Dipper; (b) south celestial pole; (c) Orion; (d) Southern Cross; (e) north celestial pole; (f) autumnal equinox; (g) nadir (point opposite the zenith); (h) Ursa Major.

6. Where on earth is it possible for the sun to be at the north point on the horizon at midnight?

7. Explain why New York has more hours of daylight on the first day of summer than does Los Angeles.

8. Suppose the obliquity of the ecliptic were only 16½°. What then would be the difference in latitude between the Arctic Circle and the Tropic of Cancer?

9. What are the approximate dates of sunrise and sunset at the South Pole? Would a lunar eclipse occuring in January be visible from there? Why or why not?

10. How many more sidereal than solar days are there in one year? Why?

11. If a star rises at 8:30 P.M. tonight, approximately what time will it rise two months from now?

12. If the local mean time is 2:30 P.M., and the Universal Time is 10:30 A.M., what is the longitude?

13. When and where on earth is it possible for the ecliptic to lie on the horizon?

14. If the earth were to speed up in its orbit so that a tropical year contained 365 days 2 hours, how often would we need to have a leap year?

15. In far northern countries the winter months tend to be so cloudy that astronomical observations are nearly impossible. Why cannot good stellar observations be made at those places during the summer months?

16. If it is 1:20 A.M. on April 13 at a longitude 160° W, what are the time and date at longitude 160° E?

17. What is the altitude of the sun at noon on December 22 as seen from a place on the Tropic of Cancer?

18. If it is June 25, 1978, according to the Gregorian calendar, what would the date be on the old Julian calendar?

A Foucault pendulum. This type of large pendulum was used in an experiment by Jean Bernard Foucault in 1891. This was the first demonstration of the rotation of the earth. *(Griffith Observatory)*

CHAPTER 5

Max Planck (1858–1947) established the quantized nature of light and other electromagnetic radiation. The multicentered *Max-Planck Institut* of West Germany is named in tribute to his contributions. *(American Institute of Physics, Niels Bohr Library)*

ENERGY FROM SPACE

The earth is constantly exposed to energy from space. In this chapter we discuss the nature and properties of some of this cosmic energy and how it interacts with matter; in the next chapter we shall describe the instruments we use to observe it. Different kinds of energy strike the earth:

1. *Electromagnetic radiation*, if we include that which we receive from the sun, is by far the most important kind of energy reaching the earth; most of this chapter and the next are concerned with it. Light is the most familiar form of electromagnetic radiation.

2. *Cosmic rays* are charged particles, mostly the nuclei of atoms, that strike the molecules of the earth's upper atmosphere, producing tremendous numbers of secondary subatomic particles that rain down to the earth's surface. The total energy we receive in the form of cosmic rays exceeds that which we get from starlight.

3. The *solar wind* is a constant flow of atomic nuclei (mostly of hydrogen) and electrons emitted into interplanetary space by the sun. Most solar wind particles do not move fast enough to be called cosmic rays. Their effects can sometimes be observed, however, in the form of the light (especially auroras) emitted by upper atmospheric molecules ionized or excited by them. The solar wind is also regularly observed with instruments carried by space vehicles and artificial satellites. We shall discuss the solar wind in Chapter 16.

4. *Neutrinos* are produced by the decay of certain subatomic particles, and also by many of the nuclear reactions that keep stars shining. Neutrinos have energy but no mass, and hence travel with the speed of light. They pass through great amounts of matter without influencing it in the slightest, and so are very difficult to detect. Nevertheless the universe must be completely bathed in the radiation of neutrinos.

5. *Gravitational radiation* (gravity waves) are predicted to be radiated by matter in motion (Chapter 18). Gravitational waves must be everywhere around us, but they are expected to be extraordinarily weak and very difficult to detect.

6. Finally, the earth, like Mars and the moon, is constantly bombarded by particles in orbit about the sun—meteoroids and micrometeoroids. They are particularly important to us for we can subject them to chemical analysis and dating. We shall take up these objects in Chapter 11.

ELECTROMAGNETIC RADIATION

We expect that astronomers will continue to investigate astronomical objects by means of electromagnetic radiation for many decades to come. However, of all the radiation from space that will be observed throughout our lifetimes, the overwhelming majority is already in space on its way to us. What is the nature of this energy approaching us from all directions, waiting to be sampled by our telescopes for the century to come—and beyond? What are the secrets it holds, and what are the revelations it will give us about those objects it left years, centuries, even thousands of millions of years ago?

All kinds of electromagnetic radiation travel through empty space in a straight line and at the same speed (the speed of light)—2.997925×10^{10} cm/s (about 300,000 km/s or 186,000 mi/s)—a speed universally denoted by the symbol c.

Most of the characteristics of electromagnetic radiation can be described adequately if it is represented as energy propagated in waves, although no medium is required to transmit the waves. Any wave motion can be characterized by a *wavelength*. Ocean waves provide an analogy. The wavelength is simply the distance separating successive wave crests. Various forms of electromagnetic energy differ from each other only in their wavelengths. Those with the longest waves, ranging up to many kilometers in length, are called *radio waves*. Forms of electromagnetic energy of successively shorter wavelengths are called, respectively, infrared radiation, light, ultraviolet radiation, X rays, and finally the very short-wave gamma rays. The shortest radio waves, of wavelength from a few millimeters to a few centimeters, are also called *microwave* radiation. All these forms of radiation are the same basic kind of energy and could be thought of as different kinds of light. Here, however, we shall reserve the term "light" to describe those wavelengths of electromagnetic radiation that, by their action upon the organs of vision, stimulate sight.

Not all the properties of electromagnetic radiation can be described as wave motion. Experiments show that radiation of any given wavelength is always absorbed or emitted in quantities of energy that are whole multiples of some basic tiny quantity of radiant energy. It is as if electromagnetic energy were composed of many discrete "packets" of energy. These units of radiant energy are called *photons*. Photons must not be thought of as particles, for, as we have said, electromagnetic energy travels with a wavelike motion. Photons can be regarded as individual wave trains propagating through space, each spreading out in all directions from its source.

Propagation of Light—Inverse-Square Law

An important property of the propagation of electromagnetic energy is the *inverse-square law*, a relation that also applies to the propagation of other kinds of energy, for example, sound. The emitted radiation spreads out and covers an ever increasing area, in proportion to the square of the distance it has traveled from its source (Figure 5.1). For example, when light from the sun reaches the earth, it is spread out over a sphere 1 AU in radius. When it has gone twice as far, to 2 AU from the sun, that same light is spread over an area four times as great, and when it reaches Saturn, 10 AU from the sun, it is spread over an area 100 times that at the earth's distance (for the surface area of a sphere is proportional to the square of its radius).

On the other hand, the apparent brightness of a light source depends on how much of its luminous flux enters the pupil of our eye, or our telescope lens, or whatever detecting device is used. Since the collecting area of the eye or telescope lens is constant, the larger the area over which the light is spread, the smaller is the fraction observed. Thus the apparent brightness of a light source varies inversely with the square of its distance. At 2 AU from the sun, it would deliver only one-fourth as much light to our telescope as at earth, and at 10 AU, only $\frac{1}{100}$ as much.

Kinds of Electromagnetic Energy

Radio waves have the longest wavelength, ranging up to several kilometers. Those used in shortwave communication and in television have wavelengths ranging from a few centimeters to a few meters. When these waves pass a conductor, such as a radio antenna, they induce in it a feeble current of electricity. The antenna conducts that current to the receiver, where it is amplified and analyzed to detect the signal encoded in the transmitted waves.

The shortest wavelengths of radio radiation, less than about one millimeter, merge into infrared radiation. Infrared radiation of wavelength less than 0.0015 mm can be photographed with special emulsions. Certain substances such as lead sulfide and indium antimonide also increase in conductivity when they absorb infrared photons. Cells containing these materials as detectors measure the intensity of infrared radiation striking them by changes in their resistance to a current.

Electromagnetic radiation with a wavelength in the range 0.0004 to 0.0007 mm is visible light. It is more convenient to express the wavelengths of visible light in *angstroms*. One angstrom (abbreviated Å) is one hundred-millionth of a centimeter. Visible light, then, has wavelengths that range from about 4000 to 7000 Å. The exact wavelength of visible light determines its *color*. Radiation with a wavelength in the range 4000 to 4500 Å is perceived by the retina of the eye as the color violet. Radiations of successively longer wavelengths are perceived as the colors blue, green, yellow, orange, and red, respectively. The array of colors of visible light is called the *spectrum*. A mixture of light of all wavelengths, in about the same relative proportions as are found in the light emitted

Dispersion of white light into a spectrum by a prism.

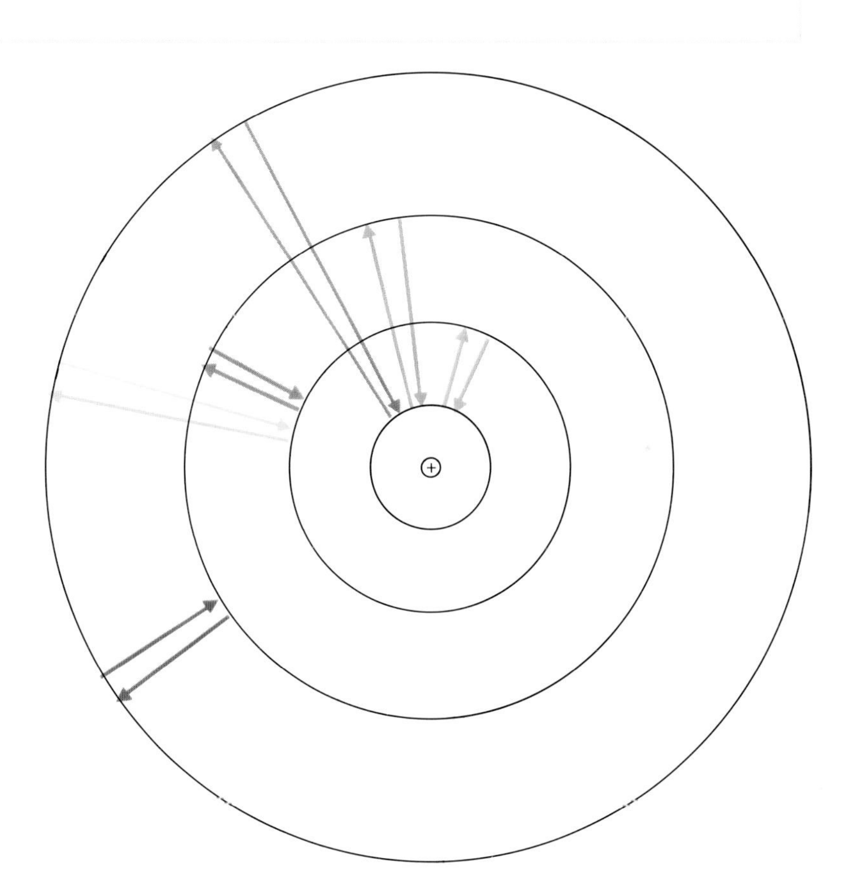

The fictitious atom Abellium and its spectrum.

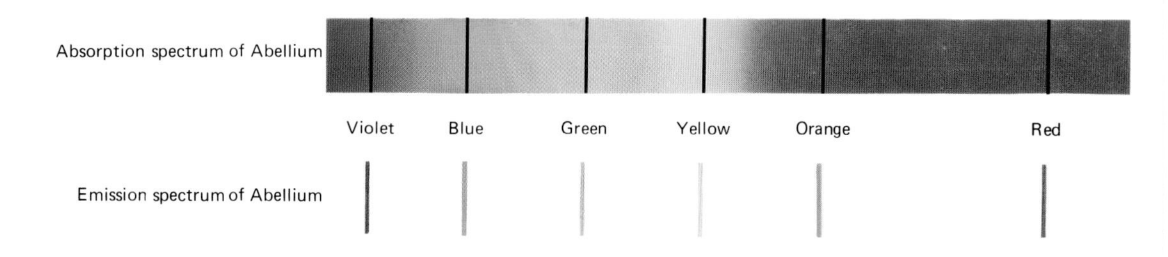

The 5-m (200-inch) telescope. *(Hale Observatories)*

The observer's cage at the prime focus of the 5-m telescope. *(Hale Observatories)*

The 122-cm (48-inch) Schmidt telescope at Palomar. *(Hale Observatories)*

The 3-m (120-inch) telescope of the Lick Observatory. *(Lick Observatory)*

The 1-m (40-inch) refracting telescope. *(Yerkes Observatory)*

The Robert McMath solar telescope. *(Kitt Peak National Observatory)*

The heliostat mirrors of the McMath solar telescope. *(Kitt Peak National Observatory)*

The Kitt Peak National Observatory. *(Kitt Peak National Observatory)*

The 90-m (300-foot) radio telescope.
(National Radio Astronomy Observatory)

Voyager spacecraft en route to Jupiter. *(NASA)*

The 4-m telescope of the Inter-American Observatory at Cerro Tololo, Chile.
(© Association of Universities for Research in Astronomy, Inc. The Cerro Tololo Inter-American Observatory)

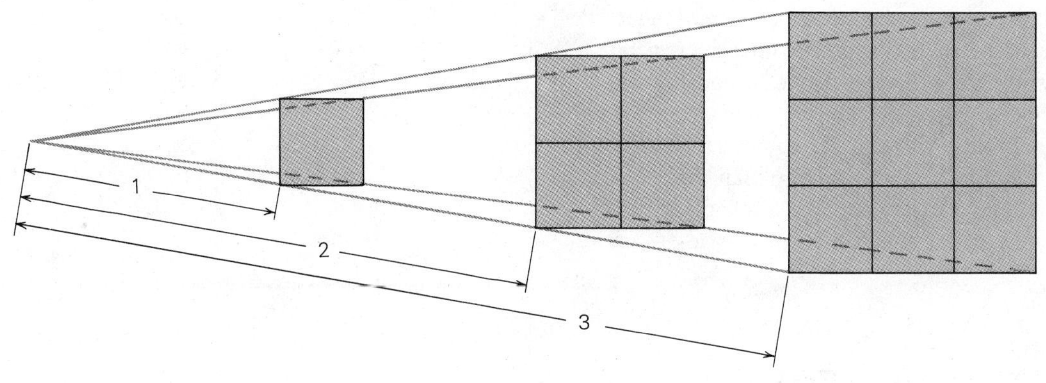

Figure 5.1
As light energy radiates away from its source, it spreads out, so that the energy passing through a unit area decreases as the square of the distance from its source.

into space by the sun, gives the impression of white light.

Radiation of wavelengths too short to be visible to the eye is called *ultraviolet*. Radiation of wavelengths less than about 200 Å are X rays. Both ultraviolet radiation and X rays, like visible light, can be detected photographically and photoelectrically.

Electromagnetic radiation of the shortest wavelength, not larger than 0.1 Å (10^{-8} mm), is called *gamma radiation*. Gamma rays are often emitted in the course of nuclear reactions and by radioactive elements. Gamma radiation is generated in the deep interiors of stars; it is gradually degraded into visible light by repeated absorption and reemission by the gases that make up the stars.

The array of radiation of all wavelengths, from radio waves to gamma rays, is called the *electromagnetic spectrum*.

The *frequency* of a wave motion is the rate at which wave crests pass a given point, that is, the number of wave crests that pass per second. Imagine a long train of waves moving to the right, past point O (Figure 5.2), at a speed c. If we measure the distance of c centimeters to the left of O, we arrive at the point P along the wave train that will just reach point O after a period of one second. The frequency f of the wave train—that is, the number of waves between P and O—times the length of each, λ, is equal to the distance c. Thus we see that for any wave motion, the speed of propagation equals the frequency times the wavelength.

The Energy of Photons

According to modern quantum theory, each photon carries a certain discrete amount of energy that depends only on the frequency of the radiation it is part of. Specifically, the energy of a photon is proportional to the frequency; the constant of proportionality, h, called *Planck's constant*, is named for Max Planck (1858–1947), the great German physicist who was one of the originators of the quantum theory. If metric units are used (that is, if energy is measured in ergs

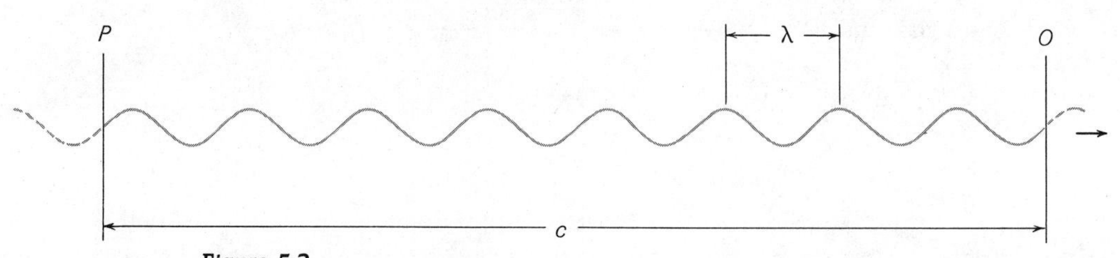

Figure 5.2
Relation between wavelength, frequency, and the speed of radiation.

and frequency in cycles or waves per second), Planck's constant has the value $h = 6.626 \times 10^{-27}$ erg•s. Since the frequency times the wavelength is equal to the speed of light, the energy of a photon is also inversely proportional to the wavelength. Photons of violet and blue light are thus of higher energy than those of red light. The highest energy photons of all are gamma rays; those of lowest energy are radio waves.

OPTICAL PROPERTIES OF LIGHT

Light obeys certain laws that are important to the design and construction of telescopes and other astronomical instruments. For example, light is *reflected* from a surface. A *normal* to a surface is simply a line or direction perpendicular to it. If light strikes a surface, its direction must make a certain angle with the normal to the surface at the point where it strikes. That angle is the *angle of incidence*. The angle that the reflected beam of light makes with the normal is called the *angle of reflection*. The law of reflection states that the angle of reflection is equal to the angle of incidence and that the reflected beam lies in the plane formed by the normal and the incident beam (see Figure 5.3).

Light is also bent or *refracted* when it passes from one kind of transparent medium into another. For a given wavelength the *index of refraction* of a transparent substance is a measure of the degree to which the speed of light is diminished while passing through it; specifically, it is the ratio of the speed of light in a vacuum to that in the substance. Usually, media of higher densities have higher indices of refraction. The index of refraction of a vacuum is, by definition, exactly 1.0. That of air at sea level is about 1.00029. Water has an index of about 1.3; crown glass and flint glass have indices of about 1.5 and 1.6, respectively. Diamond has the high index of refraction of 2.4.

When light passes from one medium into a second one of a different index of refraction, the angle the

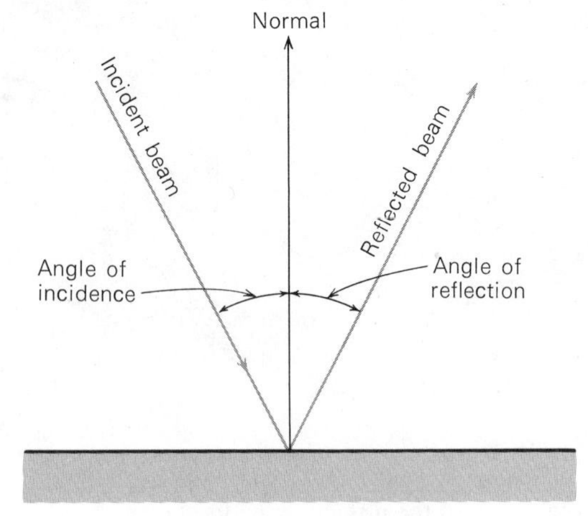

Figure 5.3
Law of reflection.

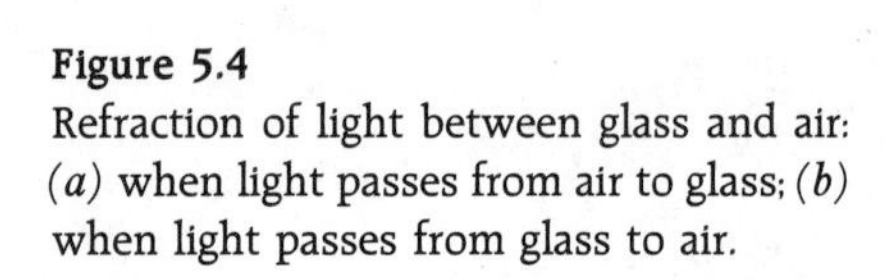

Figure 5.4
Refraction of light between glass and air: *(a)* when light passes from air to glass; *(b)* when light passes from glass to air.

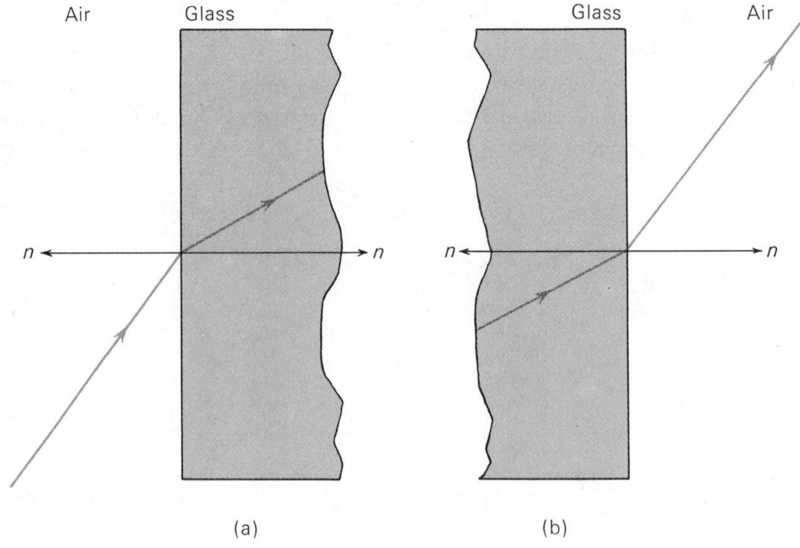

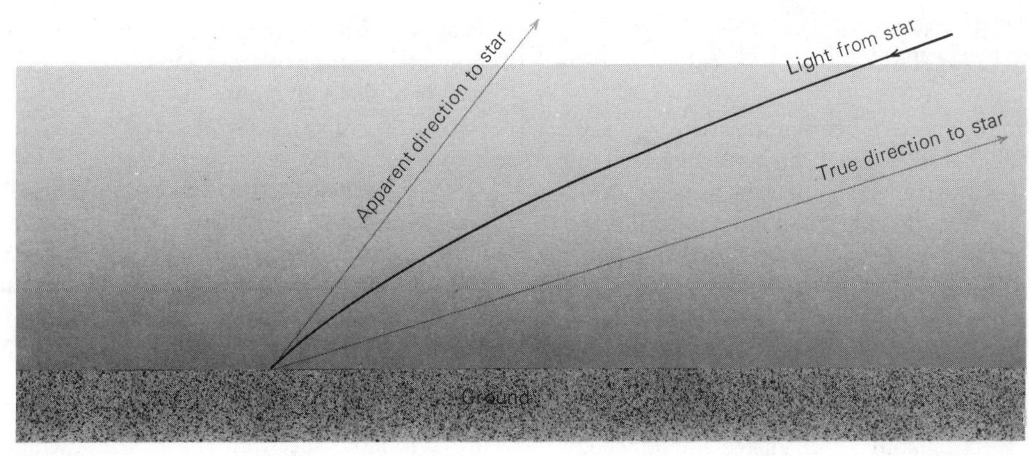

Figure 5.5
Atmospheric refraction. Light is bent upon entering the earth's atmosphere in such a way as to make stars appear at a higher altitude than they actually are. (The effect is grossly exaggerated in this figure.)

light beam makes with the normal to the interface between the two substances is always *less* in the medium of higher index. Thus, if light goes from air into glass or water, it is bent *toward* the normal to the interface, while if it goes from water or glass into air, it is bent *away* from the normal. For example, light from stars, planets, the sun, and the moon, is bent, upon entering the earth's atmosphere, in such a way as to make the object appear to be at a greater altitude above the horizon than it actually is. Atmospheric refraction is greatest for objects near the horizon, and it raises the apparent altitude of objects on the horizon by about ½°. The refraction of light passing through glass and the earth's atmosphere is illustrated in Figures 5.4 and 5.5, respectively.

The index of refraction of a transparent medium, however, is greater for light of shorter wavelengths. Thus, whenever light is refracted in passing from one medium into another, the violet and blue light, of shorter wavelengths, are bent more than are the orange and red light of longer wavelengths. This phenomenon is called *dispersion*.

Figure 5.6 shows how light can be separated into different colors with a prism, a piece of glass of triangular cross section. Upon entering one face of the prism, light is refracted once, the violet light more than the red, and upon leaving the opposite face, the light is bent again, and so is further dispersed. Even greater dispersion can be obtained by passing the light through a series of prisms. If the light leaving a prism is focused upon a screen, the different wavelengths or colors that compose white light are lined up side by side, as in a rainbow, formed by the dispersion of light through raindrops.

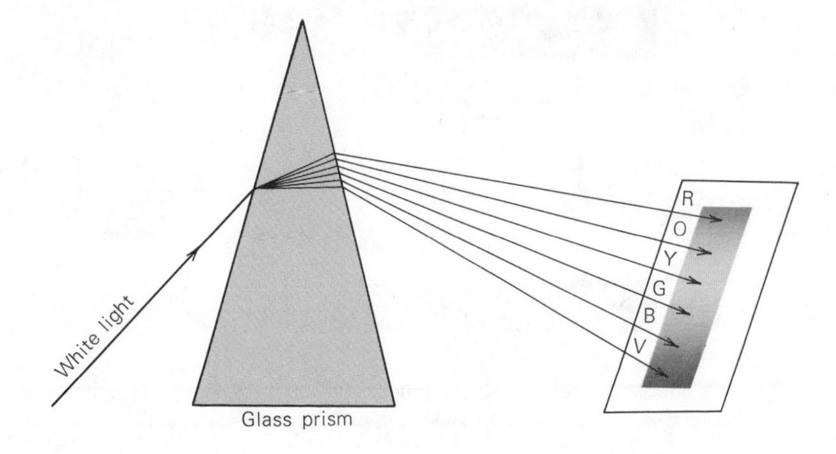

Figure 5.6
Dispersion by a prism.

SPECTROSCOPY IN ASTRONOMY

The light from a star or other luminous astronomical body, as well as from the sun, can be decomposed into its constituent wavelengths, producing a spectrum. White light is often described as a mixture of all the colors of the rainbow. Actually, white light is that particular mixture of spectral wavelengths (colors) in the light emitted from the sun. A device used to observe the spectrum of a light source visually is a *spectroscope;* one used to photograph a spectrum is a *spectrograph*. In Chapter 6 we will explain how a spectroscope (or a spectrograph) is constructed and attached to a telescope so that we may study the spectra of astronomical objects.

The Value of Stellar Spectra

If the spectrum of the white light from the sun and stars were simply a continuous rainbow of colors, astronomers would have far less interest in the study of stellar spectra. To Newton, the solar spectrum did appear as just a continuous band of colors. However, in 1802, William Wollaston (1766–1828) observed several dark lines running across the solar spectrum. He attributed these lines to natural boundaries between the colors. Later, in 1814 and 1815, the German physicist Joseph Fraunhofer (1787–1826), upon a more careful examination of the solar spectrum, found about 600 such dark lines.

Subsequently, it was found that such dark spectral lines could be produced in the spectra of artificial light sources by passing their light through various transparent substances or gases. On the other hand, the spectra of the light emitted by certain glowing gases were observed to consist of several separate bright lines.

We distinguish, then, among three types of spectra (Figure 5.7). A *continuous* spectrum is an array of all wavelengths or colors of the rainbow. A *bright line* or *emission* spectrum appears as a pattern or series of bright lines; it is formed from light in which only certain discrete wavelengths are present. A *dark line* or *absorption spectrum* consists of a series or pattern of dark lines — missing wavelengths — superposed upon the continuous spectrum of a source of white light.

Each particular chemical element or compound, when in the *gaseous form*, produces its own characteristic pattern of dark or bright lines. In other words, each particular gas can absorb or emit only certain wavelengths of light, peculiar to that gas. The presence of a particular pattern of dark (or bright) lines characteristic of a certain element is evidence of the presence of that element somewhere along the path of the light whose spectrum has been analyzed.

Thus the dark lines (Fraunhofer lines) in the solar spectrum give evidence of certain chemical elements between us and the sun, absorbing those wavelengths of light. It is easy to show that most of the lines must originate from gases in the outer part of the sun itself.

The wavelengths of the lines produced by various elements are determined by laboratory experiment. Most of the thousands of Fraunhofer lines in the sun's spectrum have now been identified with more than 60 of the known chemical elements.

Dark lines are also found in the spectra of stars

Figure 5.7
Three kinds of spectra: *(a)* continuous; *(b)* bright line; *(c)* dark line. The spectra are shown as they appear on photographic negatives, from which astronomers generally work.

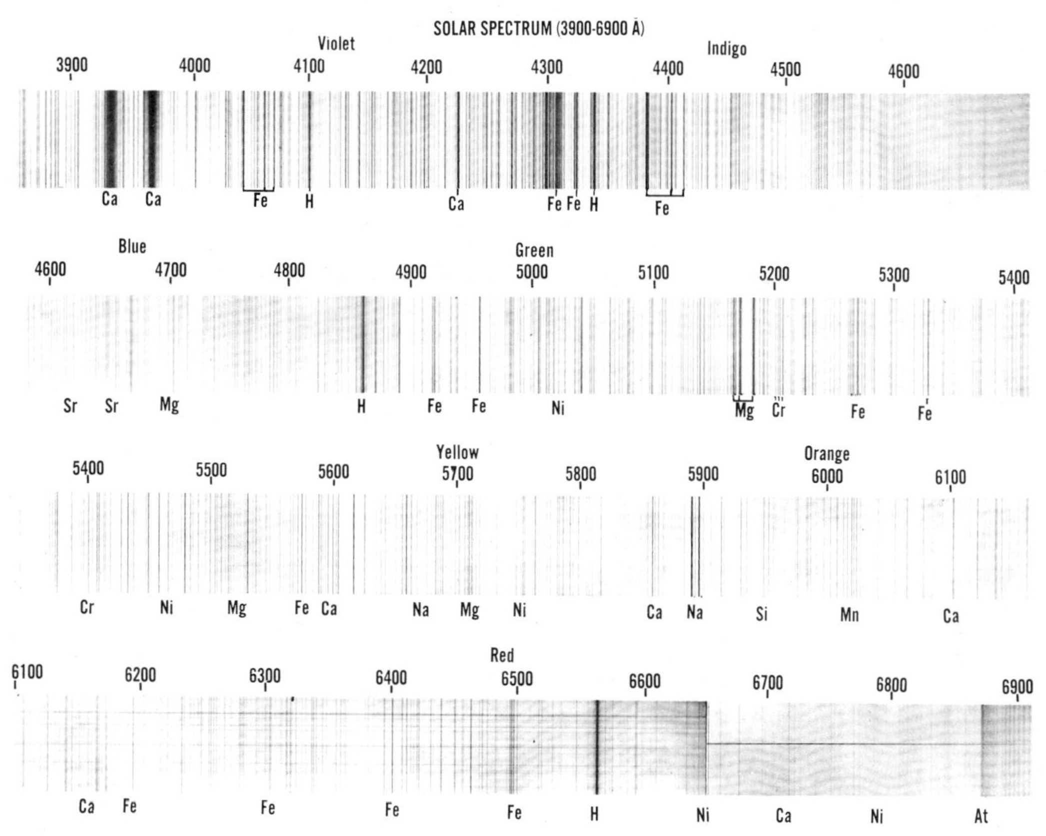

Figure 5.8
The solar spectrum. Labels indicate the elements in the sun's photosphere that cause some of the dark lines. The wavelengths in angstroms and the colors of the different parts of the spectrum are also labeled. *(Hale Observatories)*

and other celestial objects. Much can be learned from a star's spectrum, in addition to evidence of the chemical elements present in its outer layers. A detailed study of its spectral lines indicates the temperature, pressure, turbulence, and physical state of the gases in that star; whether or not magnetic and electric fields are present, and the strengths of those fields; how fast the star is approaching or receding from us; and many other data. Much additional information about a star can be obtained by studying its continuous spectrum. The study of the spectra of celestial objects is one of the most powerful means at the astronomer's disposal for obtaining data about the universe.

The Doppler Effect

In 1842 Christian Doppler (1803–1853) pointed out that if a light source is approaching or receding from the observer, the light waves will be, respectively, crowded closer together or spread out. This principle, known as the *Doppler principle* or *Doppler effect,* is illustrated in Figure 5.9. In *(a)* the light source is stationary with respect to the observer. As successive wave crests 1, 2, 3, and 4 are emitted, they spread out evenly in all directions, like the ripples from a splash in a pond. They approach the observer at a distance λ behind each other, where λ is the wavelength of the light. On the other hand, if the source is moving with respect to the observer, as in *(b)*, the successive wave crests are emitted with the source at different positions, S_1, S_2, S_3, and S_4, respectively. Thus, to observer A, the waves seem to follow each other by a distance *less* than λ, whereas to observer C they follow each other by a distance greater than λ. The wavelength of the radiation received by A is shortened; the wavelength of the radiation received by C is lengthened. Because the wave crests arrive at A following on each other more closely than they do from a stationary source, they are observed by A at a higher than normal frequency, while C receives the light at a diminished

Figure 5.9
The Doppler effect.

(a) To observer

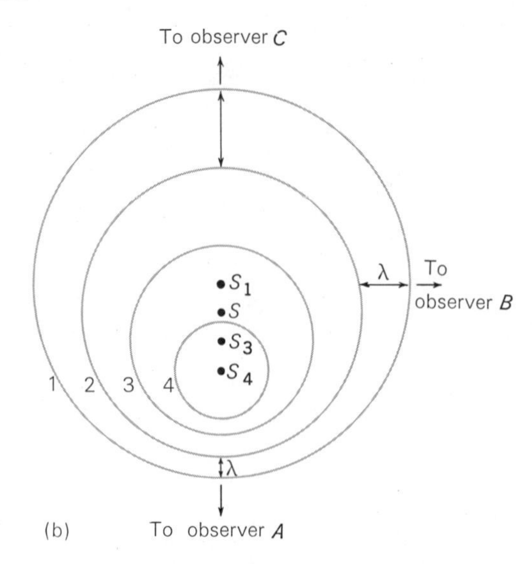

frequency. To observer B, in a direction at right angles to the motion of the source, no effect is observed. The effect is produced only by a motion *toward* or *away* from the observer, a motion called *radial velocity*. Observers between A and B and B and C would observe some shortening or lengthening of the light waves, respectively, for a component (or part) of the motion of the source is in their line of sight.

The Doppler effect is also observed in sound. Most of us have heard the higher than normal pitch of the whistle of an approaching train and the lower than normal pitch of a receding one. The precise amount by which the wavelengths of light from a source appear lengthened or shortened (the *Doppler shift*) provides us a means of determining how fast that source is moving in the line of sight—that is, its radial velocity. If its speed is small compared to the speed of light, the relative shift of the wavelengths of its light is proportional to its radial velocity divided by the speed of light.

If a star approaches or recedes from us, the wavelengths of light in its continuous spectrum appear shortened or lengthened, as well as those of the dark lines. However, unless its speed is tens of thousands of kilometers per second, the star does not appear noticeably bluer or redder than normal. The Doppler shift is thus not easily detected in a continuous spectrum (except for very remote galaxies—see Chapter 21) and cannot be measured accurately in such a spectrum. On the other hand, the wavelengths of the absorption lines can be measured accurately, and their Doppler shift is relatively simple to detect. Generally, when the spectrum of a star or other object is photographed at the telescope, sometime during the exposure the light from an iron arc or some other emission-line source is allowed to pass into the same spectrograph, and the spectrum of the arc is then photographed just beside that of the star. The known wavelengths of the bright lines in the spectrum of the arc (or other laboratory source) serve as standards against which the wavelengths of the dark lines in the star's spectrum can be accurately measured. Further illustrations of the Doppler effect are given in later chapters.

RADIATION LAWS

If a body intercepts electromagnetic radiation, it generally reflects some of it, transmits some, and absorbs the rest. Except for transparent objects the transmitted radiation is negligible, if even measurable. The energy associated with the absorbed radiation heats the body, and would continue to heat it indefinitely if the body did not begin to radiate that energy away. A body (for example, Mars) that is exposed to an approximately constant flow of radiation (sunlight) will eventually reach an equilibrium temperature such that the average rate at which it reradiates energy just equals the rate at which it absorbs it. Therefore, an opaque body (Mars, or some other planet) may be observed both by the radiation it reflects and by the reradiated energy it previously absorbed.

The quality, or distribution of energy with wavelength (called the *spectral energy distribution*), of the reflected radiation depends on the absorbing properties of the body. If, for example, it is exposed to white light and absorbs the same fraction of electromagnetic energy of all wavelengths, it appears, in reflected light, to be white. If it preferentially absorbs short wavelengths, the light it reflects is dominated by long wavelengths and it appears red. Thus the colors of objects depend on those objects' absorbing properties.

Many complicating factors determine the quality of the energy reemitted by the body, but the most important is its temperature when the energy it reemits is in equilibrium with that which it absorbs. If a body with no internal source of energy or stored heat were perfectly reflecting, it would absorb and reemit nothing and its temperature would be at absolute zero. Planets typically have temperatures of tens or hundreds of degrees absolute,[1] and reradiate most of the solar energy they absorb at infrared and radio wavelengths. At room temperature on earth, almost all objects (including people) are emitting infrared radiation, with a maximum intensity near 0.01 mm wavelength.

Perfect Radiators

Of particular interest is a hypothetical *perfect radiator*, also called a *blackbody*. This is an idealized body that absorbs all the electromagnetic energy incident on it. Its temperature then depends only on the total radiant energy striking it each second, and the spectral energy distribution of the energy it reradiates can be precisely predicted from the *radiation laws*.

A perfect radiator or blackbody is black in the sense that it is completely opaque to all wavelengths, but of course it need not look black. At room temperature the radiated energy is mostly in the invisible infrared and a blackbody does look black, but a perfect radiator with a temperature of thousands of degrees appears very bright indeed. We shall see (Chapter 13) that stars approximate perfect radiators; thus, their light has nearly the properties of blackbodies.

[1] In astronomy, temperature is almost always expressed on the Kelvin or absolute scale—that is, in centigrade degrees above absolute zero (−273 °C); see Appendix 5.

Planck's Radiation Law

The energy emitted from blackbodies had been studied experimentally during the last century. It was only about the beginning of the twentieth century, however, that the great German physicist Max Planck found a theoretical interpretation for blackbody radiation. He succeeded in this by adopting the hypothesis that light energy is *quantized* in discrete "packets" (that is, photons). An equation derived by Planck gives the radiant energy emitted per second at each wavelength from each cm^2 of the surface of a blackbody of a given temperature.[2] In Figure 5.10, this energy distribution, computed from Planck's formula, is plotted against wavelength for several different temperatures.

We notice in Figure 5.10 that there is a particular wavelength, λ_{max}, at which a perfect radiator emits its

[2] If T is the absolute temperature, λ the wavelength in centimeters, and k Boltzmann's constant (1.38×10^{-16}), $E(\lambda, T)$, the energy in ergs emitted per unit wavelength interval per second per square centimeter and into unit solid angle, is

$$E(\lambda, T) = \frac{2hc^2}{\lambda^5} \frac{1}{e^{hc/\lambda kT} - 1} .$$

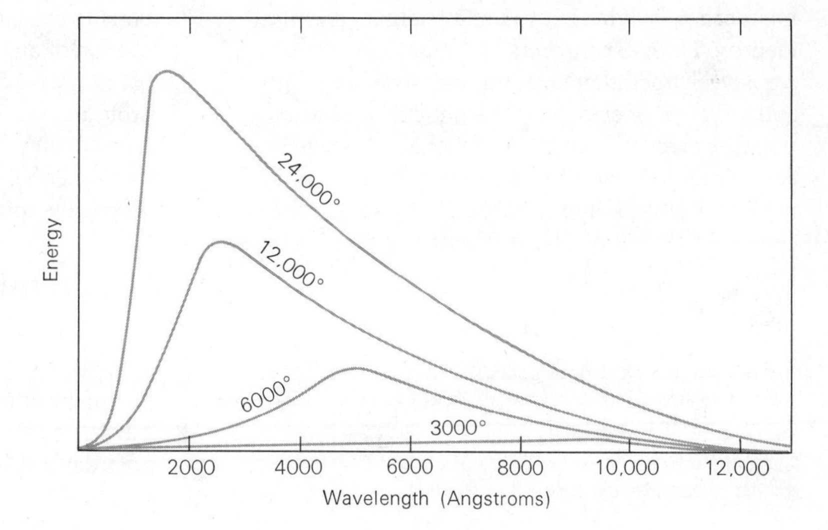

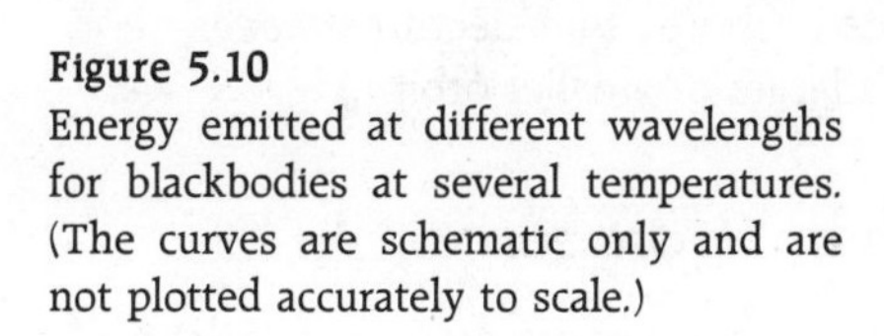

Figure 5.10
Energy emitted at different wavelengths for blackbodies at several temperatures. (The curves are schematic only and are not plotted accurately to scale.)

maximum light. The higher the temperature of the body, the shorter is λ_{max}; in fact, λ_{max} is inversely proportional to the temperature. This relation is known as *Wien's law.*[3]

If we sum up the contributions from all parts of the spectrum, we obtain the total energy emitted by a blackbody over all wavelengths. That total energy, emitted per second per square centimeter by a blackbody at a temperature T, is proportional to the fourth power of its absolute temperature. The relation is known as the *Stefan-Boltzmann law.*[4]

We shall see in coming chapters that the radiation laws are extremely useful in helping us derive properties of stars and other astronomical objects.

ABSORPTION AND EMISSION OF LIGHT BY ATOMS

The radiation from stars shows that even their outermost, coolest layers have temperatures of thousands of degrees. Stellar temperatures are above the boiling points of all chemical substances. Moreover, except in the atmospheres of the cooler stars, the temperatures are too high for chemical compounds or even for simple molecules to exist. Consequently, the gases of stars are in the atomic state. The relative abundances of those atoms of different kinds, and the physical states they are in, determine the appearance of the spectrum of the light from a star.

Structure of Atoms

Atoms are the "building blocks" of matter; they are the smallest particles into which a chemical element can be subdivided and still retain its chemical identity. From experiments in the physics laboratory, we have learned that an atom consists of a nucleus and a system of electrons. The nucleus of an atom contains practically all the atom's mass. The nucleus also carries a positive electric charge. The charge on a nucleus is always an integral (whole) multiple of a certain small unit of charge. The atoms of the different chemical elements differ from each other in the number of these charge units on their nuclei. The number of charge units on the nucleus of an atom is called its *atomic number*. An atom of hydrogen (the simplest kind of atom) has an atomic number of 1. The atomic numbers of helium and lithium are 2 and 3, respectively, of oxygen, 8, and of uranium, 92. Atoms of still higher atomic number have been produced in the nuclear physics laboratory, but most of them are unstable and in relatively short times they spontaneously decay into simpler atoms.

The mass of an atomic nucleus is also nearly equal to an integral multiple of a basic unit, called the *atomic mass unit* (amu). The atomic mass unit is about 1.66×10^{-24} g. The mass of a nucleus, in amu, is called its *atomic weight*. Most hydrogen nuclei have an atomic weight of about 1; a few, however, have atomic weights of 2 or even 3. Atomic nuclei of the same atomic number but different masses are said to compose the different *isotopes* of that element. For example, an isotope of hydrogen in which the atomic nuclei have an atomic weight of 2 is called *deuterium*, or sometimes "heavy hydrogen."

Nuclei can be decomposed into smaller particles. The most important of these are *protons*, which have a charge of one unit and a mass of approximately one unit (the nucleus of the most common isotope of hydrogen *is* a proton), and *neutrons*, which have a mass of about one unit, but are electrically neutral (have no charge).

An electron has a mass of $1/1835$ amu and carries a *negative* charge that is numerically equal to the positive charge on a single proton. Under ordinary circumstances, an atom contains just as many electrons as its nucleus has positive units of charge, and so is electrically *neutral*. The electrons of an atom are clustered outside the nucleus and revolve about it. The orbital motion of an electron corresponds to a certain amount of energy possessed by the atom, and the larger the orbit, the greater the energy. When an atom absorbs or emits energy (for example, in the form of electromagnetic energy), one or more of its electrons moves, respectively, into a larger or smaller orbit.

Spectrum of Hydrogen

A clue to the structure of atoms came from the study of the spectrum of hydrogen, whose atoms are the simplest in nature. The dark lines of hydrogen that can be observed in the spectra of many stars occur in an orderly spaced series of wavelengths. The bright

---

[3] Wien's law can be derived from Planck's formula by finding the wavelength at which the derivative of Planck's formula equals zero.

[4] By integrating Planck's formula. The total energy emitted is equivalent to the areas under the curves in Figure 5.10.

lines of hydrogen that are observed in the laboratory spectrum of glowing hydrogen are observed in the same series of wavelengths. The Swiss physicist Johann Balmer (1825–1898) found a simple formula that gives the wavelengths of this series of lines, now called the *Balmer series* of hydrogen.

The Danish physicist Niels Bohr (1885–1962) suggested that the hydrogen spectrum can be explained if it is assumed that only orbits of certain sizes are possible for the electron in the hydrogen atom. By specifying those permissible sizes for the electron orbits, Bohr was able to compute the values of energy, corresponding to the orbital motion of the electron, that are possible for an individual atom. He assumed that an atom can change from one allowed state of energy to another state of higher energy if its electron moves from a smaller to a larger allowed orbit. Conversely, according to the hypothesis, if the electron moves from a larger to a smaller orbit, the atom changes from a higher to a lower state of energy. One way in which an atom can gain or lose energy is by absorbing or emitting light. Since light is composed of *photons* whose energies depend on their wavelengths the only wavelengths of light that could be absorbed or emitted are those corresponding to photons possessing energies equal to differences between various allowed energy states of the hydrogen atom.

For example, suppose a beam of white light (which consists of photons of all wavelengths) is passed through a gas of atomic hydrogen. A photon of wavelength 6563 Å has the right energy to raise an electron in a hydrogen atom from the second to the third orbit, and can be absorbed by those hydrogen atoms that are in their second to lowest energy states. Thus, the hydrogen atoms absorb light only at certain wavelengths, and produce the spectral *lines*. Conversely, hydrogen atoms in which electrons move from larger to smaller orbits emit light—but again only light of those energies or wavelengths that correspond to the energy differences between permissible orbits. The transfer of electrons giving rise to spectral lines is shown in Figure 5.11.

A similar picture can be drawn for kinds of atoms other than hydrogen. However, since they ordinarily have more than one electron each, the energies of the orbits of their electrons are much more complicated, and the problem of their spectra is much more difficult to handle theoretically.

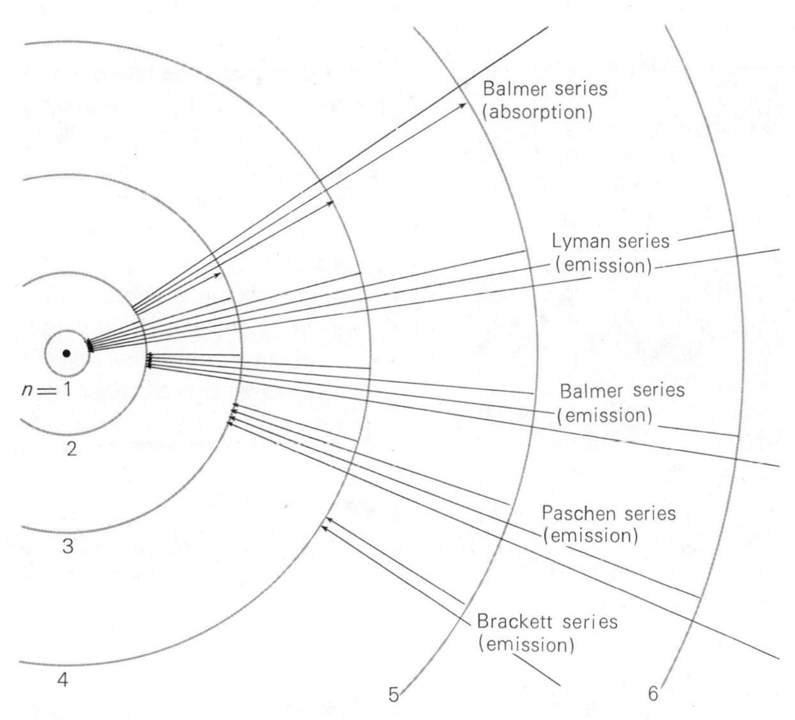

Figure 5.11
Emission and absorption of light by the hydrogen atom according to the Bohr model.

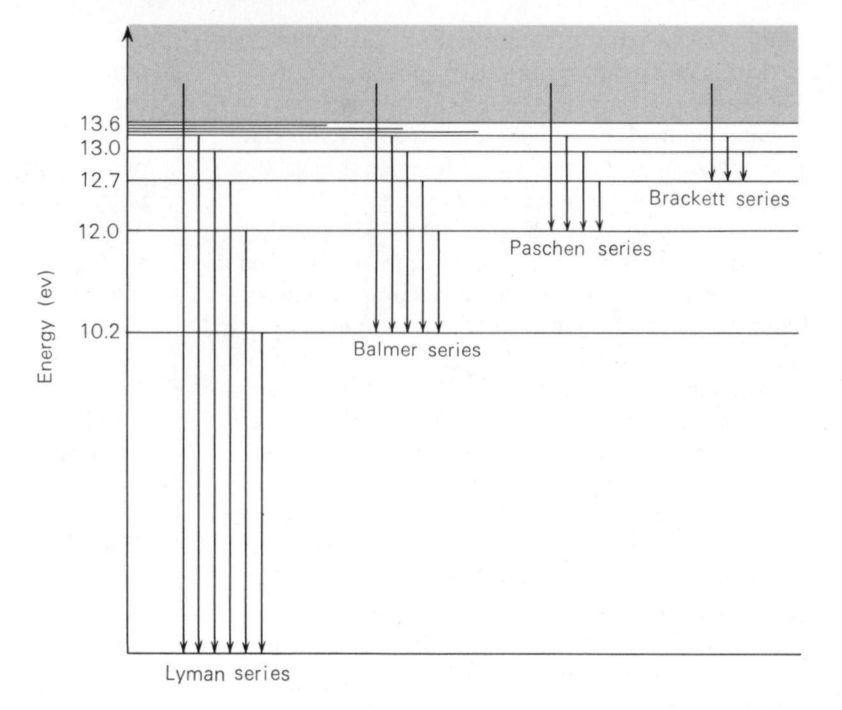

Figure 5.12
Energy-level diagram for hydrogen. The shaded region represents energies at which the atom is ionized.

Energy Levels of Atoms and Excitation

Bohr's model of the hydrogen atom was one of the beginnings of quantum theory, and was a great step forward in the development of modern physics and in our understanding of the atom. However, we know today that atoms cannot be represented by quite so simple a picture as the Bohr model. Even the concept of discrete orbits of electrons must be abandoned, because according to the modern quantum theory it is impossible at any instant to state the exact position and the exact velocity of an electron in an atom simultaneously. Nevertheless, we still retain the concept that only certain discrete energies are allowable for an atom. These energies, called *energy levels*, can be thought of as representing certain mean or average distances of an electron from the atomic nucleus.

Ordinarily, the atom is in the state of lowest possible energy, its *ground state*, which, in the Bohr model, would correspond to the electron being in the innermost orbit. However, an atom can absorb energy which raises it to a higher energy level (corresponding, in the Bohr picture, to the movement of an electron to a larger orbit). The atom is then said to be in an *excited state*. Generally, an atom remains excited only for a very brief time; after a short interval, typically a hundred-millionth of a second or so, it drops back down to its ground state, with the simultaneous emission of light, unless it chances to absorb another photon first and go to a still higher state. (In the Bohr model, this corresponds to a jump by the electron back to the innermost orbit.) The atom may return to its lowest state in one jump, or it may make the transition in steps of two or more jumps, stopping at intermediate levels on the way down. With each jump, it emits a photon of the wavelength that corresponds to the energy difference between the levels at the beginning and end of that jump. An energy-level diagram for a hydrogen atom and several possible *atomic transitions* are shown in Figure 5.12; compare this figure with the Bohr model, shown in Figure 5.11.[5]

Because atoms that have absorbed light and have thus become excited generally deexcite themselves and emit that light again, we might wonder why dark lines

[5] Actually, according to the modern quantum theory, atomic energy levels are not *perfectly* discrete (that is, "sharp"), but are, rather, the *most probable* energies of an atom. The vast majority of atoms have energies that lie in the immediate neighborhood of one of the allowed levels. At any time, however, a few atoms possess energies that deviate somewhat from one of those values. Spectral lines caused when atoms jump from one energy level to another, therefore, are not perfectly sharp, but spread over a small, but finite, range of wavelengths. The effect is called *natural line broadening*.

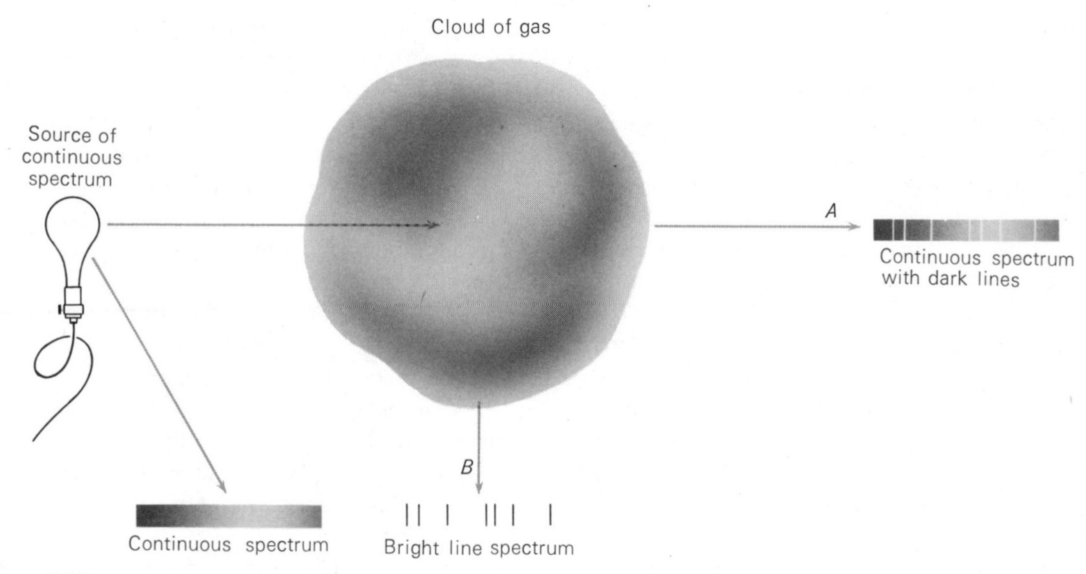

Figure 5.13
The atoms in the gas cloud produce absorption lines in the continuous spectrum of the white light source when viewed from direction A, but produce emission lines (of the light they reemit) when viewed from direction B. The spectra are shown as they appear on the photographic negatives.

are ever produced in stellar spectra. In other words, why doesn't this reemitted light "fill in" the absorption lines? Some of the reemitted light actually *is* received by us and this light does partially fill in the absorption lines, but only to a slight extent. The reason is that the atoms reemit the light they absorb in random directions. We can observe the reemitted light as emission lines only if we can view the absorbing atoms from a direction from which no light with a continuous spectrum is coming—as we do, for example, when we look at gaseous nebulae (Chapter 15). Figure 5.13 illustrates the situation.

Atoms in a gas are moving at high speeds and continually colliding with each other and with electrons. They can be excited and deexcited, therefore, by these collisions as well as by absorbing and emitting light. The mean velocity of atoms in a gas depends on its temperature (actually, *defines* the temperature), and if we know the temperature of the gas, we are able to calculate what fraction of its atoms, at any given time, are excited to any given energy level.

Ionization

If enough energy is absorbed by an atom, one or more of its electrons can be removed completely. The atom is then said to be *ionized* (it is then called an *ion*). An atom that has become ionized has lost a negative charge—that carried away by the electron—and thus is left with a net positive charge. It has, therefore, a strong affinity for a free electron (because opposite electrical charges attract each other), and eventually will capture one and become neutral (or ionized to one less degree) again. During the capture process, the atom emits one or more photons, depending on whether the electron is captured at once to the state corresponding to the lowest energy level of the atom, or whether it stops at one or more intermediate levels on the "way in."

Just as the excitation of an atom can result from a collision with another atom, ion, or electron (collisions with electrons are usually most important), so also can ionization. The rate at which such collisional ionizations occur depends on the atomic velocities and hence on the temperature of the gas. From a knowledge of the temperature and density of a gas, it is possible to calculate the fraction of atoms that have been ionized once, ionized twice, and so on. In the outer layer of the sun, for example, we find that most of the hydrogen and helium atoms are neutral, whereas most of the atoms of calcium, as well as many other metals, are once ionized.

The energy levels of an ionized atom are entirely different from those of the same atom when it is neutral. In each degree of ionization, the energy levels of the ion, and thus the wavelengths of the spectral lines it can produce, have their own characteristic values. In the sun, therefore, we find lines of *neutral* hydrogen and helium, but of *ionized* calcium. Ionized hydrogen, of course, having no electron, can produce no absorption lines.

There is an additional mechanism by which ionized atoms can absorb and emit light in the continuous spectrum. When an electron passes near an ion, it is attracted by that ion's positive charge. If the electron is not captured, it will pass the ion on a hyperbolic orbit, much like two stars passing in interstellar space. While the electron is passing, however, the ion can absorb or emit a photon which accompanies a corresponding increase or decrease in the kinetic energy of the hyperbolic motion of the electron. The process is called *free-free* absorption or emission (because the electron is "free" of the ion both before and after the encounter), or *bremsstrahlung*. Photons of any wavelength can be absorbed or emitted in bremsstrahlung.

COSMIC RAYS

Our bodies are constantly being subjected to a rain of invisible high-energy particles passing through them. These bullets of radiation, undetected by our senses, result from the entrance of some 10^{18} atomic nuclei into the earth's atmosphere each second at speeds near that of light. The total rate of influx of energy to the earth from these particles is comparable to the rate at which the earth receives energy that comes from starlight.

The physicist has learned, through the study of this phenomenon, of kinds of subatomic particles (for example, *muons* and *positrons*) *hitherto* unobserved. The incoming particles of extraterrestrial origin interest the astronomer because most are believed to come from beyond the solar system, and the discovery of the origin of these cosmic rays may provide us with new clues about the nature of the universe.

At first, cosmic rays were believed to be very high-energy photons, that is, electromagnetic energy of wavelengths even less than those of gamma rays. In 1927, however, the Dutch physicist Jacob Clay (1882–1955) found that the intensity of this radiation (cosmic rays) varies with latitude, being least near the geomagnetic equator (the circle halfway between the geomagnetic poles), and increasing as the geomagnetic poles are approached.[6] Clay's observations have been confirmed with many subsequent experiments. Photons, which have no electrical charge, could not be affected by the magnetic field of the earth as they approach it. Cosmic rays, therefore, must consist of charged particles striking the earth from outer space.

The realization only a few decades ago that the ionization noticed in the earth's atmosphere is due to charged particles striking the earth from space was one of the important discoveries of the twentieth century. Investigation of these particles became, in the 1940s, a major effort of modern physics.

Analysis of cosmic-ray particles striking the upper atmosphere shows that most of them are high-speed protons (nuclei of hydrogen atoms), that most of the rest are alpha particles (nuclei of helium atoms), and that a few are nuclei of the still heavier atoms. A primary particle traverses, on the average, only about one-tenth of the earth's atmospheric gases, however, before colliding with the nucleus of an air molecule.

When such a collision occurs, the nucleus in the air molecule breaks into several smaller subatomic particles. If the primary particle has high energy, each of these secondary particles is also given considerable energy. Each, in turn, collides with still another nucleus in an air molecule, producing more secondary particles. In this way an original primary particle moving with high speed dissipates its energy in a great many secondary particles, producing many of the particles that are recorded at intermediate and low altitudes in the atmosphere. A large proliferation of particles by successive collisions following the impingement of a primary particle of very high energy is called a *shower*.

Perhaps it is well that we do not see these primary and secondary particles or feel them as they strike us. Otherwise, it might be discomforting to see repeated skyrocketlike bursts high above our heads, and then observe many of the burst particles passing into one side of our bodies and out through the other. Those few which are intercepted by the atoms of our bodies contribute to the natural radiation our bodies are constantly receiving, and are responsible for some mutations and probably even cancer.

---

[6] The *geomagnetic poles* are in line with the ends of a hypothetical ideal bar magnet whose magnetic field most nearly matches that of the earth. The actual field of the earth, however, is somewhat irregular, and the earth's *magnetic poles*, where the actual lines of force are perpendicular to the surface, deviate by some hundreds of kilometers from the idealized geomagnetic poles.

SUMMARY REVIEW

Kinds of radiation striking earth: electromagnetic radiation; cosmic rays; solar wind; neutrinos; gravitational waves; meteoroids and micrometeoroids

Electromagnetic radiation: speed; wavelength; frequency; photons; inverse-square law of propagation; radio radiation; infrared radiation; light; ultraviolet radiation; X rays; gamma rays; Planck's constant

Optics laws: reflection; refraction; index of refraction; dispersion

Spectroscopy: spectroscope; spectrograph; continuous spectrum; Fraunhofer lines; dark line (absorption) spectrum; bright line (emission) spectrum; Doppler effect; radial velocity

Radiation laws: spectral energy distribution; perfect radiator (blackbody); Planck's law; Wien's law; Stefan–Boltzmann law

Structure of atoms: nucleus; electron; proton; neutron; atomic number; atomic weight; amu; isotope

Atomic spectra: Balmer series; Bohr atom; energy levels; absorption and emission of light; excitation; ionization; free-free transitions (bremsstrahlung)

Cosmic rays: geomagnetic poles and equator; primary and secondary particles; shower

EXERCISES

1. What kind of electromagnetic energy consists of photons (a) of the highest energy? (b) of the lowest energy? (c) that affect the organs of vision?

2. How many times brighter or fainter would a star appear if it were moved to (a) twice its present distance? (b) ten times its present distance? (c) half its present distance?

3. "Tidal waves," or *tsunamis*, are waves of seismic origin that travel rapidly through the ocean. If tsunamis traveled at the speed of 600 km/hr, and approached a shore at the rate of one wave crest every 15 minutes, what would be the distance between those wave crests at sea?

4. Because of refraction, the sun appears to rise before it is above the geometrical horizon and to set after it has dropped below the geometrical horizon. By how much does atmospheric refraction increase the hours of sunshine in a typical day?
Answer: About four minutes or more.

5. Give some examples in nature of (a) reflection of light; (b) refraction of light; (c) dispersion of light.

6. How could you measure the rotation rate of the sun by photographing the spectrum of light coming from various parts of the sun's disk?

7. How could you measure the earth's orbital speed by photographing the spectrum of a star at various times throughout the year?

8. Stars are fairly good approximations to blackbodies. Explain why they do not look black.

9. What color is the sun?

10. Most hydrogen atoms in the sun are in their lowest state of energy. What series of absorption lines would hydrogen produce most strongly in the sun?

11. How many times can each of the following kinds of atoms be ionized: (a) helium? (b) oxygen? (c) uranium? (d) hydrogen? (e) lithium?

12. Where on earth would you like to live to have the best chance of avoiding radiation resulting from primary cosmic-ray particles striking the earth?

CHAPTER 6

Grote Reber (1911–   ), an amateur astronomer and electronics expert, built in his own back yard, out of wooden two-by-four's and galvanized iron, the first radio telescope specifically designed to observe radio waves from space. From 1937 until after World War II, he was the world's *only* active radio astronomer. *(Ohio State University)*

ASTRONOMICAL OBSERVATIONS

Electromagnetic radiation from space comes at all wavelengths, from gamma rays to radio waves. Unfortunately for astronomical observations (but fortunately for biological organisms) much of the electromagnetic spectrum is filtered out by the terrestrial atmosphere. There are two spectral windows in the atmospheric filter through which we can observe celestial radiation (Figure 6.1). One of these is the *optical window*, which allows passage of the near ultraviolet (wavelengths longer than about 3000 Å), visible light, and portions of the infrared (with wavelengths up to about 0.03 mm). The other is the *radio window*, which lets in radio waves of length from about a millimeter to about 20 m (the long wavelength cutoff depends somewhat on the variable conditions of the ionosphere).

We turn first to the optical window.

OPTICAL INSTRUMENTS

The simplest kind of optical detector would do no more than reveal the presence of radiation striking it. But for most purposes it is more useful to form an *image* of an object of known direction in the sky. The image can then be detected, measured, reproduced,

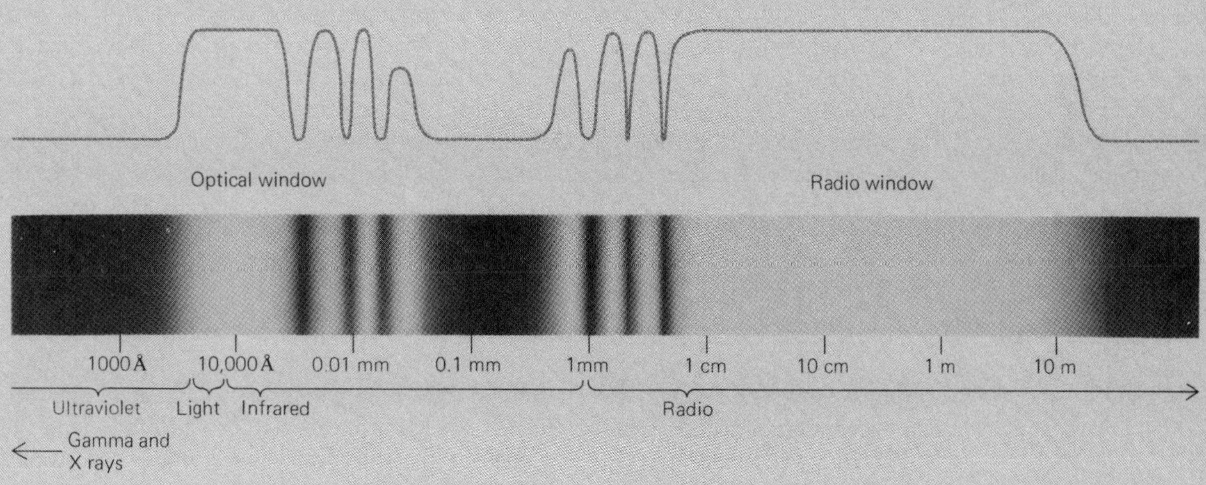

Figure 6.1
A portion of the electromagnetic spectrum, showing those regions (windows) to which the earth's atmosphere is transparent. The dark regions are those to which the atmosphere is opaque. The upper graph is a plot of the transparency of the atmosphere.

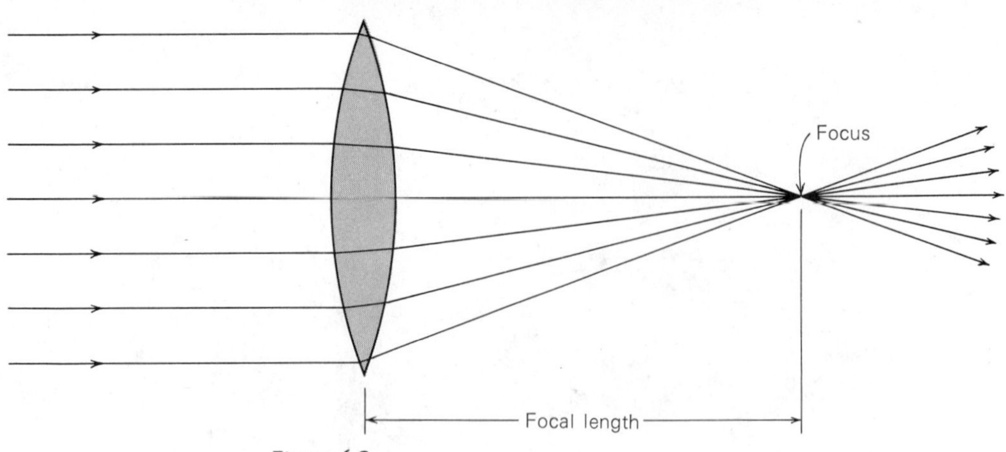

Figure 6.2
Formation of an image by a simple convex lens.

and analyzed in a host of ways. One image-forming device is the eye. It consists of a *lens*, which focuses light into an optical image on the *retina* at the back of the eye. There the image is detected by the light-sensitive retinal nerve endings, which transmit the information (through the optic nerve) to the brain for analysis. The retinal nerves can sense only that kind of electromagnetic energy we call light, but astronomers have instruments that can make images from other electromagnetic radiation as well. Ground-based optical astronomy is concerned with the forming of images in the near ultraviolet, light, and infrared, but the image-making principles are the same for even shorter and longer wavelengths.

The most important part of a telescope is the objective—the part that forms the images. In a telescope, the objective is either a lens or a mirror. The "size" of a telescope usually refers to the diameter of its objective. Thus a 6-inch telescope has a lens or mirror 6 inches in diameter, and the 5-m (200-inch) telescope has a mirror with a diameter of 5 meters.

Formation of an Image by a Lens or a Mirror

Figure 6.2 shows how a simple convex lens forms an image. Parallel light from a distant star or other point light source strikes the lens from the left. A *convex* lens is thicker in the middle than at the edges. If the curvatures of the surfaces of the lens are just right, light passing through the lens is refracted in such a way that it converges toward a point. The point where light rays come together is called the *focus* of the lens. At the focus, an image of the light source appears. The distance of the focus, or image, behind the lens is called the *focal length* of the lens.

The image of a star is just a point of light. Figure 6.3 shows how an image is formed of an extended source, for example, the moon. The moon is so distant (as are all astronomical objects) that, from it, light rays approach the lens along parallel lines. However, from different parts of the moon, the parallel rays of light approach the lens from different directions. The light from each point on the moon strikes all parts of the lens (or *fills* the lens); these rays of light are focused at a point at a distance behind the lens equal to the focal length of the lens. If a screen, such as a white card, is placed at this distance behind the lens, a bright spot of light appears, representing that point on the moon. Light from other points on the moon similarly focuses at other points, producing bright spots on the card in different places. Thus an entire image of the moon is built up at the focus of the lens. The plane in which the image is formed is called the *focal plane*.

Rays of light can also be focused to form an image with a *concave* mirror—one hollowed out in the middle (Figure 6.4), and coated with silver or aluminum to make it highly reflecting. Each ray of light is reflected according to the law of reflection (Chapter 5). If the mirror has the correct concave shape, all parallel rays are reflected back through the same point, the focus of the mirror. Thus images are produced by a mirror exactly as they are by a lens. The principal difference between image formation by a lens and by a

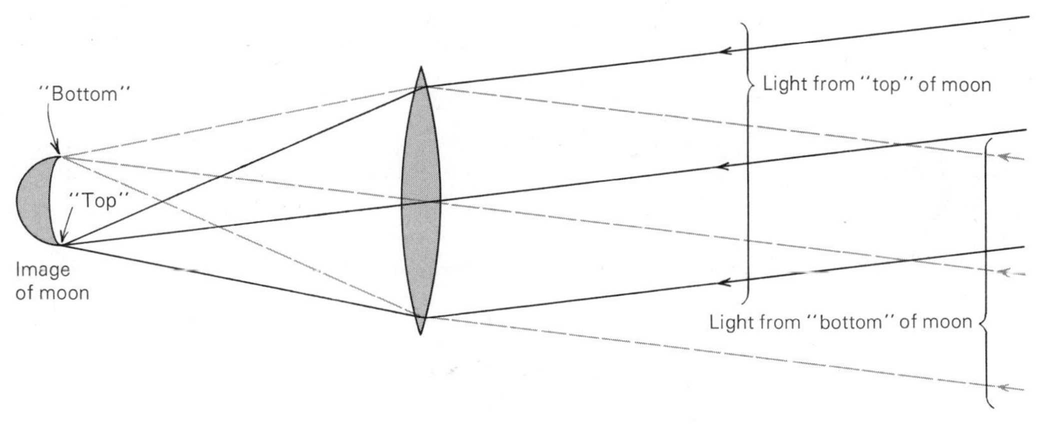

Figure 6.3
Formation of an image of an extended object.

mirror is that the mirror reflects the light back into the general direction from which it came, so that the image forms in front of the mirror.

The image formed by a lens or a mirror is always inverted and reversed (upside down and left to right) with respect to the object. The eye lens forms inverted images on the retina, but the brain interprets the image so that it appears upright.

Images have certain properties that depend on the diameter and focal length of the objective. One property is the size of an image; the size is proportional to the angular size of the object in the sky and also to the focal length of the lens or mirror. For example, the moon has an angular diameter of ½° in the sky. A lens of focal length 100 inches produces an image of the moon just under one inch across. The 5-m (200-inch) mirror of the Hale telescope on Palomar Mountain has a focal length of 660 inches and produces a lunar image about six inches across.

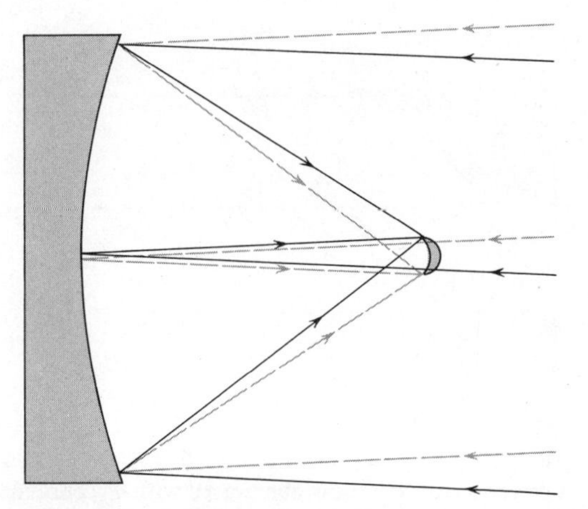

Figure 6.4
Formation of an image by a concave mirror. The dashed lines are from the "top" of the moon.

The *brightness* of an image is a measure of the amount of light energy that is concentrated into a unit area, say, a square millimeter, of the image. The brightness of an image determines whether it is above the threshold of visibility or, alternatively, how long a period of time is required to record the image photographically. The brightness of the image is greater the greater the amount of light focused into it by the objective, and is *less* when the area of the image over which the light must be spread is greater. (To be precise, the brightness of the image of an extended object is proportional to the area of the objective divided by the square of its focal length.) The image of a star, however, is a point, so its brightness depends only on the area of the objective.

Resolution refers to the fineness of detail inherently present in the image. Even if the lens or mirror is of perfect optical quality, it cannot produce perfectly sharp and detailed images. Because of a phenomenon of light called *diffraction*, a point source does not form an image as a true point but as a minute spot of light surrounded by faint, concentric, evenly spaced rings. Therefore we do not see the geometrical image of a star itself but only a *diffraction pattern* that the telescope lens produces with the star's light. If a star is viewed telescopically under good conditions, the diffraction pattern, consisting of the bright central disk and faint surrounding rings, is clearly visible.

In the image of an extended source, the diffraction patterns of the various parts of the image, all over-

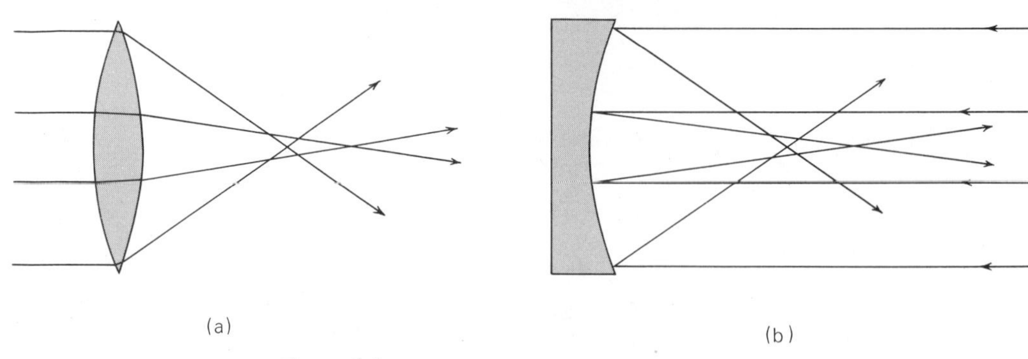

Figure 6.5
Spherical aberration: *(a)* in a lens; *(b)* in a mirror.

lapping each other, wash out the finest details. The ability of an optical system (lens or mirror) to distinguish fine detail in an image it produces, or to produce separate images of two close stars, is called its *resolving power*. In astronomical practice, the resolving power of a lens or mirror is described in terms of the smallest angle between two stars for which separate recognizable images are produced.

In an optically perfect system, the resolving power is proportional to the diameter of the objective but inversely proportional to the wavelength of the light or other radiation observed. The human eye has a resolving power (in visible light) of about one minute of arc (1′), and a 12-cm (5-inch) lens or mirror of about one second of arc (1″). The 5-m mirror of the Hale telescope can resolve about ¹⁄₅₀″ in visible light. If it were used at a radio wavelength of 5 cm, however, it could barely resolve a body of the angular size of the moon.

An image produced by any optical system always has imperfections, called *aberrations*. One kind, *chromatic*, or *color, aberration*, is a consequence of dispersion. In a simple lens the shorter wavelengths (violet and blue light) are bent the most, and focus nearest the lens, while the longer wavelengths of orange and red light focus farther from it. The effect of this chromatic aberration is to produce color fringes in the image. An image that is produced by a mirror does not suffer color aberration.

Another imperfection is called *spherical aberration* (Figure 6.5). Spherical surfaces are the most convenient to grind and polish, whether they be the concave or convex surfaces of a lens or mirror. Unfortunately, however, light striking nearer the periphery focuses closer to the lens or mirror, and light striking near the center focuses farther away. In a lens, spherical aberration and chromatic aberration can be corrected, or greatly reduced, by constructing the lens of two pieces of glass, or *elements*, of different indices of refraction. The elements are so designed that the aberrations introduced by one are at least partially canceled out by the other. Most lenses designed for astronomical purposes thus consist of two elements.

For a mirror, spherical aberration can be eliminated by grinding and polishing the surface not to a spherical shape, but to a *paraboloid of revolution*, that is, to a surface whose cross section is a parabola (Figure 6.6). A paraboloid has the property that parallel light rays striking all parts of the surface are reflected to the same focus. Similarly, light leaving the focus of a paraboloid is reflected at the surface into a parallel beam. An automobile headlight or searchlight, for example, has a parabolic reflector with the light source at the focus. Mirrors designed for astronomical telescopes generally have parabolic surfaces.

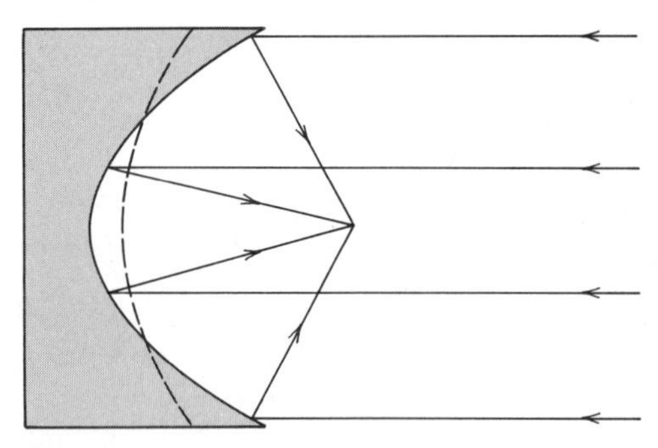

Figure 6.6
Correction of spherical aberration with a concave parabolic mirror. The dashed line shows where the surface of a spherical mirror of the same focal length would lie.

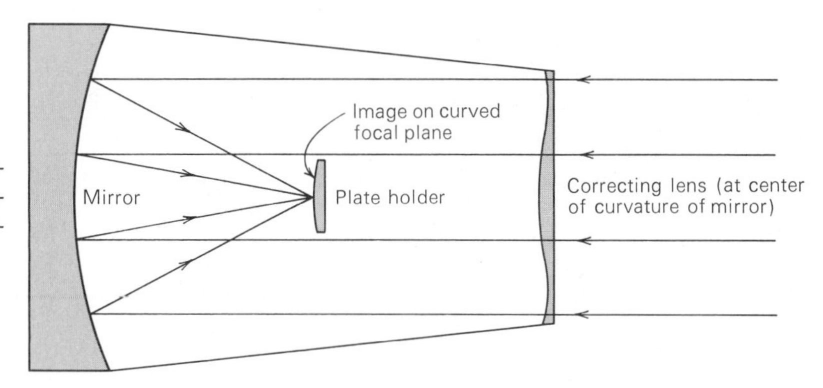

Figure 6.7
Schmidt optical system. The mirror is generally larger than the lens, so images of unreduced intensity can be produced over an appreciable field of view.

Still other aberrations exist, and their correction sometimes requires the design of very complicated objectives consisting of several elements. We have said enough, though, to give a general idea of what is involved in the design of optical systems.

The Schmidt Optical System

An ingenious optical system that utilizes both a mirror and a lens was invented by the Estonian-born optician Bernhard Schmidt (1879–1935) of the Hamburg Observatory. To avoid spherical aberration, parabolic rather than spherical mirrors are used. The principal disadvantage of the parabolic mirror is that it produces good images over only a relatively small field of view, that is, for light that approaches the mirror very nearly "square-on" (*on axis*).

On the other hand, for a spherical surface, any line reaching the surface through its center of curvature (that is, through the center of the sphere of which the surface is a part) is perpendicular, that is, square on to the surface, and hence is on axis. The Schmidt optical system, utilizing this principle, employs a spherical mirror that is allowed to receive light only through an opening located at its center of curvature (Figure 6.7). Thus there can be no off-axis aberration. The only trouble is that a spherical mirror, suffering as it does from spherical aberration, produces generally poor images for light coming from any direction. Schmidt solved this problem by introducing a thin correcting lens at the aperture at the center of curvature of the mirror. The lens is of the proper shape to correct the spherical aberration introduced by the spherical mirror but not thick enough to introduce appreciable aberrations of its own. Thus the Schmidt optical system produces excellent images over a large angular field.

The Complete Telescope

There are two general kinds of optical astronomical telescopes in use. They are (1) *refracting* telescopes, which utilize lenses to produce images; and (2) *reflecting* telescopes, which utilize mirrors to produce images. The refracting telescope is the most familiar. This is the kind of telescope that we can literally "look through." Ordinary binoculars are two refracting telescopes mounted side by side. A lens, generally consisting of two or more elements, is usually mounted at the front end of an enclosed tube. The tube is not really essential—its purpose is merely to block out scattered light; an open framework would suffice. In a refracting telescope, the objective, the optical part that produces the principal image, is the lens at the front of the tube. The image is formed at the rear of the tube, where various devices can be used to inspect, photograph, or otherwise utilize it.

The reflecting telescope was first conceived by James Gregory in 1663, and the first successful model was built by Newton in 1668. Here a concave mirror (usually a paraboloid) is used as an objective. The mirror is placed at the *bottom* of a tube or open framework. The mirror reflects the light back up the tube to form an image near the front end. The place where the image is formed by the mirror, is the *prime focus*. The image may be observed at the prime focus, or, alternatively, various systems of auxiliary mirrors can be used to intercept the light and bring it to focus at more convenient locations. Several such arrangements are shown in Figure 6.8.

The telescope tube must be mounted in such a way that it can turn to any direction in the sky. Nearly all astronomical optical telescopes have *equatorial mounts*. An equatorial mount (see Figure 6.9) allows the telescope to turn to the north and south about one axis and to the east and west about another. The axis

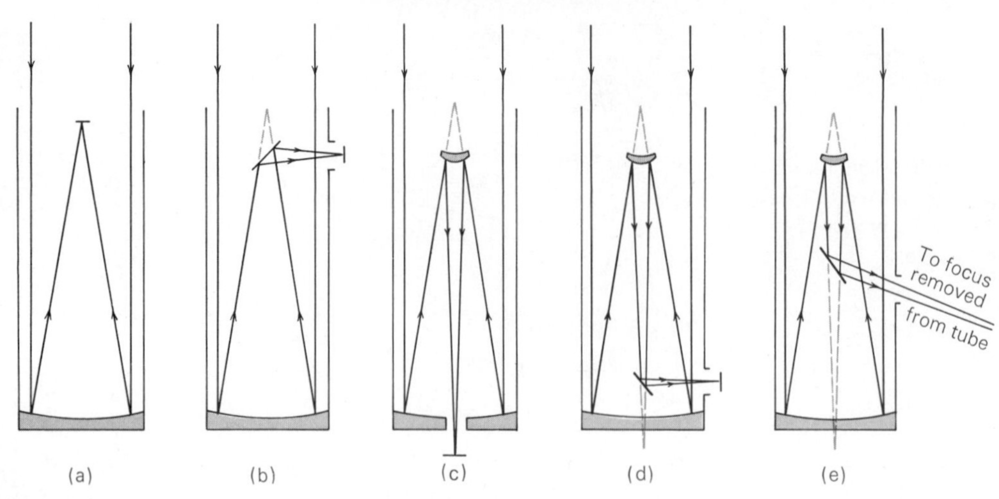

Figure 6.8
Various focus arrangements for reflecting telescopes: *(a)* prime focus; *(b)* Newtonian focus; *(c, d)* two types of Cassegrain focus; *(e)* Coudé focus.

for the east-west motion of the telescope is parallel to the axis of the earth's rotation.

Modern telescopes are equipped with setting circles or displays that indicate the direction or coordinates in the sky toward which the telescope is pointing. An important advantage of the equatorial mount is that a simple slow motion of the telescope about its axis parallel to the earth's axis—the *polar axis* of the telescope—is sufficient to compensate for the apparent motions of the stars across the sky that result from the earth's rotation.

Limitations on Telescope Size

We might wonder why astronomers do not simply build larger and larger telescopes in an attempt to see more of the universe.

A limit to the size of refracting telescopes is dictated by the sheer weight of the convex lens, which must be supported along the rim. A lens that is too large sags enough to blur the image. Moreover, for a refractor at least four optical surfaces must be ground and polished.

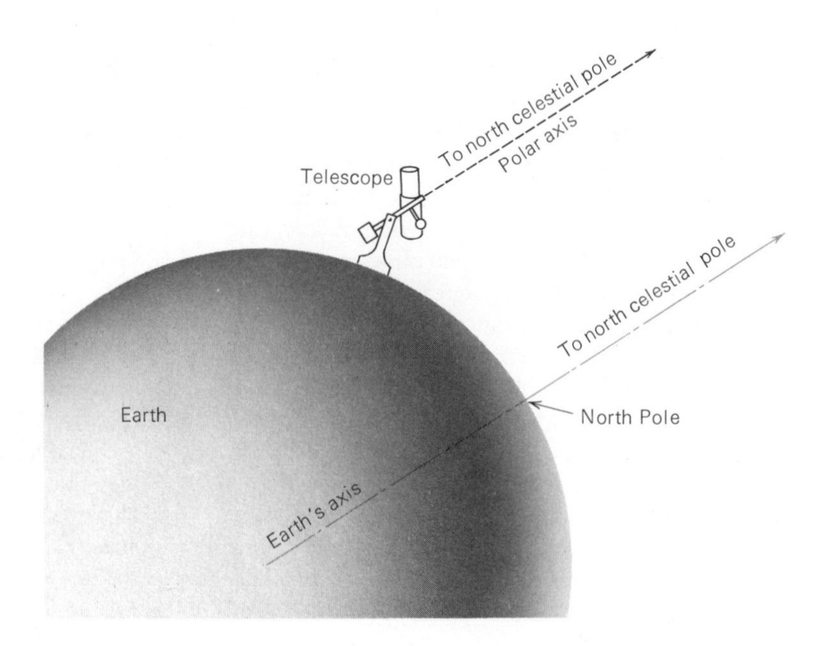

Figure 6.9
Equatorial mount.

Figure 6.10
The Special Astrophysical Observatory of the Academy of Sciences of the U.S.S.R., on Mount Pastukhov in the Caucasus, home of the 6-m telescope, the world's largest optical telescope *(right)*. *(Fotokhronica TASS—Sovfoto)*

Reflecting telescopes, on the other hand, can be built of much greater aperture, not only because a reflecting mirror has only one optical surface, but also because the mirror can be supported at all points along its back. Even so, the support of very large mirrors, especially those on telescopes with equatorial mounts, is a difficult engineering feat.

Some Famous Optical Telescopes

The largest existing refractor is the 102-cm (40-inch) telescope at the University of Chicago's Yerkes Observatory, at Williams Bay, Wisconsin.

The world's largest optical-quality telescope is the 6-m reflector at Mount Pastukhov in the Caucasus, U.S.S.R. (Figure 6.10). To support the 42-ton mirror, an altitude-azimuth mounting was used, one in which the telescope rotates about a single vertical axis and tips about a horizontal axis. This avoids the complex problem of supporting the mirror inherent in the more conventional equatorial mounting, but makes tracking stars enormously more difficult. Consequently, the tracking motion of the 6-m telescope is controlled by a sophisticated electronic computer. Because of its innovative design and the large size of the mirror, several years have been required to make the 6-m telescope fully operational.

Until the 6-m telescope went into operation in the mid-1970s, the largest telescope was the 5-m (200-inch) Hale reflector on Palomar Mountain, California. At the prime focus of the Hale telescope is a cage

TABLE 6.1 Some Major Optical Observatories

Observatory	Location	Sizes of Largest Telescopes
Astrophysical Observatory	Caucasus, U.S.S.R.	6-m (236-inch) reflector
Hale Observatories	Palomar Mountain, California	5.1-m (200-inch) reflector
		1.5-m (60-inch) reflector
		1.2-m (48-inch) Schmidt
	Mount Wilson, California	2.5-m (100-inch) reflector
		1.5-m (60-inch) reflector
	Cerro las Campanas, Chile	2.6-m (101-inch) reflector
National Observatory	Kitt Peak, Arizona	4-m (158-inch) reflector
		2.1-m (84-inch) reflector
Inter-American Observatory	Cerro Tololo, Chile	4-m (158-inch) reflector
		1.5-m (60-inch) reflector
Anglo-Australian Observatory	Siding Spring, N.S.W., Australia	3.9-m (153-inch) reflector
		1.2-m (48-inch) Schmidt
European Southern Observatory	Andes, Chile	3.6-m (142-inch) reflector
		0.9-m (36-inch) Schmidt
Lick Observatory	Mount Hamilton, California	3.0-m (120-inch) reflector
		0.9-m (36-inch) reflector
		0.9-m (36-inch) refractor
McDonald Observatory	Davis Mountains, Texas	2.7-m (107-inch) reflector
		2.1-m (82-inch) reflector
Burakan Astrophysical Observatory	Armenia, U.S.S.R.	2.6-m (104-inch) reflector
Royal Greenwich Observatory	Herstmonceux, England	2.5-m (98-inch) reflector
Mauna Kea Observatory	Mauna Kea, Hawaii	2.2-m (88-inch) reflector

1.8 m across in which the observer can ride while he makes his observations. The Hale telescope is one of the few existing telescopes in which the observer is carried "inside" the telescope. Also in California is the 2.5-m (100-inch) telescope at the Mount Wilson Observatory. The Hale Observatories—the observatories at Palomar Mountain and Mount Wilson—are operated jointly by the California Institute of Technology and the Carnegie Institution of Washington; they are under a single director and astronomical staff.

Other large telescopes, at some of the world's leading observatories, are listed in Table 6.1. In addition, several very large reflectors of nonoptical quality have been designed for observations in the long-wavelength infrared radiation.

One of the largest Schmidt telescopes is the 124-cm Schmidt of the Palomar Observatory.[1] It can photograph an area of the sky 6°.6 square (the size of the bowl of the Big Dipper) on a single 35-cm (14-inch) square plate of thin glass, and is thus ideal for surveying the sky. In 1949 the Schmidt was used to produce a comprehensive photographic atlas of that part of the sky observable from Palomar. This survey, financed by the National Geographic Society, and called the National Geographic Society–Palomar Observatory Sky Survey, took seven years to complete. Large reflectors like the Hale can probe deeper into space, but only in small regions at a time. It would take the Hale telescope at least 10,000 years to complete a comparable survey. New Schmidt telescopes in Chile and Australia are now supplementing the Palomar Survey with a similar one for the southern hemisphere.

An experimental system of telescopes has recently been put into operation by the University of Arizona and the Smithsonian Astrophysical Observatory on Mount Hopkins, about 64 km south of Tucson. The multiple-mirror telescope (MMT) utilizes six 1.8-m reflectors clustered around a common axis so that they point together to the same place in the sky. By multiple reflections, the images from the six telescopes are

[1] The universal telescope of the Tautenburg Observatory, in East Germany, can be used as a Schmidt by inserting a removable 135-cm correcting lens; it then becomes the world's largest Schmidt telescope.

combined to a single focus for analysis. It is easier to build six small telescopes than it is to build a single large one, yet the MMT has light-gathering power equivalent to a single 4.5-m telescope. Although the multiple-mirror system does not achieve the resolution of a single large telescope, for many purposes it is an admirable substitute.

OPTICAL OBSERVATIONS

Most astronomers do not live at observatories but near the universities or laboratories where they work. A typical astronomer might spend a total of only a few weeks a year observing at the telescope, and the rest of his time measuring or analyzing his data. Many astronomers work only with radio telescopes or with space experiments. Still others work at purely theoretical problems and never observe at a telescope of any kind. Even optical astronomers seldom inspect telescopic images visually except to center the telescope on a desired region of the sky or to make adjustments. On the contrary, the image is generally utilized in one of many other ways.

Telescopic Photography

If a screen or card is placed at the telescope focus, the image can be seen on the screen. To photograph the image, the screen is replaced by a photographic plate or film. The image then forms on a light-sensitive coating which, when developed, provides a permanent record of the image—one that can be measured, studied, enlarged, published, and inspected by many individuals. When used for photography, a telescope becomes nothing more than a large camera; the lens or mirror of the telescope serves as the camera lens.

The most important advantage in using a telescope as a camera is that photographic emulsions can accumulate luminous energy and build up an image during a long exposure. Most astronomical objects of interest are remote, hence the light we receive from them is feeble. However, long time exposures can be made of objects that are too dim to be seen by the human eye. Until electronic image intensifiers became available, astronomical exposures often would run hours in length, and occasionally over several successive nights. The longer the exposure, the more faint light gradually accumulates to help build up the photographic image. Objects can be photographed that are about a hundred times too faint to see by just looking through a telescope. What one can see visually is nothing compared to the spectacular photographs, such as those reproduced in this book, that result from long time exposures.

Visual Inspection with an Eyepiece

Occasionally it is desirable to inspect a telescopic image visually, doing what is commonly called "looking through" a telescope. To best inspect detail in the image, it is customary to view it with a magnifying lens. A common hand magnifier could be used for this purpose. A telescopic *eyepiece* or *ocular* is simply a high-quality magnifying lens that is used to view the image. Figure 6.11 shows how an eyepiece is used in a refracting telescope and in a Newtonian reflector.

When an extended celestial object is viewed through a telescope equipped with an eyepiece, it appears enlarged, that is, closer, than when viewed naturally. The factor by which an object appears larger (or nearer) is called the *magnifying power* of the telescope. For example, the moon appears to subtend an angle of ½° when viewed with the naked eye. If, when viewed through a particular telescope, the moon appears to subtend 10°, the *magnifying power* of the telescope is 20.

The value of the magnifying power can easily be calculated by dividing the focal length of the objective of the telescope by the focal length of the eyepiece. In principle, any desired magnification can be obtained if an eyepiece of sufficiently short focal length is used. Therefore, it does not make sense to ask an astronomer what "power" his telescope is. The power can be changed at will by using different eyepieces. Every observatory has a collection of eyepieces of different focal lengths that can be interchanged for various magnifications.

Spectroscopy

Spectroscopy occupies at least half of the available observing time of most large telescopes. We saw in Chapter 5 that white light is a mixture of all wavelengths and that these can be separated by passing the white light through a prism, producing a spectrum. A spectrum can also be formed by passing light through, or reflecting it from, a *diffraction grating*. The grating separates light into its constituent wavelengths by

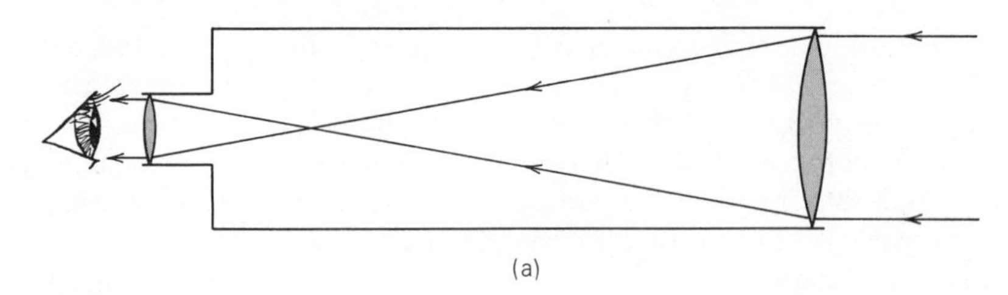

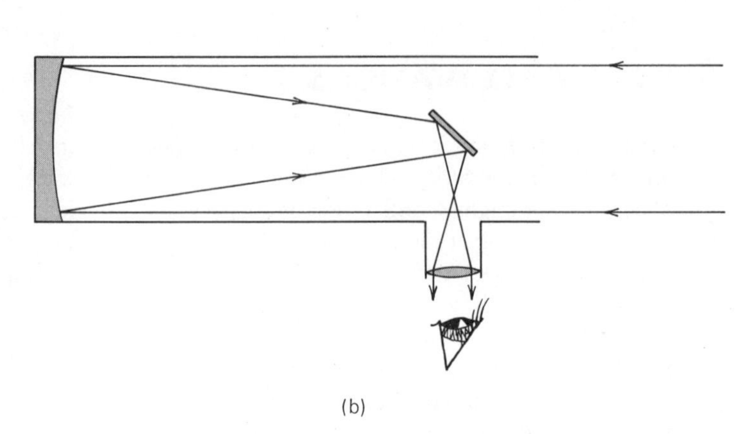

Figure 6.11
Use of an eyepiece: *(a)* with a refracting telescope; *(b)* with a reflecting telescope.

means of the phenomenon of diffraction (see the references to optics in Appendix 1).

A *spectrograph* is a device with which the spectrum of a light source can be photographed. Attached to a telescope, the spectrograph can be used to photograph the spectrum of the light from a particular star. The construction of a simple spectrograph is illustrated in Figure 6.12. Light from the source enters the spectrograph through a narrow slit and is then collimated (made into a beam of parallel rays) by a lens. In Figure 6.12 the collimated light is shown entering a prism, but it can just as well pass through or be reflected

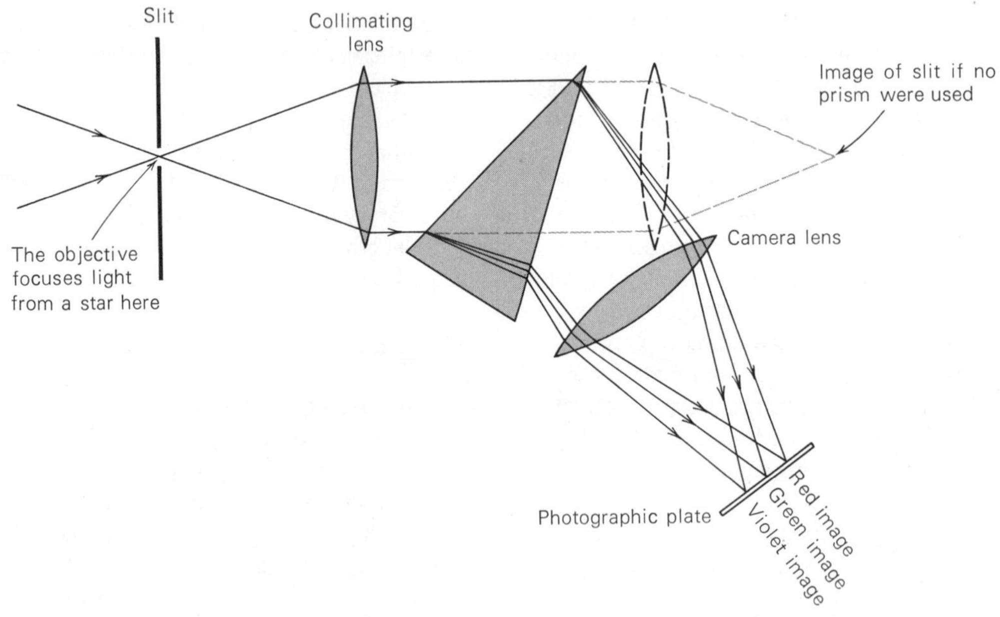

Figure 6.12
Construction of a simple prism spectrograph.

from a grating. Different wavelengths of light leave the prism (or grating) in different directions, because of dispersion (or diffraction). A second lens placed behind the prism (or grating) focuses the multiplicity of different slit images of different wavelengths lined up on the photographic emulsion. This is the desired spectrum. If the photographic plate is removed and an eyepiece is used to inspect the spectrum, the whole instrument becomes a *spectroscope*.

All of the parts of a spectrograph — slit, two lenses, prism (or grating), and photographic plate — must be attached to a rigid framework so that they do not shift during long exposures. The whole spectrograph may weigh hundreds of pounds and require a large space behind the focus.

Photoelectric Photometry

An important piece of information about a star is the amount of light we receive from it, expressed by its *magnitude*. (This matter will be explained in Chapter 13.) At present, the most accurate device for measuring stellar radiation received at the telescope is the *photomultiplier*, a light-sensitive electron tube placed behind the focus of the telescope. Light from a star is gathered by the telescope and is focused on a light-sensitive surface just inside the glass envelope of the tube. A photon striking this *photocathode* dislodges an electron, which is attracted to a positively charged element of the tube, thus creating a feeble current. This current is amplifed at a number of successive stages within the tube and finally in an external amplifier. A measure of the amount of current provides a measure of the amount of light striking the photocathode — that is, the star's brightness.

Electronic Image Intensifiers

In the past the photographic emulsion was almost universally used to record the telescopic images of astronomical objects as well as of their spectra. Photographic emulsions, however, are inefficient; it takes 100 or more photons to render a grain in the emulsion developable. Consequently, much of the information contained in starlight is wasted.

Photoemissive surfaces are coated with a substance that emits electrons when exposed to photons. In such a surface a single photon is responsible for the emission of each electron. Only a few photons need strike a sensitive photoemissive surface to dislodge one electron. Thus much higher efficiency is possible than with a photographic plate. Electronic image intensifiers utilize such sensitive photoemissive surfaces.

A large variety of devices take advantage of the efficiency of photoemissive surfaces as detectors, including *image tubes*, *image converters*, *electronographic cameras*, and *digicons*. They are now in common use at most major observatories, and improved systems are continually being invented. These electronic image intensifiers have revolutionized optical astronomy. Standard exposures of faint star fields or spectra that formerly required hours with unaided photographic plates can now be made in minutes or sometimes even in seconds. Rapid variations in the intensity of light from interesting objects (such as *pulsars* — see Chapter 17) that would have been impossible to detect photographically can now be measured. Data converted to electric currents can be recorded on magnetic tape and analyzed later in computers. Not only do astronomers use existing telescopes far more efficiently than before, but they can obtain new types of data, thereby attacking and solving new problems.

The photoemissive surface of the image intensifier or tube is placed at the focus of the telescope so that the image of the object observed, or of its spectrum, falls on that surface. From it the photons making up the image dislodge many electrons. As in a television tube, these are accelerated through a vacuum behind the photoemissive surface by an electric field and focused, often by a magnetic field on a detector. An example of a detector is a phosphor screen which emits up to hundreds of photons when struck by a single electron, thereby converting the image back into light, which is now more intense than the incident light. The phosphor screen can be photographed, or the light produced can be further amplified by passing it through a second image tube.

RADIO TELESCOPES

In 1931 K. G. Jansky of the Bell Telephone Laboratories was experimenting with antennas for long-range radio communication when he encountered interference in the form of radio radiation coming from an unknown source. He discovered that this radiation came in strongest about four minutes earlier on each successive day, and correctly concluded that since the earth's sidereal rotation period is four minutes shorter

Figure 6.13
Grote Reber's original radio telescope. *(National Radio Astronomy Observatory)*

than a solar day, the radiation must be originating from some region of the celestial sphere. Subsequent investigation showed that the source of the radiation was part of the Milky Way.

In 1936 Grote Reber, an amateur astronomer and radio ham, built the first antenna specifically designed to receive these cosmic radio waves. Reber made pioneering surveys of the sky for celestial radio sources with this first radio telescope, and remained active in radio astronomy for more than 30 years. During the first decade he worked practically alone, for professional astronomers had not yet recognized the vast potential of radio astronomy. Many of the objects that Reber discovered became subjects of intensive investigation years later. His original telescope is on display at the National Radio Astronomy Observatory (Figure 6.13).

In 1942 radio radiation was first received from the sun at radar stations in England. After World War II, the technique of making astronomical observations at radio wavelengths developed rapidly, especially in Australia, the Netherlands, England, and later in the United States.

Radio waves have now been received from many astronomical objects—the sun, moon, planets, gas clouds in our galaxy, other galaxies, and various other objects. The technique of radio astronomy is now an extremely important tool in observational astronomy.

Detection of Radio Energy from Space

It is important to understand that radio waves are not "heard"; they have nothing whatever to do with sound. Although in commercial radio broadcasting, radio waves are modulated or "coded" to carry sound information, the sound itself is not transmitted. The radio waves merely carry the information that a radio receiver must "decode" and convert into sound by means of a loudspeaker or earphones. Sound is a physical vibration of matter; radio waves, like light, are a form of electromagnetic radiation. We can also code visible light to carry sound information, as is done, for example, by the sound track on a movie film.

Many astronomical objects emit all forms of electromagnetic radiation—radio waves as well as light, infrared and ultraviolet radiation, and so on. The radio waves we can receive from space are those that can penetrate through the ionized layers of the earth's atmosphere—those with wavelengths in the range from a few millimeters to about 20 m. The human eye and photographic emulsions are not sensitive to radio waves; we must detect this form of radiation by different means. Radio waves induce a current in conductors of electricity. An antenna is such a conductor; it intercepts radio waves which induce a feeble current in it. The current is then amplified in a radio receiver until it is strong enough to measure or record.

Radio Reflecting Telescopes

Radio waves are reflected by conducting surfaces just as light is reflected from an optically shiny surface, and according to the same law of reflection. A radio reflecting telescope consists of a parabolic reflector, analogous to a telescope mirror. The reflecting surface can be solid metal, or a fine mesh such as chicken wire. In the professional jargon, the reflecting paraboloid is called a "dish." Radio dishes are usually mounted so that they can be steered to point to any direction in the sky, and gather radio waves just as an optical reflecting telescope can be turned in any direction to gather light. The radio waves collected by the dish are reflected to the focus of the paraboloid, where they form a radio image. The radio waves are focused on an antenna or some other kind of detector, and induce in it a current. This current is conducted to a receiver, not unlike ordinary home radio receivers in principle, where the current is amplified. As the optical astronomer chooses the type of photographic emulsion

Figure 6.14
The twin 90-ft radio telescopes of the Radio Observatory of the California Institute of Technology. *(California Institute of Technology)*

that is sensitive to the wavelengths of light he wants to detect, the radio astronomer tunes his receiver to amplify the specific wavelengths he wants to receive from space.

The ability of a radio telescope to gather radiation depends on its size. The radio energy received from most familiar astronomical bodies is very small compared to the energy in the optical part of the electromagnetic spectrum. Radio dishes are therefore usually built in large sizes; few are under 6 m across.

One difficulty with radio telescopes is that the wavelengths of radio radiation are far greater than those of visible light, so the resolving power of a radio telescope of a given size is correspondingly less than that of an optical telescope. Radio waves of 20 cm, for example, are some 400,000 times as long as waves of visible light, so to resolve the same angle, a radio telescope must be 400,000 times larger than an optical telescope. To resolve 1″ at 20-cm wavelength, a radio telescope would have to be nearly 65 km across. The largest steerable radio telescopes that are in use today are only about 100 m in diameter.

The solution to this limitation is provided by the use of the radio interferometer. Here two radio dishes are placed far apart. The radio waves from a source strike one antenna a brief instant before the other, so that the two antennas receive the same waves at slightly different times and thus become "out of phase" with each other. The difference in phase between the waves detected at the two antennas can be measured electronically, and the direction to the source can thus be calculated.

The farther apart the components of an interferometer are placed (the longer the *baseline*), the more accurately we can pinpoint the direction of the source. The newest extensive radio interferometer is the National Radio Astronomy Observatory's Very Large Array (VLA) in Socorro, New Mexico. It consists of an array of 27 movable radio telescopes of aperture 13 to 25 m each, and spread over an area 35 km across (Figure 6.15).

Now that international time standards can be coordinated to high precision, we can extend the interferometer principle to very long baseline interferometry (VLBI). Two different radio telescopes, thousands of kilometers apart, can simultaneously observe the radio waves from the same source and record them on tape, along with marks from a very accurate time standard (like the "ticks" of a very accurate clock). Later, these two tapes can be analyzed with a computer to find the phase difference between the radio radiation at the two stations, and hence the direction of the source. Baselines as long as from California to Parkes (in Australia) and from Greenbank, West Virginia, to the Crimea have been used. The resulting angular resolution of the sources observed is as great as a few ten-thousandths of a second of arc—far surpassing the angular resolution of optical telescopes.

Radar Astronomy

Radar is the technique of transmitting radio waves to an object and then detecting the radio radiation that the object reflects back to the transmitter. The time required for the radio waves to make the round trip can be measured electronically, and because they travel with the known speed of light the distance of the object is determined. The value of radar in navi-

Figure 6.15
Part of the Y-shaped Very Large Array (VLA), near Socorro, New Mexico. *(National Radio Astronomy Observatory)*

gation, whereby surrounding objects can be detected and their presence displayed on a screen, is well known.

In recent decades the radar technique has been applied to the investigation of the solar system. Radar observations of the moon and most of the planets have yielded our best knowledge of the distances of those worlds. In addition, as will be discussed in later chapters, radar observations have determined the rotation periods of Venus and Mercury, provided information about their surfaces, and furnished rather detailed surface contour maps.

Some Famous Radio Telescopes

Radio astronomy is young compared to optical astronomy, and large radio telescopes are still being built at a fast pace. Among the famous large steerable telescopes (that is, those that can be pointed to various positions in the sky) are the 76-m (250-foot) dish at Jodrell Bank, England, the 64-m Parkes dish in Australia, and the 91-m (300-foot) dish at the National Radio Astronomy Observatory in Greenbank, West Virginia.

Several important radio telescopes are large "bowls" carved out of the ground and lined with reflecting metal meshes. These bowls observe radio sources as they pass overhead, although some directional flexibility is provided by moving the detector about in the vicinity of the focus of the bowl. One is the 305-m bowl at Arecibo, Puerto Rico (Figure 6.16), a facility of the National Astronomy and Ionosphere Center, operated by Cornell University and funded by the National Science Foundation. Dr. Frank Drake, Cornell astronomer and director of the Center, has pointed out that the Arecibo bowl has a volume roughly equal to the world's annual beer consumption. There are other large radio telescopes in West Germany, the Soviet Union, the Netherlands, Australia, Japan, and elsewhere.

OBSERVATIONS OUTSIDE THE EARTH'S ATMOSPHERE

The earth's atmosphere absorbs many wavelengths of electromagnetic radiation, blocking them completely from passing through it. Even those wavelengths that do succeed in passing through the atmosphere are disturbed by constant convection currents in the air, which blur the images formed by optical systems. This blurring of images by the ever-turbulent air is an effect called astronomical *seeing*. It means that, even under the best conditions, images formed by a large telescope are not perfectly sharp. Theoretically, the 5-m telescope on Palomar Mountain can resolve an angle as small as 0".02, but most often the seeing limits resolution to about 1", and it is very rare indeed that seeing allows a resolution of 0".25 to be realized.

Moreover, in addition to absorbing much of the electromagnetic spectrum and blurring images of radiation that does pass through it, the atmosphere adds insult to injury by emitting light of its own. Part of this *light of the night sky* is scattered starlight, and part of it is light scattered from lights of cities (and it is becoming increasingly difficult to find observatory sites completely free of the latter), but most of it is *airglow*—light emitted by atoms in the upper atmosphere that are ionized by ultraviolet radiation from the sun, as well as high-energy particles from space. This illumination of the night sky puts a limit on the faintness of objects that can be identified through its glow.

Many areas of astronomy, therefore, have been revolutionized in recent decades by observations made from high-altitude balloons, rockets, satellites, and space probes above most or all of the atmosphere.

Ultraviolet, X-Ray, and Gamma-Ray Astronomy

In 1946 the United States Naval Research Laboratory launched a captured German V2 rocket to observe the far-ultraviolet radiation from the sun. Subsequently many rockets were launched to make X-ray and ultraviolet observations of the sun, and later of some other stars as well. Ultraviolet photographs were recovered from instrument packages that had been parachuted back to the ground. Both ultraviolet and X-ray radiation produce images on film, but because of their high energy, X rays are easily absorbed by most conventional optical systems. The early X-ray detectors were

Figure 6.16
The 305-m (1000-ft) dish at the National Astronomy and Ionosphere Center, Arecibo, Puerto Rico, operated by Cornell University and sponsored by the National Science Foundation. *(Cornell University)*

just gas chambers in which the X rays produced ionization.

Since the 1960s earth satellites have been launched to carry out astronomical observations. These have included the orbiting solar observatories (OSOs), the orbiting astronomical observatories (OAOs), and the high-energy astronomy observatories (HEAOs). The OAO2, launched December 7, 1968, and the satellite observatory now called *Copernicus*, launched in 1973, were particularly successful in obtaining ultraviolet photographs and spectra of stars, galaxies, and gas clouds. Since its launching in December 1970 from Kenya, the X-ray satellite *Uhuru* (Swahili for "freedom") has greatly increased our knowledge of X-ray sources. More than 100 sources of X rays had been found by the early 1970s, as well as what appeared to be an unresolved diffuse background of X rays coming from all directions in space. Many X-ray sources were observed to emit even in the gamma-ray region of the electromagnetic spectrum.

Special satellites have been launched expressly to detect very high energy gamma rays. The *Vela* satellite system, under the direction of the Atomic Energy Commission, conducts worldwide surveillance for possible nuclear bomb explosions. But since 1967 the Velas have been observing cosmic gamma-ray sources as well. Particularly interesting, and as yet unexplained, are some dozens of brief bursts of gamma radiation from various directions in the sky. The bursts are sometimes single and sometimes complex and multiple, and each burst or burst sequence lasts anywhere from about one-tenth of a second to several seconds.

X-Ray astronomy received a truly spectacular boost with the launching of HEAO2, now called *Einstein*, in November 1978. The Einstein X-ray telescope is the culmination of a nearly 20-year dream of Harvard–Smithsonian scientist Riccardo Giacconi, a pioneer in X-ray astronomy and current director of the orbiting observatory.

Although X rays are easily absorbed in ordinary optical systems, they can be reflected from polished surfaces that they strike at a grazing angle—like stones skipping across water. The Einstein satellite has an X-ray telescope consisting of a complex set of concentric parabolic and hyperbolic cylindrical surfaces that use the grazing reflection principle to focus X rays into an actual X-ray image that can be detected electronically and transmitted to earth. The size of the telescope aperture is 58 cm (23 inches), but because of the grazing angles of the reflecting surfaces to the incoming X-ray photons, the actual mirror surface is equivalent to that of the 2.5-m (100-inch) telescope on Mount Wilson. The Einstein telescope is designed to record X rays of wavelengths from 3 to 50 Å, and has a field of view of about 1°. In the first few months of its operation it was clear that it was an unqualified and spectacular success, with a sensitivity for detecting weak sources 1000 times as great as anything that preceded it. This is equivalent to changing from a small amateur telescope to the 200-inch Hale reflector on Palomar. In later chapters we shall have much to say about the thousands of X-ray sources discovered with the Einstein telescope and what they mean.

Observations from Space Probes

Most of the U.S. and Soviet space probes have carried instruments to gather scientific information about objects in our solar system. The moon was thoroughly mapped by television observations by Ranger landers and lunar orbiters long before the Apollo landings. The Mariner probes have obtained spectacular photographs (as well as other kinds of measurements) of Mars, Venus, and Mercury. Viking orbiters and landers have increased our knowledge about Mars incredibly, and have sent back photographs of the scenery from the Martian surface. Pioneer Venus has mapped the cloud layers of the planet Venus with optical observations and has mapped its surface with radar, and the Soviet Venera probes have landed on the planet and transmitted to earth photographs of its surface. Pioneer Jupiter and Saturn probes sent closeup pictures of the giant planets of the solar system, and the Voyagers even far more spectacular views not only of Jupiter but of its natural satellites as well. Space probes have, in a sense, carried our eyes to the vicinities of other worlds in the solar system for closeup exploration.

Space probes and manned explorations have also provided observations of other celestial objects. Mariner 10, for example, observed Comet Kohoutek, astronauts on the moon have left ultraviolet cameras there to take automatic photographs of stars, and scientist astronauts in Skylab carried out extensive ultraviolet observations of the sun and other stars.

The Space Telescope

Encouraged by the phenomenal success of our observations from space, astronomers are now excited over the prospects of a large optical telescope in space—the *Space Telescope*, now under construction, and planned for launch from the Space Shuttle in late 1983.

The Space Telescope will be a Cassegrain reflector (see Figure 6.8) with a mirror of aperture 2.4 m (94 inches). At an altitude of 500 km it will be above essentially all of the atmosphere. Unhindered by atmospheric seeing, it is expected to obtain a resolution of 0″.05, and because of the darkness of space it is expected to detect objects 50 times fainter than is possible with any existing telescope on the earth's surface. Teams of astronomers from America and Europe are planning special experiments to be carried out with the Space Telescope, making use of its television cameras, spectrographs, and photometers, which operate not only in the visible spectrum but in the near ultraviolet and near infrared as well (from 1150 to 11,000 Å). The Space Telescope will not put ground-based astronomy out of business but, if fully successful, should be the most effective optical observatory ever built.

SUMMARY REVIEW

Atmospheric transmission of electromagnetic radiation: optical and radio windows

Image formation: objective; lens; mirror; focus; focal length; focal plane

Image properties: size; brightness; resolution; aberrations; chromatic (color) and spherical aberrations

Telescopes: refractors; reflectors; Schmidt telescopes; prime focus; equatorial mounts; polar axis; major telescopes of the world

Observing techniques: photography; observation with an eyepiece; magnifying power; spectroscopy; spectrograph; prism; diffraction grating; photoelectric photometer; photometry; image intensifiers; photoemissive surfaces

Radio telescopes: detection of radio energy from space; antenna; radio dish; radio interferometer; Very Large Array (VLA); very long baseline interferometry (VLBI); radar astronomy; famous radio telescopes

Space observations: seeing; night sky glow; rocket and satellite observations; orbiting solar observatories (OSOs); orbiting astronomy observatories (OAOs); high-energy astronomy observatories (HEAOs); Copernicus satellite; Einstein telescope; Uhuru; Vela satellites; observations from planetary space probes; Mariner, Pioneer, Viking, and Voyager probes; Space Telescope; Space Shuttle

EXERCISES

1. Prepare a table, based on Figure 6.1, listing the wavelengths of radiation that can reach the earth's surface.

2. How might you plan an observing program to survey the sky in wavelengths of (a) X-ray radiation; (b) ultraviolet radiation; (c) light; (d) radio waves.

3. What happens to the image produced by a lens if it is *stopped down* with an iris diaphragm—a device that covers its periphery.

4. Most cameras have adjustable iris diaphragms that can be changed at will. Describe how the human eye is an analogy. Consider what happens to the eye in bright sunlight and in very dim light, and why.

5. What kind of telescope would you use to take

a color photograph that is entirely free of chromatic aberration? Why?

6. How would you design the reflector in a traffic signal lamp in order to send the beam of light as far as possible?

7. Telescopes with equatorial mounts are generally equipped with automatic drives that turn the telescopes about their polar axes at just the right rate to compensate for the turning of the earth, so that stars can be followed across the sky. At what rate would an equatorial telescope on the moon have to rotate about its polar axis to track the stars? (*Hint:* The moon keeps one face to the earth.)

8. Telescopes are usually housed in domes to protect them from the weather and daytime sunlight. A telescope dome has a vertical slot or window which can be opened to allow the telescope to view the sky outside. Describe how the dome must be able to move to enable the telescope to observe in all parts of the sky.

9. An eyepiece of ½-inch focal length is used with a telescope whose lens has a focal length of 100 inches. What is the magnifying power of this telescope?

10. What kind of telescope would you use for the following purposes:

(a) To photograph the spectra of many faint stars simultaneously?
(b) To measure the precise separation of two stars in a double-star system?
(c) To photograph a large field of the Milky Way at once?
(d) To make a photographic survey of the sky?
(e) To record radiation of 21-cm wavelength from a remote gas cloud?
(f) To photograph the sun in X-ray radiation?
(g) To search for ships off the coast of New England?

11. If the atmosphere is never steady enough to allow a ground-based telescope to achieve its theoretical resolution, what advantage does the 5-m (200-inch) telescope have over one of, say, 2.5-m aperture?

12. "Light pollution" of the atmosphere by illumination from city lights and advertising signs is a special problem to optical astronomy. What man-made problems do you suppose plague radio astronomy?

13. It is, admittedly, a complicated engineering task to design a space telescope. Nevertheless there are certain simplifications provided by a telescope in free-fall orbit about the earth that make certain parts of the design easier than for a large ground-based telescope. Can you think of some of these advantages?

Radio telescopes at the Goldstone Tracking Station. *(NASA/JPL)*

CHAPTER 7

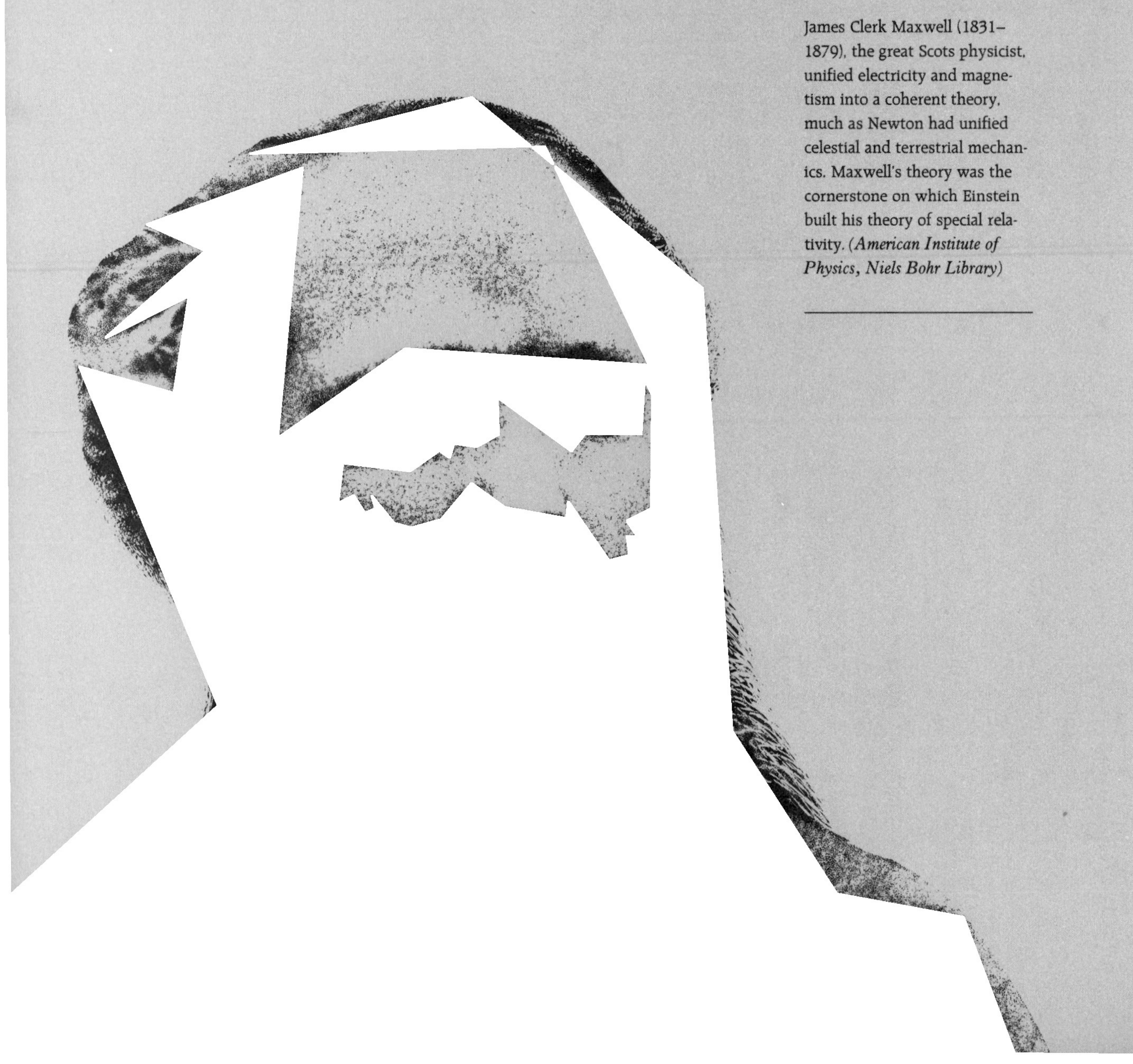

James Clerk Maxwell (1831–1879), the great Scots physicist, unified electricity and magnetism into a coherent theory, much as Newton had unified celestial and terrestrial mechanics. Maxwell's theory was the cornerstone on which Einstein built his theory of special relativity. *(American Institute of Physics, Niels Bohr Library)*

THE SPECIAL THEORY OF RELATIVITY

Newton unified science in 1687, and his theories of mechanics and gravitation dominated scientific thought for nearly two centuries. But by the middle of the nineteenth century, a new branch of physics was forming—the study of electricity and magnetism. The unifier of this new science was the Scottish physicist, James Clerk Maxwell (1831–1879), who showed that electricity and magnetism are different manifestations of the same thing. In the vicinity of an electric charge, another charge feels a force of attraction or repulsion, depending on whether the two charges have the opposite or the same sign, respectively. But in the vicinity of a *moving* charge another charge is acted on by a *magnetic* force. Maxwell spoke of an *electric field* around a static charge, and a *magnetic field* around a moving charge (as in an electric current).

Maxwell's theory of electromagnetism was codified in 1873 in his four famous equations, which describe electric and magnetic fields, and also how a change in one always induces a change in the other. If a charge moves back and forth, for example, as in an alternating current, the magnetic field set up is continually breaking down, and reforming with the opposite polarity. These changes in the magnetic field induce a constantly changing electric field, which in turn induces a changing magnetic field, and so on. These rapidly alternating fields are in the form of a disturbance that propagates away from the moving charge. Maxwell's equations also predict that this disturbance should move with a very definite speed, equal to the ratio of the electromagnetic and electrostatic units of electricity. Maxwell recognized that this speed is remarkably similar to the speed that had been measured for light, and he suggested that light must be one form of this electromagnetic radiation. He even suspected that there were other forms of electromagnetic radiation, and of course he was right, but it was more than 20 years before radio waves were generated for the first time by the German physicist Heinrich Hertz.

THE SPEED OF LIGHT

Long before Maxwell's time it was known that light travels with a finite speed, and this speed had even been rather accurately measured. The speed of light is found by measuring the time required for it to travel an accurately known distance.

Galileo suggested a way to measure the speed of light with two experimenters, separated by a mile or more, and each equipped with a lantern that can be covered. The first opens his lantern, and the second, on seeing the light from the first, uncovers his. The time that elapses between the time that the first experimenter opens his lantern and the time that he sees the light of his associate's lantern, after correction for the human reaction time, is how long light spends making the round trip. It is not clear whether Galileo actually conducted this experiment, but he correctly concluded that the speed of light is too great to be measured by so crude a technique.

Roemer's Demonstration of the Finite Speed of Light

The first demonstration that light travels at a finite speed was provided by observations of the Danish astronomer Olaus Roemer (1644–1710) in 1675. Jupiter's inner satellites are regularly eclipsed when they

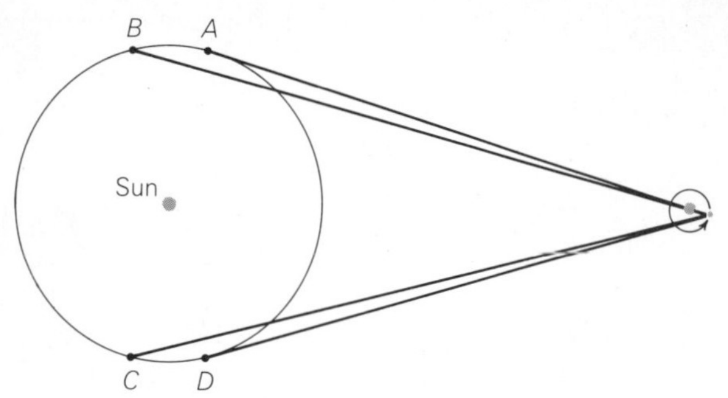

Figure 7.1
Roemer's method of demonstrating the finite speed of light.

pass into the shadow of the planet. Thus a convenient way of timing the period of revolution of a satellite is to note the interval between successive eclipses. Roemer was doing just this when he noted that the period of a satellite seems longer when the earth, in its orbit, is moving away from Jupiter than it is six months later when the earth is approaching Jupiter. Roemer correctly attributed the effect to the time it takes light to travel through space.

Suppose the earth is at A (Figure 7.1) when we note an eclipse, indicating that a cycle of revolution of the satellite has begun. Before the revolution is completed the earth has moved somewhat further from Jupiter, and light from Jupiter has farther to travel to catch up with the earth, at B, to bring us news of the next eclipse. Six months later the situation is reversed; the earth then approaches Jupiter, and thus advances to meet the oncoming light. If the first eclipse is then observed with the earth at C, the second is observed at D a little ahead of the time it would have been seen had the earth remained at C. Observations like Roemer's indicate that light takes about 16½ minutes to cross the orbit of the earth. Because the earth revolves about the circumference of its orbit in one year, it is easy to see that light must have a speed about 10,000 times that of the earth. Later, when the distance from the sun to the earth, and hence the speed of the earth, was well determined, the speed of light could be deduced in km/s.

Measuring the Speed of Light

By the middle of the nineteenth century physicists had devised means of measuring the travel time of light over measured distances on the earth. One method for doing so was invented by the French physicist Jean Foucault (1819–1868). The principle of Foucault's method is shown in Figure 7.2. Light from a source S, is reflected from a rapidly rotating mirror M. Each time that the mirror turns through the correct orientation, the reflected beam strikes a stationary mirror B some distance away. A second reflection sends the beam from B back to M, but while light is making that round trip, M, because of its rapid rotation, has turned a bit (dashed line in Figure 7.2), so that the light does not reflect directly back to S, but off to the side, to E, where it can be observed with an eyepiece. From the geometry of the setup, it is easy to calculate the angle through which the mirror M must rotate to send the light into the eyepiece at E. Thus by adjusting the rotation speed of the mirror until he could see the final reflected beam at E, Foucault could determine how long it took the mirror to turn through the requisite angle while light was making the round trip from M to B and back.

Foucault's experiment was performed entirely within the laboratory, and his total light path was only about 20 m; even so, he found the speed of light to an accuracy better than 1%. An adaptation of Foucault's method was later applied by Albert A. Michelson (1852-1931), the first American physicist to win the Nobel prize. Michelson measured the speed of light a number of times between 1878 and 1926. His most accurate determinations (1924 to 1926) were made by passing the light beam from a rotating eight-sided mirror on Mount Wilson, Southern California, to a stationary mirror on Mount San Antonio, 35 km away. Michelson arranged to have the distance between the two peaks surveyed with the help of the United States Navy and the National Bureau of Standards to a precision of better than 1 cm; it was, at that time, the most precise land survey ever attempted.

Michelson's best result at Mount Wilson gave, for

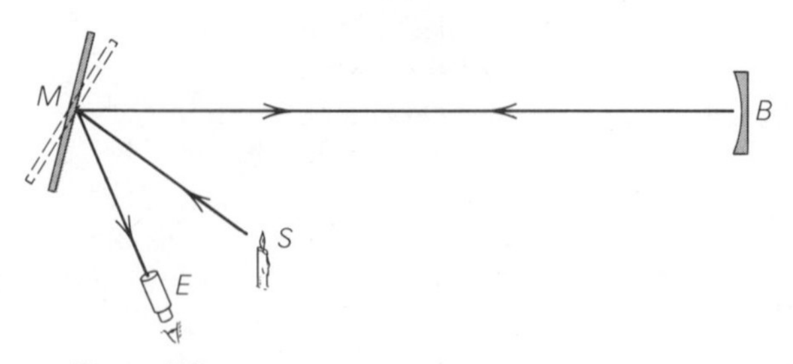

Figure 7.2
Principle of Foucault's method of measuring the speed of light.

the speed of light in air, the value 299,729 km/s. But light is slowed in passing through transparent media; in air it is retarded by nearly 70 km/s. Michelson began the construction of a mile-long vacuum tube on the Irvine Ranch in Southern California, through which he had hoped to measure the speed of light in empty space. Unfortunately he died before the experiment could be completed, and without his expert touch the results were not as good as had been hoped.

Today electronic timing techniques can measure the speed of light with high accuracy completely in the laboratory, eliminating the necessity of sending light over long distances. The modern accepted value for the speed of light in a vacuum, c, is 299,792.458 km/s.

The Speed of Light Is Something Special

The speed of light is very unlike other speeds that we normally encounter. First, of course, it is very great. Indeed, the speed of light is an absolute barrier. Nothing can go faster than light; in fact, no material body can ever travel quite at the speed of light. But in addition to these properties, the speed of light is always the same for all observers, irrespective of how they may be moving with respect to each other or to the source of the light. This absoluteness of the speed of light (and other electromagnetic radiation) is predicted by Maxwell's equations. But it is not something one would be likely to expect from "common sense." It is, indeed, a most remarkable thing.

Consider, for example, a material body, such as a bullet fired from a pistol. There is a certain muzzle velocity of the bullet, with respect to the gun, but the speed of the bullet with respect to an observer also depends on the speed with which the gun is moving when it is fired. If the bullet is discharged from a moving car, for example, its speed would certainly have the speed of the car added to its own speed. Alternately, one can, in principle, catch up with a speeding bullet, or even outrun it. (Astronauts in orbit, for example, travel much faster than bullets.) None of this is so for light! No matter how fast you approach or recede from the source, the speed of light, with respect to you, is always the same as if you and the source had no relative motion at all. Even if you could race from the earth in a spaceship at 99% the speed of light, and a colleague on earth were to send you a light signal, when that light caught up with your ship, and entered the rear port, its speed, with respect to the ship, would still be c. The speed of light depends in no way on the speed of the source.

But neither does the speed of sound, which is a compressional wave that travels through the atmosphere with a (rather slow) speed that depends on the temperature and other characteristics of the air. Similarly, ocean waves travel across the surface of the water as a transverse wave—the water level rising and falling vertically while the wave moves forward—and their speed depends on the wind and water conditions. But you can outrun sound waves, as is done in a supersonic airplane, and you can swim into ocean waves, increasing their speed with respect to you. Not so for light! No matter how fast you move, or in what direction, light waves approach you with that same speed—c. You can race forward to meet the waves of light, like the swimmer in the ocean, and, to be sure, that light will reach you sooner than if you were stationary, just as Roemer found for the times of eclipses of Jupiter's satellites, but the speed of that light when it reaches you is nevertheless the same, with respect to you, as if you were not moving; the *speed* of the light from Jupiter is the same all year round.

Light (and other electromagnetic radiation) behaving as predicted by Maxwell's theory, therefore, does not act the way one would expect from considering only the science of Newton. Does light point out a conflict between Newtonian mechanics and Maxwellian electrodynamics?

The existence of just such a contradiction was realized in 1895 by a 16-year-old schoolboy in the Luitpold Gymnasium in Munich. The boy was regarded as backward and indifferent by his elders, and became what we would today call a high school dropout. But his thought experiment at that time directed his thinking along lines that 10 years later would lead him to his special theory of relativity.

Albert Einstein (1879–1955) reasoned that it should be possible to catch up with any uniformly moving object, after which the relative velocity of the two would be zero. But if you could catch up with a light beam—an electromagnetic wave—you would find it still oscillating back and forth in time, and varying in intensity in space, but not moving! But not only is no such electromagnetic field known, it is impossible according to Maxwell's equations, which say it must be moving with a speed c. Here is surely a contradiction; either Maxwell's equations must be wrong, or our fundamental Newtonian concepts must be wrong. Yet all the predictions of Maxwell's theory that could be tested turned out to be correct. The electronic technology available to us today certainly attests to the power and success of electromagnetic theory.

But to young Einstein there was also a strong philosophical reason for suspecting that Maxwell was right: the *principle of relativity*.

THE RELATIVITY PRINCIPLE

The principle of relativity states that there is no physical experiment by which one can detect his state of uniform relative motion. What this means is that if two observers, moving uniformly with respect to each other, perform the identical experiment in their own moving environment they will obtain identical results, so neither can say, from anything the experiment told him, that he was or was not moving, or how fast he was moving.

For example, two people standing in the aisle of an airliner going 600 mi/hr can play catch exactly as they would on the ground. On that same airplane you can drop a heavy and light object together, and they will hit the cabin floor at the same time, and they fall at the same rate as they would if you had dropped them on the ground (provided that the airplane is moving uniformly—in a straight line at a constant speed). You can play table tennis quite normally on a moving ship on a calm sea. You can swing pendulums in an automobile (so long as it is not turning or accelerating in some other way) and they will swing in the same way, with the same periods, obeying the same pendulum laws as do pendulums in the laboratory. When you are moving uniformly, you experience no physical sensation of speed, or any other sensation that will tell you that you are in motion. You can, of course, look out the window and see the ground moving by, but if you were stationary and the *ground* were moving you would feel the same and see the same thing. It is common to sit in a train in a station, and momentarily wonder whether it is your train or the one on the next track that starts to move. For that matter, none of us can feel the motion of the earth carrying us about the sun with its orbital speed of 30 km/s; so emphatically do we not feel it, that scarcely three centuries have elapsed since it has been generally accepted that, indeed, the earth *does* move.

But does this principle of relativity apply only to mechanics, or does it apply to electromagnetic phenomena as well? All experiments indicate that it does apply to electromagnetic experiments, and that the principle of relativity is quite general. Do not radios and tape recorders work the same on an airplane as in the house? You can pick up iron filings with a magnet just as easily in an automobile as in a classroom.

On the other hand, if the speed measured for light depended in any way on the velocity of the observer, then relatively simple experiments could be performed that would reveal the observer's motion. All such experiments invariably fail. Like Einstein, we are forced to the conclusion that you cannot catch up with a light beam; light will always have the same speed (in a vacuum)—c—and the prediction of Maxwell's electromagnetic theory is correct. The principle of relativity holds, and there is no experiment—whether mechanical, or involving electricity and magnetism, or light, or anything else—by which we can detect our state of uniform motion.

The principle is profound, for if it is impossible to detect uniform motion, the idea of *absolute motion*—motion with respect to absolute space, as envisioned by Newton—can have no meaning. There can be no absolute reference frame or coordinate system that is guaranteed to be at rest in the universe, and with respect to which other motion can be referred. All we can define is *relative* motion with respect to something else. Thus the earth moves 30 km/s with respect to the sun; the sun moves 20 km/s with respect to the average of its stellar neighbors; these nearby stars all move at about 300 km/s with respect to the center of our galaxy; and so on. But there is no way to know how fast we are "really" moving, with respect to absolute space.

But it is a bizarre thing that two observers, one stationary with respect to a light source, and one moving rapidly away from (or toward) it will still measure the light from it to be approaching them with the same speed. All the seemingly strange results of special relativity come about because of that bizarre fact, and once we can swallow it, everything else in relativity makes perfect sense. Before proceeding, therefore, let's make sure we all understand that it is the real world we are talking about, and not fantasy. We describe briefly, therefore, a few (of very many) observations and experiments that demonstrate the absoluteness of the speed of light.

c Cannot Depend on the Motion of the Source

A simple experiment to show that the speed of light cannot depend on its source is provided by nature herself, and was pointed out by the Dutch astronomer Willem de Sitter (1872–1934) early in the century. Many stars are found in double-star systems—in which

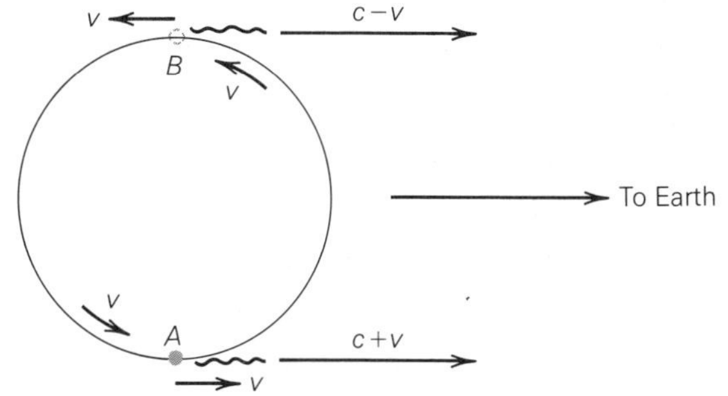

Figure 7.3
If the speed of light depended on the speed of the source, the light emitted by a binary star in a binary system would have a speed toward the earth that depended on the location of the star in its orbit.

the two stars revolve about their common center of mass. Take one star, say, the brighter, in such a system whose orbit lies roughly edge on to our line of sight (Figure 7.3). Suppose the orbital speed of that star about the center of mass of the system is v, and consider what would happen if the speed of the light it emits included the speed of the star itself, just as the speed of a bullet fired from a moving automobile has the speed of that car added to it. Then when the star approaches us, at point A in Figure 7.3, light from it should be traveling toward the earth at a speed of $c+v$, and when the star is moving away from us in its orbit, at B in the figure, its light should approach earth with a speed of $c-v$. To be sure, the speeds of the stars in binary systems (v) are very small compared to the speed of light (c), but the stars are very far away and over the many years it takes their light to reach us, the faster beam, traveling at $c+v$, can gain considerably over the slower beam, traveling only at $c-v$. If the distance to the binary system were just right, we could be receiving light from the star at position A at the same time as the light sent to us at a slower speed at an earlier time, when the star was at position B. A little thought will show that under some circumstances we could be seeing the same star in a double-star system at many different places in its orbit at once, and analysis of the orbit would end in hopeless confusion. But we have actually analyzed the orbital motions of stars in thousands of double-star systems with distances ranging from a few light years to many hundreds of light years, and the orbital motions are all well-behaved, with the stars moving in accordance with Newton's laws. The speed of light from them therefore cannot include the speeds of the stars themselves.

Further proof that the speed of light is independent of the speed of its source comes from the nuclear physics laboratory. In nuclear accelerators, subatomic particles moving at nearly the speed of light are often observed to change form (*decay*) and emit photons, but these photons are always observed to move with the normal speed of light, c, with respect to the laboratory.

c Cannot Depend on the Motion of the Observer

The most famous experiment showing that the speed of light does not depend on the motion of the observer was performed in 1887 in Ohio by A. A. Michelson and E. W. Morley, but its results have been confirmed by hundreds if not thousands of other experiments since then. Michelson and Morley, unaware at that early date of the principle of relativity, were attempting to measure the absolute speed of the earth through space by measuring the speed of light in two different directions.

To illustrate the idea of the Michelson-Morley experiment, consider three hypothetical astronauts, Able, Baker, and Charley, as shown in Figure 7.4(a), all stationary in space. Suppose, further, that the speed of light is constant through absolute space. Baker and Charley are each 4 light years (LY) away from Able, but in directions at right angles to each other. Able sends radio signals (which travel with the speed of light) to Baker and Charley at the same time, and those signals reach their destinations 4 years later; immediately Baker and Charley respond, and Able receives their answers simultaneously, 8 years after his original transmission.

Now suppose the three astronauts maintain their relative positions, but all three are moving at 60% the speed of light, and in the direction from Able toward Charley. As before, Able sends out the two messages, and, by our supposition, those signals move at the same speed, c, but with respect to stationary absolute space. This time, however, Charley is moving away from the point where the signal was emitted at $0.6c$ [Figure 7.4(b)], so waves from the signal approach Charley at only $0.4c$, and take 2.5 times as long to reach him as before—that is, 10 years. On the other hand, Able is approaching the point where Charley sends back his response, at $0.6c$, so Able moves forward to meet Charley's transmission at a relative speed of $1.6c$; thus those waves take only 2.5 years to span the 4 LY, and reach Able 12.5 years after his original transmission.

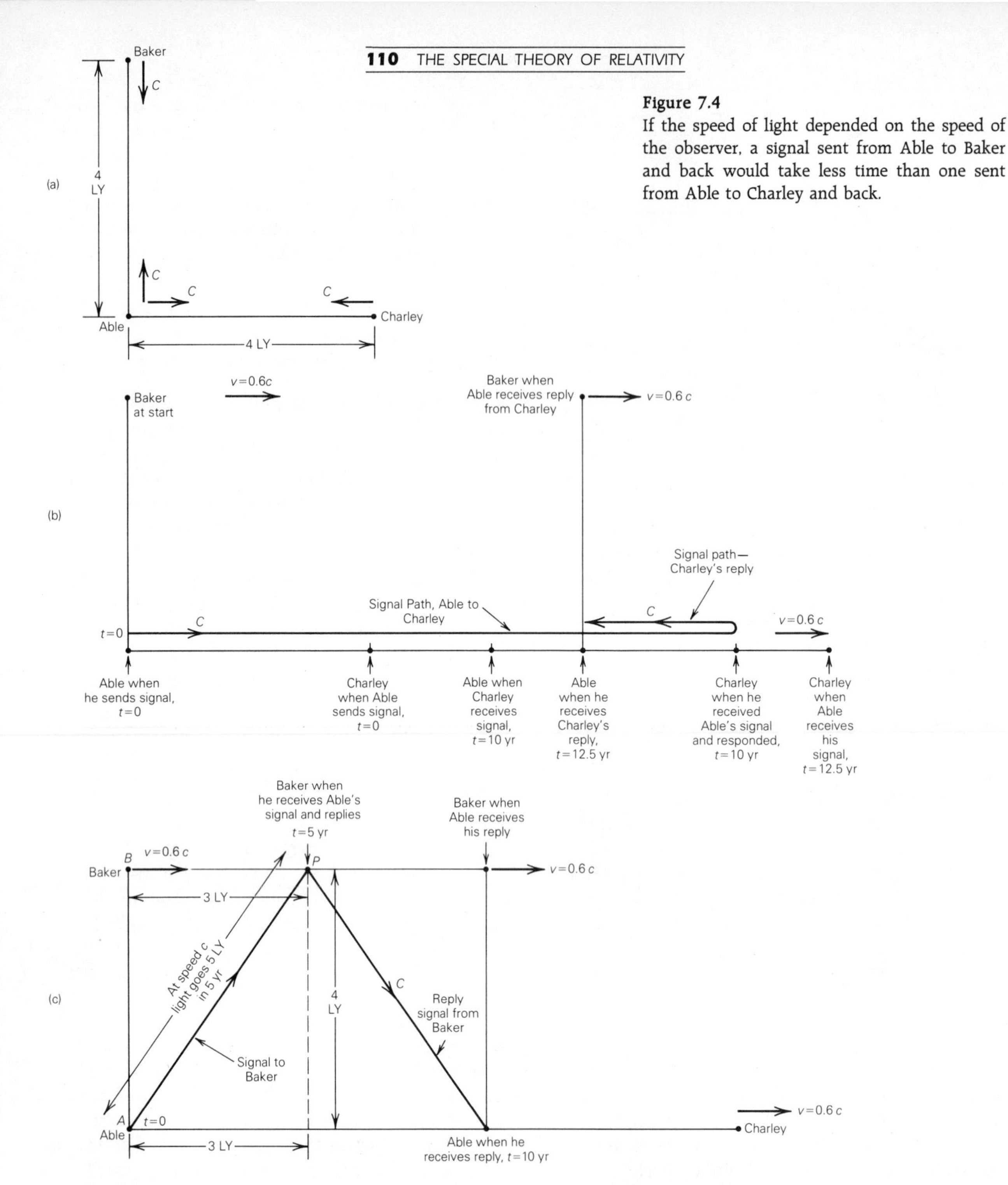

Figure 7.4
If the speed of light depended on the speed of the observer, a signal sent from Able to Baker and back would take less time than one sent from Able to Charley and back.

But how about the message going from Able to Baker and back? The radio waves that reach Baker from Able must be directed *ahead* of Baker's position (B) as seen from Able (A) at transmission time, just as a hunter must "lead" his running prey [see Figure 7.4(c)]. Thus the message that reaches Baker from Able travels an oblique path, and since Baker's speed is 60% that of the radio waves, Baker has traveled 0.6 times as far as the message has when it catches him at P. The three points—A, B, and P—make a right triangle. Since BP and AP are in the ratio 3 to 5, the theorem of Pythagoras tells us that AB must be 4/5 of AP; but AB is 4 LY, so the radio message, traveling at the speed of light, took 5 years to reach Baker. The same geometry holds for the return message, so Able receives Baker's reply 10 years after his original transmission, and 2.5 years ahead of hearing from Charley!

Thus if Able had *not* known that his speed through space was 60% of that of light, he could have deduced the fact from this experiment. Michelson and Morley performed their experiment in a basement laboratory, but the principle is the same. They hoped to determine the absolute speed of the earth through space by measuring the difference in times required for light to travel across distances in the laboratory that were at right angles to each other. The two light paths were set up by multiple reflections between mirrors on the horizontal surface of a heavy granite slab. In advance, of course, Michelson and Morley had no way of knowing that the earth, at the moment of the experiment, would be moving in a direction parallel to one of the light paths (as our astronauts were moving along the line from Able to Charley), but by rotating the granite slab through all possible directions, they reasoned that the difference between the light travel times would have to change.

It did not! Rotating the slab made absolutely no difference in the light travel time along the two beams at right angles to each other. Their experiment was accurate enough to detect a small part of the earth's orbital motion, but there was no difference in light travel times. It was as if the earth were absolutely stationary; but Copernicus could not be wrong, for the success of gravitational theory shows that the earth has to be moving. Michelson and Morley thought their experimental setup was at fault, so they repeated the experiment with even greater accuracy. Still no result, and there has never been any in all the many, many times this and comparable experiments have been repeated. The only conclusion is that the speed of light does *not* depend on the motion of the observer; it is always c with respect to him.

Strange as this result seems, however, it is completely consistent with Maxwell's theory and with the principle of relativity.

THE SPECIAL THEORY OF RELATIVITY

How can we understand the bizarre properties of the propagation of light? Einstein gave the solution in his special theory of relativity. He showed that different observers in uniform relative motion (moving at constant velocity with respect to each other) perceive space and time differently. There are two assumptions on which the special theory is based: the principle of relativity and, embodied within it, the absolute constancy of the speed of light.

Time Dilation

Let us imagine the construction of an ideal clock. Of many possible designs, we shall choose a clock consisting of two parallel mirrors, and a pulse of light reflecting back and forth perpendicularly between them. Each time the pulse passes from one mirror to the other, we shall count it as a "tick" of the clock. Because light travels at an absolutely constant rate, by carefully standardizing the spacing of the mirrors, we can agree that all such clocks should keep identical time.

On the other hand, what if an observer is moving very rapidly to the right with respect to us, carrying his two-mirror clock with him? Further, suppose his direction of motion with respect to us is parallel to the surfaces of the mirrors (see Figure 7.5). As far as he is concerned, his clock, in his own system, is at rest, for there is no experiment by which he can detect his own motion; consequently, as far as he is concerned his clock is operating normally, with the light pulse reflecting perpendicularly back and forth between the mirrors. But as *we* see the situation, because of his clock's rapid motion to the right, the light pulse is not bouncing simply back and forth along a single line, but is following a slanting path. In other words, we see the moving observer's light pulse traveling farther between ticks than he sees it traveling. But he and we agree on the *speed* of the light pulse, so *we* must conclude that the interval between *his* pulses is longer than it is between ours; that is to say, his seconds are too long, and his clock is running slowly. On the other hand, he, aware of no motion on his part, argues that it is *we* who are moving to the left, and that it is in *our* clock that light travels obliquely, and that it is *our* clock that runs slowly. Each of us insists that the other's clock is slow.

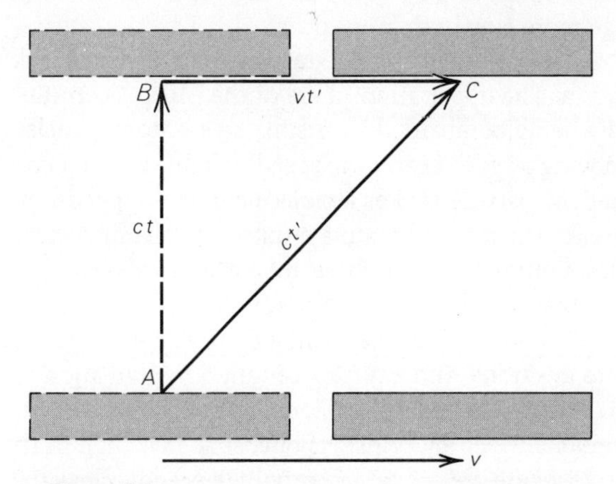

Figure 7.5
The light path in a moving observer's "ideal clock."

By isolating a triangle in Figure 7.5, with the most elementary algebra we can see by how much we disagree on the rate of passage of time. As far as our moving friend is concerned, he is stationary, and the light pulse has traveled vertically from A to B at a speed c, and the time it has taken to do so is t; since distance equals rate times time, the distance from A to B must be ct. But we see the light taking the slanting path AC, and requiring, at the same speed c, a longer time, t', to do so; thus we say that the pulse traveled a distance ct'. Meanwhile, our friend with his moving clock has gone from B to C, and if his speed relative to us is v, that distance must be vt'. The theorem of Pythagoras tells us that $c^2t^2 = c^2t'^2 - v^2t'^2$, from which we find, upon solving for t',

$$t' = \frac{t}{\sqrt{1 - v^2/c^2}} .$$

Thus what the moving observer thinks is an interval t, we see to be a longer interval t', and it is longer by the factor $1/\sqrt{1 - v^2/c^2}$. He, of course, regards his time intervals as normal and *ours* as too long by the same factor.

Which of us is right? We both are. Time really *does* move at different rates in two different systems in uniform relative motion. We simply perceive time differently: Time is not absolute; each of us has his own private time.

Reality of the Time Dilation

We must not think of this time dilation (the stretching out of time) between observers in uniform relative motion as some artifact of the clock we choose to construct. It is a very real thing. All processes slow down in moving systems; moving observers actually age more slowly than we do.

Nature provides a spectacular example of time dilation. The upper atmosphere of the earth is continually being bombarded by cosmic rays—atomic nuclei moving at very nearly the speed of light. When a cosmic ray particle strikes a molecule in the upper air, it breaks the molecule into a number of subatomic particles. Common among these are particles called *muons*. A muon is rather like an electron, but has about 200 times the mass. Muons spontaneously *decay*, turning into electrons, and emitting certain other radiation, in an average time of 1.5 millionths of a second. The muons formed by collisions of cosmic rays high in the atmosphere are moving at very high speeds, close to that of light. But even if they moved *at* the speed of light, they could travel, on the average, only about a half kilometer before decaying. Now the muons are formed at altitudes of 10 to 20 km; yet they rain down to the surface of the earth in enormous numbers; in fact, the muon is the principal kind of cosmic ray particle observed at ground level. If they are formed 15 km above the ground and decay before having time to travel as far as 1 km, how can we observe them at the bottom of the atmosphere?

As far as the muons are concerned, they *do* decay on the average in 1.5 millionths of a second, but because of their high speeds, as we observe them their time has slowed down, so that they have time to go very much farther than 0.5 km before decaying, and, in fact, most of them survive all the way to the earth's surface. Muons are also observed to live very much longer when they are accelerated to high speeds in the nuclear physics laboratory. In 1976 at CERN, the international nuclear physics laboratory in Geneva, muons were accelerated to a speed of $0.9994c$; the formula for time dilation predicts that their time should slow down by a factor of 30 at that speed. Indeed, the average lifetime of those high-speed muons was 44 millionths of a second, 30 times the 1.5 millionths of a second they survive at rest. Note that we are not speaking here of light pulses bouncing between mirrors, but of muons waiting to disintegrate; time dilation is not just a strange property of our light clock, but a funadmental property of time itself!

Would people live longer if they were rapidly moving? Not to their own way of thinking, of course, for they would have no sensation of moving. But relative to *us* they most certainly would age more slowly. In principle, long space trips could be made by astronauts if they were moving near enough to the speed of light. If we were to send a manned spaceship at a speed of $0.98c$ on a round trip to a star 100 LY away, the return journey would take just over 200 years of our time, but time for the astronauts would slow down by a factor of about 5, and on their return they would be only 40 years older than when they left. (Although such relativistic space travel is theoretically possible, the virtually prohibitive energy requirements make it unfeasible in practice—see Chapter 19.)

Contraction of Length and Distance

Let us return to those astronauts traveling to a star 100 LY away at 98% the speed of light. If they make the trip in 20 years of their time (40 years round trip) does that mean that they have traveled at 5 times the speed of light? No, for lengths (and distances) as perceived by different observers in uniform relative motion are also different. As perceived by the astronauts moving $0.98c$ with respect to the earth, the earth and star are moving at that same speed in the opposite direction. The astronauts see the separation of the earth and star to be very much less than as perceived by earthlings; in fact, they find the distance from earth to star to be just under 20 LY. If a system is in uniform motion with respect to us, we see all dimensions in that system that lie along the direction of relative motion to be *shorter* than as perceived by an observer in that moving system. He, on the other hand, sees lengths in *our* system (that lie parallel to the direction of relative motion) to be shorter than we see them. All objects in a moving system, in other words, appear foreshortened in the direction of motion.

We can see how this foreshortening must come about by reconsidering the astronauts Able, Baker, and Charley, who are lined up at right angles to each other. Suppose, as before, they are moving with respect to us along the direction from Able to Charley. We have already seen that there is no way that they can detect their own motion; thus if Able sends signals to Baker and Charley and receives simultaneous replies, he must conclude that they are equidistant from him. But not so for us, for we see Charley moving away from Able's signal until it catches up with him, and then Able rushing forward to meet the return signal from Charley. And we see the signal from Able to Baker, and Baker's return signal, traveling on slanting paths, as shown in Figure 7.4. As we found before, if Baker and Charley are equidistant from Able, we should see Able receive Baker's reply first. Thus if we see the two signals return to Able at the same time, we must conclude that Charley is *closer* to him than Baker is. In the moving system, Baker and Charley are the same distance from Able, but in our system the moving system of astronauts is foreshortened in the direction of motion. Only a bit of algebra is needed to show that the factor of foreshortening is just the same factor by which time intervals in the moving system are too long. That is, a distance in the moving system, along the direction of motion, that the moving observer would say is D, we would say is only $D\sqrt{1 - v^2/c^2}$. Of course the moving observer sees *our* distances as foreshortened, not his own. *Length* is just as private a matter as time is!

Increase in Mass

If different observers in uniform relative motion disagree on length and time, they must also disagree on velocity, which is distance covered in a given time. Thus they must, in turn, disagree on such things as momentum and energy, which depend on velocity. But they do agree on the laws of physics and the results of physical experiments — such as the conservation of momentum.

Suppose Jane and Mary are astronauts in space, moving together so that their relative velocity is zero [Figure 7.6(a)]. At a given instant each fires an elastic missile, such as a billiard ball, toward the other. The two balls are identical and are fired at identical speeds. They meet halfway between the spaceships at C, rebound, and return to the ships from which they were launched. The balls had equal but opposite momentum before the impact (since they were moving in opposite directions), and since each was turned about, they had equal and opposite momenta after the collision, so the total momentum is conserved, as it must be.

Now suppose that Jane and Mary are moving with equal speeds (with respect to us) but in opposite directions. As before, Jane and Mary discharge missiles toward each other, but because of their relative motion they fire the balls at J and M, respectively, and in directions perpendicular to their relative velocity. Because the balls move forward with the spaceships, they follow the dashed paths shown in Figure 7.6(b), meet at C, rebound, and return to their own ships at J' and M', respectively. Again, each is reversed in a symmetrical way and momentum is conserved.

But let's look at the last experiment from the point of view of Mary [Figure 7.6(c)]. Now Mary is stationary (in her own system), so her missile moves straight out perpendicular to the path of Jane's ship. But Jane's missile is released when she was way back at J. As before, the two missiles meet at C, rebound, and Mary's missile returns to her ship, while Jane's returns to hers at J', as must happen, since it is the identical experiment we described in the last paragraph. Both Mary and Jane must agree that momentum is conserved (if not, one of them would be able to detect something about her own motion).

But now there is a problem, because Jane is moving rapidly with respect to the stationary Mary, hence Jane's time passes more slowly. Similarly, all physical processes in Jane's system must slow down, including the component of velocity with which Jane's missile is fired toward Mary, perpendicular to the direction of

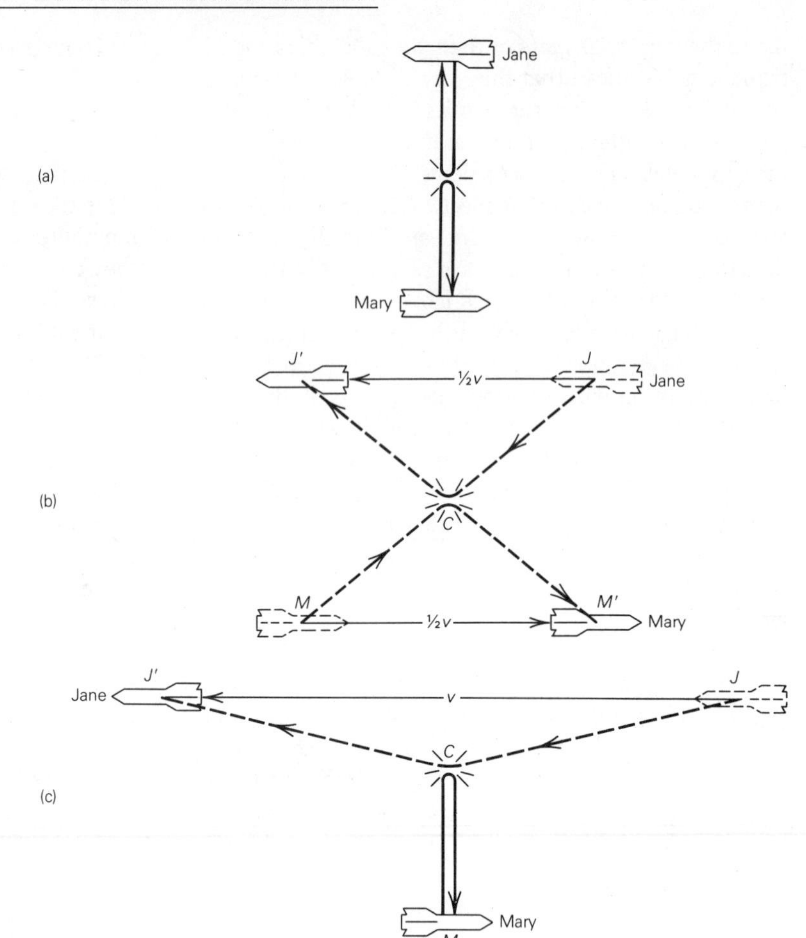

Figure 7.6
A hypothetical experiment involving collisions of elastic missiles, as seen from different perspectives. In all cases, momentum is conserved.

their relative motion. But if Jane's missile is moving more slowly than Mary's, we would expect it to have less momentum as well, since the balls are of the same mass. But then how can each turn the other around, conserving momentum? We would expect Mary's missile, with the greater momentum, to suffer less change in the impact, and not return to Mary's ship. But from our own vantage point [Figure 7.6(b)], we saw that it *did* return, and that momentum *is* conserved. The only explanation is that Jane's missile must have greater mass to compensate for its lower velocity.

If two observers are in uniform relative motion, each will say that the masses of objects in the other's system are greater than they would be if they were at rest. The factor by which mass is increased is exactly the same as the factor by which time is slowed. If an object has a mass m_0 when it is at rest, when it is moving with a speed v its effective mass is $m_0 / \sqrt{1 - v^2/c^2}$. The quantity m_0 is called the *rest mass* of the object.

The increase in mass of rapidly moving objects is not illusory; it is real. We observe it commonly in nuclear accelerators. As subatomic particles are sped up to nearly the speed of light, their masses increase manyfold, and enormously more power is required to provide them additional acceleration. Note that if the speed of a body were equal to the speed of light, v/c would be 1, and $\sqrt{1 - v^2/c^2}$ would be zero. Anything divided by zero is infinite, so the mass of a particle moving with the speed of light would be infinite, which, of course, is impossible. Thus no material body (a body with nonzero rest mass) can ever travel at quite the speed of light. Here is the physical explanation of the fact that the speed of light is an absolute barrier that no body can cross. To accelerate a body of appreciable mass to a speed even very close to that of light would require absolutely tremendous amounts of energy; so far we have succeeded in making only objects of the mass of subatomic particles reach speeds close to that of light.

Mass and Energy

All material bodies in motion possess energy of motion called *kinetic energy*.[1] With a little algebra, it can be shown that the increase in mass of an object caused by its motion is its kinetic energy divided by the square of the speed of light, or equivalently, its kinetic energy equals its mass increase times c^2. Thus there is an equivalence between the mass and energy of a moving body. Einstein postulated that even when a body is at rest there is an energy equivalence to its rest mass, so that its total energy is equal to its total mass times c^2, a concept made famous by that equation that is the hallmark of special relativity,

$$E = mc^2.$$

The equivalence of mass and energy stated in the above equation suggests that matter can be converted into energy and vice versa. Indeed, conversions in both directions are commonly observed in experiments with subatomic particles. For example, the electron has a twin called a *positron*, which has the opposite charge of the electron but identical mass. When a positron and electron come into contact, they mutually annihilate each other, turning into two photons of energy equal to the combined mass of the positron and electron times the square of the speed of light. Energetic photons can also combine to produce a positron and electron pair.

Because c^2 is a very large quantity, the conversion of even a small amount of mass results in a very great amount of energy. For example, the mutual annihilation of one gram of electrons and one gram of positrons (about 1/14 ounce in all) would produce as much energy as 30,000 barrels of oil. Here is the source of nuclear energy. Commercial nuclear power plants do not, however, involve the complete conversion of the nuclear fuel, but only a small fraction of it. In the hoped-for hydrogen reactor of the future, hydrogen is converted to helium with the destruction of a little under half of one percent of the original hydrogen. Still, the conversion of only 15 kg (about 33 lb) of hydrogen into helium per hour annihilates enough matter to produce energy at the rate of the current United States oil consumption. We are still a long way from the technology to accomplish this, but the sun and stars derive their energy by a similar process, as we shall describe in Chapter 16.

The fact that mass can be converted into energy and vice versa means that the old concepts of conservation of mass and conservation of energy are not strictly correct. However, the total of mass and energy equivalence *is* conserved; that is, if we calculate the energy equivalence of all mass by multiplying that mass by c^2, the resulting figure, added to the total energy, is conserved. Of course, we could also divide the total energy by c^2 and add that to the total mass to obtain a quantity that is conserved.

Faster Than Light?

It has been speculated that there could exist particles that always must move faster than light, and can never slow down to the speed c. Such hypothetical particles have been called tachyons, and experiments have been performed to search for them, to date with negative results. But as Nobel Laureate physicist Julian Schwinger (1918–) has pointed out, there is exellent reason for believing that tachyons cannot exist. If there *were* particles that could travel faster than light, then we could, at least in principle, use them to transmit signals at a faster rate than light can. But then we could build an ideal clock that ticks at more nearly the same rate for different observers in uniform relative motion, and all the special relativity effects we have described would not be correct; in fact if we could communicate with infinite speed (instantaneously), there would be no special relativity at all. Michelson and Morley's experiment would have given a positive result, muons would not arrive at the ground, electrons and protons would not gain mass in accelerators, and all of the many thousands of extremely accurate tests of realtivity would not have turned out the way they did. In particular, E would *not* equal mc^2, and we would not have nuclear bombs and reactors. One could, of course, hypothesize that tachyons exist but are totally unobservable, but then we can never know of them nor detect their existence, directly or indirectly, and their existence would have no practical significance on the real world. Most physicists now discard the tachyon hypothesis.

But this does not stop many people from feeling that somehow science and technology will somehow find a way to "break the light barrier." Perhaps they read the wrong science fiction authors. Anyway, irrespective of Captain Kirk's taking the *Enterprise* to "warp II" it is impossible for a material body to ever

[1] In Newtonian mechanics, the kinetic energy of a body of mass m and speed v is $\frac{1}{2}mv^2$.

reach the speed of light. It is not a technological problem, but a fundamental principle of nature. As we have seen, the mass of such a body would become infinite; it would become the entire universe itself.

Nor is there any need to travel faster than light, for, at least in principle, a person can travel at a speed as close to that of light as he wishes (given enough energy), and the closer his speed is to light's, the smaller all distances around him become. Our hypothetical astronauts going only 98% the speed of light could reach a star 100 LY away in 20 years, but by going even closer to the speed c they could make the trip in a far shorter time. As one approaches c, his time slows and distances shrink so that he can go anywhere in as short a time as he likes.

If one *could* travel at the speed of light, riding on a photon as it were, his time would stand still and he would be everywhere in the universe at once. Of course only a photon or other body of no rest mass can really do that—a body of pure energy. Massless bodies—photons and neutrinos[2]—can only travel at the speed of light; material bodies may approach that speed but never attain it.

CONCLUSION

Realm of the Universe is intended for the reader without mathematical training, and is virtually without formal mathematics. Yet in this chapter we have dipped into a little bit of algebra. The reason is that special relativity is an extremely important and extremely fascinating subject, one that is mysterious to most people, yet one whose essence can be understood with only a tiny bit of mathematics. It would have seemed such a shame to cheat the reader out of that wealth of knowledge that can be attained with so little extra effort.

[2] Neutrinos are massless particles released in certain nuclear reactions. They carry energy and have other properties, and can (with difficulty) be detected.

TABLE 7.1 Gamma (γ), the Factor $1/\sqrt{1 - v^2/c^2}$

v/c	Gamma (γ)
0.10	1.005
0.50	1.155
0.75	1.512
0.90	2.294
0.98	5.025
0.99	7.089
0.999	22.37
0.9999	70.71
0.999999	707.1
0.999999999	22360.7

At least so little effort in mathematics. Yet the concepts are very difficult to grasp. The mathematics is easy enough, but the ideas are totally alien to our experience, and present no easy conceptual hurdle. Why is this so? Because we have all grown up in a world where speeds around us are very small compared to the speed of light. All of the relativistic effects we have discussed depend on that factor $1/\sqrt{1 - v^2/c^2}$, a factor often denoted by the Greek letter gamma (γ). Values of gamma corresponding to several values of v/c are given in Table 7.1. Values in the table, show by how much masses increase, lengths shrink, and clocks slow in moving systems. Until v/c is a pretty good-sized fraction, gamma is essentially equal to 1. In such low-velocity systems, Newton's laws of motion apply with admirable precision. Even the earth's speed about the sun—30 km/s—is only $0.0001c$, and gamma is equal to unity within one part in a hundred million. We have become used to the low-velocity world, and it has prejudiced our ideas of "common sense."

On the other hand, how about a hypothetical civilization living on another world in an environment where speeds close to that of light are commonplace. Relativity would not seem strange to them. They, like us, given enough time, would discover the laws of physics, but not in the same order. As Schwinger has put it, "They would have their Maxwell and their Einstein, but alas, no Newton."

SUMMARY REVIEW

Maxwell's theory of electromagnetism: Maxwell's equations; electric and magnetic fields; propagation of electromagnetic radiation

The speed of light: Roemer's observations of the periods of Jupiter's satellites; Foucault's method of measuring the speed of light; Michelson's measurement; the currently accepted value for c

The absoluteness of c: independence of the velocity of the source; observations of binary stars; independence of the velocity of the observer; Michelson-Morley experiment

The principle of relativity: relevance to Maxwell's theory and the speed of light; abandonment of the concept of absolute space

The special theory of relativity: an ideal clock; time as perceived by observers in uniform relative motion; time dilation; muons; space travel in a human lifetime; length and distance as perceived by observers in uniform relative motion; increase of mass with velocity; equivalence of mass and energy; nuclear energy; annihilation of matter; production of pairs of particles from radiation (photons); the positron; tachyons; why one cannot go faster than light and why it is not necessary; gamma

EXERCISES

1. Do you think that two observers on systems that are rotating with respect to each other would find that all physical laws are the same in their two systems? Could they tell which observer was rotating more rapidly? If so, how?

2. Suppose a ball is thrown forward at 60 km/s from an automobile moving at 100 km/s. How fast is the ball moving with respect to an observer on the roadside? What if the ball is thrown toward the rear of the car with the same speed?

3. Compare the ways in which different observers compare the speed of the ball in the last exercise with how they compare the speed of light.

4. From the fact that light takes 16½ minutes to cross the orbit of the earth, show that the speed of light is about 10,000 times that of the earth.

5. Prepare a pendulum consisting of a small weight at the end of a string exactly 40 cm long. Start the pendulum swinging and time it for 10 complete oscillations (one oscillation is to and fro). What is your result? Now take the pendulum into an automobile, try to arrange that the car drive as smoothly and at a constant speed as possible, and repeat the experiment. Now what is your result?

6. Refer to Table 7.1. What would we measure for the mass of a 100-kg body moving past us with a speed of 90% that of light?

7. Suppose an astronaut visited a remote star and returned to earth, moving all the way at a speed so close to that of light that he aged only slightly compared with people on earth. What if he claimed that his ship was actually stationary, and that it was the earth that did the moving? Then he should appear aged and people on earth younger. How can this paradox (a famous one) be resolved? That is, how can the astronaut realize that it was, in fact, he who did the space traveling, and hence that he must have aged less?

8. According to Newton's laws the ordinary kinetic energy of a body of mass m moving at a speed v is $\frac{1}{2}mv^2$. Calculate the kinetic energy of a body of mass 1 g moving with a speed of 10^6 cm/s (about one-third the orbital speed of the earth). Now calculate the energy associated with the rest mass of the same body. How do the two energies compare?

9. By what factor does time slow for an astronaut moving 99.99% the speed of light? How long would it take an astronaut going that fast to make a round trip to a star 100 LY away: (a) according to people who stayed behind on earth? (b) according to his own time?

10. By what factor is the mass of an object increased if it moves: (a) 3/5 the speed of light? (b) 99.99% the speed of light?

CHAPTER 8

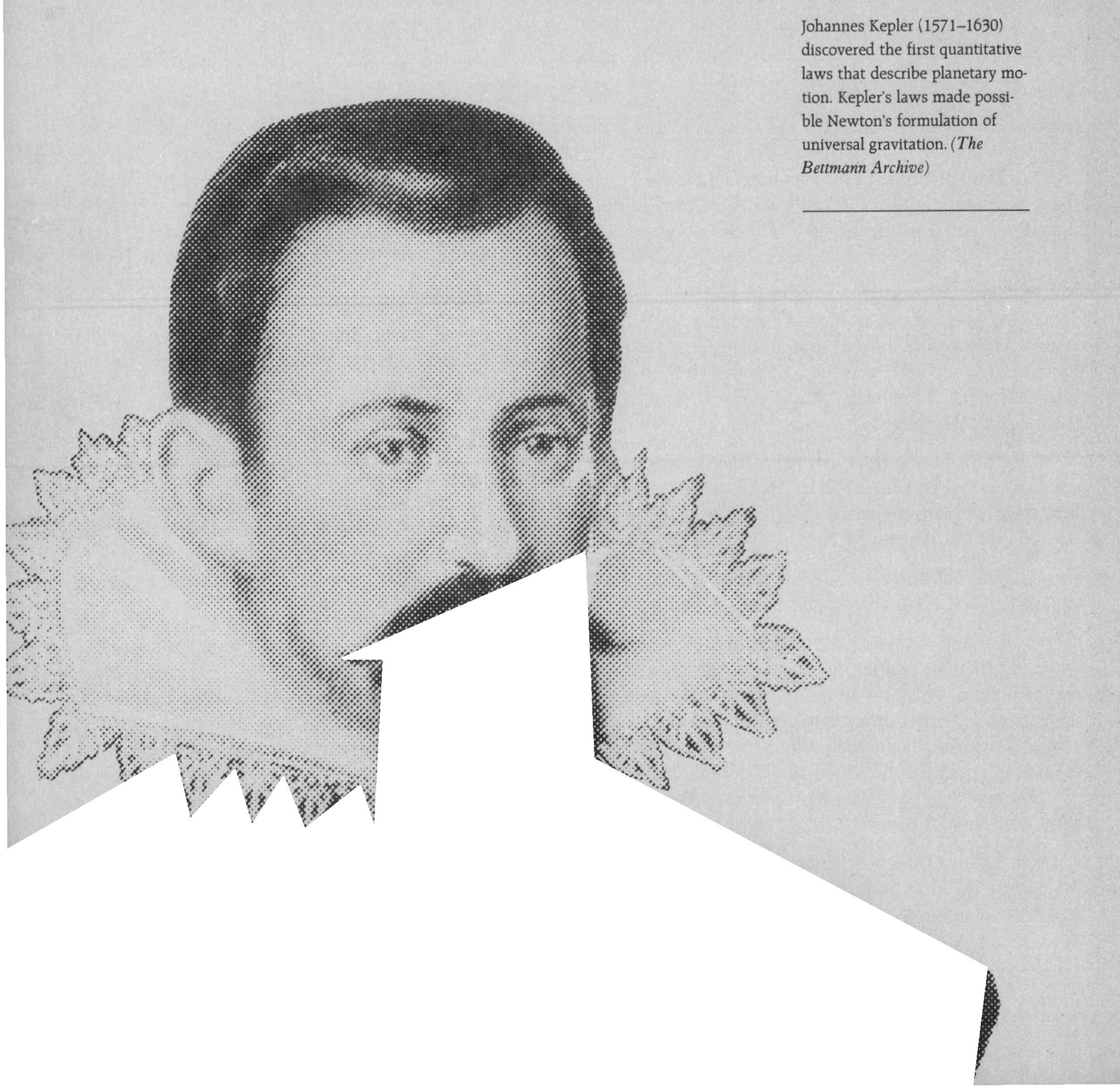

Johannes Kepler (1571–1630) discovered the first quantitative laws that describe planetary motion. Kepler's laws made possible Newton's formulation of universal gravitation. *(The Bettmann Archive)*

THE SOLAR SYSTEM

The ancient observer, who considered the earth to be central and dominant in the universe, regarded the sun, moon, and planets as luminous orbs that moved about on the celestial sphere through the zodiac.

Our solar system is indeed dominated by one body, but it is the sun, not the earth. Our sun, so important to us, is merely an ordinary, "garden-variety" star. Only careful scrutiny at close range would reveal the tiny planets to an interstellar visitor. First Jupiter, the largest, would be seen; then Saturn; and perhaps only with the greatest difficulty, the earth and other planets. Almost 99.9 percent of the matter in the system *is* the sun itself; the planets comprise most of what is left—the earth scarcely counts among them. The countless thousands of millions of other objects in the solar system, mostly unknown to the ancients, would probably remain unnoticed by a casual traveler passing through the solar neighborhood.

ORIGIN OF THE SOLAR SYSTEM

Analysis of rocks in the earth and moon (Chapter 9) and of meteorites (Chapter 11) shows that the oldest of them all have ages of about 4600 million years. Theoretical studies of the early evolution of the sun suggest that its age may be about the same (Chapter 17). We conclude that the solar system must have formed slightly less than 5000 million years ago. On the other hand, our present theory of stellar evolution indicates that the oldest systems of stars in our Galaxy, and hence the Galaxy itself, have an age at least twice as great. Thus the formation of the sun and solar system took place long after that of some of the stars surrounding us in the sky; this should not surprise us, for as we shall see, we have strong evidence that stars are forming in the Galaxy even today.

Important hints about how the solar system formed are provided by the organization of the planets in the solar system, which is very orderly. The planets' orbits lie in nearly the same plane, their orbits are nearly circular and are rather regularly spaced, they all revolve in the same direction—from west to east (the same as the direction of the sun's rotation), and most rotate and most of the satellites of planets revolve from west to east as well. The four innermost planets—Mercury, Venus, Earth, and Mars—are small, rocky, and earthlike; these are the *terrestrial* planets. The four outermost planets (except for Pluto)—Jupiter, Saturn, Uranus, and Neptune—are giant, composed largely of light elements (hydrogen and helium) and are Jupiter-like; they are the *Jovian* planets. Tiny, remote Pluto is more like a terrestrial planet. Its orbit is the most eccentric, and it may have had an origin different from that of the other outer planets.

It is generally accepted that the sun and planets formed together from the same original cloud of interstellar gas and dust (Chapter 15). The original cloud would have to have had a diameter hundreds or even thousands of times that of the orbit of the most distant present planet. It would also have had to have some original net rotation, probably because of the rotation of the Galaxy itself.

We shall see (Chapter 17) that some stars eject matter into space. Thus the interstellar medium (from which our solar system condensed) was not simply leftover matter that did not condense into stars when the Galaxy first formed, but also contained a good deal of matter that was formerly parts of other stars. Most stars derive their energy by the thermonuclear conversion of light atoms into heavier ones, thereby slowly

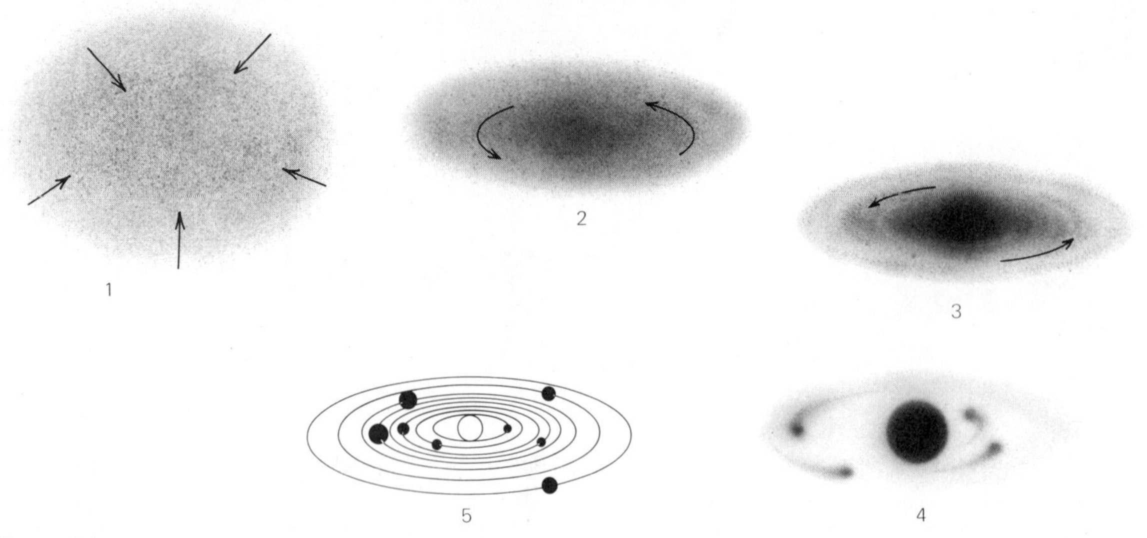

Figure 8.1
Schematic representation of the formation of the solar system. (1) The solar nebula condenses from the interstellar medium and contracts. (2) As the nebula shrinks, its rotation causes it to flatten, until (3) the nebula is a disk of matter with a concentration near the center, which (4) becomes the primordial sun. Meanwhile, solid particles condense in the inner solar nebula. These (5) accrete to form the terrestrial planets. The pressure of radiation and the wind of corpuscular radiation from the primordial sun blow the solar system clean of most of the matter in the disk that did not form into planets. The five drawings are not to the same scale; the original solar nebula had to contract greatly before its rotation produced appreciable flattening.

changing their chemical composition. The proto-solar system cloud, therefore, contained many atoms that had been built up by nuclear reactions in the interiors of earlier generation stars. It is interesting to contemplate that our own bodies are made up of atoms, many or most of which were formed in former stars.

Perhaps our cloud represented a fluctuation of density, so that it was very slightly more dense than the gas in the interstellar medium surrounding it. In any case, this *solar nebula* began to contract under its own gravity. As it contracted, however, it had to conserve its angular momentum (Chapter 3). Thus it had to spin faster and faster as the gas came together, getting closer and closer to the axis of rotation. What was very leisurely rotation of the original gas cloud became relatively rapid spinning of the much smaller ball of gas. Presently its equatorial regions were moving fast enough to stay in circular orbit, and they could fall in no farther. As the nebula kept shrinking, it left behind more and more of its equatorial matter in a rotating disk, each part moving in an approximately circular orbit about the center. Eventually, then, the solar nebula became a rotating disk, from which the planets formed, with a large blob of gas at the center, to form the sun.

We know all stars don't form in just this way. Often the original protostar fissions into two condensations, to become a double star, and the original angular momentum is conserved in the orbital motions of the two stars rather than in one star with its system of planets. Other times the clouds break up into clusters of stars. We don't know how often planetary systems can form. At most it's only half the time, because about half the stars around us are members of binary-star systems. Some recent studies of the incidence of duplicity among stars have suggested that the formation of planetary systems might be relatively rare. Alas, we have no direct evidence of a single other planetary system, but as yet we know of no way to detect planets like the earth in revolution about other stars. At least it happened here, and most astronomers expect that planetary systems have formed very many times in the Galaxy.

The infalling atoms picked up speed as they fell, and when the gas density became high enough for them to collide with each other, the kinetic energy was distributed among the atoms, becoming heat. Most of this heat was radiated away from the disk, but in the central condensation—to become the sun—the density grew until the gases of the protosun became opaque. This opacity trapped the heat inside and the pressure produced by the heat slowed down the contraction. The shrinking nebula had become a great globe of hot gas that could contract only very gradually as it was able slowly to radiate away the heat trapped in its interior. Thus a star (the sun) was born at the center, containing perhaps half of the material of the original cloud. The rest of the nebula was in the form of a relatively cold rotating disk, from which the planets and their satellites formed.

In the inner part of the solar nebula the high luminosity of the young sun would have evaporated the grains that were composed of volatile substances. Thus particles of water ice and frozen carbon dioxide could exist only far out in the disk. Rocky and metallic grains, on the other hand, could survive throughout the disk. In all parts of that rotating disk, though, the orbiting particles were constantly colliding and often sticking together, and many began to grow by accretion. A few began to get big enough to gravitationally affect those which came near. Sometimes smaller particles would pass close enough to bigger ones to bump into them and stick. But if they didn't pass close enough to hit, they could be gravitationally deflected to another part of the disk, or even out of the solar system altogether.

In this way, a few large chunks gradually won out over their neighbors, either capturing them or getting rid of them, thereby sweeping out ring-shaped swaths in the solar nebula, all centered on the sun. They became the planets. In the final stages of accretion the young planets swept up the last of the solid chunks remaining in the disk. There must have been many crater-producing explosions as these chunks smashed home. On the planets without dense atmospheres, we can still observe the heavy cratering produced, we think, in this period.

Those planets in the inner part of the solar system—the terrestrial planets—were built up of rocky and metallic particles. They could attract and hold on to none of the gases in the solar nebula. Their present atmospheres have outgassed from the rocks beneath their surfaces; Mercury, however, is too small to retain even this kind of atmosphere.

Far out in the nebula, on the other hand, it was cool enough for icy grains to exist, as well as rocky and metallic ones. The planets that accreted out there—the Jovian planets—thus formed out of lots of ices as well as rocks and metals. Jupiter and Saturn were large enough to even attract and hold a large amount of the gas in the solar nebula. Jupiter, in particular, has a present composition almost like the sun's; it is mostly hydrogen and helium, although in Jupiter those gases are compressed to the solid or liquid state.

The favored theory of the moon's origin is that it and the earth formed together, the moon accreting from material in orbit about the primordial earth. Some theorists, however, argue that the moon and earth could have formed independently in different parts of the solar nebula, but at about the same distance from the sun. In this case the earth and moon would have to have been trapped in each other's mutual gravitational fields at a later time. Most of the other planets have satellites, and many of them are believed to have formed by accretion from material in orbit about their parent planets. Some satellites, however, such as the outer ones of Jupiter, have eccentric orbits and even revolve from east to west (opposite to the revolution of the planets and most satellites); these were probably captured subsequent to their formation.

The minor planets may simply be objects that never accreted to a single large one, perhaps because there was too little mass in that part of the solar system to begin with, or perhaps because of the tidal influence of Jupiter. They may, however, be still fragmenting by collisions, and have started from a much smaller original number of bodies. We discuss the possible origins of the meteoroids and comets in Chapter 11.

The planets and minor members of the solar system contribute only minutely to its total mass. The original mass of the disk may have been much greater, for it is doubtful if practically all of the solar nebula could have condensed into the sun itself. However, when it first begins shining, a star is temporarily much more luminous than when it is fully developed. In that period of the sun's existence a large flux of energy from it, both in the form of photons and corpuscular radiation (atomic nuclei and electrons) may have interacted with the uncondensed gases and the tiny unaccreted particles and "blown" them from the solar system. Such an early solar wind of corpuscular radiation could also have carried away most of the sun's original angular momentum; today its angular momentum comprises only about 2 percent of that of the solar system, despite the fact that the sun has more than 99.8 percent of the system's present mass.

The reader is urged to regard the foregoing account of the solar system's beginnings with caution. The general picture is probably close to what actually took place, but future research can be expected to fill in many details and show others to be incorrect.

INVENTORY OF THE SOLAR SYSTEM

Today the solar system consists of the sun, the planets, their satellites, the comets, the minor planets or asteroids, the meteoroids, and an interplanetary medium of very sparse gas and microscopic solid particles. The relative prominence of the various kinds of members of the solar system can be seen from Table 8.1, which lists the approximate distribution of mass among the bodies of the solar system. The last four entries in the table are order of magnitude guesses only.

TABLE 8.1 Distribution of Mass in the Solar System

Object	Percentage of Mass
Sun	99.86
Planets	0.135
Satellites	0.00004
Comets	0.0003 (?)
Minor Planets	0.0000003 (?)
Meteoroids	0.0000003 (?)
Interplanetary Medium	<0.0000001 (?)

The Sun

The sun is a typical star—a great sphere of luminous gas. It is composed of the same chemical elements that compose the earth and other objects of the universe, but in the sun (and other stars) these elements are heated to the gaseous state. Tremendous pressure is produced by the great weight of the sun's layers. The high temperature of its interior and the consequent thermonuclear reactions keep the entire sun gaseous. There is no distinct "surface" to the sun; the apparent surface we observe is optical only—the layer in the sun at which its gases become opaque, preventing us from seeing deeper into its interior. The temperature of that region is about 6000 K. Relatively sparse outer gases of the sun extend for millions of kilometers into space in all directions. The visible part of the sun is 1,390,000 km across, which is 109 times the diameter of the earth. Its volume is 1⅓ million times that of the earth. Its mass of 2×10^{33} g exceeds that of the earth by 333,000 times. The sun's energy output of 4×10^{33} ergs/sec, or about 5×10^{23} hp, provides all the light and heat for the rest of the solar system. The sun derives this energy from thermonuclear reactions deep in its interior, where temperatures exceed 14 million K. We shall describe the sun, a typical star, in Chapter 16.

The Planets and Their Satellites

Most of the material of the solar system that is not part of the sun itself is concentrated in the planets. In contrast to the sun, the planets are small, are relatively cool, and are solid or liquid. They are not self-luminous at visible wavelengths, and shine by reflected sunlight.

The masses of the planets, in terms of the mass of the earth, range from 0.002 (Pluto) to 318 (Jupiter). The mass of Jupiter is greater than that of all the other planets combined. In diameter, the planets range from about 3000 km (Pluto) to 143,000 km (Jupiter). Most, but not all, of the planets are surrounded by gaseous atmospheres. All but two of the planets are known to have natural satellites; Jupiter leads with at least 14 moons.

The planets all rotate as they revolve about the sun. By *rotation* is meant a turning of an object on an axis running through it, as distinguished from *revolution*, which refers to a motion of the object as a whole about another object or point. The Jovian planets are all rapid rotators; Jupiter rotates most rapidly, in a period of 9^h50^m. Mercury rotates 1½ times during its 88-day revolution about the sun. The solid ball of Venus rotates still more slowly, in 243 days, but from east to west, reverse to the rotation of most of the planets (that is, *retrograde*). Some of the planets, especially Jupiter and Saturn, show marked oblateness or flattening because of their rapid rotation.

Only Mercury and Venus do not have known satellites. Jupiter has 14, Saturn has 10, Uranus has five, Neptune and Mars two each, and earth and Pluto one each. Six of the 35 known natural satellites of the solar system are about as large as or larger than our moon, although only Pluto's is as large as the moon in comparison with its primary planet. (The earth-moon system is sometimes referred to as a "double planet.") Only *Titan,* Saturn's largest satellite, is known to have an appreciable atmosphere. Most of the satellites revolve about their planets from west to east, and most have orbits that are approximately in the equatorial planes of their primary planets. (The moon is an exception, with an orbit nearly in the ecliptic plane.)

Minor Members of the Solar System

Comets are chunks of frozen gases with solid particles (nuclei) that revolve about the sun in very elongated elliptical orbits. Comets spend most of their time in those parts of their orbits that are very far from the sun, where they receive negligible radiant energy. However, as a comet moves in closer to the sun, it warms up and some of the volatile materials vaporize to form a cloud of gas and intermixed dust, the *coma,* around the solid nucleus. The particles and surrounding coma make up the *head* of a comet. When a comet is only a few astronomical units from the sun, the pressure of the sun's radiation and solar corpuscular radiation sometimes forces particles and gases away from the head to form a *tail.* The heads of comets are usually from 10,000 to 200,000 km across, and tails sometimes grow to lengths of many millions of kilometers. The entire mass of a typical comet, however, is less than one-millionth—and most are less than one-thousand-millionth—that of the earth. A comet is, therefore, a rather trivial entity. Unlike the planets, which move in orbits that are all nearly coplanar, comets approach the sun along orbits that are inclined at all angles to the plane of the earth's orbit and approach from a random distribution of directions.

The *minor planets,* or *asteroids* as they are often called, are small planets, differing from the terrestrial planets primarily only in size. Most revolve about the sun in orbits that lie between the orbits of Mars and Jupiter. There are probably tens of thousands of them large enough to observe with existing telescopes. Ceres, the largest minor planet, has a diameter of about 1000 km. Only a few hundred minor planets are over 50 km across; most are only a few kilometers or less in diameter.

The small solid objects revolving about the sun that are too small to observe with telescopes, the *meteoroids,* become known only when they collide with the earth, and plunging through the earth's atmosphere, heat with friction until they vaporize. The luminous vapors that are produced look like stars moving quickly across the sky and are popularly known as "shooting stars." Such phenomena are correctly called *meteors.* On a typical clear dark night, about half a dozen meteors can be seen per hour from any given place on earth. The total number of meteoroids that collide with the earth's atmosphere during a 24-hour period is estimated at 200 million.

Rarely, a meteoroid survives its flight through the earth's atmosphere and lands on the ground. It is then called a *meteorite.* A number of such fallen meteorites can be inspected in various museums. The largest known meteorites have masses of about 50 tons. Most are the size of pebbles. Chemical analysis reveals them to be formed of the same chemical elements that exist on earth and elsewhere in the cosmos.

On a dark clear night a faint band of light can be seen circling the sky along the ecliptic. This band of light is generally brightest near the sun and is best seen in the west within a few hours after sunset or in the east within a few hours before sunrise. Sometimes, however, it can be seen as a complete band across the sky. Because this light is confined to the region of the ecliptic or zodiac, it is called the *zodiacal light.* Spectrographic analysis of the zodiacal light shows it to be reflected sunlight. It is presumed to be due to reflection of sunlight from microscopic solid particles.

There is also a tenuous distribution of gas through the solar system. The clearest evidence of interplanetary gas comes from space probes, whose instruments have recorded rapidly moving atoms and charged atomic particles. High-altitude rockets, carrying cameras that photograph in the far ultraviolet, have recorded a faint illumination that is apparently light emitted by hydrogen gas, present either high in the earth's atmosphere or in interplanetary space. Practically all of the interplanetary gas consists of ions and electrons ejected into space from the sun. This flow of corpuscular radiation from the sun is called the *solar wind.*

We conclude that the region of interplanetary space contains minute, widely spread particles, and very sparse gas. In the neighborhood of the earth, there are only a few ions per cubic centimeter. This is a far better vacuum than can be produced in any terrestrial laboratory.

THE PLANETS

The other planets of the solar system are of special interest, for they, like the earth, are worlds that revolve about the sun and derive their light and warmth from it. The planets are considered individually in Chapter 10. Here we shall summarize some of their general properties and how certain information about them can be obtained.

Some Basic Characteristics

We determine the distances of planets from the earth, and from the sun, by essentially the same method used by Kepler, although today we employ more sophisticated mathematical techniques. The method involves geometrically surveying the distance to the planet by sighting it from different places in the earth's orbit. This procedure, of course, gives the distance to the planet only in terms of the size of the earth's orbit—that is, in astronomical units. The evaluation of the astronomical unit in, say, kilometers—that is, the determination of the scale of the solar system—is a far different problem; it is discussed in Chapter 12.

The mass of a planet must be determined by measuring the gravitational acceleration that it produces on other objects. Four different methods have been used: (1) observing the acceleration a planet produces upon one of its satellites (Chapter 3), (2) observing the perturbations a planet produces upon the motions of other planets, (3) observing the perturbations a planet produces upon the motion of a close-approaching minor planet, and (4) observing the effects a planet produces on man-made space probes. The last technique has given the best results for all but the outermost planets.

The diameter of a planet is found from its angular diameter and its distance, exactly as the diameter of the moon is calculated. The procedure is described in the next chapter.

Atmospheres

All the planets except Mercury and probably Pluto are surrounded by appreciable gaseous atmospheres. Among the planets possessing atmospheres, only Mars has an atmosphere thin enough so that we can see through it to examine features on the surface of the planet. In most cases, opaque clouds in the atmosphere reflect light brilliantly.

Spectrographic analysis of sunlight reflected from a planet can reveal some of the gases that compose its atmosphere. We have seen (Chapter 5) how atoms of gas abstract certain wavelengths from the light that passes through the solar atmosphere, leaving dark lines in the solar spectrum that are characteristic of those atoms. Sunlight reflected from the surface or cloud layers of a planet must pass through some of the outer gases in its atmosphere as well. By carefully comparing the spectrum of direct sunlight and the spectrum of sunlight reflected from a planet, the astronomer can ascertain which of the multitude of dark lines in the latter spectrum must originate in that planet's atmosphere.

Not all gases in the atmosphere of a planet can be detected spectrographically. Any oxygen, water vapor, carbon dioxide, methane, or ammonia that is abundant in the planet's atmosphere produces lines that are easy to detect. On the other hand, at the prevailing temperatures of planets, the gases helium and nitrogen do not produce conspicuous lines in the observable spectrum—that is, at wavelengths at which radiation can penetrate the earth's atmosphere.

Spectrographs are also carried, of course, on planetary space probes, and our best information on the atmospheric constituents of Venus, Mars, and Jupiter has come from these experiments. For the more remote planets, we must still rely on ground-based observations.

Which gases are present in a planet's atmosphere must depend, at least in part, on how that planet was formed and on its subsequent history. Most of the gases in the earth's atmosphere, for example, have escaped (that is, *outgassed*) from the earth's crustal rocks, and have changed subsequently due to photosynthesis and other chemical activity, whereas the constituents of Jupiter's atmosphere may reflect the original composition of the material from which Jupiter was formed.

On the other hand, the kind of atmosphere a planet has also depends on its ability to hold the various gases. The molecules of a gas are always in rapid motion; if their speeds exceed the velocity of escape of a planet (Chapter 3), that kind of gas can gradually "evaporate" from the planet into space.

At sea level on the earth there are some 10^{19} such molecules bouncing about in each cubic centimeter of air. The *kinetic energy* (or energy of motion) of a moving object depends on its mass and speed. Ordinary temperature, ***kinetic temperature***, is a measure of the mean (or average) kinetic energy of molecules. Thus the mean speed of molecules in any particular gas depends on the temperature of that gas, and on the mass of a single molecule of that gas. The higher the temperature, the faster the molecules move; at a given temperature, molecules of greater mass move slower, on the average, than those of smaller mass.

The temperature is measured from absolute zero; at $T = 0$, the mean energy of the molecules is zero. Here is the meaning of absolute zero: as gases are cooled, their molecules move more and more slowly, and at absolute zero all molecular motion ceases. This occurs at $-273°$ C ($-459°$ F). Absolute, or Kelvin, temperature is measured in units called Kelvins (K). A temperature difference of 1 K is the same as a temperature difference of 1° C; absolute temperature is thus the Celsius temperature + 273°. (Conversion from one temperature scale to another is described in Appendix 5.)

At any given time, some molecules move at less than the average speed, and others move faster. A few are moving at several times the average speed. An individual molecule, suffering frequent collisions, constantly exchanges energy with other molecules. Sometimes it moves relatively slowly; at other times it may get a good jolt in a collision and move far faster than the average. From kinetic theory we can calculate the relative numbers of molecules moving at various speeds if we know the kind of gas (and hence the mass of each molecule) and what its temperature is.

The effective temperature of a planet is the temperature of a blackbody (Chapter 5) of the same size as the planet that emits the same amount of radiation that the planet does. For Mars, whose atmosphere is nearly transparent, the effective temperature is approximately that of its solid surface. Venus and the Jovian planets, on the other hand, have opaque atmospheres, and their effective temperatures correspond to kinetic temperatures high in their atmospheres, from which is emitted the radiation we observe from them. The energy emitted from a blackbody is proportional to the fourth power of its temperature (Chapter 5). Thus we can calculate the effective temperature of a planet if we know how much energy it radiates. The source of the energy emitted by most of the planets is primarily the energy they absorb from incident sunlight. Now, just outside the earth's atmosphere, 1.36×10^6 erg/s of solar radiation is incident on each square centimeter. This is called the *solar constant*. The earth's distance from the sun is 1 AU. From the inverse-square law of propagation of electromagnetic radiation (Chapter 5), we can calculate the solar radiation falling on any other planet. Now that planet absorbs some of the energy and reflects the rest. The fraction of incident energy reflected is called the planet's *albedo*. Thus, if we measure the albedo of a planet from its brightness in reflected sunlight, we can calculate how much energy it absorbs, and hence radiates back into space. Such calculations predict that planets generally have lower temperatures the farther they are from the sun.

We can check these predictions, however, by calculating the temperature from the energy the planet is directly observed to radiate. Most of the energy emitted by planets is in the infrared. We can observe this energy at the telescope by using infrared detecting devices. The temperatures found are usually close to those predicted. However, a substantial fraction of the energy radiated by Jupiter, and to a lesser extent by Saturn, evidently comes from heat stored in their interiors, so their effective temperatures are slightly higher than their albedos and distances from the sun would lead us to expect.

Planetary temperatures derived by these methods, of course, hold only in an average sense. A planet receives all its solar energy on its daylit hemisphere. If it does not rotate rapidly it may not reemit the energy it absorbs uniformly in all directions. Mercury and the moon, for example, are hot on their sunlit sides and cold on their night sides. Moreover, it must be remembered that the effective temperatures do not correspond to temperatures at the solid surfaces of planets with opaque atmospheres. Thus, Venus is much hotter at its surface than it is high in its atmosphere, from which most of its radiated energy escapes into space (Chapter 10).

Knowing the approximate temperatures of planets, we can calculate the mean speeds of various kinds of atmospheric molecules. Those molecules at high enough levels in the atmosphere that are moving in the right direction with enough speed escape. Because some molecules are moving with much higher than average speeds, it is not necessary for the mean molecular speed in a gas to be equal to the escape velocity of a planet for that gas to be lost to space. The high level in a planetary atmosphere from which molecules can escape is called the *exosphere*. Calculations show that if the mean molecular speed is as much as one-third the velocity of escape, gases in the exosphere will escape in a few weeks. Molecules of a particular species in the lower atmosphere gradually diffuse upward into the exosphere, and if they find themselves in an environment of high enough temperature there, they will escape also. To be conservative, we shall suppose that for a planet to hold a particular gas in its

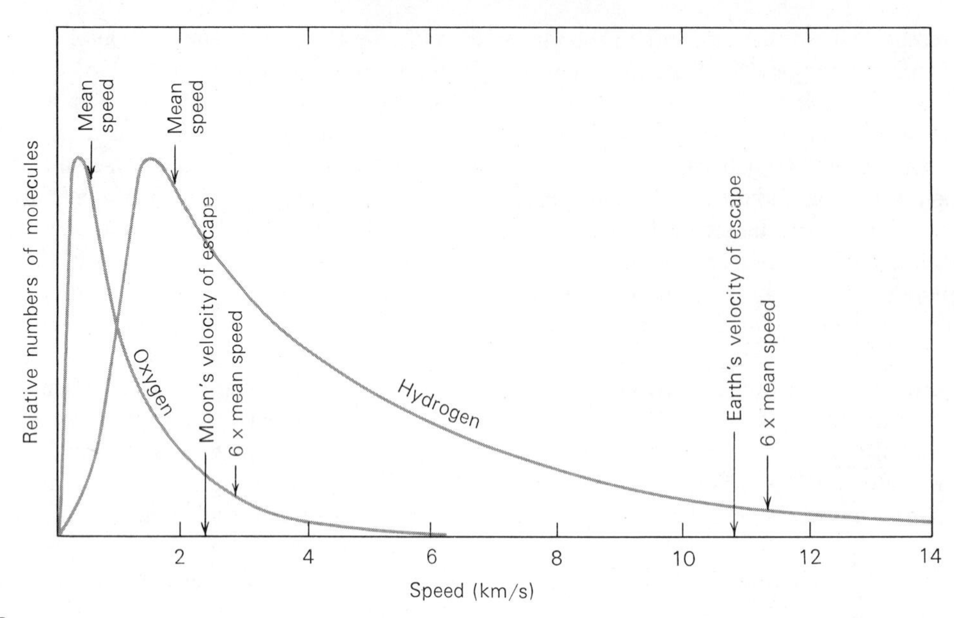

Figure 8.2
Distribution of speeds of oxygen and hydrogen molecules at a temperature appropriate for the earth and moon.

TABLE 8.2 Some Approximate Planetary Data

	Mercury	Venus	Earth	Mars	Jupiter	Saturn	Uranus	Neptune	Pluto
Distance from sun (AU)	⅖	¾	1	1½	5	10	20	30	40
Mass in terms of earth's mass	1/20	⅘	1	1/9	318	95	15	17	1/500
Radius in terms of earth's radius	⅖	1	1	½	11	9	4	4	¼
Rotation rate	59 days	243 days retrograde	24 hours	24½ hours	10 hours	10 hours	20 hours retrograde	16 hours	6 days
Velocity of escape (km/s)	4	10	11	5	60	35	20	22	1
Approx. effective temperature* (K)	450	235	240	220	100	75	50	40	40
Important gases observed in atmosphere	—	CO_2, H_2O, N_2	N_2, O_2, A, CO_2, H_2O, etc.	CO_2, H_2O, N_2	H_2, CH_4 NH_3	H_2, CH_4	H_2, CH_4	H_2, CH_4	—
Number of known satellites	0	0	1	2	14	10	5	2	1

* Temperatures on the sunlit sides of planets, or at their surfaces, may be much higher. The sunlit side of Mercury, for example, has a temperature of over 700 K and the surface temperature of Venus is also near 700 K (Chapter 10).

atmosphere for several thousand million years, its velocity of escape must be at least six times the mean molecular speed for that gas.

As an illustration, Figure 8.2 shows, roughly, the relative distributions of molecular speeds of oxygen and hydrogen that would be expected at a temperature of about 300 K—which would be representative of the earth and moon if they both had appreciable atmospheres of these gases. The solid curves show the relative numbers of molecules of each gas that would be moving at various speeds. In each case, the mean speed and six times the mean speed are indicated. It is seen that neither earth nor moon should be able to hold hydrogen, for six times its mean molecular speed would exceed the escape velocity of both. Oxygen, however, would be retained by the earth but not by the moon.

Mercury, the minor planets, and the moon would not be expected to retain any of the common gases, even if they ever did possess atmospheres. The other terrestrial planets could hold atmospheres of the heavier gases (such as nitrogen and carbon dioxide), and the Jovian planets can hold all gases in their atmospheres.

Interiors of Planets

As far as their internal structures are concerned, the planets fall naturally into two groups: the terrestrial planets, Mercury, Venus, Earth, and Mars (sometimes the moon and Pluto are also included); and the Jovian planets, Jupiter, Saturn, Uranus, and Neptune. The terrestrial planets are relatively small and dense and are believed to be composed mostly of rocky and metallic material. The Jovian planets are relatively large and of low mean density (indeed, Saturn has a mean density less than that of water). The chemical composition of the Jovian planets is probably more nearly typical of the general cosmic abundances of the various chemical elements—for example, the relative abundances found in the sun. Hydrogen and helium, the most abundant elements in the universe, probably dominate in the Jovian planets. Jupiter and Saturn, at least, are almost certainly composed mostly of hydrogen compressed to a liquid or solid state (see Chapter 10).

Summary

In Appendices 9 and 10 are tabulated various data pertaining to the planets. However, many precise figures are often more confusing than helpful to the nonscientific reader. Therefore, in Table 8.2 we have summarized, very roughly, a few of the most important data concerning the planets. It is easier to remember that Jupiter is about 5 AU from the sun than to remember the figure 778,730,000 km. The relative sizes of the planets and sun are shown in Figure 8.3.

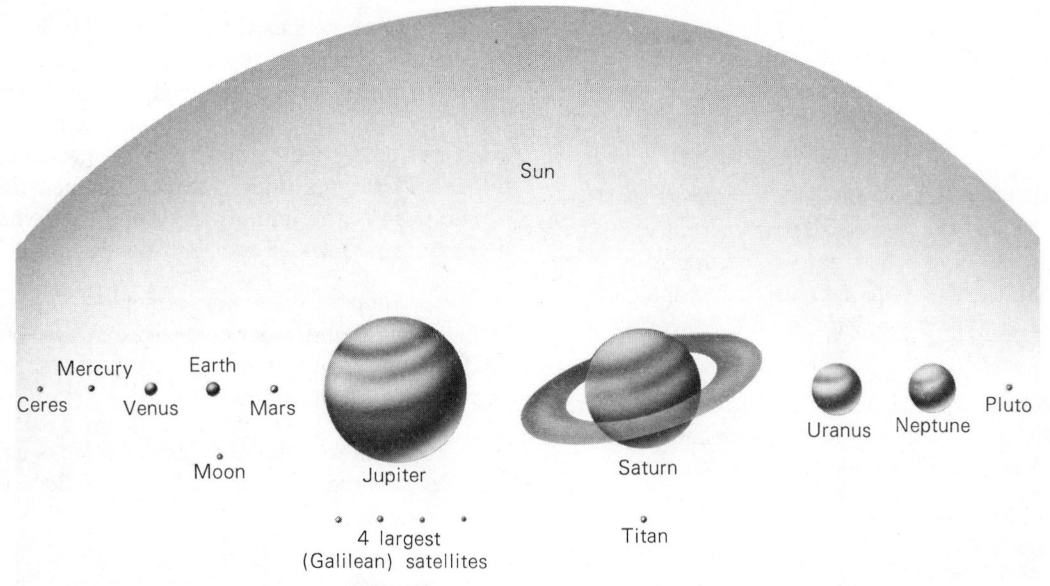

Figure 8.3
Relative sizes of the sun and planets.

SUMMARY REVIEW

Origin of the solar system: solar nebula; rotation and flattening of the solar nebula; accretion of planets; formation of the moon and other satellites; formation of minor planets; terrestrial planets; Jovian planets

Inventory of the solar system: mass of the solar system accounted for by the sun, planets, satellites, comets, minor planets, meteoroids, interplanetary medium; meteors, meteoroids, and meteorites; zodiacal light; solar wind; coma and tails of comets

Properties of planets: determination of masses and sizes; atmospheres of planets; composition of planetary atmospheres; temperature scales; effective temperature of a planet; solar constant; velocity of escape of planets; escape and retention of planetary atmospheres; interiors of planets

EXERCISES

1. Can you think of any practical applications of the principles governing rotation and the conservation of angular momentum in which rotation leads to the formation of a disk? (*Hint:* Think of a pizza parlor.)

2. The terrestrial planets formed by accretion of solid particles. The last particles to be left in the inner solar system before the accretion process was complete were large ones. Do we see any evidence of these particles being swept up in the final accretion process? Explain.

3. A number of years ago some investigators had suggested that the rings of Saturn are not a surprising phenomenon; rather, that it is strange that other Jovian planets do not have rings. What did they have in mind?

4. A theory proposed in a book published in 1950 was that Venus was formed a few thousand years ago from a comet that boiled out of the atmosphere of Jupiter. On the basis of data in this chapter, what arguments can you suggest to refute the theory?

5. It was once suggested that the minor planets and meteoroids may have originated from a planet that once broke up. In what way would that planet have had to differ from the other major planets?

6. A double planet has a period of mutual revolution of about $27\frac{1}{3}$ days, and the two bodies are separated by about 385,000 km. What is the combined mass of the two? If they are in the solar system, what are their names?

7. Can you think of any ways it might be possible to distinguish telluric lines (those due to absorption of light by the earth's atmosphere) from the absorption lines produced in the atmosphere of a planet being observed? (*Hint:* Consider the Doppler effect.)

8. Observations of both the visible and the infrared radiation respectively reflected from and emitted by a minor planet, and a knowledge of how distant it is from the sun, can enable us to calculate both its albedo and its approximate size. Can you suggest how this is done? (*Hint:* Consider what determines the optical brightness of the minor planet, and also what the intensity of its infrared radiation depends on.)

9. Suppose there were a planet 100 AU from the sun, of mass and radius like the earth's. Would it be expected to have helium in its atmosphere? How about ammonia? Explain in each case.

10. The velocity of escape of Mars is only a little greater than that of Mercury. Why then does Mars have an appreciable atmosphere while Mercury does not?

Mosaic of photographs of the planet Mercury, taken by Mariner 10 on March 29, 1974. *(NASA/JPL)*

CHAPTER 9

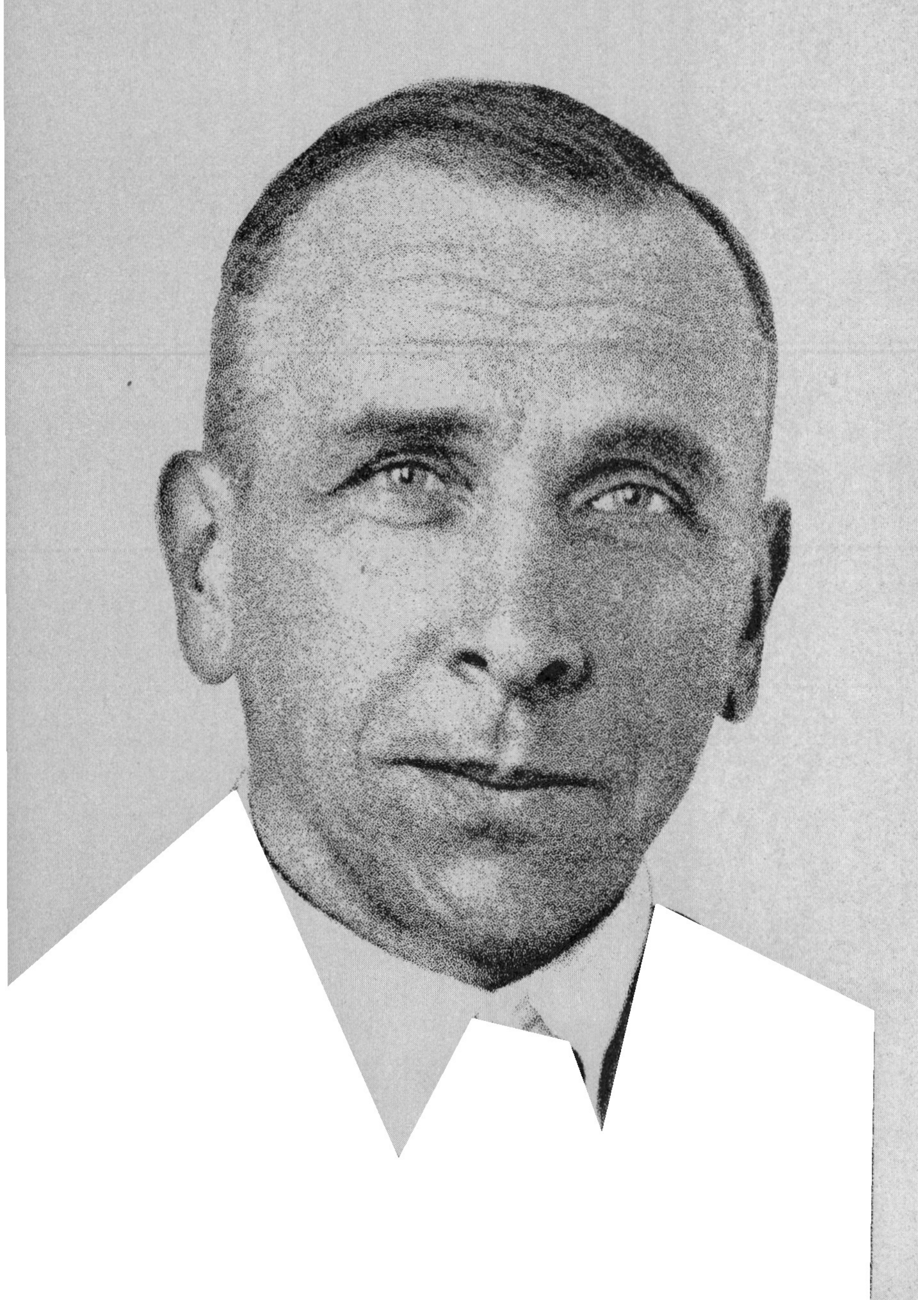

Alfred Wegener (1880–1930), German meteorologist, suggested the idea of "continental drift" in 1912. His arguments, although not accepted then, are now regarded as precursors of the theory of plate tectonics—the most important area of geophysics in the mid-twentieth century.

THE DOUBLE PLANET

Several of the satellites in the solar system are larger than the moon, but except for the recently discovered satellite of tiny Pluto, the moon is the largest in comparison to its primary; thus the earth-moon system is sometimes referred to as the "double planet."

ASPECTS OF THE MOON

The moon, because of its proximity, appears to move more rapidly in the sky than any other natural astronomical object, except meteors, which are within the earth's atmosphere. As it travels about the earth each month, the moon displays different parts of its daylight hemisphere to our view and progresses through its cycle of phases.

Phases of the Moon

When the moon is in the same general direction from earth as the sun (position A in Figure 9.1), its daylight side is turned away from the earth. Because its night side—the side turned toward us—is dark, we cannot see the moon in that position. The phase of the moon is then *new*.

A few days after new moon, the moon reaches position B, and from the earth we see a small part of its daylight hemisphere. The illuminated crescent increases in size on successive days as the moon moves farther and farther around the sky away from the direction of the sun. During these days the moon is in the *waxing crescent* phase. About a week after new moon, the moon is one quarter of the way around the sky from the sun (position C) and is at the *first quarter* phase. Here the line from the earth to the moon is at right angles to the line from the earth to the sun and half of the moon's daylight side is visible—it appears as a half moon.

During the week after the first quarter phase we see more and more of the moon's illuminated hemisphere, and the moon is in the *waxing gibbous* phase (position D). Finally, about two weeks after new moon, the moon (at E) and the sun are opposite each other in the sky; the side of the moon turned toward the sun is also turned toward the earth; we have *full moon*. During the next two weeks the moon goes through the same phases again in reverse order—through *waning gibbous*, *third* (or *last*) *quarter*, and *waning crescent*.

If you have difficulty picturing the phases of the moon from this verbal account, stand about six feet in front of a bright electric light outdoors at night and hold in your hand a small round object such as a tennis ball or an orange. If the object is then viewed from various sides, the portions of its illuminated hemisphere that are visible will represent the analogous phases of the moon. With this experiment, you should also be able to figure out what time of day or night the moon must rise or set when it is at various phases.

Brightness of the Moon

Despite the brilliance of the full moon, it shines with less than 1/400,000 the light of the sun. Even if the entire visible hemisphere of the sky were packed with full moons, the illumination would be only about one-fifth or less of that in bright sunlight.

Because the moon shines by reflected sunlight, we can calculate the moon's reflecting power from its apparent brightness. Calculation shows that if all sunlight were reflected back into space, the full moon

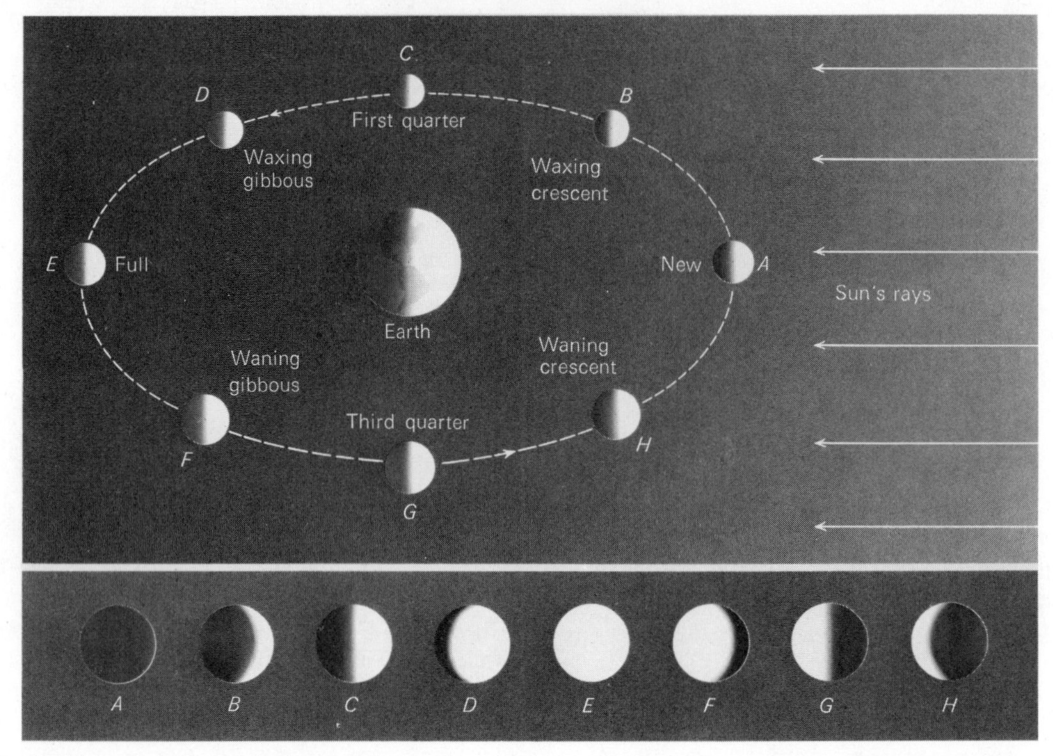

Figure 9.1
Phases of the moon. The moon's orbit is viewed obliquely. *(Below)* The appearance of the moon from the earth.

would appear about 14 times as bright as it actually is. The fraction of incident light that is reflected by a body is called its *albedo*. The average albedo of the moon is thus about 100/14, or 0.07. The moon absorbs most of the sunlight that falls upon it; its surface is quite dull. The absorbed energy heats up the surface of the moon until the energy is radiated away again as infrared radiation.

When the bright part of the moon appears as only a thin crescent, the "night" side of the moon often appears faintly illuminated. Leonardo da Vinci (1452–1519) first explained this illumination as *earthshine*, light reflected by the earth back to the night side of the moon, just as moonlight often illuminates the night side of the earth.

The Moon's Revolution and Rotation

The moon's *sidereal period*, that is, the period of its revolution about the earth with respect to the stars, is $27^d7^h43^m11^s.5$ (27.32166 days). However, during this period of the moon's sidereal revolution, the earth and the moon together revolve about 1/13 the way around the sun, or about 27°. The sun, therefore, appears to move 27° to the east on the celestial sphere during the period of the moon's sidereal revolution. In other words, the moon would not, in its sidereal period, have completed a revolution about the sky with respect to the sun, and consequently would not have completed a cycle of phases. To complete a revolution with respect to the sun, the moon requires, on the average, $29^d12^h44^m2^s.8$ (29.530588 days). We have, then, two kinds of month: the *sidereal month*, the period of revolution of the moon with respect to the stars, and the *synodic month*, the period with respect to the sun (Figure 9.2).

The moon changes its position on the celestial sphere rather rapidly, moving, on the average, about 13° to the east per day. Even during a single evening the moon creeps visibly eastward among the stars. The delay in moonrise from one day to the next, caused by this eastward motion of the moon, averages about 50 minutes.

The moon's apparent path around the sky is inclined to the sun's path (ecliptic) at about 5°, and

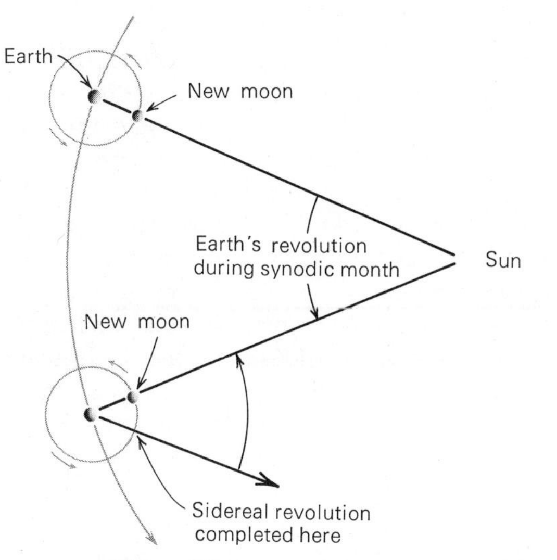

Figure 9.2
Sidereal and synodic months.

crosses it at two points on opposite sides of the celestial sphere. These points are called the *nodes* of the moon's orbit. The sun's gravitation perturbs the moon's motion, however, causing its orbit to change continually. One of the effects of these perturbations is that the nodes slide westward along the ecliptic, completing one trip around the celestial sphere in about 18.6 years. This motion is called the moon's *regression of the nodes*.

The moon's orbit changes so rapidly that if the moon's positions over a month, relative to the center of the earth, are plotted carefully, even on a sheet of standard typing paper, the orbit is seen to not close on itself.

The moon rotates on its axis with exactly the same period at which it revolves about the earth. As a consequence the moon keeps turning the same face toward the earth. This coincidence of the periods of the moon's rotation and revolution can hardly be accidental. It is believed to have come about as a result of the earth's tidal forces on the moon (Chapter 3).

We often hear the back side of the moon (the side we do not see) called the "dark side." Of course, the back side is dark no more frequently than the front side. Since the moon rotates, the sun rises and sets on all sides of the moon. The back side of the moon is receiving full daylight at new moon; the dark side is then turned toward the earth.

Mass and Dimensions of the Earth-Moon System

The distances to the moon and to some of the planets have been accurately determined by radar. In this technique, radio waves, focused into a beam by a powerful broadcasting antenna, are transmitted to the object, which reflects some of the energy back to the earth. Since radio waves are a form of electromagnetic energy, they travel with the speed of light, and half the round-trip travel time, multiplied by the speed of light, gives the distance. The time intervals involved can now be measured electronically to better than one-millionth of a second.

The distance from the center of the earth to the center of the moon, known by the ancients to be about 30 times the earth's diameter, is found by radar to be 384,404 km (238,858 mi), with an uncertainty of about 0.5 km.

If an object of known distance subtends a measurable angle in the sky, we can easily find its linear (true) size. The moon's angular diameter is about 31′5″. Notice, in Figure 9.3, that the moon's angular size is relatively small, so its linear diameter is essentially a small arc of a circle, with the observer as center and with a radius equal to the moon's distance. Obviously, the moon's diameter is the same fraction of a complete circle as the angle subtended by the moon is of 360°. A complete circle contains 1,296,000″ (there are 60″ per minute, 60′ per degree, and 360° in a circle). As seen

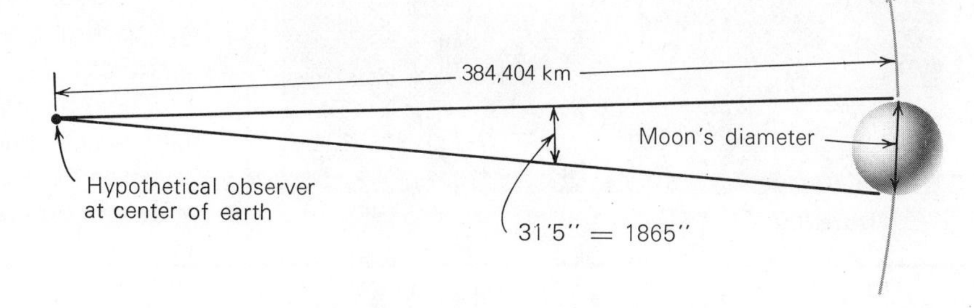

Figure 9.3
Measuring the moon's diameter.

from the center of the earth, the moon's mean angular diameter of 31′5″, or 1865″, is thus 1/695 of a circle. The moon's diameter, therefore, is 1/695 of the circumference of a circle of radius 384,404 km. Since the circumference of a circle is 2π times its radius, the diameter of the moon is $2\pi \times$ (384,404 km) divided by 695, or 3475 km (2160 mi). The equatorial diameter of the earth is 12,756 km (7927 mi).

One body does not strictly revolve about the other. Rather, the two bodies mutually revolve about their center of mass, or the *barycenter* (Chapter 3). It is the barycenter of the earth-moon system that revolves annually in an elliptical orbit about the sun, while the earth and moon simultaneously revolve about the barycenter in a shorter period—the sidereal month.

The elliptical orbit of the center of the earth about the barycenter constitutes an independent motion of the earth. The motion can be detected by careful observations of the nearer planets, or better yet, of near-approaching minor planets (Chapter 11). The motion of Mars, for example, shows monthly oscillations, carrying it a little ahead and then a little behind its regular orbital motion. This oscillation is only apparent and results from the motion of the earth, carrying us first to one side and then to the other side of the barycenter. When Mars is at its closest, the apparent displacements caused by the orbital motion of the earth around the barycenter amount to about 17″. The corresponding mean distance of the center of the earth from the barycenter is 4672 km. Thus, the earth and moon jointly revolve about a point approximately 1707 km (about 1000 mi) below the surface of the earth.

Figure 9.4
The first view of the earth taken from the vicinity of the moon; Lunar Orbiter I photograph.

One way of finding the mass of the moon compared to that of the earth uses the relative distances of the earth and moon from the barycenter. Today, however, the moon's mass is determined accurately by measuring the accelerations the moon produces on space probes—either those sent to the moon itself, or those that pass it by on interplanetary missions. For example, analysis of the motion of the Mariner V probe, which passed about 10,000 km from Venus on October 19, 1967, gives an earth-moon mass ratio of 81.3004. The earth's mass is found from laboratory measures to be 5.98×10^{21} metric tons; the moon's mass is thus 7.35×10^{19} metric tons.

ATMOSPHERES AND MAGNETOSPHERES OF THE EARTH AND MOON

We live at the bottom of the ocean of air that envelops our planet. The chemical composition of the earth's atmosphere is 78 percent nitrogen, 21 percent oxygen, and 1 percent argon, with traces of water (in the gaseous form), carbon dioxide, and other gases. At lower altitudes, variable amounts of dust particles and water droplets are also found suspended in the air.

The earth's present atmosphere, however, is not primordial. As explained in the last chapter, the earth is believed to have accreted from smaller solid particles. These particles contained hydrates, carbonates, and nitrogen compounds. Many of these compounds were evidently trapped in the outer layers of the earth, where they were subjected to heating by the decay of certain radioactive elements (such as uranium and thorium). The heating caused these compounds to lose water (in the gaseous state), carbon dioxide, and probably nitrogen. These gases escaped through the crustal layer to form the earth's atmosphere—a process called *outgassing*.

The principal means of outgassing is volcanism. Water was by far the most plentiful substance to outgas; when released from the earth's crust it cooled and condensed into the oceans. Carbon dioxide is next most important, and what did not dissolve into the oceans reacted chemically with crustal rocks. Chemical

The earth and moon, photographed by Voyager I on its way to Jupiter. *(NASA)*

The full moon, photographed by Apollo 17 astronauts. *(NASA)*

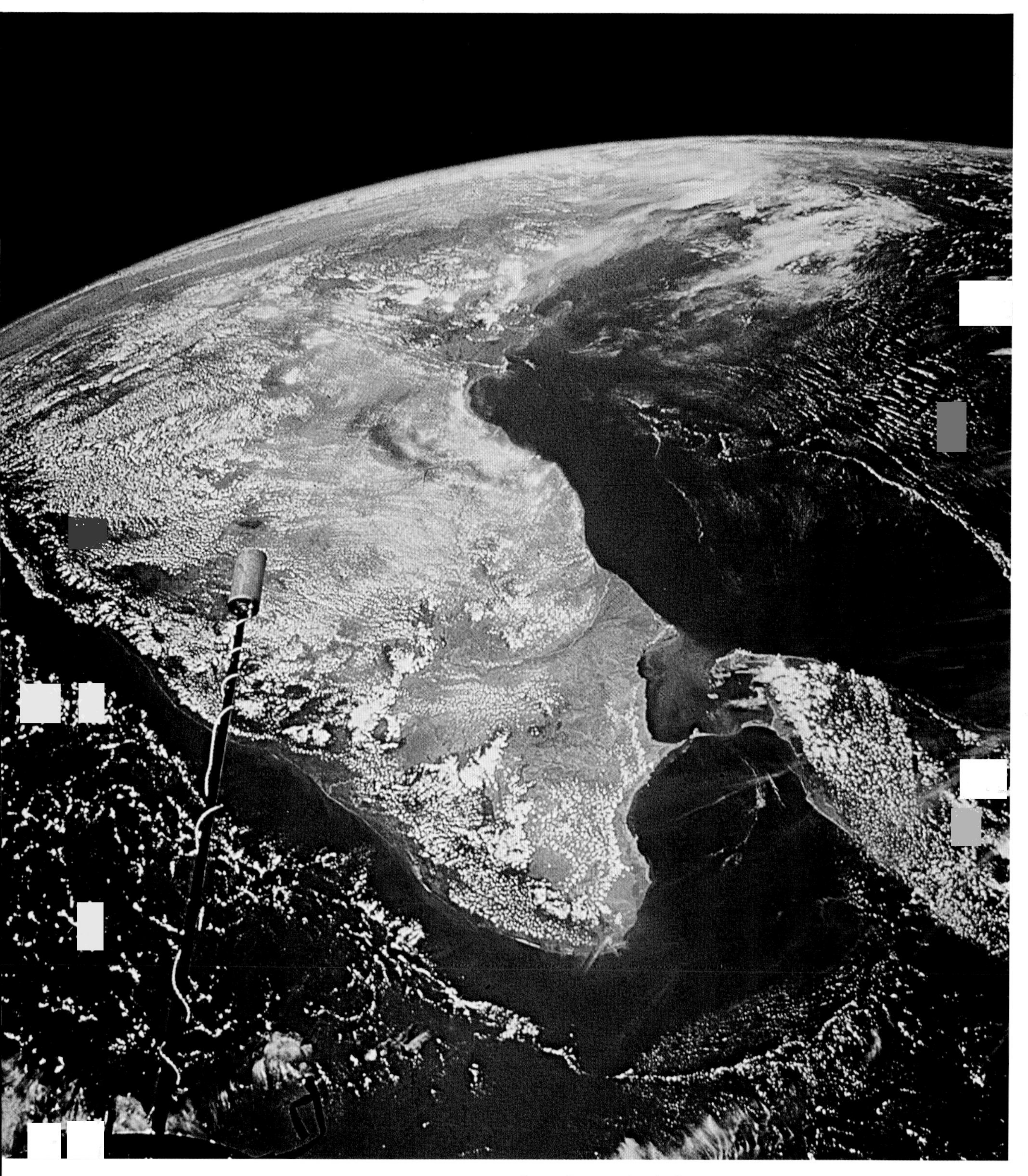

India and Sri Lanka, looking north with the Bay of Bengal on the right. Gemini XI photograph taken from an altitude of 410 nautical miles. *(NASA)*

Astronaut Harrison Schmitt working beside a large boulder on the moon. The lunar roving vehicle is at the left. *(NASA)*

activity at the earth's surface may also have produced considerable concentrations of methane and ammonia and some hydrogen in the early atmosphere. These three gases, if ever present, either escaped or dissociated relatively quickly. Nitrogen was far less abundant than carbon dioxide, but is rather inert chemically, so it has remained to be the principal component of the atmosphere today. At least some of the present nitrogen may have had a biological origin. Most of the argon (the isotope argon-40) is believed to have resulted from the radioactive decay of the isotope potassium-40.

Among the planets the earth is unique in having an appreciable amount of free oxygen in its atmosphere. The earth's oxygen is believed to have resulted from photosynthesis by green plant life (mostly in the oceans). It is estimated that oxygen increased from about 1 percent to the present 21 percent of the atmospheric content during the past 600 million years, by a very slight imbalance of photosynthesis over those factors that remove oxygen from the air—decay of vegetation and animal matter, respiration of animals, and combustion. We shall return to this subject in Chapter 19.

Levels of the Atmosphere

The atmosphere, weighing down on the surface of the earth under the force of gravitation, exerts a pressure which at sea level amounts to that produced by the weight of 1.03 kg over each square centimeter (or about 14.7 lb/in^2). If the mass of the air over 1 cm^2 is 1.03 kg, the total mass of the atmosphere may be found by multiplying this figure by the surface area of the earth in square centimeters. We find, thus, that the total mass of the atmosphere is about 5×10^{15} tons, or about a millionth that of the earth. The density of the air drops rapidly at higher and higher elevations. In fact, half of the total atmosphere is packed down within 6 km of the earth's surface. The pressure exerted by the rapid motions of the molecules that comprise air prevents it from collapsing completely and supports a small fraction of the atmosphere even to great heights. By observing auroras we know that thin vestiges of the atmosphere extend to heights of at least 1000 km. Analysis of drag on earth satellites reveals evidence for some atmosphere at even greater heights—1600 to 2000 km.

The demise of Skylab on July 11, 1979 is well-publicized evidence of the effects of atmospheric drag on near-earth satellites. Skylab had an altitude of

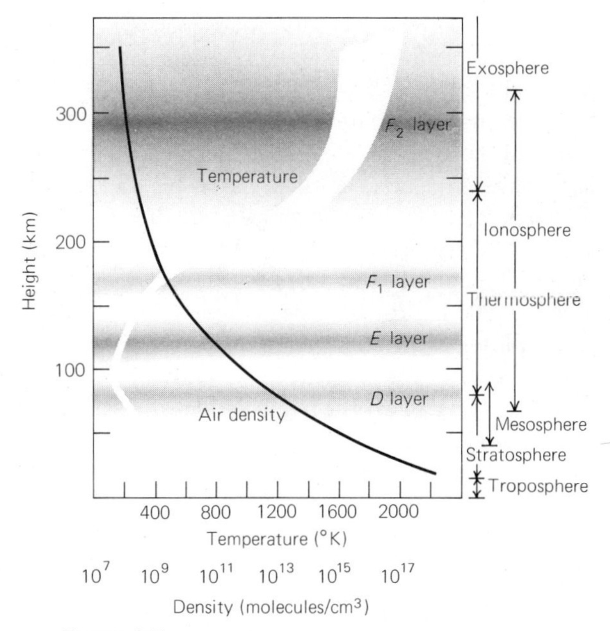

Figure 9.5
The structure of the earth's atmosphere. The broad dark band shows the ranges of temperature measured and estimated at various altitudes in the atmosphere. The single smooth line indicates the approximate density variation with altitude. The shaded regions indicate the zones in the atmosphere where the relative ionization of the atmospheric gases is highest.

about 400 km when it was launched in 1972, and served admirably for the astronauts who lived there and carried out scientific experiments for months at a time. It took only 7 years for the atmosphere to drag it down into the Indian Ocean and Western Australia.

In the lower part of the atmosphere, the temperature also drops rapidly with increasing elevation. It is this region, the *troposphere*, where all weather occurs.

Above the troposphere, and extending from about 15 to 80 km, is the *stratosphere*. The temperature through most of the stratosphere is fairly constant—a little lower than 50° C below zero. Between 40 and 65 km up, however, there is a warm zone; in 1947, V-2 rockets launched at White Sands, New Mexico, first recorded above-freezing temperatures in this region.

The hot layer is due to the presence of ozone in that level of the atmosphere. Ozone is a heavy form of oxygen, having three atoms per molecule instead of the usual two. It has the property of being a good absorber of ultraviolet light. In absorbing the sun's short-wavelength ultraviolet light, the ozone is heated up and warms the parts of the atmosphere where it is present. Incidentally, the protective ozone layer helps to prevent some of the sun's dangerous ultraviolet radiation from penetrating to the earth's surface. The region of the stratosphere extending upward from the ozone layer is sometimes called the *mesosphere*.

From 65 to 80 km, the temperature drops to below −50° C again. Above 80 km the temperature rises rapidly through a region called the *thermosphere*, and at 400 to 500 km reaches values above 1000° C. The highest layer of the atmosphere, above about 400 km, is called the *exosphere*. In the upper atmosphere, the thermosphere and exosphere, molecules of oxygen and nitrogen break up into individual atoms of those elements. Ultraviolet radiation from the sun ionizes many of these atoms. Therefore, part of the upper part of the atmosphere (above 50 km) is also called the *ionosphere*.

Lack of a Lunar Atmosphere and Water

It is very different on the moon, which could retain only gases of rather heavy atoms. The common gases that would be expected to outgas would, of course, escape. Extensive lunar orbiter probes and Apollo studies of the moon confirm that the moon has no appreciable atmosphere.

In the absence of air, the moon can have no liquid water on its surface. It is well known that water boils more easily (at a lower temperature) at high altitudes in the mountains than at sea level. If all air pressure could be removed above water, it would boil away. On the moon, therefore, if there ever were any liquid water, it would have evaporated and then, as a gas, dispersed into space with the rest of the moon's atmosphere.

In the absence of air and water, there can be no weather on the moon. There are no clouds, winds, rain, snow, or even smog, which accounts for many of the differences between the moon and earth. Weather on the earth and the running of water over its surface have been major sources of erosion that have washed and worn away entire mountain ranges, and even the faces of continents many times over during geologic time. On the moon, where air and water cannot contribute to erosion, features are more nearly permanent. There are formations of all ages on the moon, standing side by side; features that were formed thousands of millions of years ago are often still intact, standing with those formed in the recent past. Among the most recent are the footprints of Apollo astronauts.

The earth's atmosphere and water cover serve to reflect and disperse much of the sun's radiation during the day and to blanket heat in near the surface at night. On the other hand, because of the moon's lack of atmosphere and oceans, its temperature range between day and night is rather extreme. Moreover, since the moon rotates with respect to the sun in about 29½ days, the sun shines on each place on the moon for about two weeks. Consequently, the moon's temperature ranges from just above the boiling point of water where the sun is shining to about 100 K (−173° C) on its dark side.

The Magnetic Fields of the Earth and Moon

The earth has a magnetic field similar to that produced by a bar magnet. Nearly everyone is familiar with the way iron filings align themselves along the lines of force that extend between the north and south poles of a magnet. The magnetic poles of the earth are located at about 78° north and south latitude, or about 1330 km from its geographical poles; the north magnetic pole is in northeast Canada. Between the magnetic poles of the earth stretch lines of force along which compass needles align.

The origin of the earth's magnetic field is believed to be in the earth's core. Fluid motions in the electrically conducting core are thought to cause it to act like a dynamo. The rotation of the earth causes the magnetic field to be aligned approximately with the rotation axis. The energy source of this dynamo, however, is not yet known. The overall strength of the magnetism at the earth's surface is fairly weak. Moreover, both the positions of the magnetic poles and the orientation of the lines of magnetic force of the earth's field gradually shift. Even molten rocks, as in volcanic lava, containing iron compounds are weakly magnetized by the earth's field. When the rocks harden, their magnetic polarity is permanently frozen in them. We measure the strength and polarity of magnetism in rocks of different ages to trace the magnetic history of the earth, and find that the earth's field has reversed polarity 171 times in the past 76 million years.

The magnetic field of the earth also extends into space around it, throughout a region called the *magnetosphere*. Magnetometers carried on artificial satellites measure the field strength in the magnetosphere, and, as expected from theory, we find that it weakens rapidly with increasing distance from the earth. Satellites have also detected a zone of rapidly moving

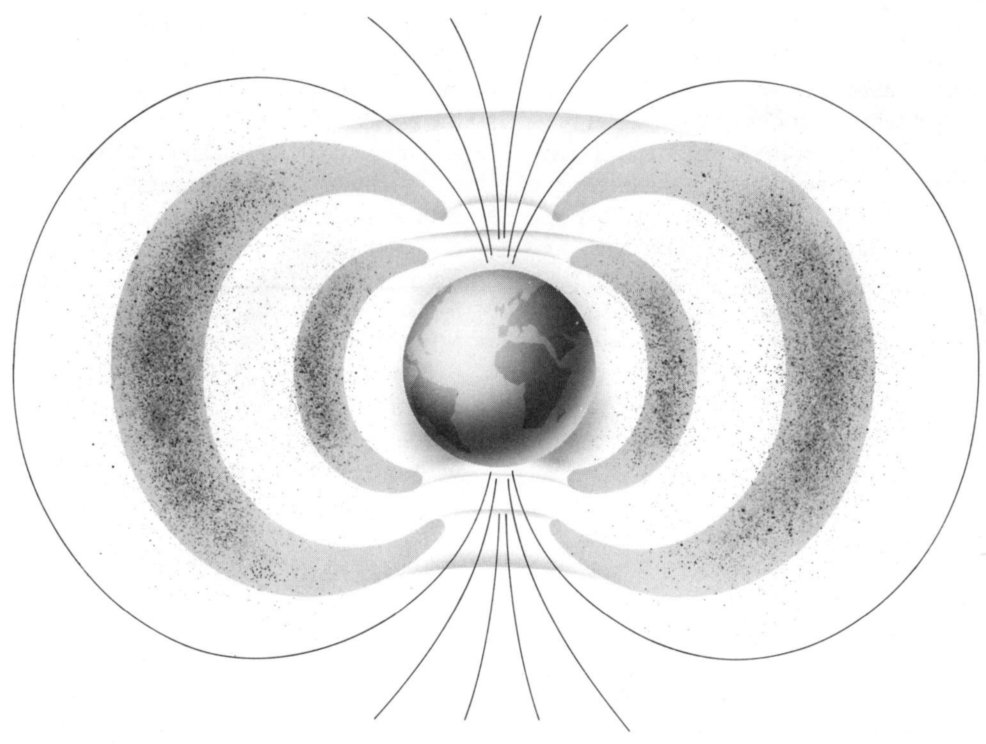

Figure 9.6
Cross section of the earth's magnetic field and the Van Allen layer.

charged particles high above the earth. This zone of radiation is called the *Van Allen layer,* in honor of the physicist James A. Van Allen (1914–), who designed the experiment that first detected it.

Figure 9.6 shows a cross section of the magnetic field surrounding the earth and the approximate location and extent of the most intense part of the Van Allen layer. It surrounds the earth, like a doughnut. The inner intense region is centered about 3000 km above the surface of the earth and has a thickness of 5000 km or more. The outer region of the Van Allen layer is about 15,000 to 20,000 km from the earth's surface and has a thickness of from 6000 to 10,000 km.

The electrically charged particles in the Van Allen radiation zone are trapped in the earth's magnetic field. Most of the fastest ones are electrons, although the inner part of the Van Allen layer contains high-energy protons as well. The origin of the trapped particles is not known with certainty. It has been suggested that the protons and electrons in the inner region may be produced in interactions between molecules of the earth's atmosphere and cosmic rays.

Magnetometers carried to the moon by Apollo astronauts, and those on earlier lunar probes and orbiters, have not detected a permanent general lunar magnetic field. There are some very weak local fields, indicating slight magnetism in some of the surface rocks. On the other hand, to be magnetized at all those rocks must have been exposed to a somewhat stronger field in the past. Possibly the rocks were magnetized in the earth's field at a time in the past when the moon was nearer the earth.

INTERNAL STRUCTURES OF THE EARTH AND MOON

The weight of the various layers of the earth bearing down on its interior causes the pressure to increase inward. Calculations indicate that the pressure at the earth's center must be close to 4 million kg/cm^2 (50 million lb/in^2). When subjected to this great pressure, matter is highly compressed and heated. We can expect the central regions of the earth, then, to be hot and dense.

Much information about the earth's interior comes from *seismology,* the study of seismic vibrations emitted during earthquakes. Stresses that build up gradually in the crust of the earth are often released by slippages along fissures or *faults*. Earthquakes result from the energy released in these sudden movements in the earth's crust. The vibrations sent out travel to all parts of the earth. Some of these waves travel along the surface; others pass directly through the interior. The seismic vibrations are picked up and recorded by delicate instruments stationed around the earth's surface.

The transmission of the waves through the earth's interior indicates that most of it is solid and rigid. This major part of the earth is called its *mantle*. The density in the mantle increases downward from about 3½ to 5½ g/cm^3. It is believed to be composed mostly of basic silicate rocks.

At the upper boundary of the mantle is the *crust*, a shell of lower density extending on the average about 30 km inward from the earth's surface under the continents. The crust consists of surface rocks such as granite and basalt, with overlying oceans and sedimentary rocks.

At the inner boundary of the mantle, about 2900 km below the surface, is the *core* of the earth. The outer part of the core acts like a liquid, for it does not transmit certain kinds of seismic waves. The innermost part of the core (about 2400 km in diameter) is extremely dense and hot (in the thousands of degrees), and is probably solid.

The temperature increases downward in the crust of the earth about 1° C for every 30 m (100 ft). As the core slowly cools, conduction gradually transmits heat from the earth's interior out through the mantle and crust. However, the conductivity of the interior is relatively low, and cooling of the hot core must occur at a very slow rate. Calculations have shown that most of the heat in the crust, perhaps over 80 percent, does not come from the deep interior but from the radioactive decay of uranium, thorium, and potassium—unstable radioactive elements in the crust itself.

The age of the earth can be estimated from the degree of decay of these radioactive elements in its crust. Atoms of uranium and thorium disintegrate spontaneously but very gradually through various elements, including radium, and end as atoms of some isotopes of lead. The exact rate of this radioactive decay from one element to another has been measured many times in the laboratory, and has been found to be constant.

Studies of the relative proportions of these elements in mineral deposits containing them indicate how long the disintegration process has been at work and give the ages of these surface rocks, from which we derive the age of the earth itself. Similar studies are made of rock samples from the moon brought back by the Apollo astronauts. The oldest rocks on the earth and moon have ages of about 4600 million years.

Analysis of the perturbations on orbits of the man-made lunar orbiting satellites gives some indication of the internal structure of the moon. As expected from its lower mass, it is found that the moon is much less centrally concentrated than the earth is. However, subsurface irregularities in the moon's mass distribution, called *mascons* (for "mass concentrations"), have been detected. These relatively dense subsurface regions are possibly submerged material that collided with the moon in the past. It is also found that the center of mass of the moon is not at its geometrical center, but is displaced about 2 km, roughly in the direction toward the earth.

The Surface of the Earth

Early in this century the German meteorologist Alfred L. Wegener (1880–1930) noted the complementarity of the east coast of South America to the west coast of Africa, and suggested that these continents may have once been contiguous, and have since drifted apart. Although Wegener presented considerable evidence in support of his ideas, his arguments were not, at the time, conclusive, and were not generally accepted. The evidence that such continental drift *actually occurs*, accumulating since the late 1950s, has led to the new field of *plate tectonics*, which has brought about a revolution in geophysics in some ways as dramatic as the Copernican revolution was in astronomy.

The Crustal Plates

The entire surface of the earth down to a depth of some 50 to 100 km consists of a horizontal mosaic of about 10 major plates, and some minor ones. These plates contain the crust and the upper mantle and make up what is called the *lithosphere*. The mantle beneath the lithosphere is called the *asthenosphere*, the upper part of which consists of a plastic region some 100 to 200 km thick, on which the lithosphere floats. For reasons not yet entirely understood, there is a general circulation of the plates, which causes them to slide about over the asthenosphere.

As the plates move about, they slide against and crunch into one another. Where the motion of one plate with respect to a neighboring one is purely lateral, they slide against each other along essentially transverse faults or fissures. An example of lateral displacement of two adjacent plates is along the famous San Andreas Fault in California. Where the plates push against one another, if the motion is not too severe, crustal folding and the building of mountain ranges on the earth's surface may result.

If two plates shove persistently into one another, one simply plows under the other and digs its way (or perhaps is drawn) into the mantle. This happens along the deep oceanic trenches, such as the Japan Trench. Where two plates separate, new molten material oozes up from the mantle and flows out in either direction to join the separating plates. Such occurs at the mid-Atlantic Ridge. The Ridge itself consists of young rocks recently hardened from the molten mantle material.

The magnetism frozen into this young crustal material on either side of the mid-Atlantic Ridge (and other oceanic ridges as well) is measured by shipboard magnetometers. The changing polarity of the earth's magnetic field is clearly exhibited in measurements made farther and farther from the Ridge. The spacing of these changes in polarity tells us how fast the plates on either side of the Ridge have been separating in the past, and comparison of the spacing in different places along the Ridge and along other oceanic ridges gives a clear picture of the motions of the plates over different parts of the earth.

The boundaries between the crustal plates (Figure 9.7) are generally the zones of higher than average seismic and volcanic activity on the earth. Stresses built up along faults where there is plate displacement release every now and then in sudden slippages with accompanying earthquakes. The epicenters of earthquakes along an oceanic trench, where one plate is diving under another one, are generally the deepest epicenters beneath the surface of the earth. The same plate boundaries are the areas of weaknesses where hot molten rock breaks through the crust to form volcanic mountains, and of course volcanic activity is particularly marked along the oceanic ridges. Iceland, on the mid-Atlantic Ridge, was formed (and *is* being formed today) by such volcanic activity. The East Pacific Rise similarly formed Easter Island.

When we learn the causes of volcanos and earthquakes, we can lose some of the irrational fear of these phenomena that many people have. Rather than fear it, we can think of the surface of the earth almost as a

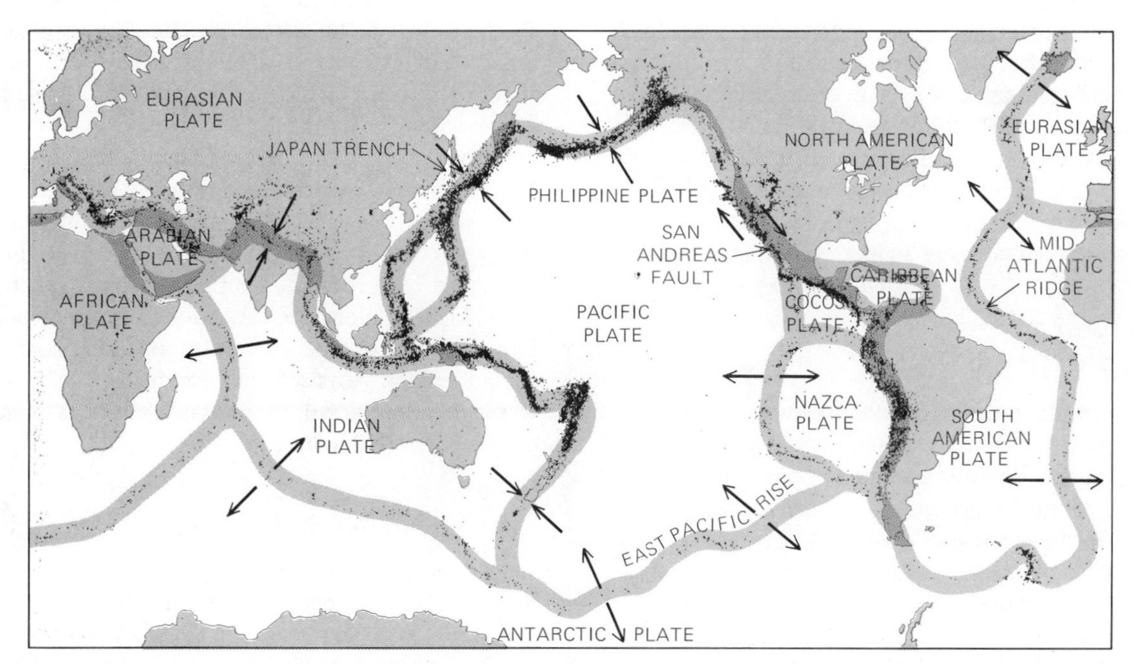

Figure 9.7
Tectonic plates on the earth. The dots indicate regions of seismic activity, where the boundaries of the plates generally lie. The major plates that have been identified are labeled, and the arrows indicate the direction of motion of the plates.

Figure 9.8
A map of the earth approximately as it appeared 150 million years ago. It is believed that at that time the continents were clustered together in a supercontinent that has been named Pangaea.

living entity, constantly churning about. If we could see a time-lapse motion picture of the earth in which 100 million years passed in a few minutes, we would find these crustal motions rather smooth and continuous. It is because we meet them on a narrow human time-scale that their effects seem abrupt and catastrophic. When we experience an earthquake, we are witnessing a minute portion of one of those grand changes in the surface of the earth by which its face is reformed, mountains are built, and continents set adrift.

Continental Drift

By tracing the plate motion backward in time and noting how the continental shelves can be matched up, it has been inferred that all of the present continents were once together in one large land mass, which has been named *Pangaea*. Pangaea broke up about 200 million years ago due to the motion of the plates, carrying the continents with them, causing the *continental drift*.

The plates that consist mostly of ocean areas tend to move the fastest—on the average about 10 cm/yr (4 in/yr). The Pacific Plate at the East Pacific Rise (at Easter Island) is separating from the Nazca Plate at about 20 cm/yr. The plates consisting mostly of continents move an average of about 2 cm/yr. This is about the speed at which Europe and North America are currently separating.

North America is also moving westward with respect to South America, but only at about 1 cm/yr. Interestingly, the Pacific Plate contains a thin slice of the west coast of North America. Its motion to the northwest relative to the North American Plate at some 3 cm/yr (along the San Andreas Fault) caused Baja California to separate from mainland Mexico some 5 million years ago, and eventually it and part of western California will be an island, with Los Angeles off the coast of San Francisco.

The Surface of the Moon

For almost a century the surface of the moon had been studied from photographs made with terrestrial telescopes. Most of those photographs have now been rendered obsolete by the observations obtained from space vehicles and the Apollo astronauts.

In 1959 a Soviet space vehicle was the first to send to earth televised photographs of the moon's far side. The Soviet Luna 9 made the first soft landing on the moon on January 31, 1966, and it transmitted close up photographs to earth for three days. Meanwhile, during the years 1964 to 1967 the spectacular United States Ranger and Surveyor lunar probes and Lunar Orbiters photographically mapped the lunar surface—both the near and far sides—in great detail in preparation for the Apollo missions. In the Apollo program 12 American astronauts were landed on the moon, beginning with Apollo 11, which landed in July 1969, and terminating with Apollo 17 in December 1972. In a period of just over a decade the moon was transformed from an astronomical object that must be viewed from afar to a world that has been intensively explored from its near environment and at first hand.

The Lunar Seas

The largest of the lunar features are the so-called seas, still called *maria* (Latin for "seas"). It is they that form the features of the "man in the moon." The maria are, of course, dry land. They are great plains, with relatively smooth, flat floors that appear darker than the surrounding regions.

The largest of the 14 lunar "seas" on the moon's earthward hemisphere is *Mare Imbrium* (the "Sea of Showers"), about 1100 km (700 mi) across. The other maria have equally fanciful names, such as Mare Nubium ("Sea of Clouds"), Mare Nectaris ("Sea of Nectar"), Mare Tranquilitatis ("Tranquil Sea"), Mare Serenitatis ("Serene Sea"), and so on. Most of the maria are roughly circular in shape, although many of them are interconnected or overlap slightly, and all have irregularities and baylike inlets, such as Sinus Iridum ("Bay of Rainbows") on the north "shore" of Mare Imbrium.

The maria do not have perfectly smooth floors, but are speckled with thousands of tiny craters resembling potholes ranging in size down to less than a few feet across, and inside some maria there are large craters. Some of the mare floors have wavelike ripples, and in a few places cliffs are found, such as the "Straight Wall" in Mare Nubium, 180 m high and 130 km long.

Apollo and Surveyor landings in several different maria found them to be remarkably similar to each other. The mare floors are covered with a layer of material that analysis shows to be basaltic in origin, although severely eroded, probably by meteoritic bombardment. The material consists mostly of fine grains with a variety of sizes, typically about 0.02 mm in diameter. However, mixed in are numerous aggregates or "clots" of fine grains and some hard rocks. The material is, on the average, very porous and compressible, but of course, it is able to support easily the weight of an astronaut. Its density is between 0.7 and 1.2 g/cm^3.

The far side of the moon (that turned away from the earth), in marked contrast to the near side, does not have large maria.

The Lunar Craters

Even from earth there can be observed on the moon some 30,000 craters, circular depressions ranging in size from about 1 km to over 200 km across. Two of the largest are *Clavius* and *Grimaldi*, both nearly 240 km (150 mi) in diameter. Following the custom started by John Riccioli in 1651, craters are generally named after famous scientists and philosophers. Other famous craters are Tycho, Copernicus, Kepler, Aristarchus, and Plato.

Although craters can be seen on the moon at almost any time, even with binoculars, the best time to view them is when the moon is near the first or last quarter phase. Then the *terminator*, the circle dividing day and night on the moon, runs about down the middle of the apparent disk of the moon in the sky. The sun strikes craters near the terminator at a glancing angle and they cast long shadows, which makes them stand out in bold relief.

In general, craters are found over those regions of the moon that are not covered by the seas. There are comparatively few craters in the maria themselves, although crater tops are barely visible in some of the mare floors. The maria are thus more recent features on the moon than most of the other craters. A great many craters are in the rugged mountainous regions of the moon. There they occur in all sizes and probably all ages as well. Frequently they overlap. An especially rugged region of overlapping craters is found in the vicinity of Tycho, near the moon's south pole.

The largest craters are often called *walled plains*. Examples are Clavius, Plato, and Ptolemaeus. A walled plain often appears to be a sunken region with little or no outside wall, sloping up to the crater rim. Some of the walled plains do not have circular walls, but are irregular in shape.

Figure 9.9
The western hemisphere of the moon. *(Lick Observatory)*

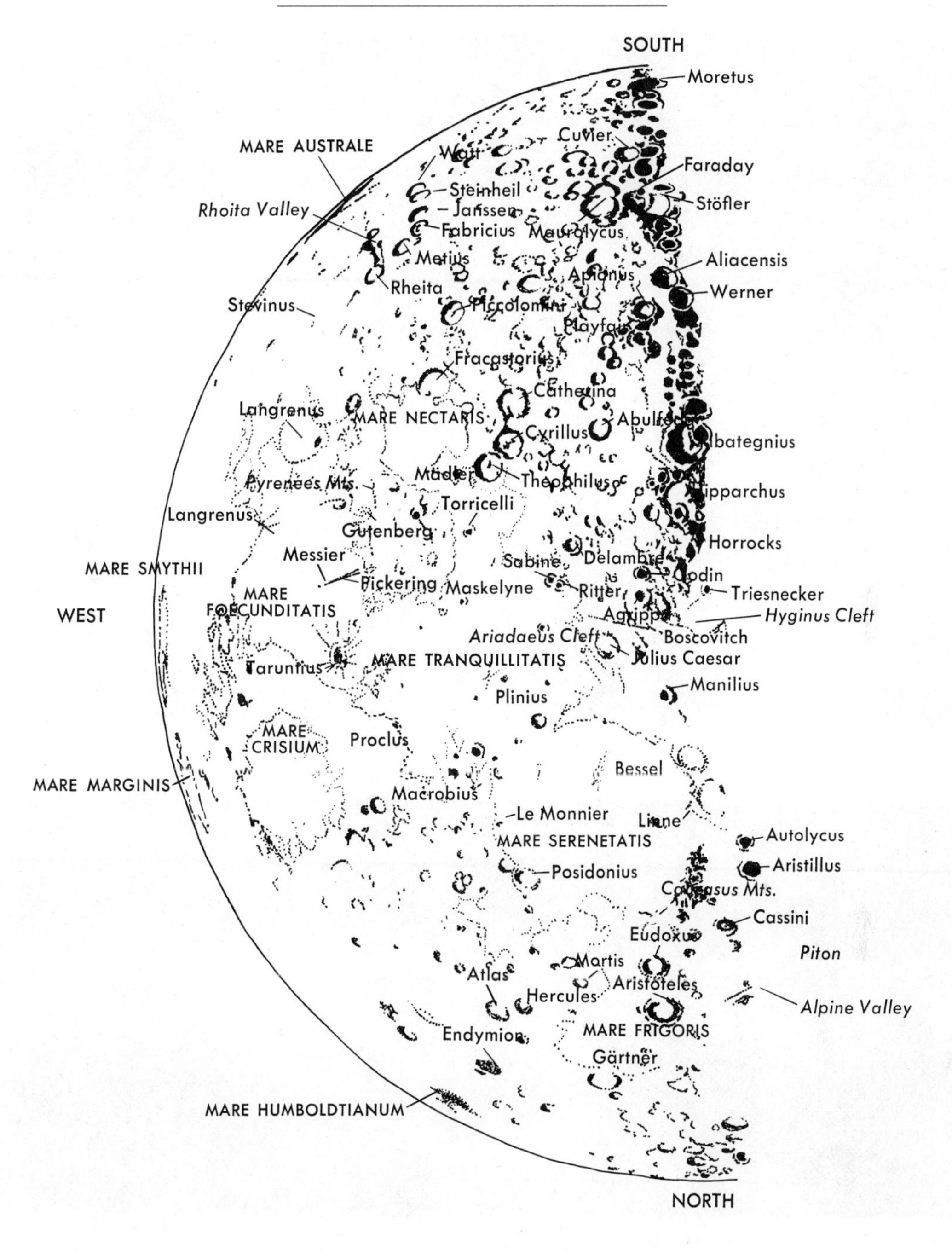
SOUTH
Moretus
Cuvier
MARE AUSTRALE
Faraday
Steinheil
Janssen
Stöfler
Rhoita Valley
Fabricius
Maurolycus
Metius
Aliacensis
Apianus
Rheita
Werner
Stevinus
Piccolomini
Playfair
Fracastorius
Catherina
Langrenus
MARE NECTARIS
Abulfeda
Cyrillus
Albategnius
Mädler
Theophilus
Pyrenees Mts.
Hipparchus
Torricelli
Langrenus
Gutenberg
Horrocks
Messier
Sabine
Delambre
MARE SMYTHII
Godin
Pickering
Maskelyne
Ritter
Triesnecker
MARE FOECUNDITATIS
WEST
Agrippa
Hyginus Cleft
Ariadaeus Cleft
Boscovitch
Julius Caesar
Taruntius
MARE TRANQUILLITATIS
Manilius
Plinius
MARE CRISIUM
Proclus
Bessel
MARE MARGINIS
Macrobius
Le Monnier
Linne
Autolycus
MARE SERENETATIS
Aristillus
Posidonius
Caucasus Mts.
Cassini
Eudoxus
Piton
Martis
Atlas
Aristoteles
Hercules
Alpine Valley
MARE FRIGORIS
Endymion
Gärtner
MARE HUMBOLDTIANUM
NORTH

Figure 9.10
The eastern hemisphere of the moon. *(Lick Observatory)*

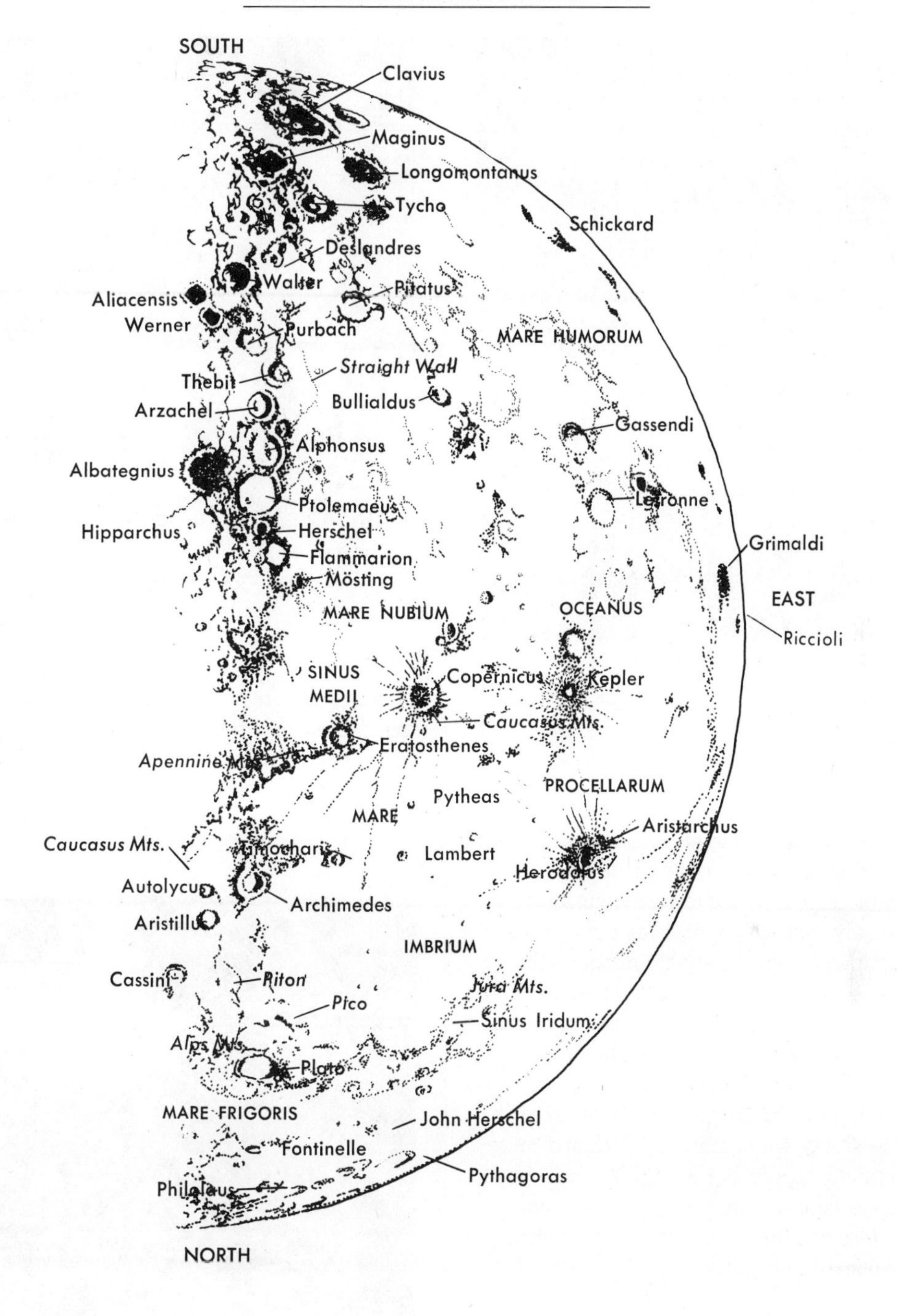
SOUTH
Clavius
Maginus
Longomontanus
Tycho
Schickard
Deslandres
Walter
Pitatus
Aliacensis
Werner
Purbach
MARE HUMORUM
Thebit
Straight Wall
Bullialdus
Arzachel
Gassendi
Alphonsus
Albategnius
Letronne
Ptolemaeus
Hipparchus
Herschel
Grimaldi
Flammarion
Mösting
EAST
MARE NUBIUM
OCEANUS
Riccioli
SINUS
MEDII
Copernicus
Kepler
Caucasus Mts.
Eratosthenes
Apennine Mts.
PROCELLARUM
Pytheas
MARE
Aristarchus
Caucasus Mts.
Timocharis
Lambert
Herodotus
Autolycus
Archimedes
Aristillus
IMBRIUM
Cassini
Piton
Jura Mts.
Pico
Sinus Iridum
Alps Mts.
Plato
MARE FRIGORIS
John Herschel
Fontinelle
Pythagoras
Philolaus
NORTH

Figure 9.11
A closeup of part of a lunar mare photographed by the Apollo 16 crew. *(NASA)*

Most of the craters of moderate size are quite circular and have outside walls. In a few cases the crater floors are higher than the surrounding landscape, but in the majority of cases they are lower. Their inside walls are almost always steeper than the outside walls and rise to heights of as much as 3000 m above the crater floors. Many of these craters have mountain peaks in their centers, and somewhat resemble craters left by bomb explosions. An excellent example is the crater Copernicus (Figures 9.10 and 9.12).

The most numerous craters are the smallest ones, those ranging in size down to craters less than a meter across. Often craters, especially the smaller ones, occur in clumps or clusters or in long lines or rows of craters. In some cases lines of craters seem to follow shallow cliffs or what appear to be faults.

The Lunar Mountains

There are several mountain ranges on the moon. Most of them bear the names of terrestrial ranges—the *Alps, Apennines, Carpathians,* and so on. The similarity of mountains on the moon and the earth ends, however, with their names. Because of the absence of water, the lunar mountain ranges are devoid of the drainage features so characteristic of our own mountain ranges, and the lack of weather erosion on the moon results in a different appearance of mountains there. Even from Earth the heights of lunar mountains above the surrounding plains can be determined by

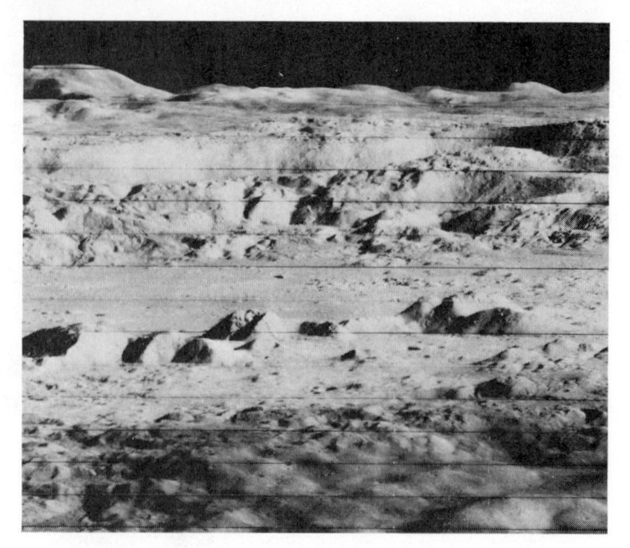

Figure 9.12
The crater Copernicus, viewed from the south, photographed by Lunar Orbiter II. *(NASA)*

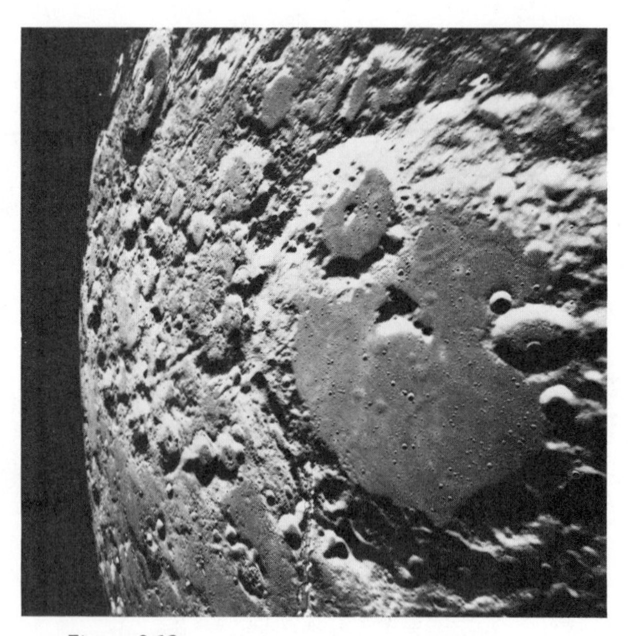

Figure 9.13
A region of dense cratering on the moon, photographed by the Apollo 16 crew. *(NASA)*

measuring the lengths of the shadows they cast. The highest lunar peaks range up to elevations of 8000 m or more, comparable to the highest peaks in the Himalayas on the earth.

Other Lunar Features

There are many other lunar features besides the seas, craters, and mountains. The *Alpine Valley,* for

Figure 9.14
The Alpine Valley, photographed by Lunar Orbiter V. *(NASA)*

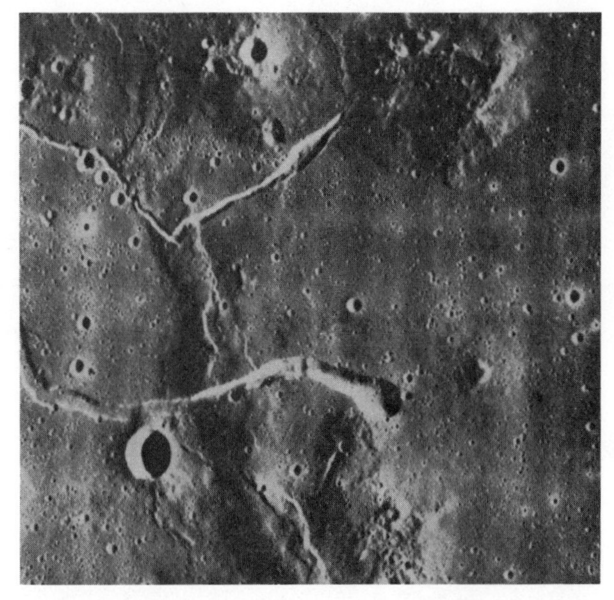

Figure 9.15
Lunar Orbiter V photograph of a lunar rile. *(NASA)*

example (Figure 9.14), is a deep straight gorge cutting through the Alps. Also on the moon are many crevasses or clefts, half a mile or so across, called *rilles*.

Especially interesting are the *rays*–bright streaks that seem to radiate from certain of the craters that appear to be of explosive origin, notably Tycho and Copernicus (see Figure 9.16). The rays have widths of 8 to 20 km or more, and many extend for distances of hundreds of kilometers. Some of the rays from Tycho seem to extend completely around the visible hemisphere of the moon. They cast no shadows and are seen best near full moon, when the sun shines down on them most directly. At full moon the ray system around Tycho gives the moon almost the appearance of a peeled orange. Ranger photographs show some of the rays to run along lines of craterlets.

Composition of the Lunar Surface

Chemical analysis of the lunar surface material has been carried out directly on lunar rocks and soil samples brought back to earth by the Apollo astronauts.

The results from widely separated places on the moon are strikingly similar. The most abundant element is oxygen, accounting for about 58 percent of the atoms present. Most of the oxygen atoms are chemically united with silicon, next most abundant, and ac-

Figure 9.16
The full moon, photographed with the 2.5-m telescope on Mount Wilson. The crater Tycho, with its many rays, is the bright feature near the bottom of the photograph. *(Hale Observatories)*

counting for about 20 percent of the atoms. Also abundant are aluminum, calcium, iron, magnesium, titanium-rich compounds, and other elements that are common on the earth. The chemical composition of the lunar material, and also the mineral structure, is similar to that of volcanic basalts on earth.

The similarity of the chemical structure of the moon and earth argues strongly that the two worlds were formed from the same reservoir of material. For example, the relative abundances of the isotopes of oxygen (O^{16}, O^{17}, and O^{18}) on the moon are identical to those on earth. On the other hand, there are subtle differences, which may tell us something about the earth. The abundance of aluminum compounds in the surface rocks of the moon is far higher than in the crust of the earth. Aluminum is a light metal whose compounds would be expected to rise to the surface of a planet with molten outer layers. Some geologists speculate that the earth's crust may once have been similarly rich in aluminum, but that the action of plate tectonics on the earth has mixed that aluminum into the earth's mantle. Perhaps a rich terrestrial supply of deeply submerged aluminum may someday be feasible to mine.

Another difference between the moon and earth is that the moon's surface layer is lower than the earth's crust in content of volatile compounds, in iron, and in such commercially valuable metals as gold, platinum, rhodium, and palladium. It has been pointed out that the moon is not a promising site for a gold mine.

Origin of the Lunar Features

From the Apollo missions, as well as the unmanned lunar probes, we have learned a great deal about the chronology of the moon's surface. Still, we have a great deal more to learn before we can begin to understand the origin of most of the moon's features. This should not surprise us, for we do not even thoroughly understand all of the features of the earth, which has been studied far more comprehensively than has the moon. Yet, what we have learned of the moon's history tells us something of the earth's early ages.

It is now known that much of the lunar surface, especially in the maria, is covered with material once molten. Thus, whether or not the moon has ever had a molten core, there has at least been some melting of its crustal rocks. Some of the necessary heat may have been produced by the decay of radioactive elements in the outer layers of the moon; most was probably produced by the energy of impact of particles accreting on the lunar surface during its formation. There are no prominent lunar volcanoes. The once solid lava that filled the mare floors has been broken down to dust by gradual erosion caused by temperature changes, cosmic rays, and meteoritic bombardment.

The large lunar craters are due to impacts, and many of the smaller ones may have been formed by secondary particles shot out in the primary impacts. Craters 80 km (50 mi) in diameter can be produced by the infall of a body with a mass of about 7×10^{12} tons or about 16 km (10 mi) in diameter. The lunar rays are thought to have been formed from secondary particles blown out over the moon's surface at the time of formation of their associated craters.

The accretion stage of the moon, during which it was constantly pelted and heated by crater-producing bodies, lasted about 600 million years. But the bombardment ceased by 3.9 to 3.8 thousand million years ago, and the moon has been essentially quiet since. The earth (and other terrestrial planets) must certainly have suffered a similar barrage, but the traces have been removed by the action of weather erosion and plate tectonics. The only craters extant on the earth today are those recently formed by chance collisions with occasional large meteoroids. The most famous example is the Barringer (or Great Meteor) Crater near Winslow, Arizona (Chapter 11).

ECLIPSES

One of the most fortunate coincidences of nature is that the two most prominent astronomical objects, the sun and the moon, have so nearly the same apparent* size in the sky. Although the sun is about 400 times as large in diameter as the moon is, it is also about 400 times as far away, so both the sun and moon subtend about the same angle—about ½°. Consequently, the moon, as seen from earth, can appear to barely cover the sun, producing the most moving and impressive event of nature. It will not always be so, however, because the slowing of the earth's rotation by tidal friction (Chapter 4) results in some of the earth's angular momentum being transferred to the moon's orbital motion, which causes the moon's distance to gradually increase. Our descendants (if there are any) millions of years hence will see the moon subtend too small a size in the sky to cover the sun. The phenomenon of a *total solar eclipse* is a privilege to those of our own geological age.

In general, an eclipse occurs whenever any part of either the earth or the moon enters the shadow of the other. When the moon's shadow strikes the earth, people on earth within that shadow see the sun covered at least partially by the moon; that is, they witness a *solar eclipse*. When the moon passes into the shadow of the earth, people on the night side of the earth see the moon darken—a *lunar eclipse*.

Eclipse Seasons

For the moon to appear to cover the sun and thus to produce a solar eclipse, it must be in the same direction as the sun in the sky; that is, it must be at the *new* phase. For the moon to enter the earth's shadow and produce a lunar eclipse, it must be opposite the sun; that is, it must be at the *full* phase. Eclipses occur, therefore, only at new moon and at full moon. If the orbit of the moon about the earth lay exactly in the plane of the earth's orbit about the sun—in the ecliptic—an eclipse of the sun would occur at every new moon and a lunar eclipse at every full moon. However, because the moon's orbit is inclined at about 5° to the ecliptic, the new moon, in most cases, is not *exactly* in line with the sun, but is a little

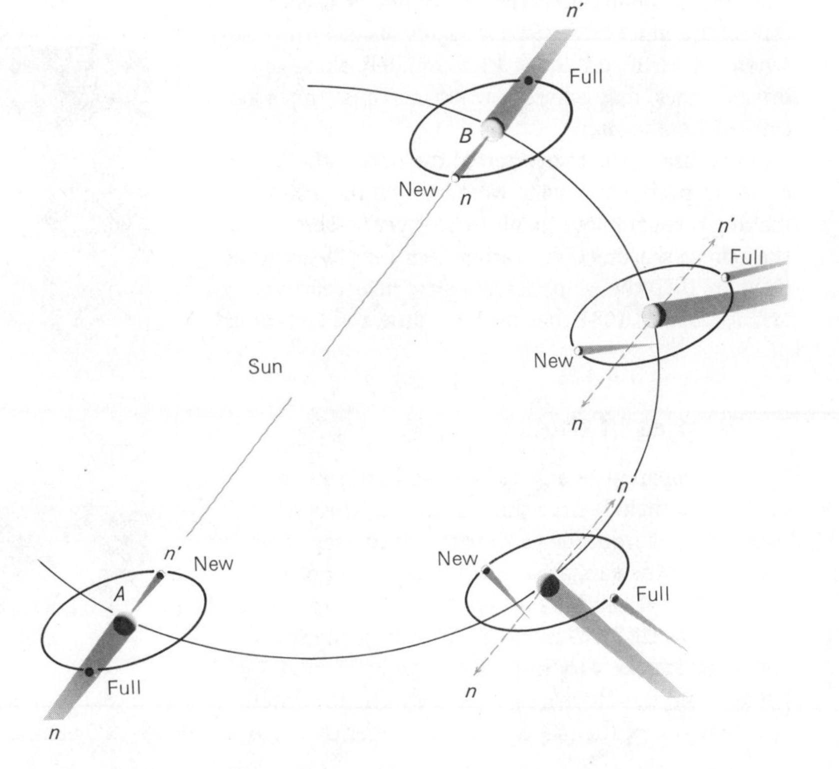

Figure 9.17
Eclipses occur only when the sun is along, or nearly along, the line of nodes.

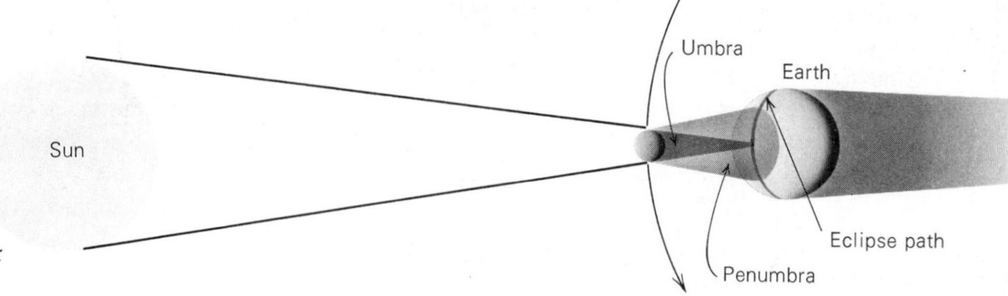

Figure 9.18
Geometry of a total solar eclipse (not to scale).

to the north or to the south of the sun in the sky. Similarly, the full moon usually passes a little south or north of the earth's shadow.

However, if full or new moon occurs when the moon is at or near one of the *nodes* of its orbit (where its orbit intercepts the ecliptic), an eclipse can occur. The line through the center of the earth that connects the nodes of the moon's orbit is called the *line of nodes*. If the direction of the sun lies along, or nearly along, the line of nodes, new or full moon occurs when the moon is near a node, and an eclipse results. The situation is illustrated in Figure 9.17. The orientation of the moon's orbit, and the line of nodes, nn', remains relatively fixed during a revolution of the earth about the sun. There are, therefore, just two places in the earth's orbit, points A and B, where the sun's direction lies along the line of nodes. It is only during the times in the year, roughly six months apart, when the earth-sun line is approximately along the line of nodes, that eclipses can occur. These times are called *eclipse seasons*.

Because of the regression of the nodes, the line of nodes is gradually moving westward on the ecliptic, making one complete circuit in 18.6 years. Therefore, the eclipse seasons occur earlier each year by about 20 days. In 1980 the eclipse seasons are near February and August; in 1982 they are near June and December.

Eclipses of the Sun

The apparent or angular sizes of both sun and moon vary slightly from time to time, as their respective distances from the earth vary. The average angular diameter of the sun (as seen from the center of the earth) is 31′59″, and the average angular diameter of the moon is slightly less, 31′5″. However, the sun's apparent size can vary from the mean by about 1.7 percent and the moon's by 7 percent. The maximum apparent size of the moon is 33′16″, larger than the sun's apparent size, even at its largest. Therefore, if an eclipse of the sun occurs when the moon is somewhat nearer than its average distance, the moon can completely hide the sun, producing a *total solar eclipse*. In other words, a total eclipse of the sun occurs whenever the dark cone (or *umbra*) of the moon's shadow reaches the surface of the earth.

The geometry of a total solar eclipse is illustrated in Figure 9.18. The earth must be at a position in its orbit such that the direction of the sun is nearly along the line of nodes of the moon's orbit. Furthermore, the moon must be at a distance from the surface of the earth that is less than the length of the moon's shadow. Then, at new moon, the moon's shadow intersects the ground at a small point on the earth's surface. Anyone on the earth within this small area covered by

Figure 9.19
Solar corona photographed during the total eclipse of June 8, 1918, at Green River, Wyoming. *(Hale Observatories)*

The earth, photographed by the Applications Technological Satellite (ATS III) from an altitude of 36,000 km. *(NASA)*

The eclipsed moon. *(Celestron International)*

The geometry of a lunar eclipse (not to scale).

The solar eclipse of
October 23, 1976,
seen from Ballaret,
Victoria, Australia.
*(Photographed by
Charles David Long)*

The solar eclipse of
March 7, 1970. *(NASA)*

A sequence of photographs
of the solar eclipse of June 20, 1974.
Photographed by Stephen Schiller.
The final frame is the totally
eclipsed sun photographed
with a special infrared film;
the different colors represent
different infrared wavelengths.

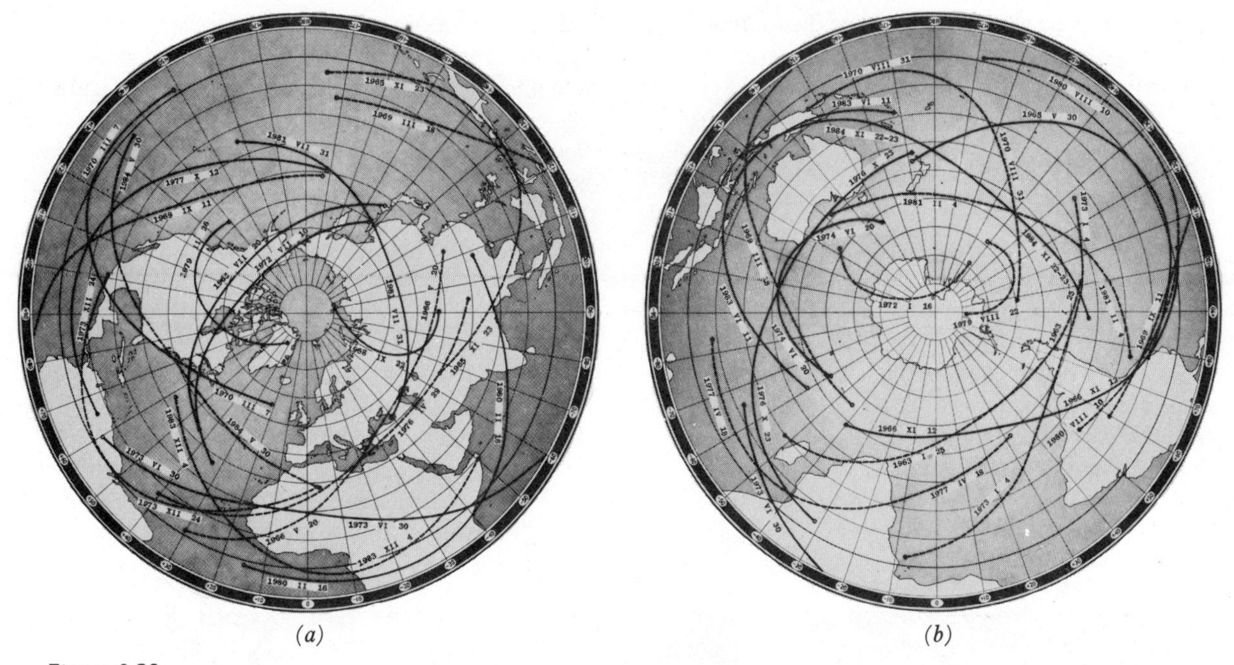

(*a*) (*b*)

Figure 9.20
Some total eclipse paths in the twentieth century: (*a*) northern hemisphere; (*b*) southern hemisphere. (*From* Canon of Solar Eclipses, *by J. Meeus, C. Grosjean, and W. Vanderleen; Pergamon Press, 1966*)

the tip of the moon's shadow will not see the sun and will witness a total eclipse. On the other hand, within a larger area of the earth's surface, one will see part but not all of the sun eclipsed by the moon—a *partial solar eclipse*. Such an observer is within the moon's partial outer shadow, called the *penumbra*.

As the moon moves eastward in its orbit, the tip of its shadow sweeps eastward at about 1000 mi/hr along a thin band across the surface of the earth, and the total solar eclipse is observed successively along this band. This path across the earth within which a total solar eclipse is visible (weather permitting) is called the *eclipse path*. Within a zone about 2000 mi on either side of the eclipse path, a partial solar eclipse is visible—the observer, inside this limit, being located in the penumbra of the shadow, sees part of the sun covered by the moon.

Because the moon's umbra just barely reaches the earth, the width of the eclipse path, within which a total eclipse can be seen, is very small. Under the most favorable conditions, the path is only 269 km (167 mi) wide in regions near the earth's equator. At far northern or southern latitudes, because the moon's shadow falls obliquely on the ground, it can cover a path somewhat more than 269 km wide.

It does not take long for the moon's umbra to

Figure 9.21
Time-lapse photograph showing the moon passing in front of the sun during a total solar eclipse. (*American Museum of Natural History*)

sweep past a given point on earth. The duration of totality may be only a brief instant. It can never exceed about 7½ minutes.

Appearance of a Total Eclipse

Almost anyone who has witnessed a *total* solar eclipse will advise you to make the effort to see one if you are anywhere near the path of totality; it is a rare and impressive event!

The very beginning of a solar eclipse is the *first contact,* when the moon just begins to silhouette itself against the edge of the sun's disk. The *partial phase* follows, during which more and more of the sun is covered by the moon. *Second contact* occurs about an hour after first contact, at the instant when the sun becomes completely hidden behind the moon. In the few minutes immediately before second contact (the beginning of totality) the sky noticeably darkens; some flowers close up, and chickens may go to roost. I saw bats appear during totality of the eclipse of March 7, 1970 as I watched it from a Mexican desert. Because the diminished light that reaches the earth must come solely from the edge of the sun's disk, and consequently from the higher layers in its atmosphere (see Chapter 16), the sky and landscape take on strange colors. In the last instant before totality, the only parts of the sun that are visible are those that shine through the lower valleys in the moon's irregular profile and line up along the periphery of the advancing edge of the moon—a phenomenon called *Baily's beads.* A final

Figure 9.22
Total solar eclipse of March 7, 1970, near a time of maximum sunspot activity. A radially symmetric, neutral density filter in the focal plane of the camera was used to compensate for the increase in brightness of the corona near the limb of the sun. Photograph by Gordon Newkirk, Jr. *(High Altitude Observatory, Boulder, Colorado, a Division of the National Center for Atmospheric Research. NACR is operated by University Corporation for Atmospheric Research under sponsorship of the National Science Foundation)*

Figure 9.23
Total solar eclipse of June 30, 1973, near a time of minimum sunspot activity. A radially graded filter was used, as in the photograph in Figure 9.22, with a technique developed by Gordon Newkirk, Jr. *(High Altitude Observatory, Boulder, Colorado, a Division of the National Center for Atmospheric Research. NCAR is operated by the University Corporation for Atmospheric Research under sponsorship of the National Science Foundation)*

flash of sunlight through a lunar valley produces a brilliant flare on the disappearing crescent of the sun—the *diamond ring.* During totality, the sky is dark enough that planets are visible, and usually the brighter stars as well.

As Baily's beads disappear and the bright disk of the sun becomes entirely hidden behind the moon, the *corona* flashes into view. The corona is the sun's outer tenuous atmosphere, consisting of sparse gases that extend for millions of miles in all directions from the apparent surface of the sun. It is ordinarily not visible because the light of the corona is feeble compared to that from the underlying layers of the sun that radiate most of the solar energy into space. Only when the brilliant glare from the sun's visible disk is blotted out by the moon during a total eclipse is the pearly white corona, the sun's outer extension, visible. It is, however, possible to photograph the inner, brighter, part of the corona with an instrument called a *coronagraph,* a telescope in which a black disk in the telescope's focal plane produces an artificial eclipse, enabling at least the brighter part of the corona to be studied at any time.

Also, during a total solar eclipse, the *chromosphere* can be observed—the layer of gases just above the sun's visible surface. *Prominences* (Chapter 16), great jets of gas extending above the sun's surface, are sometimes seen. Large prominences were especially conspicuous (even to the naked eye) during the eclipse of February 26, 1979, visible in parts of Washington, Oregon, Idaho, Montana, and Southern Canada. That was the last total solar eclipse that will be visible for the mainland United States and Canada in the twentieth century.

The total phase of the eclipse ends, as abruptly as it began, with *third contact,* when the moon begins to

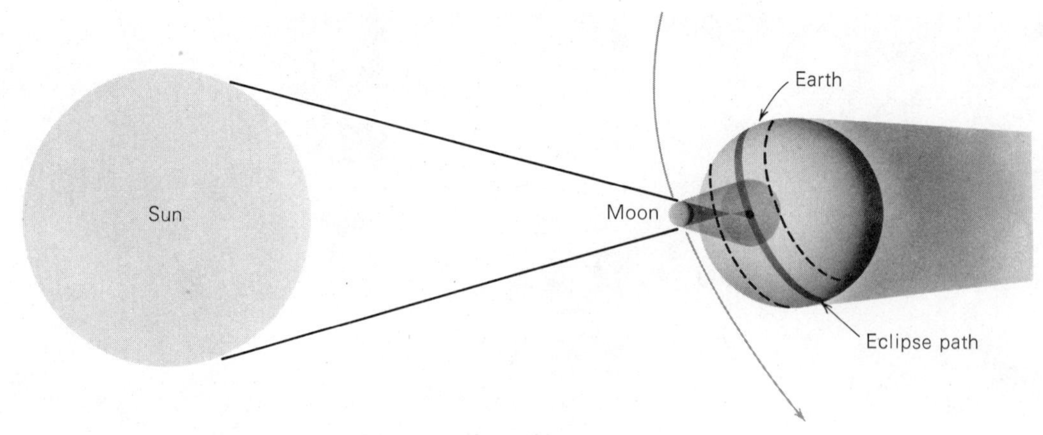

Figure 9.24
Geometry of an annular eclipse (not to scale).

uncover the sun. Gradually the partial phases of the eclipse repeat themselves, in reverse order. At *last contact* the moon has completely uncovered the sun.

Value of Total Solar Eclipses

In addition to being extraordinarily inspiring to experience, total eclipses of the sun have considerable astronomical value. Many data are obtained during eclipses that are otherwise not accessible. For example, during an eclipse we can determine the exact relative positions of the sun and moon by timing the instants of the four contacts. We can take direct photographs and make spectrographic observations of the sun's outer atmosphere and prominences. We can measure the light and heat emitted by the corona. We can determine how meteorological conditions are affected by solar eclipses and can learn something about the light-scattering properties of the earth's atmosphere. Historically, one of the most important scientific observations at a solar eclipse was the verification of a prediction of Einstein's general theory of relatively (see Chapter 18).

The important total solar eclipses that occur in the next half century are listed in Appendix 12.

Annular Eclipses of the Sun

More than half the time the moon does not appear large enough in the sky to cover the sun completely, which means that the umbra of its shadow does not reach all the way to the surface of the earth. The geometry of the situation is illustrated in Figure 9.24. When the moon's shadow cone does not reach the earth, a total eclipse is not possible. However, if the boundaries of the umbra of the shadow are extended until they intersect the earth's surface, they define a region on the ground within which the moon can be seen completely silhouetted against the sun's disk, with a ring of sunlight showing around the moon. This kind of eclipse is called an *annular eclipse*, from the Latin word *annulus*, meaning "ring." The extension of the moon's umbra, within which an annular

Figure 9.25
An annular eclipse of the sun. Bright spots are sunlight streaming through lunar valleys. *(Lick Observatory)*

eclipse is visible, sweeps across the ground in a path much like the path of totality of a total eclipse. As is true for total eclipses, a partial eclipse is visible within a region of 2000 mi or more on either side of the annular eclipse path.

An annular eclipse begins and ends like a total eclipse. However, because the sun is never completely covered by the moon, the corona is not visible, and although the sky may darken somewhat, it does not get dark enough for stars to be seen. An annular eclipse is not so spectacular, nor has it the scientific value of a total eclipse.

Partial Eclipses of the Sun

A *partial eclipse* of the sun is one in which only the penumbra of the moon's shadow strikes the earth. During such an eclipse, the moon's umbra passes north or south of the earth, and from nowhere can the sun appear to be covered completely by the moon. Also, a total or annular eclipse appears partial from regions outside the eclipse path but within the zone of the earth that is intercepted by the moon's penumbra.

Few people have seen total or annular solar eclipses, but most have had the opportunity to see the sun partially eclipsed. The moon seems to "skim" across the northern or southern part of the sun. How much of the sun can appear covered depends, of course, on how close the observer is to the path of totality or annularity. Partial eclipses are interesting but not spectacular. Only if the observer is within a few hundred kilometers of the eclipse path will he see the sky darken appreciably.

Figure 9.26
How to watch the partial phases of a solar eclipse safely.

How to Observe Eclipses

The progress of an eclipse can be observed safely by holding a card with a small (1 mm) hole punched in it several feet above a white surface, such as a concrete sidewalk. The hole in the cardboard produces a pinhole camera image of the sun (Figure 9.26).

Although there are safe filters through which one can safely look at the sun directly, many people have suffered permanent eye damage by looking at the sun through improper filters (or no filter at all!). In particular, neutral-density photographic filters are not safe, for they transmit infrared radiation that can cause severe damage to the retina.

Common sense (and pain) prevents most of us from looking at the sun directly on an ordinary day for more than a brief glance. Of course there is nothing about an eclipse that makes the radiation from the sun more dangerous than it is any other time; on the contrary, we receive less radiation from the sun when it is partly hidden by the moon. It is *never* safe, however, to look at the sun directly when it is still in *partial* eclipse; even the thin crescent of sunlight visible a few minutes before totality has a surface brightness great enough to burn and permanently destroy part of the retina. Unless you have a filter prepared especially for viewing the sun, it is best to watch the partial phases with a pinhole camera device, as described above.

It is *perfectly safe*, however, to look at the sun directly when it is *totally eclipsed*, even through binoculars or telescopes. Unfortunately, unnecessary panic has often been created by uninformed public officials, acting with the best of intentions. I have witnessed two marvelous total eclipses in Australia, during which townspeople held newspapers over their heads for protection, and school children cowered indoors, with

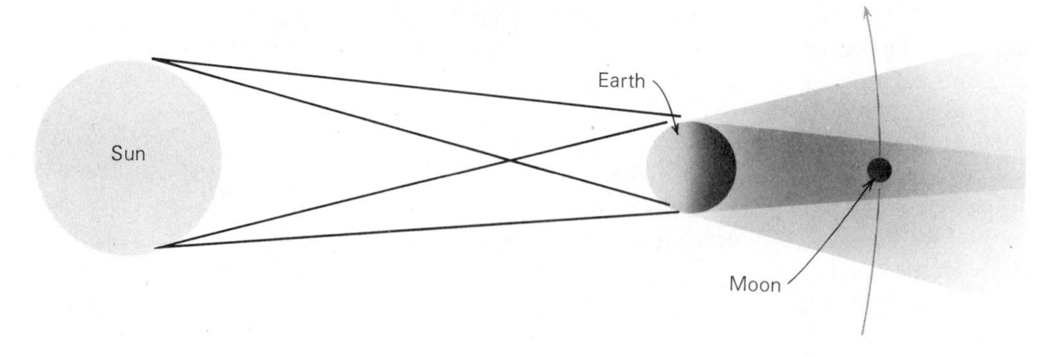

Figure 9.27
Geometry of a lunar eclipse (not to scale).

their heads under their desks. What a cheat to those people to have missed what would have been one of the most memorable experiences of their lifetimes! During totality, by all means look at the sun.

Nor should you be terrified of accidently catching a glimpse of the sun outside totality. How many times have you glanced at the sun on ordinary days while driving a car or playing ball or tennis? Common sense made you look away at once. Do the same if you inadvertantly glimpse the sun directly while it is partially eclipsed.

Eclipses of the Moon

A lunar eclipse occurs when the moon, at the full phase, enters the shadow of the earth. There are three kinds of lunar eclipses: *total*, *partial*, and *penumbral*. The geometry of lunar eclipses is shown in Figures 9.27 and 9.28. Unlike a solar eclipse, which is visible only in certain local areas on the earth, a lunar eclipse is visible to everyone who can see the moon. Weather permitting, a lunar eclipse can be seen from the entire night side of the earth, including those sections of the earth that are carried into the earth's umbra while the eclipse is in progress. Lunar eclipses, therefore, are observed far more frequently from a given place on earth than are solar eclipses.

In Figure 9.28 four of the many possible paths of the moon through the earth's shadow are shown. A total lunar eclipse occurs when the moon passes completely into the umbra (path A). A *partial eclipse* occurs if only part of the moon skims through the umbra (path B), and a *penumbral eclipse* occurs if the moon passes through the penumbra, or partially through the penumbra, but does not come into contact with the umbra (paths C and D).

Appearance of Lunar Eclipses

Penumbral eclipses usually go unnoticed even by astronomers. Only within about 1100 km of the umbra is the penumbra dark enough to produce a noticeable

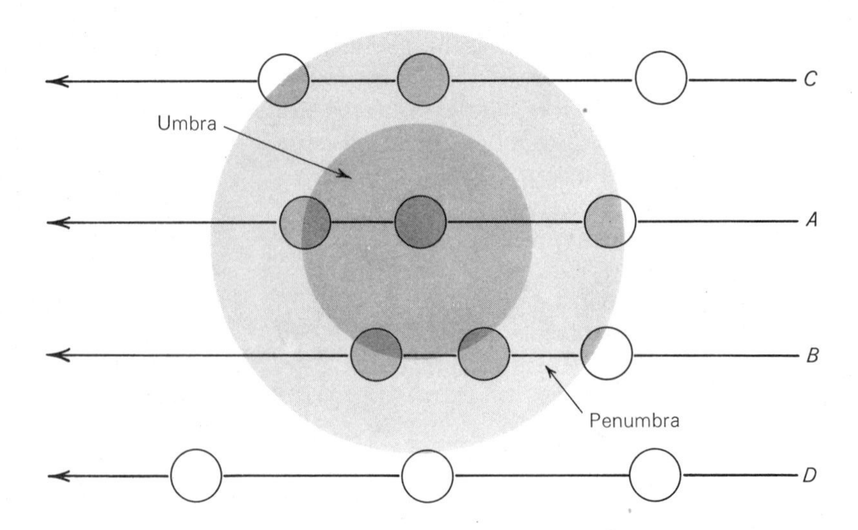

Figure 9.28
Different kinds of lunar eclipses. The penumbra pales from the full dark of the umbra at its inner boundary to full light at its outer boundary.

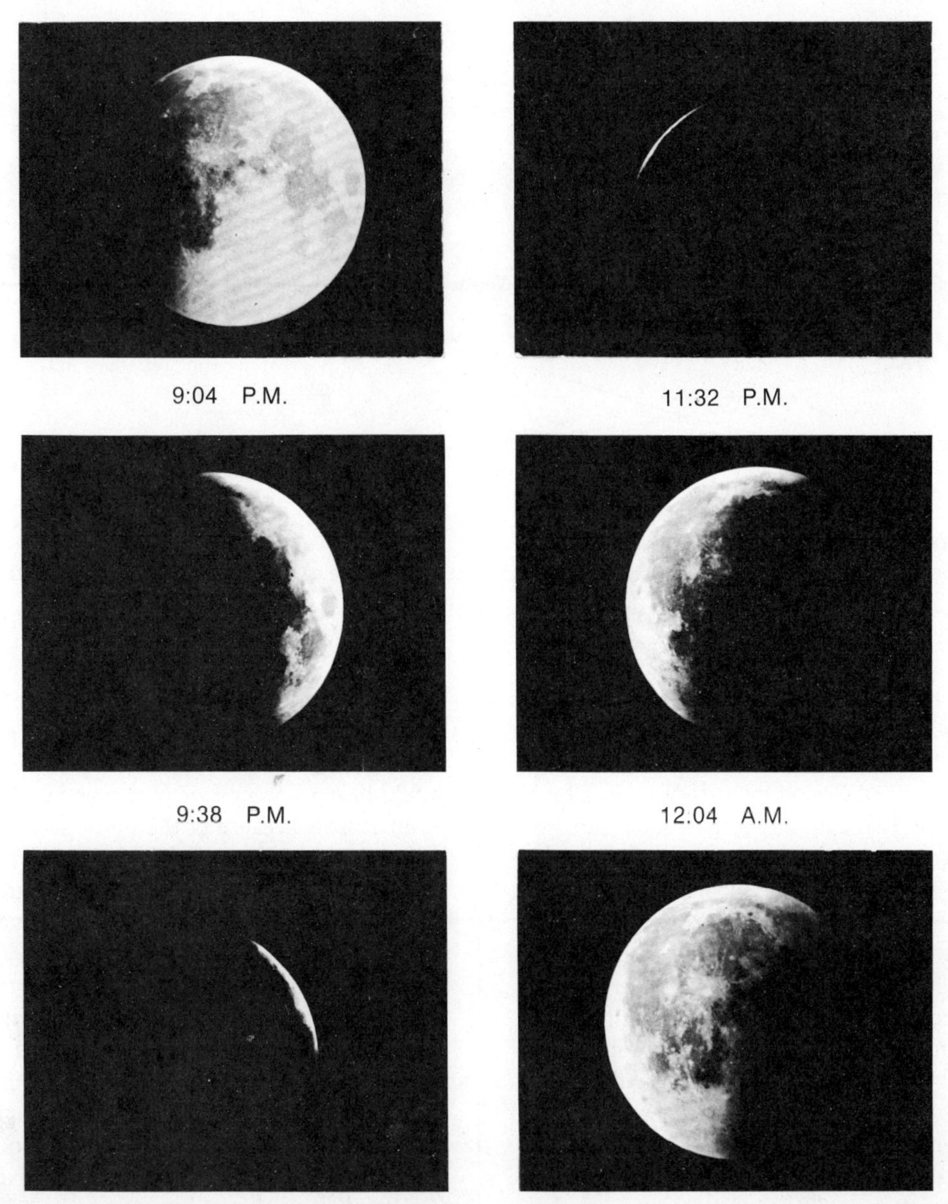

Figure 9.29
Sequence of photographs of the total lunar eclipse of November 17/18, 1956. *(Photographed by Paul Roques, Griffith Observatory)*

darkening on the moon. However, the diminished illumination on the moon's surface can be detected by photometric measurements.

Every total or partial lunar eclipse must begin with a penumbral phase. About 20 minutes or so before the moon reaches the shadow cone of the earth, the side nearest the umbra begins to darken somewhat. At the moment called *first contact*, the limb of the moon (the "edge" of its apparent disk in the sky) begins to dip into the umbra of the earth. As the moon moves farther and farther into the umbra, the curved shape of the earth's shadow upon it is very apparent. In fact, Aristotle listed the round shape of the earth's shadow as one of the earliest proofs of the fact that

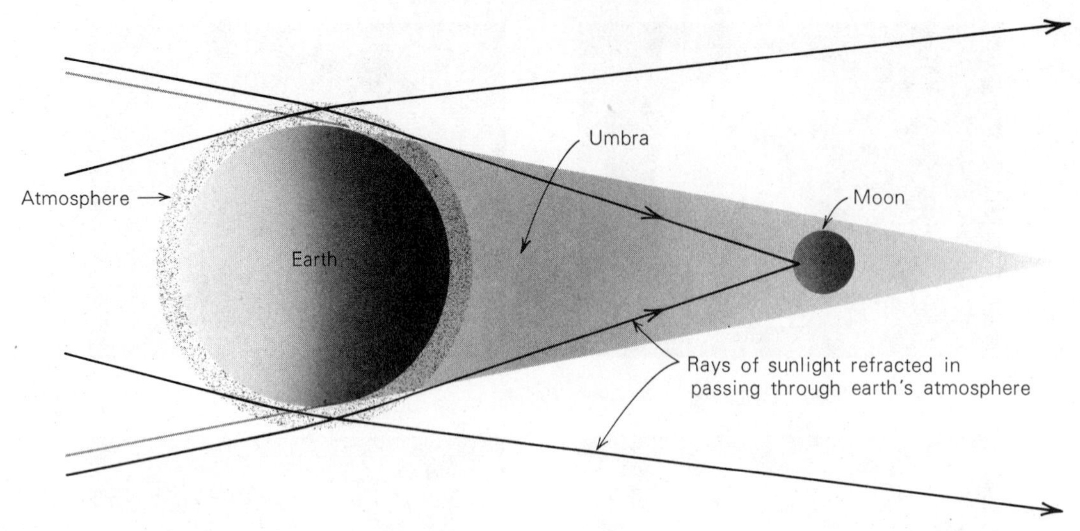

Figure 9.30
Illumination of the moon during a total eclipse by sunlight refracted by the earth's atmosphere into the earth's shadow.

the earth is spherical. At *last contact* the moon emerges from the umbra. If the eclipse is a partial one, the moon never gets completely into the umbra of the earth's shadow, but passes on by, part of it remaining in the penumbra, where it still receives some sunlight.

On the other hand, if the eclipse is a total one, at the instant of *second contact* the moon is completely inside the umbra, and the total phase of the eclipse begins. Even when totally eclipsed, the moon is still faintly visible, usually appearing a dull coppery red. The illumination on the eclipsed moon is sunlight that has passed through the earth's atmosphere and has been refracted by the air into the earth's shadow (Figure 9.30).

Totality ends at *third contact*, when the moon begins to leave the umbra. It passes through its partial phases to *last contact*, and finally emerges completely from the penumbra. The total duration of the eclipse depends on how closely the moon's path approaches the axis of the shadow during the eclipse. The penumbral phases at the beginning and end of the eclipse last about one hour each, and each partial phase consumes at least one hour. The total phase can last as long as 1 hour 40 minutes if the eclipse is central.

Ecliptic Limits

We have seen that solar eclipses occur only when the moon is new and lunar eclipses when the moon is full. Furthermore, eclipses occur only when the new or full moon is near a node. If the sun, the earth, and the moon were geometrical points, the new or full moon would have to occur *exactly* at a node if there were to be an eclipse. Because of the finite sizes of the three bodies, however, part of the sun may appear covered by part of the moon for observers at certain places on earth even though at new moon the moon is not located *exactly* at the node. Similarly, at full moon, it is possible, because of the large size of the earth's shadow, that the full moon can pass partially through it or even pass completely into it, even if the full moon is not exactly at the node. Thus, the requirement that the moon be at a node can be relaxed slightly; there is some leeway within which eclipses can occur. The limits of this leeway are called *ecliptic limits*.

It turns out that at least one solar eclipse of some kind must occur at each eclipse season. At least one lunar eclipse must also occur, but it may be penumbral, and go unnoticed. Thus the total number of observable solar eclipses over a period of time outnumbers the observable lunar eclipses by nearly three to two. Lunar eclipses are more common at any one place, however, for they can be viewed from more than half of the globe, whereas solar eclipses, even partial ones, are visible only in limited areas. Total solar eclipses occur on an average about once every 1½ years, but they are visible only within narrow eclipse paths; their rate or recurrence at any one place is highly variable, but averages about once every 360 years.

Figure 9.31
Series of photographs *(left to right)* showing the emergence of Jupiter and three of its satellites after their occultation by the moon. *(Photographed by Paul Roques, Griffith Observatory)*

The maximum number of all kinds of eclipses (solar and lunar) in any one calendar year is seven.

Phenomena Related to Eclipses

There are phenomena with geometrical properties similar to those of eclipses that involve other celestial bodies. Examples are *occultations* and *transits*.

The moon often passes between the earth and a star; the phenomenon is called an *occultation*. During a lunar occultation, a star suddenly disappears as the eastern limb of the moon crosses the line between the star and observer. If the moon is at a phase between new and full, the eastern limb is not illuminated and the star may appear to vanish mysteriously as the dark edge of the moon covers it. Geometrically, occultations are equivalent to total solar eclipses, except that they are total eclipses of stars other than the sun. Lunar occultations of the brighter stars are listed in advance in various astronomical publications. The times and durations of the occultations and the places on earth from which they are visible are given. Also listed are the comparatively rare occultations of planets by the moon, and of stars by planets.

A *transit* is a passage of an inferior planet (Mercury or Venus) across the front of the sun's disk when it passes between the earth and sun—inferior conjunction. Usually each of these planets appears to pass north or south of the sun, but it can pass in front of the sun if inferior conjunction occurs when the planet is near one of the nodes of its orbit—the points where its orbit crosses the ecliptic.

Transits are analogous to annular eclipses of the sun. The planets Venus and Mercury, seen from earth, are far too small to cover the sun completely; their shadow cones fall far short of reaching the surface of the earth. The appearance of a transit is that of a black dot slowly crossing the disk of the sun from east to west. The silhouette of Mercury against the sun is too small to see without a telescope. That of Venus can be barely observed, without optical aid, if the sun is properly viewed by projection, or through dense filters to protect the eye.

SUMMARY REVIEW

Aspects of the moon: phases; moonlight; earthshine on the moon; the moon's revolution and rotation; sidereal and synodic periods; regression of the nodes

Dimensions and masses of the earth and moon: radar distance determination; finding the size of a body from its angular size and distance; location of the barycenter; mass of the moon from perturbations of space vehicles

Atmospheres of the earth and moon: origin of earth's atmosphere; outgassing; composition of the atmosphere; levels in the atmosphere—troposphere, stratosphere, mesosphere, thermosphere, exosphere, ionosphere; lack of a lunar atmosphere

Magnetic fields on the earth and moon: magnetosphere; Van Allen layers; the very weak lunar magnetism

Interiors of the earth and moon: seismological studies; core; mantle; crust; the lunar interior; mascons; ages of the earth and moon

Surface of the earth: plate tectonics; continental drift; lithosphere; asthenosphere; Pangaea

Surface of the moon: studies from space and Apollo; maria; craters; lunar mountains; rilles; composition of lunar surface; origin of lunar features; impact cratering on the earth and moon

Eclipses: eclipse seasons; solar eclipses—total, annular, partial; phenomena of solar eclipse—contacts, Baily's beads, diamond ring, prominences, solar corona; value of eclipse observations; how to view an eclipse; lunar eclipses—total (umbral), partial, penumbral; ecliptic limits

Occultations and transits

EXERCISES

1. When earthshine is brightest on the moon, what must be the phase of the earth, as seen from the moon?

2. If the moon revolved from east to west rather than from west to east, would a synodic month be longer or shorter than a sidereal month? Why?

3. About what time does the moon rise when it is at each of the following phases:
(a) new?
(b) full?
(c) third quarter?
(d) two days past new?
(e) three days past full?
(f) first quarter?

4. What is the phase of the moon if (a) it rises at 3:00 P.M.? (b) it is on the meridian at 7:00 A.M.? (c) it sets at 10:00 A.M.?

5. What time does (a) the first quarter moon cross the meridian? (b) the third quarter moon set? (c) the new moon rise?

6. Describe the phases of the earth as seen from the moon. At what phase of the moon (as seen from the earth) would the earth (as seen from the moon) be a waning gibbous?

7. Suppose you lived on a plain on the moon near the center of the moon's disk as seen from the earth.
(a) How often would you see the sun rise?
(b) How often would you see the earth set?
(c) Over what fraction of time would the vernal equinox be above the horizon?

8. Describe the phases that the earth would appear to go through as seen from (a) Venus; (b) Mars; (c) the moon. In each case, would a telescope be needed to observe the phase of the earth?

9. If 100 times as much water as nitrogen has outgassed from the crust of the earth, what fraction of the earth's mass is in its oceans?

10. Give two reasons why the discovery of free oxygen in a lunar atmosphere would be a surprise.

11. Imagine a baseball game on the moon. Describe such phenomena as:
(a) How far a ball might be hit.
(b) The desirable distance between the bases.
(c) The likelihood of rain checks.
(d) The necessity of calling the game because of darkness.
(e) The difficulty of bawling out the umpire (consider how sound travels on the moon).

12. Consult Figure 9.7 and select several regions of the earth that you would expect to be relatively free of seismic and volcanic activity.

13. If Europe and North America were separating from each other at a constant rate of 2 cm/yr, how long a time would have been required for them to have moved from contact to their present separation of about 4000 km?

14. In the photographs in this chapter identify some craters submerged in lunar maria, and also some cliffs. Cite examples by figure number and location on the figures.

15. Draw a diagram illustrating how you might measure the height of a lunar mountain from the length of its shadow. Describe your procedure, and in each case explain what must be known or assumed.

16. Make up a table listing the major surface features of the earth and moon. Comment on the differences between corresponding features on the two worlds. Explain which features are found only on one body, and why.

17. The red army staged a surprise attack on the blue army on the night of February 10 during the relative darkness of a total eclipse of the moon. If you were a general of the blue army, when would you plan a return attack if you wanted to be sure of a complete night of no moonlight?

18. If a solar eclipse were annular over part of the eclipse path and total over part, would the total eclipse be of long or short duration? Why?

19. Does the longest duration of a solar eclipse occur when
(a) the sun is at its nearest and the moon at its nearest?
(b) the sun is at its farthest and the moon at its nearest?
(c) the sun is at its nearest and the moon at its farthest?
(d) the sun is at its farthest and the moon at its farthest?

20. Describe what an observer at the crater Copernicus on the moon would observe during what would be a total solar eclipse as viewed from the earth.

21. Describe the phenomenon observed by a spectator on the moon while the moon is being eclipsed.

22. Which planets can the moon never occult while at the full phase? Why?

23. If the earth and the moon had their present distance from each other and their present period of mutual revolution, but if they were removed to the distance of Jupiter from the sun, would total solar eclipses be more common or less common at any one place? Explain.

24. Draw a diagram showing the geometry of a transit of the sun by an inferior planet. Compare your diagram to Figure 9.24.

CHAPTER 10

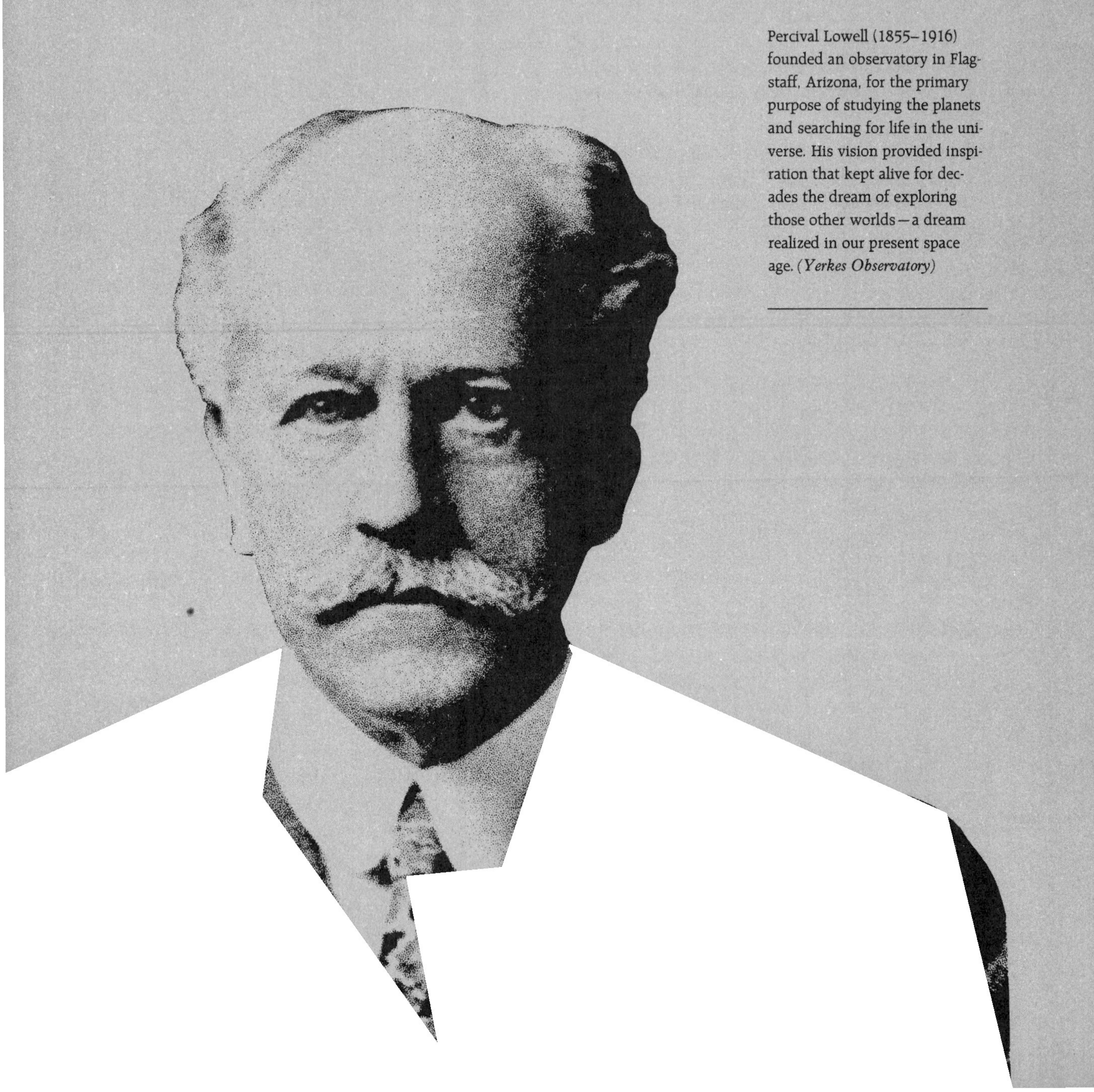

Percival Lowell (1855–1916) founded an observatory in Flagstaff, Arizona, for the primary purpose of studying the planets and searching for life in the universe. His vision provided inspiration that kept alive for decades the dream of exploring those other worlds—a dream realized in our present space age. *(Yerkes Observatory)*

THE OTHER WORLDS

The ancients named the planets for their gods, and associated them with those gods. Even in modern America tens of millions of people still believe in the Greek religion of astrology—that the planet-gods somehow control or at least influence our lives. In contrast to this 2000-year-old belief, with the technology made possible by modern science we have actually explored at close range all of the planets known in antiquity. What we have seen is fabulous indeed, and far more stimulating to the human intellect than the man-invented gods could ever have been.

The program of space exploration is less than two decades old, but in that short time its spectacular achievements have accomplished what hundreds of years of viewing from afar, and highly imperfect viewing from the bottom of our own ocean of air at that, could not even lead us to imagine. Our speculations and dreams have been replaced by hard data and

TABLE 8.1 Some of the More Significant Planetary Probes (All Unmanned)

Planet	Spacecraft	Arrival Date	Comments
Mercury	Mariner 10 (U.S.)	1974	Three flybys
Venus	Venera 7 (U.S.S.R.)	1970	Landed; broadcast for 23 min
	Venera 8 (U.S.S.R.)	1972	Landed; broadcast for 50 min
	Venera 9 (U.S.S.R.)	1975	Landed; sent photo
	Venera 10 (U.S.S.R.)	1975	Landed; sent photo
	Venera 11 (U.S.S.R.)	1978	
	Venera 12 (U.S.S.R.)	1978	
	Pioneer Venus 1 (U.S.)	1978	Orbiter
	Pioneer Venus 2 (U.S.)	1978	Multiprobe lander
Mars	Mariner 4 (U.S.)	1965	Flyby; 22 photos sent back
	Mariner 6 (U.S.)	1969	Flyby; many photos
	Mariner 7 (U.S.)	1969	Flyby; many photos
	Mariner 9 (U.S.)	1971	Orbiter; 7329 photos
	Viking 1 (U.S.)	1976	Lander; photos and life experiments
	Viking 2 (U.S.)	1976	Lander; photos and life experiments
Jupiter	Pioneer 10 (U.S.)	1973	Flyby; photos
	Pioneer 11 (U.S.)	1974	Flyby; photos
	Voyager 1 (U.S.)	1979	Flyby; photos of Jupiter and satellites
	Voyager 2 (U.S.)	1979	Flyby; photos of Jupiter and satellites
Saturn	Pioneer 11 (U.S.)	1979	Flyby; photos
	Voyager 1	1980	
	Voyager 2	1981	

close-up pictures. Yet, it was thousands of years of watching the planets move in the sky that led to the discovery of the gravitational theory that made our space explorations possible in the first place.

Some of the more significant planetary probes are listed in Table 10.1 (the table is *not* a complete list of all U.S. and Soviet missions). All of these probes carried scientific instruments to obtain data on such things as magnetic fields, temperature, and atmospheric constituents. Many also carried television systems to transmit photographs back to earth. The Soviet Union has sent a long series of probes to each of Venus and Mars; their Venera missions to Venus have been the more successful. All of these probes, of course, were unmanned.

MERCURY

At its brightest, Mercury is inferior in brilliance only to the sun, the moon, the planets Venus, Mars, and Jupiter, and the star Sirius. Yet, most people—including even Copernicus, it is said—have never seen Mercury. The planet's elusiveness is due to its proximity to the sun. Its orbit is only about 40 percent the size of the earth's; it can never appear further from the sun than about 28°. It is visible to the unaided eye, for a period of only about one week, at times when it is near eastern elongation, appearing above the western horizon just after sunset, and also when it is near western elongation, rising in the east shortly before sunrise. The intervals of its visibility as an "evening star" after sunset, and as a "morning star" before sunrise, occur about three times a year. However, Mercury sets so soon after the sun (or rises so shortly before the sun), that only rarely can it appear above the horizon when the sky is completely dark; generally, one must look for it in twilight.

Mercury is the nearest planet to the sun, and except for Pluto is also the smallest planet, having a diameter less than half that of the earth. Its mass, best determined from its gravitational influence on the Mariner 10 space probe, is 1/18 that of the earth. Its mean density is about 5.4 times that of water—about the same as the mean density of the earth. If Mercury were identical in chemical composition to the earth, it should not have compressed itself to so high a density. We believe it has a higher relative abundance of iron than the earth does, probably in the form of a sizable iron core. Heavier elements like iron must have contributed more to the material near the sun, where Mercury formed, than to that at the earth's position in the solar nebula.

Figure 10.1
The planet Mercury in a mosaic of photographs made by Mariner 10 during its approach to the planet on March 29, 1974. *(NASA/JPL)*

Mercury, like Venus, presents different portions of its illuminated hemisphere to us as it revolves about the sun and hence goes through phases similar to those of the moon. The alternate crescent and gibbous shape of Mercury is its only conspicuous telescopic characteristic.

Radar observations of Mercury in the mid-1960s

showed it to rotate with respect to the sun. Its sidereal period of rotation (that is, with respect to the distant stars) is 59 days. It is believed that this particular period of rotation results from the tidal force of the sun on Mercury. If the planet is slightly deformed, the tidal force should cause its longest diameter to align with the sun every time Mercury passes perihelion. One way this can happen is for it to rotate in 58.65 days, which is exactly two-thirds of its revolution period.

The best temperature measurements of Mercury come from the Mariner 10 flyby in 1974. The daylight temperature on the surface ranges up to about 700 K at noontime. Just after sunset, however, the temperature drops quickly to about 150 K, and then slowly descends to about 100 K at midnight. Extrapolation of the Mariner data leads to a temperature estimate of about 90 K just before dawn. The range in temperature on Mercury is thus over 600 K, more than on any other planet.

Mercury was not expected to have an atmosphere because of its high daylight surface temperature and low velocity of escape. As expected, no evidence for a permanent atmosphere was found by Mariner 10. There is a very tenuous hydrogen cloud around the planet, but the gas pressure near the surface is completely negligible—less than 10^{-11} that of the earth's atmosphere. This sparse hydrogen may be due to the interaction of the planet with the solar wind.

Mariner 10 found Mercury to have a weak magnetic field, with a strength at the surface of 1 to 2×10^{-3} gauss. This field is hundreds of times weaker than the earth's, but it can influence the motions of ions in the solar wind.

The Surface of Mercury

The first close-up look at Mercury came on March 29, 1974, when the Jet Propulsion Laboratory's Mariner 10 passed 9500 km from the surface of the planet at a speed of about 11 km/s. On that date, and for several days before and after the encounter, Mariner

Figure 10.2
Cratered terrain of Mercury photographed by Mariner 10. The large flat-floored crater at the right is about 100 km in diameter—about the size of the lunar crater Copernicus. Many of the smaller craters are probably caused by particles thrown out by the impact that formed the large crater. *(NASA/JPL)*

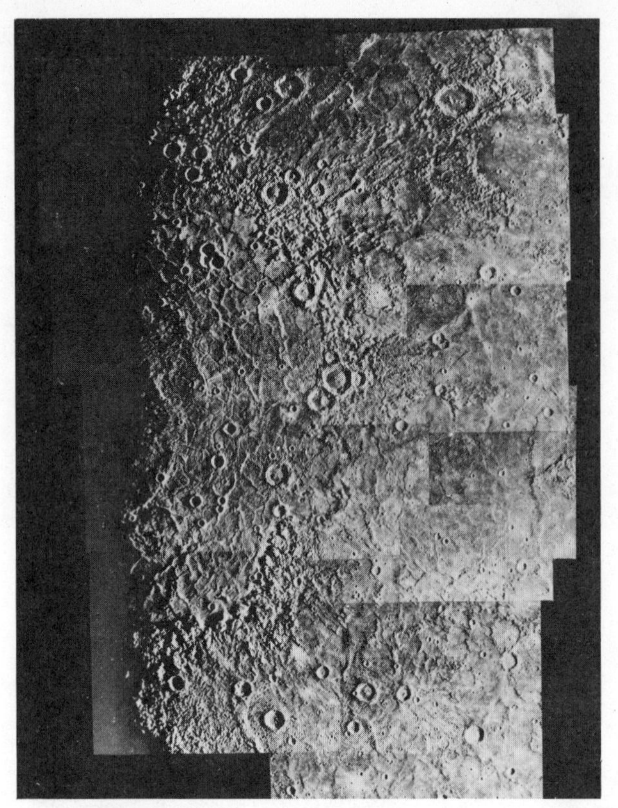

Figure 10.3
The left half of this mosaic shows the ring basin Caloris, 1300 km in diameter. This is the largest structural feature on Mercury seen by Mariner 10 in its first pass of the planet in late March 1974. It is so named because it is believed to be the place on Mercury where the temperature reaches its highest value. *(NASA/JPL)*

10 televised more than 2000 photographs to earth, revealing details with a resolution down to 150 m. Mariner 10, in a planned elliptical orbit about the sun, passed Mercury again on September 21, 1974, and finally on March 16, 1975, obtaining over 2000 more pictures.

Mercury strongly resembles the moon in appearance. It is covered with thousands of craters up to hundreds of kilometers across, and larger basins up to 1300 km in diameter. There are also scarps (cliffs) over a kilometer high and hundreds of kilometers long, as well as ridges and plains. Some of the brighter craters are rayed, like Tycho and Copernicus on the moon, and many have central mountain peaks.

The larger basins resemble the lunar maria, both in size and appearance. They show evidence of much flooding, as do the maria, from lava flows, evidently released during the impacts that produced the features. The fact that Mercury is as cratered as the moon is strong evidence that the craters of those worlds were formed near the ends of their accretion processes, as they swept up the remaining large objects in their regions of the solar system. Many of the smaller craters on Mercury are secondary ones, formed by debris thrown out by the impacts that caused the main craters.

VENUS

We might expect that Mercury, being nearest the sun, would be the hottest planet in the solar system. Not so; that honor goes to Venus, with a surface temperature of more than 700 K.

The high temperature of Venus is due to its dense atmosphere. That atmosphere has been thoroughly explored by various Soviet and U.S. missions, and its structure has been especially well documented with measures by the Pioneer Venus probes. Pioneer Venus 1 arrived at its destination and went into orbit about the planet on December 4, 1978. Pioneer Venus 2 was a multiprobe vehicle that on December 9, 1978, sent five separate instrument packages into the Venerian atmosphere and on down to the surface. These consisted of a large probe, three small ones known as North, Day, and Night (for the areas of the planet they explored), and the Bus—the vehicle on which the other four were mounted before deployment.

From the earth (and even from the Pioneer orbiter) we see light reflected from upper cloud layers about 60 km above the surface. It is very cold up there—230 to 240 K. Those upper clouds form two distinct layers, which together are about 12 km thick; they consist mostly of drops of sulfuric acid. Below them is a dark opaque cloud some 4 or 5 km thick, composed largely of solid and liquid sulfur with intermixed sulfur dioxide, which has shrouded the planet in mystery until the eyes of the Soviet Venera and U.S. Pioneer probes penetrated it. Below this main cloud a haze extends down to an altitude of about 32 km, but below that the atmosphere is clear—right down to the surface.

However, both the Soviet and U.S. probes detected frequent and almost constant displays of lightning in the middle atmosphere of Venus. These are not caused by thunderstorms of the earthly variety, but result

The planets as seen from earth:

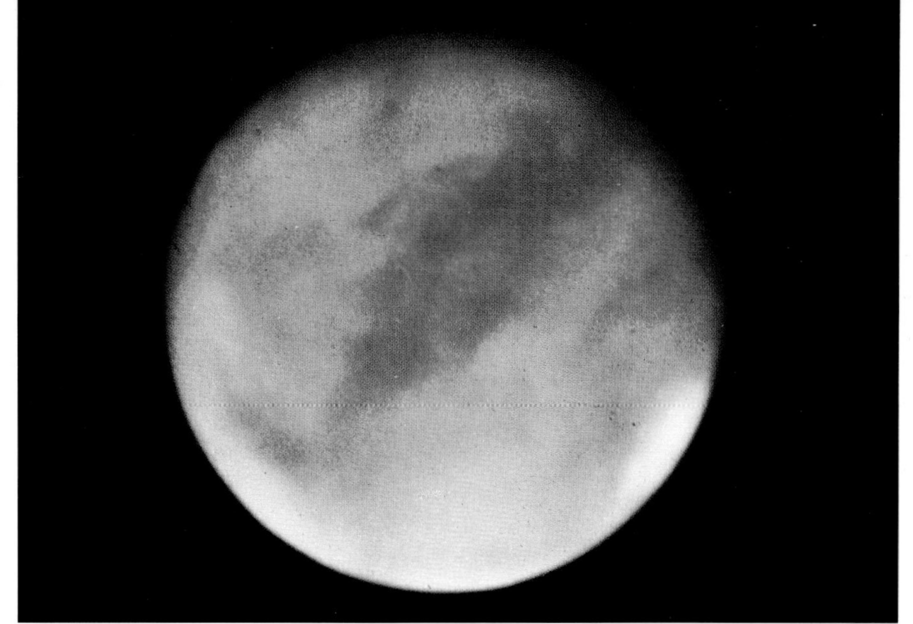

Mars, photographed with the 5-m telescope on Palomar Mountain. (*Hale Observatories*)

Saturn, photographed with the 3-m telescope of the Lick Observatory. (*Lick Observatory*)

Jupiter, photographed with the Palomar 5-m telescope. (*Hale Observatories*)

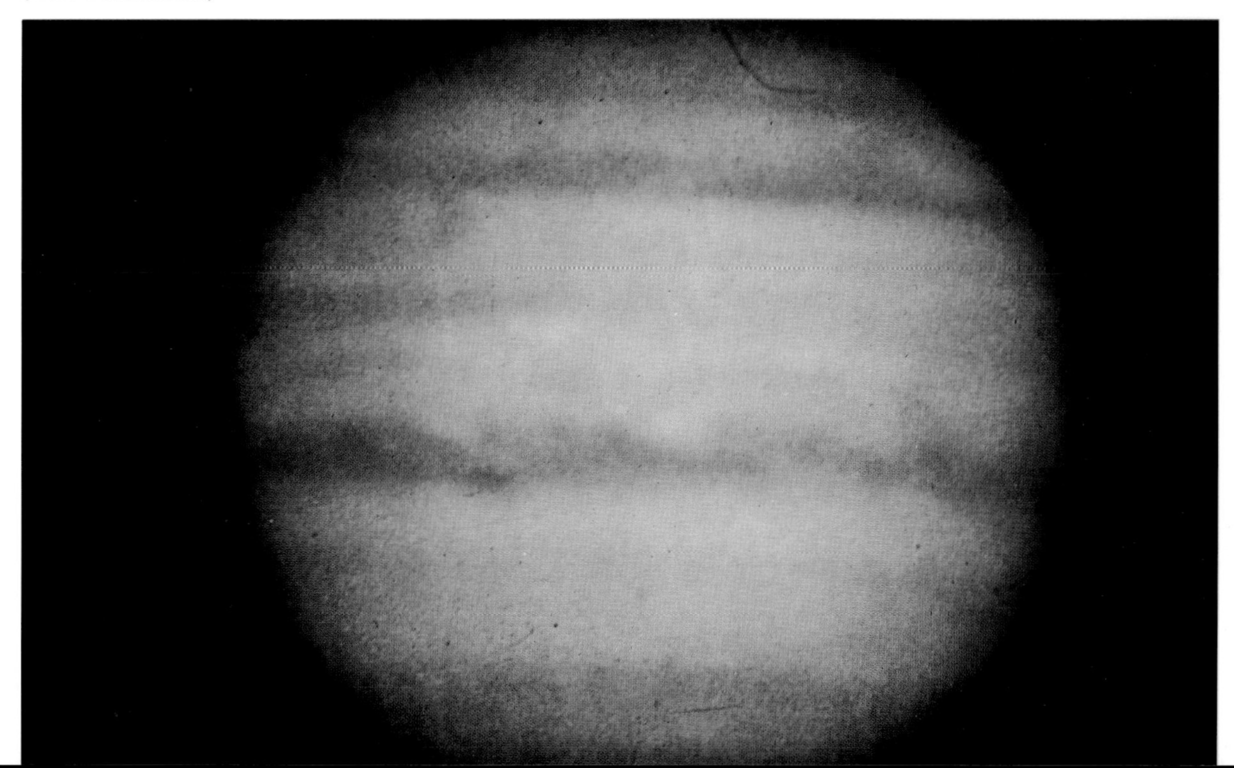

Mars, photographed from Viking 1 spacecraft in 1976. (*NASA*)

Mars, photographed from
Viking 2 lander in 1976.
(NASA)

Looking south from the Viking 2 lander on September 6, 1976. *(NASA)*

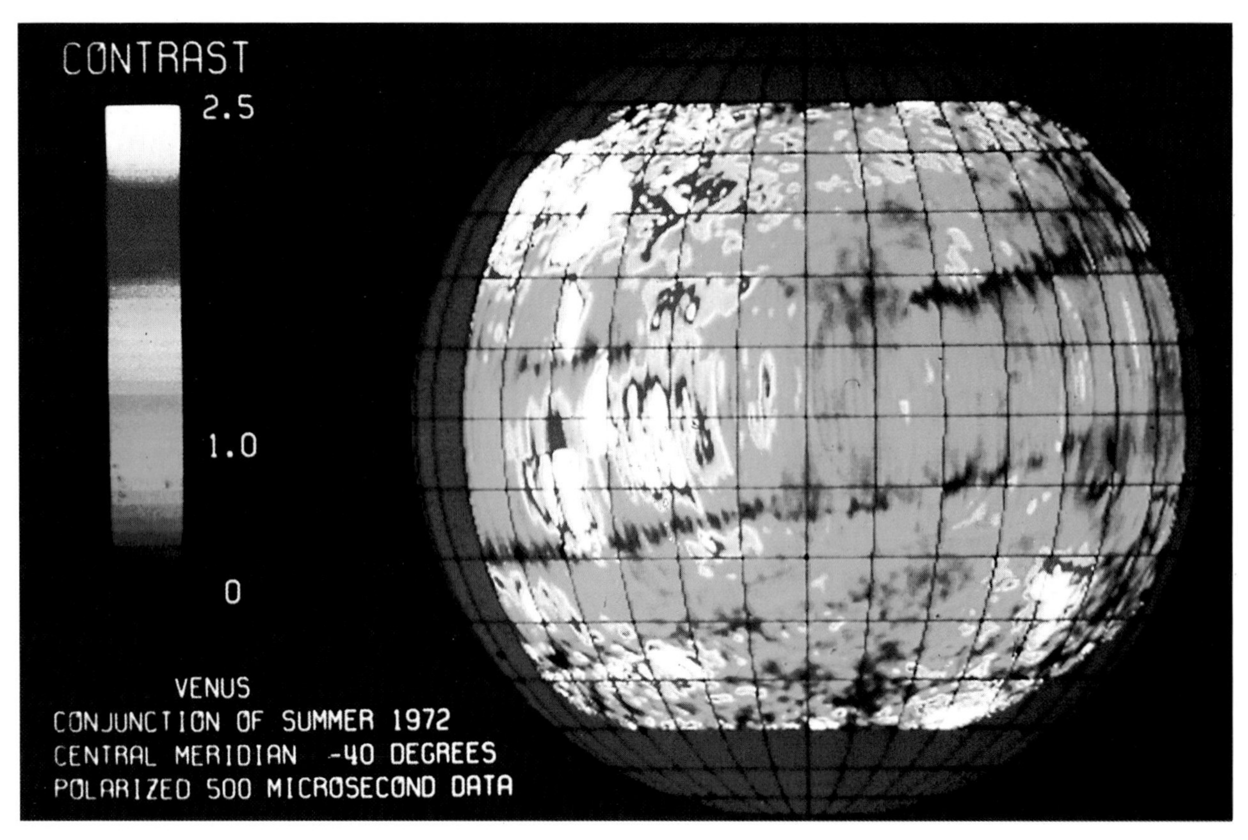

A radar map of Venus, prepared by the National Astronomy and Ionospheric Center, Arecibo, Puerto Rico. *(Cornell University)*

An ultraviolet photograph of Venus made by Mariner 10. *(NASA)*

Saturn, photographed by Pioneer 11 on September 1, 1979. *(NASA/Ames Research Center)*

Figure 10.4
Pioneer Venus 2 Multiprobe *(foreground)* and Pioneer Venus 1 Orbiter. The entry probes can be seen mounted on the Bus, which itself carried instruments into the Venerian atmosphere. *(NASA)*

from electrical or chemical activity not yet thoroughly understood.

Even the earlier space vehicles, such as Mariner 10, revealed high speed winds — 100 to 200 km/hr — in the upper Venerian atmosphere, showing evidence of atmospheric circulation. The Pioneer probes found that at the north pole of the planet there is a clear area in the clouds about 1100 km in diameter. Evidently, the winds circulate poleward and down toward the surface at the poles, depressing the clouds to lower elevations, where they evaporate. In the lower atmosphere of Venus, however, there is no sign of weather whatsoever.

The first gas to be detected in the Venerian atmosphere, from earthbound observations in 1932, was carbon dioxide. Now that the chemical composition of the atmosphere has been measured accurately by the Pioneer probes, we can confirm that Venus' air is 96 percent carbon dioxide. Next in importance is nitrogen, at 3.4 percent. Water vapor accounts for 0.1 to 0.5 percent, and there are traces of argon, oxygen, neon, and sulfur dioxide.

The argon in the Venus atmosphere presents a puzzle. On the earth, the argon is mostly the isotope argon-40, which arises from the radioactive decay of potassium-40, and then outgases. But in Venus there is a fairly large proportion of argon-36 — very rare on earth. We expect argon-36 to be primordial; that is, it was present in the solar nebula from which the planets formed. But then why is it not more plentiful on earth? Unless it is locked in the earth's interior, we have no immediate explanation; its presence on Venus was one of the unexpected findings by Pioneer.

We expect that the atmosphere of Venus, like that of the earth, originated by outgassing from the crust. The total amount of nitrogen on Venus is about the same as on earth, and so is the carbon dioxide, except that on Venus it is in the gaseous state in the atmosphere, while on the cooler earth it has reunited with surface rocks. But the major compound to outgas on earth was water, which exists only in small quantities on Venus. Where is the Venerian water? The favored guess is that it circulated to the top of the atmosphere, where solar ultraviolet radiation decomposed it to hydrogen, which would easily escape, and oxygen, which

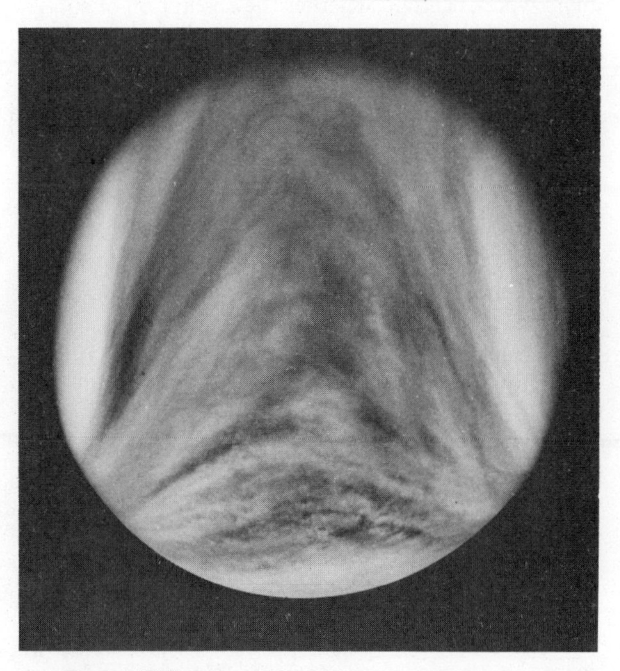

Figure 10.5
Venus, photographed by Pioneer Venus Orbiter at a distance of 58,000 km. *(NASA)*

Figure 10.6
The surface of Venus, showing fairly rough rocks, photographed by the landing apparatus of Venera 9. *(Novoste from SOVFOTO)*

is chemically active enough to form compounds with other elements. It is possible, though, that hydrates (chemical compounds that contain water) were not common in the stuff from which Venus accreted, in which case there may never have been any water there.

At least our new knowledge of the composition of the Venerian atmosphere has completely settled the question of the high surface temperature of the planet. Earth-based radio observations of Venus provided the first evidence of the planet's high temperature, and the various space probes confirmed it, but some astronomers were uncertain whether the absorption of sunlight that penetrates to the lower atmosphere and surface could heat up the planet so much. They realized that the planet's reradiated energy, in the form of infrared radiation, would be partly trapped by the carbon dioxide to keep the temperature high—the so-called *greenhouse effect*. But more information was needed to make sure the numbers work out right; that information was provided by the Pioneer probes. There is, indeed, enough carbon dioxide and water vapor, combined, to account for the observed temperature.

Except for the sun and moon, Venus is the brightest object regularly seen in the sky. It can cast a shadow on occasion, and can even be seen in broad daylight when it is at its brightest (you must know where to look for it). It often appears as a truly beautiful object in the morning or evening sky, and it is not uncommon, when it happens to be in the evening sky near Christmas time, for people to call observatories and inquire whether it is the "Christmas Star." Venus certainly gives a romantic illusion, and it must have seemed quite appropriate to name it for the goddess of love and beauty. Venus is also the planet most like the earth in size and mass, and has long been called our "sister planet." Our modern exploration of the planet, however, shows it to be very different from the earth, and hardly well named for the love goddess; Venus is an absolute hellhole!

Imagine life at the surface of that world. The massive stagnant atmosphere of carbon dioxide weighs down with a pressure averaging 91 times that of the earth's atmosphere. The temperature is between 721 and 732 K (about 850° F). Since Venus' upper clouds reflect about 70 percent of the incident sunlight back into space (which is what makes Venus appear so bright), not much passes through the clouds. In fact, only about 2 or 3 percent reaches the surface, and most of that is reddish light of longer wavelength. So the sky would be dark and red and gloomy, with a visibility of only a couple of kilometers. Moreover, it is dry and dusty. Small wonder that the early entry probes did not survive to the ground. Even the hardiest of them have lasted on the ground only an hour or so. Still, the Soviet Venera 9 and 10 landers did manage to transmit photographs of the surface rocks back to earth before ceasing to operate (Figures 10.6 and 10.7).

Except for the changing illumination, day and night on Venus are much the same, with virtually no change in temperature. Venus rotates very slowly; its rotation period, first detected by radar observations from earth, is 243 of our days. The Venus Pioneer 1 orbiter was designed to operate for this period of time to watch Venus through one of its complete days. Because of its slow rotation (the slowest of any planet in the solar system), Venus has no measurable magnetic field of its own.

In addition to the Soviet Venera photographs, we can learn about the surface features of Venus with radar observations, both from earth and from the Pio-

Figure 10.7
The surface of Venus, photographed by the landing apparatus of Venera 10. The vertical stripes in this figure and in Figure 10.6 occur where the transmission carried nonpicture information for other scientific experiments. *(TASS from SOVFOTO)*

neer orbiter. The planet has a topography not unlike that of the earth, and it even shows signs of faults, suggesting that plate tectonics may once have occurred or may still be going on. There are highlands and low plains. The Great Northern Plateau is a continent-sized region, and rising above the plateau is a rugged highland area named Maxwell; Maxwell rises about 13 km above the surrounding plains—considerably higher than Mt. Everest above sea level. There is a line of volcanic peaks larger than the Hawaiian chain. Most impressive of all is a great rift valley, the widest such feature in the solar system. The valley is 1400 km long, 280 km wide, and 5 km deep. There are also many large old craters in the lowland plains—remnants, presumably, of the final accretion phases of Venus' formation.

Despite its scenery, Venus would not be a nice place to live. It would not even be a nice place to visit! Let's see what Mars is like.

THE MARTIAN DREAM

The most significant new thing we have learned about Mars is that apparently it has no life. It's a pity, because Mars seemed to provide the best hope for finding life elsewhere in the solar system. It has excited the interest and imagination of man for centuries. In some ways it is like the earth. Its day lasts 24 hours 37 minutes. Its obliquity is 24°, compared to 23½° for the earth; thus Mars has seasons similar to ours, although each season lasts almost six of our months. For more than 100 years we have been watching seasonal changes on Mars, the most conspicuous being the growth and recession of its polar caps. Seasonal color changes were also reported in the temperate zones of the planet; some observers thought those regions turned green in spring and summer and brown in fall. It is actually less a change in color, however, than a change in the visibility of various surface markings, as viewed through earthbound telescopes.

Mars also was known for a long time to have an atmosphere. Vast yellow dust storms were sometimes seen to cover the planet. In fact, when Mariner 9 reached Mars and went into orbit around it in 1971, the entire planet was hidden by such blowing dust during the first weeks of observation. White clouds on Mars have long been observed from earth, too. We know now that some of these are quite high—about 45 km, and are probably carbon dioxide crystals. Other clouds are lower, 15 to 30 km, and are believed to be of ice crystals, like our cirrus clouds.

The Martian atmosphere, however, is now very thin; the surface pressure is somewhat less than 1 percent sea-level pressure on earth. The best determination of the abundances of the various gases present there is from data gathered by the Viking landers in 1976:

Carbon dioxide (CO_2)	95%
Nitrogen (N_2)	2 to 3%
Argon (A)	1 to 2%
Oxygen (O_2)	0.1 to 0.4%
Water (H_2O)	0.01 to 0.1%
Krypton (Kr)	less than 0.0001%
Xenon (Xe)	less than 0.0001%

The abundance of the argon is a good indication of how much total gas outgassed, and hence what Mars' atmosphere may have been like in the past. A best guess at the time of this writing is that the Martian

Figure 10.8
Artist's conception of the Great Rift Valley on Venus. *(NASA)*

atmosphere may have once had a pressure of about 10 percent that of the earth's.

The maximum equatorial temperatures of Mars are about 30° C. The sunrise and sunset temperatures at the equator are much lower, around −40 and −20° C, respectively. At night, the temperature drops to nearly −75° C. The temperature at the south polar cap was measured by Mariner 7 and was found to range down to about −150° C, which is near the frost point for carbon dioxide.

Before the space age, all astronomers (amateurs, too) who spent much time looking at Mars through a telescope agreed that it abounds in surface detail. The problem is making out what that detail is. Under the best conditions, Mars shows telescopically about the same amount of resolution that the full moon does when seen with the unaided eye. Those of you who have looked at Mars through a telescope probably saw mainly a shimmering red or orange ball, with perhaps a slight spot of white at one or the other pole. The shimmering is, of course, caused by the earth's atmosphere, a phenomenon called *seeing*. Seasoned observers, though, waiting for the steadiest (terrestrial) atmospheric conditions, occasionally see much more, especially if they wait for rare moments when sharp features seem to flash into view for a brief instant. I have had the experience myself several times, and can fully understand how the professional Mars observers can have become terribly excited at times.

The trouble is that the eye tends to simplify complex patterns that are seen indistinctly. What some observers called many fine specks, others saw as connected blobs. So it was that in 1877 the Italian astronomer Giovanni Schiaparelli thought he saw fine straight lines on Mars. He called them *canali,* or "channels." Then lots of other people saw them too. Most famous among them was Percival Lowell, who devoted most of his life to studying Mars. In fact, the main reason he established his Flagstaff, Arizona, observatory was to study that mysterious red planet.

Lowell mapped hundreds of canals. He saw them connected at spots he called *oases*. He believed they were real water canals, built by intelligent Martians to carry the waters from the melting polar caps across the desert to irrigate their fields. So grew the Martian dream.

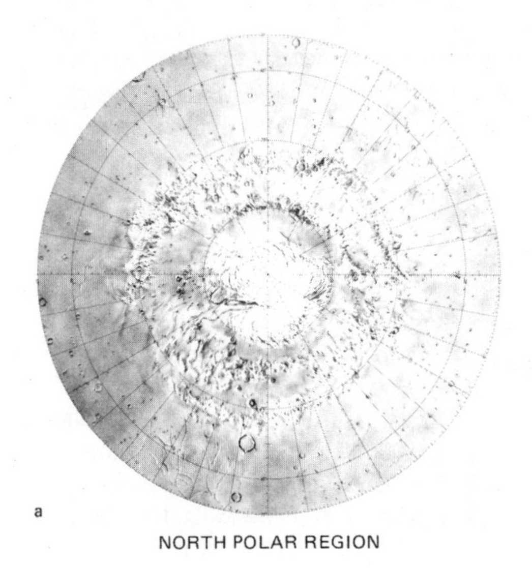

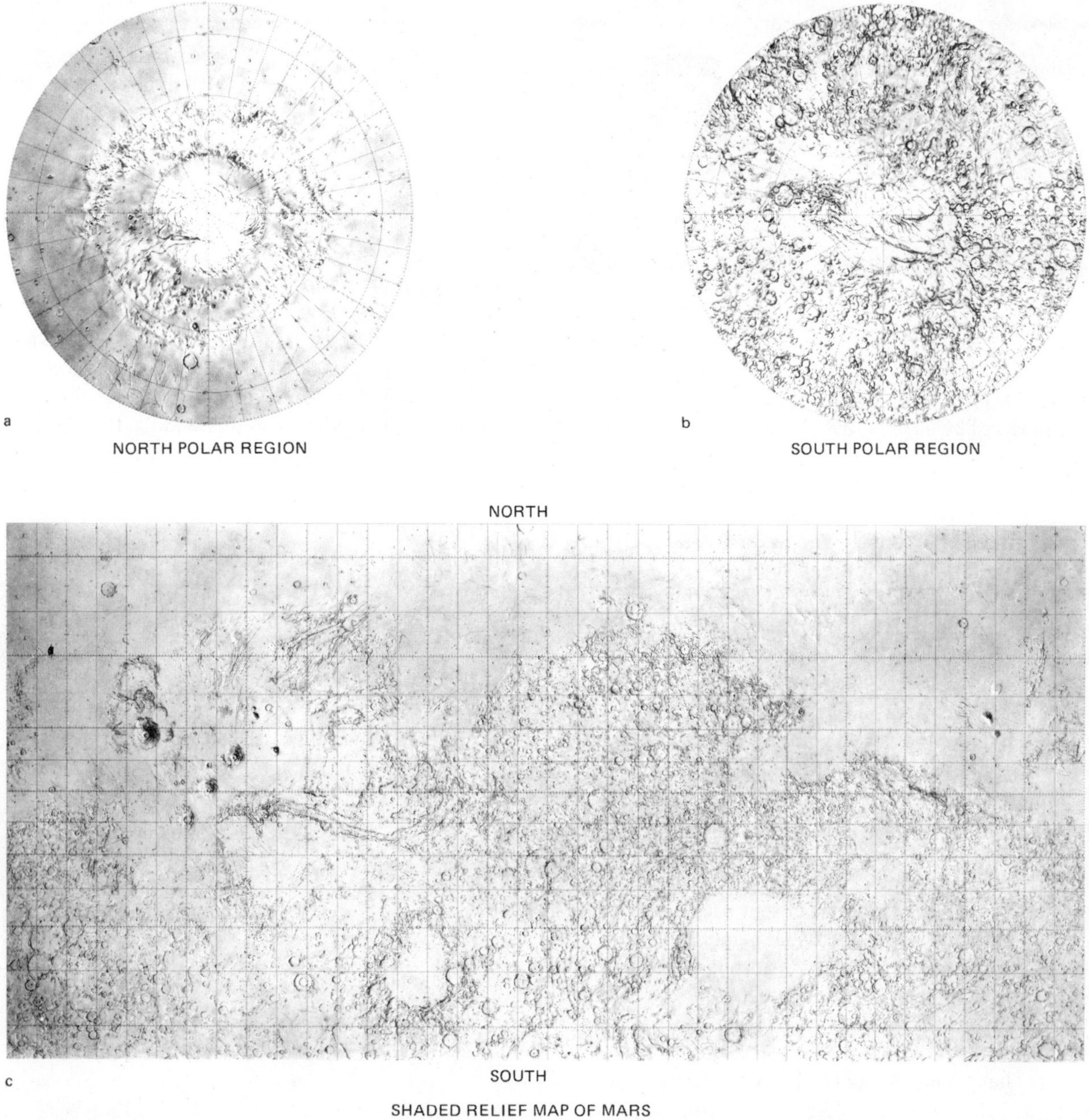

a

NORTH POLAR REGION

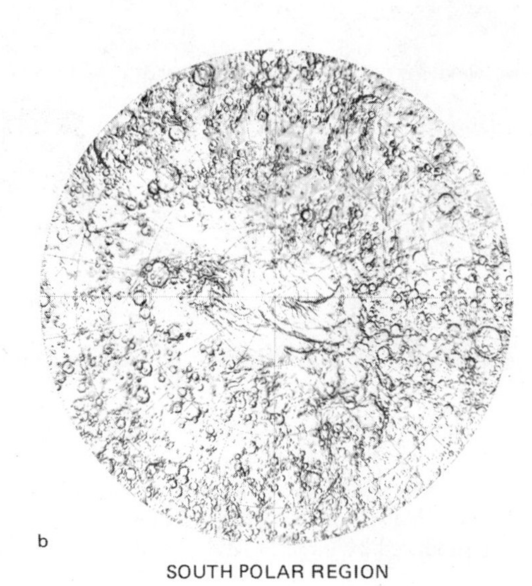

b

SOUTH POLAR REGION

NORTH

SOUTH

c

SHADED RELIEF MAP OF MARS

Figure 10.9
A topographical map of the planet Mars: *(a)* the north polar region; *(b)* the south polar region; *(c)* the equatorial and temperate regions. *(NASA/JPL)*

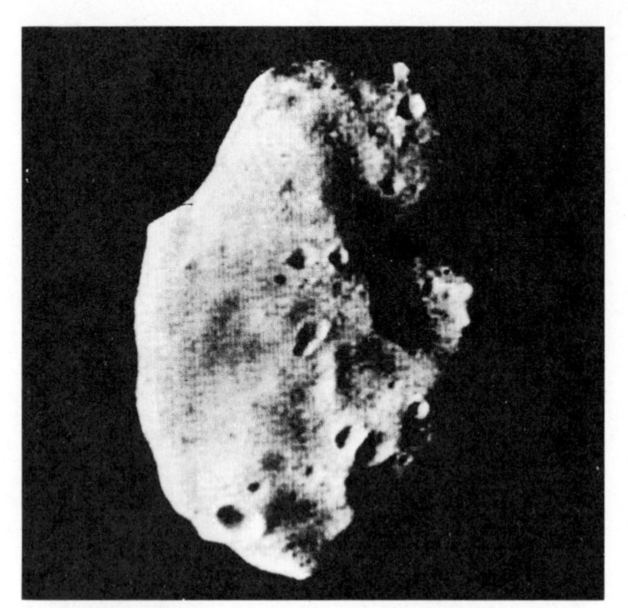

Figure 10.10
Phobos, Mars' larger satellite, photographed by Mariner 9 in November 1971. The sun's illumination is from the left, and the right (night) portion of the satellite is hidden in shadow. Note the many craters produced by impacts. *(NASA/JPL)*

It wasn't that observers like Schiaparelli and Lowell were seeing things. The features were there all right. It's just that they were misinterpreted, because of the limited resolution permitted by earthbound telescopes, due especially to the turbulence of the earth's atmosphere. The first close-up view of Mars came in 1965, when Mariner 4 flew past the planet and televised 22 pictures back to earth. The photographs showed a bleak and barren planet, covered with craters, more like the moon than the earth. But then Mariners 6 and 7 flew by in 1969, and sent back a couple of hundred more pictures, including 33 of the south polar cap. But no picture showed canals.

The real coup came in 1971, when Mariner 9 went into orbit around the planet and sent back 7329 photographs that have mapped Mars far better than we knew the face of the earth 200 years ago. Mariner 9 even photographed those little satellites, Phobos and Deimos (which mean "Fear" and "Panic," companions of the god of war), 25 and 13 km in diameter, respectively. The culmination of the U.S. exploration of Mars came in 1976 when the Viking 1 and 2 landers settled on the surface of Mars and proceeded to send back pictures and to chemically analyze the surface material.

The Martian polar caps are now well understood. They change rapidly with the seasons. The south cap, in particular, reaches halfway to the Martian equator in midwinter, and often disappears entirely in summer. The main stuff of those white caps is frozen carbon dioxide (like dry ice). But not all! Part of the caps is water. At least in the north, there is a permanent cap of water ice. It doesn't go away in summer, and it is probably hundreds of meters thick. If it could all be melted, it is estimated that it could cover the planet with water to a depth of one-half meter. Furthermore, a lot of water could be frozen as permafrost under the Martian surface; we don't know, of course, for sure.

There is also volcanism, and on a large scale. At least 12 huge volcanos exist on Mars, and their eruptions have flooded a large part of the planet with newly formed lava rock. Most spectacular among them is Olympus Mons, 25 km high and 600 km across and displaying a giant caldera 70 km across at its summit. It is by far the largest volcanic pile ever seen by man.

Canyons also are present, especially a great canyon (Valles Marineris) 5000 km long, 75 km wide (on the average), and 6 km deep. Although there is what appears to have been lifting and subsiding, there is no evidence for *horizontal* plate tectonics, as on earth.

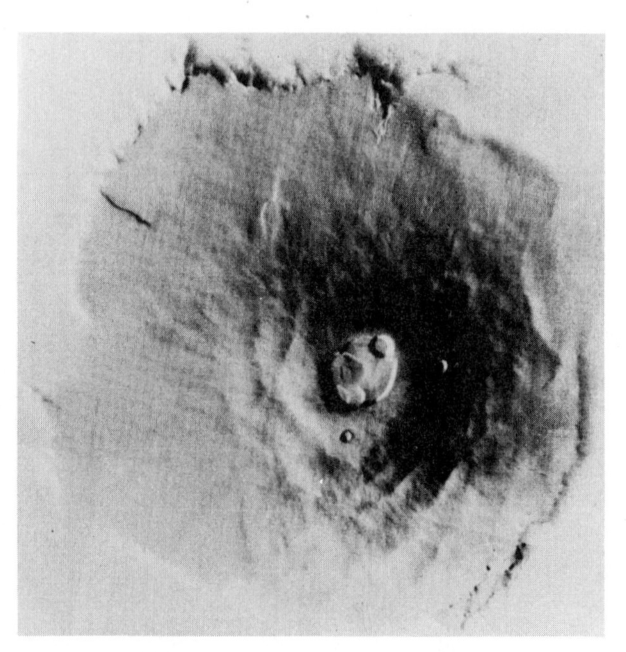

Figure 10.11
Olympus Mons, a gigantic volcanic mountain on Mars, photographed by Mariner 9. *(NASA/JPL)*

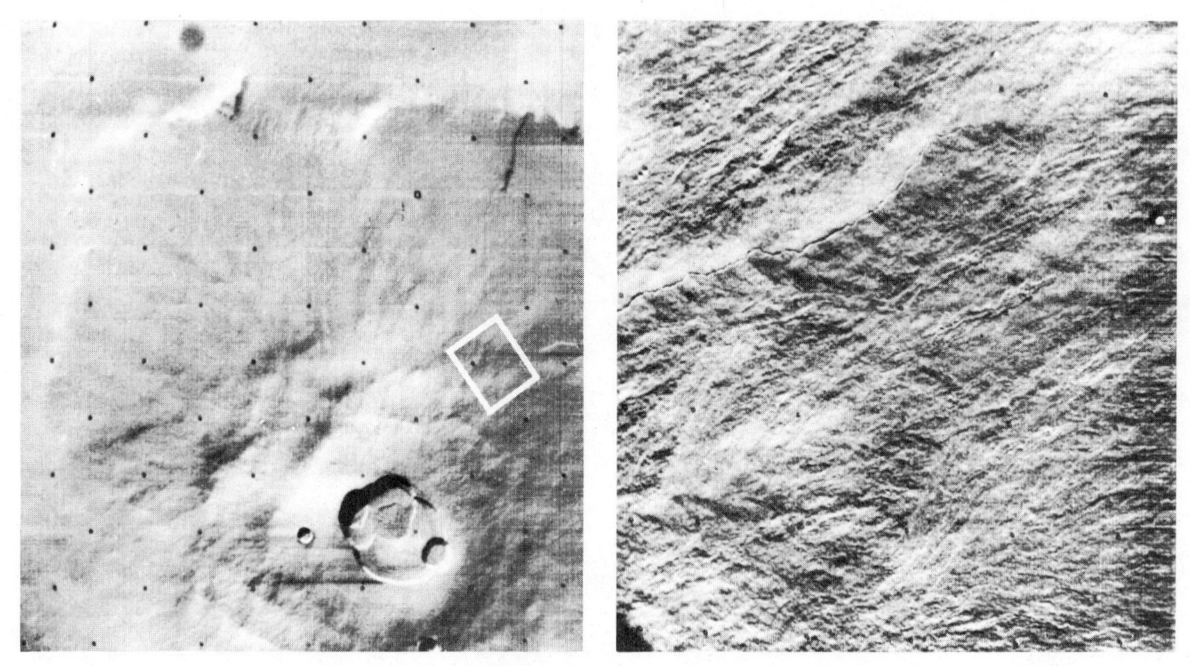

Figure 10.12
Olympus Mons region: *(left)* summit crater of the volcanic mountain; *(right)* detail of white rectangular region of left photograph, showing lava flow. Mariner 9 photographs. *(NASA/JPL)*

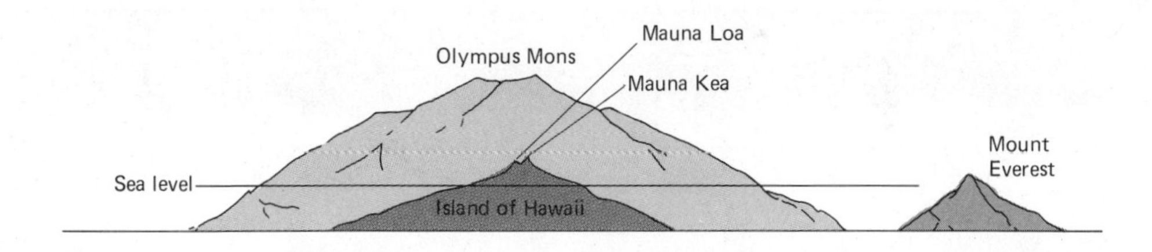

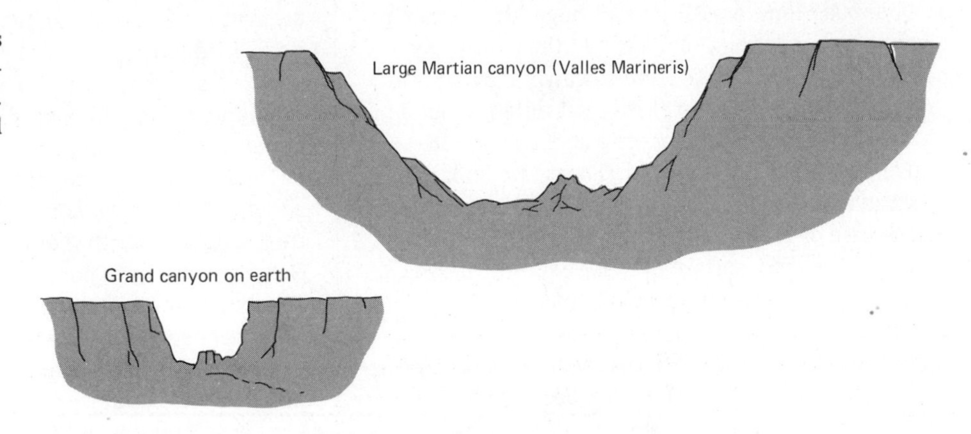

Figure 10.13
The relative sizes of Olympus Mons, Mauna Loa, Valles Marineris, and the Grand Canyon. The vertical scale is enlarged 5 times.

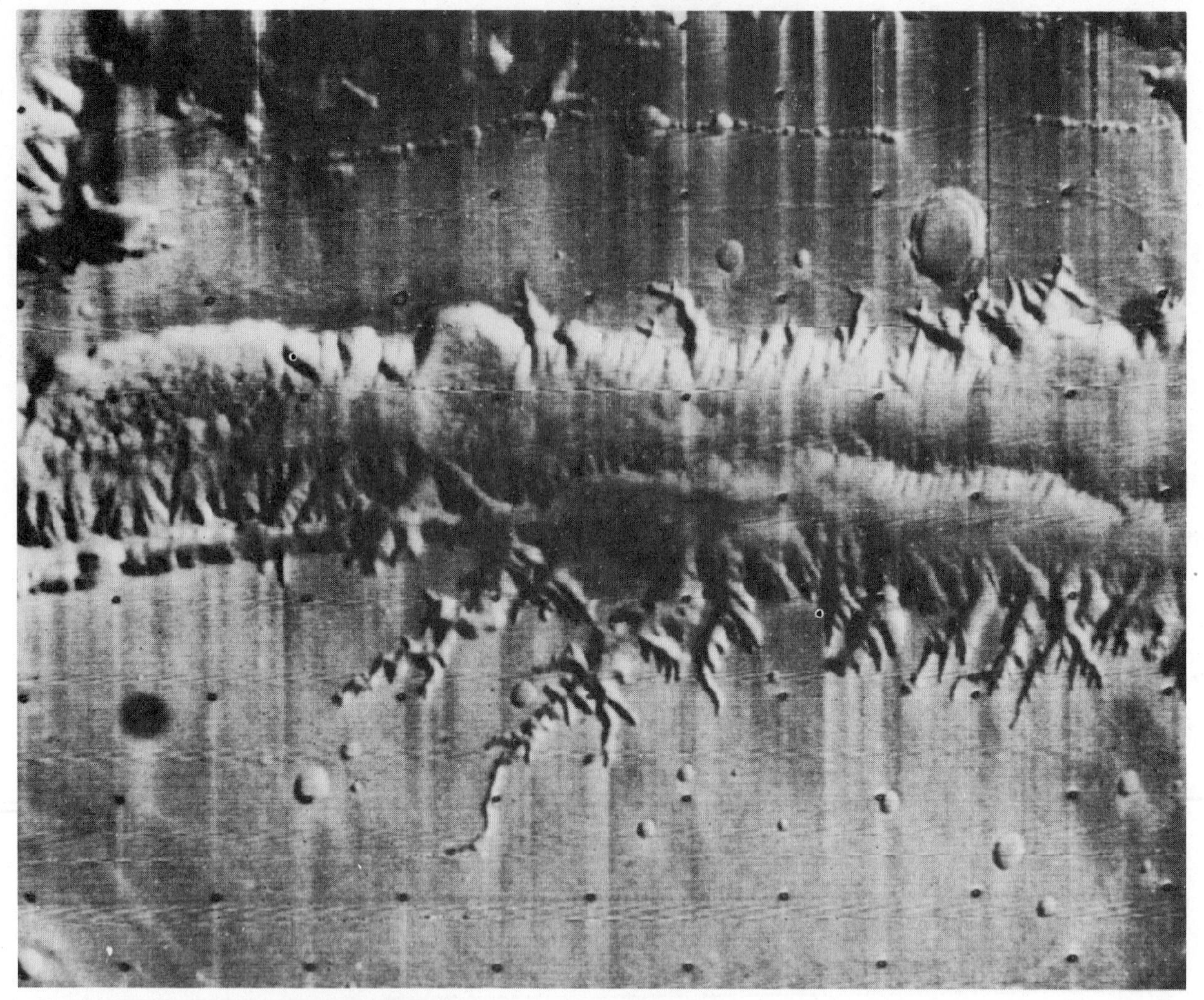

Figure 10.14
Valles Marineris, a huge Martian canyon photographed by Mariner 9. *(NASA/JPL)*

This can explain why certain Martian volcanos, such as Olympus Mons, have grown so huge. On the earth, a volcano grows over a "hot spot" in the mantle beneath it. Now, if a crustal plate is slowly moving over that hot spot, the volcano emerges at different places on the surface of the crustal plate as time goes on. Thus island chains, like that of Hawaii, are built up over the tens of millions of years, while a moving plate (in the case of Hawaii, the Pacific plate) carries new parts of the earth's surface over the exuding magma.

The most interesting things found on Mars are dry river beds. Apparently all experts now agree that there is no question that running water was once present on Mars. Braided streams have been found, and even ancient islands in dry stream beds. Certainly there has been erosion by some liquid, and water seems to be the only candidate. Where is that water now? Evidently frozen, either as permafrost or in those polar caps, or both. Or perhaps most of it has evaporated into space, although a considerable amount of frozen water appears to be left.

Then why does the water not flow now? In part, perhaps, because Mars may be in a glacial period, like the ice ages on earth. But there is another problem. Geophysicists have pointed out that liquid water can simply not survive in the low-pressure atmosphere Mars has today. Various experts estimate that the atmospheric pressure would have to be from 5 to 50 times as great as at present to allow liquid water to flow over the surface of the planet, and not immedi-

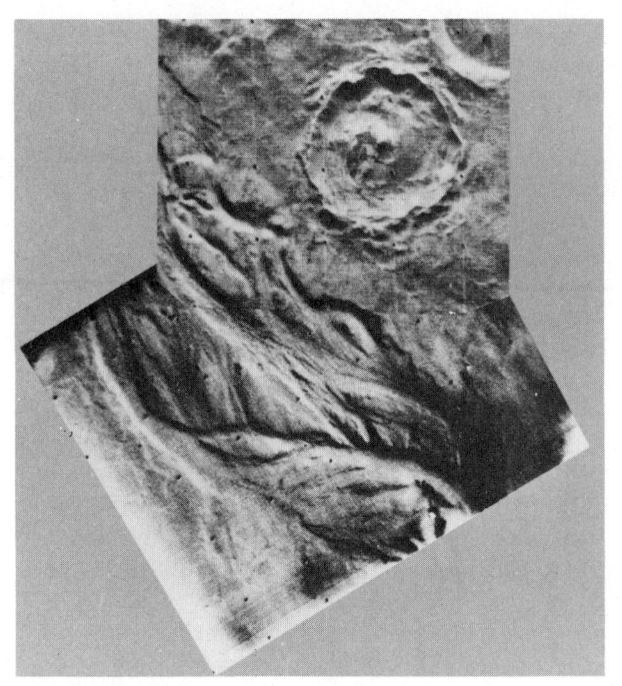

Figure 10.15
A dry Martian river bed, showing the braided pattern characteristic of many terrestrial stream beds. *(NASA/JPL)*

ately evaporate. Yet, the abundance of argon suggests that the requisite atmospheric pressure may have existed in the past, perhaps more than 1000 million years ago.

Does this mean that Mars may once have been a planet of flowing water, like the earth? If so, has that water evaporated and dispersed into space, along with Mars' gradually dispersing atmosphere, or perhaps frozen out, never to flow again? There is, alas, some reason to think that such a fate may, indeed, have fallen upon Mars. The authorities disagree on the ages of the youngest great volcanos; some say a few hundred million years, and others say over a thousand million years. In any case, those youngest looking volcanic lava slopes do *not* have water channels eroded into them, which strongly suggest that they came after the last running water.

If there was running water on Mars in the past, perhaps there was life as well; perhaps life, in some form, or its remains, still exists in the Martian soil. It was this possibility, however unlikely, that provided the main purpose of the Viking landers.

The Viking landers—in addition to carrying experiments to measure weather, seismic events, and other things, as well as television cameras to see any Martians who might be walking up—had long arms that went out a few meters to scoop up soil and bring it back into the laboratory for analysis. There were four biological experiments. Three were to test for respiration of living animals, absorption of nutrients by any living organisms, and exchange of gases between the Martian soil and the laboratory environment for any reason whatsoever. In the various experiments the soil samples were isolated and incubated in contact with various gases, radioactive isotopes, and nutrients to see what would happen. The fourth experiment pulverized the soil sample, and analyzed it carefully to see what organic material it contains.

The Viking experiments were sensitive enough that had one of the probes landed anywhere on earth, with the possible exception of Antarctica, it would easily have detected life. Those experiments that tested for absorption of nutrients and gas exchange did show activity, but this was caused by the chemically active soil, not living or dead organisms. The organic chemistry experiments showed no trace whatsoever of any organic material. The possibility of biological organisms somewhere else on Mars has not been completely eliminated, but most experts consider the chance of any life on that planet to be negligible.

Compared to Venus, Mars would not be so bad a place to visit, but nobody seems to be home.

Figure 10.16
The Martian landscape, photographed by Viking 1. *(NASA)*

JUPITER

Next to the sun, Jupiter is the largest and most massive object in the solar system. It has at least 14 satellites, four of which are themselves the sizes of small planets. In a sense, Jupiter is almost a miniature solar system in its own right.

Jupiter is 318 times as massive as the earth, a value which is very close to 0.001 the mass of the sun. Because of the tremendous gravitational attraction of Jupiter for its constituent parts, it would most certainly be compressed to a far greater mean density than it is unless it is composed almost entirely of hydrogen and helium—the lightest and most abundant elements in the universe. Even these elements must be compressed to a liquid metallic state throughout most of Jupiter's interior, although there is probably a small rocky core, perhaps about the size of the earth, at the planet's center. We think the overall chemical composition of Jupiter is very close to that of the sun and the primordial solar nebula.

It is interesting to note that Jupiter has very nearly the maximum possible size for a body of "cold" hydrogen—that is, one that is not generating energy. Less massive bodies than Jupiter would occupy a smaller volume. More massive bodies, by virtue of their greater gravitation, would also be compressed to a smaller volume than Jupiter's.

A remarkable thing about Jupiter is that it radiates more than twice as much energy into space (at infrared wavelengths) as it receives from the sun. The source of most of this energy is probably stored heat trapped in the planet's interior at the time of its formation. Jupiter must be slowly radiating away this energy as it cools. It might also be deriving some energy from a slow gravitational contraction.

The first close-up look at Jupiter was provided by Pioneers 10 and 11 in December 1973 and December 1974, respectively. But the really good look came in 1979 when Voyagers 1 and 2 reached Jupiter, and transmitted more than 33,000 pictures of the planet and its satellites back to earth. All four of these probes were flybys, but Pioneer 11 and both Voyagers were placed on trajectories such that Jupiter's gravitational deflection on them sent them on toward Saturn. Pioneer 11 passed Saturn in 1979, and the Voyagers in 1980 and 1981.

The surface of Jupiter is crossed with alternate

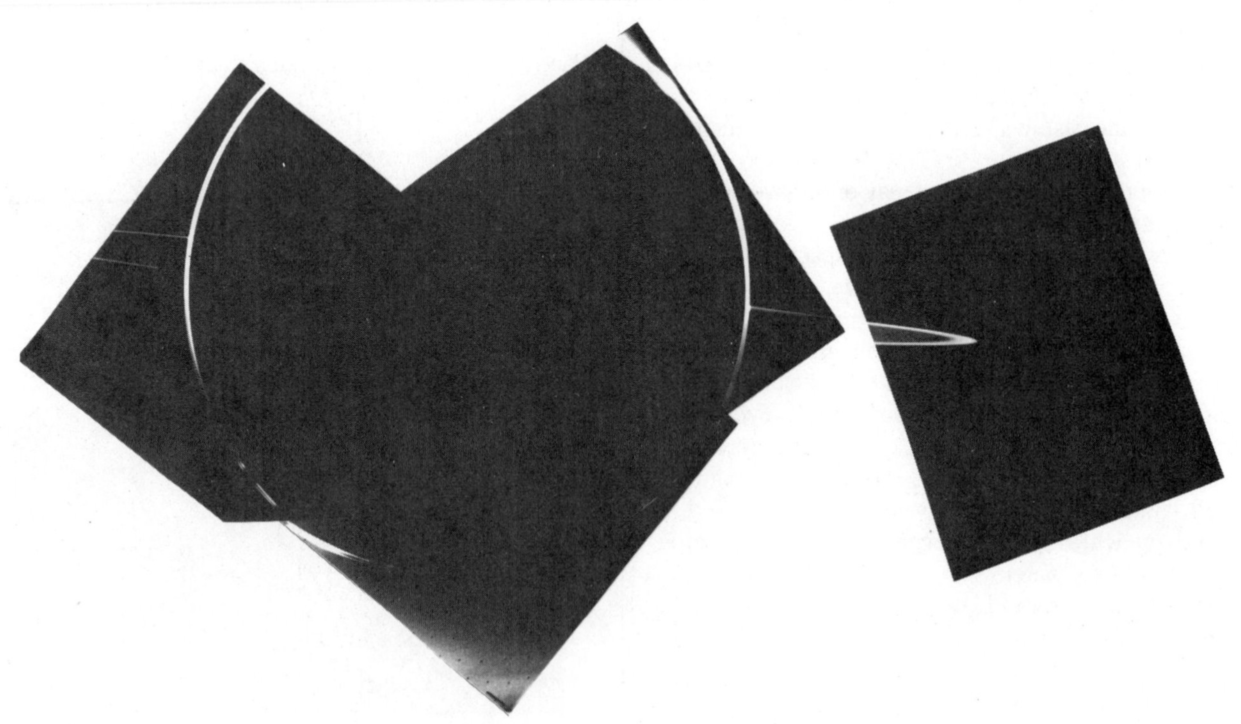

Figure 10.17
A mosaic of photographs obtained by Voyager 2, showing the night side of Jupiter, with the ring silhouetted against the dark sky. *(NASA)*

light and dark, brightly colored bands parallel to its equator. The bands abound in detail and as seen telescopically from earth exhibit gradual changes. Time-lapse photographs from the Voyagers, however, show them to be enormously active and complex features. Jupiter is the most rapidly rotating planet, turning once in just under ten hours. But its atmosphere does not rotate as a solid unit; different latitudes have slightly different velocities. As a result the cloud bands, especially at their boundaries, show almost every conceivable kind of current and flow.

Even more striking is the Great Red Spot. The spot was first seen telescopically from earth in 1831. It has changed slightly in size and shape since then, as well as in intensity of color, but on the whole it has persisted. It has been as large as 50,000 km across, and has always been far bigger than the earth. The Pioneer data suggested that it was some kind of a long-lived storm system in the planet's atmosphere. The Voyagers showed it to be a massive eddy with enormously complex, and changing, small eddies in gas streams flowing around it.

When Voyager 1 flew past Jupiter and turned back to photograph its night side, it sent us a surprise: Jupiter has a thin faint ring in its equatorial plane. The ring has an outer diameter of 256,000 km, and a width of at least 6000 km, although it may extend very faintly all the way in to the planet's surface. The ring is very thin, no more than 30 km thick, and is composed of very tiny particles. The famous rings on Saturn are not only far more substantial, but contain sizable chunks. Analysis of the scattering of light from the Jovian ring shows the particles making it up to have sizes of only 8 to 10 microns (a micron is one-millionth of a meter). It is doubtful that particles that small can remain in stable orbits in the presence of the various perturbing forces about Jupiter, which suggests that they are not permanently in that ring. Perhaps the ring itself is a temporary phenomenon, or perhaps it results from a continual flow of tiny particles swept up by Jupiter from interplanetary space. After its discovery by Voyager, the Jovian ring has been confirmed by observations from earth.

Jupiter has a very strong magnetic field. The field was first suspected when earth-based radio observations showed Jupiter to be a strong source of radio waves emitted by electrons moving in a magnetic field. The magnetic field strength was measured directly by the Pioneer probes and, of course, by Voyager. It is 20 to 30 times the strength of the earth's field and also extends over far vaster a volume. High velocity ions are trapped in Jupiter's magnetic field much as they are in the Van Allen radiation belts around the earth. But the numbers and energies of charged particles around Jupiter are enormously greater; the space probes were subjected to doses of radiation from the impact of these particles that are hundreds of times the lethal dose for man. The region occupied by Jupiter's magnetic field and the trapped charged particles is called its *magnetosphere*. The Jovian magnetosphere far exceeds the sun in volume; if we could see it with the eye from the earth, it would appear the size of the full moon.

Jupiter's 14 known satellites, (a fifteenth is suspected but not confirmed) fall into three groups. The outer satellites have eccentric orbits at large inclinations to Jupiter's equator, and revolve about the planet from east to west—backward with respect to most motions in the solar system. They are probably objects (minor planets?) captured by Jupiter long after its formation. The satellites in the middle group have direct orbital revolution, but their paths are also eccentric and highly inclined; they, too, may be captured minor planets. The inner six satellites, on the other hand, all move in the usual west-east direction, and all have nearly circular orbits lying in Jupiter's equatorial plane; the best guess is that they were formed from a disk of

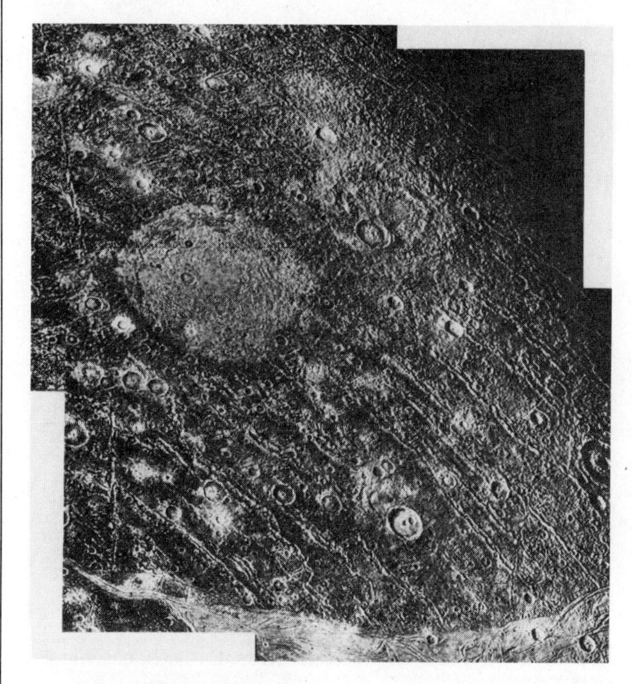

Figure 10.18
Cratered terrain on Jupiter's largest satellite, Ganymede. *(NASA)*

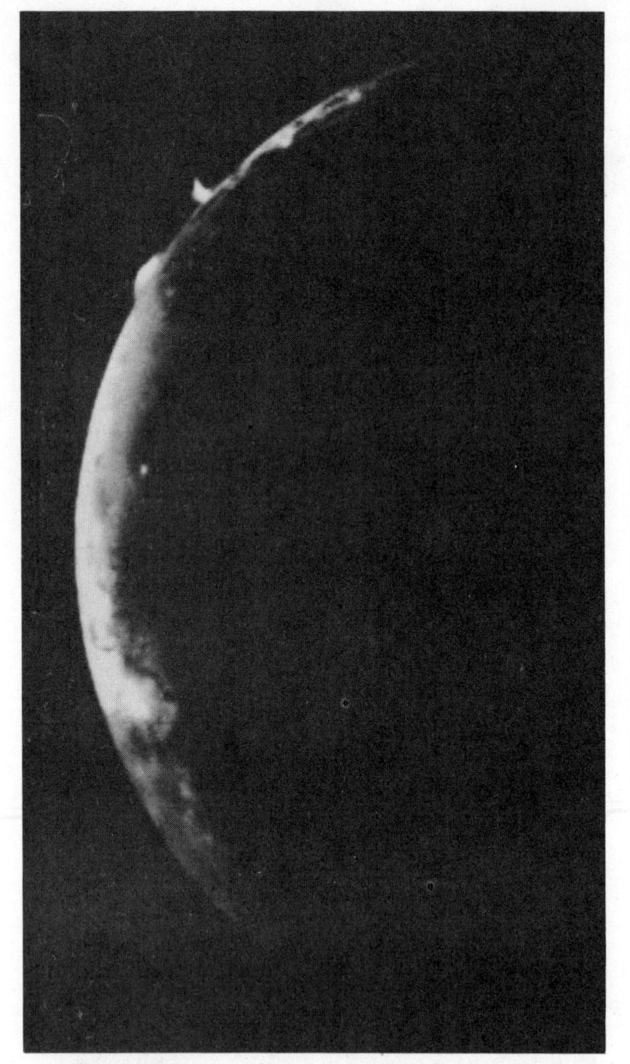

Figure 10.19
One of Jupiter's satellites, Io, showing two plumes from erupting volcanoes. Photographed from the night side by Voyager 2. *(NASA)*

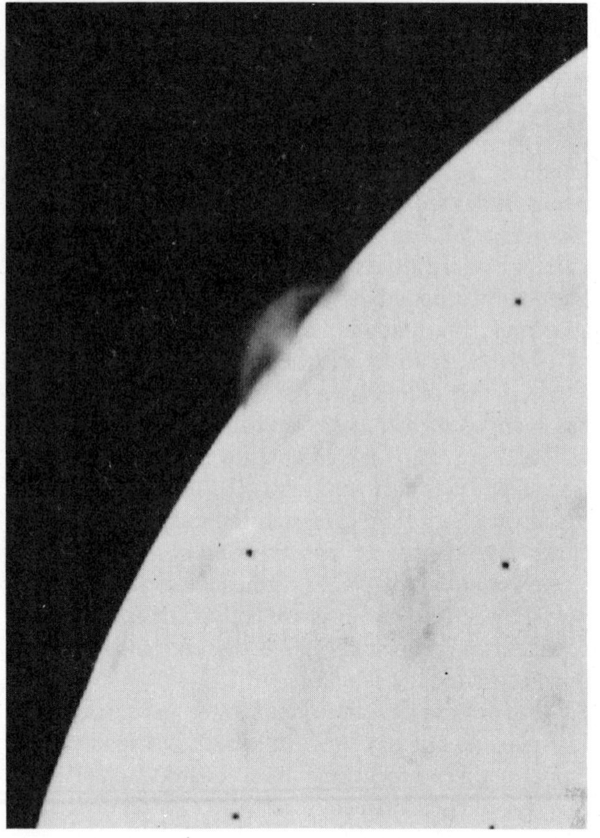

Figure 10.20
A portion of the limb of Io, showing a particularly striking plume from an erupting volcano. *(NASA)*

matter revolving about the primordial Jupiter much as the planets are thought to have formed from the solar nebula.

Of the inner six Jovian moons, the nearest to the planet is a very tiny one, discovered by Voyager 2. Next out is Almathea, about the size of California. The other four, all larger than our moon, are the satellites discovered by Galileo, and are called the *Galilean* satellites. They are amazing little worlds. Callisto, the most remote of the four, is completely covered with craters, so evidently has suffered very little erosion since its accretion days. Ganymede, the next in and the largest, is also cratered, but less so than Callisto, and shows many scars of cracks or faults. Europa, still further in, has relatively few craters, but many more cracks. Io, the innermost of the big four, has no sign of craters or cracks, but a brightly colored dusty material covers its entire surface. That material is probably mostly sulfur.

Ganymede and Callisto have low densities and are probably about 50 percent water ice, with the rest of their mass being made up of rocky material. Io and Europa have higher densities, and are at most only 20 percent ice. Io is now believed to be mostly molten inside a solid crust, its core kept hot and fluid by the strong tidal forces exerted on it by nearby Jupiter. Io also has hot spots on its surface—up to 150° warmer than the surrounding landscape.

But the most striking phenomena on Io, and perhaps the greatest surprise found by Voyagers, are its active volcanos. Seven volcanos were observed, but when account is taken of the fact that we cannot see all sides of Io at once, we predict that it may have a dozen or more. These volcanos are erupting almost continually, spewing clouds of sulfur and gases into graceful plumes hundreds of kilometers above its surface. The solid particles fall back, which is believed to result in the moon being slowly buried in its own sulfur. Fresh sulfur must be depositing at an average rate of a millimeter or so per year, which over a period of millions of years is enough to hide all evidence of any old erosion.

The gases belched out by Io's volcanos, thought to be largely sulfur dioxide, would stay as an atmosphere of the moon were it not for the fact that the molecules are dissociated and ionized by the charged particles in Jupiter's magnetosphere. Once ionized, they are swept up in Jupiter's magnetic field and drawn away from Io into a large doughnut or torus lying roughly in the orbit of the satellite. This *plasma torus* of Io is another discovery of Voyager 1.

So our new exploration of Jupiter and its family has proved very exciting. Recall that before Voyager, only five solid planetlike objects had ever been examined: Mercury, Venus, the earth and moon, and Mars. The close-up looks at the Galilean satellites has nearly doubled our number of studied worlds, and has greatly increased the variety and diversity of known planetary phenomena.

SATURN

Saturn is the solar system's second largest planet. Its ring system, once thought to be unique, makes it one of the most impressive of telescopic objects. In addition to its ring system, Saturn has 10 known satellites. Several of them are easily visible with small telescopes. The largest, Titan, is larger than our moon and is known to possess an appreciable atmosphere; methane was identified in its spectrum in 1944. The last satellite to be discovered, and the innermost one, is Janus, discovered in 1966.

Galileo first saw the rings of Saturn, but he was unable to discern them clearly with his crude telescopes. It was not until half a century later, in 1655, that Huygens described their true form. The rings have approximately the appearance of the brim of a straw hat surrounding the planet in its equatorial plane. There are three concentric portions of the rings. The brightest and broadest is the central or *bright* ring. Surrounding this is the *outer* ring with an outside diameter of 275,000 km. The faintest is the inner or *crape* ring, whose inside diameter of about 142,000 km allows only a 13,000 km gap between it and the ball of the planet itself.

In 1979 Pioneer 11 passed very close to the plane of the rings. Care was taken not to send the probe *through* the ring system because an impact of one of the centimeter-sized particles on the spacecraft traveling 100,000 km/hr would be devastating to it. But to everyone's surprise, a new, very narrow ring was found by Pioneer lying beyond the outer known ring. Moreover, there is a hint of still another ring, much further out yet.

The rings are composed of many thousands of millions of minute solid particles, most of which are probably pebble-sized or smaller. However, radar observations of the ring system show that a few of the particles must be over 1 m in diameter. Sunlight reflected from the countless myriads of particles gives the illusion of solid rings.

The rings lie in the equatorial plane of Saturn, whose orientation remains constant during the planet's revolution about the sun. This plane is inclined at about 28° to the ecliptic, so that during part of Saturn's orbital revolution we see one face of the rings, and when the planet is on the opposite side of the sun, we see the other face. At intermediate points, the rings may appear edge on to our line of sight; in that condition they virtually disappear (Figure 10.21), which shows that they are very thin.

Saturn has a mass equal to 95 earth masses, and a diameter 9 times that of the earth. These figures give it an average density (in terms of water) of only 0.7—the lowest of any planet in the solar system. In fact, Saturn would be light enough to float, if an ocean existed large enough to launch it. The internal structure of Saturn is very similar to that of Jupiter. Except for its rings, Saturn looks very much like Jupiter. It, too, has parallel alternately dark and light cloud bands, although the details, color, and irregularities in them are much less distinct than in Jupiter's clouds. Small light spots are occasionally seen on the planet.

The rotation period of the planet, as determined both from the Doppler shift in its spectrum and from the apparent motions of the spots on its disk, is just over 10 hours at the equator. Like Jupiter, however, Saturn rotates more slowly at latitudes away from the equatorial regions. The mean rotation period for most of the planet is near 10^h38^m. Because of its rapid rotation, Saturn is the most oblate of all the planets; its equatorial diameter is about 10 percent greater than that through its poles.

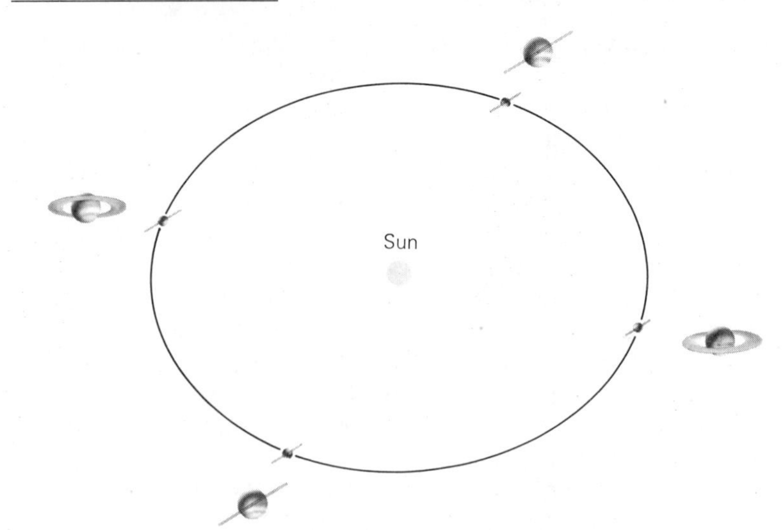

Figure 10.21
Orientation of the rings as seen from the sun with Saturn at different places in its orbit.

Pioneer 11 showed Saturn to have a magnetic field, but one weaker than Jupiter's. More important, though, there are far fewer trapped ions and electrons. Investigators think that those charged particles that do become trapped in Saturn's magnetosphere are absorbed by collision with ring particles.

Measures of the radiant heat from Saturn indicate a temperature of 120 to 130 K, or near −150° C. This

Figure 10.22
Saturn, photographed by Pioneer 11. One of Saturn's satellites, Rhea, is seen as a speck of light below the planet. *(NASA/Ames Research Center)*

is some 20° C higher than might be expected for a planet of Saturn's albedo and distance from the sun. Saturn, like Jupiter, seems to emit more energy than it receives from the sun. The structure of Saturn's atmosphere is thought to be similar to Jupiter's. The gaseous region probably merges into a hydrogen "ocean" a few hundred kilometers below the cloud layers. The ocean may extend a little deeper than in Jupiter, but probably eventually becomes liquid metallic hydrogen.

URANUS

Uranus is a planet that must surely have been seen by the ancients and yet was unknown to them. It was discovered on March 13, 1781, by the German-English astronomer, William Herschel (1738–1822), who was making a routine survey of the sky in the constellation of Gemini. Herschel noted that through his telescope the planet did not appear as a stellar point but seemed to present a small disk. He believed it to be a comet and followed its motion for some weeks. Several months later, a preliminary solution for its orbit was computed and it was found to be a nearly circular one, lying beyond that of Saturn; the object was unquestionably a new planet.

Uranus can be seen by the unaided eye on a dark clear night, but is near enough to the limit of visibility so that it is indistinguishable from a very faint star. It is so inconspicuous that its motion escaped notice until after its telescopic discovery. However, it turned out that Uranus had been plotted as a star on charts of the sky on at least 20 previous occasions since the year 1690. These earlier observations were later of use in the determination of how perturbations were altering the planet's orbit.

Herschel proposed to name the newly discovered planet *Georgium Sidus*, in honor of George III, England's reigning king. Others suggested the name Herschel; the name finally adopted, in keeping with the tradition of naming planets for gods of Greek mythology, was Uranus, father of the Titans, and grandfather of Jupiter.

Uranus has five known satellites; the last to be discovered, and the faintest, was found by Kuiper in 1948. Herschel himself found the two brightest moons of the planet. None of the satellites is probably much over 1000 km in diameter. Their distances range from 123,000 to 586,000 km from the center of Uranus.

Until 1977, only one planet, Saturn, was known to have rings. In that year, however, a small ring system about Uranus was discovered. The Voyager discovery of a ring around Jupiter brings to a total of three the number of ringed planets.

As in the case for Saturn, the particles in the rings of Uranus revolve about that planet in its equatorial plane. The five small rings, however, reflect too little sunlight to be seen directly; they were discovered when Uranus passed in front of a star on March 10, 1977. The rings are not solid and did not occult the star completely, but the star did dim somewhat as each ring portion passed by. The outer diameter of Saturn's bright rings is 275,000 km, 2.31 times that planet's diameter. The outer ring of Uranus has a maximum diameter of 102,000 km, 1.83 times that of Uranus itself.

Uranus, like Jupiter and Saturn, is surrounded by a cloud layer. Spectrographic studies reveal conspicuous methane (it is too cold for ammonia), but hydrogen and helium are the main gases present in its atmosphere. The internal structure of Uranus differs from that of Jupiter and Saturn in that it contains considerable ice, and has a smaller proportion of hydrogen than do the two giant planets.

Uranus appears as a greenish disk when seen through the telescope. The green color is probably due to its atmospheric methane. A few observers report faint markings, but these are too indefinite to indicate the rotation. The rotation period, therefore, must be obtained from the Doppler shift in the spectrum of

Figure 10.23
Uranus and its satellites, photographed with the 3.05-m telescope. *(Lick Observatory)*

light from different parts of its disk. A recent value, somewhat uncertain, is 20^h to 25^h.

A unique feature about Uranus is that its axis of rotation lies almost in the plane of its orbit; during some parts of its revolution, it is so oriented that we look almost directly at one or the other of its poles. The actual inclination of its equatorial plane to that of its orbit is 82°. Its direction of rotation is the same as that of the revolution of its satellites, and both are in the *reverse* direction from the rotation of all the other planets except Venus. Its direction of orbital revolution, however, is normal—that is, from west to east.

NEPTUNE

Whereas the discovery of Uranus was quite unexpected, Neptune was found as the result of mathematical prediction. The discoveries of the two planets could hardly have been made under more different circumstances, yet in other respects Uranus and Neptune are more alike than any two other worlds in the solar system.

By 1790, an orbit had been calculated for Uranus (first by Delambre) on the basis of observations of its motion in the decades following its discovery. Even after allowance was made for the perturbating effects of Jupiter and Saturn, however, it was found that Uranus did not move on an orbit that fitted exactly the earlier observations of it made since 1690, before it was known as a planet. By 1840, the discrepancy between the positions observed for Uranus and those predicted from its computed orbit amounted to about 2′—an angle barely discernible to the unaided eye but still much larger than the probable errors in the orbital calculations. In other words, Uranus did not seem to move on an orbit that would have been predicted from Newtonian theory.

Figure 10.24
Neptune and a satellite, photographed with the 3.05-m telescope. *(Lick Observatory)*

In 1843 John Couch Adams, a young Englishman who had just completed his work at Cambridge, began an analysis of the irregularities in the motion of Uranus to see whether they could be produced by the perturbative action of an unknown planet. His calculations indicated the existence of a planet more distant than Uranus from the sun. In October 1845 Adams sent his results to Sir George Airy, the Astronomer Royal, informing him where in the sky he should look to find the new planet. Adams' predicted position for the unknown body was correct to within 2°.

Meanwhile, Leverrier, a French mathematician, unaware of Adams or his work, attacked the same problem, and published its solution in June 1846. Airy, noting that Leverrier's predicted position for the unknown planet agreed to within 1° with that of Adams', suggested to Challis, director of the Cambridge Observatory, that he begin a search for the new object. The Cambridge astronomer, having no up-to-date star charts of the region of the sky in *Aquarius* where the planet was predicted to be, proceeded by plotting all the faint stars he could observe with his telescope in that location. It was Challis' plan to repeat such plots at intervals of several days, in the hope that the planet would reveal its presence and distinguish itself from a star by its motion. Unfortunately, he was negligent in examining his observations; although he had actually seen the planet, he did not recognize it.

About one month later, Leverrier suggested to Galle, an astronomer at the Berlin Observatory, that he look for the planet. Galle received Leverrier's letter on September 23, 1846, and, possessing new charts of the *Aquarius* region, he found and identified the planet that very night. It was only 52′ from the position Leverrier predicted.

The discovery of the eighth planet, now known as Neptune (named for the god of the sea), was a major

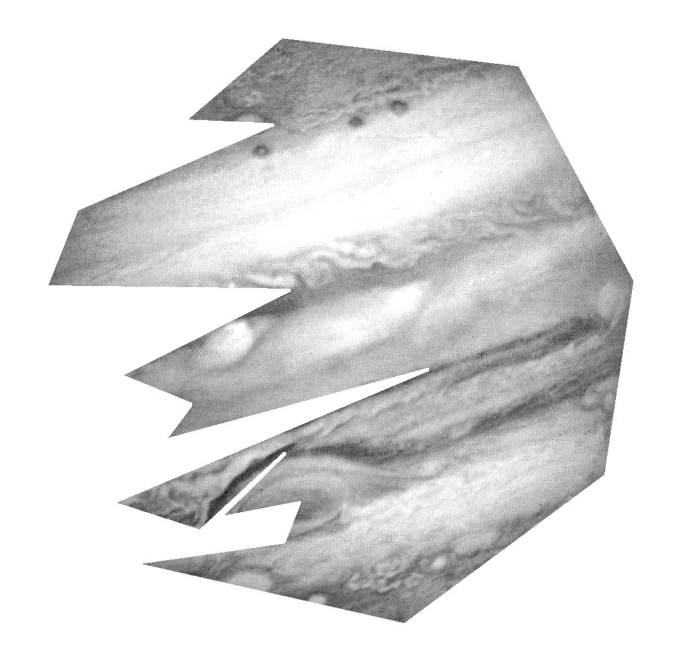

Jupiter, photographed by Voyager 1 at a distance of 33,000,000 km. *(NASA/JPL)*

Closeup of Jupiter, photographed by Voyager 1 at a distance of 5 million km. The famous red spot is at the upper right. Note the turbulence to the left (west). *(NASA/JPL)*

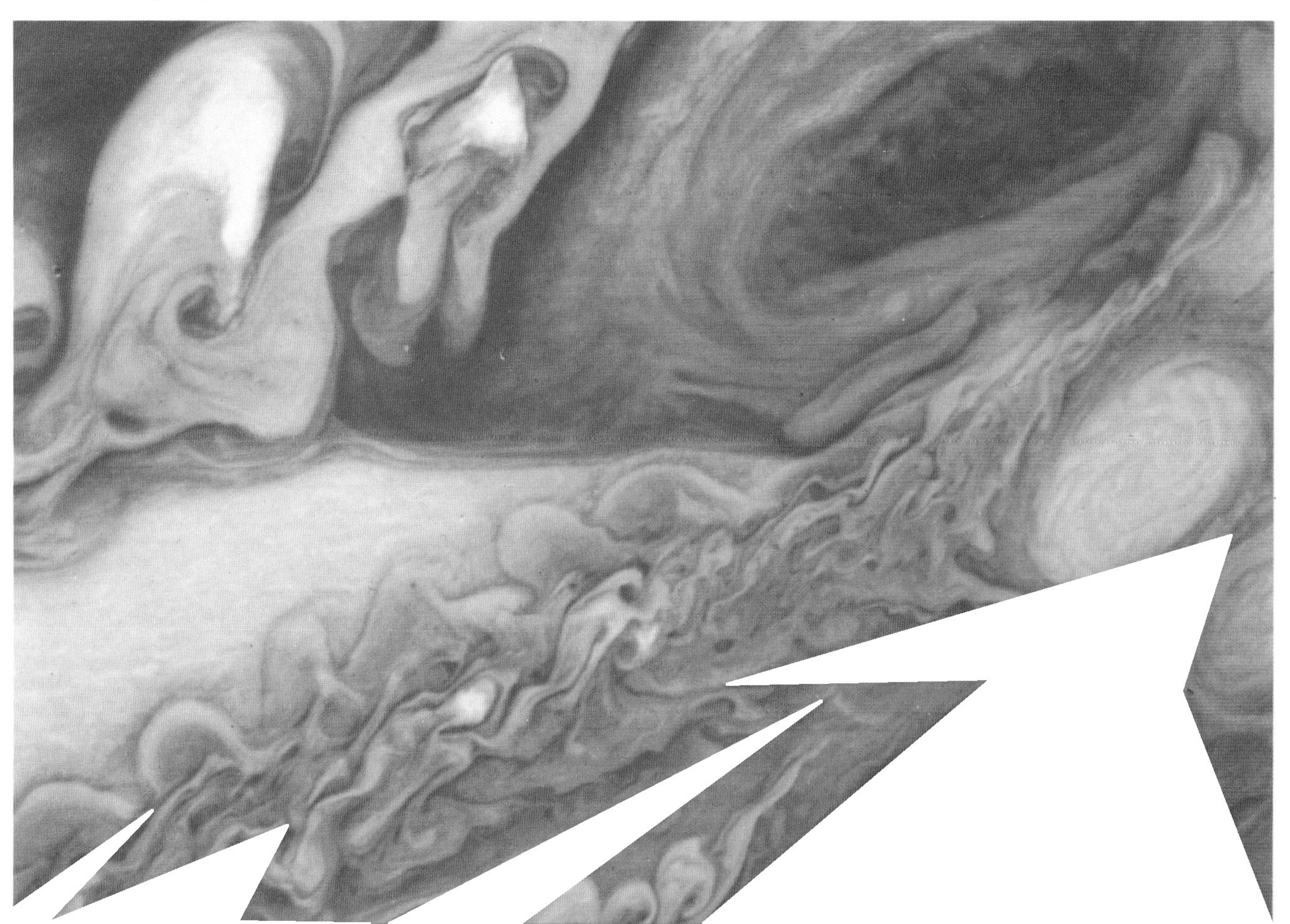

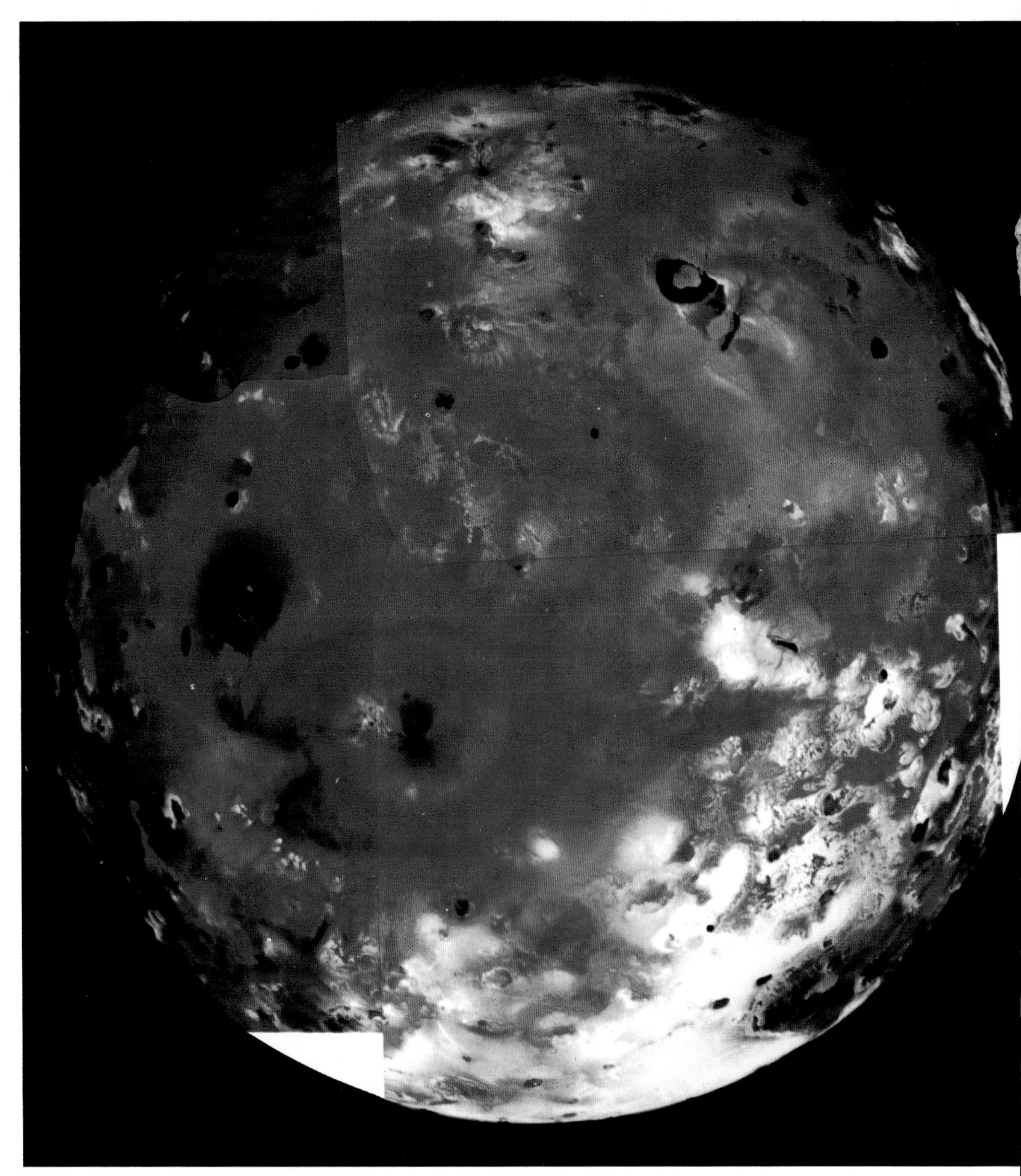

Io, Jupiter's innermost Galilean satellite, photographed by Voyager 1 at a distance of 377,000 km on March 4, 1979. Many of the features are believed to be volcanic features, and the soil is thought to consist of sulfur and various salts. *(NASA/JPL)*

Europa, the second Galilean satellite from Jupiter, photographed March 4, 1979, by Voyager 1. *(NASA/JPL)*

Closeup of Europa, showing some of the many fractures in its icy crust. *(NASA/JPL)*

Jupiter's largest satellite, Ganymede, photographed by Voyager 1 from 2.6 million km. *(NASA/JPL)*

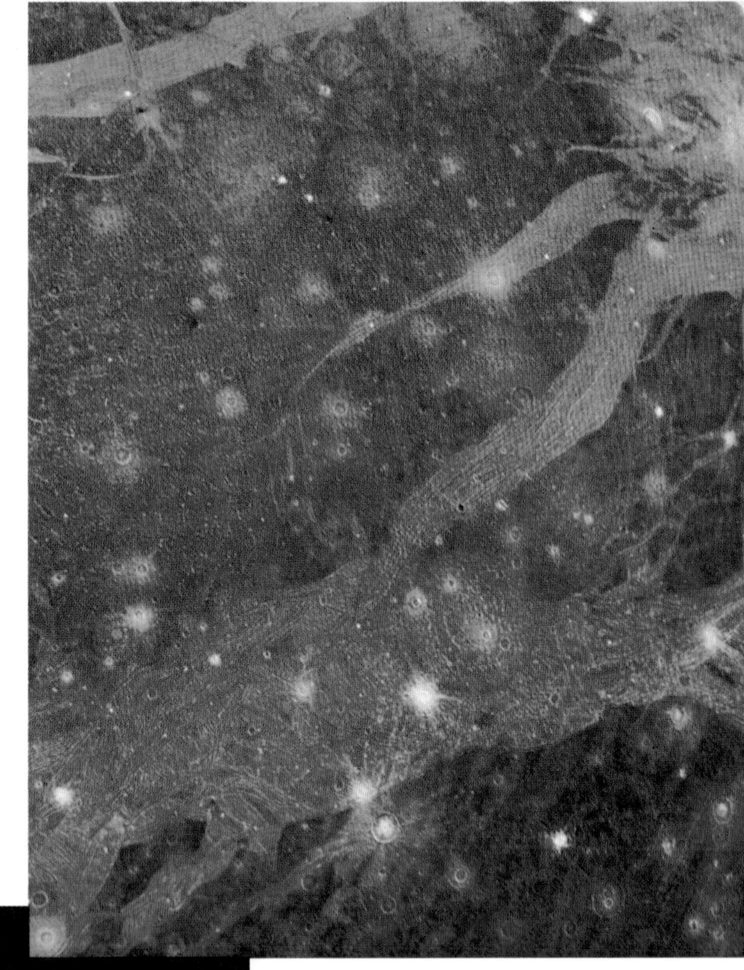

Closeup of Ganymede, photographed by Voyager 1 from a distance of 312,000 km. Ganymede has both craters and grooved terrain, suggesting the existence of faulting. *(NASA/JPL)*

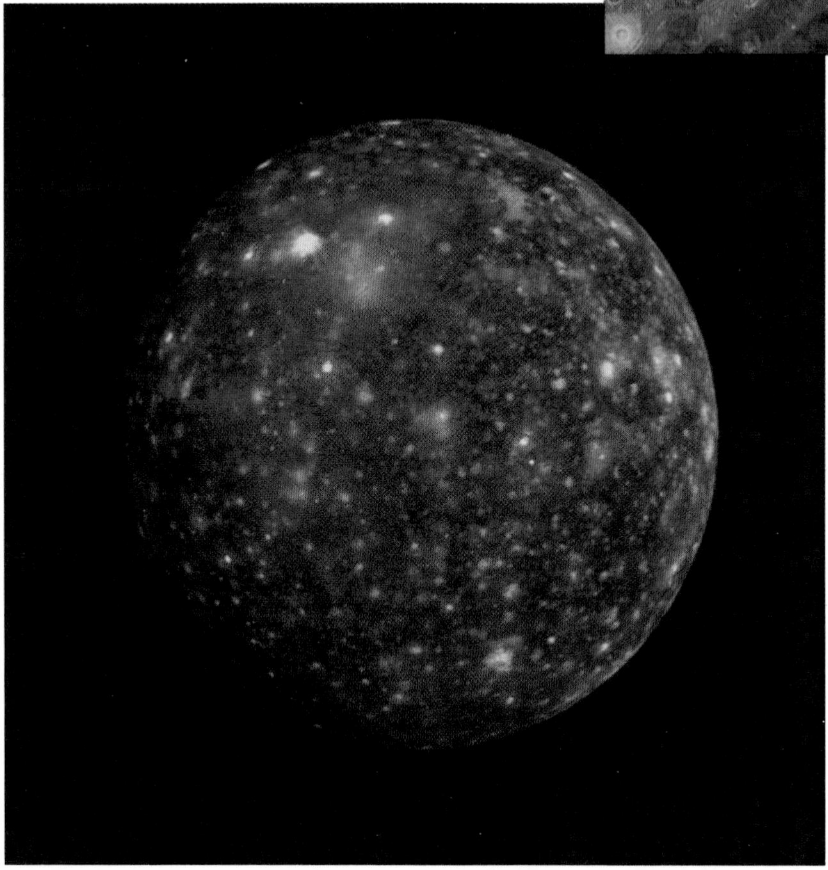

The outermost of Jupiter's Galilean satellites, Callisto, photographed by Voyager 2 on July 7, 1979, at a range of 1 million km. Callisto is the most cratered of the Galilean satellites. *(NASA/JPL)*

triumph for gravitational theory and ranks as one of the great scientific achievements. The honor for the discovery is quite properly shared by the two mathematicians, Adams and Leverrier.

Although a great triumph, the discovery of Neptune was not a complete surprise to astronomers, who had long suspected the existence of the planet. On September 10, 1846, Sir John Herschel, son of the discoverer of Uranus, remarked: "We see it as Columbus saw America from the shores of Spain. Its movements have been felt trembling along the far-reaching line of our analysis with a certainty hardly inferior to ocular demonstration."

Neptune is almost exactly 30 AU from the sun. Traveling with an orbital speed of only 5.4 km/s, it requires 165 years to complete one revolution around the sun. It will not have completed its first revolution since its discovery until 2011. Its axial rotation period, determined from the Doppler shift in its spectrum, is 15 to 20 hours; the value is uncertain.

Neptune has two satellites. *Triton*, the larger, is somewhat greater in diameter than our moon. Its distance from the center of Neptune is only about 353,000 km, and its motion is backward—that is, east to west. Neptune's other satellite, discovered in 1949, is smaller, and has the most eccentric orbit of any satellite in the solar system; its distance from Neptune varies from 2 to 10 million km.

Neptune's mass is 17.2 times that of the earth. Since it subtends an angle of only about 2″, its diameter is difficult to measure, but it is in the range 45,000 to 50,000 km. Its mean density is somewhat more than twice that of water—higher than the density of Uranus. Its internal structure is believed to resemble that of Uranus.

Neptune appears telescopically as a small greenish disk. There are no conspicuous markings in its atmosphere. Both methane and hydrogen have been detected spectrographically, as in Uranus. Helium is probably also present but cannot be observed in the spectrum. Ammonia appears to be absent in the gaseous state, as we would expect, since the temperature of Neptune must be near 60 K.

PLUTO

After the perturbative effects of Neptune on Uranus had been taken into account, the discrepancies between the observed and predicted positions of Uranus were reduced to only 1/60 of what they had been. Today those remaining discrepancies are realized to be smaller than the errors of the original observations and thus are almost certainly not real. Yet several investigators made attempts to account for these remaining deviations by the gravitational influence of a ninth planet, beyond the orbit of Neptune. Among them were Gaillot, W. H. Pickering, and Percival Lowell. It was Lowell's solution to the problem that led to the discovery of Pluto.

By the beginning of the twentieth century, when Lowell made his calculations, Neptune had moved such a short distance in its orbit that a knowledge of the perturbations upon it was not yet available. Therefore Lowell, as did the others, based his calculations entirely on the minute remaining irregularities of the motion of Uranus. His computations indicated two places where a perturbing planet could be, the more likely of the two being in the constellation of *Gemini*. Lowell searched for the unknown planet at his Arizona Observatory from 1906 until his death in 1916, without success. Subsequently, Lowell's brother do-

Figure 10.25
Two photographs of Pluto, showing its motion among the stars in a 24-hour period; photographed with the 5.08-m telescope. *(Hale Observatories)*

nated to the observatory a 33-cm photographic telescope that could record a 12° by 14° area of the sky on a single photograph. The new camera went into operation in 1929, and the search was continued for the ninth planet.

Unfortunately, Gemini lies near the Milky Way, and some 300,000 star images were recorded on each exposure. It was an immense task to compare all the star images on each of two or more photographs of the same field in the hope of finding one image that changed position with respect to the rest, revealing itself as the new planet. The job was facilitated by the invention of the ***blink microscope***, a device in which are placed two different photographs of the same region of the sky. The operator's vision is automatically shifted back and forth between corresponding parts of the two photographs. If the star patterns are the same on the two plates, the observer sees a constant, although flickering picture. However, if one object has moved slightly in the interval between the times the two plates were taken, the image of that object appears to jump back and forth as the view is transferred from one photograph to the other. In this way, moving objects can quickly be picked out from among the many thousands of star images. The blink microscope is also a very useful instrument for locating stars that vary in brightness.

In February 1930, Clyde Tombaugh, comparing photographs made on January 23 and 29 of that year, found an object whose motion appeared to be about right for a planet far beyond the orbit of Neptune. It was within 6° of the position Lowell predicted for the unknown planet; subsequent investigation of the object showed its orbit to have elements very similar to those Lowell had calculated. Announcement of the discovery was made on March 13, 1930. The new planet was named for Pluto, the god of the underworld. (Appropriately, the first two letters of Pluto are the initials of Percival Lowell.)

Pluto's discovery was based on alleged residuals in the positions of Uranus after account was taken of the perturbations that could have been produced on Uranus' motion by Neptune and other known planets. These residuals, only about 4″ to 5″, were in observations of the positions of Uranus made in the early eighteenth century before it had been recognized as a planet. They were, in fact, smaller than the errors expected in observations made that long ago, and have now been generally discredited by the experts as having no significance. Thus Lowell's calculations, although mathematically correct, must have been irrelevant. The discovery of Pluto, in other words, was accidental!

The fact that Pluto *was* discovered, with so nearly the orbital elements that Lowell predicted, is one of the startling coincidences in the history of astronomy. Be that as it may, it was Lowell's faith and enthusiasm that led to our knowledge of Pluto, and he has justly earned the honor for its discovery.

Pluto's median distance from the sun is 39.52 AU, or 5896 million km, but its aphelion distance is over 7000 million km and its perihelion distance under 4500 million km. Part of its orbit is closer to the sun than the orbit of Neptune. There is no danger of collision between the two planets, however; because of its high inclination, Pluto's orbit clears that of Neptune by 385 million km. Pluto completes its orbital revolution in a period of 248.4 years.

Pluto varies slightly in light with a period of 6.3867 days. The variation is interpreted as arising from the rotation of Pluto, alternately turning to our view hemispheres of greater and lesser albedo. Evidently, the surface is not uniform.

In 1978 James W. Christy of the United States Naval Observatory noticed a peculiarity on a routine photograph of Pluto taken with the Observatory's 155-cm astrometric telescope at its Flagstaff Station: the image of Pluto appeared elongated but those of the stars did not. Christy then checked other photographs on file and found that some of them showed the same effect, but that most did not. It occurred to him that Pluto may have a satellite that is almost but not quite resolved as a separate object from Pluto on photographs taken when the two had their maximum separation as seen from earth. Christy followed up on his hunch and found that the image was indeed that of a satellite revolving about Pluto in a period of 6.387 days—the same as Pluto's period of rotation. The satellite has been named Charon.

From the radius of Charon's orbit (about 20,000 km) and its period of revolution, the combined mass of Pluto and Charon can be found from Kepler's third law (as reformulated by Newton—Chapter 3); it turns out to be only about 0.002 times the mass of the earth. From the relative brightnesses of the two, it is concluded that Charon has only about half the diameter of Pluto, and probably about a tenth the mass. If both reflect 50 percent of the incident sunlight on them, we find the diameters of Pluto and Charon to be about 3000 km and 1500 km, respectively. Pluto's small size and mass makes it the smallest planet in the solar system.

Pluto is several thousand times too faint to see with the unaided eye. Even under the best conditions a telescope of aperture 15 cm or more is required to see it, and even then it appears only as a star. No gases are revealed in its spectrum, but at its expected temperature of only about 40 K all common gases except neon, hydrogen, and helium would be frozen, and of these all but neon would probably escape the planet. None of the gases that would remain unfrozen would be easy to observe spectrographically. Probably Pluto has no appreciable atmosphere.

From Pluto, the solar system must appear a bleak and empty place. The sun would appear as a bright star, although it would still provide Pluto with 250 times as much light as we on earth receive from our full moon. The earth as seen from Pluto would scarcely ever be at more than 1° angular separation from the sun and would be slightly too faint to be seen with the unaided eye. Even Jupiter could only be 7° from the sun and would appear only as a medium-bright star.

ARE THERE UNKNOWN PLANETS?

The possibility of undiscovered planets exists. However, a careful search by Tombaugh after his discovery of Pluto failed to reveal any other trans-Neptunian planet. He should have picked up any object as large as Neptune within a distance of 270 AU.

SUMMARY REVIEW

Planetary space probes

Mercury: appearance in the sky; physical characteristics; phases; surface; Mariner 10 observations

Venus: high temperature; atmospheric structure; clouds; composition of the atmosphere; greenhouse effect; pressure of the atmosphere; surface conditions; topography

Mars: seasons; atmosphere; temperatures; "canals"; polar caps; vulcanism; topography; old rivers; search for life; Viking experiments

Jupiter: physical characteristics; structure; surface markings; Great Red Spot; ring; magnetic field and magnetosphere; satellites; Io and its volcanos

Saturn: rings; satellites; structure; physical properties; temperature

Uranus: discovery; physical nature; rings and satellites; inclination of its axis

Neptune: discovery; satellites; physical properties

Pluto: discovery; physical properties; discovery of satellite; the solar system from Pluto

Search for unknown planets

EXERCISES

1. Explain why Mercury is visible in the west after sunset when it is at eastern elongation, and in the east before sunrise when it is at western elongation. Draw a diagram.

2. Give several reasons why Mercury would be a particularly unpleasant place to live.

3. Compare Mercury, Mars, and the moon. What do these worlds have in common? How do they differ? Explain why in each case.

4. Venus requires 440 days to move from greatest western to greatest eastern elongation but only 144 to move from greatest eastern to greatest western elongation. Explain why. A diagram will help.

5. At its nearest, Venus comes within about 40 million km of the earth. How distant is it at its farthest?

6. Why isn't Venus always *exactly behind* the sun at superior conjunction?

7. What problems would be encountered by scientists planning a station on the surface of Venus?

8. When might it be possible to observe both Mercury and Venus, at the same time, when they are at inferior conjunction? Why would this be an extraordinarily rare event?

9. Which satellite would have the greatest period, one 1 million km from the center of Jupiter or one 1 million km from the center of earth? Why?

10. Describe at least three respects in which Mars resembles the earth, and three others in which it differs from the earth.

11. Compare and contrast the seasons on earth, Mars, and Uranus.

12. How often are Saturn's rings turned edge-on to us? Assume, for this exercise, that Saturn's orbit is in the plane of the ecliptic. What is the relevance of this assumption?

13. In the solar system there are three pairs of planets in which the members of each pair have certain rather similar characteristics. Identify these three pairs, and discuss the similarities of the members of each.

14. Compare and contrast the discoveries of Uranus, Neptune, and Pluto.

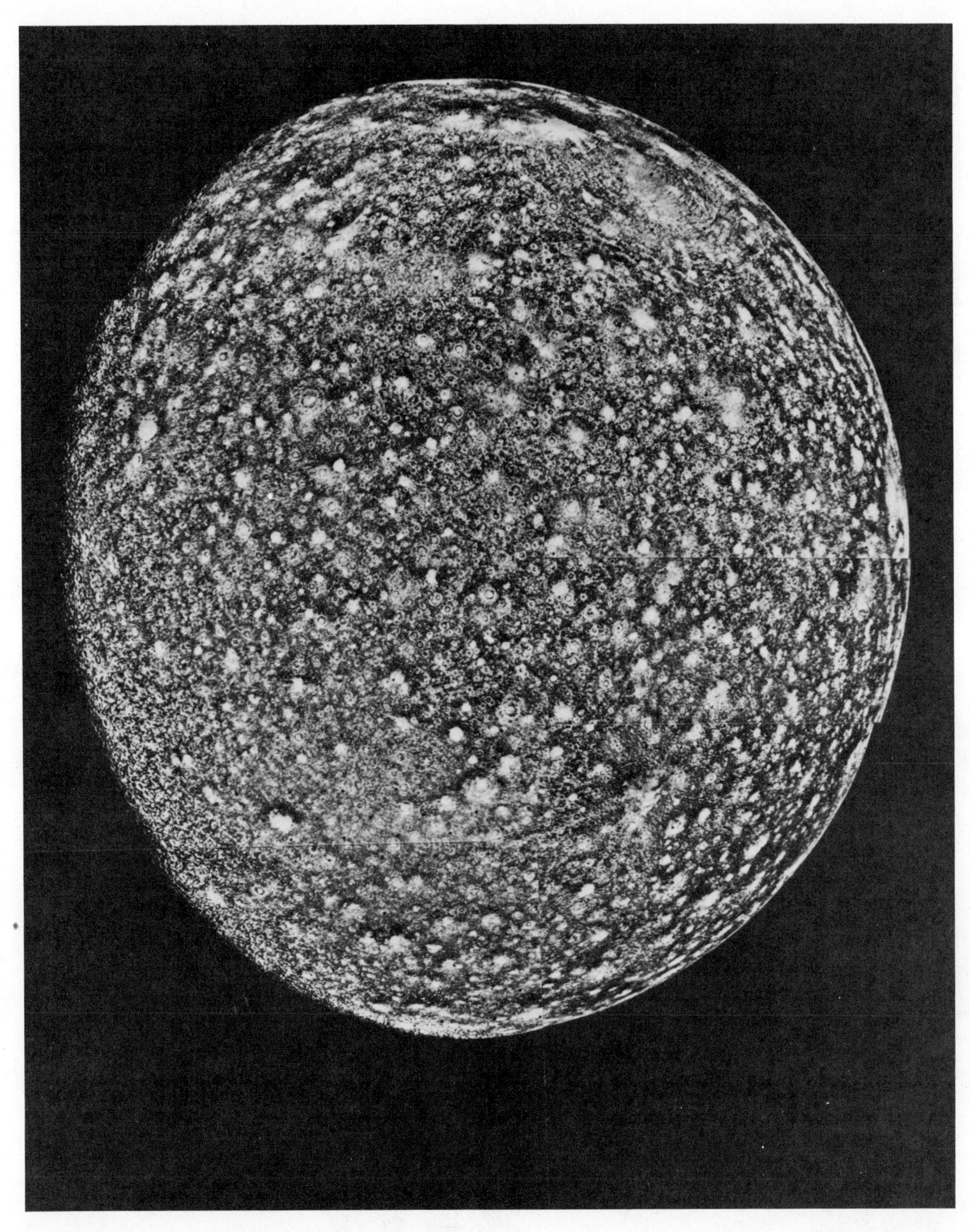

Photomosaic of Jupiter's heavily cratered satellite Callisto, taken by Voyager 2 on July 8, 1979. *(NASA/JPL)*

CHAPTER 11

Karl Friedrich Gauss (1777–1855), prince of mathematics, physicist, and astronomer, developed the mathematics that later led to the geometry Einstein needed to describe curved spacetime. Among his many other accomplishments, he invented a method for calculating the orbit of an object moving about the sun from three observations of that object, spaced over only a few weeks. *(Yerkes Observatory)*

DEBRIS OF THE SOLAR SYSTEM

The sun, the nine major planets, and their satellites make up practically all of the mass of the solar system. By number, however, they scarcely count among the billions of far smaller objects in orbit about the sun. We now examine briefly these minor members of our solar system.

THE MINOR PLANETS

Of the many thousands of minor planets, the vast majority appear telescopically like stars—that is, as small points of light. The term *asteroid* (meaning "star-like") is therefore often applied to these tiny worlds. Another synonym is *planetoid*. The term *minor planet*, preferred by the International Astronomical Union, is used in this text.

The orbits of most of the minor planets, including that of Ceres, the largest and first to be discovered, lie between the orbits of Mars and Jupiter. From the time of Kepler it was recognized that this region of the solar system constituted a considerable gap in the spacing of the planetary orbits; consequently, when Ceres was found there in 1801 it occasioned no great surprise.

Moreover, the median distance of Ceres from the sun fitted well into a sequence of numbers, discovered by Titius of Wittenberg in 1766, that gives the approximate distances of the planets from the sun in astronomical units. The progression is generally known as *Bode's law*, after J. E. Bode, director of the Berlin Observatory, who published it in 1772. Titius' progression is obtained by writing down the numbers: 0, 3, 6, 12, . . . , each succeeding number in the sequence (after the first two) being obtained by doubling the preceding one. If 4 is now added to each of the numbers, and in each case the sum is divided by 10, the numbers obtained give the approximate distances from the sun (in astronomical units) of the planets known in 1766. Bode's law predicts a planet with a semimajor axis of 2.8 AU, where no major planet exists. Thus a search for the "missing planet" was organized.

The discovery of the minor planets was independent of this search, but they fell nicely into the gap at 2.8 AU; the orbit of Ceres has a semimajor axis of 2.77 AU. Today, Bode's law is regarded as a useful scheme for remembering the distances of some of the planets from the sun but it is not a precise law. On the other hand, it does call attention to the rather orderly spacing of the planetary orbits, which, in turn, gives theo-

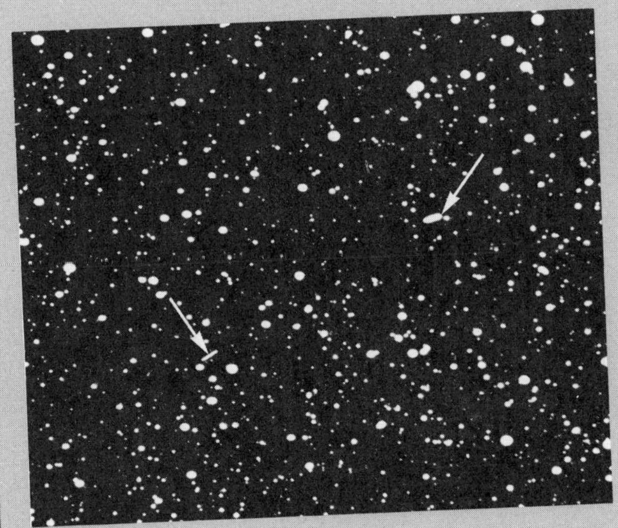

Figure 11.1
Time exposure showing trails left by two minor planets (marked by arrows). *(Yerkes Observatory)*

reticians some limits about the conditions under which the solar system formed.

Discovery of the Minor Planets

The Sicilian astronomer Giuseppe Piazzi (1746–1826) was engaged in mapping a region of the sky in Taurus, when on January 1, 1801 (the first night of the nineteenth century) he observed an uncharted "star." During the next two nights, he noted that the new object had shifted its position slightly. Piazzi continued to observe the object until February 11, after which he was forced to interrupt his work because of illness.

In mid-January Piazzi wrote of his discovery to Bode, at the Berlin Observatory. Unfortunately, his letter did not reach Bode until March 20, when the object was too nearly in the direction of the sun to be observed. However, the brilliant young German mathematician, Karl Friedrich Gauss (1777–1855), had devised new mathematical techniques that could be applied to orbit calculations. By November, Gauss had successfully calculated the orbit of the object from Piazzi's short series of observations. The object was found very near the position Gauss predicted in the constellation of Virgo by von Zach, on New Year's Eve, the last night of the year of its first discovery. At Piazzi's request, the new planet was named for *Ceres*, goddess of agriculture and the protecting goddess of Sicily.

Ceres was widely assumed to be the missing planet predicted by "Bode's law." It came as a complete surprise, therefore, when in March 1802 Heinrich Olbers discovered a second moving starlike object—the minor planet to be named *Pallas*. It was a natural speculation that if there were room for *two* minor planets, there could be room for others as well, and a search for such objects began in earnest. The discovery of *Juno* followed in 1804 and of *Vesta* in 1807. It was 1845 before Karl Hencke discovered the fifth minor planet after 15 years of search. Subsequently, new ones were found with increasing frequency, until by 1890 more than 300 were known.

Today, minor planets are frequently recorded on telescopic photographs. The angular motion of a minor planet is large enough (especially if it is near opposition) so that during a long time exposure its image forms a trail on the emulsion. The object appears on the photograph, therefore, as a short dash rather than a starlike point image. Special photographic searches for minor planets have recorded images of thousands of them. The total number bright enough to detect this way is estimated to be 100,000.

Orbits of the Minor Planets

A standard reference in which the orbital elements and ephemerides of minor planets are published is the Soviet *Minor Planet Ephemerides*. The 1979 edition gives data for 2042 minor planets.

The minor planets all revolve about the sun in the same direction as the principal planets (from west to east), and most of them have orbits that lie near the plane of the earth's orbit, and have semimajor axes that lie in the range 2.3 to 3.3 AU. The corresponding sidereal periods are from 3.5 to 6 years. Icarus has the smallest orbit—with a semimajor axis equal to 1.0777 AU. It also has the orbit of highest eccentricity among the known minor planets (0.83), and a perihelion within the orbit of Mercury. Icarus belongs to a small family called the *Apollo* minor planets that come closer to the sun than the earth. The Apollos, however, are not the only minor planets that are not in the minor planet "belt" between Mars and Jupiter.

Chiron

On November 1, 1977, Charles Kowal of the Hale Observatories discovered an object on photographs he had taken during the previous month with the 48-inch Schmidt telescope at Palomar. Kowal's new object, which he has named *Chiron*, has a period of about 50 years, and moves on an orbit that takes it from 8.5 to 20 AU's from the sun, and thus spends most of its time beyond the orbit of Saturn. If Chiron is considered to be a minor planet, it is by far the most distant one known. On the other hand, it may have had an origin very different from that of the typical minor planets and be a completely unrelated object. It has been suggested that it could be a comet (next section), but its diameter of 300 km is enormous compared to the nuclei of typical comets. In any case, it is a minor planet-sized object out among the Jovian planets. Perhaps there are many such worlds awaiting discovery.

Physical Nature of the Minor Planets

When the minor planet Vesta is at opposition while it is also at perihelion, it is faintly visible to the unaided eye. With this single exception, the minor planets are visible only telescopically. A few of the brightest, and presumably largest, show measurable disks, but these can be measured only with considerable difficulty. We can, however, calculate the size of any minor planet of known distance from the amount of sunlight it reflects if we guess its *albedo* (reflecting power). If a minor planet can be observed in both visible light and infrared radiation, the guesswork is removed, because the infrared radiation it emits is, on the average, equal to the energy from the sun that it *absorbs*, while its brightness in visible light is equal to the sunlight it *reflects*. Comparison of its reemitted energy (in the infrared) and its brightness in reflected sunlight thus tells us what its albedo has to be (see Exercise 4).

Approximate sizes for some of the larger minor planets are given in Table 11.1. There are probably only a few dozen minor planets with diameters greater than 160 km. There may be hundreds that are more than 40 km across. Most observable minor planets must have diameters not much greater than 1 km. Some minor planets have elongated irregular shapes, so that in different orientations they reflect different amounts of light to us. We can sometimes detect rotation periods of these objects from the periodic variations of brightness they exhibit.

The mass of Ceres, the largest minor planet, is probably only about 1/8000 that of the earth. The velocity of escape from its surface must be only about ½ km/s. Even Ceres, therefore, could not retain an atmosphere. From the smaller minor planets, a good pitcher could easily pitch a baseball into space. These small ones cannot even have enough gravity to pull themselves into a spherical shape. The combined mass of all minor planets is not known, but a guess is 1/20 that of the moon.

Satellites of Minor Planets

Occasionally minor planets occult (pass in front of) distant stars. The times of these occultations are calculated and announced in advance by the United States Naval Observatory. An active group of amateur astronomers cooperates with professionals by observing such events. Thus on June 7, 1978, J. H. McMahon observed an occultation of the star SAO 120774 by the minor planet 532 Herculina. The star disappeared, as expected, for 20.6 s while the 220-km-diameter Herculina passed in front of it. But in addition, McMahon observed the star to dim six other times within 2 min

TABLE 11.1 Sizes of Minor Planets*

Number	Name	Diameter (km)
1	Ceres	1025
2	Pallas	583
4	Vesta	555
10	Hygeia	443
704	Interamnia	338
511	Davida	335
65	Cybele	311

* Courtesy T. Gehrels, University of Arizona.

of the main occultation, for periods that ranged from 0.5 to 4 s.

The longest secondary dimming was confirmed by photoelectric records obtained by astronomers at the Lowell Observatory in Arizona. It is interpreted as due to a 50-km-diameter satellite at a distance of about 1000 km from 532 Herculina.

Calculations by R. P. Binzel and T. C. van Flandern show that satellites can have stable orbits up to a distance of 15,000 km from the centers of the larger minor planets, and those astronomers suggest that satellites of minor planets might be quite common. They have therefore checked through old records for indications of secondary dimmings associated with other occultations that are similar to the dimmings observed by McMahon, and have compiled a list of seven additional minor planets with a dozen probable satellites among them, and 10 other suggested satellites. It is possible that many minor planets have one or more satellites in revolution about them.

Naming Minor Planets

By modern custom, after a newly found minor planet has had its orbit calculated, and has been observed again after another circuit of the sun since its first discovery, it is given a name and a number. The number is a running index that indicates the order of discovery among the minor planets. The discoverer is customarily given the honor of supplying the name. The full designation of the minor planet contains both number and name, with the number preceding the name, thus: 1 Ceres, 2 Pallas, 433 Eros, 1566 Icarus, and so on.

Originally, the names were chosen from gods in Greek and Roman mythology. These, however, were soon used up. After exhausting the names of heroines from Wagnerian operas, discoverers chose names of wives, friends, flowers, cities, colleges, pets, and even favorite desserts. One is named for a computer (NORC). The thousandth minor planet to be discovered was named Piazzia, and number 1001 is Gaussia. Other minor planets bear the names Washingtonia, Hooveria, and Rockefellia.

Origin of the Minor Planets

Most authorities regard it as probable that the minor planets were formed from the same material that formed the principal planets, and at about the same time. Most likely they formed from material that could not form a major planet, perhaps because too little material was there to begin with, or perhaps also because the disruptive tidal effects of Jupiter prevented such an object from forming.

On the other hand, there is some evidence that many of the minor planets may have originated from the breakup of several somewhat larger bodies. A number of the minor planets fall into "families" or groups of similar orbital characteristics. It is hypothesized that each family may have resulted from the collision of two bodies. Slight differences in the initial velocities given the fragments of the collision would have resulted in the relatively small differences now observed among the orbits of the different minor planets in a given family.

Collisions among a relatively few larger objects would be extremely rare. If they once occurred, however, subsequent collisions would be more likely among the larger number of smaller bodies that resulted. If this collision and fragmentation hypothesis for the origin of the minor planets should be shown to be correct, it would be inferred that the fragmentation process is still going on.

If minor planets are colliding with each other, and if some come close to the earth, we might wonder about the possibility of one colliding with us. The earth may have been struck many times, but erosion would have erased most of the evidence. From the number of closely approaching minor planets we might expect collisions at the rate of one every million years or so. A minor planet of 1-km diameter striking the earth with the minimum possible speed would still hit with the energy of 20,000 megaton hydrogen bombs, and an actual collision could involve much greater energy. Such an object striking a populated area would, needless to say, be catastrophic.

COMETS

Comets have been observed from the earliest times. Accounts of spectacular comets are found in the histories of virtually all ancient civilizations. Yet, until comparatively recently, comets were not generally regarded as celestial objects.

A typical comet that is bright enough to be conspicuous to the unaided eye has the appearance of a rather faint, diffuse spot of light, somewhat smaller than the full moon and many times less brilliant. There may be a very faint nebulous tail, extending for a length of several degrees away from the main body of the comet. Like the moon and planets, comets slowly shift their positions in the sky from night to night, remaining visible for periods that range from a few days to a few months. Unlike the planets, however, most comets appear at unpredictable times. In medieval Europe, comets were usually regarded as poisonous vapors in the earth's atmosphere and as bad omens. More fear and superstition have been attached to comets than to any other astronomical objects. Even in 1973 there were dire predictions of disaster when Comet Kohoutek passed near the sun, and numerous books about comets written by cranks and charlatans were on sale in bookstores.

Early Investigations

When Newton applied his law of gravitation to the motions of the planets, he wondered whether comets might similarly be gravitationally accelerated by the sun. If so, their orbits should be conic sections. If comets, like planets, had nearly circular orbits, they should be visible at regular and frequent intervals. On the other hand, if a comet moved in an elongated elliptical orbit of large size, it would necessarily be visible during only that relatively brief period of time during which it passed near the sun (perihelion); over most of its orbit the comet would be so far from the sun as to be invisible. Furthermore, the periods of comets, moving in such orbits, would be very great. A comet that had been seen for a short time, and then was seen again many tens or hundreds of years later when it next appeared near perihelion, would quite naturally be mistaken for a new object.

Edmond Halley greatly extended Newton's studies of the motions of comets. In 1705 he published calculations relating to 24 cometary orbits. In particular, he noted that the elements of the orbits of the bright comets of 1531, 1607, and 1682 were so similar that the three could well be the same comet, returning to perihelion at average intervals of 76 years. If so, he predicted that the object should return about 1758.

The comet was first sighted by an amateur astronomer, George Palitzsch, on Christmas night, 1758. It has been named *Halley's comet,* in honor of the man who first recognized it to be a permanent member of the solar system. Subsequent investigation has shown that Halley's comet has been observed and recorded on every passage near the sun at intervals from 74 to 79 years since 239 B.C. The period varies somewhat because of perturbations upon its orbit produced by the Jovian planets. It last appeared in 1910, and is due again about the spring of 1986.

Discovery

Today, 5 to 10 new comets are usually discovered each year. Some of these are found accidentally on astronomical photographs taken for other purposes. Many are discovered by amateur astronomers.

Most of the new comets found each year never become conspicuous, and are visible only on photographs made with large telescopes. On an average of every few years, however, a comet appears that is bright enough to be seen easily with the unaided eye. About two or three times each century, there appear spectacular comets that reach naked-eye visibility even in daylight. The first "daylight comet" to appear since the brilliant comet of 1910 (that preceded the more famous Halley's comet by a few months), was Comet Ikeya-Seki in 1965.

One of the best studied comets was discovered by the Czech astronomer, Lubos Kohoutek, in 1973. The great importance of Comet Kohoutek is that it not only passed very close to the sun (about 0.14 AU) on December 28, but also that it was discovered when it was still 5.2 AU from the sun, more than nine months before its perihelion passage, which enabled astronomers to plan detailed observing programs. It was studied from ground-based observatories all over the world, from an airplane, from rockets, from NASA's Skylab, and even with instruments on the Venus-Mercury probe, Mariner 10. Moreover, it was the first comet for which radio observations were obtained. Comet Kohoutek was barely visible to the unaided eye in late December and early January, 1974, and it did not become as bright as some early predictions had led many to expect. Its visual faintness, however, in no way detracted from its astronomical importance.

Orbits of Comets

The orbit of a comet, like that of a planet, can be determined from three or more fairly well spaced observations of its position in the sky among the stars. Most comet orbits are nearly indistinguishable from parabolas. However, a parabolic orbit would be possible only for a body that, before being attracted toward the sun, was moving through space in almost exactly the same direction and at the same speed as the sun, that is, which was *at rest with respect to the sun.* Such a body would have to have been associated with the sun before approaching it. Comets certainly cannot be interstellar interlopers; otherwise their orbits would be hyperbolas.

It is generally accepted, therefore, that comets are members of the solar system, and that they approach the sun on elliptical orbits, most of which are extremely long and have eccentricities near unity. A comet can be observed from the earth only when it traverses that end of its orbit where it is near perihelion. If the orbit has an eccentricity greater than about 0.99, that part of it which can be observed is often indistinguishable from a parabola. Thus "parabolic" orbits observed for comets are really the ends of very long elliptical orbits that bring comets from the farthest reaches of the solar system, tens of thousands of astronomical units away. Such comets have periods that range up to millions of years.

Calculations show that observed comets with such highly elongated orbits must be approaching the sun for the first time. To have an orbit so nearly parabolic, a comet must, at each point in its orbit, travel almost exactly with its velocity of escape from the solar system. If, however, it comes within 5 or 10 AU from the sun, the perturbative effects of Jupiter are sufficient to change its speed enough that it subsequently speeds up slightly and escapes the solar system on a hyperbolic orbit, or slows down and switches to a much smaller orbit.

Indeed, some comets—a few hundred—have orbits of low enough eccentricities to determine their semimajor axes from the relatively small portions of the orbits that are actually observed. These comets, whose periods can be well determined, are called *periodic comets.* The comets of longest definitely established periods are Pons-Brooks (71 years), Halley (76 years), and Rigollet (151 years). The comet of the shortest known period is Comet Encke (3.3 years).

Comets are very flimsy objects and they often suffer severe damage as the result of tidal forces produced

Figure 11.2
Halley's comet. *(Yerkes Observatory)*

by the sun when they pass perihelion. Some have been observed to split into two separate comets. Comet Biela, for example, with a period of seven years, was discovered in 1772. During its approach to the sun in 1846, it was observed to break into two separate comets. Both comets returned on about the same orbit in 1852, but neither has been seen since. However, spectacular meteor showers (see below) were observed on November 27, 1872, and on November 27, 1885, on which occasions the earth passed through the orbit of Comet Biela and encountered swarms of particles traveling in the path of the disintegrated comet. The orbit of the swarm has subsequently been altered by Jupiter; consequently, spectacular displays of Bielid meteors have not been observed in the present century. The nuclei of Comets Ikeya-Seki (in 1965) and West (in 1975) also broke up into several pieces within a few days of passing perihelion.

We shall see that the gases in comets result from solar heat that evaporates their nuclei. It is estimated that most comets would completely evaporate in this manner after a few hundred or at most a few thousand perihelion passages. Thus, because of their high mortality rates, periodic comets cannot have been periodic for long but must originally have been comets whose orbits did not bring them near the sun. Within the recent past (perhaps the last few thousand years) they must have approached the sun for the first time and have had their orbits altered to their present relatively small size by perturbations produced by the planets, especially Jupiter. Those occasional comets that are highly spectacular, and hence cannot have suffered appreciable disintegration, usually have nearly parabolic orbits and are on their first approach to the sun.

Physical Nature of Comets

All comets have one characteristic in common—the *coma,* which appears as a round, diffuse, nebulous glow. As a comet approaches the sun, the coma usually grows in size and brightness. Typically it reaches a maximum size as large as or larger than the planet Jupiter. Often, but not always, the small *nucleus* is visible in the middle of the coma. Together, the coma and nucleus constitute the *head* of the comet. Many comets, as they approach the sun, develop *tails* of luminous material that extend for millions of kilometers away from the head.

Far from the sun, where a typical comet spends practically all of its time, it is very cold and all of its material is frozen into the nucleus. Fred Whipple (1906–) describes it as a "dirty iceberg." Half or less of the nucleus is dusty, stony, or metallic material, and the remainder is substance that vaporizes at terrestrial temperatures. If a comet approaches within a few astronomical units of the sun the surface of the nucleus is warmed and begins to evaporate. The evaporated molecules, carrying many of the small solid particles along with them, then begin to produce a coma of gas and dust. Far from the sun, the nucleus can be seen from earth only by the sunlight it reflects. When a coma develops, however, the dust reflects still more sunlight, and the gas in the coma, absorbing ultraviolet solar radiation, begins to fluoresce. If the comet comes close enough to the sun (less than 5 AU), the light that the gas atoms reemit (by fluorescence) is usually more intense than the sunlight the solid matter reflects.

From the way they reflect sunlight, we find that the nuclei of comets (those dirty icebergs) turn out to be only 1 to 10 km in diameter. Cometary nuclei are expected to have densities about like that of water, and certainly less than 2 g/cm^3. Comet masses, therefore, are only from 10^{-10} to 10^{-12} that of the earth; a value near 10^{-11} earth masses is probably typical for bright comets.

Figure 11.3
Comet Kohoutek in a mosaic of two photographs made with the Palomar Schmidt telescope on January 15, 1974. *(Hale Observatories)*

The gas atoms in the coma have speeds of up to about 1 km/s. At these speeds they easily escape the nucleus, which has only a feeble gravitational field. The coma thus expands to an enormous size as the atoms disperse into space, typically to a diameter of 100,000 km or more. The head of a comet is, on the average, a high vacuum by laboratory standards. In 1910, stars were observed at full brilliance through some 70,000 km of the head of Halley's comet, yet starlight is appreciably dimmed by only a few kilometers of the earth's atmosphere.

The spectra of the light from cometary comas show that they contain the common gases water (H_2O), methane (CH_4), and ammonia (NH_3). Solar ultraviolet radiation dissociates molecules of these compounds into carbon (C_2), cyanogen (CN), hydroxyl (OH), NH, and NH_2, and bright emission lines of these radicals account for most of the light from bright comets. A great deal of hydrogen is released as well. It was first detected around Comet Bennett in 1970 with the Orbiting Astronomical Observatory (OAO). Mariner 10 instruments detected radiation from hydrogen from a region around Comet Kohoutek at least 40 million km across. Space observations of Comet Kohoutek also showed that water is dissociating into hydrogen and the OH radical within the inner 15,000 km around the nucleus, and that OH is further dissociating into hydrogen and oxygen beyond 45,000 km from the nucleus.

Kohoutek was the first comet to be observed at radio wavelengths. In the microwave region radio spectral lines of the molecules hydrogen cyanide (HCN) and methyl cyanide (CH_3CN) were observed. This observation is particularly significant, because these are molecules also observed in interstellar space (Chapter 15), but not in the solar system. These data give support to the hypothesis that comets were not formed when the planets formed from the solar nebula, but rather from fragments of the interstellar cloud before it contracted down to the planetary disk.

Every time a comet passes by the sun and forms a coma, it loses part of its material. Calculations based on the brightness of emission lines in the coma and on the radiation from the surrounding hydrogen indicate that something like 10^{28} to 10^{29} molecules boil off the surface of the nucleus each second. At this rate it works out that a typical comet loses the outer few meters of its nucleus each perihelion passage. After at most some 1000 perihelions, a periodic comet uses itself up. This is why periodic comets must be relative newcomers to the inner solar system.

How much light is emitted by the coma depends on how many atoms it contains and of what kind (which in turn depends on how rapidly they evaporate

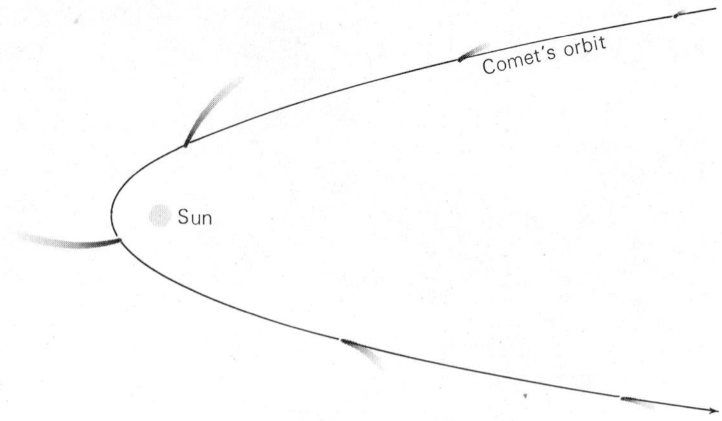

Comet's orbit

Sun

Figure 11.4
Shape of a typical comet tail as the comet passes perihelion.

from the nucleus) and how much ultraviolet solar radiation it absorbs. As comets approach the sun and heat up, not only do the nuclei evaporate faster but more solar radiation is absorbed and reemitted by the gas; consequently comets sometimes brighten many times more rapidly than if they were merely reflecting sunlight. Comets vary among each other, however, in size, in content of dust, and probably in chemical composition as well. There are so many variables that it is not possible to know in advance exactly how a comet will brighten, and predictions are risky. Many people, for example, were disappointed that Comet Kohoutek brightened more slowly near the sun than a typical comet of its size, and did not become the spectacle it was expected to be.

The steady flow of cometary material from the nucleus is picked up by solar repulsive forces and driven radially away from the sun, producing the comet's tail. The orbit followed by the tail material depends on the relative magnitudes of the forces acting on it, but in any case is different from that of the parent comet; thus this material is lost permanently to the comet and, usually, to the solar system.

The repulsive forces from the sun that produce comet tails are of two kinds: the pressure of radiation pushing on small dust particles and some atoms and molecules, and the force of corpuscular radiation comprising the *solar wind*—a continuous rain of atomic nuclei ejected from the sun into space. The solar wind, acting on charged atoms (ions) in the comet, is the greater force, and produces the straightest comet tails.

Tails generally grow in size as comets near the sun; some comet tails have reached lengths of more than 150 million km. Once the tail material has left the vicinity of the comet's head, the only forces acting upon it are radial forces toward and away from the sun (gravity and the force of repulsion). As the tail material recedes farther from the sun, it slows down, in accordance with Kepler's third law, and lags behind the head of the comet. Thus, in general, comet tails lie in the plane of the comet's orbit, pointing more or less away from the sun, but curving somewhat backward, away from the direction of the comet's motion (Figure 11.4). Most comets have two tails: a nearly straight one of ions driven by the solar wind, and a more diffuse tail of dust particles driven by solar radiation pressure. The acceleration produced by the force of

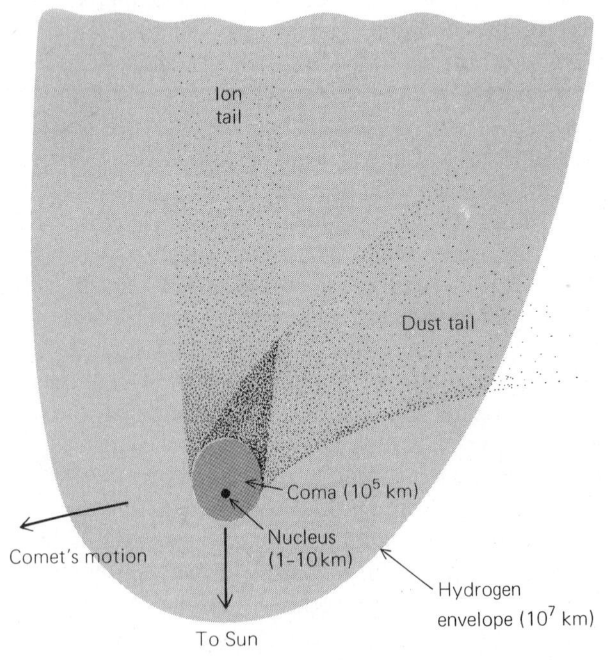

Figure 11.5
Parts of a typical comet.

repulsion from the sun varies according to the size of the particles it acts on and according to whether the particles are gas molecules or solid grains. Material of different types, therefore, is accelerated away from the sun by different amounts. If the repulsive force exceeds the gravitational force on a certain kind of particle by 20 or 30 times, particles of that type are driven outward from the sun so rapidly that they form a tail that is nearly straight. On the other hand, if the repulsive force is only 2 or 3 times the gravitational force, the particles move outward more slowly; the difference between the orbital motions of the comet and tail material becomes apparent before the latter has receded very far from the comet's head, and the tail appears noticeably curved.

Origin of Comets

The most widely accepted hypothesis for the origin of comets is that of the Dutch astronomer, Jan H. Oort (1900–). He postulates a huge cloud of comets revolving about the sun with orbits whose aphelia are 50,000 AU or more from the sun. Their orbits cannot have originally brought them close to the sun, or, as we have already explained, Jupiter would have transformed them to short-period comets, or have ejected them from the solar system.

On the other hand, the outer fringe of the cloud of comets cannot extend beyond about 60 percent of the distance of the nearest star (Alpha Centauri) because beyond that point their orbits would be unstable. Even so, perturbations produced by those stars that pass, every few thousand years, within a few light-years of the sun must alter many of the primeval cometary orbits. Some comets are lost to the solar system as the result of such stellar perturbations, but others, retarded in their motion by the star's attraction, fall in toward the sun. Thus, Oort accounts for the steady influx of comets into the central part of the solar system.

Those comets that are so deflected toward the sun move subsequently on orbits of very high eccentricity, until, by perihelion passage, further perturbations by Jupiter either convert them into "periodic" comets, or deflect them into interstellar space on hyperbolic orbits. In the former case, they are eventually decomposed by evaporation or by tidal forces; in the latter, they are permanently lost to the solar system.

Although the supply of comets in the "cloud" must be gradually diminishing, the remaining number of comets may still be very large. Calculations show that there must be at least 100 thousand million. Their combined mass is probably comparable to that of the earth.

Collisions of the Earth and Comets

The chance of a collision between the earth and a typical comet is remote in our lifetimes, but over the age of the solar system it may well have occurred even several times. The coma of a comet is so tenuous that the earth would pass through it with probably no more effect than a spectacular meteor shower caused by the small particles in the coma. The only damage likely to result would be from direct impact of the nucleus, which has about the mass of a modest minor planet. A city or a small nation could be devastated by such a blow, for the energy of the impact could be that of a million 100-megaton hydrogen bombs. Most likely, the comet would strike an ocean, and loss of life might be minimal although great tsunamis (tidal waves) could also be devastating; there would be no question, though, that we would know it had happened.

The earth very probably did pass through the tail of Halley's comet in 1910. There was no detectable evidence of the encounter, although it is reported that at least one enterprising person made his fortune selling "comet pills."

METEOROIDS, METEORITES, AND METEORS

Although the layman often confuses comets and meteors, these two phenomena could hardly be more different. Comets can be seen when they are many millions of miles away from the earth, and may be visible in the sky for weeks, or even months, slowly shifting their positions from day to day. They rise and set with the stars, and during a single night appear motionless to the casual glance. Meteors, on the other hand, appear only when small solid particles enter the earth's atmosphere from interplanetary space. Since they move at speeds of many kilometers per second, they vaporize as a result of the high friction they encounter with the air. The light caused by the luminous vapors formed in such an encounter appears like a star moving rapidly across the sky, fading out within a few seconds. Meteors are commonly called "shooting stars."

On rare occasions, an exceptionally large particle may survive its flight through the earth's atmosphere and land on the ground. Such an occurrence is called a

Figure 11.6
Photograph of the trail of a bright meteor that happened to cross the field of view of the telescope photographing the Andromeda galaxy. *(F. Klepesta)*

meteorite fall, and if the particle is later recovered, it is known as a fallen *meteorite*. Frequently, the term "meteorite" is reserved for the fallen particle, and the particle, when still in space, is called a *meteor*. In this book, however, we shall use the terminology recommended by the International Astronomical Union in 1961. The particle, when it is in space, is called a *meteoroid;* the luminous phenomenon caused when the particle vaporizes in the earth's atmosphere is a *meteor;* and if it survives and lands on the ground, the particle is a *meteorite*.

Phenomenon of a Meteor

On a typical dark moonless night an alert observer can see half a dozen or more meteors per hour. To be visible, a meteor must be within 150 to 200 km of the observer; over the entire earth, the total number of meteors bright enough to be visible must total about 25 million per day. Faint meteors are far more numerous than bright ones; the number of meteors that are potentially visible with binoculars or through telescopes, and that range down to a hundred times fainter than naked-eye visibility, must number near 8 thousand million per day.

There are certain times when meteors can be seen with much greater than average frequency—up to 60 or more per hour. These unusual meteor displays are called *showers*. Meteor showers occur when the earth collides with swarms of meteoroids moving together in space. A distinction is made between shower meteors and nonshower, or *sporadic,* meteors.

The mass of a meteoroid producing a typical bright meteor is about ¼ g. A small fraction, however, have masses greater than a few grams. Many visible meteors are produced by particles with masses of only a few milligrams. Most meteoroids, therefore, are much smaller than pebbles.

The rate at which the earth accumulates meteoritic material is not accurately known. However, the total mass of meteor-producing meteoroids that enters the atmosphere each day is estimated to be from 10 to 100 tons.

Occasionally an exceptionally bright meteor is reported by many observers. These bright meteors are called *fireballs,* or *bolides*. It is estimated that tens or hundreds of thousands of fireballs appear every day over the entire earth, most of them over oceans or uninhabited regions. Some fireballs are visible in broad daylight; many are as bright as the full moon, and a few are as much as 100 times as bright. Sometimes they break up in midair with explosions that are audible from the ground.

Orbits of Meteoroids

The paths of meteors come from the special photographic patrols and radar observations. In the former, two specially designed, high-speed, wide-angle *meteor cameras* are placed many kilometers apart, and are directed to the same region of the sky. When trails of the same meteor are identified on photographs obtained simultaneously by the two cameras, the meteor's elevation and direction of motion through the atmosphere can be determined.

It is found that meteoroids produce meteors at an average height of 95 km (60 mi). Nearly all meteoroids completely disintegrate, and their meteors disappear, by the time they reach altitudes of 80 km (50 mi). The vast majority of meteoroids have speeds between 12 and 72 km/s. Knowing the speeds and directions of meteors in the earth's atmosphere, we can find the orbits of the meteoroids in space before they encountered the earth.

The orbits of some of those meteoroids producing fireballs are found to lie close to the plane of the ecliptic and have moderate eccentricities, and the revolution of the particles is *direct*—that is, from west to east, as is the revolution of the planets. These meteoroids are thought to have orbits similar to those unusual minor planets whose orbits bring them within 1 AU of the sun and possibly origins connected with those bodies.

For the vast majority of meteoroids, however—virtually all the thousands of nonshower objects whose orbits have been determined by photographic or radar methods—there is no preference for the ecliptic plane. Like the comets, they approach the earth from all directions. Also, like the comets, most of them are found to be moving at speeds very near the velocity of escape from the solar system; that is, they travel on near-parabolic heliocentric orbits. They are thus thought to be debris from old comets.

Meteor Showers

Meteor showers occur when the earth encounters swarms of particles moving together through space. Many such swarms of particles that the earth intercepts at regular intervals are known. Showers produced by them are predictable. On rare occasions the earth unexpectedly encounters swarms of particles that produce spectacular meteor displays.

Unlike sporadic meteors, which seem to come from any direction, meteors belonging to a shower all seem to radiate or diverge away from a single point on the celestial sphere; that point is called the *radiant* of the shower. Recurrent showers are named for the constellation within which the radiant lies or for a bright star near the radiant.

The seeming divergence of shower meteors from a common point is easily explained. The meteoroids producing a meteor shower are members of a swarm; they are all traveling together in closely spaced parallel orbits about the sun. When the earth passes through such a swarm, it is struck by many meteoroids, all approaching it from the same direction. As we, on the ground, look toward the direction from which the particles are coming, they all seem to diverge from it. Similarly, if we look along railway tracks, those tracks, although parallel to each other, seem to diverge away from a point in the distance (Figure 11.7).

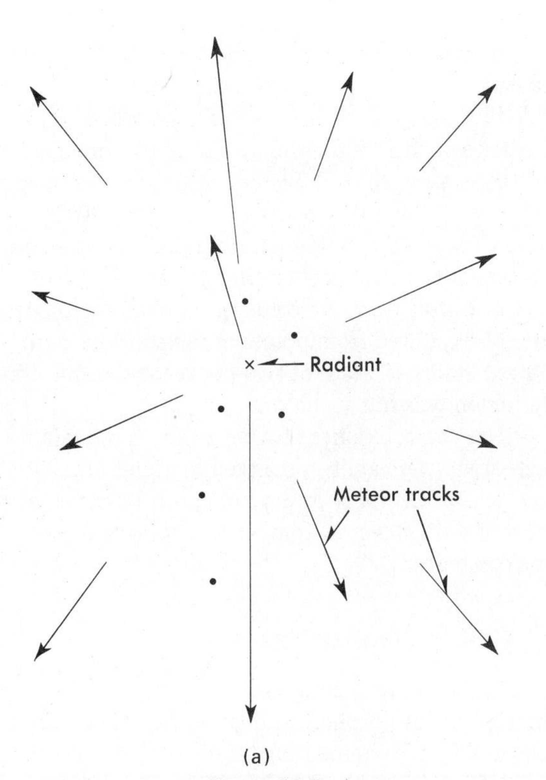

(a)

(b)

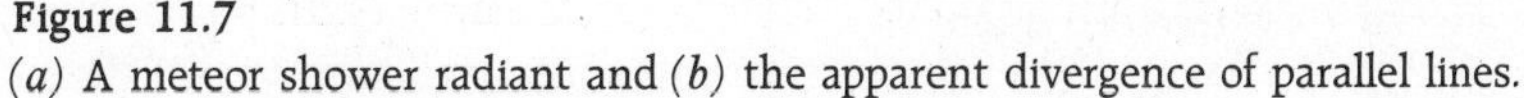
Figure 11.7
(a) A meteor shower radiant and *(b)* the apparent divergence of parallel lines.

TABLE 11.2 Characteristics of Some Meteor Showers

Shower	Date of Maximum Display	Velocity (km/s)	Associated Comet	Period of Comet (yr)
Quadrantid	Jan 3	43	—	7.0
Lyrid	Apr 21	48	1861 I	415.0
Eta Aquarid	May 4	59	Halley	76.0
Delta Aquarid	Jul 30	43	—	3.6
Perseid	Aug 11	61	1862 III	105.0
Draconid	Oct 9	24	Giacobini-Zinner	6.6
Orionid	Oct 20	66	Halley	76.0
Taurid	Oct 31	30	Encke	3.3
Andromedid	Nov 14	16	Biela	6.6
Leonid	Nov 16	72	1866 I	33.0
Geminid	Dec 13	37	—	1.6

On about August 11 of each year, the earth passes through a swarm of particles that approach from the direction of Perseus. In 1866 it was observed that the particles producing this *Perseid shower* travel in an orbit that is almost identical to that of Comet 1862 III. It was then realized that those meteoroids encountered each August are debris from the comet that has spread out along the comet's orbit.

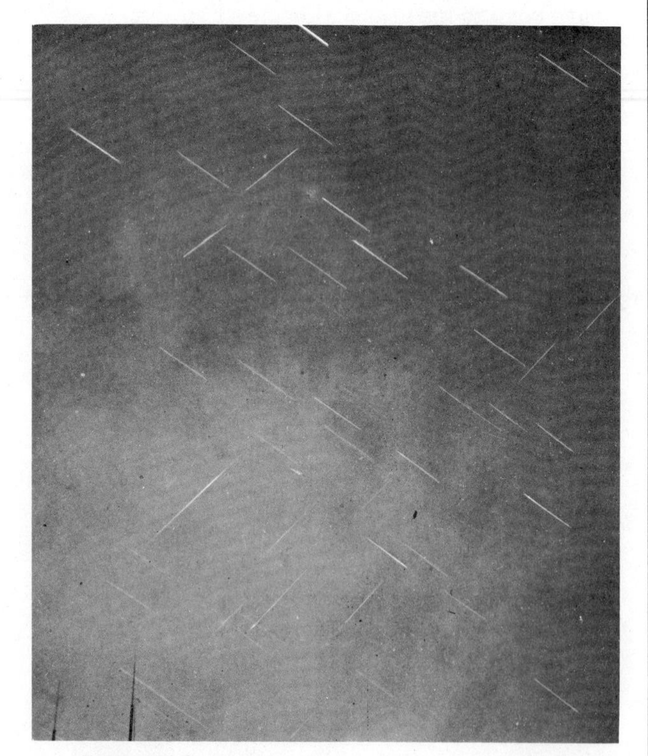

Figure 11.8
Photograph showing trails of many meteors during the shower of October 1946. Note the divergence of the trails from a radiant (off the field to the upper right). Other streaks are star trails recorded during the time exposure.

Subsequently, it has been found that the elements of the orbits of many other meteoroid swarms are similar to those of the orbits of known comets. Not all meteor showers have yet been identified with individual comets, but it is presumed that all showers have had a cometary origin. These swarms of debris, provided by the gradual disintegration of comets, give further evidence of the flimsy nature of comets.

One spectacular shower of recent decades was the *Draconid* shower that reached maximum display on October 9, 1946. On that date the earth reached the point that Comet Giacobini-Zinner had passed 15 days earlier. Debris from the comet produced meteors that could be counted from points in the southwestern United States at a rate of two per second, even though the moon was full at the time.

The characteristics of some of the more famous meteor showers are summarized in Table 11.2. Other spectacular meteor showers can occur, however, at almost any time, just as some bright comets appear unexpectedly.

Fallen Meteorites

Occasionally, a meteoroid survives its flight through the atmosphere and lands on the ground; this happens with extreme rarity in any one locality, but over the entire earth probably more than a thousand meteorites fall each year. Fallen meteorites have been found for centuries; it was nearly 1800, however, before their association with "shooting stars" began to be appreciated. The general acceptance that indeed "stones fall from the sky" occurred after the French physicist J. B. Biot described the circumstances of a fall in Orne on April 26, 1803, in which many witnesses observed the explosion of a bolide, after which many meteoritic stones were found, reportedly still warm, on the ground.

Today, fallen meteorites are found in two ways. First, sometimes bright fireballs are observed to penetrate the atmosphere to very low altitudes. A search of the area beneath the point where the fireball was observed to burn out may reveal one or more remnants of the meteoroid. *Observed falls*, in other words, may lead to discoveries of fallen meteorites.

Secondly, unusual-looking "rocks" are occasionally discovered that turn out to be meteoritic. These are termed "finds." Now that the public has become "meteorite conscious," many suspected meteorites are sent to experts each year. The late F. C. Leonard, a specialist in the field, referred to these objects as "meteorites" and "meteorwrongs." Genuine meteorites are turned up at an average rate of 25 per year.

Meteorites can be classified in three general groups:

1. *Irons*–alloys of metals. From 85 to 95 percent of their mass is iron; the rest is mostly nickel.
2. *Stony irons*–relatively rare meteorites that are about half iron and half "stony" silicates.
3. *Stones*–composed mostly of silicates and other stony materials. Stones contain about 10 or 15 percent iron and nickel in metallic flakes.

Tektites are somewhat rounded glassy bodies that have been found in Indonesia, Australia, and elsewhere. A few investigators still suspect them to be of extraterrestrial origin, but most favor the hypothesis that they are formed of terrestrial material that is melted and thrown out by the explosion produced when a large meteoroid strikes the earth. Tektites show evidence of having been molten at high pressure, and having undergone rapid cooling. They contain small spheres of iron and nickel, and schreibersite, a mineral known otherwise only in meteorites.

Stones are the most common kind of fallen meteorites. Ninety-three percent of meteorites obtained from observed falls are stones. However, after erosive action, they resemble terrestrial rocks, and generally only experts can ascertain their meteoritic nature from the iron pieces embedded in them. Consequently, about 58 percent of the total number of finds are irons. (Irons are seldom found, however, in those regions of the world where for centuries man has made use of iron for tools.) When irons are cut, polished, and etched, about 80 percent show a characteristic crystalline structure *(Widmanstätten figures)* that makes their identification as meteorites certain.

The chemical composition of many meteorites has been determined by laboratory analysis. It is found that meteorites contain the same common chemical elements that we find in the crust of the earth. When allowance is made for the fact that stones are at least nine times as numerous as irons, the following abundances (by weight) of elements are derived: iron, 30 to 40 percent; oxygen, 30 percent; silicon, 15 percent; magnesium, 12 percent; sulfur and nickel, 2 percent each; calcium, 1 percent; and small amounts of carbon, sodium, aluminum, lead, chlorine, potassium, titanium, chromium, manganese, cobalt, copper, and other elements.

Natural radioactivity provides a means of determining the ages of meteorites just as it provides a means of determining the ages of rocks in the earth's crust (Chapter 9). A derived age, of course, applies only to the solid particle in its present mineral state. The ages of meteorites are found to be about 4.6×10^9 years—the same as that of the earth.

The largest meteorite ever found on the earth is *Hoba West*, near Grootfontein, Namibia. It has a volume of about 7 cubic meters and an estimated mass of more than 45 tons. The largest meteorite on display in a museum has a mass of 31 tons; it was found by Peary in Greenland in 1897 and is now at the American Museum of Natural History in New York. The largest fallen meteorite found in a single piece in the United States was discovered in a forest near Willamette, Oregon, in 1902; it has a mass of 13 tons. (The discoverer spent about three months hauling the meteorite to his own property, where he put it on display for an admission price. Among those interested in the new exhibit were the attorneys of the Oregon Iron and Steel Company, owner of the land on which the meteorite was found. After litigation, it was decided that the company was the rightful owner of the object.)

Meteorite Falls

A total of 1800 meteorite falls and finds are listed in the *Hey Catalogue of Meteorites*.

Figure 11.9
The Barringer Meteorite Crater in Arizona. *(American Meteorite Museum)*

A spectacular meteoroid collision took place in Tunguska, Siberia, on June 30, 1908. A brilliant fireball was seen in broad daylight, and the impact produced shock waves that were registered on seismographs in distant Europe. Trees were seared of their branches and were felled from the impact over an area more than 30 km in radius. About 1500 reindeer were killed, and a man standing on the porch of his home 80 km away was knocked down. No pieces of the original particle have been recovered and no crater is found in the area; evidently the impacting body exploded in the air. Its original mass before it entered the earth's atmosphere is estimated to have been about 10^5 tons. Because the mass completely disintegrated it has been hypothesized that it was the nucleus of a small comet; but the object would have been only about 60 m or less in diameter—many times smaller than the nuclei of comets that become visible in the sky or on photographs.

Another spectacular Siberian fall occurred on February 12, 1947, near Vladivostok. The approaching fireball was described as "bright as the sun." The impact produced 106 craters and impact holes ranging in size up to 28 m across. Trees were felled radially around each of the large craters. The entire region covers nearly 5 square kilometers. More than 23 tons of iron meteorite fragments have been recovered from the area.

It is estimated that in each century several meteorites strike the earth with enough force to produce craters more than 10 m across. A large meteorite produces a crater larger than itself, for when it strikes the ground, its kinetic energy is dissipated with explosive violence.

The first crater to be discovered, and the most famous definitely known to be meteoritic, is the Barringer Meteorite Crater, near Winslow, Arizona. The crater was shown to be meteoritic as a result of research instigated by the Barringers (owners of the land) early in the twentieth century. The crater is 1300 m (nearly a mile) across and 180 m deep, and its rim rises 45 m above the level of the surrounding ground. In the area over 25 tons of iron meteorite fragments have been found, some buried near the crater and many more scattered at distances up to 7 km. The meteorite blew up completely into small pieces when it exploded, forming the crater. The age of the crater is estimated at about 22,000 years.

A larger crater is the New Quebec Crater (formerly Chubb Crater) in Quebec. It was discovered in 1950 on aerial photographs of the region. It is similar in appearance to the Barringer Crater but is about twice as large, being more than 3 km in diameter. It is currently filled with water and forms a lake in solid granite. No trace of meteoritic fragments has been found in the area, but it is almost certain that the crater is meteoritic. There are more than a dozen other craters about the world that are believed to be meteoritic. Those craters that were formed more than a few tens of thousands of years ago, of course, have long since eroded away.

THE INTERPLANETARY MATERIAL

Many tiny meteorites also strike the earth each day. Those that are only a few microns in diameter (1 micron is 10^{-4} cm) are slowed in the air before they have a chance to heat up; they eventually settle to the ground. These particles, too small to make meteors, are called *micrometeorites*. Their annual accretion is estimated at 10^5 tons. We see, then, that the space between the planets contains a vast number of micrometeorites. These particles comprise a distribution of *interplanetary dust*. Micrometeorites are a few microns or more in diameter. Particles less than 1 micron in size are "blown" out of the solar system by radiation pressure from the sun, just as small particles are blown out of comets in the form of comet tails. Particles the size of micrometeorites are not blown away but revolve about the sun as tiny planets. In addition to micrometeorites, or interplanetary dust, space vehicles have found evidence for interplanetary gas. Large numbers of charged atoms (ions) have been encountered. It is now well established that these ions have been expelled from the sun, and this outflow of gas is known as the *solar wind*. In the neighborhood of the earth, the ion density is from 1 to 10 particles per cubic centimeter. Further evidence of the solar wind is provided by its effect on comets in producing the straight tails.

There is additional evidence for the interplanetary dust (micrometeoritic material): the *zodiacal light* and the *gegenschein*. The zodiacal light is a faint glow of light along the zodiac (or ecliptic). It is brightest along those parts of the ecliptic nearest the sun and is best seen in the west in the few hours after sunset or in the east before sunrise. Under the most favorable circumstances, the zodiacal light rivals the Milky Way in brilliance. It is sometimes called the "false dawn" because of its visibility in the morning hours before twilight actually begins. The zodiacal light has the same spectrum as the sun, which shows it to be reflected sunlight. The present interpretation of the zodiacal light is that the interplanetary dust is concentrated most heavily in the plane of the ecliptic. The dust reflects enough sunlight to produce the faint glow along the zodiac.

The *gegenschein*, which means "counterglow," is a faint glow of light, centered on the ecliptic, which is exactly opposite the sun in the sky. Its angular size is from 8° to 10° by 5° to 7°. It is much more difficult to see than the zodiacal light, but it can be measured photoelectrically, and it has also been photographed. Like the zodiacal light, the gegenschein appears to be reflected sunlight. We believe it is simply that portion of the zodiacal light opposite the sun where the particles reflecting sunlight are at the "full phase," and hence at the most favorable illumination angle.

SUMMARY REVIEW

Minor planets: minor planet, asteroid, and planetoid; Bode's law; discovery of Ceres; computation of Ceres' orbit by Gauss and rediscovery; subsequent discoveries; total number of minor planets; orbits of minor planets; *Minor Planet Ephemeris;* Apollo minor planets; Chiron and its orbit; sizes of minor planets and how they are determined; masses of minor planets; origin of minor planets; satellites of minor planets; fragmentation of minor planets

Comets: early ideas about them; Halley's investigations; Halley's Comet; daylight comets; nucleus, coma, and tail of a comet; orbits of comets; membership in the solar system; periodic comets; disintegration of comets; sizes and masses of comets; chemical composition of comets; formation of comet tails—solar wind and radiation pressure; origin of comets; Oort cloud of comets; deflection of comets by stars; deflection of comets by Jupiter; evolution of cometary orbits; collisions of comets with earth

Meteoroids, meteorites, and meteors: "shooting stars"; shower and sporadic meteors; observations of meteors; fireballs; bolides; meteorite falls; orbits and origins of meteoroids; shower radiants; recognition of meteorites as extraterrestrial; types of meteorites; composition of meteorites; largest and most famous meteorites; meteorite collisions with the earth; meteorite craters; Barringer Crater; Tunguska event; fossile craters

Interplanetary material: micrometeorites; solar wind; zodiacal light; gegenschein

EXERCISES

1. Calculate the distances of the planets from the sun as predicted by Bode's law (Titius progression). List these values in a table, and compare them with the actual distances of the planets from the sun. Identify each distance predicted by Bode's law with the planet who's orbital semimajor axis most nearly coincides with it. How do the planets Uranus, Neptune, and Pluto fit the law? What do you think of Bode's law as a law of nature?

2. Minor planets can be discovered by the trails they leave on astronomical photographs. Fainter objects could be recorded, however, if their images were points rather than trails. Explain how you might plan a photographic search for faint minor planets.

3. At times Eros fluctuates in brightness by a factor of about five. At other times its light fluctuates by a much smaller amount. Can you offer an explanation for this phenomenon?

4. What would be the albedo of a minor planet if it were observed to emit no infrared radiation at all? What if the energy it emits in the infrared is equal to that in visible light that it reflects?

5. Years ago it was suggested that the minor planets may have resulted from the explosion of a single planet that once had an orbit with a semimajor axis of 2.8 AU. What objections can you think of to this theory?

6. Why do most comets move in their orbits at speeds that are very nearly their velocities of escape from the solar system? (Chapter 3)

7. How many typical comets would it take to have enough matter to make up a planet like the earth or Venus?

8. When can the tail of a comet lie in the direction toward which the comet is moving?

9. Why do you suppose that most comets that are very bright are most conspicuous in the hours after sunset or before sunrise?

10. If Oort's hypothesis for the origin of comets is correct, why do comets that come close to the sun for the first time always have aphelion points in the region of the cloud from which they originate?

11. Several years ago, a book published by a reputable publishing house advocated the theory that Venus was formed a few thousand years ago by a comet that boiled out of Jupiter and even stopped the earth from rotating for a brief time. Describe your reasons for supporting or discounting this theory.

12. Contrast comets to planets, typical satellites, meteoroids, and the sun, as regards their (a) sizes; (b) masses; (c) densities.

13. Suppose meteoroids all moved in nearly circular orbits. What then would be the range of velocities we should observe for them, relative to the earth?

14. The earth intercepts the orbit of Halley's comet twice, and we observe two meteor showers each year associated with this comet. Show, by a diagram, how this is possible.

15. Comets that have been associated with meteor showers are all periodic comets. Why do you suppose showers have not been identified with comets having near-parabolic orbits?

16. Why must the perihelion of the orbit of a meteoroid swarm never be greater than 1 AU from the sun for us to see a shower produced by that swarm, and why are the perihelia of such swarms usually closer to the sun than 1 AU?

17. Why is it that a particular shower may be seen only at certain hours of the night, for example, early evening or the hours before sunrise?

Comet Ikeye-Seki. *(Hale Observatories)*

CHAPTER 12

Friedrich Wilhelm Bessel (1784–1846) made the first authenticated measurement of the distance to a star (61 Cygni) in 1838, a feat that had eluded many dedicated astronomers for almost a century. *(Yerkes Observatory)*

CELESTIAL SURVEYING—DISTANCES AND MOTIONS OF STARS

The distance to the moon had been measured in the second century B.C., and the distances to the planets in relation to that of the sun were known to Copernicus. The goal of measuring distances to the stars, on the other hand, was not realized until the nineteenth century. The determination of astronomical distances is central to man's concept of the universe.

SURVEYING THE HEAVENS

In principle, we survey distances to planets and stars by the same technique that the civil engineer employs to survey the distance to an inaccessible mountain or tree (see Figure 12.1). The method is that of *triangulation*. Two observing stations are set up some distance apart. That distance (AB in Figure 12.1) is called the *baseline*. Now the direction to the remote object (C in the figure) in relation to the baseline is observed from each station. Note that C appears in different directions from the two stations. This apparent change in direction of the remote object due to a change of vantage point of the observer is called *parallax*. The parallax is also the angle at C between A and B—that is, the angle *subtended* by the baseline. A knowledge of the angles at A and B and the length of the baseline, AB, allows the triangle ABC to be solved for any of its dimensions, say, the distance AC or BC. The solution could be accomplished by constructing a scale drawing or by numerical calculation with the technique of trigonometry.

Depth perception is an example of the same principle. Our eyes are separated by a baseline of a few inches, so our two eyes see an object in front of us from slightly different directions. The brain, like an electronic computer, solves the triangle and gives us an impression of the distance of the object. The greater the parallax, the nearer the object. Hold a pencil a few inches in front of your face and look at it first with one eye and then with the other; note the large shift in direction (parallax) against the more distant wall across the room. Now hold the pencil at arm's length and note how the parallax is less. If an object is fairly distant, the parallax is too small to notice with the eyes; thus depth perception fails for objects more than a few tens of meters away. It would take a larger

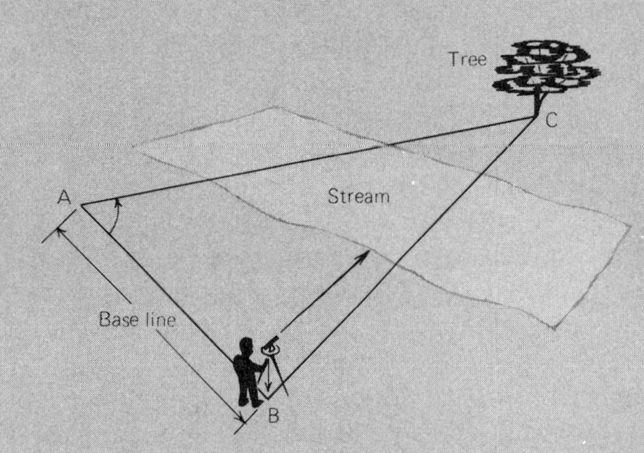

Figure 12.1
Triangulation of an inaccessible object.

baseline than the distance between the eyes to see the parallax of an object, say, 500 m distant.

Nearly all astronomical objects are very far away, and either a very large baseline must be found or highly precise angular measurements must be made, or both. The moon is the only object near enough that its distance can be found fairly accurately with measurements made without a telescope. Ptolemy had determined the distance to the moon correctly to within a few percent. He used the earth itself as a baseline; its rotation carries the observer from an observing station on one side of the earth to one thousands of kilometers away in a few hours.

Distances in the Solar System

The planets and other bodies that revolve about the sun in the solar system are so far away that at any time each one is seen in almost the same direction by all terrestrial observers. Thus, the parallaxes of these bodies, although measurable, are very small and are difficult to measure accurately; a larger baseline than the diameter of the earth is required. As the earth moves about the sun, however, it carries us across a baseline that can be as large as the diameter of the earth's orbit—about 300 million km. Thus, we can determine accurate distances to the other members of the solar system by observing them at times when the earth is in two different places in its orbit. We must, however, take into account the motion of the object during the interval between the sightings of it. There are various mathematical techniques (which will not be gone into here) for unscrambling the effects of the combined motions of the body and the earth.

Unfortunately, the measures described above—measures within the solar system—are not obtained directly in kilometers; they are rather found in terms of the *astronomical unit*, the semimajor axis of the earth's orbit. The foregoing procedure for surveying the distances to the planets provides us with an accurate map of the solar system, but does not give us the absolute scale of the map. On a map of the United States that is correct proportionally, the relative distances between cities, for example, can be determined simply by measuring distances on the drawing itself. Distances so obtained cannot be converted to kilometers, however, until the scale of the map is established—that is, one cm equals so many kilometers or other units of distance. The scale can be determined if the actual distance in kilometers between two places (perhaps New York and Philadelphia) is known. Similarly, to find the scale of the solar system—that is, to evaluate the astronomical unit—we must find the distance to some object that revolves about the sun, both in astronomical units and in kilometers or other units.

The modern accurate technique for evaluating the astronomical unit is by means of radar. A strong beam of radio energy is sent toward a planet, and some of the radiation reflected by it can be detected back on earth. Now radio waves travel with the speed of light, an accurately known quantity; the time between their transmission and when the return echo is received, multiplied by the speed of light, thus gives the round-trip distance to that planet. Radar observations of several planets, especially Venus, are now regularly carried out to study these worlds, and the length of the astronomical unit is thus known very accurately. The Jet Propulsion Laboratory finds 1 AU = 149,597,870.7 km, with an uncertainty of about 100 m.

Surveying Distances to Stars

Aristotle argued that the earth could not revolve about the sun, or we would observe parallaxes of stars when they are observed from different sides of the earth's orbit. Tycho Brahe advanced the same argument nearly two millennia later. By the eighteenth century, when there was no longer serious doubt of the earth's revolution, it was realized that the stars must be extremely distant and their parallaxes exceedingly tiny.

Many astronomers attempted to detect stellar parallaxes. Among them was J. Bradley in 1729. He attached a telescope rigidly to his chimney and hoped to observe stars passing through its field of view (as the rotating earth carried them through the zenith) in slightly different places during the year because of parallax. What he found, to his surprise, was that *all* stars shifted back and forth during the year, but by exactly the same amount—20″.5. Moreover, the shift was always in the direction of the motion of the earth, not the kind of shift that parallax would produce.

What Bradley had observed was the *aberration of starlight*. The stars always appear displaced in the direction of the earth's motion because the earth is moving forward into the light coming from them. Similarly, vertically falling raindrops appear to approach us slightly if we walk or run through the rain. For example, if you walk with an upright hollow stove pipe in the rain, you must tilt the top of the pipe forward slightly in order that drops falling into it pass completely through it. Bradley's observation of aberration demonstrated both the motion of the earth and the finite speed of light. This important discovery, made while looking for something quite different, is an example of *serendipity*.

Another serendipitous discovery made in the quest of stellar parallaxes is of the orbital motions of stars in binary star systems. The great German-English astronomer William Herschel, late in the eighteenth century, made systematic telescopic surveys of the sky. He noted many examples of pairs of stars, usually one brighter than the other. He thought the fainter star in each pair was more distant, and that the nearer, brighter one should therefore appear to shift back and forth with respect to it during the year as the earth revolves about the sun. Thus he cataloged many such pairs, thinking they would be good candidates for detecting stellar parallax.

Subsequently it was found that the stars did, indeed, move with respect to each other, but not with a yearly period. Rather, each star in such a pair was seen to be revolving about the other, and in an elliptical orbit. Herschel had discovered that double stars were common, and that in such binary star systems the stars revolved about each other in accord with Newton's and Kepler's laws. The discovery that these laws apply to stars far beyond the solar system is of immense importance, for it gives us reason to believe that the whole universe is governed by the same natural laws.

Eventually, the true parallaxes of some of the nearer stars were observed. The first successful detections were in the year 1838, when Friedrich Bessel (Germany), Thomas Henderson (Cape of Good Hope), and Friedrich Struve (Russia) measured the parallaxes of the stars 61 Cygni, Alpha Centauri, and Vega, respectively. However, even the nearest star, Alpha Centauri, showed a total displacement of only about 1″.5 during the year. Small wonder that Tycho Brahe was unable to observe the stellar parallaxes and concluded that the earth was stationary.

As the earth moves about its orbit, the place from which we observe the stars continually changes. Consequently, the positions of the comparatively near stars, projected against the more remote ones, are also continually changing. If a star is in the direction of the ecliptic (in the plane of the earth's orbit), it seems merely to shift back and forth in a straight line as the earth passes from one side of the sun to the other. A star that is at the pole of the ecliptic (90° from the ecliptic) seems to move about in a small circle against the background of more distant stars, as we view it from different positions in our nearly circular orbit. A star whose direction is intermediate between the ecliptic and the ecliptic pole seems to shift its position along a small elliptical path during the year. The eccentricity of the ellipse ranges from that of the earth's orbit (nearly a circle) for a star that is at the ecliptic pole to unity (a straight line) for a star on the ecliptic (see Figure 12.2). This small ellipse is called the *parallactic ellipse*.

Units of Stellar Distances

The angular semimajor axis of the parallactic ellipse is called the *stellar parallax* of the star. Since the major axis of the ellipse is the maximum apparent angular deflection of the star as viewed from opposite ends of a diameter of the earth's orbit, the stellar parallax, the semimajor axis of this ellipse, is the angle, at the star's distance, subtended by 1 AU perpendicular to the line of sight.

A convenient unit of distance for stars is the *parsec* (*pc*). One parsec is the distance of a hypothetical star (none exists) with a stellar parallax of one second. Recall that the parallax is smaller, the more distant the star. In fact, the distance of a star, in parsecs, is equal to the reciprocal of its parallax. Thus a star with a parallax of 0″.1 would be at a distance of 10 pc; one with a parallax of 0″.05 would be 20 pc away. It takes light 3.26 years to travel one parsec; thus 1 pc = 3.26 *light years* (LY). One parsec contains 3.086×10^{13} km, and one light year 9.47×10^{12} km (about 6×10^{12} mi).

No known star (other than the sun) is near enough to have a parallax of one second of arc. The nearest stellar neighbors to the sun are three stars that make up a multiple system. To the naked eye the system appears as a single bright star, Alpha Centauri, which is only 30° from the south celestial pole and hence is not visible from the mainland United States. Alpha Centauri itself is a double star—two stars in mutual revolution, too close together to be separated by the naked eye. Nearby is the third member of the system, a faint star known as *Proxima Centauri*. Proxima is slightly closer to us than the other two stars in the system; it has a parallax of 0″.763 and a distance of

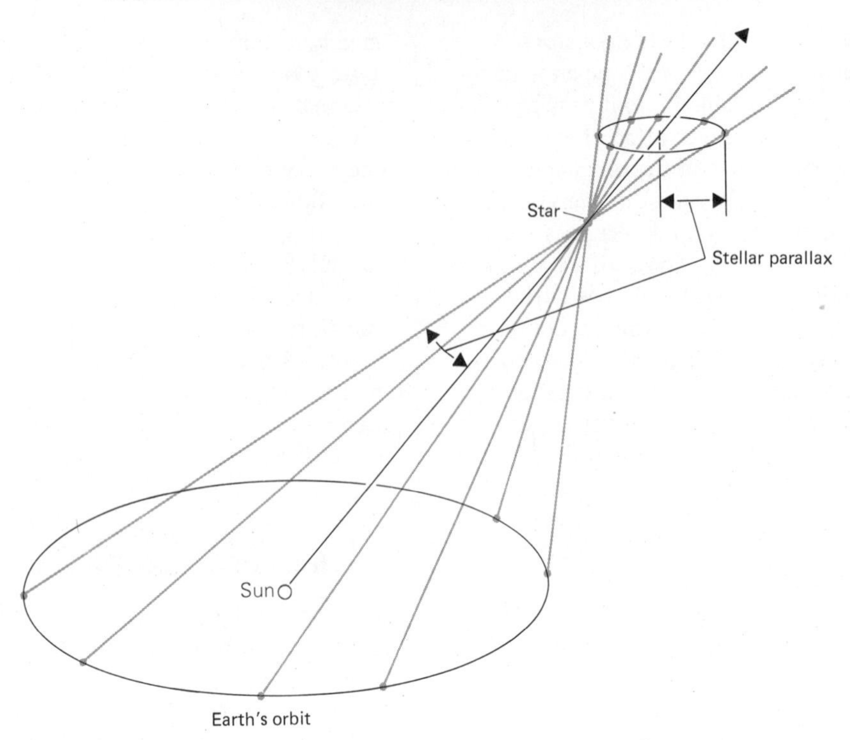

Figure 12.2
The apparent motion of a nearby star relative to the more remote stars. This motion results from the orbital motion of the earth, and appears as a small ellipse in the sky called the paralactic ellipse.

1/0.763 = 1.31 pc (4.3 LY). The two stars that make up Alpha Centauri have a parallax of 0".74 and a distance of 1.35 pc. The nearest star visible to the naked eye from most parts of the United States is the brightest appearing of all the stars, *Sirius*. Sirius has a distance of 2.6 pc, or about 8 LY. It is interesting to note that light reaches us from the sun in eight minutes and from Sirius in eight years.

Parallaxes have been measured for about 7000 stars. Only for a fraction of them, however, are the parallaxes large enough (about 0".05 or more) to be measured with a precision of 10 percent or better. The 1969 edition of Gliese's catalog of nearby stars lists 1049 within 20 pc, but the total number of such stars, including those not yet discovered, may be near 4000. Of those stars within about 20 pc, most are invisible to the unaided eye and actually are intrinsically less luminous than the sun. Most of the stars visible to the unaided eye, on the other hand, have distances of hundreds or even thousands of parsecs and are visible not because they are relatively close, but because they are intrinsically very luminous. The nearer stars are described more fully in Chapter 14.

Until now we have implied that the star whose parallax is observed is motionless with respect to the sun. Actually, the stars are all moving at many kilometers per second. The effects of the relative motion of the star and the sun must be separated from the effects of the earth's motion (the star's parallax) by observing the star's direction at the same time of the year several years apart. Any observed change in its direction then must be due to its own motion relative to the sun. This change indicates the corrections that must be applied to observed total changes in the direction of the star in order to obtain its true stellar parallax.

MOTIONS OF STARS

The ancients distinguished between the "wandering stars" (planets) and the "fixed stars," which seemed to maintain permanent patterns with one another in the sky. The stars are, indeed, so nearly fixed on the celestial sphere that the apparent groupings they seem to form—the constellations—look today much as they did when they were first named, more than 2000 years ago. Yet the stars are moving with respect to the sun, most of them with speeds of many kilometers per second. Their motions are not apparent to the unaided eye in the course of a single human lifetime, but if an ancient observer who knew the sky well—Hipparchus, for example—could return to life today he would find that several of the stars had noticeably changed their positions relative to the others. After some 50,000 years or so, terrestrial observers will find the handle of the Big Dipper unmistakably "bent" more than it is now. Changes in the positions of the nearer stars can be measured with telescopes after an interval of only a few years.

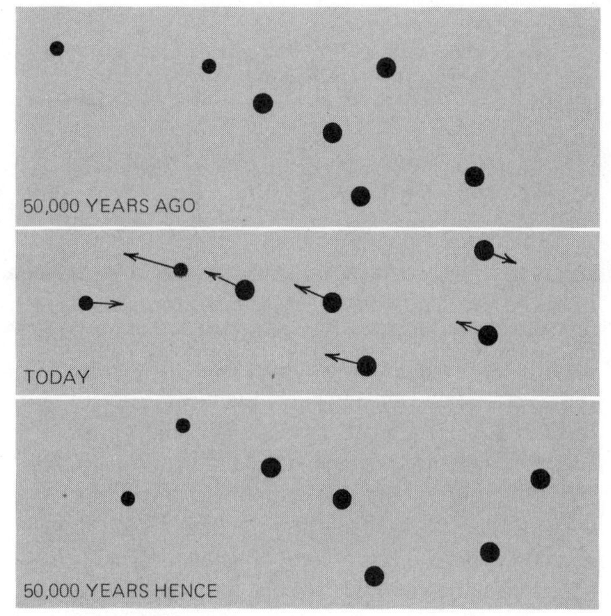

Figure 12.3
Appearance of the Big Dipper over 100,000 years.

Proper Motion

The *proper motion* of a star is the rate at which its direction in the sky changes, usually expressed in seconds of arc per year. It is almost always an angle that is too small to measure with much precision in a single year; in an interval of 20 to 50 years, on the other hand, many stars change their directions by easily detectable amounts. The modern procedure for determining proper motions is to compare the positions of the star images on two different photographs of the same region of the sky taken at least two decades apart. Most of the star images on such photographs do not appear to have changed their positions measurably; these are statistically, the more distant stars that are relatively "fixed," even over the time interval separating the two photographs. With respect to these "background" stars, the motions of a few comparatively nearby stars can be observed.

The star of largest proper motion is *Barnard's star*, whose direction changes by 10″.34 each year. This large proper motion is partially due to the star's relatively high velocity with respect to the sun but is mostly the result of its relative proximity. Barnard's star is the nearest known star beyond the triple system containing Alpha Centauri; its distance is only 1.8 pc, but it emits only about 1/2000 as much light as the sun and is 25 times too faint to see with the unaided eye. A few hundred stars have proper motions

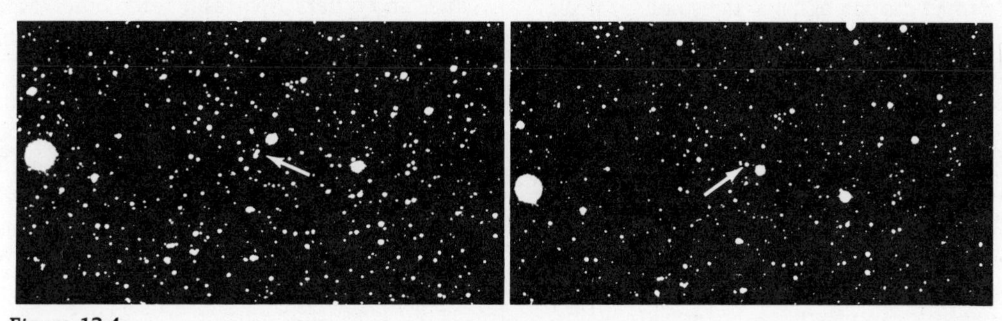

Figure 12.4
Two photographs of Barnard's star, showing its motion over a period of 22 years. *(Yerkes Observatory)*

as great as 1″.0. The mean proper motion for all naked-eye stars is less than 0″.1; nevertheless, the proper motions of most stars are larger than their stellar parallaxes.

Radial Velocity

The *radial* velocity (or line-of-sight velocity) of a star is the speed with which it approaches or recedes from the sun. This can be determined from the Doppler shift of the lines in its spectrum (Chapter 5). Unlike the proper motion, which is observable only for the comparatively nearby stars, the radial velocity can be measured for any star that is bright enough for its spectrum to be photographed. The radial velocity of a star, of course, is only that component of its actual velocity that is projected along the line of sight—that is, that carries the star toward or away from the sun. Radial velocity is usually expressed in kilometers per second, and is counted as *positive* if the star is moving *away* from the sun, and *negative* if the star is moving *toward* the sun. Since motion of either the star or the observer (or both) produces a Doppler shift in the spectral lines, a knowledge of the radial velocity alone does not enable us to decide whether it is the star or the sun that is "doing the moving" (indeed, as we saw in Chapter 7, it does not even make sense to ask which is moving). What we really measure, therefore, is the speed with which the distance between the star and sun is increasing or decreasing—that is, the star's radial velocity *with respect to the sun.*

Space Velocity

Radial velocity is a motion of a star along the line of sight, while proper motion is produced by the star's motion *across*, or at right angles to, the line of sight. Whereas the radial velocity is known in kilometers per second, the proper motion of a star does not, by itself, give the star's actual *speed* at right angles to the line of sight. The latter is called the *tangential* or *transverse* velocity. To find the tangential velocity of a star, we must know both its proper motion and *distance*. A star with a proper motion of 1″.0, for example, might have a relatively low tangential velocity and be nearby, or a high tangential velocity and be far away.

If, however, we know the distance of a star of a particular proper motion, we can calculate what its tangential velocity must be to produce the observed proper motion. If, in addition, we know its radial velocity—how fast it is moving in the line of sight—we can easily calculate how fast and in what direction it is moving in space with respect to the sun. This is called its *space velocity*.

The Solar Motion

The sun, a typical star, is in motion, just as the other stars are. In fact, it is a member of our Galaxy, a system of a hundred thousand million stars. The Galaxy is flat, like a pancake, and is rotating. The sun, partaking of this general rotation of the Galaxy, moves with a speed of from 250 to 300 km/s to complete its orbit about the galactic center in a period of about 200 million years. At first thought it might seem that the galactic center is the natural reference point with which to refer the stellar motions. However, our observations of the proper motions and radial velocities of the stars that surround the sun in space, all in our own so-called local "neighborhood" of the Galaxy, do not give us directly the motions of these stars about the galactic center. The reason is that the stars' orbits around the galactic center and their orbital velocities are both nearly the same as those of the sun. The motions we observe are merely small differences between the orbital motions of these stars in the Galaxy and that of the sun. These small residual motions arise because our neighboring stars' orbits about the galactic center are not absolutely identical to our own. We are overtaking and passing some stars, while others are passing us; the slightly different eccentricities and inclinations of our respective orbits bring us closer to some stars and carry us farther from others. We can study these residual motions without knowing anything about the actual motions of stars around the center of the Galaxy. Our situation is analogous to that of a man driving an automobile on a busy highway. All the cars around him are going the same direction and at roughly the same speed, but some are changing lanes and others are passing each other. More or less like the highway traffic, the residual motions of the stars around us seem to be helter-skelter.

We deduce the motion of the sun with respect to our neighboring stars by analyzing the proper motions and radial velocities of the stars around us. The easiest way to understand how the sun's motion is found is to consider the effect it has on the apparent motions of the other stars.

First, consider the radial velocities of stars with respect to the sun. If we look in the direction *toward* which the sun is moving, we find that most of the stars are approaching us, because, of course, we are moving forward to meet them. The only stars in that direction that are receding from us are those that are moving in the same direction we are going, but at a faster rate, so that they are pulling away from us. The observed radial velocities of all the stars in the direction toward which the sun is moving do not average to zero, but to −20 km/s, showing that we are moving toward them at about 20 km/s. Similarly, stars in the opposite direction have an average radial velocity of about +20 km/s, because we are pulling away from them at that speed.

The sun's motion also affects the observed proper motions of stars. If the stars were at rest, they would all show a backward drift due to our forward motion. As it is, the stars have motions of their own, but only those moving in the same direction we are and at a faster rate appear to have "forward" proper motions—the rest, by far the majority, *do* appear to drift backward.

William Herschel was the first to attempt to detect the direction of the solar motion from the proper motions of stars. In 1783 he analyzed the proper motions of 14 stars and deduced that the sun was moving in a direction toward the constellation Hercules—a nearly correct result.

Modern analysis of the proper motions and radial velocities of the stars around the sun has shown that the sun is moving approximately toward the direction now occupied by the bright star Vega in the constellation of Lyra. The value found for the sun's speed depends somewhat on what stars are observed to determine it. Analysis of most of the stars in the standard catalogs gives the *standard solar motion*, which is 19.5 km/s (4.14 AU/yr). The direction in the sky toward which the sun is moving is called the *apex* of solar motion,[1] and the opposite direction, away from which the sun is moving, is called the *antapex*.

Distances from Stellar Motions

The proper motions of stars can be expected to be largest, statistically, for the nearest stars. If, for example, a star is only a few parsecs away, its proper motion will almost certainly be observable after a few years. The proper motion of a very distant star, on the other hand, may be detectable only after a long time, and then only if the star has a very great space velocity. Searches for nearby stars, therefore, are usually conducted by searching for stars of large proper motion. Conversely, remote stars can be identified by their lack of observable proper motions; they serve as standards against which we can measure the parallaxes of the nearby stars to determine their distances. The proper motion of an individual star does not in itself indicate its distance uniquely. However, proper motions and distances of stars are inversely correlated, and investigations of such motions do give statistical information about stellar distances.

[1] The equatorial coordinates of the *solar apex* are (1950) $\alpha = 18^h4^m \pm 7^m$, $\delta = +30° \pm 1°$.

SUMMARY REVIEW

Triangulation: baseline; parallax; depth perception

Surveying in the solar system: the earth's orbit as a baseline; the length of the astronomical unit; use of radar in finding distances

Stellar parallaxes: serendipitous discoveries—aberration of starlight and binary stars; parallactic ellipse; the parsec; light year; nearest stars—Proxima Centauri and Alpha Centauri; Sirius; Gleise catalog

Stellar motions: proper motion; Barnard's star; radial velocity; space velocity; tangential or transverse velocity

Solar motion: sun's motion in the Galaxy; solar motion from radial velocities and proper motions of surrounding stars; standard solar motion; solar apex and solar antapex

Distances determined from statistical studies of stellar motions

EXERCISES

1. A radar astronomer claims that he beamed radio waves to Jupiter and received an echo exactly 48 minutes later. Do you believe him? Why?

2. How would observed parallaxes of stars as measured by a hypothetical observer on Saturn (10 AU from the sun) compare with parallaxes measured from the earth?

3. Would the aberration of starlight be greater or less on Saturn than on the earth? Why?

4. Give the distances to stars having the following parallaxes: (a) 0".1; (b) 0".5; (c) 0".005; (d) 0".001.

5. Give parallaxes of stars having the following distances: (a) 10 pc; (b) 3.26 LY; (c) 326 LY; (d) 10,000 pc.

6. The value of the astronomical unit can be obtained from radar observations of a planet, such as Venus, not only from the time required for the radio waves to travel the round trip to Venus and back, but also from the difference between the frequencies of the waves beamed to Venus and those received waves that are reflected back to earth. Explain how this is possible.

7. Make up a table relating the following units of astronomical distance: kilometer, earth radius, astronomical unit, light year, parsec.

8. Since we observe stars from the earth, does the Doppler shift we measure in the spectrum of a star indicate directly the radial velocity of that star with respect to the sun? If not, what kind of correction must be applied?

9. Following are data on five stars:

Star	Distance	Parallax	Proper Motion	Radial Velocity
1	1000 pc		0.05	−80 km/s
2		0.0001	0.1	+75 km/s
3	40 pc		0.25	+20 km/s
4		0.1	0.01	− 7 km/s
5	100 pc		1.0	−18 km/s

(a) Which star has the largest space motion?
(b) Which star has the smallest space motion?
(c) Which is most distant?
(d) Which is nearest?
(e) Which approaches us the fastest?
(f) Which has a parallax of 0".01?

10. In 50 years a star is seen to change its direction by 1′40″. What is its proper motion?

11. Suppose a star at a distance of 10 pc has a radial velocity of 150 km/s. By what percentage does its distance change in 100 years?

12. Show by a diagram how two stars can have the same radial velocity and proper motion but different space motions.

Comet Ikeya-Seki, photographed by J. B. Irwin in Chile, October 1965.

Comet Humason (1961a), photographed with the Palomar Schmidt telescope. *(Hale Observatories)*

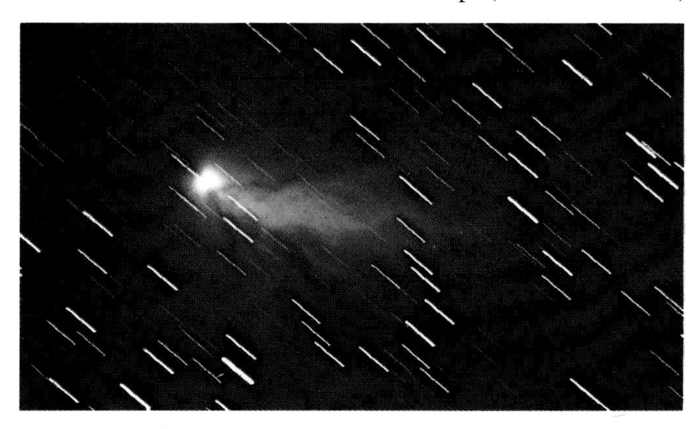

The trail of a meteor in the Perseid shower, photographed by Ronald Oriti. *(Courtesy, Ronald Oriti)*

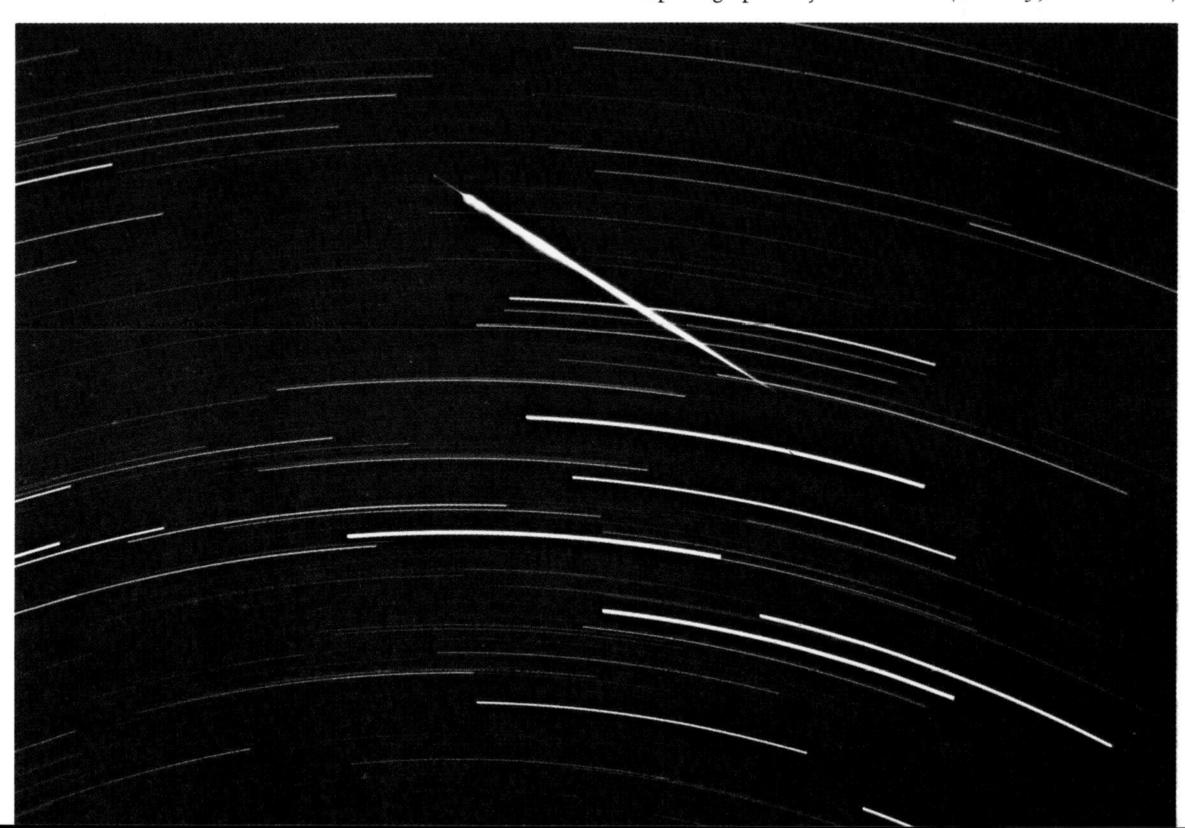

Slice of the Kamkas iron meteorite, which has been polished and then etched with a dilute nitric acid solution to show the criss-cross Widmanstätten figures. *(Photograph by Ivan Dryer)*

Stony-iron meteorite, Glorieta Mountains, New Mexico. The specimen has been polished and etched to show the metallic structure. *(Photograph by Ivan Dryer)*

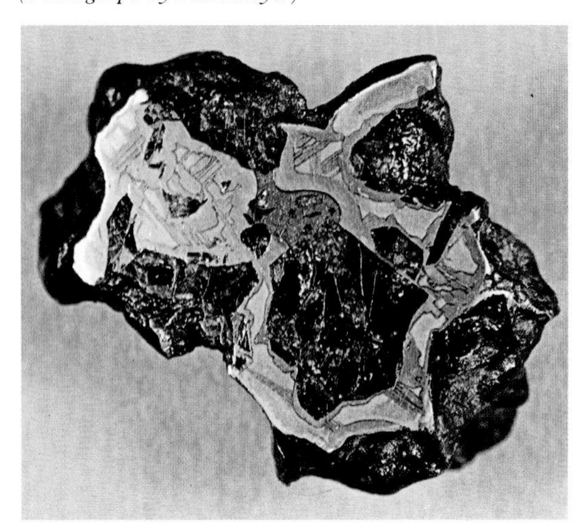

Polished slice of the Albin, Wyoming, stony meteorite. This type of meteorite consists of nickel-iron metal with inclusions of the green mineral olivine. *(Photograph by Ivan Dryer. All meteorite specimens on this page are from the collection of Ronald Oriti, and are reproduced by the kind permission of Mr. Oriti.)*

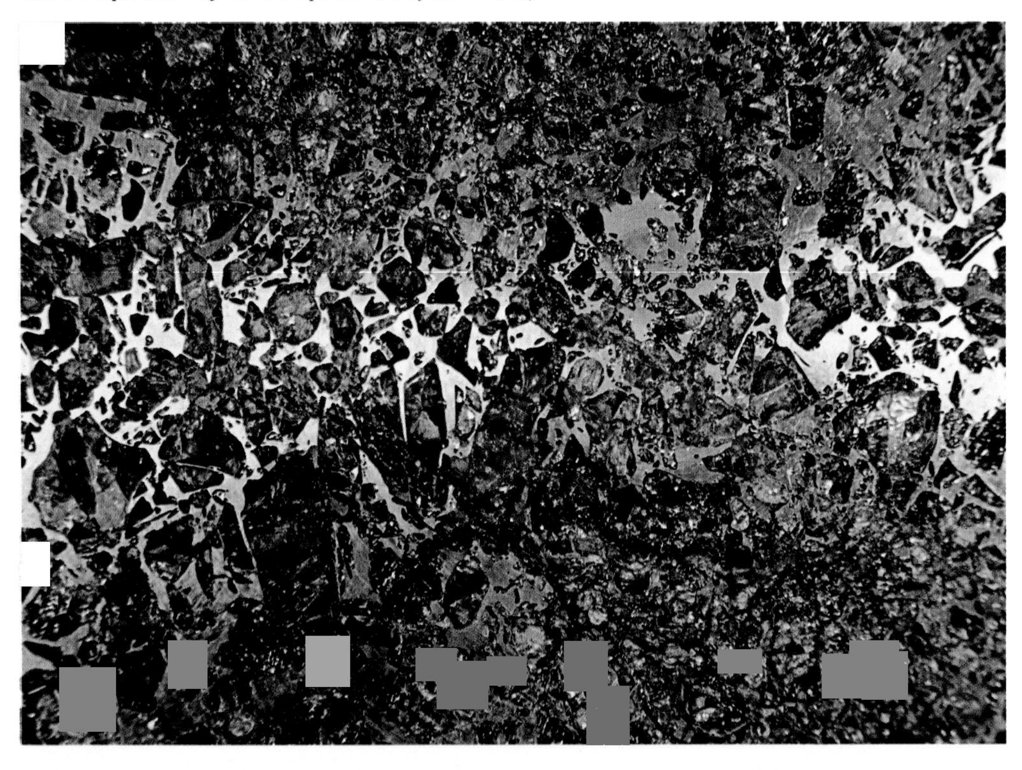

A transit circle installed at the U.S. Naval Observatory during the late nineteenth century and modified by them for their use. Instruments of this nature were, and still are, used for making precise measurements of the locations of objects in the sky. *(Courtesy, Paul M. Routly, U.S. Naval Observatory)*

CHAPTER 13

Sir William Herschel (1738–1822), a German musician, emigrated to England to avoid service in the Seven Years War. While composing and giving music lessons, he built the first large reflecting telescopes, surveyed the sky, and attempted the first quantitative measurements of star brightnesses. *(Yerkes Observatory)*

ANALYZING STARLIGHT

In the second century B.C., Hipparchus compiled a catalog of about a thousand stars (Chapter 2). He classified these stars into six categories of brightness, which are now called *magnitudes*. The brightest-appearing stars in his catalog are of the *first magnitude;* the faintest naked-eye stars are of the *sixth magnitude*. The other stars are assigned intermediate magnitudes. This system of stellar magnitudes, which began in ancient Greece, has survived to the present time, with the improvement that today magnitudes are based on precise measurements of apparent or total luminosity rather than arbitrary and uncertain eye estimates of star brightness.

MEASURING STARLIGHT—PHOTOMETRY

That branch of observational astronomy, which deals with the measurement of the amount of light energy, or *luminous flux*, from stars is called *photometry*. One method of comparing the fluxes of light received from stars is *photographic photometry*, in which the degrees of blackness and sizes of star images on photographic negatives are measured.

The modern method of stellar photometry is *photoelectric photometry*. A metal plate with a small hole is placed in the focal plane of the telescope. The light from a star coming to focus on the plate is allowed to pass through the hole and thence onto the photosensitive surface of a photomultiplier (Chapter 6). The electric current generated in the photomultiplier is amplified and recorded and provides an accurate measure of the light passing through the hole. Because even the darkest night sky is not completely dark, however, not all the light striking the photomultiplier comes from the star; some is provided by the light of the night sky, which is, of course, also gathered by the telescope and passed through the hole. Consequently, the hole is next moved to one side of the star image, so that only

Figure 13.1
Effect of telescope aperture and exposure time. *(Top left)* 36 cm, 1 min. Stars to twelfth magnitude. *(Top right)* 152 cm, 1 min. Stars to fifteenth magnitude. *(Lower left)* 152 cm, 27 min. Stars to eighteenth magnitude. *(Lower right)* 152 cm, 4 hr. Stars to twentieth magnitude. *(Hale Observatories)*

light from the sky passes through and is recorded; the difference between the first and second readings is a measure of the star's light entering the telescope.

The Magnitude Scale

In 1856 Norman R. Pogson proposed the quantitative scale of stellar magnitudes that is now generally adopted. He noted, as had William Herschel before him, that we receive about 100 times as much light from a star of the first magnitude as from one of the sixth, and that, therefore, a difference of five magnitudes corresponds to a ratio in luminous flux of 100:1. Now it is a widely accepted assumption in the physiology of sense perception that what appear to be equal intervals of brightness are really equal *ratios* of luminous energy. Pogson proposed, therefore, that the ratio of light flux corresponding to a step of one magnitude be the fifth root of 100, which is about 2.512. Thus, a fifth-magnitude star gives us 2.512 times as much light as one of sixth magnitude, and a fourth-magnitude star, 2.512 times as much light as a fifth, or 2.512 × 2.512 times as much as a sixth-magnitude star. From stars of third, second, and first magnitude, we receive 2.512^3, 2.512^4, and 2.512^5 (= 100) times as much light as from a sixth-magnitude star. By assigning a magnitude of 1.0 to the bright stars Aldebaran and Altair, Pogson's new scale gave magnitudes that agreed roughly with those in current use at the time.

Table 13.1 gives the approximate ratios of light flux corresponding to several selected magnitude differences. Note that a given ratio of light flux, whether between bright or faint stars, always corresponds to

TABLE 13.1 Magnitude Differences and Light Ratios

Difference in Magnitude	Ratio of Light
0.0	1:1
0.5	1.6:1
0.75	2:1
1.0	2.5:1
1.5	4:1
2.0	6.3:1
2.5	10:1
3.0	16:1
4.0	40:1
5.0	100:1
6.0	251:1
10.0	10,000:1
15.0	1,000,000:1
20.0	100,000,000:1
25.0	10,000,000,000:1

TABLE 13.2 Some Magnitude Data

Object	Magnitude
Sun	−26.5
Full moon	−12.5
Venus (at brightest)	−4
Jupiter, Mars (at brightest)	−2
Sirius	−1.5
Aldebaran, Altair	1.0
Naked-eye limit	6.5
Binocular limit	10
15-cm telescope limit	13
5.08-m (visual) limit	20
5.08-m photographic limit	24

the same magnitude interval. Further, note that the numerically *smaller* magnitudes are associated with the *brighter* stars; a numerically *large* magnitude, therefore, refers to a faint star.[1]

With optical aid, stars can be seen that are beyond the reach of the naked eye. The *limiting magnitude* of a telescope is the magnitude of the faintest stars that can be seen with that telescope under ideal conditions. A 15-cm telescope, for example, has a limiting magnitude of about 13. The *photographic limiting magnitude* of a telescope is the magnitude of the faintest stars that can be photographed with it. The photographic limiting magnitude of the 5.08-m telescope on Palomar Mountain is about 24. The space telescope (Chapter 6) is expected to be able to record the light of stars at least 3 magnitudes fainter.

The so-called first-magnitude stars are not all of the same apparent brightness. The brightest-appearing star, Sirius, sends us about 10 times as much light as the average star of first-magnitude, and so has a magnitude of 1.0 − 2.5 (see Table 13.1), or of about −1.5. Several of the planets appear even brighter; Venus, at its brightest, is of magnitude −4. The sun has a magnitude of −26.5. Some magnitude data are given in Table 13.2. It is of interest to note that the brightness of the sun and Sirius differ by 25 magnitudes—a factor of 10,000 million (10^{10}) in light energy, or flux—and that we also receive 10^{10} times as much light from Sirius as from the faintest stars that can be photo-

[1] If m_1 and m_2 are the magnitudes corresponding to stars from which we receive light flux in the amounts l_1 and l_2, the difference between m_1 and m_2 is defined by

$$m_1 - m_2 = 2.5 \log \frac{l_2}{l_1} .$$

graphed with the 5.08-m telescope. The entire range of light flux represented in Table 13.2 covers a ratio of about 10^{20} to 1.

We have seen that the magnitude differences between objects indicate the *relative* amounts of luminous flux received from them. To set the scale unambiguously, accurate photoelectric measures have been made of a large number of stars distributed over the sky. These stars serve as modern standards, with respect to which other stars are compared. The scale has been adjusted so that the brightest-appearing stars average about the first magnitude, in keeping with astronomical tradition.

The "Real" Brightnesses of Stars

Even if all stars were identical, and if interstellar space were entirely free of absorbing matter, stars would not all appear to have the same brightness, because they are at different distances from us, and the light that we receive from a star is inversely proportional to the square of its distance (Chapter 5). The apparent brightnesses of stars therefore do not provide a basis for comparing the amounts of light that they actually emit into space. To make such a comparison we would first have to calculate how much light we would receive from each star if all stars were at the same distance from us.

The sun, for example, gives us thousands of millions of times as much light as any of the other stars, but on the other hand, it is hundreds of thousands of times as close to us as any other star is. To compare the intrinsic luminous outputs of the sun and of other stars, we first have to determine what magnitude the sun would have if it were at a specified distance, typical of the distances of the other stars. Suppose we choose 10 pc as a more or less representative distance of the nearer stars. Since 1 parsec is about 200,000 AU, the sun would be 2,000,000 times as distant as it is now if it were removed to a distance of 10 pc; consequently, it would deliver to us $(1/2{,}000{,}000)^2$ or $1/(4 \times 10^{12})$ of the light it now does. A factor of 4×10^{12} corresponds to about 31½ magnitudes (which can be verified by raising 2.512 to the 31.5 power). The sun, therefore, if removed to a distance of 10 pc, would appear fainter by some 31½ magnitudes than its present magnitude of -26.5; that is, the sun would then appear as a faint star of the fifth magnitude.

Similarly, we can use the inverse-square law of light to calculate how luminous all other stars of known distance would appear if they were 10 pc away. Suppose, for example, that a tenth-magnitude star has a distance of 100 pc. If it were only 10 pc away, it would be only one-tenth as far away, and hence 100 times as bright—a difference of five magnitudes. At 10 pc, therefore, it too would appear as a fifth-magnitude star.

We define the *absolute magnitude* of a star as the magnitude that star would have if it were at the standard distance of 10 pc (about 32.6 LY). The absolute magnitude of the sun is about $+5$. Most stars have absolute magnitudes that lie in the range 0 to $+15$. The extreme range of absolute magnitudes observed for normal stars is -10 to $+19$, a range of a factor of nearly 10^{12} in intrinsic light output. The absolute magnitudes of stars are measures of how bright they really are; they provide a basis for comparing the actual amount of light emitted by stars.

Distances of Stars from Their Magnitudes

The absolute magnitude of a star, of course, is independent of its distance. On the other hand, the magnitude of a star (sometimes called the *apparent magnitude* to avoid confusion with *absolute magnitude*) is a measure of how bright the star *appears* to be, and this depends on both the star's actual rate of light output and its distance. The *difference* between the star's apparent magnitude, symbolized m, and absolute magnitude, symbolized M, can be calculated from the inverse-square law of light and from a knowledge of how much greater or less than 10 pc the star's distance actually is. The difference $m - M$ therefore depends only on the distance of the star.[2]

---

[2] Let $l(r)$ be the observed light of a star at its actual distance, r, and $l(10)$ the amount of light we would receive from it if it were a distance of 10 pc. From the definition of magnitudes, we have

$$m - M = 2.5 \log \frac{l(10)}{l(r)} .$$

and from the inverse-square law of light,

$$\frac{l(10)}{l(r)} = \left(\frac{r}{10}\right)^2 .$$

Combining the above equations, we obtain

$$m - M = 5 \log \frac{r}{10} .$$

The quantity, $5 \log (r/10)$, is called the *distance modulus*.

For example, suppose the difference, $m - M$, for some star is 10 magnitudes. Ten magnitudes (see Table 13.1) corresponds to a ratio of 10,000:1 in light. Thus, we actually receive from the star $1/10{,}000$ of the light that we would receive if it were 10 pc away; it must, therefore, be 100 times as distant as 10 pc, or at a distance of 1000 pc. In the next chapter we shall see that the absolute magnitude of a star can often be inferred from its spectrum. Since the apparent magnitude of a star can be observed, a knowledge of its absolute magnitude is equivalent to a knowledge of its distance.

Most of us make use of this same principle, subconsciously, in everyday life. Every experienced motorist has an intuitive notion of the actual brightness of a stop light. If, while driving down the highway at night, he sees a stop light, he judges its distance from its apparent faintness. In other words, the difference between the light's *apparent* and *real* brightness indicates its distance. The computation of a star's distance from its apparent and absolute magnitudes is analogous.

Colors of Stars

Every device for detecting light has a particular color or spectral sensitivity. The human eye, for example, is most sensitive to green and yellow light; it has a lower sensitivity to the shorter wavelengths of blue and violet light and to the longer wavelengths of orange and red light. It does not respond at all to ultraviolet or to infrared radiation. The eye, in fact, responds roughly to the same kind of light that the sun emits most intensely; this coincidence is probably not accidental—the eye may have evolved to respond to the kind of light most available on earth.

Another detecting device is the photographic plate (or film). The early photographic emulsions, before the development of yellow- and red-sensitive and panchromatic emulsions, were sensitive only to violet and blue light and did not respond to light of wavelengths longer than about 5000 Å (in the blue-green). The basic photographic emulsion is still sensitive to violet and blue; dyes must be added to the basic emulsion to make it sensitive to longer wavelengths.

Suppose, now, that the total amount of light energy entering a telescope from each of two stars is exactly the same if light of all wavelengths is considered, but that one star emits most of its light in the blue spectral region and the other in the yellow spectral region. If these stars are observed visually (that is, by looking at them through the telescope), the yellow one will appear brighter, that is, will have a numerically smaller magnitude, because the eye is less sensitive to most of the light emitted by the blue star. If the stars are photographed on a blue-sensitive photographic plate, however, the blue star will produce the more conspicuous image; measures of the photographic images will show the blue star appearing brighter and having the smaller magnitude. Consequently, when a magnitude system is defined, it is necessary also to specify how the magnitudes are to be measured—that is, what detecting device is to be used.

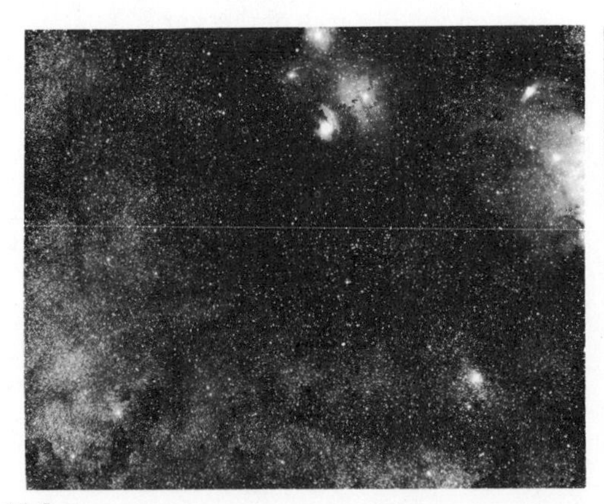

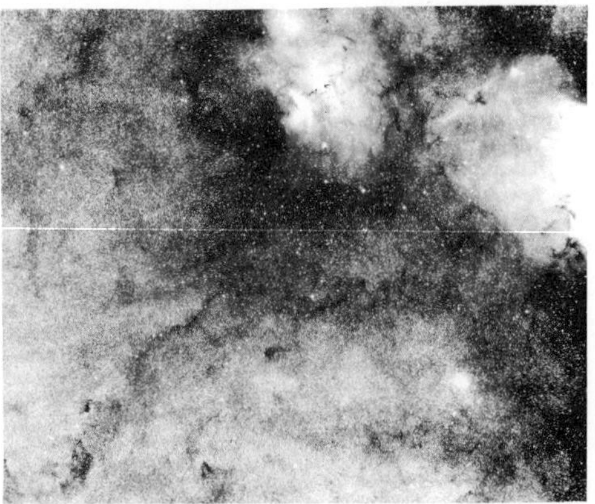

Figure 13.2
Two photographs of the same region of the Milky Way taken with the 124-cm Schmidt telescope: *(left)* on a photographic emulsion sensitive to blue light; *(right)* through a red filter and on a photographic emulsion sensitive to red light. Note the difference with which the stars and nebulae show up in the different colors. *(National Geographic Society–Palomar Observatory Sky Survey; reproduced by permission from the Hale Observatories)*

Figure 13.3
Two photographs of the same region of the sky (in Coma Berenices). The upper is in blue light, and the lower in red; note how the colors of the stars are apparent from comparison of the two. These are negative prints, the stars showing as black dots on a white sky. *(National Geographic Society–Palomar Observatory Sky Survey; reproduced by permission from the Hale Observatories)*

Magnitudes, whether apparent or absolute, that are based on stellar brightness as they are observed with the human eye are called *visual magnitudes* or *absolute visual magnitudes*. Today photographic plates or photomultipliers can be used in conjunction with many kinds of color filters to produce a great variety of different magnitude systems—for example, red magnitudes, ultraviolet magnitudes, infrared magnitudes, and so on. A certain few spectral bands have become more or less standard, however, and are now widely used to define magnitudes.

Since about 1960 one commonly used system of magnitudes has been U (ultraviolet), B (blue), and V (visual). The U and B magnitudes are obtained from measures of the flux from stars through certain standardized ultraviolet and blue filters with a common type of photomultiplier. The visual magnitude (V) is measured with the same photocell through a filter that approximates the response of the human eye. The *difference* between any two of these magnitudes, say, between blue and visual magnitudes ($B - V$), is called a *color index*. Since the inverse-square law of light applies equally to all wavelengths, the color index of a star would not change if the star's distance were changed.

A very blue star appears brighter through a blue filter than through a yellow one; its blue magnitude, therefore, is algebraically *less* than its visual magnitude, and its $B - V$ color index is *negative*. A yellow or red star, on the other hand, has a brighter (smaller) visual magnitude, and a *positive* color index. Color indices, therefore, provide measures of the *colors* of stars. Colors, in turn, indicate the temperatures of stars. Ultraviolet, blue, and visual magnitudes are adjusted to be equal to each other, so that they give a color index of zero to a star with a temperature of about 10,000 K. The $B - V$ color indices of stars range from -0.4 for the bluest to more than $+2.0$ for the reddest.

If a star field is photographed on emulsions and through filters chosen to isolate the blue and visual spectral regions, the blue and visual magnitudes and color indices so obtained comprise a powerful tool for determining the temperatures of all the stars appearing on the photographs.

Bolometric Magnitudes and Luminosities

A magnitude system based on *all* the electromagnetic energy reaching the earth from the stars, rather than just from some specified wavelength range, would seem to be most fundamental. Magnitudes so based are called *bolometric magnitudes*, m_{bol}, and the bolometric magnitudes that stars would have at a distance of 10 pc are *absolute bolometric magnitudes*, M_{bol}. Unfortunately, bolometric magnitudes are difficult to observe directly because some wavelengths of electromagnetic energy do not penetrate the earth's atmosphere. Although most of the radiation from stars like the sun does reach the earth's surface, a large part of the energy from stars that are substantially hotter or cooler than the sun lies in the far ultraviolet or in the infrared, which is blocked by the earth's atmosphere and cannot be observed from the ground; for those stars, bolometric magnitudes can only be estimated or calculated from theoretical considerations unless the stars are observed from rockets or satellites. The scale is set so that the bolometric magnitude of a star like the sun is almost the same as its visual magnitude.

The absolute bolometric magnitude of a star is a measure of the rate of its entire output of radiant energy. The rate at which a star pours radiant energy into space, usually expressed in ergs per second, is called its *luminosity*.

We find the luminosity of the sun by measuring the rate at which its radiation falls on the earth. It is found that a surface area of 1 cm^2 just outside the atmosphere and oriented perpendicular to the direction of the sun receives from the sun 1.36×10^6 erg/s. This value is known as the *solar constant*.

The total energy that leaves the sun during an interval of one second diverges outward, away from the sun, in all directions. Since one astronomical unit is 1.49×10^{13} cm, the area of the spherical surface over which the solar radiation has spread by the time it reaches the earth's distance from the sun (about eight minutes later) is 2.8×10^{27} cm^2. The solar constant of 1.36×10^6 ergs/s/cm^2 is the energy that crosses just one of those square centimeters. The total energy that leaves the sun in one second—its luminosity—is thus $1.36 \times 10^6 \times 2.8 \times 10^{27} = 3.8 \times 10^{33}$ ergs/s. The corresponding absolute bolometric magnitude of the sun is $+4.6$.

A dramatic illustration of the magnitude of that amount of energy is obtained by imagining a bridge of ice 3 km wide and 1.5 km thick and extending over the 150-million-km span from the earth to the sun. If all the sun's radiation could be directed along the bridge, it would be enough to melt the entire column of ice in one second.

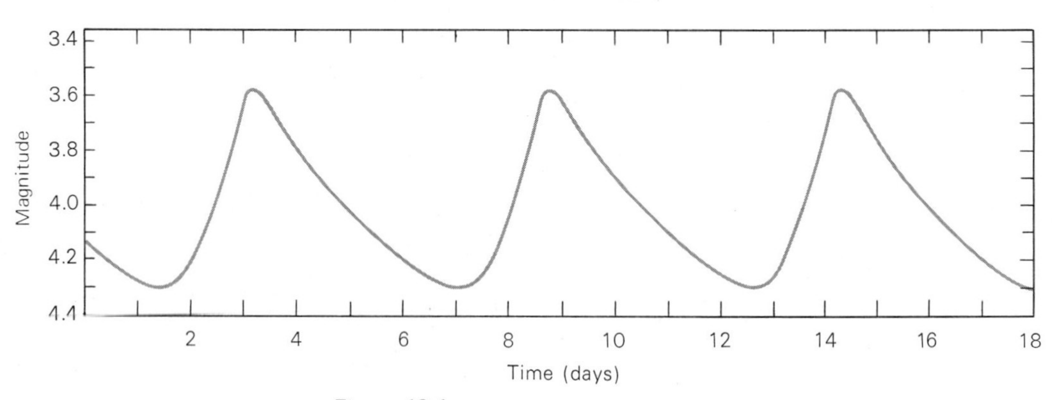

Figure 13.4
Light curve of a typical cepheid variable.

STARS THAT VARY IN LIGHT

Most stars shine with constant light. A minority, however, are variable in magnitude. The standard international index of stars that vary in light is the Soviet *General Catalogue of Variable Stars*. The 1968 edition of this catalogue (the most recent at the time of writing) lists 20,448 known variable stars in our Galaxy, but supplements of this catalog increase the number yearly.

A variable star is studied by analyzing its spectrum and by measuring the variation of its light with lapse of time. Some stars show light variations that are apparent to the unaided eye. Generally, however, the apparent brightness of a variable star is determined by telescopic observation.

A graph that shows how the magnitude of a variable star changes with time is called a *light curve* of that star. An example is given in Figure 13.4. The *maximum* is the point on the light curve where the maximum amount of light is received from the star; the *minimum* is the point where the least amount of light is received. If the light variations of a variable star repeat themselves periodically, the interval between successive maxima is called the *period* of the star. The *median light* of a variable star is the amount of light it emits when it is halfway between its maximum and minimum brightness. The *amplitude* is the difference in light (usually expressed in magnitudes) between the maximum and the minimum. The amplitudes of variable stars range from less than 0.1 to several magnitudes.

Types of Variable Stars

The *General Catalogue of Variable Stars* lists three types of variable stars: (1) pulsating variables, (2) eruptive variables, and (3) eclipsing variables. *Pulsating variables* are stars that periodically expand and contract, pulsating in size as well as in light. *Eruptive variables* (including novae and supernovae—see Chapter 17) are stars that show sudden, usually unpredictable, outbursts of light, or, in some cases, diminutions of light. *Eclipsing variables* (or eclipsing binaries) are "close" binary (double) stars whose orbits of mutual revolution lie nearly edge-on to our line of sight and which periodically eclipse each other. Eclipsing variables are not, of course, true variable stars; they will be discussed in the next chapter. The cataloged numbers of different kinds of variable stars (in 1968) are summarized in Table 13.3. More complete data are given in Appendices 15 and 16.

We shall return to the physical properties of variable stars and to the causes of their pulsations later (Chapter 17). Some characteristic and easily recognized variable stars, however, have well-determined absolute

TABLE 13.3 Numbers of Variable Stars

Type	Number
Pulsating	13,782
Eruptive	1,618
Eclipsing	4,062
Unclassified or unstudied	986
All kinds	20,448

magnitudes at median light; thus when they are identified in remote clusters or systems of stars, they provide a means of finding the distances to those systems. We describe here the more important of these objects.

Long-Period Variables

The largest group of pulsating stars consists of the *red variables*, or *Mira-type* stars; these are named for their prototype, Mira, in the constellation of Cetus. They are giant stars that pulsate in very long and somewhat irregular periods of months or years. Because they are not highly predictable, an important service is provided by amateur astronomers who keep track of the magnitudes of these stars. It would require far too much of the time of professional astronomers to maintain constant vigil on all of them. The *American Association of Variable Star Observers* has a well-planned program of careful surveillance of these long-period and irregular variables and has been gathering valuable data on them for years.

Cepheid Variables

Although relatively rare, the cepheid variables are very important in astronomy. They are large yellow stars named for the prototype and first known star of the group, δ Cephei. The variability of δ Cephei was discovered in 1784 by the young English astronomer John Goodricke just two years before his death at the age of 21. The magnitude of δ Cephei varies between 3.6 and 4.3 in a period of 5.4 days. The star rises rather rapidly to maximum light and then falls more slowly to minimum light (see Figure 13.4).

More than 700 cepheid variables are known in our galaxy. Most cepheids have periods in the range 3 to 50 days and absolute magnitudes (at median light) from -1.5 to -5. The amplitudes of cepheids range from 0.1 to 2 magnitudes. Polaris, the *North Star*, is a small-amplitude cepheid variable that varies between magnitudes 2.5 and 2.6 in a period of just under four days.

The importance of cepheid variables lies in the fact that a relation exists between their periods of pulsation (or light variation) and their median luminosities, that is, their absolute magnitudes at median light. The relation was discovered in 1912 by Henrietta Leavitt, an astronomer of the Harvard College Observatory, when she studied the cepheid variables revealed on photographs of the Large and Small Magellanic Clouds, two great stellar systems that are actually neighboring galaxies (although they were not known to be galaxies in 1912—see Chapter 20). Miss Leavitt found that the brighter-appearing cepheids always have the longer periods of light variation.

Harlow Shapley was one of the astronomers who recognized the importance of cepheids as distance indicators, and he pioneered the work of determining distances to some of them in our Galaxy. His work was extended by others, and today we have a fair knowledge of the absolute magnitudes of the cepheids of different pulsation periods. When these stars are recognized in remote stellar systems, we can determine their absolute magnitudes from their periods. Then, by observing their apparent magnitudes we can find their distances, and hence the distances of the stellar systems in which those cepheids lie (see Chapter 20).

There are actually two kinds of cepheids. The most common, those of type I, are the ones described above. Cepheids of the other type, type II, are rare, and many of those known are found in the globular star clusters (Chapter 15). Type II cepheids obey a period-luminosity law too, although they are about four times fainter than type I cepheids of the same periods.

RR Lyrae Stars

Next to the long-period variables, the most common variable stars are the *RR Lyrae* stars, named for RR Lyrae, best known member of the group. Nearly 4500 of these variables are known in our Galaxy. Almost all of them are found in the nucleus or the corona of our Galaxy (Chapter 15) or in globular clusters. In fact, nearly all globular clusters contain at least a few RR Lyrae variables, and some contain hundreds; these stars, therefore, are sometimes called *cluster-type* variables.

The periods of RR Lyrae stars are less than one day; most periods fall in the range 0.3 to 0.7 day. Their amplitudes never exceed two magnitudes, and most RR Lyrae stars have amplitudes less than one magnitude. Several subclasses of RR Lyrae stars are recognized, but the differences between these subclasses are small and need not be considered here.

It is observed that the RR Lyrae stars occurring in any particular globular cluster all have about the same median apparent magnitude. Since they are all at approximately the same distance, it follows that they must also have nearly the same absolute magnitude. Because the RR Lyrae stars in different clusters are all similar to each other in observable characteristics, it is reasonable to assume that *all* RR Lyrae stars have

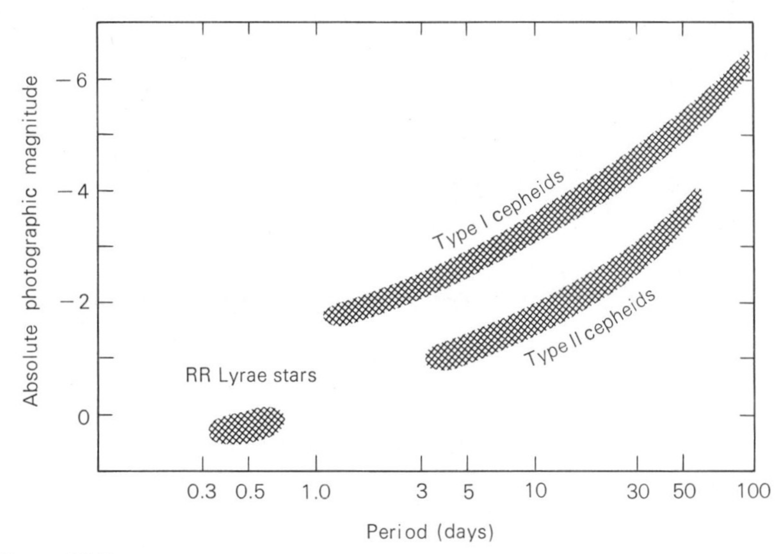

Figure 13.5
Period-luminosity relation for type I cepheids, type II cepheids, and RR Lyrae stars.

about the same absolute magnitude. Recent work shows that RR Lyrae stars average between 0 and +1 in absolute magnitude. Figure 13.5 displays the ranges of periods and absolute magnitudes for the two types of cepheids and the RR Lyrae stars.

SPECTRA OF STARS

About 1665 Newton showed (perhaps not for the first time) that white sunlight is really a composite of all colors of the rainbow (Chapter 5), and that the various colors, or wavelengths, of light could be separated by passing the light through a glass prism. William Wollaston first observed dark lines in the solar spectrum, and Joseph Fraunhofer cataloged about 600 such dark lines. As early as 1823, Fraunhofer observed that stars, like the sun, also have spectra that are characterized by dark lines crossing a continuous band of colors. Sir William Huggins, in 1864, first identified some of the lines in stellar spectra with those of known terrestrial elements.

The Spectral Sequence

When the spectra of different stars were observed, it was found that they differed greatly among themselves. In 1863 the Jesuit astronomer Angelo Secchi classified stars into four groups according to the general arrangement of the dark lines in their spectra. Secchi's scheme subsequently was modified and augmented, until today we recognize seven such principal *spectral classes*.

As we have seen (Chapter 5), each dark line in a stellar spectrum is due to the presence of a particular chemical element in the atmosphere of the star observed. It might seem, therefore, that stellar spectra differ from each other because of differences in the chemical makeup of the stars. Actually, the principal differences in stellar spectra are caused by the widely differing temperatures in the outer layers of the various stars. Hydrogen, for example, is by far the most abundant element in all stars (except those at advanced stages of evolution—Chapter 17). In the atmospheres of the hottest stars, however, hydrogen atoms are completely ionized, and can thus produce no absorption lines. In the atmospheres of the coolest stars hydrogen is neutral and can produce absorption lines, but in these stars practically all of the hydrogen atoms are in the lowest energy state (unexcited), and can absorb only those photons that can lift them from that first energy level to higher ones; the photons so absorbed produce the *Lyman series* of absorption lines (Chapter 5), which lies in the unobservable ultraviolet part of the spectrum. In a stellar atmosphere with a temperature of about 10,000 K, many hydrogen atoms are not ionized; nevertheless, an appreciable number of

TABLE 13.4 Spectral Sequence

Spectral Class	Color	Approximate Temperature (K)	Principal Features	Stellar Examples
O	Blue	>25,000	Relatively few absorption lines in observable spectrum. Lines of ionized helium, doubly ionized nitrogen, triply ionized silicon, and other lines of highly ionized atoms. Hydrogen lines appear only weakly.	10 Lacertae
B	Blue	11,000–25,000	Lines of neutral helium, singly and doubly ionized silicon, singly ionized oxygen and magnesium. Hydrogen lines more pronounced than in O-type stars.	Rigel Spica
A	Blue	7,500–11,000	Strong lines of hydrogen. Also lines of singly ionized magnesium, silicon, iron, titanium, calcium, and others. Lines of some neutral metals show weakly.	Sirius Vega
F	Blue to white	6,000–7,500	Hydrogen lines are weaker than in A-type stars but are still conspicuous. Lines of singly ionized calcium, iron, and chromium, and also lines of neutral iron and chromium are present, as are lines of other neutral metals.	Canopus Procyon
G	White to yellow	5,000–6,000	Lines of ionized calcium are the most conspicuous spectral features. Many lines of ionized and neutral metals are present. Hydrogen lines are weaker even than in F-type stars. Bands of CH, the hydrogen radical, are strong.	Sun Capella
K	Orange to red	3,500–5,000	Lines of neutral metals predominate. The CH bands are still present.	Arcturus Aldebaran
M	Red	<3,500	Strong lines of neutral metals and molecular bands of titanium oxide dominate.	Betelgeuse Antares

them are excited to the second energy level, from which they can absorb additional photons and rise to still higher levels of excitation. These photons correspond to the wavelengths of the *Balmer series,* which is in the part of the spectrum that is readily observable. Absorption lines due to hydrogen, therefore, are strongest in the spectra of stars whose atmospheres have temperatures near 10,000 K, and they are less conspicuous in the spectra of both hotter and cooler stars, even though hydrogen is, roughly, equally abundant in all the stars. Similarly, every other chemical element, in each of its possible stages of ionization, has a characteristic temperature at which it is most effective in producing absorption lines in the observable part of the spectrum.

Once we have ascertained how the temperature of a star can determine the physical states of the gases in its outer layers, and thus their ability to produce absorption lines, we need only to observe what patterns of absorption lines are present in the spectrum of a star to learn its temperature. We can therefore arrange the seven classes of stellar spectra in a continuous sequence in order of decreasing temperature. In the hottest stars (temperatures over 25,000 K) only lines of ionized helium and highly ionized atoms of other elements are conspicuous. Hydrogen lines are strongest in stars with atmospheric temperatures of about 10,000 K. Ionized metals provide the most conspicuous lines in stars with temperatures from 6000 to 8000 K. Lines of neutral metals are the strongest in somewhat cooler stars. In the coolest stars (below 4000 K), bands of some molecules are very strong. The most important among the molecular bands are those due to titanium oxide, a tenacious chemical compound which can exist at the temperatures of the cooler stars. The sequence of spectral types is summarized in Table 13.4 and Figure 13.6.

Spectrum Analysis

The design of a spectroscope or spectrograph is described in Chapter 6. Stellar spectra were first observed by means of a spectroscope placed at the focus of a telescope. With a magnifying glass (eyepiece) the observer viewed the spectrum of the light from a star

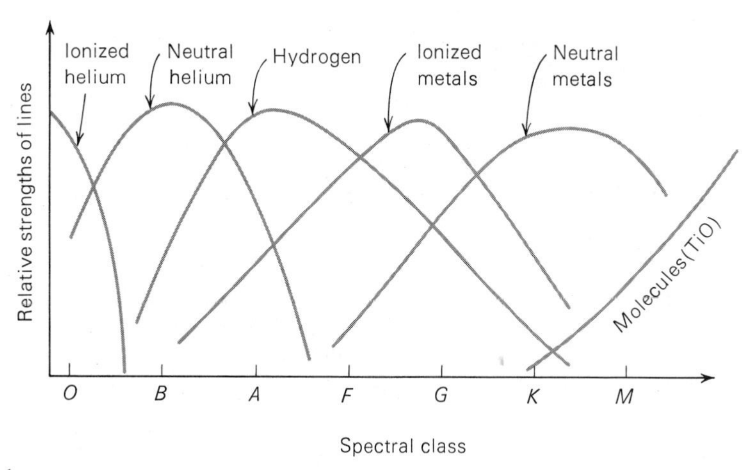

Figure 13.6
Relative intensities of different absorption lines in stars at various places in the spectral sequence.

gathered by the telescope and passed through the apparatus. Today, however, stellar spectra are usually photographed or scanned photoelectrically.

In practice, the spectra of the star and a laboratory source are both photographed on the same negative. The laboratory source is often iron vaporized in an electric arc; the light emitted from the glowing iron vapor is passed through portions of the slit of the spectrograph adjacent to the portion through which the starlight is passed. The spectrum of the iron arc consists of many bright emission lines whose wavelengths have been measured accurately in the labora-

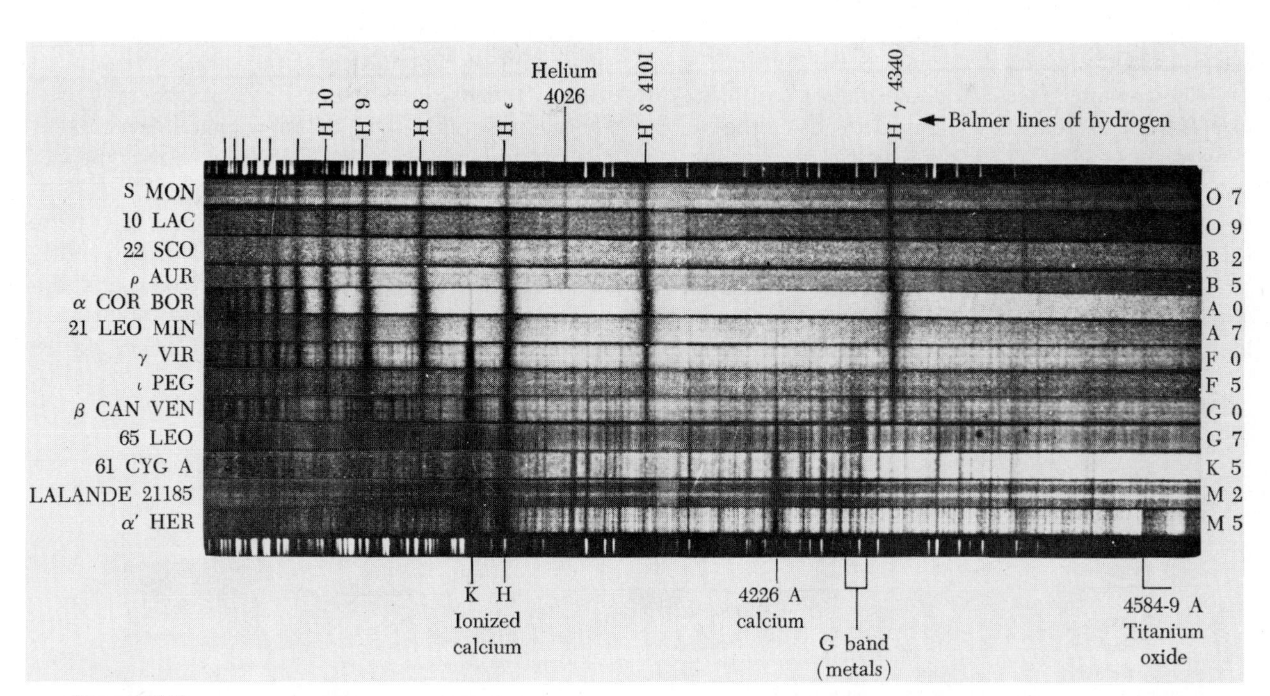

Figure 13.7
Spectra of several stars of representative spectral classes. A comparison spectrum is shown at top and bottom. *(UCLA Observatory)*

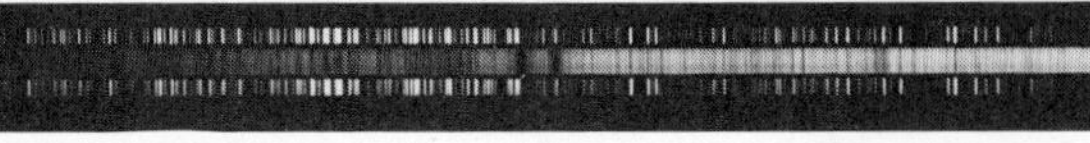

Figure 13.8
A typical stellar spectrogram. The bright streak along the middle, crossed by dark lines, is the spectrum of the star. The bright lines flanking it on either side form the comparison (emission) spectrum of iron.

tory (Figure 13.8). They serve as a convenient standard with which the absorption or emission lines in the star's spectrum can be compared, so that the wavelengths of the stellar lines can be determined. In this way, the chemical elements that give rise to the lines can be identified, and the Doppler shift (Chapter 5) of those lines (due to the radial velocity of the star) can be measured.

As we have seen, the spectrum of a star is influenced mostly by its temperature. Now the radiation we observe from a star emerges from a range of depths in its outer layers. Thus the temperature we derive from its spectrum corresponds to an average of the temperatures of gases at various depths in the star's atmosphere. For the sun, those levels from which light escapes and is radiated into space comprise a zone about 200 km thick. This region of the sun (or of any star) is called its *photosphere*.

The pressure in a stellar photosphere also affects its spectrum, because the pressure depends on the density of the photospheric gases as well as on their temperature. At the very low densities expected for the very extended tenuous photospheres of giant stars the pressures are also low, and ionized atoms recombine with electrons more slowly. These low density gases, therefore, maintain a higher average degree of ionization than high density gas of the same temperature. Subtle details in the spectrum of a star of high photospheric pressure thus enable us to distinguish it from the spectrum of a giant star of the same temperature but of low photospheric pressure.

Dark lines of a majority of the known chemical elements have now been identified in the spectra of the sun and stars. The lines of *all* elements are not observable in the spectrum of a star, nor are lines of any one element visible in the spectra of *all* stars. As we have seen, because of variations among the stars in temperature and pressure of the photosphere, only certain of the prevailing kinds of atoms are able to produce absorption lines in any one star. The absence of the lines of a particular element, therefore, does not necessarily imply that that element is not present. Only if the physical conditions in the photosphere of a particular star are such that lines of an element *should* be visible were the element present in reasonable abundance in that star, can we conclude that the absence of observable spectral lines implies low abundance of the element. On the other hand, spectral lines of an element, in the neutral state or in one of its ionized states, certainly imply the presence of that element in the star.

Once due allowance has been made for the prevailing conditions of temperature and pressure in a star's photosphere, analyses of the strengths of absorption lines in its spectrum can yield information regarding the relative abundances of the various chemical elements whose lines appear. It is found that the relative abundances of the different chemical elements in the sun and in most stars (as well as in most other regions of space that have been investigated) are approximately the same. Hydrogen comprises from 60 to 80 percent of the mass of most stars. Hydrogen and helium together comprise from 96 to 99 percent of the mass; in some stars they comprise more than 99.9 percent. Among the 4 percent or less of "heavy ele-

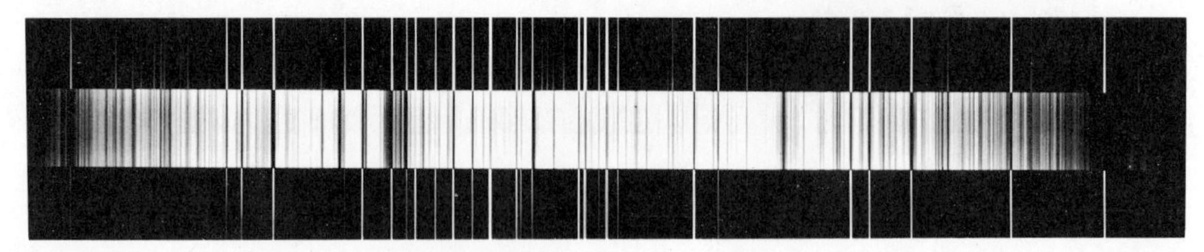

Figure 13.9
A portion of the solar spectrum and, flanking it, a comparison spectrum of iron photographed with the same spectrograph. Notice that many of the dark lines in the solar spectrum are matched by the bright lines in the comparison spectrum, showing that iron is present in the sun. *(Lick Observatory)*

ments," neon, oxygen, nitrogen, carbon, magnesium, argon, silicon, sulfur, iron, and chlorine are among the most abundant. Generally, but not invariably, the elements of lower atomic weight are more abundant than those of higher atomic weight.

The radial or "line-of-sight" velocity of a star can be determined from the Doppler shift of the lines in its spectrum (Chapter 5). W. Huggins made the first radial-velocity determination of a star in 1868. He observed the Doppler shift in one of the hydrogen lines in the spectrum of Sirius and found that the star is approaching the solar system.

If a star is rotating, unless its axis of rotation happens to be directed exactly toward the sun, one of its limbs approaches us and the other recedes from us, relative to the star as a whole. For the sun or a planet, we can observe the light from one limb or the other and measure directly the Doppler shifts that arise from the rotation. A star, however, appears as a point of light, and we are obliged to analyze the light from its entire disk at once. Nevertheless, if the star is rotating, part of the light from it, including the spectral lines, is shifted to shorter wavelengths and part is shifted to longer wavelengths. Each spectral line of the star is a composite of spectral lines originating from different parts of the star's disk, all of which are moving at different speeds with respect to us. The effect produced by a rapidly rotating star is that all its spectral lines are broadened to a characteristic "dish" shape. The amount of this rotational broadening of the spectral lines, if observable, can be measured, and a lower limit to the rate of rotation of the star can be calculated.

In the presence of a magnetic field each energy level of an atom splits up into several levels. The result is that the spectral lines formed by atomic transitions from those levels are separated into several lines each, the spacing of which depends on the strength of the magnetic field. The splitting of spectral lines of atoms due to this so-called *Zeeman effect* is generally so slight as to go unnoticed. However, in some stars the magnetic fields are strong enough that the incipient splitting of their spectral lines makes them appear broader than normal. This characteristic *Zeeman broadening* enables us to detect the strength of those fields of stars which have a high degree of magnetism in their photospheres.

In the spectra of some stars, absorption lines are observed that do not appear to originate in the photospheres. Sometimes such lines can be associated with shells or rings of material ejected by the star. If enough material is ejected, and if the star has radiation of high enough energy to excite or ionize the gas, the latter produces emission lines instead of absorption lines, superposed on the stellar spectrum. Examples are Wolf-Rayet stars and other emission-line stars.

SUMMARY REVIEW

Photometry: magnitudes; luminous flux; photographic photometry; photoelectric photometry; Pogson scale; limiting magnitude; photographic limiting magnitude of a telescope; magnitude the sun would have at a distance of 10 pc; absolute magnitude; apparent magnitude; calculation of distances of stars from the differences between their apparent and absolute magnitudes; detecting devices—the eye, photographic emulsions, etc.; colors of stars; visual magnitudes (and visual absolute magnitudes); U, B, V systems of magnitudes; color index; color index and stellar temperatures; bolometric magnitudes; absolute bolometric magnitudes; stellar luminosities; solar constant; luminosity of the sun; absolute bolometric magnitude of the sun

Variable stars: *General Catalogue of Variable Stars;* light curve; period of a variable star; median light; amplitude; types of variables—pulsating, eruptive, and eclipsing; long period variables (red variables, Mira stars); American Association of Variable Star Observers (AAVSO); cepheid variables; period luminosity relation; RR Lyrae stars

Spectra of stars: Fraunhofer lines; spectral classes; the spectral sequence; role of temperature in determining the spectral class of a star; spectrum analysis; comparison spectrum; stellar or solar photosphere; role of pressure in the spectrum of a star; finding chemical composition of stars; distribution of different kinds of elements in stars; radial velocity measures of stars; stellar rotation; stellar magnitude fields; Zeeman effect; Zeeman broadening; circumstellar shells and rings; Wolf-Rayet stars and other emission-line stars

EXERCISES

1. Suppose that star A is just barely visible through a 15-cm telescope and that star B is just barely visible through a 30-cm telescope. Which star gives us more light and by what factor? What is the approximate difference in magnitude of the stars? (See Table 13.1.)

2. What magnitude would be assigned to an object from which we receive:
(a) 100 times as much light as from Venus when it is at its brightest?
(b) 1/100 as much light as a star at the naked-eye limit?
(c) no light at all?

3. Here are data on five stars:

Star	m	M
1	1.5	5.0
2	8.4	7.7
3	13.9	15.0
4	3.7	3.7
5	16.7	−7.0

(a) Which is nearest?
(b) Which is most distant?
(c) Which is 10 pc away?
(d) Which is brightest appearing?
(e) Which is faintest appearing?
(f) Which is intrinsically most luminous?
(g) Which is intrinsically least luminous?
(h) Which has a luminosity most like the sun's?
(i) Which of the stars would be visible without the aid of a telescope?

4. If a star has a color index of $B - V = 2.5$, how many times brighter in visual light does it appear than in blue light?

5. Suppose a type I cepheid variable is observed in a remote stellar system. The cepheid has a period of 50 days and an apparent magnitude of +20. What is the distance to the stellar system? (See Figure 13.5.)

6. Draw a diagram showing how two stars of equal bolometric magnitude, one of which is blue and the other red, would appear on photographs sensitive to blue and to yellow light.

7. Saturn is about 10 AU from the sun. Approximately what is the sun's apparent magnitude as seen from Saturn?

8. Consult Figure 13.5. How many times as bright is a cepheid of period 50 days as is a typical RR Lyrae star?

9. How many times farther away can a cepheid like that in Exercise 8 be than an RR Lyrae star and still be observed with the same telescope?

10. Explain why we can only find a lower limit to the rotation rate of a star from the rotational broadening of its spectral lines.

11. Explain (with a diagram) how stellar rotation broadens spectral lines.

12. Star A has lines of ionized helium in its spectrum, and star B has bands of titanium oxide. Which is the hotter? Why?

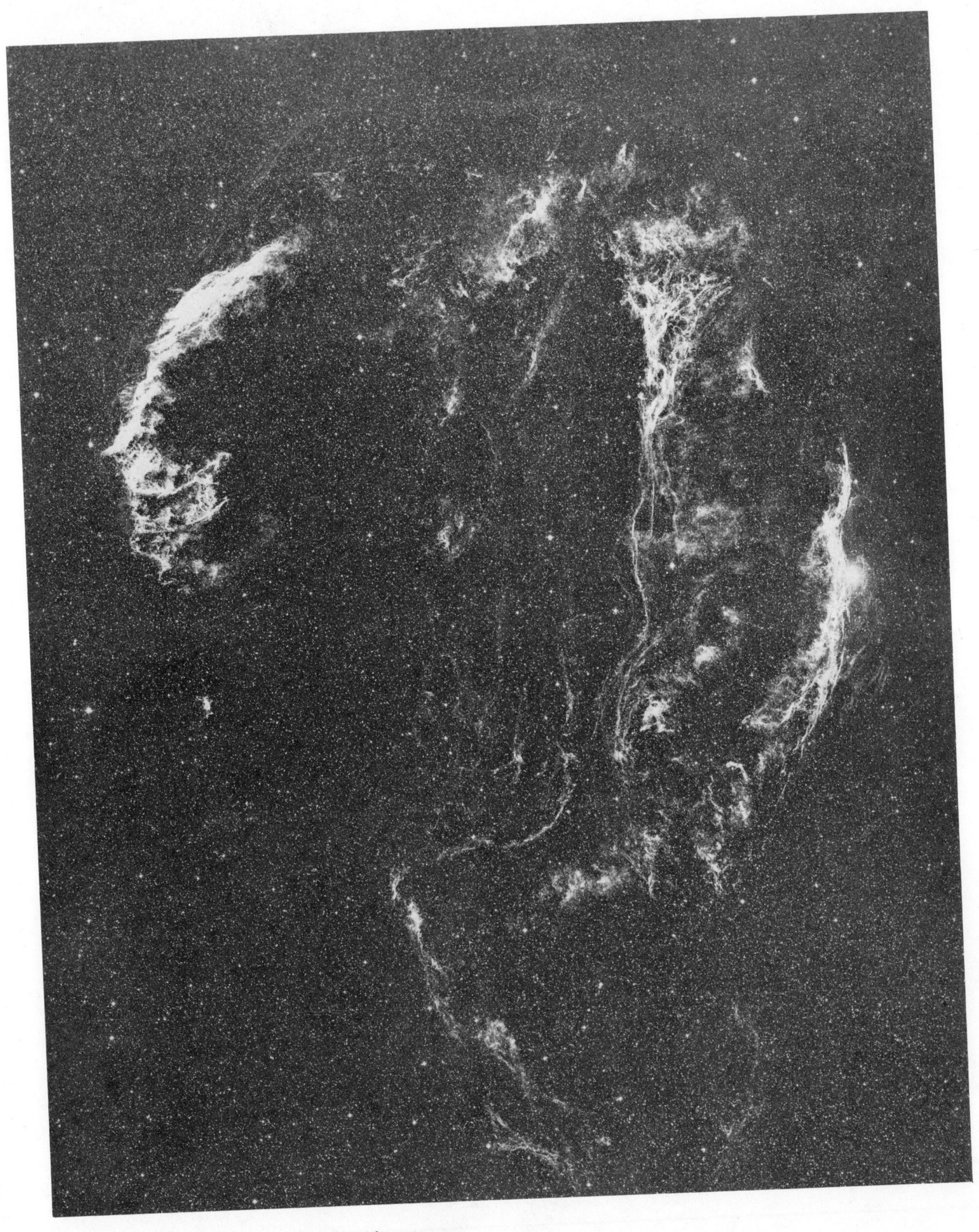

The Cygnus "Loop" nebula, the remnant of a prehistoric supernova (exploding star), photographed with the 48-inch Schmidt telescope. *(Hale Observatories)*

CHAPTER 14

Henry Norris Russell (1877–1957), famous American astronomer, is best remembered for his discovery (independently of Ejnar Hertzsprung) of the main sequence. *(Princeton University Archives)*

PROPERTIES OF STARS

Let's look at our stellar neighbors. The nearest and the brightest-appearing stars in the sky are listed in Appendices 13 and 14, respectively. Many of these are double or triple star systems; in such cases data are given for each component.

THE NEAREST AND THE BRIGHTEST STARS

The most striking thing about the brightest-appearing stars is that they are bright not because they are nearby, but because they are actually of high intrinsic luminosity. Of the 20 brightest stars listed in Appendix 14, only six are within 10 pc of the sun. Remember that the absolute magnitude of a star is the apparent magnitude it would have if it were at a distance of 10 pc. Since the 20 brightest stars are of apparent magnitude 1.5 or brighter, the 14 of them that are more distant than 10 pc must have absolute magnitudes *less* (that is, brighter) than 1.5. Even among the approximately 3000 stars with apparent magnitudes less than 6.0, only about 60 are within 10 pc. Most naked-eye stars are tens or even hundreds of parsecs away and are many times as luminous as the sun. Figure 14.1 is a histogram showing the distribution among various absolute visual magnitudes of the 30 brightest-appearing stars (the absolute visual magnitude of the sun is +4.8).

From Appendix 14 or Figure 14.1 we might gain the impression that the sun is far below average among stars in luminosity. Not so. Most stars are really much less luminous than the sun is. They are too faint, in fact, to be conspicuous unless they are nearby. Appendix 13 lists the 39 known stars within 5 pc of the sun, according to Gliese's 1969 *Catalogue of Nearby*

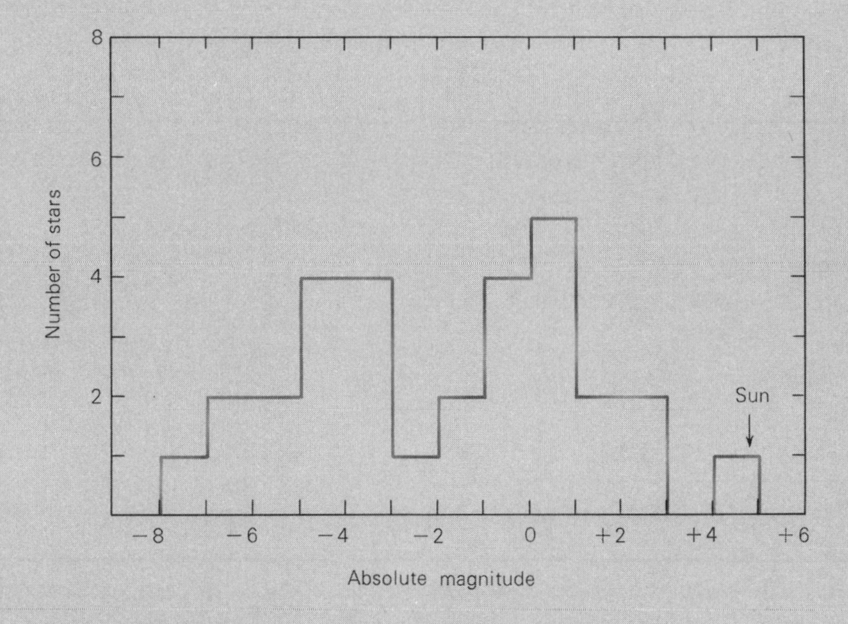

Figure 14.1
Distribution among absolute magnitudes of the 30 brightest-appearing stars. The units are the numbers of stars per unit of absolute magnitude.

Stars. (Note that most of these stars have large proper motion. As we saw in Chapter 12, nearby stars are often discovered because of their large proper motions.)

The Most Common Stars

Note that 11 of the 39 stars listed in Appendix 13 are binary, or multiple, star systems. Counting these companions, the table contains a total of 51 stars. Now only three of these 51 stars are among the 20 brightest-appearing stars: Sirius, Alpha Centauri, and Procyon. We see, then, that most of the nearest stars are intrinsically faint. Only eight of those 51 nearest stars are visible to the unaided eye. Only three are as intrinsically luminous as the sun; 37 have absolute magnitudes fainter than +10. If the stars in our immediate stellar neighborhoods are representative of the stellar population in general, we must conclude that the most numerous stars are those of low luminosity. Stars of high luminosity are rare—so rare that the chance of finding one within a small volume of space, say, within 10 pc of the sun, is very slight.

We can clarify this point with the help of some examples. The sun, whose absolute visual magnitude is +4.8, would appear as a very faint star to the naked eye if it were 10 pc away. Stars much less luminous than the sun would not be visible at all at that distance. Stars with absolute magnitudes in the range +10 to +15 are very common, but a star of absolute magnitude +10 would have to be within 1.6 pc to be visible to the naked eye. Only Alpha Centauri is closer than this. The intrinsically faintest star observed has an absolute magnitude of about +19. For this star to be visible to the naked eye, it would have to be within 0.025 pc, or 5200 AU. The star could not be photographed even with the 5.08-m telescope if it were more distant than 100 pc. It is clear, then, that the vast majority of nearby stars, those less luminous than the sun, do not send enough light across interstellar distances to be seen without optical aid.

In contrast, consider the highly luminious stars. Stars with absolute magnitudes of 0 have luminosities of about 100 times that of the sun. They are far less common than stars less luminous than the sun, but they are visible to the naked eye even out to a distance of 160 pc. A star with an absolute magnitude of −5 (10,000 times the sun's luminosity) can be seen without a telescope to a distance of 1600 pc (if there is no dimming of light by interstellar dust—see Chapter 15). Such stars are very rare, and we would not expect to find one within a distance of only 10 pc; the volume of space included within a distance of 1600 pc, however, is about 4 million times that included within a distance of only 10 pc. Hence many stars of high luminosity are visible to the unaided eye.

The Density of Stars in Space

There are at least 52 stars within 5 pc (counting the members of binary and multiple star systems and the sun). A sphere of radius 5 pc has a volume of

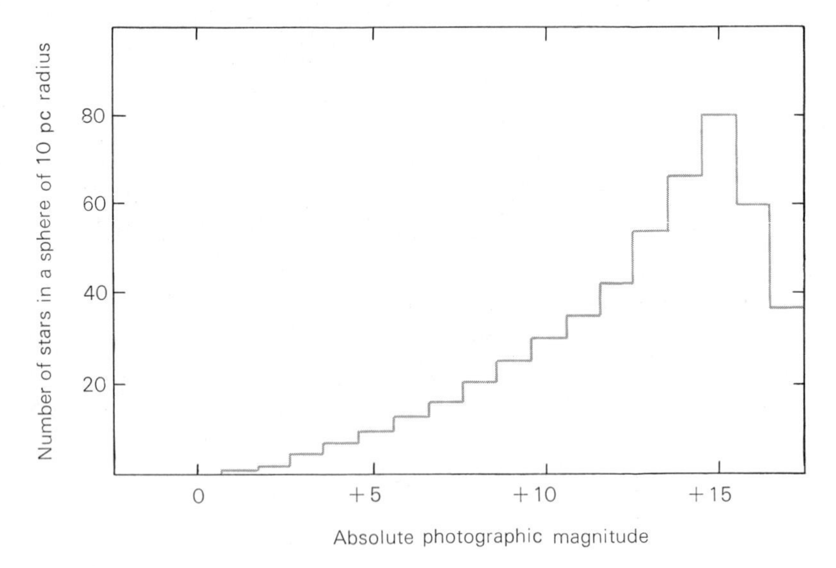

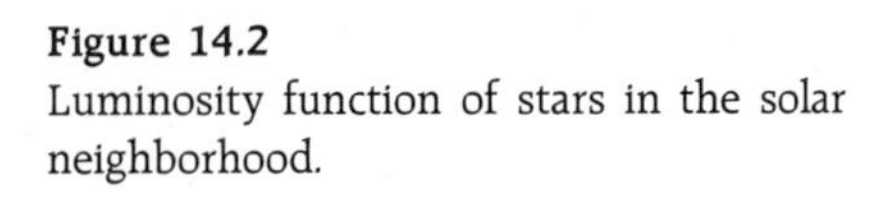
Figure 14.2
Luminosity function of stars in the solar neighborhood.

$^4/_3\pi(5)^3$, or about 520 pc^3. Since this volume of space contains at least 52 stars, the density of stars in space in the neighborhood of the sun is at least one star for every 10 pc^3; the actual stellar density, of course, can be greater than this figure if there are undiscovered stars with 5 pc. At this density we expect a total of between 3000 and 4000 stars within 20 pc. The mean separation between stars is the cube root of 10, or about 2.1 pc. If the matter contained in stars could be spread out evenly over space, and if a typical star has a mass of 0.4 times that of the sun, the mean density of matter in the solar neighborhood would be about 3×10^{-24} g/cm^3.

The Luminosity Function

Once the numbers of stars of various intrinsic luminosities have been found, the relative numbers of stars in successive intervals of absolute magnitude within any given volume of space can be established. This relationship is called the *luminosity function*. Figure 14.2 shows the luminosity function for stars in the solar neighborhood, as it has been determined by W. J. Luyten. Compare Figure 14.2 with Figure 14.1.

The sun, we see, is more luminous than the vast majority of stars. Most of the stellar mass is contributed by stars that are fainter than the sun. On the other hand, the relatively few stars of higher luminosity than the sun compensate for their small numbers by their high rate of energy output. It takes only ten stars of absolute magnitude 0 to outshine 1000 stars fainter than the sun, and only one star of absolute magnitude -5 to outshine 10,000 stars fainter than the sun. Most of the starlight from our part of space, it turns out, comes from the relatively few stars that are more luminous than the sun.

BINARY (DOUBLE) STARS

We have seen that among the 52 nearest stars, 23, or roughly one-half, are members of systems containing more than one star. The circumstance is fortunate, because analyses of binary systems provide us with our best means of learning stellar masses and sizes.

Discovery of Binary Stars

In 1650, less than half a century after Galileo turned a telescope to the sky, the Italian astronomer John Baptiste Riccioli observed that the star Mizar, in the middle of the handle of the Big Dipper, appeared through his telescope as two stars; Mizar was the first *double star* to be discovered. In the century and a half that followed, many other closely separated pairs of stars were discovered telescopically.

Usually, one star of a pair is brighter than the other. We have already seen (Chapter 12) how William Herschel, assuming that the fainter star was the more distant, thought that it might be possible to measure the parallax of the nearer star with respect to it. Accordingly, he began a search for such pairs, and between 1782 and 1821 he published three catalogs, listing more than 800 double stars. Actually, only rarely does a double star consist of one nearby and one distant star; the vast majority of these systems found by Herschel are *physical pairs* of stars, *revolving about each other*.

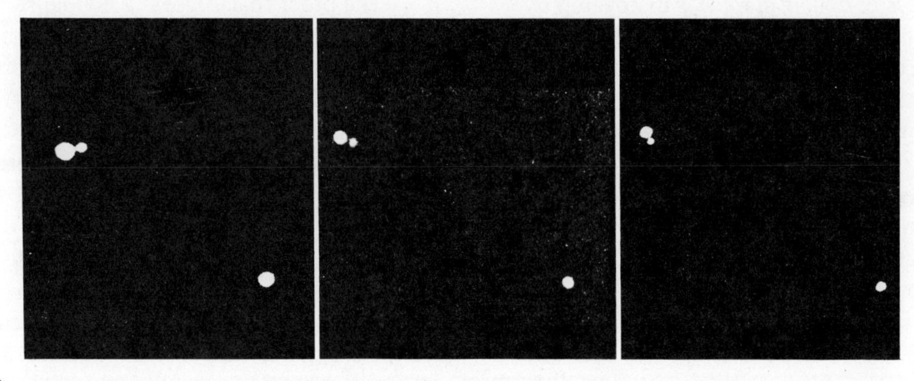

Figure 14.3
Three photographs, covering a period of about 12 years, which show the mutual revolution of the components of the double star Kruger 60. *(Yerkes Observatory)*

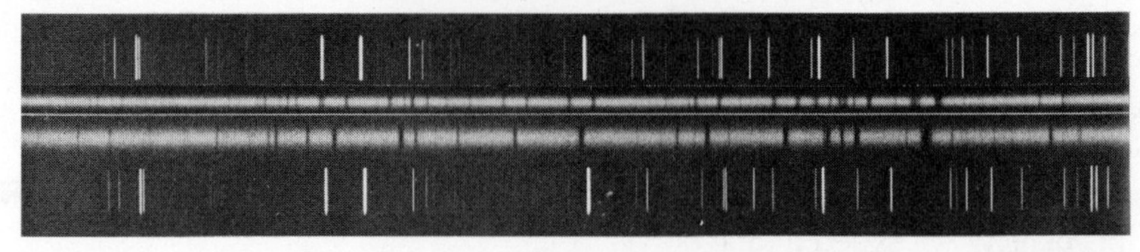

Figure 14.4
Two spectra of the spectroscopic binary κ Arietis. When the components are moving at right angles to the line of sight *(bottom)*, the lines are single. When one star is approaching us and the other receding *(top)*, the spectral lines of the two stars are separated by the Doppler shift. *(Lick Observatory)*

One famous double star is Castor, in Gemini. The telescope reveals Castor to be two stars separated by an angle of about 5″. By 1804 Herschel had noted that the fainter component of Castor had changed, slightly, its direction from the brighter component. Here, finally, was observational evidence that one star was moving about another; it was the first evidence that gravitational influences exist outside the solar system. Herschel had failed in his program to facilitate parallax determinations but had found something of far greater interest. As he put it, he was like Saul, who had gone out to seek his father's asses and had found a kingdom. His son, John Herschel, continued the search for double stars, and prepared a catalog (published posthumously) of more than 10,000 systems of two, three, or more stars.

If the gravitational forces between stars are like those in the solar system, the orbit of one star about the other must be an *ellipse*. The first to show that such is the case was Felix Savary, who in 1827 showed that the relative orbit of the two stars in the double system ξ *Ursae Majoris* is an ellipse, the stars completing one mutual revolution in a period of 60 years.

Another class of double stars was discovered by E. C. Pickering, at Harvard, in 1889. He found that the lines in the spectrum of the brighter component of Mizar (the first double star to be discovered) are usually *double*, but that the spacing of the components of the lines varies periodically, and at times the lines even become single. He correctly deduced that the brighter component of Mizar (Mizar A) itself is really *two* stars that revolve about each other in a period of 104 days. When one star is approaching us, relative to the center of mass of the two, the other star is receding from us; the radial velocities of the two stars, and therefore the Doppler shifts of their spectral lines, are different, so that when the composite spectrum of the two stars is observed, each line appears double. When the two stars are both moving across our line of sight, however, they both have the same *radial* velocity (that of the center of mass of the pair), and hence the spectral lines of the two stars coalesce.

Stars like Mizar A, which appear as single stars when photographed or observed visually through the telescope, but which the spectroscope shows really to be double stars, are called *spectroscopic binaries;* systems that can be observed visually as double stars are called *visual binaries*. In 1908 Frost found that the fainter component of Mizar, Mizar B, is also a spectroscopic binary.

Almost immediately following Pickering's discovery of the duplicity of Mizar A, Vogel discovered that the star Algol, in Perseus, is a spectroscopic binary. The spectral lines of Algol were not observed to be double, because the fainter star of the pair gives off too little light compared to the brighter for its lines to be conspicuous in the composite spectrum. Nevertheless, the periodic shifting back and forth of the lines of the brighter star gave evidence that it was revolving about an unseen companion; the lines of both components need not be visible in order for a star to be recognized as a spectroscopic binary.

The proof that Algol is a double star is significant for another reason. In 1669 Montonari had noted that the star varied in brightness; in 1783 John Goodricke established the nature of the variation. Normally, Algol is a second-magnitude star, but at intervals of $2^d20^h49^m$ it fades to one-third of its regular brightness; after a few hours, it brightens to normal again. Goodricke suggested that the variations might be due to large dark spots on the star, turned to our view periodically by its rotation, or that the star might be eclipsed regularly by an invisible companion. Vogel's discovery that Algol is a spectroscopic binary verified the latter hypothesis. The plane in which the stars revolve is turned nearly edgewise to our line of sight, and each star is eclipsed once by the other during every revolution. The eclipse of the fainter star is not very noticeable because the part of it that is covered contributes little to the total light of the system; this second eclipse can, however, be observed. A binary such as Algol, in which the orbit is nearly edge on to the earth so that the stars eclipse each other, is called an *eclipsing binary*.

Classes of Binary Stars

A *visual binary* is a gravitationally associated pair of stars; the members are either so near the sun, or so widely separated from each other (usually, both), that they can be observed visually (in a telescope) as two stars. Typical separations for the two stars in a visual binary system are hundreds of astronomical units; thus the orbital speeds of the stars are usually quite small and their orbital motion may not be apparent over a few decades of observation. Nevertheless, two closely separated stars are generally assumed to comprise a visual binary system if there is no reason to doubt that they are at the same distance from us and if they have the same proper motion and radial velocity, indicating that they are moving together through space. Over 64,000 such systems have been cataloged.

Sometimes one member of what would otherwise be a visual binary system is too faint to be observed; its presence may be detected, however, by the "wavy" motion of its companion, revolving about the center of mass of the two stars as they move through space. In 1844 Bessel discovered that the bright star Sirius displays such a sinusoidal motion with a period of 50 years. Sirius remained such an *astrometric binary* until 1862, when Alvan G. Clark found its faint companion—a member of the class of stars known as *white dwarfs* (discussed later in the chapter).

When the binary nature of a star is known only from the variations of its radial velocity (or of both radial velocities if the spectral lines of both stars are visible), it is said to be a *spectroscopic binary*. Over 700 systems have been analyzed.

If the orbit of a binary system is turned nearly edge on to us, so that the stars eclipse each other, it is called an *eclipsing binary*. More than 4000 have been cataloged.

The different kinds of binaries are not mutually exclusive. An eclipsing binary, for example, may *also* be a spectroscopic binary, if it is bright enough that its spectrum can be photographed, and if its radial velocity variations have been observed. Also, a small number of relatively nearby spectroscopic binaries can also be observed as visual binaries. Two stars, in nearly the same line of sight, of which one is far more distant than the other, are said to comprise an *optical double;* these are not true binary stars, and are not discussed further here.

DYNAMICS OF BINARY STAR SYSTEMS

We learn the masses of stars from their gravitational influences on other bodies. For example, the earth is accelerated into a nearly circular orbit by the sun's gravitational force on it. Now the circular (or centripetal) acceleration on the earth depends on its speed and its distance from the sun (Chapter 3), while the gravitational attraction the sun exerts on the earth depends on its distance and on the sun's mass. Thus we can calculate what the sun's mass must be for it to have a strong enough gravitational pull on the earth to produce the observed acceleration. The calculation shows the sun's mass to be 2×10^{33} g, or about 333,000 times that of the earth. We do not directly observe planets revolving about other stars, but similar calculations can be performed on the members of double star systems.

Mass Determinations of Binary Stars

We find the masses of double star systems most conveniently with Newton's reformulation of Kepler's third law. We recall (Chapter 3) that if two objects are in mutual revolution, the square of the period with which they go around each other is proportional to the cube of the semimajor axis of the orbit of one with respect to the other divided by their combined mass. The semimajor axis is simply the mean of the maximum and minimum separations of the objects. Thus if we can observe the size of the orbit and the period of mutual revolution of the stars in a binary system, we can calculate the sum of their masses.

The two stars of a binary pair revolve mutually about their common center of mass (or barycenter), which in turn moves in a straight line among the neighboring stars. Each star, therefore, describes a wavy path around the course followed by the barycenter. With careful observations it is possible to determine these individual motions of the member stars in a visual binary system. It is far more convenient, how-

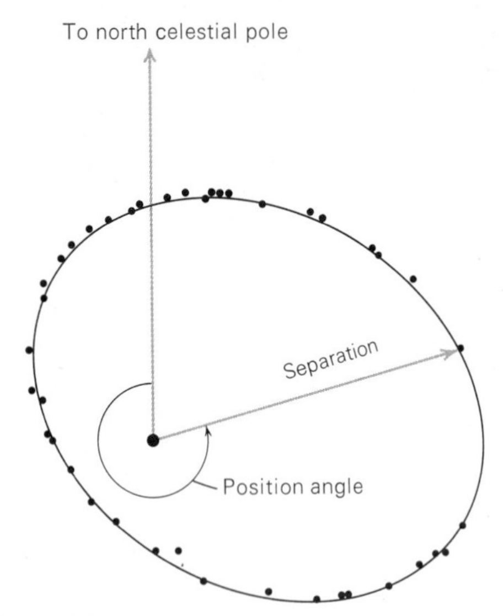

Figure 14.5
Separation and position angle in a visual-binary star.

ever, simply to observe the motion of one star (by convention the fainter), about the other (the brighter). The observed motion shows the *apparent relative orbit*. The periods of mutual revolution for visual binaries range from a few years to thousands of years, but only for those systems with periods less than a few hundred years can the apparent relative orbits be determined with much precision; even then, a long series of observations covering a number of decades is usually necessary. The data observed are the angular separation of the stars and the ***position angle***, which is the direction, reckoned from the north, around toward the east, of the fainter star from the brighter one. These data can be measured on a photograph if the separation of the stars is not too small; in any case, they can be measured directly at the telescope. A typical apparent relative orbit, assembled from many observations of separation and position angle, is shown in Figure 14.5.

The true orbit of the binary system does not, in general, happen to lie exactly perpendicular to the line of sight, so we do not see it face-on. Consequently, the apparent relative orbit is merely a *projection* of the *true relative orbit*. Now it is easy to show that when an ellipse is viewed obliquely, it still appears as an ellipse. However, the foci of the original ellipse are not the foci of the ellipse viewed obliquely. Therefore, the brighter star, although it is located at one focus of the *true* relative orbit, is *not* at a focus of the *apparent* relative orbit. This circumstance makes it possible to determine the inclination of the true orbit to the line of sight. There are several techniques for solving this geometry problem, so if the distance to the system is known, the shape and size of the orbit can be found. The period of mutual revolution, of course, is observed directly. Finally, with this knowledge of the period of revolution and the size of the semimajor axis of the orbit, we can calculate the sum of the masses of the stars from Newton's formula. To find what share of the total mass belongs to each star, it is necessary to investigate the individual motions of the stars with respect to the center of mass of the system. The distance of each star from the barycenter is inversely proportional to its own mass.

If the two stars of a binary system have a small linear separation, that is, if their relative orbit is small, there is little chance that they will be resolved as a visual binary pair. On the other hand, they have a shorter period, and their orbital velocities are relatively high, as compared with the stars of a visual binary system; unless the plane of orbital revolution is almost face-on to our line of sight, there is a good chance that we will be able to observe radial velocity variations of the stars due to their orbital motions. In other words, they comprise a spectroscopic binary system.

Most spectroscopic binaries have periods in the range from a few days to a few months; the mean separations of their member stars are usually less than 1 AU. If the two stars of a spectroscopic binary are not too different in luminosity, the spectrum of the system displays the lines of both stars, each set of lines oscillating in the period of mutual revolution. More often, lines of only one star are observed. A graph showing the radial velocity of a member of a binary star system plotted against time is called a ***radial velocity curve***, or simply, a ***velocity curve*** (Figure 14.6).

The actual analysis of the velocity curve of a spectroscopic binary is complex, but in principle the idea is simple. Suppose the orbit is circular and edge-on to our line of sight, and that spectral lines of both stars are observed. The radial velocities of the stars then tell us their speeds in their orbits. These speeds, multiplied by the time for a complete revolution (or a complete cycle of radial velocity variations), are the distances the stars actually move around their orbits. Thus (for such a hypothetical edge-on system) we can find the size of the relative orbit—that is, its semimajor axis. As with visual binaries, therefore, we find the sum of the masses of the stars. The relative speeds of the two stars tell us how much of the mass sum each star has

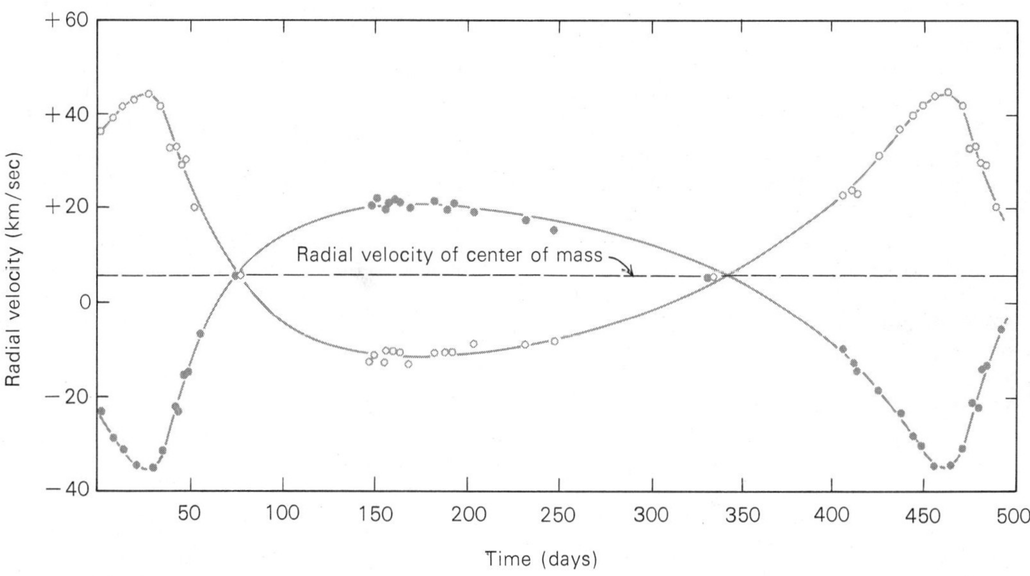

Figure 14.6
Radial velocity curves for the spectroscopic binary system ϕ Cygni. *(Adapted from Rach and Herbig)*

(the more massive star, closer to the barycenter, has a smaller orbit and hence moves more slowly to get around in the same time).

The fact that spectroscopic binaries generally have elliptical rather than precisely circular orbits complicates the problem, but we have geometrical techniques to handle it. We do not, however, have any way of knowing the true inclination of the orbit to our line of sight (we cannot observe the orbit directly as in a visual binary). Thus the radial velocities we observe are only a part of the stars' true orbital speeds, and the true orbit can be larger than the size we derive in the manner just described. Analyses of spectroscopic binaries, therefore, give us only *lower limits* to stellar masses. Their real masses can be larger by an unknown amount.

The exception to this limitation occurs if the system is also an *eclipsing binary,* for then we know that the orbit is almost edge-on to our line of sight (or there would be no eclipses) and that we are observing the true orbital velocities. (Actually, the system need not be *exactly* edge on; it can be off somewhat and still produce eclipses if the stars are not too far apart compared to their sizes. However, analysis of the way their light drops during the eclipses does tell us what the inclination is.) Thus we can find the masses of stars in spectroscopic binary systems if the spectral lines of both stars show up, and if the system is also an eclipsing binary.

The Mass-Luminosity Relation

Rather complete analyses have been carried out for a few dozen visual binary systems and a few dozen eclipsing binary systems. Thus, studies of binary stars have provided a fairly accurate knowledge of the masses of several dozen individual stars. When we compare the masses and luminosities of those stars for which both of these quantities are well determined, it is found that, in general, the more massive stars are also the more luminous. This relation, known as the *mass-luminosity relation,* is shown graphically in Figure 14.7. Each point represents a star of known mass and luminosity; its horizontal position (abscissa) indicates its mass, given in units of the sun's mass, and its vertical position (ordinate) indicates its luminosity in units of the sun's luminosity.

Most stars fall along a narrow sequence running from the lower left (low mass, low luminosity) corner of the diagram to the upper right (high mass, high luminosity) corner. The relation between the mass and luminosity of a star is not accidental or mysterious but results from the fundamental laws that govern the internal structures of stars; we shall return to this matter in Chapter 17. It is estimated that about 90 percent of all stars obey the mass-luminosity relation.

In particular, it should be noted how very much greater the range of stellar luminosities is than the

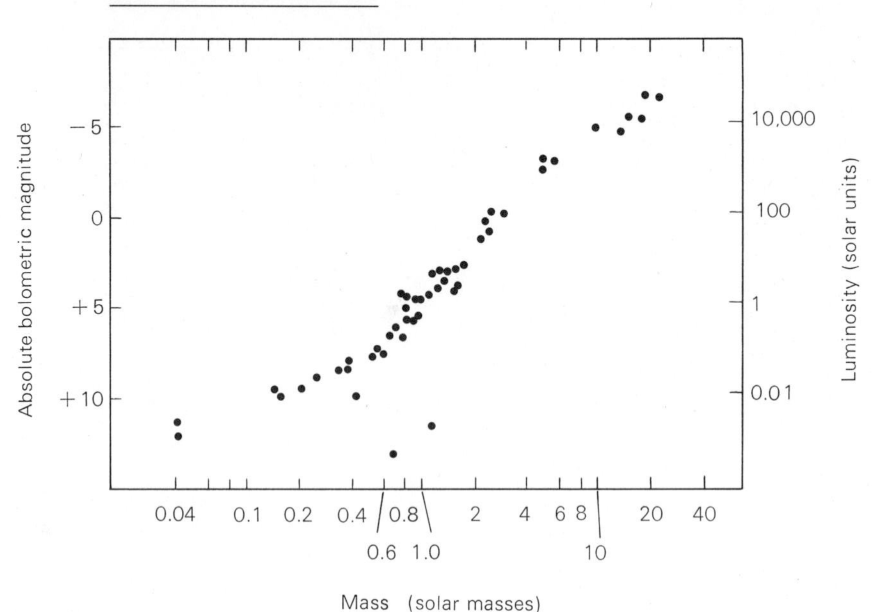

Figure 14.7
Mass-luminosity relation. The three points lying below the sequence represent white dwarf stars, which do not conform to the relation.

range of stellar masses. Luminosities of stars are roughly proportional to their masses raised to the 3.5 power. Most stars have masses between one-tenth and 50 times that of the sun; according to the mass-luminosity relation, however, the corresponding luminosities of stars at either end of the range are respectively less than 0.01 and about 10^6 solar luminosities. The intrinsically faintest known star has a luminosity near 10^{-6} that of the sun; its mass is probably greater than 0.01 of the sun's. If two stars differ in mass by a factor of 2, their luminosities would then be expected to differ by a factor of 10.

Mass Exchange in Close Binaries

We shall see (Chapter 17) that as stars age, the generation of nuclear energy in their interiors causes them to distend their outer layers greatly, so that those stars become giants. If such a star is a member of a close binary system, the atoms in its expanding outer layers may reach and pass through the point between the stars where the gravitational attraction toward each of them is equal. Thus the matter from the expanding star can flow to the other star.

Mass exchange is believed to occur between some stars in close binary systems. This exchange of mass can have profound effects on the evolution of the stars in a system. The problem was first studied by Plavec in Czechoslovakia, and independently by Kippenhahn in West Germany, and today it is an important part of the study of stellar evolution. Not only may matter stream from one star to another in a close binary system, but it can also form a large circumstellar disk or ring of material around the binary system. Mass exchange between stars is now thought to be involved in the creation of novae, supernovae, neutron stars, and black holes (Chapter 17).

The Search for Planets

The barycenter of the sun-Jupiter system is about 0.001 of the distance from the center of the sun to the center of Jupiter, which puts it outside the surface of the sun. Thus every 12 years the sun revolves about a tiny orbit with a radius of 0.005 AU. An observer on a hypothetical planet about even a nearby star would probably not be able to detect such a motion of the sun with equipment comparable to that available to our astronomers. Yet if the sun were a somewhat less massive star, and if Jupiter were a little more massive, the sun's motion might just be barely detectable.

Peter van de Kamp of the Sproul Observatory has been attempting to detect just such orbital motions of nearby stars about unseen companions that could be classed as large planets. There are several suspects, including a possible companion (or pair of them) of Barnard's star, but no confirmed discoveries of extraterrestrial planets exist at this writing (1980). Of course this does not mean that there are no other planets. Even the influence of a planet like Jupiter would be difficult to detect on even nearby stars, and there is as yet no suggestion of how we might hope to detect observationally a planet like the earth revolving about another star. But we know of no reason today why such planets could *not* exist, or even be plentiful (see Chapter 19).

SIZES OF STARS

The sun presents an observable angular diameter to us. Thus we can calculate the sun's true (linear) diameter by the same techniques by which we find the sizes of the moon and planets (Chapter 9). The sun's diameter is 1.39 million km (865,000 mi), or about 109 times the diameter of the earth.

The sun is the only star whose angular size can be resolved optically and whose diameter can be calculated simply. There are a few other stars, however, whose angular sizes are only slightly beyond the limit of resolution of the largest telescopes and which can be measured with a device known as the *stellar interferometer*.

The Stellar Interferometer

The stellar interferometer, invented by the physicist A. E. Michelson, can be used to effectively increase the resolving power of a telescope. In the use of a stellar interferometer, light from a star is gathered by two mirrors separated some distance from each other on a long beam. These mirrors bring the star's light together and direct it into a telescope. The telescopic image produced, however, is not the usual point image of a star, but an interference pattern—a spot of light crossed with alternating bright and dark fringes. The interference is caused by the waves of light from the two separated mirrors interacting with each other. The spacing of the fringes in the interference pattern depends on the precise geometry of the mirror system and on the angular diameter of the star being observed. Thus analysis of the pattern can lead to a determination of the angular size of the light source. In 1920 Michelson and Pease mounted an interferometer with a 20-ft beam on the front of the 100-inch telescope at Mount Wilson. With the apparatus, they were able to measure the angular sizes of seven giant stars that are large enough and near enough so that their disks, although too small to photograph or to see directly, are still just large enough to measure with the interferometer. The data for these stars are given in Table 14.1.

TABLE 14.1 Stars Measured with the Stellar Interferometer

Star	Angular Diameter	Distance (in pc)	Linear Diameter (in terms of sun's)
Betelgeuse (α Orionis)	0".034*	150	500
	0.042		750
Aldebaran (α Tauri)	0.020	16	34
Arcturus (α Bootis)	0.020	11	23
Antares (α Scorpii)	0.040	120	510
Scheat (β Pegasi)	0.021	50	110
Ras Algethi (α Herculis)	0.030	150	500
Mira (o Ceti)	0.056	70	420

* Variable in size.

An electronic analog of the stellar interferometer has been applied more recently to the measurement of star diameters. Separate optical telescopes, placed up to hundreds of meters apart and equipped with photomultiplier tubes, are used to observe the same star simultaneously. The electric currents generated in the two tubes are brought together to a single amplifier. The electrical impulses carry information that can be made to interfere in a way somewhat analogous to the production of fringes in the optical interferometer. The radio astronomers Brown and Twiss first used the device to measure the angular diameter of the star Sirius.

The Brown-Twiss type of interferometer is known as an *intensity interferometer*. The largest model, assembled in Australia, uses two telescopes that can be separated up to 188 m. Each uses a mosaic mirror 6.7 m in diameter, which in turn consists of 251 separate small mirrors, all mounted to reflect the light from a star to one point. These special-purpose mirrors are not of high optical quality and could not be used for ordinary telescopic observations. They are adequate, however, to focus starlight on the light-sensitive surface of a photomultiplier. This large intensity interferometer can achieve a resolution of 5×10^{-4} seconds of arc and (at the time of writing) has been used to measure the angular diameters of 32 stars. These range from Canopus, whose angular diameter is 0".0066 to Zeta Puppis with an angular diameter of only 0".00042.

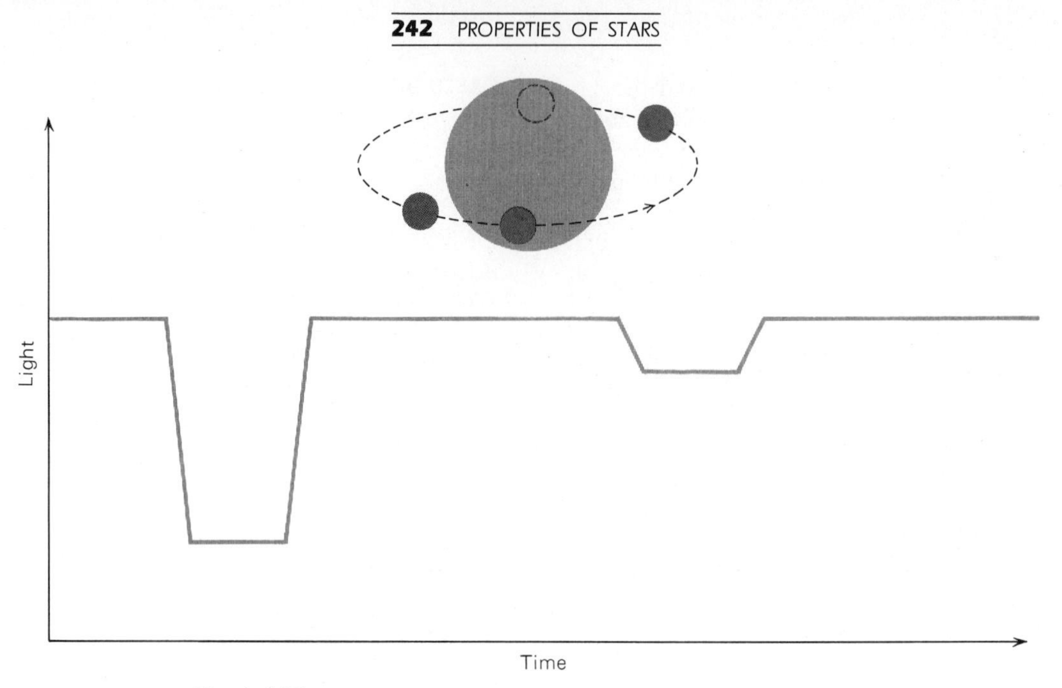

Figure 14.8
Light curve of a hypothetical eclipsing binary star with total eclipses.

The vast majority of stars, however, are too remote for even sophisticated interferometers to be able to measure their angular sizes directly.

Stellar Radii from Analysis of Eclipsing Binaries

During the period of revolution of an eclipsing binary, there are two times when the light from the system diminishes—once when the smaller star passes behind the larger one and is eclipsed, and once when the smaller star passes in front of the larger one and eclipses part of it. If the smaller star goes completely behind the larger one, that eclipse is *total* and the other eclipse half a period later is *annular* (see Figure 14.8). If the smaller star is never completely hidden behind the larger star, both eclipses are partial. Each interval during an eclipse when the light from the system is farthest below normal is called a *minimum*. Both minima are not, in general, equally low in light. The relative amount of light drop at each minimum, depends on the relative surface brightnesses of the two stars, and hence on their temperatures. *Primary minimum* occurs when the hotter star is eclipsed (whether it is a total, an annular, or a partial eclipse), and *secondary minimum* occurs when the cooler star is eclipsed. A graph of the light from an eclipsing binary system, plotted against time through a complete period, is called a *light curve*.

To illustrate how the sizes of the stars are related to the light curve, we may consider a hypothetical eclipsing binary in which the stars are very different in size, and in which the orbit is exactly edge-on, so that the eclipses are *central* (Figure 14.9). When the small star is at point *a* (*first contact*), and is just beginning to pass behind the large star, the light curve begins to drop. At point *b* (*second contact*), the small star has gone entirely behind the large one and the total phase of the eclipse begins. At *c* (*third contact*) it begins to emerge, and when the small star has reached *d* (*last contact*) the eclipse is over. During the time interval between first and second contact (or between third and last contacts) the small star has moved a distance equal to its own diameter. During the time interval from first to third contacts (or from second to last contacts) the small star has moved a distance equal to the diameter of the large star. If the lines of both stars are visible in the composite spectrum of the binary, the speed of the small star with respect to the large one is also known. This speed, multiplied by the time intervals from first to second contacts and from first to third contacts, gives, respectively, the diameters of the small and large stars.

In actuality the orbits are not, generally, exactly edge-on, and the eclipses are not central. However, it is

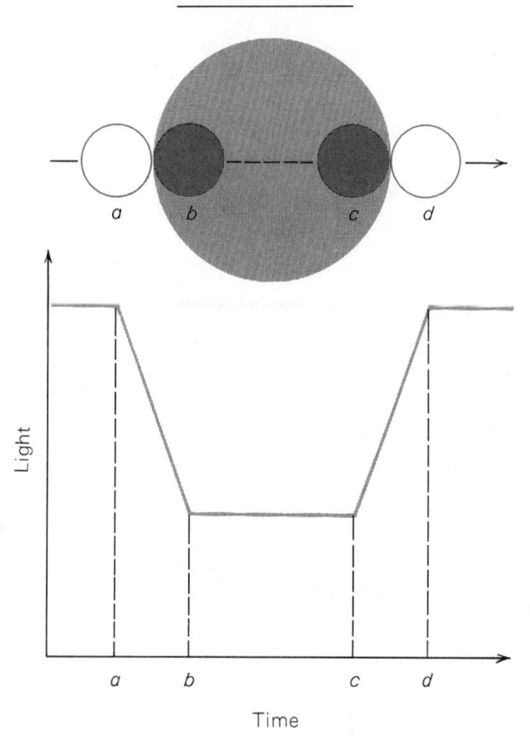

Figure 14.9
Contacts in the light curve of a hypothetical eclipsing binary with central eclipses.

a relatively simple geometry problem, at least in principle, to sort out the effects, and from the depths of the minima and the exact instants of the various contacts to calculate both the inclination of the orbit and the sizes of the stars in relation to their separation. If the eclipses are partial, the analysis is far more difficult, but even so it can be accomplished.

Among the thousands of known eclipsing binaries, however, there are only a few dozen that are so favorably disposed for observation that all the necessary data can be obtained. Only for these few binaries have complete analyses led to fairly reliable values of the radii of their member stars.

Stellar Radii from Radiation Laws

For most stars we must use an indirect method, by which we can calculate their radii from theory. The theory involved is the Stefan-Boltzmann law (Chapter 5); we calculate the radius of a spherical perfect radiator that has the same luminosity and temperature that a star does.

The luminosity of a star can be obtained by the procedure discussed in the last chapter, and the temperature of a star can be obtained in various ways, as from its color or its spectrum. Now stars are fairly good approximations to perfect radiators, so we can apply the Stefan-Boltzmann law to them. The energy emitted per unit area of a star is proportional to the fourth power of its temperature. This energy emitted per unit area, multiplied by the total area of the surface of the star, must be equal to its total luminosity. Since the surface area of a sphere of radius R is $4\pi R^2$, the luminosity of a star is proportional to the square of its radius multiplied by the fourth power of its temperature.[1] A knowledge of a star's temperature and luminosity, therefore, enables us to calculate its size. We give some illustrations in the next section.

The validity of this indirect method of obtaining stellar radii is verified by noting that it gives approxi-

[1] The equation is

$$L = 4\pi R^2 \sigma T^4,$$

where L is the luminosity, R the radius, and T the temperature of the star, and σ is the *Stefan-Boltzmann* constant (see Appendix 6).

mately correct answers for those stars whose sizes can also be determined by geometrical means, such as by the interferometer or from analysis of eclipsing binary systems.

THE HERTZSPRUNG–RUSSELL DIAGRAM

In 1911 the Danish astronomer E. Hertzsprung (1873–1967) compared stars within several clusters by plotting their magnitudes against their colors. In 1913 the American astronomer Henry Norris Russell (1877–1957) undertook a similar investigation of stars in the solar neighborhood by plotting the absolute magnitudes of stars of known distance against their spectral classes. These investigations by Hertzsprung and by Russell led to an extremely important discovery concerning the relation between the luminosities and surface temperatures of stars. The discovery is exhibited graphically on a diagram named in honor of the two astronomers—the *Hertzsprung–Russell* or *H–R diagram*.

Features of the H–R Diagram

Two easily derived characteristics of stars of known distances are their absolute magnitudes (or luminosities) and their surface temperatures. The absolute magnitudes can be found from the known distances and the observed apparent magnitudes. The surface temperature of a star is indicated either by its color or its spectral class.

If the absolute magnitudes of stars are plotted against their temperatures (or spectral classes, or color indices), an H–R diagram like that of Figure 14.10 is obtained. The most significant feature of the H–R diagram is that the stars are not distributed over it at random, exhibiting all combinations of absolute magnitude and temperature, but rather cluster into certain parts of the diagram. The majority of stars are aligned along a narrow sequence running from the upper left (hot, highly luminous) part of the diagram to the lower right (cool, less luminous) part. This band of points is called the *main sequence*. A substantial number of stars, however, lie above the main sequence on the H–R diagram, in the upper right (cool, high luminosity) region. These are called *giants*. At the top part of the diagram are stars of even higher luminosity, called *supergiants*. Finally, there are stars in the lower left (hot, low luminosity) corner known as *white dwarfs*. To say that a star lies "on" or "off" the main sequence does not refer to its position in space, but only to the point that represents its luminosity and temperature on the H–R diagram.

An H–R diagram, such as Figure 14.10, that is plotted for stars of known distance does not show the

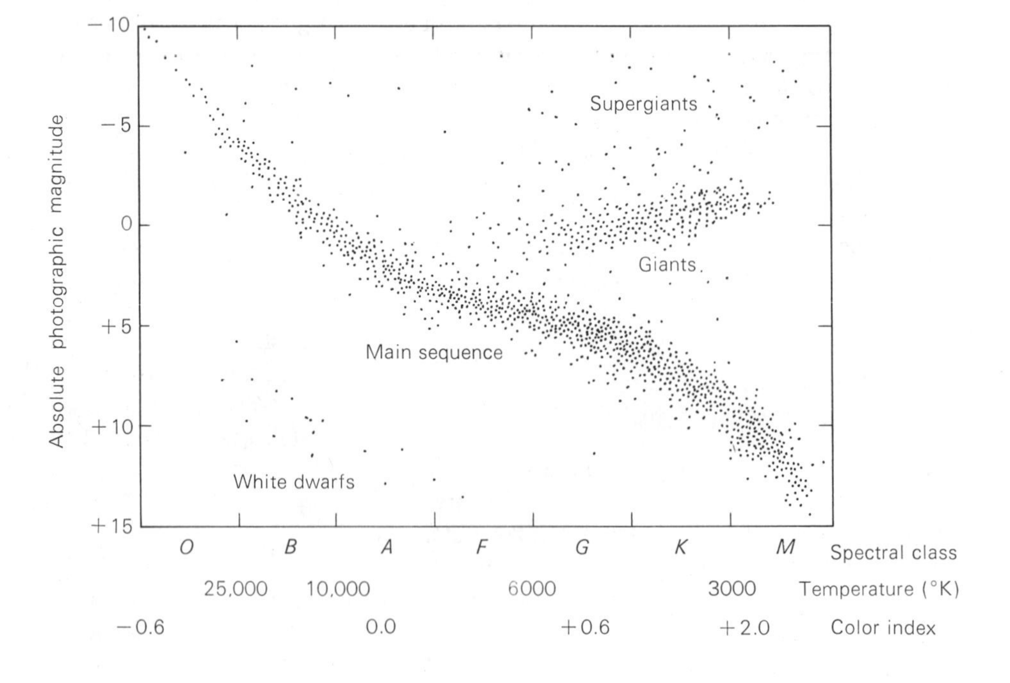

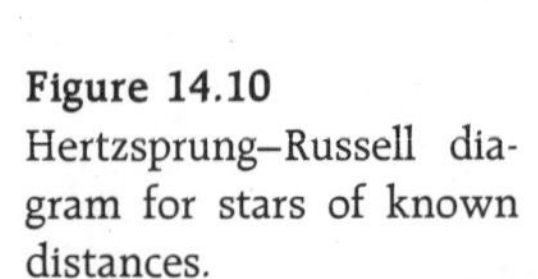

Figure 14.10
Hertzsprung–Russell diagram for stars of known distances.

relative proportions of various kinds of stars, because only the nearest of the intrinsically faint stars can be observed. To be truly representative of the stellar population, an H–R diagram should be plotted for all stars within a certain distance (see Exercise 11). Unfortunately, our knowledge is reasonably complete only for stars within a few parsecs of the sun, among which there are no giants or supergiants. It is estimated that about 90 percent of the stars in our part of space are main-sequence stars and about 10 percent are white dwarfs. Less than 1 percent are giants or supergiants.

Stellar Distances from Their Spectra

Examination of the H–R diagram reveals a very important method for determining stellar distances. Suppose, for example, that a star is known to be a spectral class G star on the main sequence. Its absolute magnitude could then be read off the H–R diagram at once; it would be about +5. From this absolute magnitude and the star's apparent magnitude, its distance can be calculated (Chapter 13).

In general, however, the spectral class alone is not enough to fix, unambiguously, the absolute magnitude of a star. The G star described in the last paragraph could have been, for example, a main-sequence star of absolute magnitude +5, a giant of absolute magnitude 0, or a supergiant of still higher luminosity. We recall, however (Chapter 13), that pressure differences in the atmospheres of stars of different sizes result in slightly different degrees of ionization for a given temperature. We will see in a moment that giant stars are larger than main-sequence stars of the same spectral class and that supergiants are larger still. It is thus possible to classify a star by its spectrum, not only according to its temperature (spectral class) but also according to whether it is a main-sequence star, a giant, or a supergiant, that is, its *luminosity class*.

With both its spectral class and luminosity class known, a star's position on the H–R diagram is uniquely determined. Its absolute magnitude, therefore, is also known, and its distance can be calculated. Distances determined this way, from the spectral and luminosity classes, are said to be obtained from the *method of spectroscopic parallaxes*.

There are some stars that do not fit into the standard classification scheme (see, especially, Chapter 17). The method of spectroscopic parallaxes does not work for them.

Extremes of Stellar Luminosities, Radii, and Densities

Let us now investigate the extremes in size, luminosity, and density found for stars. The most massive stars are the most luminous ones, at least for main-sequence stars. These stars have absolute magnitudes of −6 to −8. A few stars are known that have absolute bolometric magnitudes of −10; they are a million times as luminous as the sun. These superluminous stars, most of which are at the upper left on the H–R diagram, are very hot spectral-type O and B stars, and are very blue. These are the stars that would be the most conspicuous at very great distances in space.

Consider now the stars at the upper right corner of the H–R diagram, both giants and supergiants. Some have surface temperatures less than half that of the sun, so each unit area of the surface of such a star must emit only $\frac{1}{16}$ or less as much light as the sun does. Yet they are at least a few hundred times as luminous as the sun (if they are giants) or some thousands of times as luminous (if they are supergiants). We see, then, how aptly the cooler stars of high luminosity are called giants or supergiants.

Consider a red, cool supergiant that has a surface temperature of 3000 K and an absolute bolometric magnitude of −5. This star has 10,000 times the sun's luminosity but only half its surface temperature. Since each unit area of the star emits only $\frac{1}{16}$ as much light as a unit area of the sun, its total surface area must be greater than the sun's by 160,000 times. Its radius, therefore, is 400 times the sun's radius. If the sun could be placed in the center of such a star, the star's surface would lie beyond the orbit of Mars. Even larger supergiants exist that are so cool that they emit nearly all of their radiation in the infrared.

Red giant stars have extremely low mean densities. The volume of the star described in the last paragraph is 64 million times that of the sun. The masses of such giant stars, however, are probably at most only 50 solar masses, and very likely much less. (Plaskett's star, a spectral-type O star with a mass of at least 50 solar masses, is one of the most massive stars known.) If we assume that the supergiant star with 64 million times the sun's volume has only 10 times its mass, we find that it has just over 1 ten-millionth the sun's mean density, or only about 2 ten-millionths the density of water; the outer parts of such a star would constitute an excellent laboratory vacuum.

In contrast, the very common red, cool stars of low luminosity at the lower end of the main sequence are much smaller and more compact than the sun. An example of such a red dwarf is the star Ross 614B, which has a surface temperature of 2700 K and an absolute bolometric magnitude of about + 13 ($\frac{1}{2300}$ of the sun's luminosity). Each unit area of this star emits only $\frac{1}{20}$ as much light as a unit area of the sun, but to have only $\frac{1}{2300}$ the sun's luminosity, the star need have only about $\frac{1}{115}$ the sun's surface area, or $\frac{1}{11}$ its radius. A star with such a low luminosity also has a low mass (Ross 614B has a mass about $\frac{1}{12}$ that of the sun), but still would have a mean density about 100 times that of the sun. Its density must be higher, in fact, than that of any known solid found on the surface of the earth.

The faint red main-sequence stars are not the stars of the most extreme densities, however. The white dwarfs, at the lower left corner of the H–R diagram, have the highest densities of the normal stars known to be common.

The White Dwarfs

The first white dwarf stars to be discovered were the companions to the stars 40 Eridani, Sirius, and Van Maanen's star. Sirius, the brightest-appearing star in the sky, is the most conspicuous star in the constellation of Canis Major (the Big Dog). It is an interesting coincidence that Procyon, the brightest star in the constellation of Canis Minor (the Little Dog), also has a white dwarf companion. Both Sirius and Procyon are visual binaries within 5 pc of the sun. One other star within 5 pc, 40 Eridani, is a multiple star system that contains a white dwarf.

The white dwarf companion of 40 Eridani, 40 Eridani B, is a good example of a typical white dwarf. Its absolute magnitude is 10.7 and its temperature is about 12,000 K; it has, therefore, 2.1 times the sun's surface temperature and $\frac{1}{275}$ its luminosity. Its surface area is only $1/(275 \times 2.1^4)$, or $\frac{1}{5450}$, of the sun's, which gives it a radius of 0.014 and a volume of 2.5×10^{-6} the sun's. Its mass, however, is 0.43 times that of the sun, so its mean density is $0.43/(2.5 \times 10^{-6})$, or about 170,000 times the mean density of the sun, and over 200,000 times the density of water.

Since white dwarfs are intrinsically faint stars, they must be relatively nearby to be observed. Today they are usually discovered by searching for faint stars of large proper motion. Some hundreds of white dwarfs have now been found, largely due to the efforts of the astronomer W. J. Luyten of the University of Minnesota.

The theory of white dwarfs (Chapter 17) predicts a relation between their masses and radii. The masses of all white dwarfs should range from 0.1 to 1.4 solar masses, and their corresponding radii should range from four times to less than half that of the earth (the more massive white dwarfs being the smaller). The predicted mean densities of these stars range from about 50,000 to over 1,000,000 times that of water and their central densities can be more than 10^7 times that of water. A teaspoonful of such material would have a mass of some 50 tons. At such densities, matter cannot exist in its usual state. Although it is still gaseous, its atoms are completely stripped of their electrons, which are obliged to move according to certain restrictive laws. The matter in white dwarfs is said to be *degenerate*.

As we shall see in Chapter 17, many or most stars are believed to become white dwarfs near the end of their evolution. Eventually, after many thousands of millions of years, white dwarfs radiate away their internal heat, cooling off to become *black dwarfs* — cold, dense stars no longer shining. White or black dwarfs, however, are not the only possible final evolutionary states for stars. Some stars, we shall see, evidently become *neutron stars* with densities 1000 million times as great as those of white dwarfs. Still others may collapse to *black holes* of even greater density. We take up these bizarre objects in later chapters.

THE DISTRIBUTION OF THE STARS IN SPACE

In the immediate neighborhood of the sun, the stars seem to be distributed more or less at random (except for their tendency to occur in binary systems and small clusters). The larger the volume of space we survey, the more stars we find, and if allowance is made for the fact that the faintest stars become invisible at larger distances, it is found that the number of stars we can count is roughly proportional to the cube of the distance to which we look. Eventually, however, the stars do thin out more rapidly in some directions than in others. The way they thin out is a clue to the nature of the stellar system to which the sun belongs. The idea that the sun is a part of a large system of stars was suggested as early as 1750 by Thomas Wright in his *Theory of the Universe*. Immanuel Kant, the great German philosopher, suggested the same hy-

North America nebula in Cygnus, NGC 7000. *(Hale Observatories)*

The Lagoon nebula in Sagittarius, M8. *(Kitt Peak National Observatory)*

The Trifid nebula in Sagittarius; the blue region on the left is starlight reflected by interstellar dust, and the red region on the right is light emitted by ionized gas. *(Kitt Peak National Observatory)*

The globular cluster M13.
(U. S. Naval Observatory)

The Orion nebula.
(Hale Observatories)

The Rosette nebula in Monoceros, NGC 2237, photographed with the Palomar Schmidt telescope.
(Hale Observatories)

pothesis five years later. It was the German-English astronomer William Herschel, however, who first demonstrated the nature of the stellar system.

Herschel's Star Gauging

Herschel sampled the distribution of stars about the sky by a procedure he called *star gauging*. He observed that in some directions he could count more stars through his telescope than in other directions. In 1785 he published the results of gauges or counts of stars that he was able to observe in 683 selected regions scattered over the sky. While in some of these fields he could see only a single star, in others he was able to count nearly 600. Herschel reasoned that in those directions in which he saw the greatest numbers of faint stars, the stars extended the farthest, and in other directions they thinned out at relatively short distances. As a result of his star gauging, Herschel arrived at the conclusion (only partially correct, as we shall see) that the sun is inside a great sidereal system, and that the system is disk-shaped, roughly like a grindstone, with the sun near the center.

Figure 14.11
The Milky Way in Sagittarius. *(Yerkes Observatory)*

The Phenomenon of the Milky Way

All of us who have looked at the sky on a moonless night away from the glare of city lights are aware of the Milky Way, a faint, luminous band of light that completely encircles the sky. Galileo solved the first mystery of the Milky Way when he turned his telescope on it and saw that it really consists of myriads of faint stars. Herschel's star gauging solved the second mystery by explaining why the Milky Way should appear as a band all the way around the sky.

It must be recalled that we view our sidereal system from the inside. Figure 14.12 shows a portion of the "grindstone," viewed edge-on. The sun's position is at O. If we look from O toward either face of the wheel, that is, in directions a or b, we see only those stars that lie between us and the nearest boundary of the stellar system. In these directions in the sky, therefore, we see only scattered stars. On the other hand, if we look edge-on through the wheel, say in directions c or d, we encounter so many stars along our line of sight that we get the illusion of a continuous band of light. Since the greatest dimensions of the grindstone extend in all directions along its flat plane, the band of light extends completely around the sky. This band of light is the Milky Way; it is simply the light from the many distant stars that appear lined up in projection when we look edge-on through our own flattened stellar system.

The Galaxy

We call our stellar system the *Galaxy*, or sometimes, the *Milky Way Galaxy*. In our modern view, the Galaxy is a vast, wheel-shaped system of some 10^{11} stars, with a diameter that probably exceeds 30,000 pc (100,000 LY). The flattened shape of the

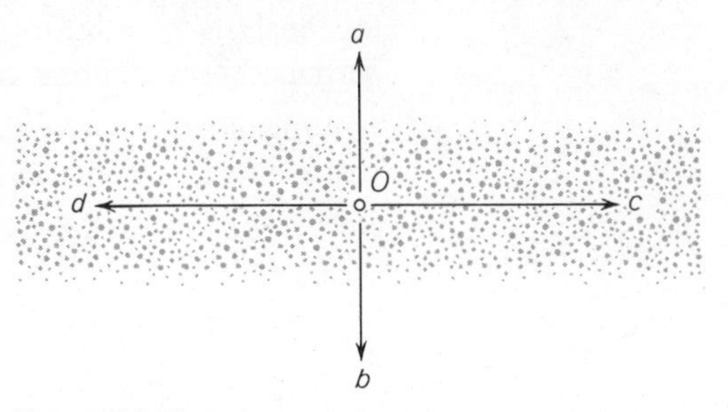

Figure 14.12
We see the Milky Way when we look from the earth, O, edge-on through our Galaxy (directions c and d).

Galaxy is a consequence of its rotation. The sun, about two-thirds of the way from the center out to the rim of the wheel, moves at a speed of from 250 to 300 km/s to complete its orbital revolution about the galactic center in some 200 million years.

In the central region of the Galaxy is a huge hub or *nuclear bulge* of stars, extending outward from the bulge, and winding through the disk of the Galaxy, like the spirals of light in a gigantic pinwheel, are the *spiral arms*. The spiral arms contain vast clouds of gas and cosmic dust—the interstellar medium. Associated with these gas and dust clouds are many young stars, a few of which are very hot and luminous because in the interstellar clouds of the spiral arms star formation is still taking place. The sun is located in or near a spiral arm.

In addition to individual stars and clouds of interstellar matter, the Galaxy contains many *star clusters*—groups of stars having a common origin and probably a common age. The most common star clusters, numbering in the thousands, are the *open* or *galactic clusters*. Typically, an open cluster consists of a few hundred stars, loosely held together by their mutual gravitation, and moving together through space. The open clusters are located in the main disk of the Galaxy and are usually in or near spiral arms. Besides the open clusters, there are over a hundred *globular clusters*—beautiful, spherically symmetrical clusters, each containing hundreds of thousands of member stars. Most of the globular clusters are scattered in a roughly spherical distribution about the main wheel of the Galaxy, grouped around it rather like bees around a flower. They form a more or less spherical *halo* or *corona* surrounding the main body of the Galaxy.

Our Galaxy, the interstellar matter, star clusters, are the subjects of the next chapter.

SUMMARY REVIEW

The brightest and nearest stars: most bright stars are intrinsically luminous; most stars are intrinsically faint; the most common stars; density of stars in space; luminosity function

Binary stars: discovery; Mizar; orbital motion; spectroscopic binaries; visual binaries; eclipsing binaries; Algol; how binary stars are observed; numbers of known binary stars of various types; astrometric binaries; companion of Sirius; optical binaries

Analysis of binary stars: mass of the sun; finding masses of stars in binary systems; apparent relative orbit; true relative orbit; separation and position angle; radial velocity (or velocity) curve; analysis of the velocity variations in a spectroscopic binary; mass-luminosity relation; ranges of stellar masses and luminosities; mass exchange in binary systems; search for planets about other stars

Sizes of stars: size of the sun; direct measures of stars with the stellar interferometer; the intensity interferometer; stellar size found from the analysis of eclipsing binaries; light curve; total, annular, and partial eclipses; minima; contacts; stellar radii from radiation laws

Hertzsprung–Russell (H–R) diagram: main sequence; giants; supergiants; white dwarfs; luminosity class; spectroscopic parallaxes

Extreme ranges of stellar properties: most luminous stars; sizes and densities of giants and supergiants; faint red main sequence stars; white dwarfs; sizes and densities of white dwarfs; observation of gravitational redshifts; mass-radius relation for white dwarfs; degenerate matter; black dwarfs

Distribution of the stars in space: Herschel's star gauging; the Milky Way; the Galaxy; sun's position in the Galaxy; nuclear bulge; spiral arms; star clusters—open or galactic and globular clusters; galactic halo (or corona)

EXERCISES

1. Describe an everyday situation that is analogous to the fact that most naked-eye stars are of far more than average stellar luminosity. (Think of living organisms in the forest.)

2. From the data in Appendix 13 (the nearest stars), plot the luminosity function for stars nearer than 5 pc. Compare your plot with Figure 14.2. Explain any differences.

3. Many eclipsing binaries can be observed which are *not observed* as spectroscopic binaries. Can you suggest an explanation?

4. A few stars are both visual binaries *and* spectroscopic binaries (their radial velocity variations can be detected). Why do you suppose such stars are rare?

5. Describe the apparent relative orbit of a visual binary whose true orbital plane is edge-on to the line of sight. Describe the apparent motions of the individual stars of the system among the background stars in the sky.

6. Why do most visual binaries have relatively long periods and most spectroscopic binaries relatively short periods? Under what circumstances could a binary with a relatively long period (over a year) be observed as a spectroscopic binary?

7. Find the combined mass of two stars in a binary system whose period of mutual revolution is two years and for which the semimajor axis of the relative orbit is two astronomical units.

8. Although the periods of known eclipsing binaries range from 4^h39^m to 27 years, the average of their periods is less than the average period of all known spectroscopic binaries. Can you suggest an explanation?

9. Most eclipsing binary stars are giant stars of high luminosity. Why do you suppose this is the case?

10. What is the radius of a star (in terms of the sun's radius) with the following characteristics:

(a) Twice the sun's temperature and four times its luminosity?

(b) Eighty-one times the sun's luminosity and three times its temperature?

11. Plot a Hertzsprung–Russell diagram for the stars within 5 pc of the sun. Use the data of Appendix 13. How does this H–R diagram differ from the one in Figure 14.10? Explain the reasons for these differences.

12. Consider the following data on five stars:

Star	m	Spectrum
1	12	G main sequence
2	8	K giant
3	12	K main sequence
4	15	O main sequence
5	5	M main sequence

(a) Which is hottest? (b) coolest? (c) most luminous? (d) least luminous? (e) nearest? (f) most distant?
In each case, give your reasoning.

13. Suppose you had data on the apparent magnitudes and colors of several hundred stars in a cluster. Explain how you could use these data to determine the distance to the cluster.

14. Why do you suppose that most visual binaries are stars of low luminosity?

15. Sometimes our Galaxy is called, simply, the "Milky Way." Why is this poor terminology? Where, exactly, *is* the Milky Way? Does the question make sense? Why?

16. Suppose the Milky Way were a band of light extending only halfway around the sky (that is, in a semicircle). What then would you conclude about the sun's location in the Galaxy? Give your reasoning.

CHAPTER 15

Harlow Shapley (1885–1972) began his career as a newspaper reporter, but returned to school to study astronomy. His investigation of the Galaxy removed the sun from its center, much as Copernicus removed the earth from the center of the solar system. *(Harvard University)*

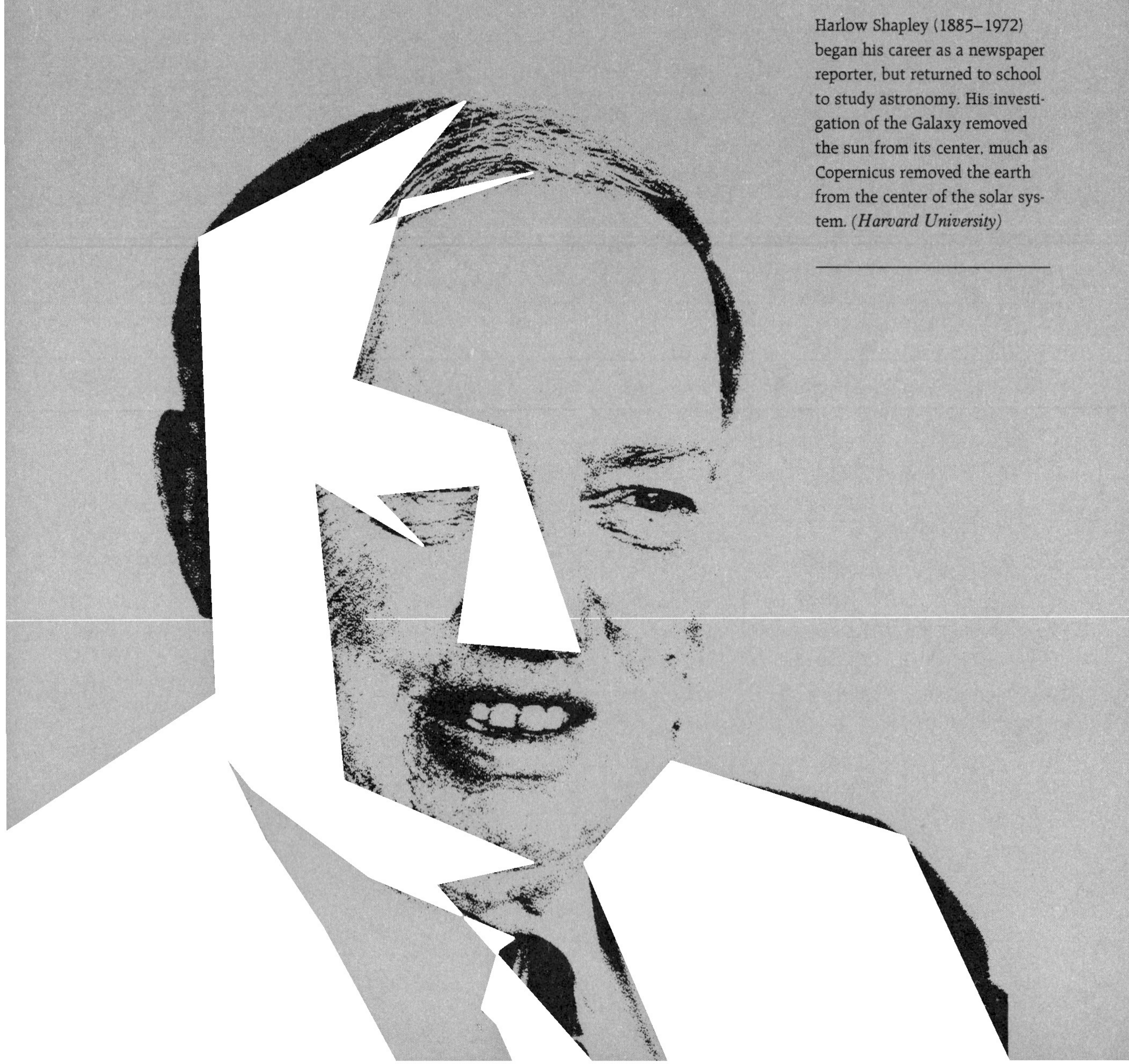

OUR MILKY WAY GALAXY

Until early in the twentieth century, the Galaxy was generally believed to be centered approximately at the sun and to extend only a few thousand light years from it. The shift from the "heliocentric" to the "galactocentric" view of our system, as well as the first knowledge of its true size, came about largely through the efforts of Harlow Shapley (1885–1972).

Shapley began his career as a crime reporter for a local newspaper, but was unsatisfied with that profession, and returned to school, where he studied astronomy. He earned his PhD degree at Princeton University in 1913, after which he took a job at the Mount Wilson Observatory. There he used the 60-inch telescope, then the world's largest, to study the distribution of globular star clusters. That investigation was to lead to one of the great revolutions in twentieth century astronomy.

SIZE OF THE GALAXY AND THE SUN'S ORBIT

Globular clusters (Chapter 14) are great symmetrical star clusters, containing tens of thousands to hundreds of thousands of stars each. Because of their brilliance, and the fact that they are not confined to the central plane of the Galaxy where they would otherwise be largely obscured by interstellar dust, they can be observed (with telescopes) to very large distances.

Distribution of the Globular Clusters

Most globular clusters contain at least a few RR Lyrae variable stars (or cluster-type variables—see Chapter 13), whose absolute magnitudes are known to lie between 0 and +1. The distance to an RR Lyrae star in a globular cluster, and hence to the cluster itself, can therefore be calculated from its observed apparent magnitude. Shapley was able to determine the distances to the closer globular clusters which contained RR Lyrae stars that could be observed individually. Distance estimates to globular clusters that do not contain RR Lyrae stars, or which are too far away to permit resolution of the variables, were obtained indirectly. Shapley measured angular diameters of globular clusters of known distance, thus obtaining their true diameters. Assuming a statistical average for the true diameter of the clusters, he was then able to obtain distance estimates for the remote ones from their observed angular diameters.

From their directions and derived distances, Shapley, in 1917, mapped out the three-dimensional distribution in space of the 93 globular clusters then known. He found that the clusters formed a spheroidal system with the highest concentration of clusters at the center. That center was not at the sun, however, but at a point in the middle of the Milky Way in the direction of Sagittarius, and at a distance of some 25,000 to 30,000 LY. Shapley then made the bold—and correct—assumption that the system of globular clusters represented the "bony frame" of the entire Galaxy; that not only is the distribution of clusters centered upon the center of the Galaxy, but, moreover, that the extent of the galactic system is indicated by the cluster distribution. Today the assumption has been verified by the observed distributions of globular clusters in other spiral galaxies. Although the sun lies far from the galactic center, the main disk of the Galaxy probably extends a nearly equal distance beyond the sun and comprises a gigantic system 100,000 LY across. The exact size of the disk is not known, however, nor is the exact distance of the sun from its center. Today, the center of the galactic nucleus is estimated to be about 10,000 pc from the sun (or about 30,000 LY), with an uncertainty of about 20 percent.

The Galactic Corona

Although the main body of the Galaxy is confined to a relatively flat disk, the globular clusters and a sparse "haze" of individual stars—not members of clusters—define a more or less spheroidal system superimposed on the disk. This haze of stars and clusters forms the galactic *corona*, or *halo*. The corona is either a spherical or spheroidal system at least 100,000 LY thick, and in the direction of the galactic plane may have a diameter of two or three times this figure, extending far beyond the "rim" of the main disk of the Galaxy. Coronas of some other galaxies have been traced to similar distances. It is possible, but not established, that the corona contains a large fraction of the total mass of the Galaxy.

Orbit of the Sun in the Galaxy

Like a gigantic solar system, the entire Galaxy is rotating. The sun, partaking of the galactic rotation, moves in a nearly circular path about the nucleus. The motion of the sun in the Galaxy is deduced from the apparent motions of objects surrounding us that do not share in the general galactic rotation. The globular clusters comprise a class of such objects. These clusters are moving, to be sure, but the fact that they are found in a spheroidal distribution, rather than being confined to the flat plane of the Galaxy, is evidence that the system of globular clusters as a whole is not rotating as rapidly as the disk of the Galaxy. By analyzing the radial velocities of the globular clusters in various directions, we can determine the motion of the sun with respect to them. In one direction, the globular clusters, on the average, seem to be approaching us, while in the opposite direction they seem to recede from us.

When the data from various sources are combined, they indicate that the sun is moving in the direction of the constellation Cygnus, with a speed that, although somewhat uncertain, probably lies in the range 200 to 300 km/s. This direction lies in the Milky Way, and is about 90° from the direction of the galactic center, which shows that the sun's orbit is probably nearly circular and lies, approximately, in the main plane of the Galaxy. As viewed from the north side of the galactic plane, the orbital motion of the sun is clockwise. The period of the sun's revolution about the nucleus, the *galactic year*, can be found by dividing the circumference of the sun's orbit by its speed; it comes out *roughly* 200 million (2×10^8) of our terrestrial years. We can observe, therefore, only a "snapshot" of the Galaxy in rotation; we do not actually see stars traverse appreciable portions of their orbits.

The majority of the stars near the sun move nearly parallel to the sun's path about the galactic nucleus, and their speeds with respect to the sun are generally less than 40 or 50 km/s.

Some stars, on the other hand, have speeds relative to the sun in excess of 80 km/s. They move along orbits of rather high eccentricity that cross the sun's orbit in the plane of the Galaxy at rather large angles. Nearby stars moving on such orbits are passing through the solar neighborhood and are only temporarily near us. Globular clusters, in particular, are believed to revolve about the nucleus of the Galaxy in orbits of high eccentricity and inclination to the galactic plane, perhaps rather like the comets revolving about the sun in the solar system. A globular cluster must pass through the plane of the Galaxy twice during each revolution. The large distances between stars, both within the cluster and within the Galaxy itself, make stellar collisions during the cluster's penetration of the galactic disk exceedingly improbable.

The Mass of the Galaxy

The mass of the Galaxy can be calculated from the gravitational influence of the system as a whole on the sun. For purposes of illustration, a few simplifying assumptions will be made which enable us to estimate the mass with the use of Kepler's third law.

The greatest spatial density of stars occurs in the region of the nucleus of the Galaxy. As a rough approximation, therefore, let us assume that it acts, gravitationally, as if its mass were concentrated at its center. Let us assume further that the sun moves on a strictly circular orbit of radius 10,000 pc and that it completes an orbital revolution about the galactic center in 2×10^8 years. Finally, we ignore the contribution of matter farther from the galactic nucleus than the orbit of the sun and assume that all of the Galaxy's mass is interior to the sun's orbit. With these assump-

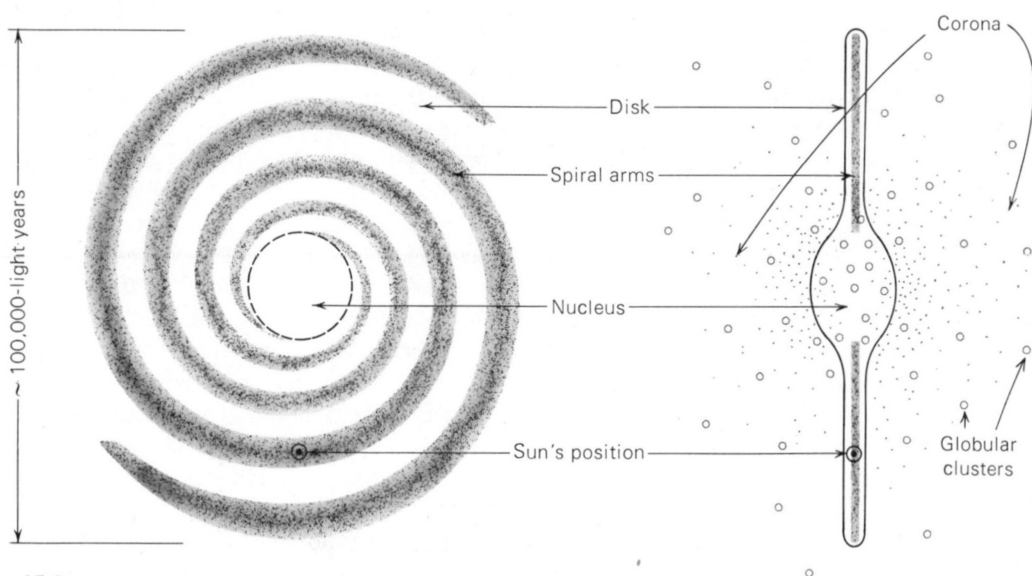

Figure 15.1
Schematic representation of the Galaxy. In the face-on view on the left (seen from the south side of the galactic plane), the sun's revolution is counterclockwise.

tions, we can regard the sun and the Galaxy as a whole as two mutually revolving bodies.

Since there are 2×10^5 AU per parsec (Chapter 12), the radius of the sun's orbit is 2×10^9 AU. Applying Kepler's third law (as corrected by Newton) and ignoring the mass of the sun in comparison to that of the Galaxy, we find that the mass of the Galaxy is 2×10^{11} times that of the sun.[1]

More sophisticated calculations that take into account the actual distribution of matter in the Galaxy (as well as it is known) give very nearly the same result. In the opinion of some investigators, however, the corona of the Galaxy may extend so far that despite its very low spatial density it might still contribute so much to the mass of the system that more than half of its mass lies farther than the sun from the nucleus. If they are correct, the mass of the Galaxy could be two or more times as great as the value found by our rough approximation.

Most of the material of the Galaxy is in the form of stars. If the mass of the sun is taken as average, we find that the Galaxy contains some hundreds of thousands of millions of stars.

THE INTERSTELLAR MEDIUM

By earthly standards the space between the stars is empty, for in no laboratory on earth can so complete a vacuum be produced. Yet throughout large regions of space this "emptiness" consists of vast clouds of gas and tiny solid particles. The greatest concentration of this interstellar material is found between the stars in the spiral arms of our own and other galaxies. The density of the interstellar matter in the arms of our Galaxy in the neighborhood of the sun, for example, is estimated to be from three to 20 times that of the interarm regions. The gas and dust are not distributed uniformly, however, but have a patchy, irregular distribution, being denser in some areas than in others, hence forming "clouds."

Sometimes, these tenuous clouds are visible, or partially so, in the form of *nebulae* (Latin for "clouds"). More often, they are invisible, and their presence must be deduced. In the spiral arms, on the average, there is about one atom of gas per cubic centimeter in interstellar space, and about 25 or 50 tiny particles or "grains," each less than a thousandth of a millimeter in diameter, per cubic kilometer. In some of the denser

---

[1] With these approximations the mass of the Galaxy, including that of the sun, is

$$M_{Galaxy} = a^3/P^2 \text{ solar masses,}$$

where a, the radius of the sun's orbit, is 2×10^9 AU, and P, the sun's period of revolution, is 2×10^8 yr.

clouds, the densities of gas and dust may exceed the average by as much as several thousand times, but even this is more nearly a vacuum than any attainable on earth. In air, for contrast, the number of molecules per cubic centimeter at sea level is of the order 10^{19}.

Absorption of Starlight by Cosmic Dust

Relatively dense clouds of the solid grains produce the *dark nebulae*, the opaque-appearing clouds that are conspicuous on any photograph of the Milky Way. Even in the densest clouds, the particles are very sparse, but the clouds extend over such vast regions (measured in parsecs) that they absorb or scatter a considerable portion of the starlight passing through them. Such concentrations of dust often have the appearance of dark curtains, greatly dimming or completely obscuring the light of stars behind them.

The "dark rift," running lengthwise down a long part of the Milky Way and appearing to split it in two, is an excellent example of a collection of such obscuring clouds. The obstruction of light from the stars located behind it is so great that less than a century ago astronomers thought that it was a sort of "tunnel" through which they could see beyond the Milky Way, into extragalactic space. In his study of the distribution of stars in the late 1780s, William Herschel described the dark rift as a "hole in the heavens."

Figure 15.2
A portion of the Milky Way in Cygnus, showing the "dark rift." *(Hale Observatories)*

In addition to the large dark clouds, many very small dark patches can be seen on Milky Way photographs, silhouetted against bright backgrounds of star fields or glowing gas clouds. Many of these patches, called *globules*, are round or oval and have angular diameters of only a few seconds of arc. The sharp contrast of their dark boundaries shows that they cannot be more distant than about 1000 pc, which allows us to calculate upper limits to their sizes. Most are less than a hundred thousand astronomical units across. The high opacity of the globules (they dim background objects by five magnitudes or more) implies that they must be very dense compared to the usual interstellar material. Astronomer Bart Bok and others propose that the globules are condensations of matter that may ultimately form into stars.

Although the distribution of the interstellar dust is spotty, and dense clouds produce conspicuous dark nebulae, some of the dust is thinly scattered more or less evenly throughout the spiral arms of the Galaxy. As a result, some absorption of starlight occurs even in regions where dark clouds are not apparent. Unfortunately, the presence of such sparse absorbing matter is not obvious, and it has been the cause of considerable difficulty in the determination of stellar distances.

In Chapter 13 we described how the distance to a star can be calculated from comparison of its apparent and absolute magnitudes. If light from a star has had to pass through interstellar dust to reach us, however, it is dimmed, much as a traffic light is dimmed by fog. We therefore underestimate the true apparent brightness of the star, and assign to it too large (that is, faint) an apparent magnitude; thus the distance we calculate for the star, corresponding to its known (or assumed) absolute magnitude, is too large. A motorist may similarly overestimate the distance to that stoplight seen through fog.

The early investigators, unaware of this absorption of light from remote stars, thought that stars thinned out at a moderate distance. They arrived at the erroneous conclusion that the Galaxy was centered on the sun, and thinned out to its "edge" at a distance of only a few thousand LY. In actual fact, we do not even see (in visible light) as far as the Galaxy's brilliant central nucleus. Were it not for the obscuring dust in space, we would be able to read at night by the light of the Milky Way.

Fortunately, the interstellar dust dims light of short wavelengths more than that of long wavelengths. Astronomers were once puzzled by the existence of stars whose spectra indicate that they are intrinsically hot and blue, of spectral type B, although they actually appear as red as cool stars of spectral type G. We know today that the light from these stars has been reddened by the interstellar absorbing material; most of their violet, blue, and green light has been obscured, leaving a greater percentage of their orange and red light, of longer wavelengths, to penetrate through the obscuring dust. This *reddening* of starlight by interstellar dust not only shows that the stars are dimmed, but also provides a means of estimating the amount of obscuration they have suffered.

We find that the more a star is dimmed by dust in space, the more it is reddened. We learn the amount of reddening corresponding to a given amount of dimming by careful comparison of two stars, one of which is seen through interstellar dust, while the other is in a direction in the sky relatively free of such obscuring matter.

Once we have ascertained this so-called *reddening law*, we can estimate the amount by which obscuring dust in space increases a star's magnitude simply by observing its color index. The star's color index is *increased* by the reddening of its light in space (the redder the star, the greater its color index—Chapter 13). The difference between the *observed* color index and the color index that the star *would have* in the absence of obscuration and reddening is called the *color excess*. The $B - V$ color excess, for example, is the amount by which the difference between the blue and visual magnitudes of a star is increased by reddening. In most directions in the Galaxy, the total absorption, in visual magnitudes, is found empirically to be about three times the $B - V$ color excess.

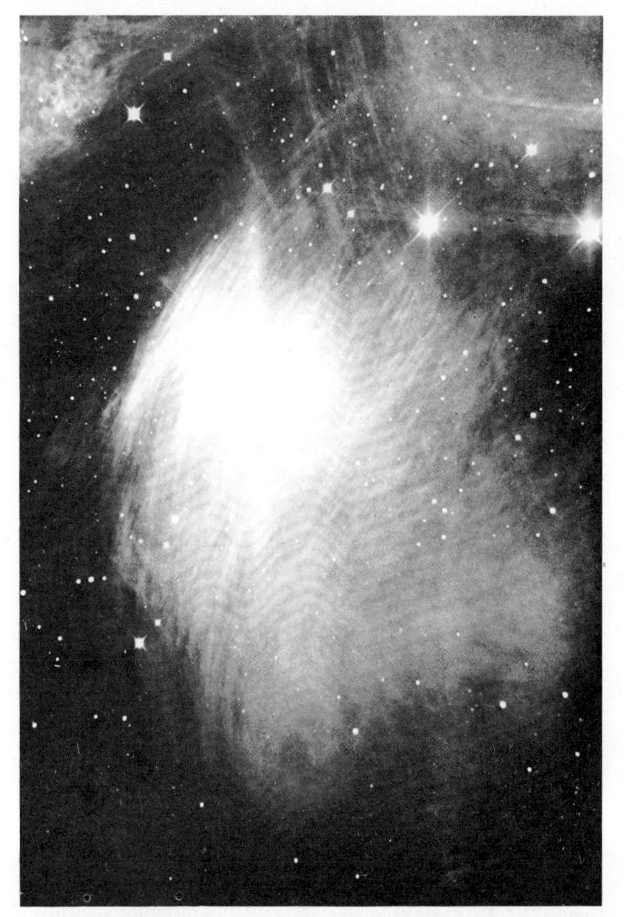

Figure 15.3
Reflection nebula about the star Merope in the Pleiades. *(Hale Observatories)*

Reflection Nebulae and Circumstellar Dust

Actually the tiny interstellar grains literally absorb only a small fraction of the starlight they intercept. Most of it they merely scatter—that is, they redirect it helter-skelter in all directions. Since the starlight that is scattered does not come directly to us, the star itself appears dimmed, just as if the light were truly absorbed. That scattering, however, illuminates the dust itself. Consequently, even the darkest dark nebulae are not completely dark but are illuminated by a faint glow of scattered starlight that can actually be measured. It is estimated that about one-third of the light of the Milky Way is diffused starlight, scattered by interstellar dust.

The light scattered by a particularly dense cloud of dust around a luminous star may be bright enough to be seen or photographed telescopically. Such a cloud of dust, illuminated by starlight, is called a *reflection nebula*. Blue light is scattered more than red by the dust. Most reflection nebulae that are conspicuous, therefore, are very blue. Examples are the blue nebulosity around each of the brightest stars in the Pleiades cluster, and the blue glow next to the Trifid nebula in Sagittarius; both are reproduced in color in this book.

Infrared surveys of the sky reveal a large number of stars that appear very much brighter at infrared wavelengths than in visible light. Many of these objects are stars surrounded by circumstellar dust clouds.

Light from such a star is greatly dimmed by the dust, but the absorbed radiation heats up the dust itself, which then reradiates the energy at infrared wavelengths. Some of these objects may be very young stars imbedded in dust that condenses from the prestellar nebulae from which they are still forming. Others may be old giant stars; we shall see that interstellar dust probably forms in the outer layers of certain red giants and is ejected from them.

Nature of the Interstellar Grains

The absorption of starlight is accomplished by *solid particles*, and not by interstellar gas. To be sure, molecules or atoms of gas also scatter short wavelengths of light more efficiently than long wavelengths (it is the scattering of sunlight by the molecules of gas in the earth's atmosphere that produces the blue daytime sky), but molecules scatter only a very tiny fraction of visible light, so gas is almost transparent. The earth's atmosphere, despite its incredibly high density as compared to interstellar gas, is so transparent as to be practically invisible. The absorbing power of the interstellar medium exceeds that of an equal mass of gas by more than 100,000 times. The quantity of gas that would be required to produce the observed absorption in space would have to be many thousands of times the amount that can possibly exist. The gravitational attraction of so great a mass of gas would produce effects upon the motions of stars that would be easily detected; such effects, however, are not observed.

Whereas gas can contribute only negligibly to absorption of light, we know from our everyday experience that tiny particles can be very efficient absorbers. Water vapor in the gaseous state in the air is quite invisible. When some of that vapor condenses into tiny water droplets, however, the resulting cloud is opaque. Dust storms, smoke, and smog furnish other familiar examples of the opacity of solid particles.

Calculations confirm that widely scattered solid particles in interstellar space are responsible for the observed dimming of starlight. Infrared observations give direct spectral evidence that they contain water ice, silicates, silicon carbide, and graphite. The complicated physical theory that deals with the scattering of light by solid particles indicates that a variety of sizes and shapes of particles probably exists. Those most responsible for the extinction of light must have diameters of the order of 10^3 Å.

Interstellar Gas; Emission Nebulae

Interstellar gas, although transparent and often invisible, is estimated to be a hundred times as abundant by mass, on the average, as the dust is. Hydrogen comprises about three-quarters of the gas, and hydrogen and helium together comprise from 96 to 99 percent of it by mass. Most of the gas is cold and nonluminous. Near very hot stars, however, it is ionized by the ultraviolet radiation from those stars. Since hydrogen is the main constituent of the gas, we often characterize a region of interstellar space according to whether its hydrogen is neutral—an "H I region"— or ionized—an "H II region."

The gas in the H II regions glows by the process of fluorescence. The light emitted from these regions of ionized gas consists largely of emission lines, so they are also called *emission nebulae*. Those emission nebulae in which the gas happens to be much denser than average (it occasionally reaches densities of 10^3 or 10^4 atoms per cubic centimeter—still an extremely high vacuum on earth) are especially conspicuous. The best known example is the Orion nebula (Figure 15.4),

Figure 15.4
Orion nebula. *(Lick Observatory)*

which is barely visible to the unaided eye, but easily seen with binoculars, in the middle of the sword of the hunter. Other famous emission nebulae are the North America nebula in Cygnus and the Lagoon nebula in Sagittarius (Figure 15.5).

All ultraviolet radiation of wavelength 912 Å or less can be absorbed by neutral hydrogen, and in the process the hydrogen is ionized (Chapter 5). An appreciable fraction of the energy emitted by the hottest stars lies at wavelengths shorter than 912 Å. If such a star is embedded in a cloud of interstellar gas, the ultraviolet radiation from that star ionizes the hydrogen in the gas, converting it into positive hydrogen ions (protons) and free electrons. Protons in the gas are continually colliding with electrons and capturing them, becoming neutral hydrogen again. As the electrons cascade down through the various energy levels of the hydrogen atoms on their way to the ground states, they emit light in the form of emission lines. The lines of the Balmer series are the ones most easily observed from the surface of the earth because they lie in the visible spectrum. Part of the invisible ultraviolet light from the star is thus transformed into visible light in the Balmer emission lines of hydrogen. After an atom has captured an electron and emitted light, it loses that electron again almost immediately by the subsequent absorption of another ultraviolet photon from the star. Thus, although neutral hydrogen absorbs and emits light in H II regions, almost all the hydrogen, at any given time, is in the ionized state.

The interstellar gas, of course, contains other elements besides hydrogen. Many of them are also ionized in the vicinity of hot stars and are capturing electrons and emitting light, just as the hydrogen does. Especially important is the emission of light by ionized oxygen and nitrogen and by helium. The light emitted by ionized oxygen and nitrogen arises from transitions to the lowest energy states of these ions from levels to which they have been excited by collisions with free electrons. These emission lines are observed only if a very great amount of gas is at a density so low that it is transparent. Hence, these lines are not observed in the laboratory. For years, their presence in gaseous nebulae was a mystery; they were, in fact, attributed to an unknown element, which was named *nebulium* (for the nebulae). The correct explanation was provided by the American physicist I. S. Bowen in 1927.

Gas and dust are generally intermixed in space, although the proportions are not everywhere exactly the same. The presence of dust is apparent on many photographs of emission nebulae. Clouds of dark material can be seen silhouetted on the Orion nebula, actually hiding a large part of the H II region from our view. Foreground dust clouds produce the "lagoon" in the Lagoon nebula, and the "Atlantic Ocean" and "Atlantic coastline" in the North America nebula. Although the dust is most conspicuous when it is in front of an emission nebula and is silhouetted against it, the dust is also intermixed with the gas. Spectra of H II regions often reveal the faint continuous spectrum (with absorption lines) of the central star, whose light is reflected to us by the dust associated with the gas. In other words, emission nebulae are generally superimposed upon *reflection nebulae*.

Figure 15.5
The Lagoon nebula in Sagittarius, photographed in red light with the 200-inch telescope. *(Hale Observatories)*

The gas in H II regions is heated in the ionization process, and usually has a temperature near 10,000 K. The gas in the surrounding H I regions is at about 100 K (below $-270°$ F), and some clouds of even very much colder gas have been found. The hot gas in an H II region tries to expand, and as it pushes into the surrounding cold material of the H I region, gas densities build up at the boundary. These higher-than-average densities often appear as bright "edges" in photographs of H II regions. Sometimes the hot gas expands around tongues of colder material, which then have the appearance of dark jets intruding into the bright nebulae. These relatively dense regions of the interstellar medium may be sites of star formation.

Interstellar Absorption Lines

The cold interstellar gas—that in the H I regions—is not visible by reflected or emitted light, nor does it appreciably dim the light of stars shining through it. Yet it often reveals its presence by leaving dark absorption lines superposed on the spectra of stars that lie beyond it.

Interstellar lines have been found of most of those elements for which observable lines would be expected. The most conspicuous interstellar lines are produced by sodium and calcium. Lines are also observed of some other common elements, as well as bands of CN, CH, and CH^+. Ultraviolet satellite observations have detected lines of additional elements and of CO (carbon monoxide). The strengths of interstellar lines lead to estimates of the relative abundances of the elements that produce them. For most elements such estimates do not differ markedly from their relative abundances in the solar and other stellar photospheres.

Recent observations from rockets and satellites have shown that molecular hydrogen (H_2) is also prevalent in interstellar space. It is now believed that a large fraction of the interstellar hydrogen—perhaps nearly one-half—is in the molecular form. However, ultraviolet radiation from stars easily dissociates the hydrogen molecule into two hydrogen atoms. Thus the H_2 is found mostly in interstellar clouds that are shielded from ultraviolet photons by interstellar dust. We shall see that the dust particles are probably necessary for the formation of molecular hydrogen as well as for its protection.

Radio Emission from Interstellar Gas

Emission of energy at radio wavelengths from the Milky Way was the first radio radiation of astronomical origin to be observed. This energy does not come from stars. The sun, to be sure, is an apparently strong source of radio waves, but the sun is very close to us. If its radio emission is typical of that of the other stars, all the stars in the Galaxy would emit less than 10^{-9} of the radio energy actually observed. The radio waves come from the interstellar gas.

Some strong sources of radio energy are individual gaseous nebulae. One example is the Crab nebula (Chapter 17); another is a group of faint gaseous filaments in Cassiopeia (known as the Cassiopeia A source). Many other discrete sources of radio radiation have been identified with supernovae remnants (Chapter 17), and some have been identified with the more conspicuous nebulae (such as the Orion nebula). The majority of discrete radio sources scattered along the Milky Way have not yet been associated with objects visible optically. Superposed on them all is the general background of radio radiation from those regions where interstellar material prevails.

A number of emission and absorption lines at radio wavelengths are also observed. The first radio line to be detected was the 21-cm line of neutral hydrogen. A hydrogen atom possesses a tiny amount of energy by virtue of the axial spin of its electron and the electron's orbital motion about the nucleus (proton). In addition, the proton has an axial spin of its own, and this spin may either add or subtract energy from that of the electron, depending on their relative orientations. If the spins of the two particles oppose each other, the atom as a whole has a very slightly lower energy than if the two spins are aligned. Ordinarily, an atom of hydrogen is in the state of lower energy. If the requisite minute amount of energy is imparted to it by a collision with another particle, however, the spins of the proton and electron can be aligned, leaving the atom in a slightly *excited state*. It radiates that same amount of energy again when the atom returns to its ground state. The amount of energy involved is that associated with a photon of 21-cm wavelength.

In 1944 the Dutch astronomer H. C. van de Hulst predicted that enough atoms of interstellar hydrogen would be radiating photons of 21-cm wavelength to make this radio emission line observable. Equipment sensitive enough to detect the line was not available until 1951. Since that time, "21-cm astronomy" has been a very active field of astronomical research, because 21-cm emission reveals where neutral hydrogen is located in the Galaxy.

Observations at 21 cm show that the neutral hydrogen in the Galaxy is confined to an extremely flat layer, most of it in a sheet less than 100 pc thick, extending throughout the plane of the Milky Way. The strength of the line indicates that neutral interstellar hydrogen makes up from 1 to 2 percent of the mass of the Galaxy. One important use of 21-cm radiation is the measurement of its Doppler shifts in various directions, which helps us to map out the motions of clouds of neutral hydrogen.

Molecules in Interstellar Space

Since the discovery of the 21-cm line, other radio spectral lines have been observed. Among them is a set of lines due to the OH radical—one oxygen and one hydrogen atom, bonded together. In 1968, radio lines of interstellar polyatomic molecules were seen for the first time. Since then, observations of many absorption and emission lines at infrared and radio wavelengths have led to the exciting new field of interstellar chemistry.

By 1979 more than 40 kinds of molecules and radicals had been identified in space. Common atoms of hydrogen, oxygen, carbon, nitrogen, and sulfur make up molecules of water, carbon monoxide, ammonia, hydrogen sulfide, and such common organic molecules as formaldehyde, hydrogen cyanide, methyl cyanide, and simple alcohols.

Like the hydrogen molecule (H_2), most of these more complex molecules are dissociated by short-wave stellar radiation. Hence they are found in relatively dense, dark interstellar clouds where dust shields the region from ultraviolet starlight. Such clouds are estimated to have a thousand to a million times the mass of the sun and to be several light years across. The molecules in the clouds are excited to various states of vibration and rotation by collisions among themselves; they subsequently transform to lower energy states with the emission of the infrared and radio radiation by which we observe them. In the process, energy is removed from the dark clouds, which causes them to cool, contract, and become denser. Particularly cold clouds, rich in interstellar molecules, are found behind the Orion nebula, near the galactic center, and elsewhere, and are believed to be prime sites of star formation (Chapter 17).

Relatively heavy molecules, such as HC_9H, are found in some cold clouds, and ethyl alcohol, C_2H_5OH, is quite plentiful—up to one molecule for every cubic meter in space. The largest of the cold clouds have enough ethyl alcohol to make 10^{28} fifths of 100-proof liquor. Wives of future interstellar astronauts, however, need not fear that their husbands will become interstellar alcoholics; even if a spaceship were equipped with a giant funnel 1 km across and could scoop through such a cloud at the speed of light, it would take about a thousand years to gather up enough alcohol for one standard martini!

The cold interstellar clouds also contain cyanoacetylene and acetaldehyde, generally regarded as starting points for the formation of amino acids necessary for living organisms. The presence of these organic molecules does not, of course, imply the existence of life in space. On the other hand, as we learn more about the processes by which they are produced, we gain an increased understanding of similar processes which must have preceded the beginnings of life on the primitive earth some thousands of millions of years ago.

Origin of the Interstellar Matter

In part, the interstellar material must represent left-over matter that did not form completely into stars when the galaxy condensed from a protogalactic cloud some 10^{10} years ago (perhaps much as the solar system formed from the solar nebula). A large part of the interstellar material, however, has certainly been ejected into space from stars, especially those near their final evolutionary states (Chapter 17). Nuclear transformations within stars, as we shall see, have altered the chemical composition of the interstellar medium—in particular, they have enriched its content of the heavier elements. New stars forming from gas and dust clouds of space—as we think the sun and planets formed 5×10^9 years ago—are thus *second generation* stars (or stars of an even higher generation). According to this picture many of the atoms of our bodies have been produced by nuclear reactions in stellar interiors!

Interstellar gas densities, however, appear to be too low to permit an appreciable amount of the gas to have condensed into dust particles. Rather, many investigators think that the dust particles condense from matter ejected from the outer layers of extreme red giants. From some of these stars matter is observed to be streaming away into space (Chapter 17). Thus it may be that as this material flows out and cools, those elements that most easily condense, such as aluminum, calcium, and titanium, begin to collect in solid grains. Condensation theory predicts that other substances, including graphite, magnesium, silicates, silicon carbide, and metallic iron also become locked up in the grains.

After the dust grains become part of the interstellar cloud, they interact with the gases in space. According to a model by G. Field, atoms such as those of hydrogen, oxygen, carbon, and nitrogen strike the grains. The last three tend to stick and unite with the impinging hydrogen atoms, building up a mantle of ice and other molecules. Some of the hydrogen forms H_2 molecules, which are easily dislodged from the grains and escape into space again as molecular hydrogen. Ultraviolet photons from stars striking the grains break up some of the molecules. These molecular fragments may unite with other molecules or fragments to form some of the more complex molecules observed in space. Further chemical changes can occur through interactions of cosmic rays with the grains or with the molecules in space. In any event, we can best understand the formation of H_2 and other molecules as taking place on the surfaces of dust grains.

Field points out that his model of an interstellar grain (Figure 15.6) is like a tiny planet about 1000 Å in diameter. There is a body of rocky and metallic material, mostly silicates and iron. Around this is a layer of ice analogous to the oceans of the earth. On the very outside there should be a thin "atmosphere" of absorbed hydrogen atoms, coming and going from interstellar space.

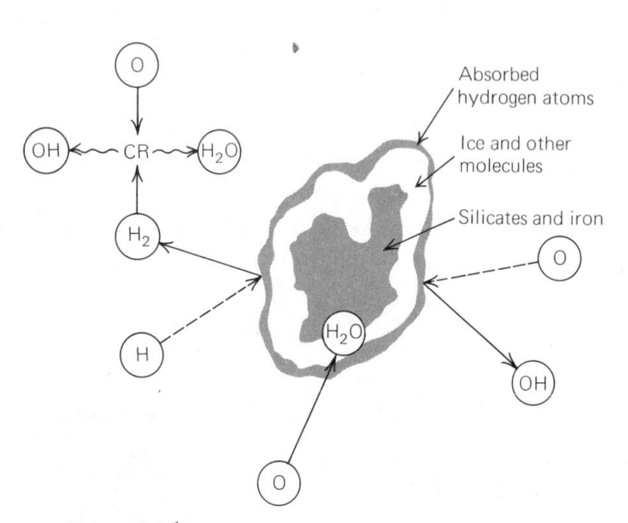

Figure 15.6
Model of a dust grain, magnified about 0.5 million times. Surface reactions with hydrogen and oxygen are shown. The H_2, aided by cosmic ray (CR) ionization, may further react with oxygen to form OH and water. Other atoms, including those of carbon and nitrogen, produce other molecules by similar processes. *(Adapted from a diagram by G. Field)*

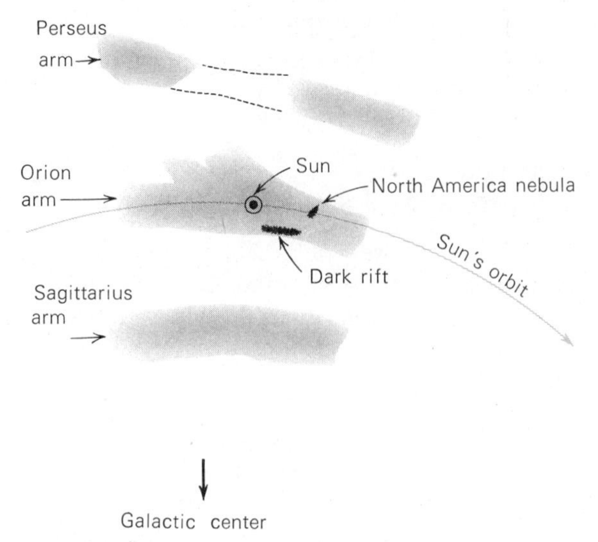

Figure 15.7
Spiral structure of the Galaxy as deduced from optical observations.

SPIRAL STRUCTURE OF THE GALAXY

In other galaxies both the interstellar matter and the most luminous resolved stars are generally concentrated in spiral arms. The association of the brightest stars and the gas and dust is not mysterious; highly luminous stars are almost certainly relatively young objects, which have recently formed from clouds of interstellar matter. One way to map out the spiral structure of our own Galaxy, therefore, is to use the gaseous nebulae and the very luminous main-sequence O and B stars as "tracers" to identify spiral arms.

Because we are surrounded by interstellar dust, however, it is difficult for us to identify even the nearby spiral arms optically. Nevertheless, short pieces of three nearby arms have been detected from the observed directions and distances of a large number of O and B stars and from the distribution of emission nebulae. The fragmentary data on the spiral structure of the Galaxy, as deduced from such optical observations, are summarized in Figure 15.7.

Another method of demonstrating spiral structure of the Galaxy is from radio observations of the 21-cm line of neutral hydrogen. The radio waves are not absorbed by interstellar dust and thus traverse the entire disk of the Galaxy. Because the Galaxy is rotating, the gas clouds in remote parts of the disk have high velocities compared to the sun; the actual radial velocity of a gas cloud depends on its distance along a given line of sight through the galactic plane.

Now because different gas clouds along such a line of sight have different radial velocities, we observe that the 21-cm radiation from the hydrogen is Doppler shifted by various amounts; that is, we do not see a single emission line at 21.11 cm, but several emission lines, all at slightly different wavelengths, each arising from a different concentration of hydrogen.

If we knew the precise orbits of the neutral hydrogen in various parts of the Galaxy, we could deduce just how far away the hydrogen concentrations would have to lie to account for the observed Doppler shifts of their 21-cm radiation. In the 1950s and 1960s great attention was paid to this problem, and various models of spiral structure were deduced on the assumption that the hydrogen moved uniformly on circular orbits about the galactic center (Figure 15.8). Today it is quite clear that the motion of the interstellar gas is considerably more complex, and specific maps of spiral structure pieced together from 21-cm observations should not be taken too seriously. These observations certainly do, however, demonstrate the existence of the Galaxy's spiral arms.

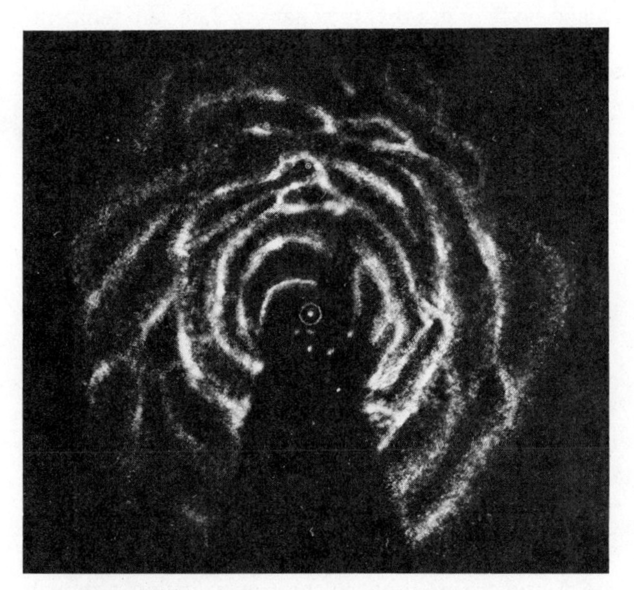

Figure 15.8
An early model of the spiral structure of the Galaxy, deduced from 21-cm observations made at Leiden, The Netherlands, and at Sydney, Australia. The large and small circles show the location of the galactic center and the sun, respectively.

Different Stellar Populations in the Galaxy

Some classes of objects are found only in regions of interstellar matter, that is, in the spiral arms of the Galaxy. Examples are bright supergiants, main-sequence stars of high luminosity (spectral classes O and B), type I cepheid variables, and young open star clusters. Other objects show no concentration to the spiral arms, but are found throughout the disk of the Galaxy, with greatest concentration toward the nucleus. These include planetary nebulae (Chapter 17), novae, and many older stars. Still others, for example, RR Lyrae stars and globular clusters, do not concentrate strongly to the disk, but extend out into the sparse galactic corona.

The stars associated with the spiral arms are sometimes said to belong to *extreme population I*. Those in the disk make up the intermediate, or *disk population*, and those in the corona are *extreme population II*.

We interpret some of the phenomena of these different stellar populations in the light of stellar evolution. As we will see in Chapter 17, only in the interstellar matter of spiral arms is star formation expected to take place. Thus population I is comprised of stars of relatively young age, including some that were recently formed or are still forming. Population II consists entirely of old stars, formed, probably, early in the history of the Galaxy, and the disk population stars have a wide range of ages.

Magnetism in the Galaxy and Synchrotron Radiation

We are familiar with the fairly strong magnetic field around the earth, and have seen (Chapter 10) that there is a much stronger field around Jupiter. In the sun (next chapter) there are very strong magnetic fields associated with the active regions around sunspots, and a general solar field of about the strength of the earth's. Some stars also have extremely strong magnetic fields, and pulsars (Chapter 17) must have incredibly strong fields. But in addition to magnetism associated with particular objects in the Galaxy, there are general fields whose force lines usually run along the spiral arms, and possibly a weak field throughout the corona as well. These general interstellar magnetic fields of the Galaxy are of low intensity, but cover

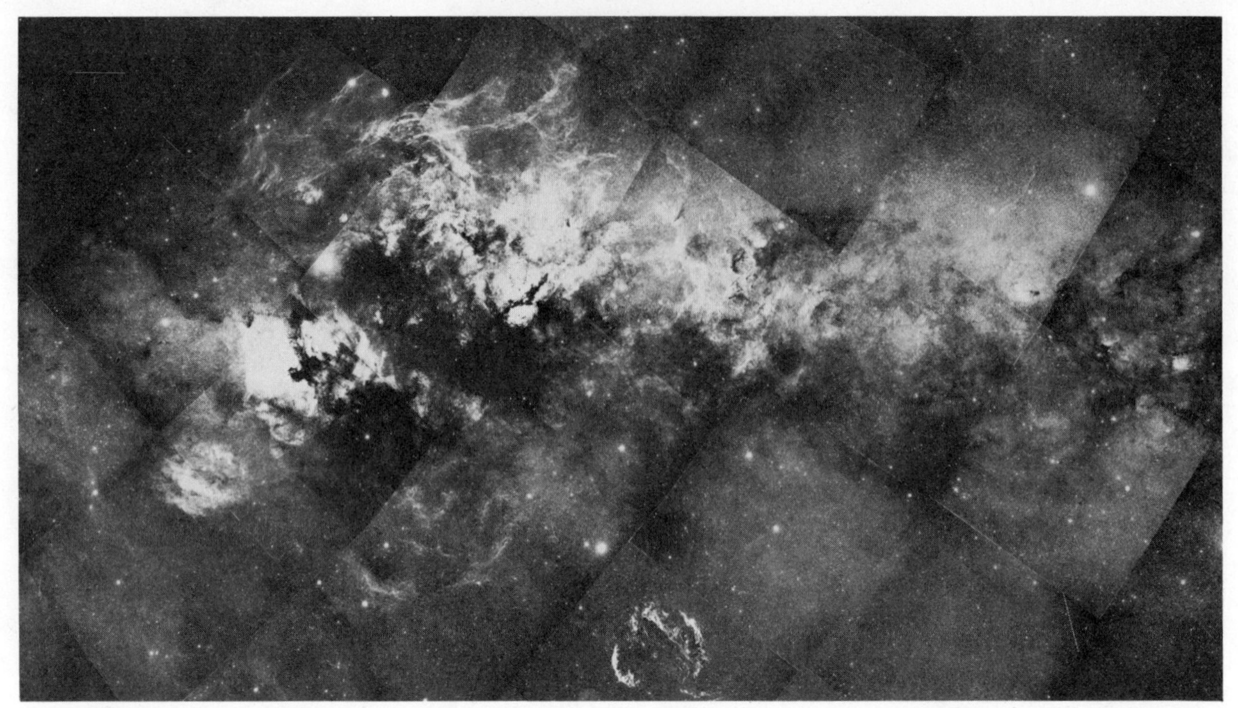

Figure 15.9
Mosaic photographs of the Milky Way in Cygnus, photographed with the 124-cm Schmidt telescope. *(National Geographic Society–Palomar Observatory Sky Survey, reproduced by permission from the Hale Observatories)*

such vast regions of space that they involve a great deal of total energy.

The galactic magnetism reveals its presence in several ways. One is the fact that when stars are dimmed by dust, their light is also slightly polarized. We can understand this phenomenon if the dust grains are slightly elongated and are aligned along magnetic field lines. Also, magnetic fields cause charged particles moving through them to follow curved paths. We believe that most of the cosmic rays striking the earth (Chapter 5) originate in the Galaxy (possibly from supernovae—see Chapter 17). These charged particles, we think, are trapped by magnetic fields forcing them to keep spiraling about inside the Galaxy.

Further evidence for the existence of magnetic fields in the Galaxy is the observation of *synchrotron radiation*. Charged particles, such as electrons and ions, radiate electromagnetic energy when they are accelerated by a magnetic field. If the speed of the particle is nearly the speed of light, the particle is said to be *relativistic*. In this case, the energy it radiates is called *synchrotron radiation*, because particles so radiate when they are accelerated to relativistic speeds in a laboratory synchrotron. Now, we have seen that any solid or liquid body or gas that is not at absolute zero temperature radiates electromagnetic energy (Chapter 5). To distinguish it from this normal *thermal* radiation from gases or bodies, synchrotron radiation is sometimes called *nonthermal* radiation.

Nonthermal radiation has properties that make it easy to recognize: it has a distinctive distribution of intensity with wavelength, it is highly polarized, and the energy radiated by a particle is primarily in the direction of the particle's instantaneous motion (see Figure 15.10). Both atomic nuclei (positive ions) and electrons radiate when accelerated, but the nuclei are thousands of times as massive as electrons, and consequently for the same energy have much lower speeds. Thus the nuclei do not generally move fast enough to emit significant synchrotron radiation.

We find many astronomical examples in which relativistic electrons are spiraling through magnetic fields and are emitting synchrotron radiation, although we do not yet, in all cases, know the origin of these energetic electrons, nor the mechanisms that give them their great speeds. Usually this radiation is most intense at radio wavelengths.

If cosmic ray particles (mostly atomic nuclei) are

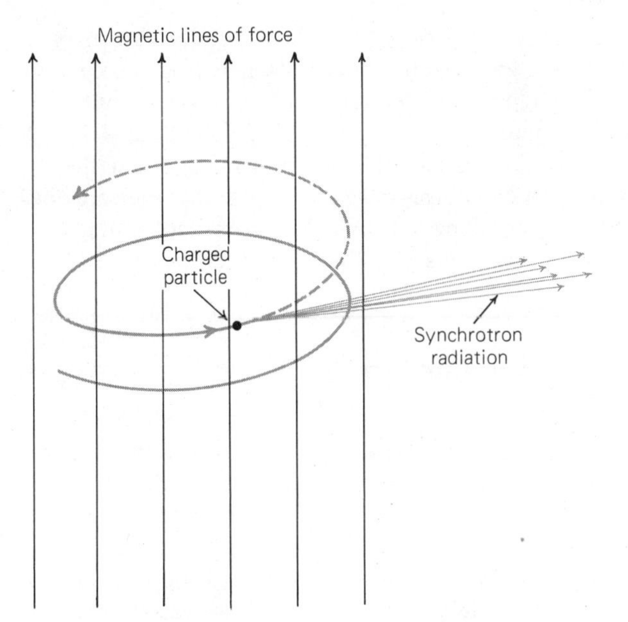

Figure 15.10
The emission of synchrotron radiation by a charged particle moving at nearly the speed of light in a magnetic field.

trapped in galactic magnetic fields, it is not surprising that electrons are present as well. The Galaxy, in fact, abounds in sources of nonthermal radiation. Some are in its very nucleus. This radiation shows us that there are vast interstellar magnetic fields in the Galaxy. In certain other galaxies (Chapter 20) we find synchrotron radiation emanating from their entire coronas, and often from invisible regions far outside the parts of those galaxies that we can observe optically. Evidently those galaxies have very extended magnetic fields that sometimes reach far beyond their visible images. Calculations of the field strengths in our Galaxy are not very precise, but they indicate an average field in the spiral arms of from 10^{-6} to 10^{-5} gauss; by comparison, the field at the surface of the earth is a little less than 1 gauss.

STAR CLUSTERS

A star cluster is a congregation of stars that have a stronger gravitational attraction for each other (because of their proximity) than do stars of the general field (that is, those *not* in clusters). Clusters range from rich aggregates of many thousands of stars to loose associations of only a few stars. The mutual gravitation of the stars in the larger clusters may hold them together more or less permanently; the small clusters may be held together so weakly that they are gradually dissipating into the field.

A number of clusters bear proper names. Some of these are names of mythological characters (the Pleiades); other clusters bear the names of the constellations in which they appear (the Double Cluster in Perseus). Most of the conspicuous clusters are listed in the early catalogs of star clusters and nebulae, especially in Sir John Herschel's *General Catalogue* (1864). The catalog designations of clusters most often referred to today are those in Messier's (1781) catalog (for example, M13—the famous globular cluster in Hercules), and in Dreyer's revisions of Herschel's catalog, the *New General Catalogue* (*NGC*) and the *Index Catalogue* (*IC*), published between 1888 and 1908. For example, in Dreyer's catalog, M13 is known as NGC 6205.

Globular Clusters

As we have already mentioned, about a hundred globular clusters are known, most of them in the corona and nuclear bulge of our Galaxy. All are very far from the sun, and some are found at distances of 60,000 LY or more from the galactic plane. A few, nevertheless, are bright enough to be seen with the naked eye; they appear as faint, fuzzy stars. One of the most famous naked-eye globular clusters is M13, in the constellation of Hercules, which passes nearly overhead on a summer evening at most places in the United States. Through a good pair of binoculars the more conspicuous globular clusters resemble tiny moth balls. A small telescope reveals their brightest stars, while a large telescope shows them to be beautiful, globe-shaped systems of stars. Visual observation, however, even through the largest telescope, does not reveal the multitude of fainter stars in globular clusters that can be recorded on telescopic photographs of long exposure.

A good photograph of a typical globular cluster shows it to be a nearly circularly symmetrical system of stars, with the highest concentration of stars near its own center (a few globular clusters, such as Omega Centauri, appear slightly flattened). Most of the stars in the central regions of the cluster are not resolved as individual points of light but appear as a nebulous glow. The brightest stars in globular clusters are red giants.

Distances to globular clusters are usually calculated from the apparent magnitudes of the RR Lyrae

Figure 15.11
The globular cluster M13. *(Hale Observatories)*

stars they usually contain (Chapter 13). From their angular sizes (typically a few minutes of arc) their actual linear diameters are found to be from 20 to 100 pc or more. In one of the nearer globular clusters more than 30,000 stars have been counted, but if those stars too faint to be observed are considered, most globular clusters must contain hundreds of thousands of member stars.

There is plenty of space between the stars, however. A bullet fired on a straight line through the center of a cluster would have far less than one chance in 10^{11} of striking a star. If the earth revolved not about the sun, but about a star in the densest part of a globular cluster, the nearest neighboring stars, light-months away, would appear as points of light. Thousands of stars, however, would be scattered uniformly over the sky. The Milky Way would be difficult, if not impossible, to see, and even on the darkest of nights the brightness of the sky would be comparable to faint moonlight.

The globular clusters do not partake of the general galactic rotation. They are believed to revolve about the nucleus of the Galaxy on orbits of high eccentricity and high inclination to the galactic plane. Obeying Kepler's second law, a cluster spends most of its time far from the nucleus; a typical cluster probably has a period of revolution of the order of 10^8 years.

X-Ray sources have been found in several globular clusters, and most of these display sudden bursts of X-ray energy. The bursts last anywhere from a second to a few tens of seconds, and recur irregularly at intervals of minutes to hours. The cause of these flashes of high-energy radiation is not known for certain; the favored hypothesis at this writing is that they are due to matter falling onto collapsed stars, and igniting in nuclear reactions. (We discuss collapsed stars in Chapter 17.)

Open Clusters

Open clusters, as mentioned previously, are found in the disk of the Galaxy, often associated with interstellar matter. Because of their locations, they are sometimes called *galactic clusters* rather than open clusters (this term should not be confused, however, with *clusters of galaxies* — Chapters 20 and 21). Over 1000 open clusters have been cataloged as of 1980, but many more are identifiable on good search photographs such as those of the Palomar Sky Survey. Yet only the nearest open clusters can be observed, because of interstellar obscuration in the Milky Way plane. We conclude, therefore, that we see only a small fraction of the open clusters that actually exist in the Galaxy; possibly tens or even hundreds of thousands of them escape detection.

Several open clusters are visible to the unaided eye. Most famous among them is the *Pleiades,* which appears as a tiny group of six stars (some people see more than six) arranged like a tiny dipper in the constellation of *Taurus;* a good pair of binoculars shows dozens of stars in the cluster, and a telescope reveals hundreds. (The Pleiades is *not* the Little Dipper; the latter is part of the constellation of *Ursa Minor,* which also contains the north star.) The *Hyades* is another famous open cluster in Taurus. To the naked eye, the cluster appears as a V-shaped group of faint stars, marking the face of the bull. Telescopes show that the Hyades actually contains more than 200 stars. The naked-eye appearance of the *Praesepe,* in Cancer, is that of a barely distinguishable patch of light; this group is often called the "Beehive" cluster, because its many stars, when viewed through a telescope, appear like a swarm of bees.

Typical open clusters contain several dozen to several hundred member stars, although a few, such as M67, contain more than a thousand. Compared to globular clusters, open clusters are small, usually having diameters of less than 10 pc. Bright supergiant stars of high luminosity in some open clusters, however, may cause them to outshine the far richer globular clusters. The RR Lyrae stars are never found in open clusters, but other kinds of variable stars, such as type I cepheids, are sometimes present.

Figure 15.12
The Pleiades. North is to the left. *(Lick Observatory)*

Associations

An association appears as a group of several (5 to 50) O stars and B stars scattered over a region of space some 30 to 200 pc in diameter. Because these stars are rare, it would be very unlikely for so many of them to exist by chance in so relatively small a volume of space. It is assumed, therefore, that the stars in an association are either physically associated or at least have had a common origin. Stars of other spectral types may also belong to associations, but these more common stars are not conspicuous against the general star field and do not attract attention to themselves as belonging to any particular group.

Often a small open cluster is found near the center of an association. It is presumed, in such cases, that the stars in the association are the outlying members of the cluster, and that they probably had a common origin with the cluster stars. There is evidence that some associations may be expanding, but as yet there is no generally accepted hypothesis to explain why the O and B stars should move rapidly away from the other stars of their parent clusters.

About 70 associations are now cataloged. Like ordinary open clusters, however, they lie in regions occupied by interstellar matter, and many others must be obscured. There are probably several thousand undiscovered associations in our Galaxy.

Summary of Clusters

The foregoing descriptions of star clusters are summarized in Table 15.1. The numbers of globular clusters, open clusters, and associations are taken from

TABLE 15.1 Characteristics of Star Clusters

	Globular Clusters	Open Clusters	Associations
Number known in Galaxy	125	1055	70
Location in Galaxy	Corona and nuclear bulge	Disk (especially spiral arms)	Spiral arms
Diameter (pc)	20 to 100	<10	30 to 200
Mass (solar masses)	10^4 to 10^5	10^2 to 10^3	10^2 to 10^3?
Number of stars	10^4 to 10^5	50 to 10^3	10 to 100?
Color of brightest stars	Red	Red or blue	Blue
Integrated absolute visual magnitude of cluster	-5 to -9	0 to -10	-6 to -10
Density of stars (solar masses per cubic parsec)	0.5 to 1000	0.1 to 10	<0.01
Examples	Hercules Cluster (M13)	Hyades, Pleiades, h and χ Persei, Praesepe	Zeta Persei, Orion

the second edition of the *Catalogue of Star Clusters and Associations*, published by the Hungarian Academy of Sciences. The *Catalogue* includes all objects reported up to 1970 and is the most comprehensive star cluster catalog available. The sizes, absolute magnitudes, and numbers of stars listed for each type of cluster are approximate only and are intended as representative values.

THE NUCLEUS OF THE GALAXY

Near the center of the Galaxy is a large concentration of stars, generally called the *nuclear bulge*. At its center, lying behind the constellation Sagittarius, is the *nucleus* of the Galaxy. We cannot see the nucleus in visible light or in the ultraviolet, because those wavelengths are absorbed by the intervening interstellar dust. High energy X rays and gamma rays, however, force their way through the interstellar medium and are recorded by instruments on rockets and satellites. Also the infrared and radio radiation, whose wavelengths are long compared to the sizes of the interstellar grains, flow around them and reach us from the center of the Galaxy. The very bright radio source in that region is known as *Sagittarius A*.

Much of the infrared radiation is believed to be from sparse dust in the nuclear bulge, and some of it to be from circumstellar dust around stars. At least part of the radiation, however, comes simply from stars themselves, and its intensity allows us to derive the approximate density of stars at various distances from the galactic center; the results are given in Table 15.2. We see that near the center of the Galaxy the stars are still light-months or more apart, but within the inner parsec the average separation of stars is only a few thousand astronomical units—only one or a few light-weeks. At the very center of the galactic nucleus, the stars are only a few hundred astronomical units apart. Yet if the earth were a planet revolving about a sun in almost any part of the galactic nucleus we would still find the neighboring stars far enough away to appear as points of light. In most parts of the nuclear bulge the night sky would appear almost as dark as it does from our present location in the Galaxy. On the other hand, in the inner few parsecs there would probably be enough nearby bright stars to illuminate our night sky as much as moderate moonlight does, and at the very center there would probably be some stars near enough to give us a sky brighter than at full moon.

TABLE 15.2 Distribution of Stars near the Galactic Center

Distance from Center (pc)	Number of Stars per Cubic pc	Total Mass Within This Distance (Solar Masses)
0.1	3×10^7	3×10^5
1.0	4×10^5	4×10^6
10.0	7×10^3	7×10^7
20.0	2×10^3	2×10^8

The Sagittarius radio source displays synchrotron radiation, which means that relativistic electrons and magnetic fields are present at the galactic center. Many of the most complex interstellar molecules observed at radio wavelengths are found in this region as well. The molecules may well have an origin associated with the circumstellar dust clouds inferred from some of the infrared radiation.

We also receive 21-cm radiation from neutral hydrogen in the region around the galactic center. The Doppler shifts of this radiation show that the neutral hydrogen is flowing away from the center at a speed of about 50 km/s. Therefore, the nucleus of our Galaxy is emitting radiation from gamma rays to radio waves, both thermal and nonthermal, and is ejecting gas as well.

Nuclear Activity

The emission of many kinds of radiation as well as the ejection of matter from the nucleus of our Galaxy leads us to compare it to the nuclei of other galaxies. There exists a class of galaxies called *Seyfert galaxies*, most of which are spiral stellar systems like our own, except that they have nuclei that emit very strongly in visible light, infrared radiation, and sometimes radio waves and X-rays. Seyfert galaxy nuclei evidently have regions of very hot gas which may be associated with violent activity. The well-known Seyfert galaxy NGC 1068 has a nucleus from which gas is also flowing at a rate of about 600 km/s. Other galaxies have nuclei of still more explosive properties. Most spectacular are the quasars (Chapter 20). Some investigators suspect that quasars are extreme examples of explosive events at the centers of remote galaxies whose main parts are too faint to observe because of their great distances.

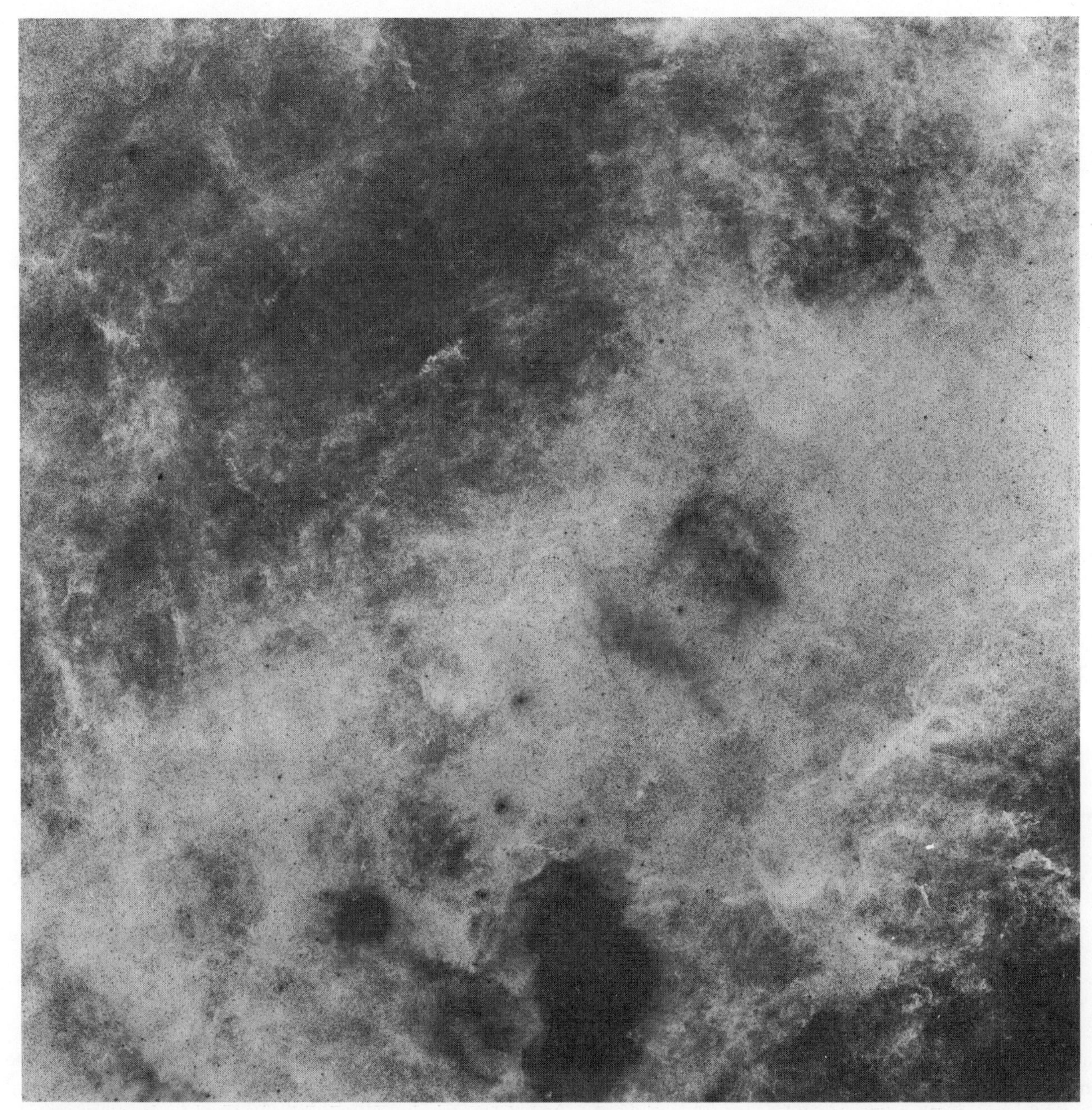

Figure 15.13
A region of the Milky Way near the galactic center. Negative print, photographed in red light with the 124-cm Schmidt telescope. The white regions are absorbing dust clouds. *(Copyright, National Geographic Society–Palomar Observatory Sky Survey, reproduced by permission of the Hale Observatory)*

There seems, in other words, to be a continuum of nuclear phenomena, ranging from the mighty quasars, through the Seyfert galaxies, down to the nuclei of apparently "normal" galaxies like our own. It is not ruled out that a violent nuclear event might occur, from time to time—or possibly at least once—in *any* galaxy. In other words, it is conceivable that our own Galaxy is in a quiescent period, before or after which it too could display the more violent characteristics of Seyfert galaxies or even quasars.

Until recently the evolution of a galaxy was thought of largely in terms of the evolution of its stellar content. Now we know that this is not all of the picture. The existence of events such as those described above shows that the life history of a galaxy may be influenced, and perhaps even largely determined, by large-scale collective phenomena involving the structure of the galaxy as a whole rather than just the evolution of its individual components.

SUMMARY REVIEW

Structure of the Galaxy: distribution of globular clusters; RR Lyrae stars; the sun's location in the Galaxy; disk; extent of disk; galactic corona (halo); sun's orbit in the Galaxy; galactic year; orbits of neighboring stars; mass of the Galaxy

Interstellar medium: *nebulae;* density of the interstellar material in space; dark nebulae; dark rift; globules; general obscuration by dust; reddening law; color excess; correction of measured magnitudes for interstellar obscuration; reflection nebulae; circumstellar dust; nature of the dust grains; interstellar gas; emission nebulae; H I and H II regions; "nebulium" lines; interstellar absorption lines; radio emission from the gas; the 21-cm line of hydrogen; interstellar molecules; cold clouds; origin of interstellar grains and molecules

Spiral structure of the Galaxy: observation of O and B stars and emission nebulae; observation of 21-cm radiation; stellar populations; galactic magnetic fields; polarization of starlight; synchrotron (nonthermal) radiation; relativistic particles

Star clusters: nomenclature and catalogs; globular clusters; distances; open (galactic) clusters; associations; properties of clusters and associations

Galactic nucleus: X-ray and infrared radio observations; Sagittarius A; star density; hydrogen streaming from the nucleus; comparison with nuclear activity in other galaxies—Seyfert galaxies and quasars

EXERCISES

1. Sketch the distribution of globular clusters about the Galaxy, and show the sun's position. Show how they would appear on a Mercator map of the sky, with the central line of the Milky Way chosen as the "equator."

2. The globular clusters probably have highly eccentric orbits, and either oscillate through the plane of the Galaxy or revolve about its nucleus. Suppose the latter is the case; where would the clusters spend most of their time? (Think of Kepler's second law.) At any given time, would you expect most globular clusters to be moving at high or low speeds with respect to the center of the Galaxy? Why?

3. Suppose the mean mass of a star in the Galaxy were only ⅓ solar mass. Using the value for the mass of the Galaxy found in the text, find how many stars the system contains. What did you assume about the total mass of interstellar matter in finding your answer?

4. Distinguish clearly between the orbital motion of the sun, toward galactic longitude 90°, and the *solar motion,* toward the *solar apex,* which was described in Chapter 12.

5. Identify several dark nebulae on photographs of the Milky Way in this book. Give the figure numbers of the photographs, and specify where the dark nebulae are to be found on them.

6. The red color of the sun, when seen close to

the horizon, and the blue color of the daytime sky, provide analogies to the reddening of starlight and the blue color of reflection nebulae. Discuss this analogy more fully, and also explain how it breaks down.

7. Suppose all stars had the same luminosity and were uniformly distributed through space. Now suppose there were a cloud of dust at such a distance that some stars could be seen in front of it, and that other stars could be seen shining through it, appearing dimmer, of course, than if the dust cloud were not there. Describe how you might detect the cloud's presence from counts of stars of different apparent magnitudes.

8. Describe the spectrum of (a) starlight reflected by dust; (b) a star behind invisible interstellar gas; (c) an emission nebula.

9. Here are five kinds of interstellar clouds: (1) a reflection nebula, (2) a dark nebula, (3) an emission nebula, (4) gas in an H I region, (5) circumstellar dust.
(a) Which is detected by the infrared radiation it emits?
(b) Which is detected by the observation of scattered starlight?
(c) Which has a bright-line spectrum?
(d) Which could be detected by the 21-cm line of hydrogen?
(e) Which is detected by the absence of observable stars?

10. Here are five kinds of objects: (1) open cluster, (2) association, (3) gaseous nebula, (4) globular cluster, (5) a group of O and B stars.
(a) Which one or ones are found only in spiral arms?
(b) Which one or ones are found only in the parts of the Galaxy that are *not* in the spiral arms?
(c) Which are thought to be very young?
(d) Which are thought to be very old?
(e) Which have stars that are of the highest temperatures?

11. Where in the Galaxy do you suppose undiscovered globular clusters may exist?

12. Table 15.1 indicates that stellar associations can emit even more light than a globular cluster. How is this possible if the associations have so few stars?

13. What is the color of a globular cluster? Why?

14. From the data of Table 15.1, estimate the average mass of the stars in each of the three different cluster types.

15. It is often possible to observe fainter main-sequence stars in open clusters than in globular clusters. Why do you suppose this is the case?

16. Cosmic ray particles strike the earth at speeds very near to that of light. At these speeds, they would leave the Galaxy entirely in less than 100,000 years. Yet we observe cosmic rays that we think formed in the Galaxy millions of years ago. How can they still be around?

17. There is a cold interstellar cloud, containing interstellar molecules, behind (that is, beyond) the Orion nebula. Why do we not observe molecules of, say, ethyl alcohol *within* the Orion nebula?

18. What if a violent event, say a great explosion, occurred at *this instant* at the center of the Galaxy. By what means might we observe it, and when? Would you be worried about your personal safety? Why?

CHAPTER 16

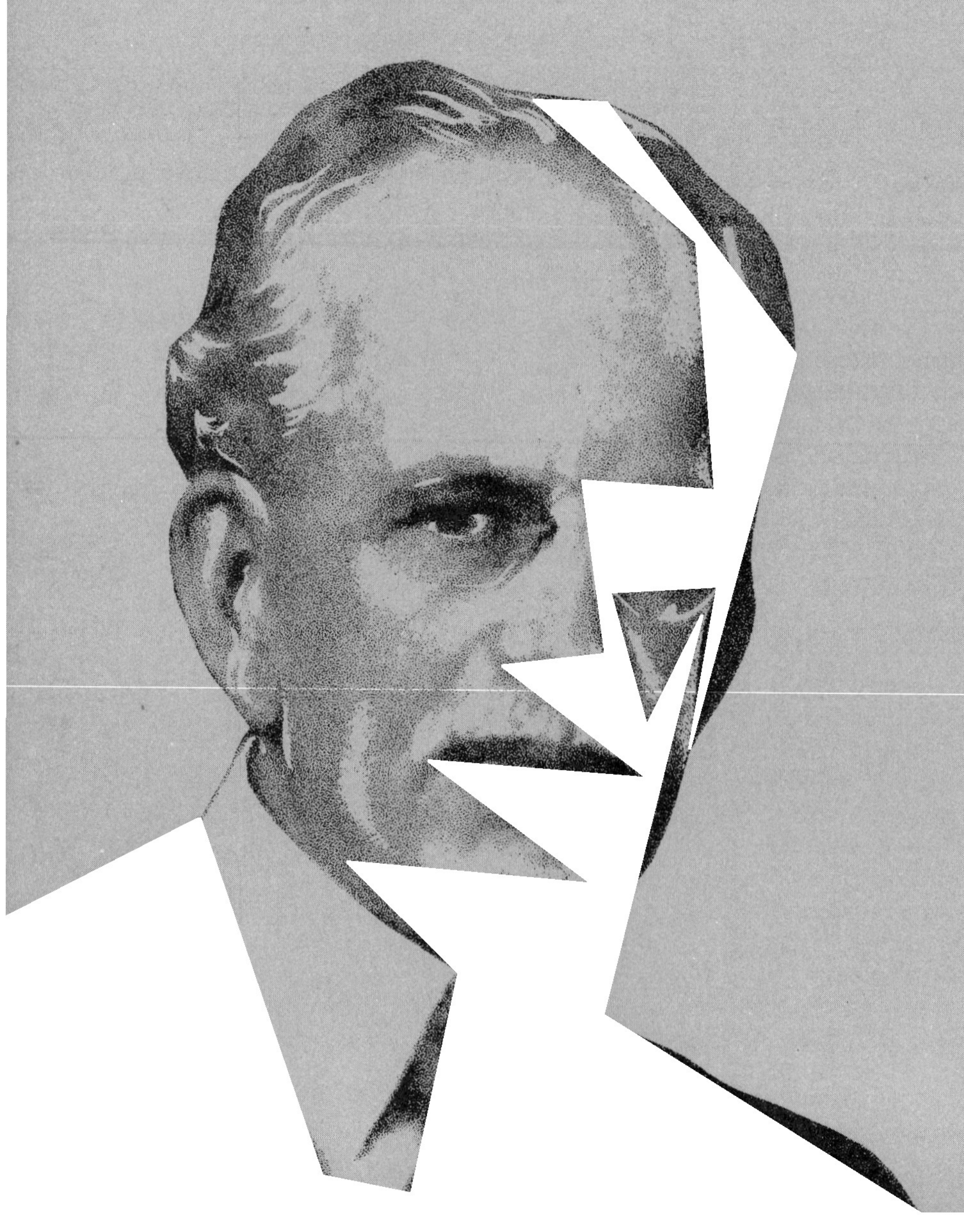

George Ellery Hale (1868–1938) had the vision and leadership to create the world's largest telescope no less than three times! The famous observatories on Mount Wilson and Palomar Mountain are named the *Hale Observatories* in his honor. He was also the foremost solar astronomer of his time. *(Hale Observatories)*

A GARDEN-VARIETY STAR

Among the hundreds of thousands of millions of stars in our Milky Way Galaxy, our sun is way above average in luminosity, size, and mass. Yet, as we saw in Chapter 14, it is roughly midway between the extremes of the properties of the normal stars. The sun is, in other words, pretty much a "garden-variety" star.

THE SOLAR ATMOSPHERE

The only parts of the sun that can be observed directly are its outer layers, collectively known as the sun's *atmosphere*. The solar atmosphere is not stratified into physically distinct layers with sharp boundaries. Yet there are three general regions, each having substantially different properties, even though there is a gradual transition from one region to the next. These are the *photosphere*, the *chromosphere*, and the *corona*.

The Solar Photosphere

What we see when we look at the sun is the solar photosphere. As stated in Chapter 13, the photosphere is not a discrete surface but covers the range of depths from which the solar radiation escapes. As we look toward the limb (the apparent "edge") of the sun, our line of sight enters the photosphere at a grazing angle, and the depth below the outer suface of the photosphere to which we can see is even less than at the center of the sun's disk. The light from the limb of the sun, therefore, comes from higher and cooler regions of the photosphere.

From analysis of this *limb darkening*, and with our knowledge of the physics of gases and the way in which atoms absorb and emit light, we can calculate how the temperature, density, and pressure of the gases vary through the photosphere. It is found that within a depth of about 250 km (160 mi) the pressure and density increase by a factor of 10, while the temperature climbs from 4500 to 6800 K. (The corresponding range of temperature on the Fahrenheit scale is 7600° to 11,700°.) At a typical point in the photo-

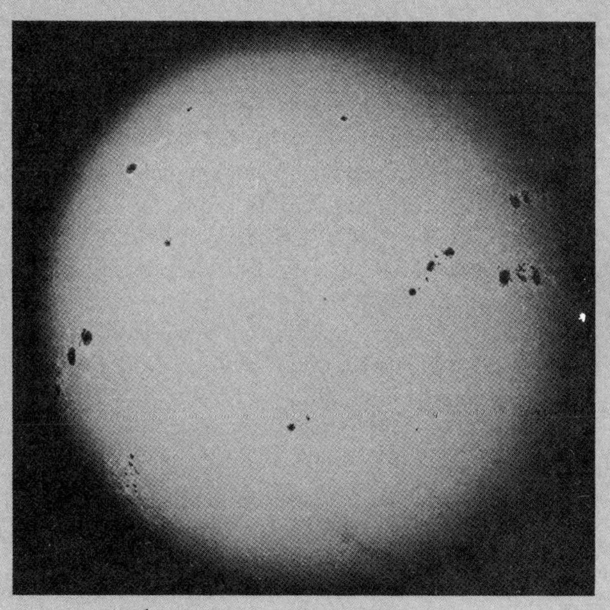

Figure 16.1
The sun, photographed under excellent conditions, showing a large number of sunspots, on September 15, 1957. *(Hale Observatories)*

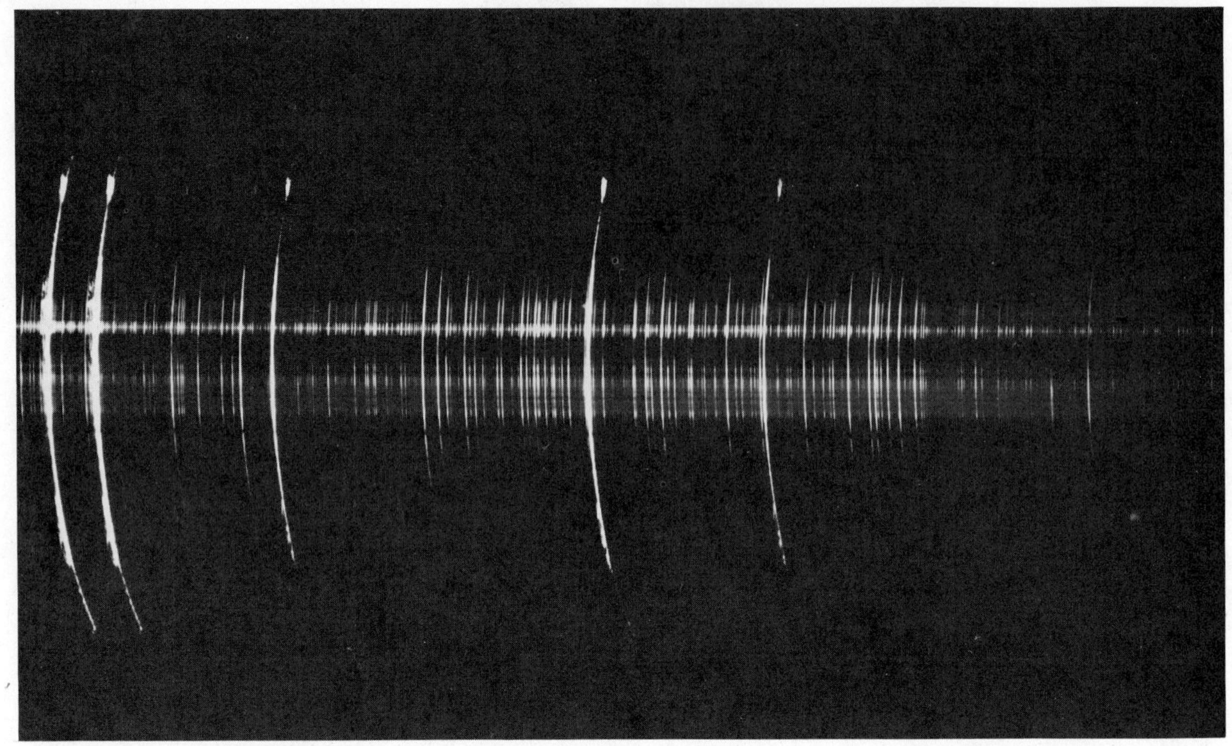

Figure 16.2
Flash spectrum of the eclipse of February 2, 1968. The bright arcs are images of the solar limb. The irregularities are prominences. *(Courtesy Sacramento Peak Observatory, Air Force Cambridge Research Laboratories)*

sphere, the pressure is only a few hundredths of sea-level pressure on the earth, and the density is about one ten-thousandth of the earth's atmospheric density at sea level.

More than 60 of the elements known on the earth have now been identified in the solar spectrum. Those that have not been identified in the sun either do not produce lines in the observable spectrum or are so rare on the earth that they cannot be expected to produce lines of observable strength on the sun unless, proportionately, they are far more abundant there. Most of the elements found in the sun are in the atomic form, but more than 18 types of molecules have been identified. Most of the molecular spectra are observed only in the light from the cooler regions of the sun, such as the sunspots.

The relative abundances of the chemical elements in the sun are similar to the relative abundances found for other stars. About three-fourths of the sun (by weight) is hydrogen; 97 to 98 percent is hydrogen and helium. The remaining few percent is made up of the other chemical elements, in approximately the amounts that are given in Appendix 19.

The Chromosphere

Gases extend far beyond the photosphere, but they are transparent to most visible radiation. The region of the sun's atmosphere that lies immediately above the photosphere is the chromosphere. Until this century the chromosphere was best observed when the photosphere was occulted by the moon during a total solar eclipse. In the seventeenth century several observers described what appeared to them as a narrow red "streak" or "fringe" around one limb of the moon during a brief instant after the sun's photosphere had been covered. Not until the careful observations of the solar eclipses of 1842, 1851, and 1860, however, was much attention paid to the chromosphere. During the eclipse of 1868 the spectrum of the chromosphere was first observed; it was found to be made up of bright lines, which showed that the chromosphere consists of hot gases that are emitting light in discrete wavelengths.

Because of the brief instant during which the chromospheric spectrum can be photographed during an eclipse, its spectrum, when so observed, is called the *flash spectrum*. The element *helium* (from *helios*, the Greek word for "sun") was discovered in the chromospheric spectrum before its discovery on earth in 1895. Today it is possible to photograph both the chromosphere and its spectrum outside eclipse with special instruments.

The chromosphere is about 2000 to 3000 km thick, but in its upper region it breaks up into a forest of jets (called *spicules*), so the position of the upper boundary of the chromosphere is somewhat arbitrary. The density of the chromospheric gases decreases upward above the photosphere, but spectrographic studies show that the temperature *increases* through the chromosphere, from 4500 K at the photosphere to 100,000 K or so at the upper chromospheric levels. Far ultraviolet observations of the chromosphere such as those made from the Skylab, have been especially helpful in revealing information about its structure.

The Corona

The chromosphere merges into the outermost part of the sun's atmosphere, the corona. Like the chromosphere, the corona was first observed only during total eclipses (Chapter 9), but unlike the chromosphere, the corona has been known for many centuries; it is referred to by Plutarch and was discussed in some detail by Kepler. Many of the early investigators regarded the corona as an optical illusion, but photography confirmed its existence in the nineteenth century. The corona extends millions of kilometers above the photosphere and emits half as much light as the full moon. Under ordinary circumstances we cannot see the corona because of the overpowering brilliance of the photosphere. Like the chromosphere, the corona can now be photographed, with special instruments, under other than eclipse conditions. The corona is also observed at radio and X-ray wavelengths.

Studies of the coronal spectrum show it to be very low in density and very hot. At the base of the corona there are about 10^9 atoms/cm^3, compared to about 10^{16}/cm^3 in the upper photosphere, and 10^{19}/cm^3 at sea level in the earth's atmosphere. The corona thins out very rapidly to ever lower density at greater heights, and corresponds to a high vacuum on laboratory standards.

The high temperature of the corona can be inferred from the fact that some of the lines observed are of highly ionized atoms of iron, nickel, argon, calcium, and other elements. For example, lines are observed of iron whose atoms have lost 13 electrons and of calcium whose atoms have lost 14 or more electrons. Such a high degree of ionization requires a temperature of at least one million kelvins. The corona, in other words, is many times as hot as the photosphere. Because of the low density of the corona, however, it does not contain much actual heat, despite its high temperature. The heating mechanism of the corona is believed to be shock waves originating from convection currents in the photosphere. Because of its high temperature, the bulk of the spectral lines emitted by the corona lie in the far ultraviolet. Skylab, for example, has mapped the corona from 300 to 1350 Å.

SOLAR ROTATION

Galileo first demonstrated that the sun rotates on its axis by recording the apparent motions of the sunspots as the turning sun carried them across its disk.

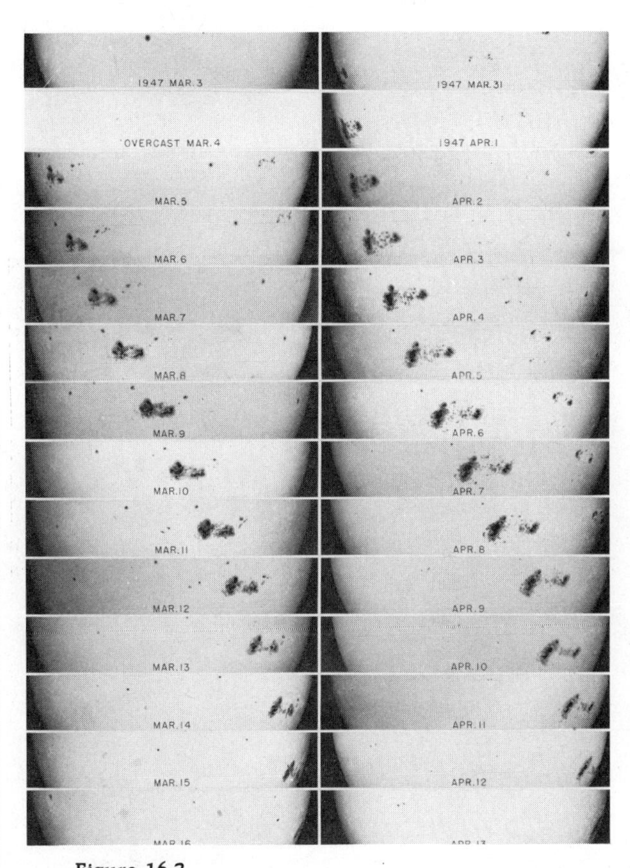

Figure 16.3
Series of photographs showing the motions of sunspots, indicating the solar rotation. *(Hale Observatories)*

He found that the rotation period of the sun is a little less than one month. The period was determined more accurately in subsequent observations by others. In 1859 Richard Carrington found that in its equatorial regions the sun rotates in about 25 days, but that at a latitude of 30° the period is about 2½ days longer. The phenomenon whereby different parts of the sun rotate at different rates is called the sun's *differential rotation*.

The sun's rotation rate can be determined also from the difference in the Doppler shifts of the light coming from the receding and approaching limbs. Its *direction* of rotation is from west to east, like the orbital revolution of the planets and (except for Uranus and Venus) like their axial rotations. It is possible for different parts of the sun to rotate at different rates because, of course, it is fluid rather than solid like the planets. The details of its differential rotation, however, are not yet completely explained.

PHENOMENA OF THE SOLAR ATMOSPHERE

In its gross characteristics the sun is very stable, but the detailed features of its atmosphere are constantly changing. Collectively, those constantly changing features of the solar atmosphere are called *solar phenomena*.

Photospheric Granulation and Supergranulation

The photosphere is not perfectly smooth but has a mottled appearance resembling rice grains—this structure of the photosphere is called *granulation*. Typically, granules are about 1000 km in diameter; the smallest

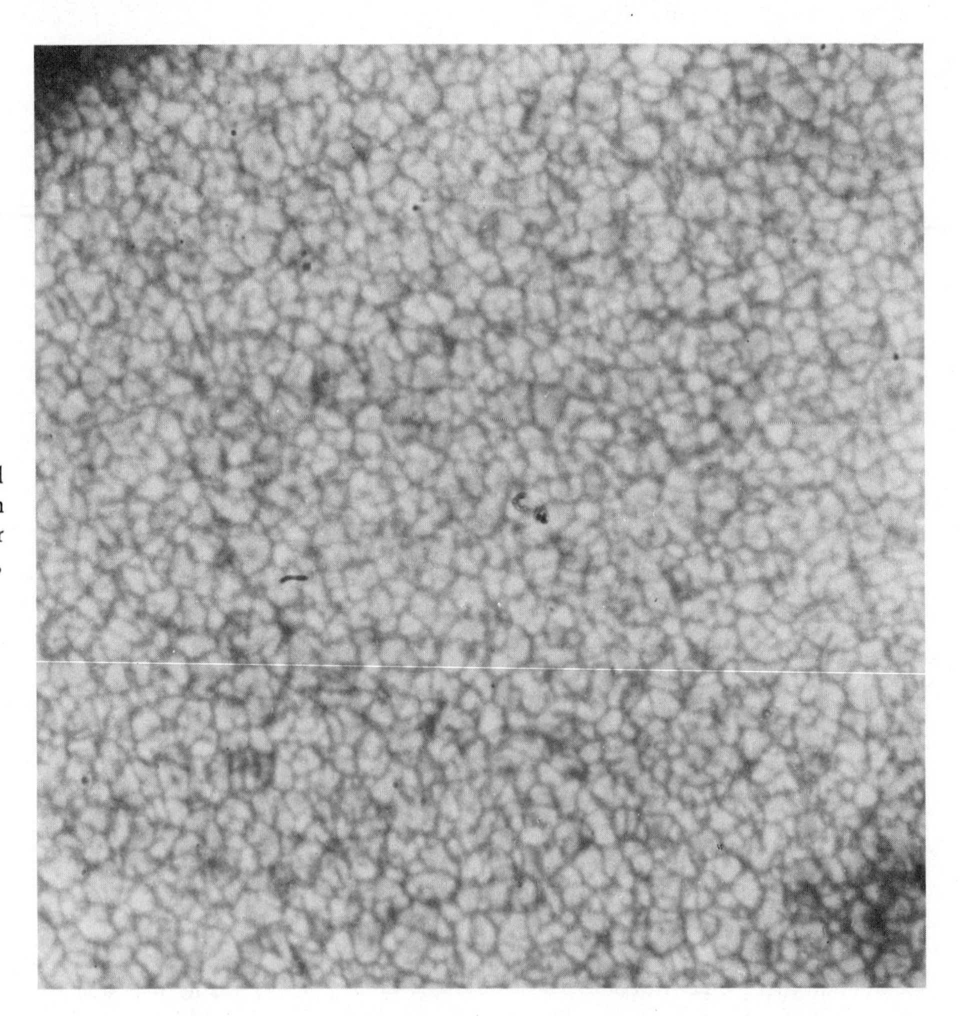

Figure 16.4
Solar granulation, revealed clearly on this photograph from a balloon at an altitude near 25,000 m. *(Project Stratoscope, Princeton University)*

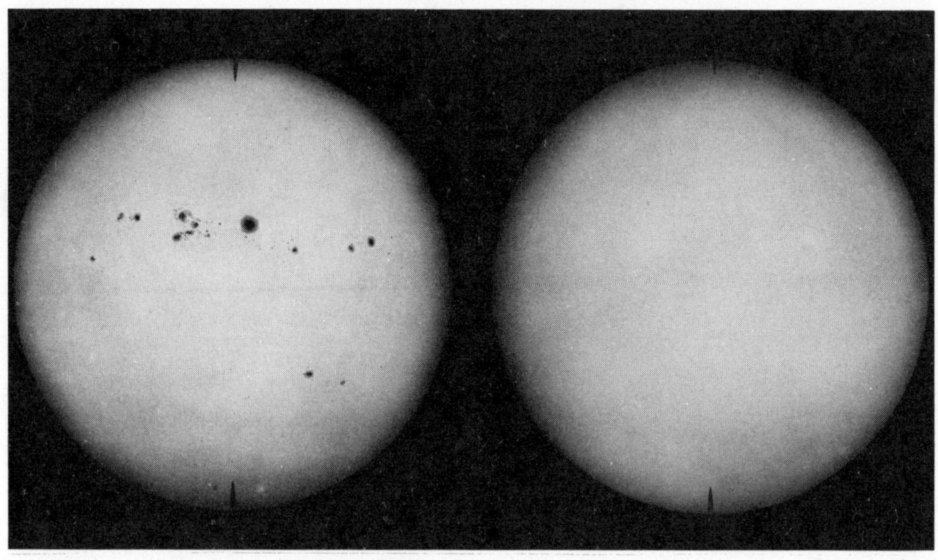

Figure 16.5
Direct photographs of the sun near the time of sunspot maximum *(left)* and near sunspot minimum *(right)*. The dark spikes show the sun's rotation axis. *(Hale Observatories)*

observed are about 300 km across. They appear as bright spots surrounded by narrow darker regions.

The motions of the granules can be studied by the Doppler shifts in the spectra of gases just above them. It is found that the granules themselves are columns of hotter gases arising from below the photosphere. As the rising gas reaches the photosphere it spreads out and sinks down again. The darker intergranular regions are the cooler gases sinking back. The centers of the granules are hotter than the intergranular regions by 50 to 100 K. The vertical motions of gases in the granules have speeds of about 2 or 3 km/s. Individual granules persist for about 8 minutes. The granules, then, are the tops of convection currents of gases seen through the photosphere.

The granules themselves form part of a structure of still larger scale called *supergranulation*. Supergranules are cells that average about 30,000 km in diameter, within which there is a flow of gases from center to edge. In addition to the vertical currents of gases in the granules and the center-toward-edge flow in the supergranules, in each region of the sun (from 3500 to 7000 km across) the gases rhythmically pulse up and down with speeds of about ⅓ km/s, taking about five minutes for a complete cycle, a pulsing called the *five-minute oscillation*.

Sunspots

The most conspicuous of the photospheric features are the *sunspots*. Occasionally, spots on the sun are large enough to be visible to the naked eye, and such spots have been observed for many centuries. It was Galileo, however, who first showed that sunspots are actually on the surface of the sun, rather than being opaque patches in the earth's atmosphere or the silhouettes of planets between the sun and earth.

The spots are photospheric regions where the gases are up to 1500 K cooler than the surrounding gases. Sunspots are nevertheless hotter than the surfaces of many stars. If they could be removed from the sun, they would be seen to shine brightly; they appear dark only by contrast with the hotter, brighter surrounding photosphere.

Individual sunspots have lifetimes that range from a few hours to a few months. They are first seen as small dark "pores" somewhat over 1500 km in diameter. Most of them disappear within a day, but a few persist for a week or occasionally much longer. If a spot lasts and develops, it is usually seen to consist of two parts: an inner darker core, the *umbra*, and a surrounding less dark region, the *penumbra*. Many spots become much larger than the earth, and a few

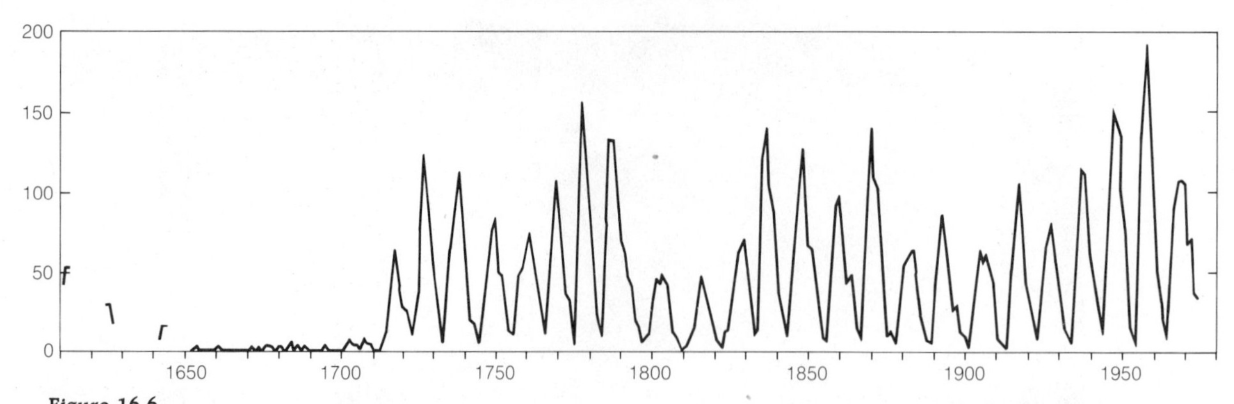

Figure 16.6
The relative numbers of sunspots, as a function of time. Note the absence of sunspots from 1645 to 1715. *(Courtesy, John Eddy, National High Altitude Observatory)*

have reached diameters of 50,000 km. Frequently spots occur in groups of from two to twenty or more. If a group contains many spots, it is likely to include two large ones, one approximately east of the other, and many smaller spots clustered around the two principal ones. The principal spot to the east is most often the largest one of the group. The largest groups are very complex and may have over a hundred spots. Like storms on the earth, sunspots move on the surface of the sun, but their individual motions are slow when compared to the solar rotation, which carries them across the disk of the sun.

In 1851 a German apothecary and amateur astronomer, Heinrich Schwabe, found that although individual sunspots are short-lived, the total number of spots visible on the sun at any one time is likely to be very much greater during certain periods, the periods of *sunspot maximum*, than at other times, the periods of *sunspot minimum*. Sunspot maxima have occurred at an average interval of 11.1 years, but the intervals between successive maxima have ranged from eight to 16 years. During sunspot maxima, more than 100 spots can often be seen on the sun at once. During sunspot minima, the sun sometimes has no visible spots.

This 11-year sunspot cycle is correlated with all of the other phenomena of the solar atmosphere discussed below. Even the corona changes with the cycle; it is large and roughly spherical during sunspot maximum, while it is smaller, compressed along the sun's rotational axis, but extended in the equatorial direction at times of sunspot minimum (see Figures 9.22 and 9.23).

The actual period of the sunspot cycle is far from constant. Although it has *averaged* 11.1 years in recent centuries, it has varied from 8 to 16 years. Moreover, there is considerable evidence that solar activity ceased entirely, or nearly entirely, from 1645 to 1715. During that 70-year interval there were no records of sunspots except for two small groups in 1705; but there are ample reports of sunspots preceding 1645. Furthermore, the solar corona that was visible during the solar eclipses that occurred in those 70 years had no streamers, and reached out only 1′ to 3′. This interval of quiescence in solar activity, recently brought to the attention of astronomers by John Eddy, of the High Altitude Observatory in Boulder, Colorado, was first noted by Gustav Spörer in 1887 and by E. W. Maunder in 1890, and is now called the *Maunder minimum*. A plot showing the incidence of sunspots over time is in Figure 16.6.

Magnetism in the Solar Cycle

The solar cycle is closely related to magnetism in the sun. The Zeeman effect (Chapter 13) in the spectra of the light from sunspot regions shows them to have strong magnetic fields, ranging from 100 to nearly 4000 gauss. There is also a much weaker general magnetic field of the whole sun itself. The general field has a strength of about 1 gauss.

Whenever sunspots are observed in pairs or in groups containing two principal spots, one of the spots usually has the magnetic polarity of a north-seeking magnetic pole, and the other has the opposite polarity. Moreover, during a given cycle, the leading spots of

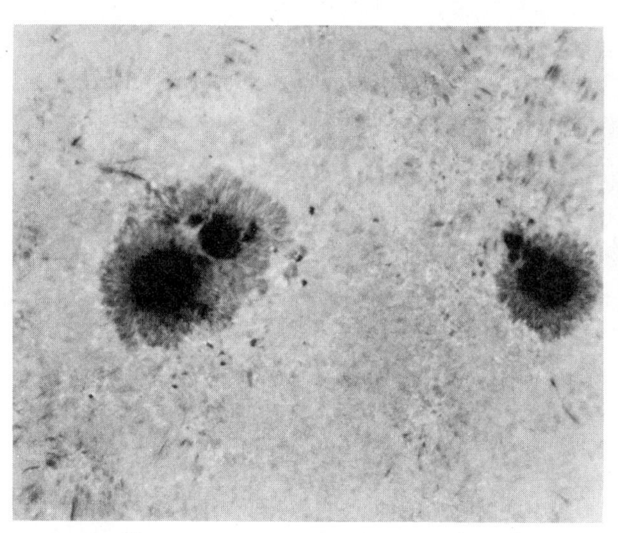
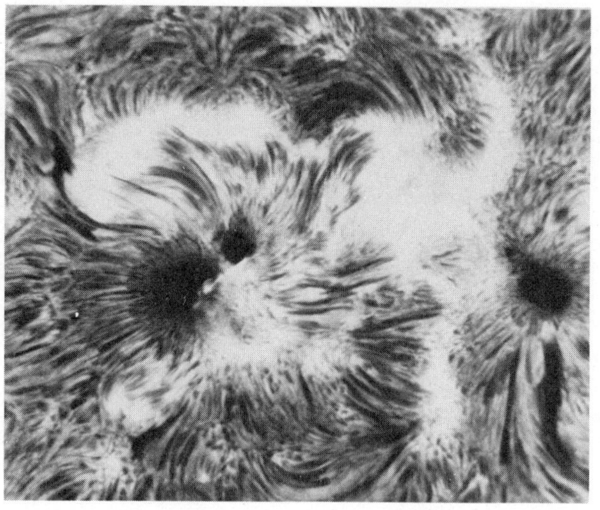

Figure 16.7
Bipolar sunspot group photographed May 21, 1972, at Big Bear Solar Observatory: *(left)* in white light; *(right)* in the red light of the hydrogen Hα line. *(Hale Observatories)*

pairs (or leading principal spots of groups) in the northern hemisphere all tend to have the same polarity, while those in the southern hemisphere all tend to have the opposite polarity. During the next sunspot cycle, however, the polarity of the leading spots is reversed in each hemisphere. For example, if during one cycle the leading spots in the northern hemisphere all had the polarity of a north-seeking pole, the leading spots in the southern hemisphere would have the polarity of a south-seeking pole; during the next cycle, the leading spots in the northern hemisphere would have south-seeking polarity and those of the southern hemisphere would have north-seeking polarity. We see, therefore, that the sunspot cycle does not repeat itself as regards magnetic polarity until *two* 11-year maxima have passed. The sunspot cycle is therefore sometimes regarded to last 22 years, rather than 11.

Phenomena Just Above the Photosphere

In order to see regions of the sun that lie directly above the photosphere, we may observe in spectral regions to which the photospheric gases are especially opaque—at the centers of strong absorption lines such as those of hydrogen and calcium.

There are special filters which pass only light in these narrow spectral regions, and now astronomers routinely photograph the sun through such *monochromatic* filters. These photographs are called *filtergrams*. Spectacular motion pictures have been taken through such monochromatic filters. Time-lapse photographs, in which frames are exposed every few seconds or every few minutes and then run through a projector at normal speed, show in a dramatic way changes that occur in the solar chromosphere.

Filtergrams in the light of calcium and hydrogen show bright "clouds" in the chromosphere in the magnetic-field regions around sunspots. These bright regions are known as *plages*. The plages are not really clouds of any particular element, but are regions where some of the atoms of the element observed are changing their states of ionization or excitation and are emitting more light than other regions. Plages of hydrogen and calcium usually occur in approximately the same projected regions at the same time.

Plages sometimes emit light at many wavelengths and can be seen in the direct image of the sun. These "white-light" plages are called *faculae* ("little torches"), and were first described by Galileo's contemporary Christopher Scheiner. Faculae are seen best near the limb of the sun where the photosphere is not so bright and the contrast is more favorable for their visibility.

The chromosphere also contains many small jet-like spikes of gas rising vertically through it. These features, called *spicules*, occur at the edges of supergranule cells; when viewed near the limb of the sun so many are seen in projection that they give the effect of a forest (Figure 16.8). They show up best when the chromosphere is viewed in the light of hydrogen. They

Figure 16.8
Solar spicules near the sun's limb, photographed in the light of Hα at Big Bear Solar Observatory. *(Hale Observatories)*

consist of gas jets moving upward at about 30 km/s and rising to heights of from 5000 to 20,000 km above the photosphere. Individual spicules last only 10 minutes or so. Through the spicules matter continually flows into the corona. They are now believed to be of fundamental importance in the energy and momentum balance in the solar atmosphere.

Among the more spectacular of coronal phenomena are the *prominences*. Prominences have been viewed telescopically during solar eclipses for centuries. They appear as red flamelike protuberances rising above the limb of the sun. Prominences can now be viewed at any time on filtergrams, and motions of prominences are exhibited in motion pictures. The gross features of some, the *quiescent* prominences, may remain nearly stable for many hours, or even days, and may extend to heights of tens of thousands of kilometers above the solar surface. Others, the more active prominences, move upward or have arches that surge slowly back and forth. The relatively rare *eruptive* prominences appear to send matter upward into the corona at speeds up to 700 km/s, and the most active *surge* prominences may move upward at speeds up to 1300 km/s. Some eruptive prominences have reached heights of over one million km above the photosphere. When seen silhouetted on the disk of the sun, prominences have the appearance of irregular dark filaments.

Superficially, prominences appear to be material ejected upward from the sun, but the motion pictures show that whereas a prominence may grow in size and rise higher and higher above the photosphere, the actual material in the prominence most often appears to move downward in graceful arcs, evidently along lines of magnetic force. Apparently, most prominences form from coronal material that cools and moves downward, even though the disturbance that characterizes the prominence may move upward. Prominences usually originate near regions of sunspot activity and lie on the boundary between regions of opposite magnetic polarity. Quiescent prominences are supported by coronal magnetic fields, and eruptive prominences evidently result from sudden changes in the magnetic fields. Prominences seem to be further symptoms of the same general disturbances that produce spots and plages, that is, local magnetic fields.

The sun in the red light of the first Balmer line of hydrogen. *(Carl Zeiss)*

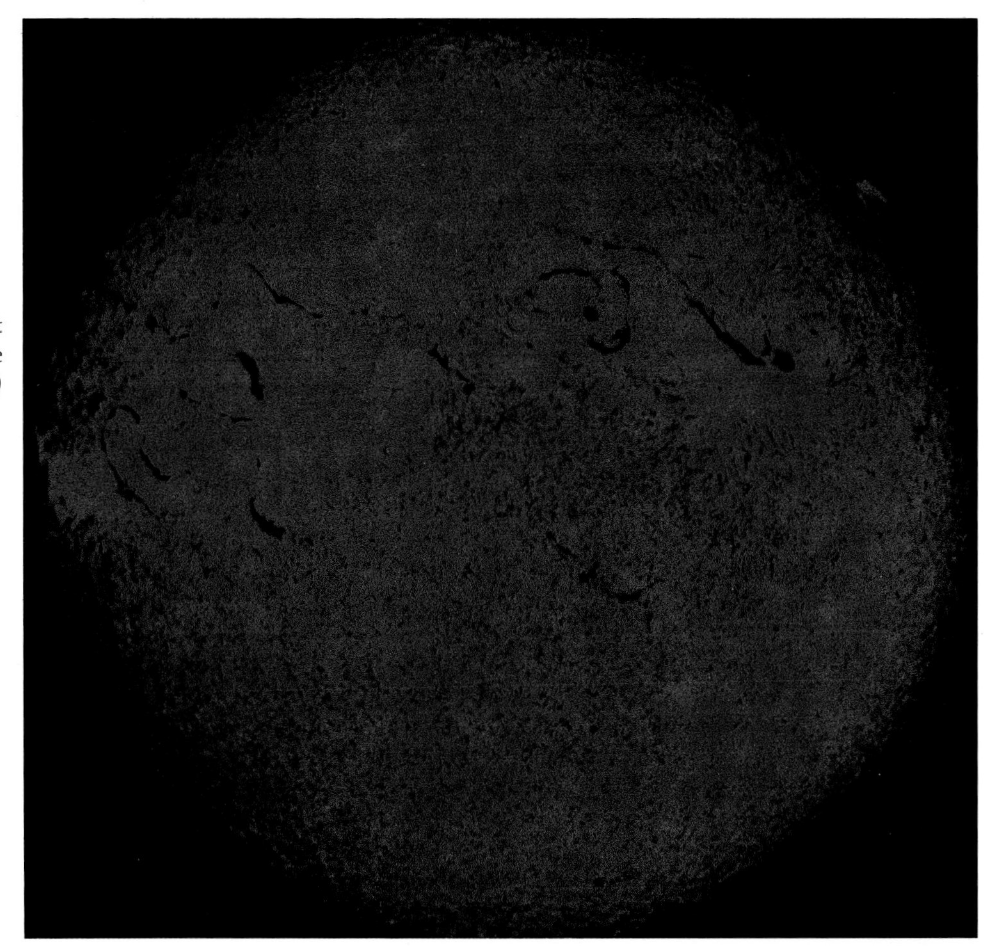

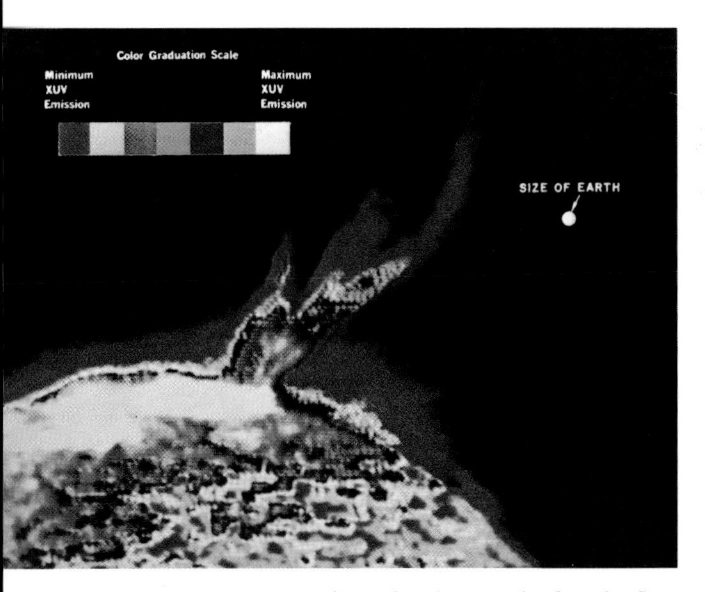

Ultraviolet photograph of a solar flare, taken by Skylab astronauts. The different colors indicate relative ultraviolet intensity. *(NASA)*

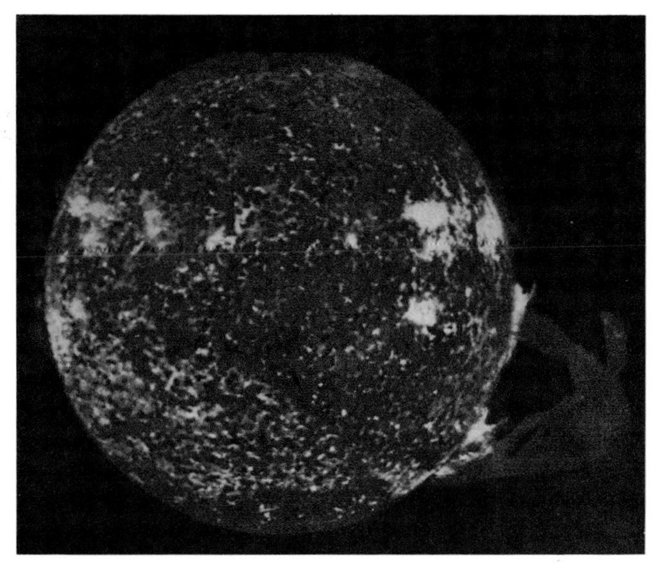

Ultraviolet photograph of the entire sun, showing several flares and a large prominence, taken by Skylab. *(NASA)*

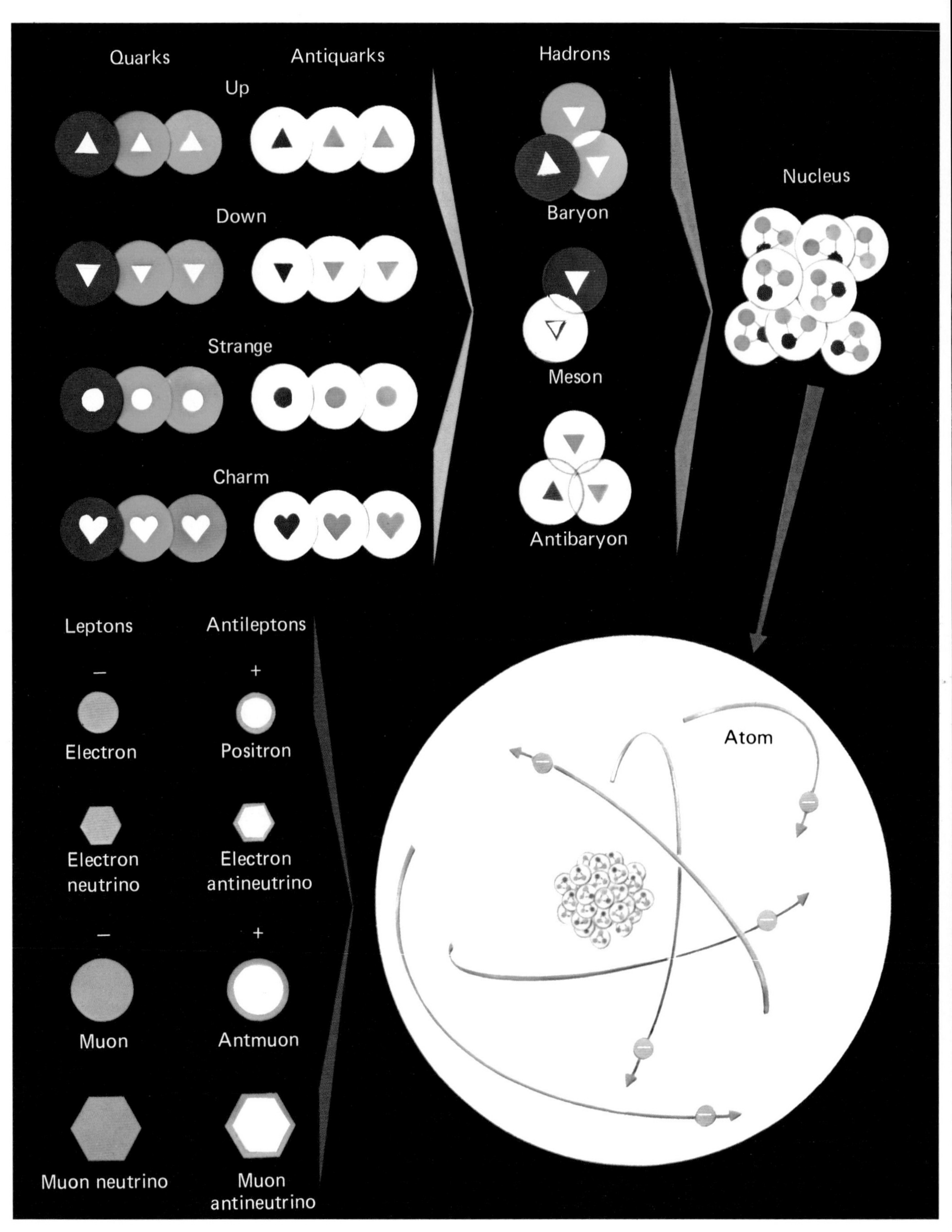

A current model of the subatomic and subnuclear particles.

Figure 16.9
A solar prominence 160,000 km high, photographed in the light of Hα at Big Bear Solar Observatory on June 12, 1972. *(Hale Observatories)*

Flares

Occasionally, the chromospheric emission lines (the Hα and ionized calcium lines in particular) in a small region of the sun brighten up to unusually high intensity. Such an occurrence is called a *flare*. A flare is usually discovered on a filtergram that is being made in the light of one of the spectral lines that brighten. It appears as an intensely bright spot on the photograph a few thousand or tens of thousands of kilometers in diameter. Very rarely, the continuous spectrum of that part of the solar surface affected also brightens during a flare. These white-light flares are among the most intense observed. A flare usually reaches maximum intensity a few minutes after its onset. It fades out slowly, and after an interval ranging from a few minutes to a few hours, it disappears. Near sunspot maximum, small flares occur several times per day and major ones may occur every few weeks.

During major flares, an enormous amount of en-

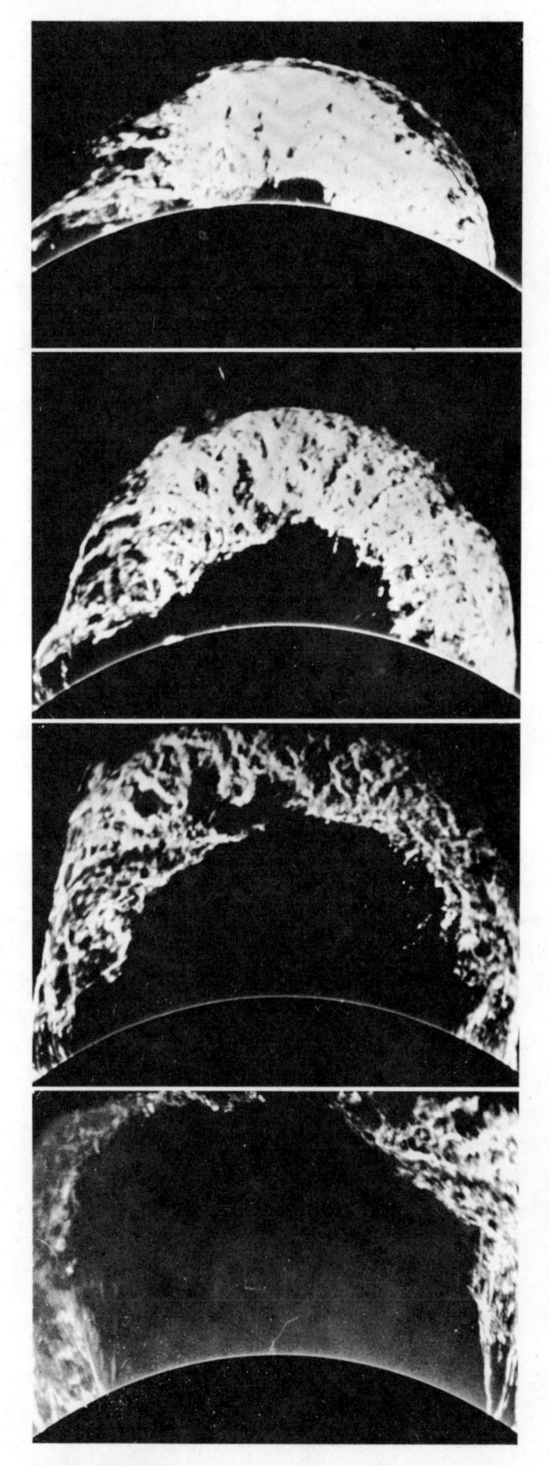

Figure 16.10
Four successive spectroheliograms of the great eruptive prominence of June 4, 1946, taken with a coronagraph. The total elapsed time between the first *(top)* and the last *(bottom)* picture was 1 hr. *(Harvard Observatory)*

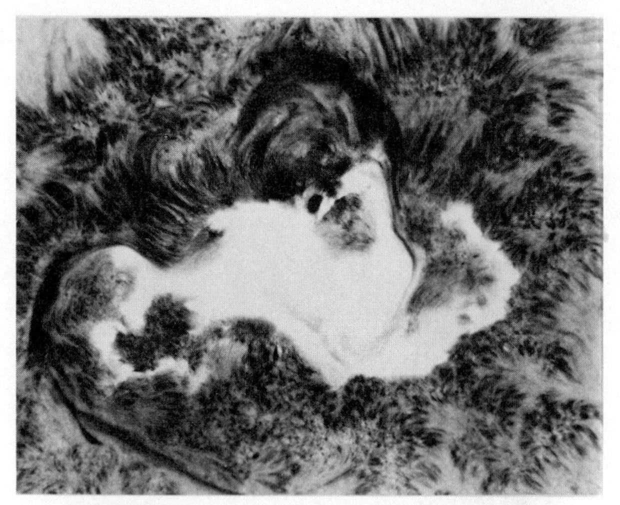

Figure 16.11
A solar flare at its peak intensity, photographed in the light of Hα on August 7, 1972, at Big Bear Solar Observatory. *(Hale Observatories)*

ergy is released. The visible light emitted is, of course, very small compared to that from the entire sun, but a flare covering only a thousandth of the solar surface can actually outshine the sun in the ultraviolet. X rays and gamma rays are emitted as well and are observed from space probes and the earth-orbiting X-ray telescopes. In addition, matter is thrown out at speeds of 500 to 1000 km/s.

Flares are most frequent in the regions of complex sunspot groups with complicated magnetic field structure. Material thrown into the corona from a flare expands and cools and, as ions and electrons recombine, it may emit light and be seen as a loop prominence over the active region of the sun. Loop prominences are the hottest of all prominences and are indicative of rather violent activity in the sun's atmosphere. As is the case with almost all solar phenomena, we do not know the energy source of flares or the mechanism that triggers them, but they are somehow connected with the magnetic fields of sunspot regions.

The Babcock Theory of Solar Activity

The favored theory to account for the solar cycle is that due to Hale Observatories astronomer H. D. Babcock. Babcock's idea is that there is a general, although rather weak, solar magnetic field of roughly the intensity of (or perhaps a little greater than) that of the earth. Since the 1950s Babcock and his father established the existence of just such a solar magnetic field. Most of the field lines are in the photosphere and are forced to follow the motions of the photospheric gases. But recall that the sun rotates differently, the equatorial parts moving faster than those regions at higher latitudes.

A look at Figure 16.12 shows what must occur. The low-latitude regions of the sun draw the field lines forward, compared to their positions in the polar regions, and after a few rotations, stretch them around the sun so that they wind up close to themselves. As the differential rotation continues, the field lines wind up tighter and tighter, and as they do so, adjacent field lines, wound even closer together, build up strong local magnetic fields. Local turbulence can cause the lines to become twisted together. The matter in such a region tries to expand and lower the field strength, and as it does so, it rises. The rising currents can carry "cables" of field lines out through the photosphere into loops above it. Sunspots, in this theory, occur where the field cables intersect the photosphere, and expanding gases in those regions cool and darken. The phenomena in the chromosphere and corona are associated with those same loops above the active sunspot re-

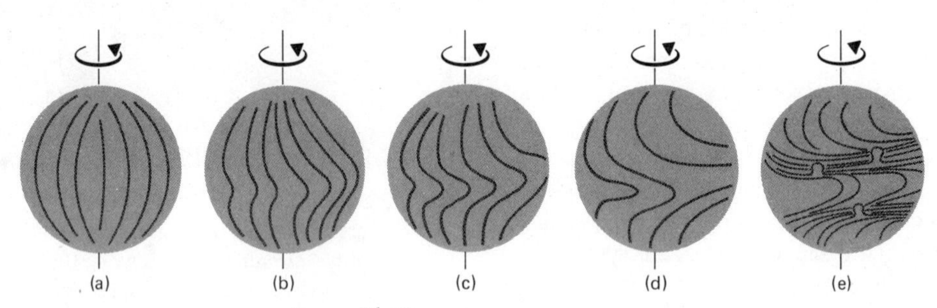

Figure 16.12
The Babcock model of the solar cycle.

gions. Eventually, as the differential rotation continues, the field lines coalesce and disperse, dissipating the strong field, and the process starts over again.

THE SOLAR WIND

More than a century ago, it was discovered that sunspot activity is correlated with magnetic storms on the earth. During geomagnetic storms, the earth's magnetic field is disturbed, and the compass needle shows fluctuations. Today we know that long-range shortwave radio interference and displays of the aurora are also correlated with geomagnetic storms and the sunspot cycle. Some investigators are of the opinion that even long-term changes in climate may be connected with solar activity.

These effects are due to the ultraviolet and X-ray radiation from the sun, and also to the *solar wind*. In addition to electromagnetic radiation, the sun emits corpuscular radiation in the form of charged atomic particles, mostly protons and electrons. The solar wind is the more or less continuous emission of these particles from the sun. They are regularly recorded by instruments carried on artificial earth satellites and space probes.

The speed of the solar wind near the earth's orbit averages about 400 km/s, and its density is usually from 2 to 10 ions/cm^3. Both the speed and density of the solar wind, however, are highly variable. During solar flares, for example, very high-speed particles are sometimes ejected from the sun into space. Some of these have energies in the low-energy cosmic ray range. Thus the sun (at times of flare activity) is an occasional source of weak cosmic rays. These, however, comprise only a tiny fraction of the total cosmic ray influx to the earth.

At the surface of the earth, we are protected from the solar wind and solar cosmic rays by the atmosphere and by the earth's magnetic field. These charged particles, along with electromagnetic radiation from the sun interacting with the upper atmosphere of the earth, do, however, disrupt the ionized layers of gas in the *ionosphere* (Chapter 9). It is these ionized layers that reflect shortwave radio waves back to the earth, making long-range radio communication between distant stations on the earth possible. When the ionospheric layers are disturbed, they may disrupt the reflection of radio waves transmitted from the ground; this results in radio "fadeouts." The rain of particles is also responsible for the aurora (northern and southern lights). As the particles strike atoms and molecules in the upper atmosphere they excite them. Radiation from the ions and atoms in the atmosphere gives rise to the auroral emission of light. The most spectacular auroras occur at elevations of 75 to 150 km.

THE INTERNAL STRUCTURES OF THE SUN AND OTHER STARS

Even at the surfaces of the sun and stars the temperatures are so high that all chemical elements are vaporized to gases. The internal temperatures of stars are many times higher than their surface temperatures; thus stars are gaseous throughout. This circumstance greatly facilitates the computation of conditions in the interiors of the sun and other stars because gases obey rather simple laws.

The particles that make up a gas are in rapid motion, frequently colliding with each other. This constant bombardment is the *pressure* of the gas. The pressure is greater, the greater the number of particles within a given volume of gas, for, of course, the combined impact of the moving particles increases with their number. The pressure, in other words, is proportional to the density of the gas. The pressure is also greater the faster the molecules or atoms are moving; since their rate of motion is determined by the temperature of the gas, the pressure is greater the higher the temperature.

These two ideas can be combined to give us the *perfect gas law* (also called the *equation of state* for a perfect gas). The perfect gas law states that the pressure is proportional to the product of the density and temperature of the gas. The gases in most stars closely approximate an ideal gas; thus, they must obey this law.

Equilibrium in Stars

Now, the mutual gravitational attraction between all the gases within a star produces tremendous forces that tend to collapse the star toward its center. Yet, since stars like the sun have remained more or less unchanged for billions of years, the gravitational force that tends to collapse a star must be exactly balanced by a force from within. Most of the internal force that supports the outer layers of a star against the downward pull of gravitation upon them is the pressure of the gases themselves. In some very luminous stars the pressure of radiation also contributes appreciably to support of the outer layers.

If the internal gas pressure in a star were not great enough to balance the weight of its outer parts, the star would collapse somewhat, contracting until the gas pressure inside built up to the point where it could support the star. Analogously, if the air in a tire or balloon cools and its pressure is thus reduced, the rubber contracts, further compressing the air until a balance is restored. On the other hand, if the gas pressure in a star were greater than the weight of the overlying layers, the star would expand, thus decreasing the internal pressure. Expansion would stop when the pressure at every point within the star again equaled the weight of the stellar layers above that point. Similarly, if the air in a balloon or tire heats up and increases in pressure, the balloon expands until the increased tension in the rubber is just enough to match the higher air pressure.

Thus, a star adjusts until it is in balance. The pressure within a star increases downward from the surface to the center; the increase of pressure through each layer of a star is exactly the right amount necessary to support the weight of that layer. This condition is called *hydrostatic equilibrium*. Stable stars are all in hydrostatic equilibrium; so are the oceans of the earth, as well as the earth's atmosphere; it is the pressure of the air that keeps it from falling to the ground.

The condition of hydrostatic equilibrium and the gas law enable us to derive minimum values for the internal pressures and temperatures of a star. It is found, for example, that for the internal gases in the sun to support their own weight, the mean pressure must be at least 500 million times the sea-level pressure of the earth's atmosphere, the central pressure must be at least 1.3 thousand million times that of the earth's atmosphere, and the mean temperature must be at least 2.3 million Kelvins. These pressures and temperatures are only lower limits; the actual values are much higher. Under such conditions all elements are in the gaseous form, and the atoms cannot even be combined into molecules. Moreover, most of the atoms are completely ionized—that is, stripped of their electrons (Chapter 5). These electrons, freed from their parent atoms, become part of the gas itself, moving about as individual particles.

Another kind of equilibrium in the sun and stars is *thermal equilibrium*, which expresses the balance of heat gain and loss from each internal region. Electromagnetic energy (mostly light in the sun's case) continually flows from the surface of a star. The second law of thermodynamics requires that heat always tries to flow from hotter to cooler regions. Therefore, as energy filters outward toward the surface of a star, it must be flowing from inner hotter regions. The outward flow of energy through a star, however, robs it of its internal heat and would result in a cooling of the interior gases were that energy not replaced. There must therefore be a source of energy within each star.

If a star is in a steady state (that is, in hydrostatic equilibrium and shining with a steady luminosity), the temperature and pressure at each point within it must remain approximately constant. If the temperature were to change suddenly at some point, the pressure would similarly change, causing the star to contract suddenly, or to expand, or to otherwise deviate from hydrostatic equilibrium. Energy must be supplied, therefore, to each layer in the star at just the right rate to balance the loss of heat in that layer as it passes energy outward toward the surface. Moreover, the rate at which energy is supplied to the star as a whole must, at least on the average, exactly balance the rate at which the whole star loses energy by radiating it into space; that is, the rate of energy production in a star is equal to its luminosity. Before we examine the source of energy of stars, however, we must digress a moment to consider atomic nuclei, for it is there that we find the powerhouses that maintain the stars.

The Atomic Nucleus

At the core of each atom lies its nucleus. The nucleus has almost all of the atom's mass, and carries a positive electric charge equal, in a neutral atom, to the sum of the negative charges of the surrounding electrons. The nucleus, however, is not an indivisible entity.

The most important particles in the nucleus are protons and neutrons (Chapter 5). Each proton carries a positive charge equal but opposite to the negative charge of the electron. Neutrons are electrically neutral and have masses minutely greater than those of protons. In most atoms, the number of neutrons is roughly equal to the number of protons, but the most common kind of atom of hydrogen contains one proton but no neutron in its nucleus. Atoms of the same chemical element (whose nuclei have the same number of protons) but with different number of neutrons are said to be different *isotopes* of that element. Thus *deuterium* is an isotope of hydrogen whose atoms contain nuclei of one proton and one neutron each, and *tritium* is an isotope of hydrogen whose atomic nuclei have one proton and two neutrons each. The *atomic weight* of an atom is the mass of its nucleus in terms of the *atomic mass unit* (amu), defined in Chapter 5. The particles of the nucleus are bound by the strong nuclear force (Chapter 1).

In the nuclear physics laboratory, nuclei bombarded by very high-energy particles are observed to break up not only into smaller nuclei, protons, and neutrons, but into other kinds of particles as well. The other particles are unstable and decay in an extremely tiny fraction of a second. After one or more decays they end up as stable protons or electrons. More than 300 subnuclear particles have been discovered, some of which are a little less massive, and some of which are more massive than the proton. They are called *hadrons;* protons and neutrons are the most familiar hadrons.

So many hadrons have been discovered that nuclear physicists have doubted for some years that they are really fundamental at all, but have thought that they represent different states of excitation of a much smaller number of nuclear particles. The physicists G. Zweig and M. Gell-Mann have theorized that hadrons are made up of various combinations and configurations of several kinds of simple particles that they have named *quarks*. The name "quark" is whimsically taken from James Joyce's *Finnegan's Wake:* "Three quarks for Muster Mark. . . ." Originally Zweig and Gell-Mann had postulated three kinds of quark. Now, however, there are thought to be at least four, and perhaps six different kinds of quark. The existence of quarks, however, is still hypothetical. The idea of quarks can explain many of the properties of the nucleus, but quarks have never been isolated in the laboratory, and some physicists are skeptical of their reality. In any event, the quark theory is a useful model, and it has resulted in progress in nuclear physics.

Even the neutron, when outside a nucleus, is not stable, but after an average interval of about 11 minutes it decays into a proton and an electron. This is an example of the weak nuclear force (Chapter 1). Whenever a weak nuclear reaction occurs in which a particle decays into an electron and a hadron (as in neutron decay) a small fraction of the energy is carried away by a particle called a *neutrino*. The neutrino has energy and momentum, but no mass, and hence travels with the speed of light. Because they react extremely weakly with other matter, neutrinos can easily pass unhindered through great bulks of matter. Neutrinos that are formed in the centers of stars almost all flow directly out of those stars with the speed of light. The universe must be completely bathed in neutrinos. Because they interact so poorly with matter, neutrinos are very difficult to absorb, and hence are almost impossible to observe. Yet careful laboratory experiments can and do detect a few of them.

Antimatter

For every nuclear particle, there appears to exist a corresponding *antiparticle*. In 1932, C. D. Anderson discovered the *positron* in the course of research on cosmic rays. The positron is an electron with positive, rather than negative charge. More recently, the antiproton—a proton with a negative charge—was produced in the laboratory. All of the hadrons have antiparticles, including the antineutron, which decays into an antiproton and a positron. Quarks too are presumed to have their corresponding kinds of *antiquarks*.

Whenever a particle comes into contact with its antiparticle, the two annihilate each other, their combined mass being converted completely into energy or into energy and less massive particles. Consequently, in our world of ordinary matter, antiparticles cannot survive long, for they soon encounter their ordinary counterparts and are destroyed. Moreover, at most only trivial amounts of antimatter could exist in our Galaxy—such as antiparticles produced by collisions between cosmic rays and interstellar gas atoms. One can speculate, however, that entire systems of galaxies composed entirely of antimatter might exist in remote parts of the universe. If such systems of antimatter should exist, they can never have come into contact with ordinary matter. We know of no way to recognize antimatter by its external appearance; stars of antimatter would emit electromagnetic radiation into space (so far as we know) just as the sun does. Some investigators have speculated that roughly equal amounts of matter and antimatter could exist in the universe, well segregated into separate parts of space. There is, however, no direct evidence to support this speculation.

Binding Energy of the Nucleus

Stars and planets are held together by their own gravitation. The prestellar material from which a star condensed (like the solar nebula—Chapter 8) originally had a great deal of gravitational potential energy, much of which has been given up as the star has shrunk to its present size. Similarly, an object dropped from a great height gives up some of its potential energy as it falls. The actual amount of potential energy released by a star is the same as the amount of energy (or work) that would be required to separate its atoms infinitely far from each other, working against their mutual gravitation. We can regard the star as bound by this gravitational energy.

Similarly, the atomic nucleus is held together by the extremely strong, but short-range, nuclear force. An enormous amount of energy (or work) is required to pull the parts of the atom apart and spread them far from each other. If those pieces fell together to form the nucleus, that same amount of potential energy would be released. This is the *binding energy* that holds the nucleus together. We must put a lot of energy into a nucleus to tear it apart, but we obtain a lot of energy from it when we build it from its parts.

The binding energy is greatest for atoms with a mass near that of the iron nucleus, and it is less both for the lighter and heavier atoms. Thus heavy nuclei, like those of atoms of lead and uranium, are held together with less energy per atomic mass unit than iron is, and so are the lighter nuclei, like carbon and lithium. In general, therefore, if light atomic nuclei come together to form a heavier one (up to iron), energy is released; this is called *nuclear fusion*. On the other hand, if heavy atomic nuclei can be broken up into lighter ones (down to iron), an increase in the total binding energy results, with the release of that much potential energy; this is called *nuclear fission*. Nuclear fission sometimes occurs spontaneously, as in natural radioactivity, whereby the nuclei of particular isotopes of certain elements, such as of uranium and radium, spontaneously break up, after a time, producing lighter atomic nuclei and radiation.

We recall (Chapter 7) that mass and energy are equivalent, and are related to each other by the equation $E = mc^2$. Indeed, we find that the mass of every nucleus (other than the simple proton nucleus of hydrogen) is less than the sum of the masses of the nuclear particles that are required to build it. This slight deficiency in mass, always only a small fraction of 1 amu, is called the *mass defect* of the nucleus. The mass defect is greatest for the nucleus of the iron atom, and is less for both more massive and less massive nuclei. A *nuclear transformation* is a buildup of a heavier nucleus from lighter ones, or a breakup of a heavier nucleus into lighter ones. In any such nuclear transformation, if the mass defect increases, the equivalent amount of energy is released. That energy, of course, is the difference in mass defect times the square of the speed of light. On the other hand, if, in the nuclear transformation, the mass defect *decreases*, a corresponding amount of energy must be put into the system.

STELLAR ENERGY

The rate at which the sun emits electromagnetic radiation into space, and thus the rate at which energy must be generated within it, is about 4×10^{33} ergs/s (Chapter 13). Moreover, the power output of the sun has been about the same throughout recorded history and, according to geological evidence, not very different since the formation of the earth thousands of millions of years ago. Our problem is now to find what sources can provide the gigantic amounts of energy required to keep the stars like the sun shining so long.

Two large stores of energy in a star are its internal heat, or *thermal energy*, and its *gravitational energy*. The heat stored in a gas is simply the energy of motion (kinetic energy) of the particles that comprise it. If the speeds of these particles decrease, the loss in kinetic energy is radiated away as heat and light. This is how a hot iron cools after it is withdrawn from a fire (except that the atoms in a solid vibrate within a crystalline structure, rather than moving freely, as in a gas).

Because a star is bound together by gravity it has gravitational potential energy. If the various parts of a star fall closer together, that is, if the star contracts, it converts part of its potential energy into heat, and it radiates away the rest of it. A century ago, the physicists H. von Helmholtz (1821–1894) and Lord W. Thompson Kelvin (1824–1907) postulated that the source of the sun's luminosity was indeed the conversion of part of its gravitational potential energy into radiant energy. They showed that because of its enormous mass, the sun need contract only extremely slowly to release enough gravitational potential energy to account for its present luminosity, and that the sun could have been shining in this way for millions of years. We know today, however, that the earth, and hence the sun, has an age of at least several thousand million years, and therefore that the sun's gravitational energy is grossly inadequate to account for the luminosity it has generated over its lifetime.

It was suggested about 1928 that the energy source of the sun and stars might be fusion of hydrogen into helium. Helium nuclei are about four times as massive as hydrogen nuclei, so it would take four nuclei of hydrogen to produce one of helium. Now the masses of hydrogen and helium nuclei are 1.00813 and 4.00389 amu, respectively. The four hydrogen nuclei, therefore, with a combined mass of $4 \times 1.00813 = 4.03252$ amu, outweigh the finished helium nuclei by 0.02863 amu. Thus about 0.71 percent of the original mass of hydrogen, when it is converted into helium, is released in the form of energy. For example, if 1 g of hydrogen is turned into helium, 0.0071 g of material is converted into energy. The velocity of light is 3×10^{10} cm/s, so the energy released is

$$E = 0.0071 \times (3 \times 10^{10})^2 = 6.4 \times 10^{18} \text{ ergs}$$

This 6×10^{18} ergs is enough energy to raise a 500-ton mass 150 km above the ground.

To produce the sun's luminosity of 4×10^{33} erg/s some 600 million tons of hydrogen must be converted to helium each second, with the simultaneous conversion of about 4 million tons of matter into energy. As large as these numbers are, the store of nuclear energy in the sun is still enormous. Suppose half of the sun's mass of 2×10^{33} g is hydrogen that can ultimately be converted into helium; then the total store of nuclear energy would be 6×10^{51} ergs. Even at the sun's current rate of energy expenditure, 10^{41} ergs/yr, the sun could survive for more than 10^{10} yr.

There is little doubt today that the principal source of energy in stars is thermonuclear reactions. Deep in the interiors of stars, where the temperatures range up to many millions of degrees, nuclei of lighter atoms are fusing into heavier ones, with an accompanying release of energy. The most important of these changes is the conversion of hydrogen to helium. Several processes are known by which energetic collisions between atomic nuclei and subsequent nuclear transformations can result in this fusion. The steps in these processes are listed in Appendix 8. The discovery that some of the energy locked up in the nuclei of atoms is released in the interiors of stars may be one of the most significant contributions of astronomy in the twentieth century.

Most of the electromagnetic radiation released in these nuclear reactions is at very short wavelengths—in the form of X rays and gamma rays. Nuclear reactions are important, however, only deep in the interior of a star. Before this released energy reaches the stellar surface, it is absorbed and reemitted by atoms a very great number of times. Photons of high energy (short wavelength) that are absorbed by atoms are often reemitted as two or more photons, each of lower energy. By the time the energy filters out to the surface of the star, therefore, it has been converted from a relatively small number of photons, each of very high energy, to a very much larger number of photons of lower energy and longer wavelength, which constitute the radiation we actually observe leaving the star.

Solar Neutrinos

The nuclear process which is thought to provide almost certainly most of the sun's energy, called the *proton-proton chain*, involves the emission of neutrinos. In fact, the neutrinos should comprise about 3 percent of the energy released. Their detection, however, because of their low interaction with matter, is extremely difficult.

Nevertheless, Raymond Davis, Jr., and his colleagues at Brookhaven National Laboratory have devised a technique by which they are detecting solar neutrinos. On rare occasions a neutrino of the energy of some of those emitted from the sun should react with the isotope chlorine-37 to transmute it to argon-37 and an electron. Davis has placed a tank containing 378,000 liters of tetrachloroethylene, C_2Cl_4 (ordinarily used as cleaning fluid), 1.5 km beneath the surface of the earth in a gold mine at Lead, South Dakota. Even though an individual neutrino is extremely unlikely to react with the chlorine in the cleaning fluid, calculations show that about one atom of argon-37 should nevertheless be produced every day or so. Because argon-37 is radioactive (about half of it decaying in 35 days) it is possible to isolate and detect most of those few argon-37 atoms from the more than 10^{30} atoms of chlorine in the tank.

At first Davis was not able to detect solar neutrinos. After refining the experimental techniques, however, he has been able to measure some flux of solar neutrinos. That flux, though, is somewhat lower than was at first expected from the theory of nuclear reactions in the solar interior. The discrepancy has led to a relook at the theory, and today some experts feel that what was once a discrepancy between theory and observation has been resolved. The matter is *not* completely resolved to the satisfaction of all authorities, however, and is an excellent example of the interplay of theory and experiment in science.

Figure 16.13
Raymond Davis, Jr.'s, 378,000-liter neutrino detector in a mine at Lead, South Dakota. *(Brookhaven National Laboratory)*

MODEL STARS

To determine the internal structure of a star, we must now combine the principles we have described: hydrostatic equilibrium, the perfect gas law, thermal equilibrium, and the rate of energy generation from nuclear processes. These physical ideas are formulated into mathematical equations which are solved to determine the march of temperature, pressure, density, and other physical variables throughout the stellar interior. The set of solutions so obtained, based on a specific set of physical assumptions, is called a theoretical model for the interior of the star in question.

The sun is the most studied of all stars, and models of its interior have been calculated for several decades. Each new model of the sun represents a refinement resulting from an improvement in our knowledge of physics or of computing methods or both. One model, based on the best physical data available in 1979, and on the assumption that the sun was originally 73 percent hydrogen and 24.5 percent helium, is presented in Table 16.1. According to this model, the temperature within the sun increases gradually toward its center and reaches a value of about 15 million degrees at the center. The density (also the pressure), on the other hand, increases very sharply near the center

TABLE 16.1 Model for the Structure of the Sun*

Fraction of Radius	Fraction of Mass	Fraction of Luminosity	Temperature (millions of kelvins)	Density (g/cm^3)	Fraction Hydrogen (by weight)
0.00	0.000	0.00	15.0	148	0.38
0.05	0.011	0.10	14.2	125	0.47
0.10	0.076	0.45	12.5	86	0.59
0.15	0.19	0.78	10.7	56	0.67
0.20	0.33	0.94	9.0	36	0.71
0.30	0.61	1.00	6.5	12	0.73
0.40	0.79	1.00	4.9	4	0.73
0.60	0.95	1.00	3.1	0.5	0.73
0.80	0.99	1.00	1.3	0.1	0.73
1.00	1.00	1.00	0.0	0.0	0.73

* Adapted from Ulrich.

of the sun, indicating a high degree of central concentration of its material, and reaches a maximum value of over 100 times the density of water. The hydrogen abundance at the very center has been reduced (by nuclear reactions) to only 38 percent, and the present age of the sun is about 4.5×10^9 years.

SUMMARY REVIEW

Solar atmosphere: photosphere; limb darkening; chemical composition; chromosphere; flash spectrum; discovery of helium in the sun; spicules; corona; high temperature of the corona

Solar rotation: determination from observation of sunspots and from Doppler shifts

Solar phenomena: granules; supergranules; five-minute oscillation; sunspots; umbra and penumbra of spots; spot groups; sunspot cycle; Maunder minimum; solar magnetism; filtergrams; plages; faculae; prominences—quiescent, eruptive, and surge prominences; flares; Babcock theory of solar activity

The solar wind and solar-terrestrial relations: geomagnetic storms and aurora; solar cosmic rays; radio fade-outs

Solar interior: perfect gas law (equation of state); hydrostatic equilibrium; internal pressures and temperatures of the sun; thermal equilibrium

Atomic nucleus: proton; neutron; isotopes; deuterium; tritium; atomic weight; atomic mass unit (amu); other nuclear particles; hadrons; quarks; neutrinos; antimatter; positron; antiproton; antiquarks; binding energy; nuclear fusion; nuclear fission; natural radioactivity; mass defect; nuclear transformation

Stellar energy: thermal energy; gravitational energy; Helmholtz-Kelvin theory of solar contraction; hydrogen fusion in the sun; store of nuclear energy; proton-proton chain; neutrinos from the sun; Davis' solar neutrino detector

Model stars: model for the sun

EXERCISES

1. Describe the principal spectral features of the sun.

2. How might you convince an ignorant friend that the sun is not hollow?

3. Give at least three good arguments that refute the view proposed by Herschel that the sun has a cool interior.

4. Draw a diagram illustrating limb darkening on the sun and its explanation.

5. If the rotation period of the sun is determined by observing the apparent motions of sunspots, must any correction be made for the orbital motion of the earth? If so, explain what the correction is and how it arises. If not, explain why the earth's orbital revolution does not affect the observations.

6. If the corona, which is outside the photosphere, has a temperature of 1,000,000 K why do we measure a temperature near 6000 K for the surface of the sun?

7. Draw a diagram showing a vertical cross section of a granule and the motions of gas in and around it.

8. Turn to Chapter 14, and compare and contrast the sun with the most common stars.

9. Give some everyday examples of hydrostatic equilibrium. It is known that the pressure in a container of water increases with depth in the container. Is this a consequence of hydrostatic equilibrium? Explain. Compare the pressure-depth relation in water with that in the earth's atmosphere. Why is the case much simpler for water?

10. If the atmospheric pressure were the same on two different days, but if the temperature were much higher on one day than on the other, what could you say about the relative density of the air on the two days?

11. If, in a vacuum chamber, the pressure is only one-millionth of sea-level pressure, how does the density of the gas in the chamber compare with the average density of air at sea level?

12. Why do you suppose so great a fraction of the sun's energy comes from its central regions? Within what fraction of the sun's radius does practically all of the sun's luminosity originate? (See Table 16.1.) Within what radius of the sun has its original hydrogen been partially used up? Discuss what relation the answers to these questions bear to each other.

13. Verify that some 600 million tons of hydrogen are converted to helium in the sun each second.

14. Stars exist that are as much as a million times as luminous as the sun. Consider a star of mass 2×10^{35} g, and luminosity 4×10^{39} ergs/s. Assume that the star is 100 percent hydrogen, all of which can be converted to helium, and calculate approximately how long it can shine at its present luminosity. There are about 3×10^{7} s in a year.

15. Perform a similar computation for a typical star less massive than the sun, such as one whose mass is 1×10^{33} g and whose luminosity is 4×10^{32} ergs/s.

16. Which of the following transformations is fusion and which is fission (see Appendix 19): the tranformation of (a) helium to carbon; (b) carbon to iron; (c) uranium to lead; (d) boron to carbon; (e) oxygen to neon.

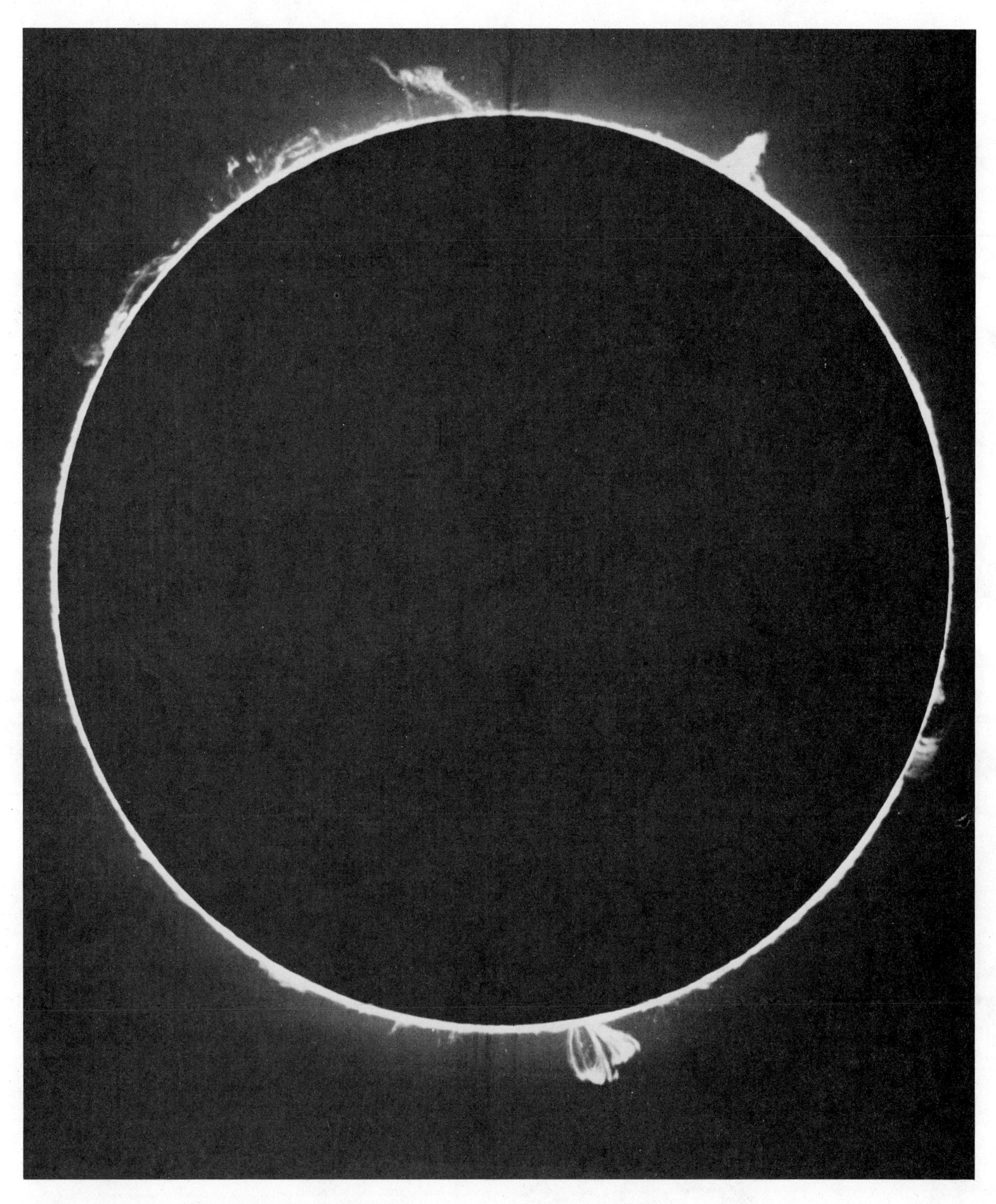

The entire limb of the sun, photographed in the light of the calcium K line. Several prominences are visible. *(Hale Observatories)*

CHAPTER 17

Sir Fred Hoyle (1915–), the well-known British astrophysicist and cosmologist, is also well known for his science fiction, and even for an opera libretto. In the 1950s, Hoyle's brilliant deduction about the nature of the carbon nucleus enabled us to understand where the atoms of our own bodies originated. *(Floyd Clark, California Institute of Technology)*

THE BIRTH AND EVOLUTION OF STARS

We need a certain minimum amount of information to determine the structure of a star, if we assume that all pertinent physical laws are perfectly understood and that infinitely accurate calculations can be performed. Theory shows that if a star is in hydrostatic and thermal equilibrium, and if it derives all its energy from nuclear reactions, then its structure is completely and uniquely determined by its total mass and by the distribution of the various chemical elements throughout its interior.

Imagine a cluster of stars forming from a cloud of interstellar material whose chemical composition is similar to the sun's. All condensations that become stars will then begin with the same chemical composition and will differ from each other only in mass. Suppose now that we were to compute a model for each of these stars for the time at which it became stable, and derived its energy from nuclear reactions, but before it had had time to alter its composition appreciably as a result of these reactions.

The models we would calculate for these stars would indicate, among other things, their luminosities and radii. From Stefan's law we known that the luminosity of a star is proportional to the product of its surface area and the fourth power of its effective temperature (Chapter 14). We can, therefore, calculate the temperature for each of the stars and plot it on the Hertzsprung–Russell diagram (Chapter 14). We would find that the most massive stars were the hottest and most luminous and would lie at the upper left corner of the diagram, while the least massive were coolest and least luminous and would lie at the lower right. The other stars would all lie along a line running diagonally across the diagram—the *main sequence*. The main sequence, then, is the locus of points on the H–R diagram representing stars of chemical composition similar to the sun's but of different mass.

Those stars that do not lie on the main sequence in the Hertzsprung–Russell diagram (for example, red giants and white dwarfs) must differ somehow from the majority in their chemical compositions, or else they are not stable or are not shining by nuclear energy alone. We have seen, however, that as stars age they convert hydrogen to helium, and so change their compositions, especially near their centers.

Most non-main-sequence stars can be interpreted either as stars that are still forming from interstellar matter and are not yet deriving all their energy from nuclear sources, or as stars that, by virtue of nuclear transformations, have altered their chemical compositions and hence their entire structures.

In the study of stellar evolution we calculate from the theory of stellar structure how stars should change, as they contract gravitationally or age, through changes in their chemical composition produced by nuclear reactions. We check the theory by observing stars or groups of stars that are at different stages in their evolution to see whether they exhibit the characteristics expected of them from the theoretical predictions.

Star clusters are the most useful objects to study to check the theory of stellar evolution, because the stars within a cluster can usually be presumed to have a common origin and age, and to have all had, originally, similar chemical compositions. For a cluster of known distance we can plot a color-magnitude or Hertzsprung–Russell diagram. It is convenient to compare the theory and observations by considering the tracks of evolution of stars on the H–R diagram. Of course the "position" of a star, or its "evolution," on the H–R diagram does not refer to its position or motion in space. Rather, these terms refer to the position and motion on a diagram of a point that represents the star's luminosity and surface temperature—they indicate indirectly changes in the structure of the star.

EARLY STAGES OF STELLAR EVOLUTION

No star that is shining today can be infinitely old, for eventually it exhausts its sources of energy. Moreover, the stars of highest known luminosity (100 thousand to 1 million times that of the sun) can continue to exist at the rate they are now expending energy for at most a few million years. Had they been formed when the sun was formed, thousands of millions of years ago, they would long since have burned themselves out. At least some stars, therefore, have formed recently (in the cosmic time scale), and there is every reason to expect that stars are still forming today. The "birthplaces" of stars must be the clouds of interstellar gas and dust. That period in a star's existence during which it condenses from interstellar matter, and contracts into an "adult" main-sequence star, may be considered its "early stages."

Theoretical Studies of Pre-Main-Sequence Stellar Evolution

Here and there, in comparatively dense regions of interstellar matter, small condensations begin to form—atoms of gas and particles of dust slowly begin to collect under the influence of their mutual gravitation. As a condensing region grows, so does its gravitational influence, and more and more material is attracted to it. Eventually, material over a large region of space falls toward the central condensation.

Many stars must have formed in the initial "collapse" or condensation of the Galaxy; the globular star clusters, in particular, are believed to be among the oldest objects in the Galaxy. We also know of several probable kinds of sites of current star formation: (1) We described the *cold clouds* in Chapter 15. These are regions where interstellar molecules, protected from stellar ultraviolet radiation by dust, become excited by molecular collisions and then radiate infrared and radio energy away from the cloud, thereby cooling it. The cool gas then contracts, building up in density. One such cool cloud, thought to be a site of star formation, lies just beyond the Orion nebula. (2) Globules, formed of pockets of neutral gas and intermixed dust compressed to relatively high density by the hot ionized gas of surrounding H II regions, are possible sites for new stars to condense. (3) Modern theory of galactic spiral structure predicts that the spiral arms are regions or "waves" of higher density in the galactic disk, through which stars and gas clouds slow down in their orbital speeds and pile up—rather like a pocket of traffic congestion in a highway at the site of an accident (even after the accident has been cleared away). In this picture, gas jamming into the trailing edges of spiral arms builds up to great enough densities for stars to condense from it.

The central regions of the forming star fall together more rapidly than the outer parts. This occurs in the center of a protostar, where the pressure first becomes great enough to stop the infall of matter, thus achieving hydrostatic equilibrium. The internal temperature is not yet high enough to support thermonuclear reactions, however, and its only sources of energy, so far, are gravitational and thermal. Thus the star gradually radiates into space the energy that it derives from a very slow shrinking, while its internal temperature and pressure continue to rise. The whole process is so gradual that hydrostatic equilibrium is never upset. This is exactly the process whereby Helmholtz and Kelvin attempted to explain the sun's source of energy more than a century ago (Chapter 16).

For stars of mass greater than about 1/12 that of the sun, after a period of some thousands or millions of years the central temperature becomes high enough to support nuclear reactions. Soon this new source of energy supplies heat to the interior of the star as fast as energy is radiated away. The central pressures and temperatures are thus maintained and the contraction of the star ceases; it is now on the main sequence. By this time the infall of matter is complete and the star is fully formed. The evolutionary tracks of these forming stars are shown in Figure 17.1.

Calculations by R. B. Larson, however, show that stars more massive than the sun would not be visible to us during most of their pre-main-sequence evolution because the energy they emit is absorbed by surrounding dust in the infalling material. Because the dust is cooler than the stellar embryos (or forming stars), when it reradiates the energy that it absorbed it emits it in long infrared wavelengths.

In general, the pre-main-sequence evolution of a star slows down with time; the numbers labeling the

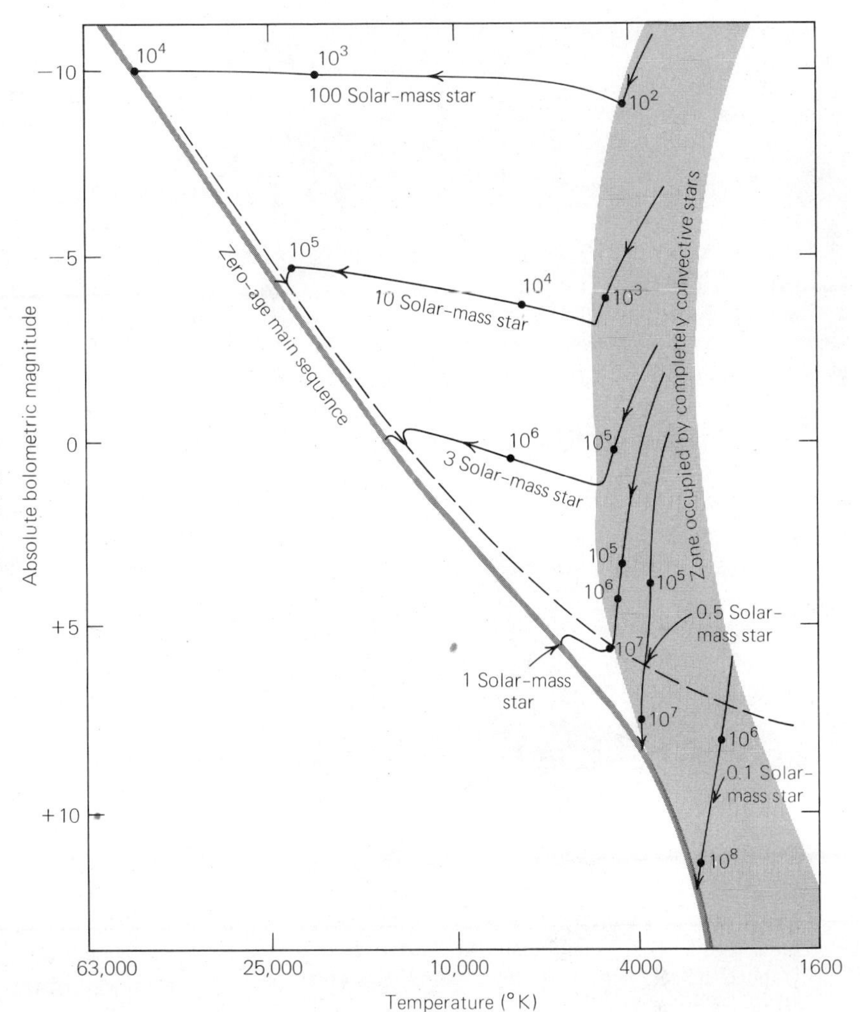

Figure 17.1
Theoretical evolutionary tracks of contracting stars or stellar embryos on the Hertzsprung–Russell diagram. According to calculations by Larson, stars or embryos lying roughly above the dashed line are still surrounded by infalling matter and would be hidden by it.

points on each evolution track in Figure 17.1 are the times, in years, required for the embryo stars to reach those stages of contraction. The time for the whole evolutionary process, however, is highly mass dependent. Stars of mass much higher than the sun's reach the main sequence in a few thousand to a million years; the sun required millions of years; tens of millions of years are required for stars to evolve to the lower main sequence. For all stars, however, we should distinguish three evolutionary time scales:

1. The initial gravitational collapse from interstellar matter is relatively quick. Once the condensation is, say, 1000 AU in diameter, the time for it to reach hydrostatic equilibrium is measured in thousands of years.
2. Pre-main-sequence gravitational contraction is much more gradual; from the onset of hydrostatic equilibrium to the main sequence requires, typically, millions of years.
3. Subsequent evolution on the main sequence is very slow, for a star changes only as thermonuclear reactions alter its chemical composition. For a star of a solar mass, this gradual process requires thousands of millions of years. All evolutionary stages are relatively faster in stars of high mass and slower in those of low mass.

Observations of Very Young Star Clusters

The theoretical calculations described above enable us to predict what a cluster of stars that is now forming from interstellar matter should be like. Within

Figure 17.2
Nebulosity in Monoceros, situated in the south outer region of the young cluster NGC 2264. Photographed in red light with the 5.08-m telescope. *(Hale Observatories)*

a few million years, the most massive stars of the cluster should complete the contraction phase of their evolution and settle on the main sequence. As time goes on, more and more stars that are less and less massive should reach the main sequence. When the contraction phase is over, all the stars in the cluster should line up on the main sequence—just as we observe in the H–R diagrams of many clusters.

Since star formation is constantly going on, among the hundreds of known star clusters we might expect to find a few that are still in the process of formation, that is, with some of their stars still in the contraction phase. The more massive and luminous cluster stars might be expected to have reached the main sequence while their less massive companions would still be "on their way in." Since 1950, several clusters have been observed that fit this description.

The first such cluster to be studied was NGC 2264, a small open cluster embedded in a cloud of gas and dust in the constellation Monoceros. The H–R diagram for this cluster (by M. Walker at the Lick Observatory) is shown in Figure 17.3. The brighter stars in the cluster are on the main sequence (solid line in the figure), while the less luminous ones (presumably also less massive) are off to the right. They are interpreted as young stars, still contracting from the interstellar material associated with the cluster.

Several other such clusters, evidently very young ones, are known, all of which are associated with interstellar matter. One of them is the cluster in the central part of the Orion nebula. It is generally ac-

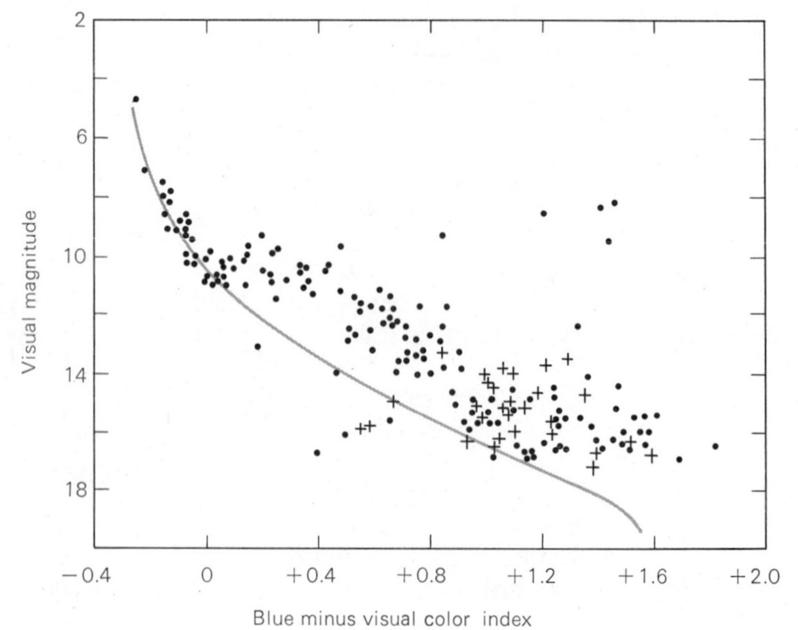

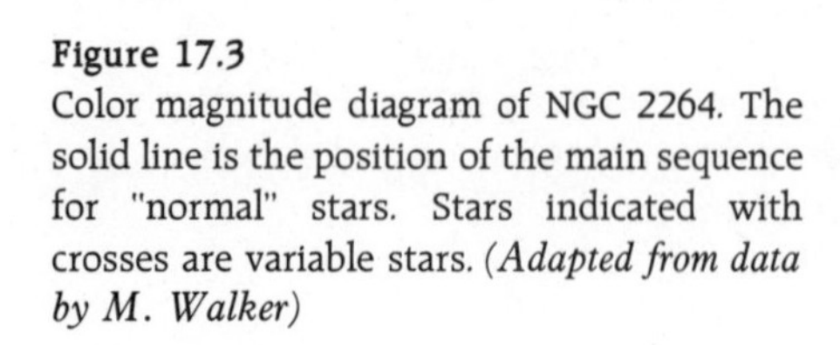

Figure 17.3
Color magnitude diagram of NGC 2264. The solid line is the position of the main sequence for "normal" stars. Stars indicated with crosses are variable stars. *(Adapted from data by M. Walker)*

cepted that in these clusters we are witnessing the early evolution of stars.

EVOLUTION FROM THE MAIN SEQUENCE TO GIANTS

As soon as a star has reached the main sequence, it derives its energy almost entirely from the thermonuclear conversion of hydrogen to helium. It remains on the main sequence for most of its "life." Since only 0.7 percent of the hydrogen used up is converted to energy, the star does not change its mass appreciably, but in its central regions, where the nuclear reactions occur, the chemical composition gradually changes as hydrogen is depleted and helium accumulates. This change of composition forces the star to change its structure, including its luminosity and size. Eventually, the point that represents it on the H–R diagram evolves away from the main sequence. The original main sequence, corresponding to stars of homogeneous chemical composition, is called the *zero-age main sequence*.

From the Main Sequence to Red Giants

As helium accumulates at the expense of hydrogen in the center of a star, calculations show that the temperature and density in that region must increase. Consequently, the rate of nuclear-energy generation increases, and the luminosity of the star slowly rises. When the hydrogen has been depleted completely in the central part of a star, a core develops containing only helium, "contaminated" by whatever small percentage of heavier elements the star had to begin with. The energy source from hydrogen fusion is now used up, and with nothing more to supply heat to the helium core, it begins again to contract gravitationally.

These changes result in a substantial and rather rapid readjustment of the star's entire structure, so that the star leaves the vicinity of the main sequence altogether. About 10 percent of a star's mass must be depleted of hydrogen before the star evolves away from the main sequence. The more luminous and massive a star, the sooner this happens, ending its term on the main sequence. Because the total rate of energy production in a star must be equal to its luminosity, the core hydrogen is used up first in the very luminous stars. The most massive stars spend less than 1 million years on the main sequence; a star of one solar mass remains there for about 10^{10} years, and a spectral-type M0 V star of about 0.4 solar mass has a main-sequence life of some 2×10^{11} years.

Calculations show that as the central core of a star contracts, the star as a whole expands to enormous proportions; all but its central parts acquire a very low density. The expansion of the outer layers causes them to cool, and the star becomes red. Meanwhile, some of the potential energy released from the contracting core heats up the hydrogen surrounding it to ever higher temperatures. In these hot regions the conversion of hydrogen to helium accelerates, causing most stars actually to increase in total luminosity. After leaving the main sequence, then, stars move to the upper right portion of the H–R diagram; they become red giants.

Figure 17.4, based on theoretical calculations by Icko Iben, shows the tracks of evolution on the H–R diagram from the main sequence to red giants for stars of several representative masses and with chemical composition similar to that of the sun. The broad band is the zero-age main sequence. The numbers along the

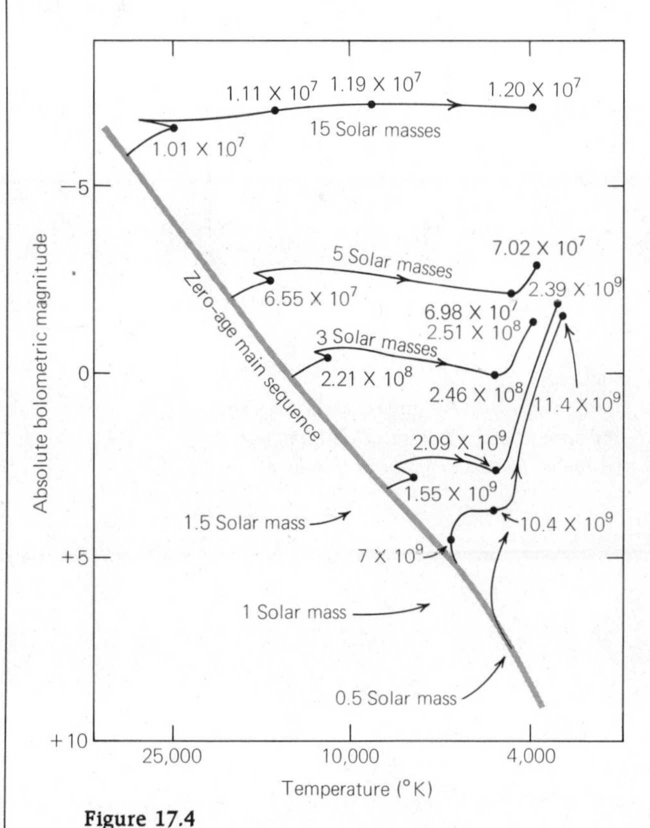

Figure 17.4
Predicted evolution of stars from the main sequence to red giants. See text for explanation. *(Based on calculations by I. Iben)*

tracks indicate the times, in years, required for the stars to reach those points in their evolution after leaving the main sequence.

Observations of Star Clusters of Different Ages

The results of theoretical studies outlined above predict certain characteristics of the H–R diagrams of star clusters of different ages. In each cluster there should be some point along the main sequence at which stars have just reached the critical age where they rapidly evolve away from it; this point is the upper termination of the main sequence for the cluster. In a young cluster, the main sequence extends to stars of high luminosity; in successively older clusters, it terminates at successively lower luminosities. As a cluster ages, its main sequence "burns down" like a candle.

Figure 17.5 shows a composite H–R diagram for several star clusters of different ages (compare with Figure 17.4). On the left side is shown the absolute visual magnitude scale. On the right side is a scale that gives the approximate ages of star clusters corresponding to the points where their main sequences terminate. These ages are based on computations of the times required for the cores of stars of various masses to become depleted of enough hydrogen to cause them to contract. We see that the clusters shown range in age from only about 1 million years to several thousand million years.

As expected, most clusters have red giants. Those few which do not may just not happen to have any stars of the proper mass to be entering that stage of their evolution. In the younger clusters the red giants have magnitudes about the same as those of the brightest main-sequence stars; for stars of those masses and compositions, the tracks of evolution from

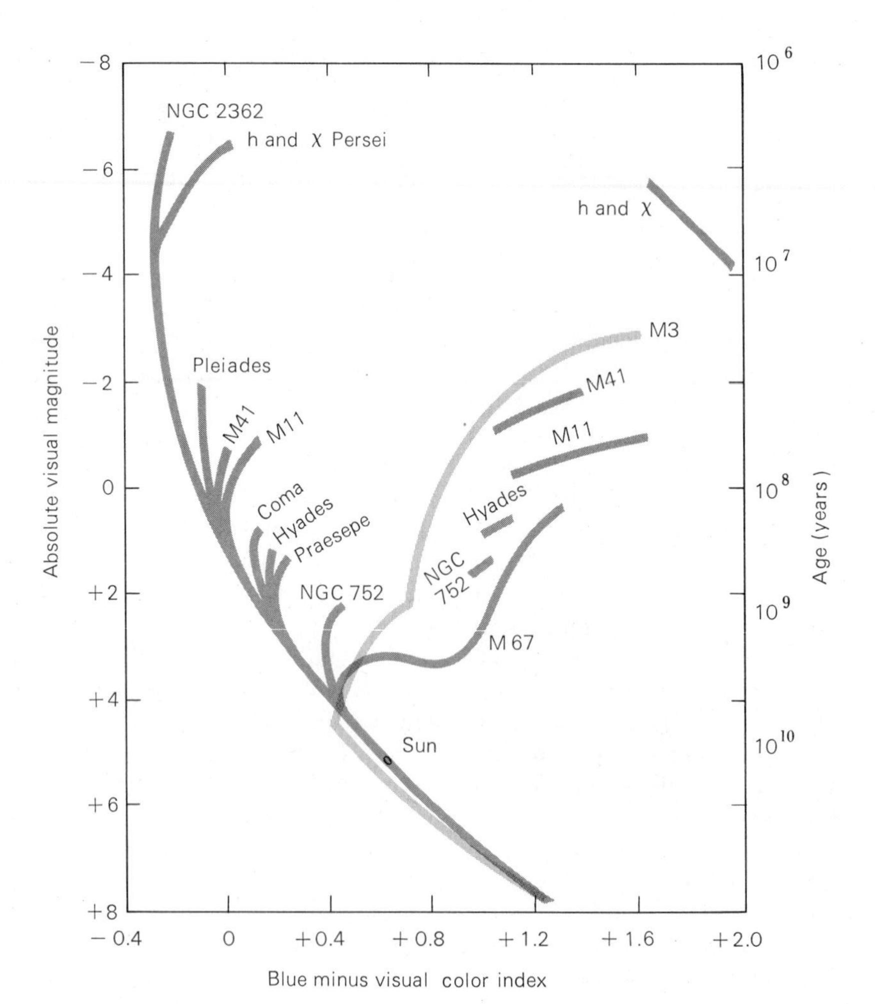

Figure 17.5
Composite Hertzsprung–Russell diagram for several star clusters of different ages. *(Adapted from a diagram by Sandage)*

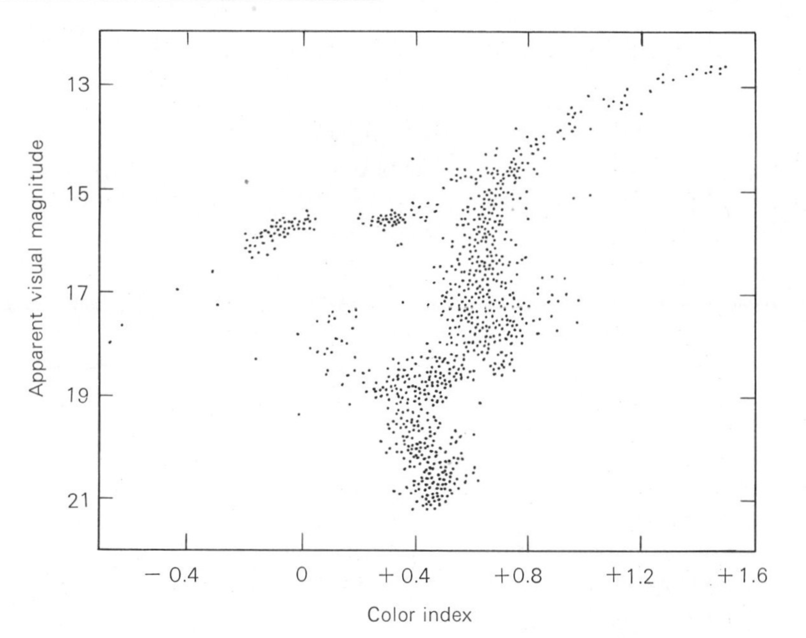

Figure 17.6
Color-magnitude diagram for the globular cluster M3. *(Adapted from data by Arp, Baum, and Sandage)*

the main sequence to the giant stages are approximately to the right across the diagram. In the older clusters, however, the giants are brighter than the brightest main-sequence stars and must, therefore, have increased in luminosity during their evolution from the main sequence.

The oldest assemblages of stars in the Galaxy are believed to be the globular clusters. The lighter colored line in Figure 17.5 outlines the H–R diagram for M3, a typical globular cluster; note that the main sequence terminates and branches into the giant sequence at a lower luminosity than for any of the other clusters. Ages for globular clusters are now estimated at from 8 to 15×10^9 years. A few open clusters (for example, NGC 188 and M67) may approach the globular clusters in age. In no cluster, however, does the main sequence terminate below the luminosity of the sun. Apparently, stars of solar mass in the Galaxy have not yet had time to evolve away from the main sequence.

In any case, the red giant stage must be a relatively brief part of a star's life. In the youngest clusters these stars are red *supergiants* of high luminosity. In the older clusters, stars *increase* their luminosities as they become red giants. In this stage of evolution, therefore, a star's nuclear fuel is consumed relatively quickly, and further evolutionary changes soon follow.

We summarize by noting that globular clusters (and, presumably, other very old systems) have main sequences that terminate at about absolute magnitude 3.5; these systems evidently have not experienced recent star formation. Open clusters, on the other hand, are often located in regions of interstellar matter, where star formation can still take place. Indeed, we find open clusters of all ages from less than 1 million to several thousand million years. The H–R diagrams of the oldest of them resemble those of globular clusters (at least as far as main-sequence and giant stars are concerned), except for differences that can be understood in terms of the differences in chemical composition of the two kinds of clusters.

Nucleosynthesis of Heavy Elements in Stars

We shall see in Chapter 21 that we expect hydrogen and helium—but no heavier elements—to have been produced in the first few minutes of the "big bang" that started the expansion of the universe. If this is correct, when our own Galaxy formed it would have had no carbon, oxygen, nitrogen, sulfur, iron, or any of the other heavy elements that make up the earth—and us. But if the atoms of which we are composed were not formed in the big bang, where did they come from? Perhaps they were synthesized by nuclear fusion of helium into those heavier elements in stars. But at the temperature of stellar interiors lithium, beryllium, and boron are unstable, and until the 1950s the known properties of the carbon nucleus made fusion into carbon impossible.

The obvious question, "How can *we* be here?" inspired the British astrophysicist Fred Hoyle (now *Sir* Fred Hoyle) to rethink the nuclear physics of carbon. Hoyle realized that part of the gravitational potential energy that is released in the contracting cores of red giants heats up those cores. By the time a star reaches the end of the red giant branch in the H–R diagram, the temperature of its central regions must exceed 100 million kelvins. At these temperatures, Hoyle reasoned, fusion of helium to carbon somehow *has* to occur. There must, therefore, be a property of the carbon nucleus that makes such fusion possible. Later, stars must eject some of that material into the interstellar medium, from which new stars can form.

Hoyle carried out the relevant calculations about the time he made a visit to the California Institute of Technology in Pasadena, California. According to Caltech nuclear physicist William Fowler, in 1953 Hoyle walked into the laboratory and urged Fowler to search for a resonance level in carbon at 7.7 megaelectronvolts (MeV). That level of excitation in the carbon nucleus was not then believed to exist, but on Hoyle's recommendation it was searched for with the help of the laboratory's particle accelerators; it was found at 7.653 MeV! Fowler reminds us that Hoyle's prediction remains to this day the most accurate one ever made in nuclear physics.

Fusion of helium into carbon can and does occur at a temperature of 10^8 K by a process called the *triple-alpha process*—so named because the nucleus of the helium atom is also called an *alpha particle*. Successive bombardments by helium nuclei can build carbon up into other, still heavier nuclei. With astrophysicists G. and E. M. Burbidge, Fowler and Hoyle have found mechanisms whereby virtually all chemical elements of weight up to that of iron can be built up by this nucleosynthesis in the centers of red giant stars, in approximately the relative abundances with which they occur in nature. It is now generally believed that a gradual buildup of the elements heavier than helium is continually going on in the hot centers of at least the more massive red giants. Fusion of elements heavier than iron, however, *require* energy, rather than release it. We think today that the most probable place where these elements originate is in nuclear reactions that occur in the outbursts of supernovae (see below).

The triple-alpha process is expected to begin abruptly in the central core of a red giant, and it causes the core to rise rapidly in temperature. With the sudden rise in temperature, the helium fusion into carbon accelerates; the phenomenon is called the *helium flash*.

The new energy released expands the core and reverses the growth of the outer parts of the red giant. The star then shrinks rapidly and increases in surface temperature, ending its red giant stage. Calculations indicate, however, that a star may actually move first to the left across the H–R diagram, and then back to be a red giant several times, each time as a consequence of the onset of new nuclear reactions or of nuclear energy released in new parts of the star. All these evolutionary stages occur in tens or hundreds of millions of years or less—a brief time compared to the stars' main-sequence lives.

PULSATING STARS

Some stars are unstable against pulsation (Chapter 13). Most types of stellar variability are now thought to represent temporary stages of evolution of stars of various masses or compositions, in which they become unstable and pulsate. For example, the upper middle part of the H–R diagram is a region in which most single stars, when their post red-giant evolution places them there, are variable.

Cause of Pulsation; Cepheid Variables

Unlike the stable sun, a pulsating star is not in hydrostatic equilibrium. Rather it is something like a spring: as the star contracts, its internal pressures build up until they surpass the weights of its outer layers. Eventually, these pressures start the star pulsing outward, but because of their inertia, the outward-moving layers overshoot the equilibrium point where their weights will just balance the internal pressures. As the star expands further, the weights of the overlying layers decrease, but the internal pressure decreases faster. Hence, the overlying layers are not supported adequately, and the star begins to contract. As it does so, it overshoots again, and this time it becomes too highly contracted. Once more the inner pressures cause the star to expand—and so the pulsation continues.

Typically, the radii of pulsating stars change by less than 10 percent. However, some giant red variables—for example, the star Mira—change in radius by 20 percent or more.

The energy needed to drive the pulsation comes, of course, from the nuclear, thermal, or gravitational energy sources that make the star shine. The mechanism that continually feeds this energy into pulsation for certain stars is a sort of valve action provided by the ionization and de-ionization of some abundant elements such as hydrogen and helium.

The pulsation of a cepheid variable, for example, is evidently produced by the ionized gases in its outer envelope cooling and becoming neutral when it is at its largest. As the gas cools it radiates its heat into space and the pressures within it drop, allowing the outer parts of the star to fall inward. As they do so, they absorb energy from the interior of the star and become ionized again. The re-ionization heats the gas and raises its pressure again, causing it to expand and cool; thus the process is repeated.

In most stars the neutral hydrogen and helium are either in too narrow a layer near the surface to be important or there is too little energy from the stellar interior to ionize it anyway. Thus this mechanism does not cause most stars to pulsate. Even the variable stars, evolving as nuclear reactions alter their chemical compositions, change their structures enough to stabilize and cease pulsating after some thousands or millions of years. Cepheids, in fact, are thought to pass through unstable conditions and pulsate several times during their evolution back and forth across the H–R diagram, each successive time becoming larger and more luminous and pulsating more slowly.

Periods of Pulsation for Different Kinds of Stars

It is interesting to ask what the pulsation period would be for an ordinary star if it were unstable. It turns out that the period is greater for a giant star of low mean density than for a smaller compact star of higher density—just as a long piano string vibrates more slowly than a short one. It can be shown that for pulsating stars of any one type, the period of a particular star is inversely proportional to the square root of its mean density.

Detailed calculation shows that if the sun were to pulsate its period would be about 0.5 hr. Thus we can estimate the periods of pulsation of other kinds of stars by comparing their mean densities to the sun's. Table 17.1 shows the results of such a comparison for stars all of one solar mass but of different radii. Because most stars have masses that differ from the sun's by less than a factor of 10, their pulsation periods would differ from the tabulated values for one solar mass stars of the same radii by only at most a factor of 3.

We see that we would expect the giant cepheids to pulsate in days or weeks, and the supergiant red (Mira) variables to require months or even years for a cycle, and this is just what we observe. Appendix 15 lists the principal kinds of pulsating variable stars. Note that most are giant stars of periods greater than one day.

TABLE 17.1 Pulsation Periods for Various Stars of One Solar Mass

Radius (solar radii)	Period	Examples
1	0.5 hr	Sun
1000	2 yrs	Red supergiants
100	1 month	Cepheid
10	0.7 day	RR Lyra star
0.1	1 min	
0.01	2 s	White dwarf
10^{-5}	6×10^{-5} s	Neutron star

The shortest period with which a star can rotate without flying apart is, within a factor of a few times, the same as the period with which it could pulsate. Note in Table 17.1 that only exceedingly tiny stars could pulsate or rotate in as short a time as a few seconds.

FINAL STAGES OF STELLAR EVOLUTION

Sooner or later a star must exhaust its store of nuclear energy. Then it can only contract and release more of its potential energy. Eventually, the shrinking star will attain an enormous density. We know of three such possible end states for stars: *white dwarfs, neutron stars,* and *black holes*. Of these, by far the most common are the extremely compact white dwarf stars (Chapter 14), whose mean densities range up to more than one million times that of water.

White Dwarfs

White dwarf stars are far more dense, of course, than any solid substance. The high density of a white dwarf is possible because the atoms that comprise the gases in a stellar interior are almost completely ionized, that is, stripped of virtually all their electrons. Most of a neutral atom is empty space; once an atom is completely ionized, it and its freed electrons can occupy a volume many times smaller than when the electrons are still revolving about the nucleus.

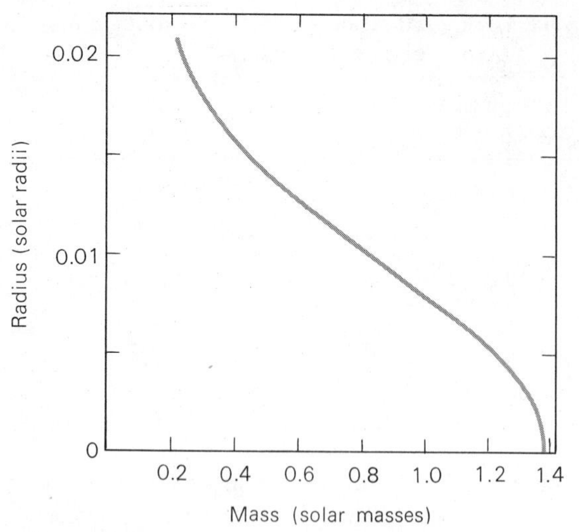

Figure 17.7
Theoretical relation between the masses and radii of white dwarf stars.

It is the free electrons in a white dwarf that dominate its structure. The quantum theory shows that electrons cannot be packed closer together than to a certain critical separation. What that critical density of electrons is depends on how fast the fastest of them can move. If a star of considerable mass contracts to a small size, it releases enough gravitational potential energy to speed up the electrons to nearly the speed of light; if they can be given ever greater momenta they will never reach a final limiting density. White dwarf stars, however, are at most only a little more massive than the sun. Their contraction cannot release enough energy, so the electrons pack up to their maximum limit and exert an enormous pressure that prevents further collapse of the stars. Electrons in this state are said to be *degenerate*.

White dwarfs are thus stars whose electrons have become degenerate. Given the mass of a star, we can calculate how far it can contract before the degenerate electrons stop it, and hence what the radius of the final white dwarf will be. A white dwarf of mass like the sun has a size about that of the earth. The theoretically calculated relation between the masses and radii of white dwarfs is shown in Figure 17.7. Note that the larger the mass of the star, the smaller is its radius. A white dwarf with a mass of about 1.4 solar masses would have a radius of zero! According to the theory, therefore, no white dwarf can have a mass quite as large as 1.4 times that of the sun. (This upper limit to the mass of a stable white dwarf is somewhat uncertain; it may be somewhat lower than 1.4 and possibly as low as 1.2 solar masses.)

Since a white dwarf is a star that has exhausted its nuclear fuel, and since it can no longer contract, its only energy source is the heat represented by the motions of the atomic nuclei (ions) in its interior that, unlike the electrons, are *not* degenerate. The light by which we see a white dwarf comes from this internal stored heat. Gradually, however, the white dwarf must radiate away its heat, and so it must slowly fade. After many hundreds or thousands of millions of years that heat will be gone, the nuclei will be still, and the white dwarf will be dark—a final cold body with the mass of a sun and the size of a planet, called a *black dwarf*.

Now, the only stars that have had time to exhaust their nuclear fuel supply and evolve to the white dwarf stage must have had original masses comparable to or greater than that limiting value of 1.4 solar masses, for those more massive stars are the very ones that use up their energy store most rapidly. On the other hand, white dwarf stars are plentiful, and they must have come from somewhere. Moreover, the number of white dwarfs is high enough, as nearly as can be determined, to account for most or all evolved stars of original mass greater than 1.4 solar masses. It is thought therefore, that most stars eventually do, in fact, become white dwarfs. Consequently they must lower their masses somehow, before reaching that stage, by ejecting matter into space.

Mass Ejection by Stars; Planetary Nebulae

We know of several mechanisms by which stars lose some of their matter. For example, many stars are surrounded by extended atmospheres of expanding gas shells. The extended atmosphere is usually revealed by the presence of emission lines or bands superposed on the continuous spectrum of the star. Sometimes the light absorbed and then reemitted by the gas shell is too feeble to be observed, but the shell may still reveal its presence by producing absorption lines that, because of their wavelengths or sharpness, cannot originate in the stellar photosphere. In a few cases, a gaseous envelope surrounding a star can be seen or photographed telescopically.

Many red giants are examples of stars with extended atmospheres, whose spectra show that the gaseous shells have been and are being ejected and are now expanding about them. Also about 4000 spectral class B stars are known whose spectra show emission lines, usually of hydrogen, and sometimes of other elements as well. These stars, known as *B emission,* or *Be, stars,* have evidently ejected material from their outer layers.

For stars of 1.2 or so solar masses—the most massive stars among red giants in globular clusters and in the galactic corona, nucleus, and disk—one of the most important mass-ejection mechanisms is the planetary nebula phenomenon. *Planetary nebulae* are shells of gas ejected from and expanding about certain extremely hot stars. They derive their name from the fact that a few bear a superficial telescopic resemblance to planets; actually they are thousands of times larger than the entire solar system, and have nothing whatever to do with planets.

The most famous example is the Ring nebula, in Lyra. It is typical of many planetaries in that, although actually a hollow shell of material emitting light, it appears as a ring. The explanation is that we are looking through the *thin* dimensions of the front and rear parts of the shell, while along its periphery our line of sight encounters a long path through the glowing material. Similarly, a soap bubble often appears to be a thin ring. Altogether, about 1000 planetary nebulae have been cataloged. Doubtless there are many distant ones that have escaped detection, so there must be some tens of thousands in the Galaxy. An appreciable amount of material is ejected in the shell of a planetary nebula. From the light emitted by the shells, we calculate that they must have masses of 10 to 20 percent that of the sun. The shells, typically, expand about their parent stars at speeds of 20 to 30 km/s.

A typical planetary has a diameter of about ½ LY to 1 LY. If it is assumed that the gas shell has always expanded at the speed with which it is now enlarging about its parent star, its age can be calculated. Most of the gas shells have been ejected within the past 50,000 years; an age of 20,000 years is more or less typical. After about 100,000 years, the shell is so enlarged that it is too thin and tenuous to be seen. When we take account of the relatively short time over which planetary nebulae exist, we find that they are very common, and that an appreciable fraction of all stars must sometime evolve through the planetary nebula phase.

The gas shells of planetary nebulae shine by the process of fluorescence. They absorb ultraviolet radiation from their central stars and reemit this energy as visible light. We can calculate the rate at which ultraviolet radiation must be leaving the star to account for the visible light coming from the gas shell; it turns out to be a far greater amount of energy than the star radiates in its observable, visible spectrum. The central star of a planetary nebula, must be many times hotter than the sun for so large a fraction of its luminosity to be in the ultraviolet (see Chapter 5). Nearly all these stars are hotter than 20,000 K, and some have temperatures well in excess of 100,000 K, which makes them among the hottest known stars.

Despite their high temperatures, the central stars of planetary nebulae do not have exceedingly high luminosities—some emit little more total energy than does the sun. They must, therefore, be stars of small size; some, in fact, appear to have the dimensions of white dwarfs. A planetary nebula may be the last ejection of matter by a red giant star before it collapses to a white dwarf.

Novae

Some stars eject mass violently. These include some of the so-called *eruptive variables.* A table of summary data on these kinds of stars is given in Appendix 16. Here we describe the most familiar, the *novae* and *supernovae.*

Nova literally means "new." Actually, a nova is an existing star that suddenly emits an outburst of light. In ancient times, when such an outburst brought a star's luminosity up to naked-eye visibility, it seemed like a new star. Novae remain bright for only a few days or weeks and then gradually fade. They seldom remain visible to the unaided eye for more than a few months. The Chinese, whose annals record novae from centuries before Christ, called them "guest stars." Only occasionally are novae visible to the naked eye, but, on the average, two or three are found telescopically each year. Many must escape detection; altogether there may be as many as two or three dozen nova outbursts per year in our Galaxy.

Novae, before and after their outbursts, are stars of very high temperature but small size, so that despite their high temperatures they have lower luminosities than equally hot main-sequence stars. A typical nova, however, flares up during its outburst to thousands or even tens of thousands of times its normal luminosity, and may reach an absolute magnitude of from -6 to -9. The rise to maximum light is very rapid, often requiring less than one day. The subsequent decline in light is much slower; the star requires years, or even decades, to return to normal. A typical nova light curve is shown in Figure 17.9. Different types of novae, however, decline at different rates. Some (possibly all) show variability during their gradual fading.

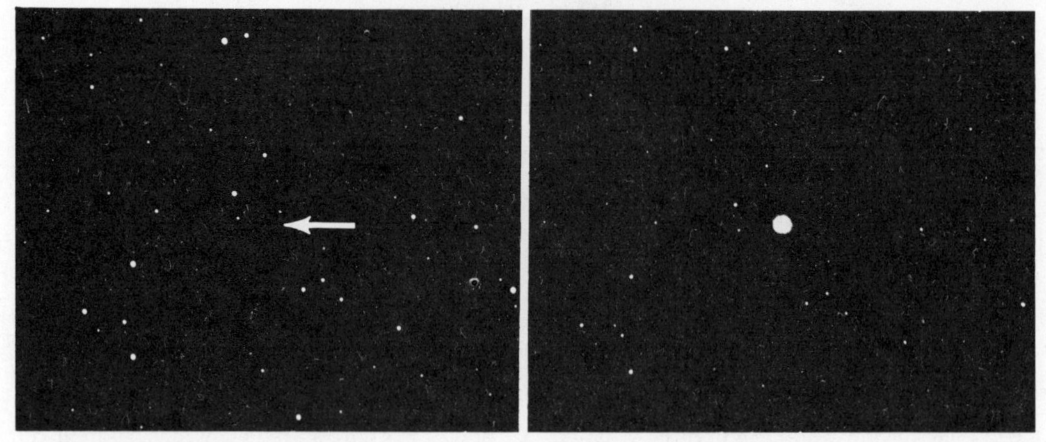

Figure 17.8
Nova Herculis, as it appeared before and after its outburst in 1934. *(Yerkes Observatory)*

Some novae have undergone more than one outburst over the period they have been under observation; these are called *recurrent novae*. There is no way of knowing, of course, whether ordinary novae may have had previous flareups before recorded history, or whether they will produce outbursts again. It is possible, in other words, that all novae may be recurrent but that those with the more spectacular outbursts wait for longer intervals between them.

The current theory of novae is that they are white dwarfs belonging to binary star systems. The companion star to the white dwarf is presumed to be evolving to a red giant, the surface of the giant becomes so large that it is attracted more strongly by the gravitation of the white dwarf than by the giant itself. Thus, some of the large star's material falls onto the dwarf. This fresh supply of matter, piling up on the compact star, is heated until nuclear reactions start combining the hydrogen in it into helium. The sudden release of nuclear energy then shoots this blanket of new matter off the white dwarf into an expanding shell, with the emission of a great amount of light. The outburst dies down, but the giant star continues to grow, spilling still more matter onto the white dwarf until, after decades or hundreds or thousands of years, the process is repeated.

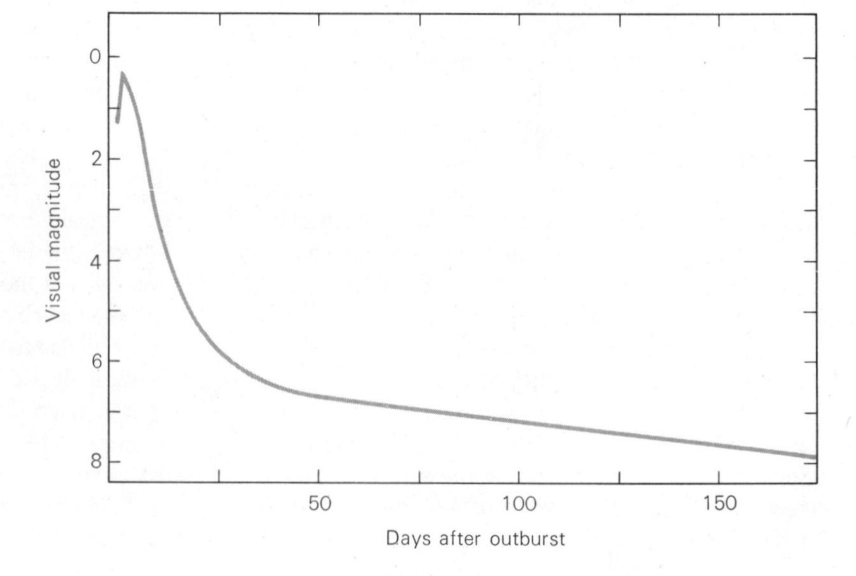

Figure 17.9
Light curve of Nova Puppis, 1942.

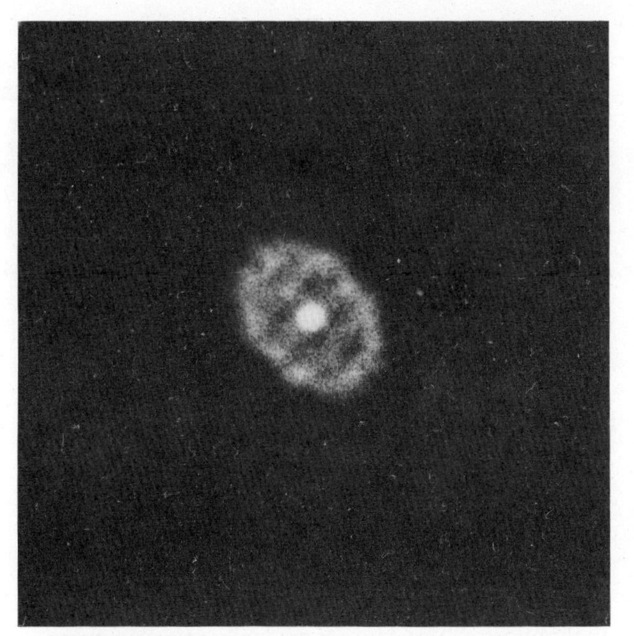

Figure 17.10
The shell of Nova Herculis photographed with the Lick Observatory's 3-m telescope in 1972. *(Courtesy H. Ford, UCLA)*

Supernovae

Among the more spectacular of the cataclysms of nature is the *supernova*. In contrast to an ordinary nova, which increases in luminosity a paltry few thousands or at most tens of thousands of times, a supernova is a star that flares up to hundreds of millions of times its former brightness. At maximum light, supernovae reach absolute magnitude -14 to -18, or possibly even -20. The three most famous supernovae to have been recorded during the last 10 centuries in our Galaxy are (1) the supernova of 1054 in Taurus (described in the *Chinese Annals*), (2) Tycho's "star" of 1572 in the constellation Cassiopeia, and (3) the supernova of 1604 in Serpens, described by both Kepler and Galileo. We also observe many gaseous remnants of what are believed to have been prehistoric supernovae. Supernovae may occur in the Galaxy at an average rate as high as one every 30 to 50 years. They are also commonly observed in other galaxies.

The light curve of a supernova is similar to that of an ordinary nova, except for the far greater luminosity of the supernova. There are several different kinds of supernovae, but they all rise to maximum light extremely quickly (in a few days or less), and for a brief time one may outshine the entire galaxy in which it appears. Just after maximum, the gradual decline sets in, and the star fades in light until it disappears from telescopic visibility within a few months or years after its outburst. Bright emission lines in the spectra of supernovae indicate that they, like ordinary novae, eject material at the time of their outbursts. The velocities of ejection are substantially greater than in ordinary novae, however; speeds of up to 10,000 km/s have been observed. Moreover, a much larger amount of material is ejected; in fact, a large fraction of the original star may go off in the expanding envelope. As mentioned above, we believe that supernovae synthesize small amounts of the heaviest elements during their outbursts, and spew those elements into the interstellar medium. Supernovae may well be responsible for such elements as copper, silver, and gold on the earth.

The Crab Nebula

An example of the remnant of a supernova explosion is the Crab nebula in Taurus, a chaotic, expanding mass of gas, visible telescopically, and spectacular on telescopic photographs. It was formed by the supernova of 1054, and in the nine centuries since that outburst the cloud has reached an angular diameter of 6′ of arc. At its estimated distance of 2000 pc, the nebula must be 3.5 pc, or more than 11 LY, across. The

Figure 17.11
The Crab nebula, photographed in red light with the 5.08-m telescope. *(Hale Observatories)*

Crab nebula is one of the most interesting and most studied objects in the sky.

The Crab nebula is a strong source of radio waves, infrared radiation, X rays, and gamma rays, as well as of light. As the first X-ray source discovered in Taurus, it is known as Tau X-1. The radio spectrum (variation of radio energy with wavelength) has characteristics that led the Soviet astrophysicist I. S. Shklovsky, in 1953, to propose that the radiation is from the synchrotron process (Chapter 15). The Crab nebula is the first astronomical object from which synchrotron radiation was recognized. Much of its visible light similarly originates from the synchrotron mechanism. The red filaments (see the color photograph) derive their light mostly from hydrogen ions recombining with electrons, but the white light, and the other radiation, from radio to gamma rays is synchrotron, showing the Crab nebula to possess strong magnetism, and a large source of relativistic electrons.

At the center of the nebula is a star that was once thought to be the white dwarf remains of the supernova explosion. In 1968, however, the light from that star was discovered to consist of pulses of energy coming at the rate of 30 per second! It is a *pulsar*.

PULSARS

Pulsars are objects now thought to be formed by supernovae. Pulsars, an acronym for *pulsating radio stars*, are radio sources emitting sharp, intense, rapid, and extremely regular pulses. The first known was discovered accidentally in 1967 by Jocelyn Bell, a research student at Cambridge University. Miss Bell, working with Anthony Hewish, was studying the twinkling of radio sources with a radio telescope whose antenna is simply a large array of wires strung horizontally a few feet above the ground. (In fact, the wires are so low that a man cannot conveniently push a lawnmower under them, so in the summer sheep are brought in to graze on the grass.)

The pulsar found by Miss Bell is in the constellation Vulpecula, and emits pulses with great regularity every 1.3373 s. In a couple of weeks the object pulses nearly a million times, and at a rate more stable than a fine clock.

At first it seemed conceivable that the pulses could be intelligently coded signals, and the source was half-jokingly dubbed "LGM" for "little green men." But within a few months the Cambridge astronomers found three additional similar sources of radio pulses, of somewhat different period, and in widely separated parts of the sky, which made it highly unlikely that they were intelligent signals from other beings.

By 1979 more than 300 pulsars were known, all more or less similar to the first few discovered. They have periods in the range from .03 to 3 s. Although the pulses from any one pulsar are extremely regular in period, they do often vary considerably from one to the next in intensity. Some pulsars are more than 1000 pc away. At such distances, the radio energy emitted in each pulse must be enormous. Moreover, because of the sharpness of the pulses, that radio energy must be coming from a region at most a few hundred kilometers in diameter; otherwise, the light-travel time across the emitting region would smear out the pulse.

One pulsar, as mentioned above, is smack in the middle of the Crab nebula. It has the shortest pulse period known (at this writing)—0.033 s, and the period is observed to be very slowly increasing, showing that pulsars evolve, pulsing gradually more slowly as they age. The Crab pulsar also emits optical (visible light) and X-ray pulses with that same 0.033-s period. About 10 to 15 percent of the X-ray radiation from the Crab nebula comes from the pulsar.

Several other pulsars are also associated with wisps of gas that are believed to be the remnants of prehistoric supernovae. One such pulsar, centered on a supernova remnant in the constellation Vela, is the only other one besides the Crab pulsar that is observed to pulse in visible light as well as in radio energy. The Vela pulsar also has a short period (0.089 s), although not so short as that of the Crab. Presumably, these pulsars are still young enough to be emitting appreciable energy at other than radio wavelengths.

In addition to these pulsating radio sources, there are dozens of X-ray sources that pulse in short regular periods. Of these, only the Crab pulsar is seen also in visible light and radio waves, but the other X-ray pulsars are believed to be objects similar to radio pulsars, but more energetic in their emission. At least some are members of binary star systems, and for four of these enough information is available to calculate masses by techniques described in Chapter 14. These four X-ray pulsars have masses in the range 1.4 to 1.8 times that of the sun.

Theory of Pulsars—Neutron Stars

The energy emitted by pulsars is not small; the Crab pulsar puts out thousands of times as much en-

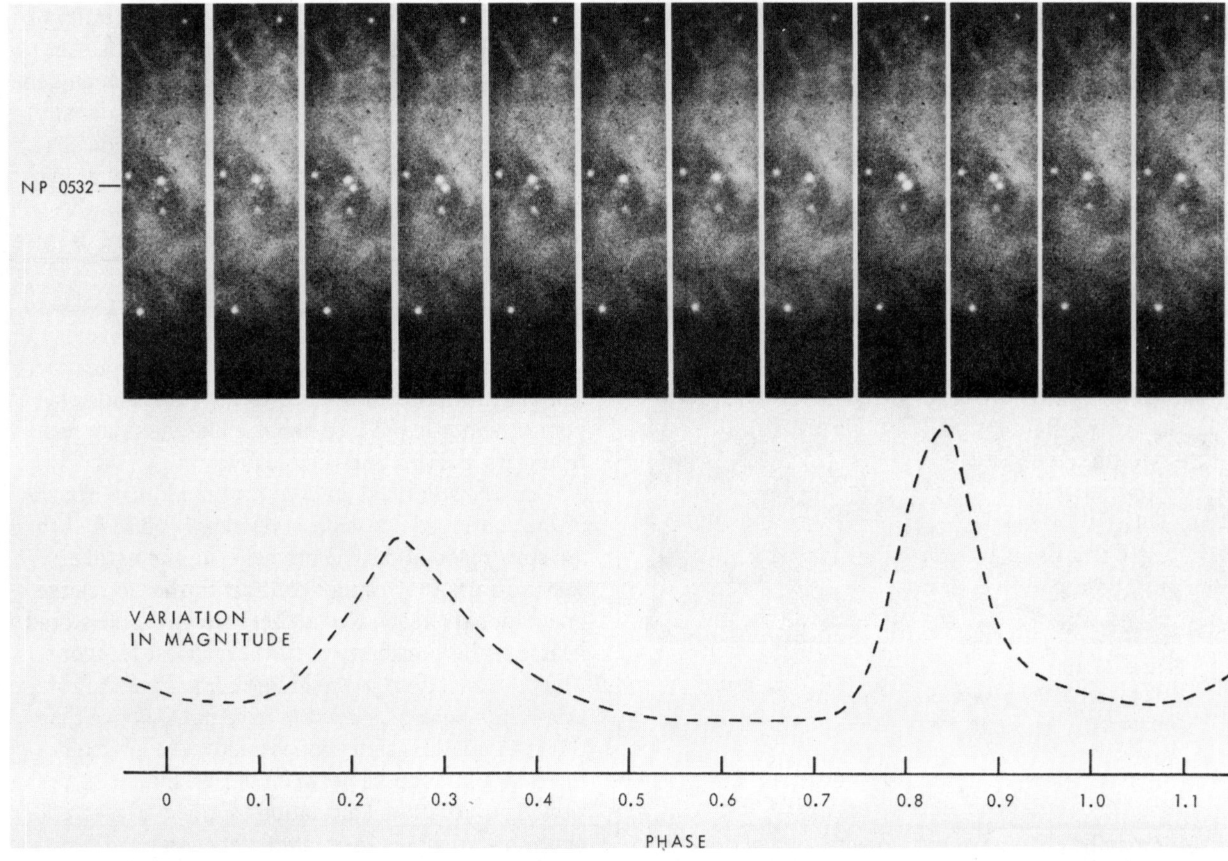

Figure 17.12
A series of photographs of the central part of the Crab nebula taken by S. P. Maran at Kitt Peak National Observatory. Note the star that seems to blink on and off; it is a pulsar which has a period of 1/30 s. *(Kitt Peak National Observatory)*

ergy as does the sun. Thus pulsars are like stars in their output of radiation. Yet they emit this energy in pulses of up to 30 per second, as if they were pulsating at such high frequency. But what kind of object can pulsate up to 30 times per second? Even a white dwarf would pulsate with a period of a few seconds or more—hardly fast enough for the pulsars. A *neutron star*, however, could pulsate with a very much shorter period, and could rotate with any period longer than about a ten-thousandth of a second. Pulsars, therefore, are generally thought to be associated with rotating neutron stars.

Neutron stars are configurations composed almost entirely of neutrons. Ordinarily, a free neutron (one not bound in an atomic nucleus) survives only about 11 minutes before decaying into a proton and an electron. Under extremely high pressures, however, a neutron is stable. Suppose, somehow, that all the electrons in a star could be forced, under tremendous pressure, into the atomic nuclei. Since stars are electrically neutral, there are just as many electrons as there are protons in the nuclei. Thus all the matter would become neutrons.

Neutrons, like electrons, become degenerate if crowded into a sufficiently small volume for a given velocity range. Thus, the structure of a neutron star is analogous to that of a white dwarf, except that neutron stars are much smaller. A neutron star of one solar mass would have a radius of only about 10 km. There exists a mass-radius relation for neutron stars, and an upper mass limit as well, although uncertainties in the theory make it difficult to say exactly what that upper mass limit is. The best guess for the limit to the mass of a neutron star is from two to three solar masses. This is greater than the mass limit for white dwarfs. Thus a star of mass greater than 1.4 solar

Figure 17.13
IC 443, probably a supernova remnant. *(Hale Observatories)*

masses may be able to contract gravitationally directly to a neutron star, missing the white dwarf configuration. It is interesting that the four X-ray pulsars of known mass have masses in the range expected for neutron stars.

We think, however, that neutron stars can also be formed in supernova explosions. We know of several mechanisms by which this may occur. One idea is that a massive red giant star may have a small dense degenerate core whose mass may approach that of the limiting value for a white dwarf. Continued conversion of hydrogen to helium in a shell around the degenerate core can add mass to it until it exceeds the white dwarf mass limit and collapses to a neutron star. In any case, gravitational energy is released from the mass collapsing to a neutron star, and that energy causes ejection of the outer layers of the star in a supernova outburst.

Any magnetic field that existed in the original star is highly compressed if the core of the star collapses to a neutron star. Thus a moderate field of the order of 1 gauss in a star the size of the sun increases to the order of 10^{10} to 10^{12} gauss around the neutron star. At the very surface of the collapsed star neutrons decay into protons and electrons. Many of these charged particles should leave the stellar surface and, in the vicinity of the magnetic poles, move out into the circumstellar magnetic field. With such intense fields and high densities many of the particles are relativistic, especially the electrons, which emit nonthermal radiation. The star's rotation turns first one and then the other magnetic pole into our view, so the radiation from the rotating magnetic field can be directed toward us once each time the star turns on its axis.

Meanwhile, the atomic nuclei are thought also to be accelerated by the magnetic field to relativistic speeds, producing cosmic rays. If this entire picture (or some modification of it) is correct, we can understand how supernovae produce nebulae like the Crab, neutron stars, pulsars, and cosmic rays.

But we still must account for the ultimate source of the continued emission of energy by pulsars. If they are stars made of degenerate neutron matter, like white dwarfs they cannot contract further to release gravitational energy. Nor is there any available stored heat, nor the possibility of further nuclear reactions. Thus the energy of pulsars must come from the rotation of the neutron stars. We have noted above that the Crab pulsar is actually observed to be gradually increasing its interval between pulses, indicating a slowing of rotation. Calculations show that such a neutron star should lose rotational energy at the rate of 10^{38} erg/s. This is just the amount of electromagnetic energy, of all wavelengths, that we observe to be emitted by the Crab nebula and its associated pulsar. Thus we predict that as pulsars age, they slow down, thereby extracting energy from the rotating neutron stars less efficiently, so that over thousands of years they fade out—first at X-ray, then visible, and eventually at radio wavelengths as well.

Another Possible Stellar Fate: the Black Hole

We have seen that most stars must end up as white dwarfs. Yet, some stars, evidently through processes connected with the supernova phenomenon, end up as neutron stars. White dwarfs must have masses less than 1.4 times that of the sun, and neutron stars probably cannot exceed three solar masses. However, what becomes of a more massive star—beyond the neutron star mass limit—that for some reason does not eject matter, yet collapses after exhaustion of its nuclear fuel? We suspect that such stars may become *black holes*.

By now, however, we are well into the realm of

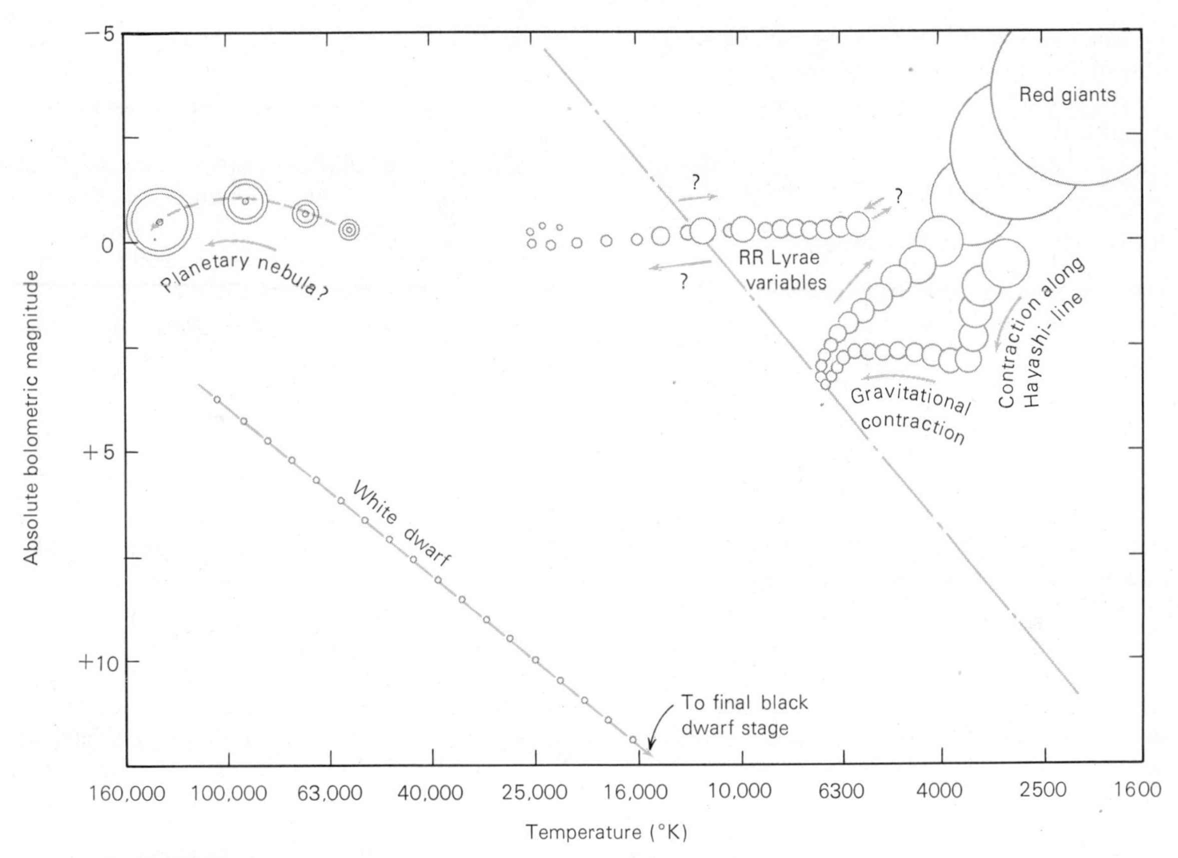

Figure 17.14
Summary of the evolutionary track on the Hertzsprung–Russell diagram for a star of 1.2 solar masses.

general relativity, so we shall defer discussion of black holes to the next chapter.

SS433

A chapter on stellar evolution is probably not the appropriate place to describe SS433, but then it is hard to know where else to describe it because no one yet knows what it is. It is quite clear, however, that this bizarre object *warrants* description.

SS433 is in a catalog of objects showing H-alpha (hydrogen) emission lines, compiled by C. B. Stephenson and N. Sanduleak in 1977. More recently it attracted attention when it was found to be a source of variable radio emission and X-rays. Moreover, it is near the center of a supernova remnant and may be related to a supernova outburst. But the really astonishing features of SS433 became known when its optical spectrum was studied in a long series of observations —especially by B. Margon and his associates at UCLA.

The spectrum shows bright emission lines of hydrogen and helium arising from three different sources. One source appears to be the main object, and the Doppler shift of its lines indicates a relatively low speed, like that expected for an object in our Galaxy. The other two sources, however, have very high speed with respect to the main body, and are evidently streams of gas ejected from it in oppositely directed jets. Analysis shows the gas to be streaming out (in both directions) at 27 percent the speed of light. Moreover, the whole system is evidently rotating in such a way that the directions of the jets describe a conical motion (suggesting precession) with a period of 164 days.

The importance of SS433 is the study of the mechanisms for the mass ejection at so high a speed and the periodic change in orientation. No complete physical theory has been suggested at this writing. However, in some respects SS433 has properties that on a small scale remind us of those of radio galaxies and quasars (Chapter 20). Perhaps our eventual understanding of SS433 will bring us the insight needed to fathom the powerhouses of those far grander objects associated with remote galaxies.

Chemical Evolution in the Universe

We have seen that stars convert hydrogen into helium and, moreover, that at least some stars in some stages of their evolution are building up helium into carbon and heavier elements. Thus, inside stars, some of the lighter elements of the universe are gradually being converted into heavier ones. As these stars eject matter into the interstellar medium, that matter is richer in heavy elements than was the material from which the stars were formed. In other words, a gradual enrichment of the heavy-element abundance in interstellar matter is taking place. The heavy-element abundance in stars that are forming now is thus higher than in those that formed in the past. The fact that the oldest known stars (those in globular clusters) are the stars with the lowest known abundance of heavy elements provides evidence for this scenario.

Current theory suggests that the original stars in our Galaxy consisted of pure hydrogen and helium. Stars like the sun, in whose outer layers heavy elements are observed spectroscopically, must then be of the second or third (or even higher) "generation." Indeed, our very bodies, we think, are made of atoms "cooked up" in the hot cores of earlier generation stars.

SOLAR EVOLUTION

Figure 17.14 summarizes our current ideas on the evolution of a star of about 1.2 solar masses on the H–R diagram. In its early stages, the star contracts and moves to the left, reaching the main sequence with a size only slightly greater than that of the sun. In its subsequent evolution to the red giant stage, it grows to a radius of tens of millions of kilometers. The further evolution is uncertain. Perhaps the star goes through stages of variability, or emits material as a planetary nebula. Its final size as a white dwarf is only about that of the earth.

From theoretical calculations we can now form a fairly clear picture of the approximate past history of the sun, and we can make at least educated guesses about its future. Since it reached the main sequence, the sun has increased somewhat in luminosity, probably by about 30 to 50 percent. During that interval of nearly 5 thousand million years, it has depleted much of the hydrogen at its very center, but a pure helium core has not yet had time to form.

All available evidence leads us to expect that sometime in the future the sun will leave the main sequence and evolve to a red giant. According to calculations of R. Ulrich, that time will occur in about 5 thousand million years, when the sun's photosphere will reach past the orbit of the earth. The earth, then well inside the sun, and exposed to temperatures of thousands of degrees, will gradually vaporize. The gases in the greatly distended outer layers of the sun will, of course, be very tenuous, but they should still offer enough resistance to the partially vaporized earth to slow it in its orbital motion. Ulrich believes that the earth will spiral inward toward the very hot interior of the sun, reaching its final end about 10 thousand years after being swallowed by the sun.

If, in that remote time, man could leave the doomed earth but remain in the solar system, he would find that most of the naked-eye stars in our twentieth-century sky would long since have exhausted their nuclear fuel and evolved to white dwarfs (or perhaps some other end). Main-sequence stars less massive than the sun would still be shining, but only those few of them passing temporarily through the solar neighborhood would be near enough to see with the unaided eye. It is doubtful, however, if by then *all* star formation in the Galaxy would have ceased. Luminous young stars might be shining in remote clouds of interstellar matter; a Milky Way might still stretch around the sky.

SUMMARY REVIEW

Factors determining the structure of a star: role of mass and chemical composition; interpretation of the main sequence

Early stages of evolution: sites of star formation (cold clouds, globules, spiral arms); initial collapse; gravitational contraction; main sequence; observations of young star clusters

Main-sequence evolution: zero-age main sequence

Evolution to red giant phase: helium core; contraction of core; lifetime on main sequence; evolutionary tracks on the H–R diagram; observations of star clusters of various ages

Synthesis of elements: theory of original hydrogen and helium composition of the Galaxy; formation of heavier elements by nucleosynthesis in the centers of red giant stars; the helium flash

Pulsating stars: stellar instability; cause of pulsation; cepheid variables; pulsation as a stage of stellar evolution; the period-density relation for pulsating stars; the sun's natural period of pulsation; the kinds of stars or objects with various periods of pulsation or maximum rotation rate

Final stages of stellar evolution: white dwarfs; electron degeneracy in white dwarfs; mass-radius relation; mass limit; mass ejection; B-emission (Be) stars; planetary nebulae; central stars of planetary nebulae; novae; recurrent novae; theory of nova outbursts; supernovae; remnants of prehistoric supernovae; the Crab nebulae; synchrotron radiation from the Crab nebula; the Crab pulsar

Pulsars: discovery; pulse periods; energy involved; association with supernovae; theory of pulsars; association with neutron stars; nature of neutron stars; formation of neutron stars in supernova outbursts; source of energy of pulses; loss of rotational energy of neutron stars; evolution of pulsars

Other fates of stars: black holes; SS433; chemical evolution of the universe (nucleosynthesis)

Past and future evolution of the sun

EXERCISES

1. The H–R diagram for field stars (that is, stars all around us in the sky) shows very luminous main-sequence stars and also various kinds of red giants and supergiants. Explain these features, and interpret the H–R diagram for field stars.

2. In the H–R diagrams for some young clusters, stars of very low and very high luminosity are off to the right of the main sequence, while those of intermediate luminosity are on the main sequence. Can you offer an explanation? Sketch an H–R diagram for such a cluster.

3. If the sun were a member of the cluster NGC 2264, would it be on the main sequence yet? Why?

4. Explain how you could decide whether red giants seen in a star cluster probably had evolved away from the main sequence or were still evolving toward the main sequence.

5. If all the stars in a cluster have the *same* age, how can clusters be useful in studying evolutionary effects?

6. Where on the H–R diagram does a star *end?*

7. Suppose a star spends 10×10^9 years on the main sequence and burns up 10 percent of its hydrogen. Then it quickly becomes a red giant with a luminosity 100 times as great as that it had while on the main sequence and remains a red giant until it burns up the rest of its hydrogen. How long a time would it be a red giant? Ignore helium burning and other nuclear reactions, and assume that the star brightens from main sequence to red giant almost instantaneously.

8. Can you think of a reason why the fact that most known variable stars are giants could be an effect of observational selection? How might we discover faint telescopic variable stars that are smaller and denser than the sun?

9. Since supernovae occur so rarely in any one galaxy, how might a search for them be conducted? (It may help to glance at Chapter 20.)

10. Compare and contrast nova shells, supernova remnants, and planetary nebulae.

11. Suppose the luminous shell of a planetary nebula is easily resolved with a telescope. Now suppose the spectrum of the nebula is photographed, with the slit of the spectrograph extending completely along one diameter of the shell. Sketch the appearance of a typical emission line in the spectrogram, and explain your sketch. It may help to look over the description of the construction of a spectrograph in Chapter 6.

12. Assume that a pulsar is 100 pc away. Suppose that no star brighter than apparent magnitude 23 shows up in that position of the sky. What is the brighter limit to the absolute magnitude that a star associated with the pulsar could have? (*Hint:* see Chapter 13.)

13. Why do you suppose masses are known from observation only for those few white dwarfs that are members of nearby visual binary systems? (*Hint:* see Chapter 14.)

14. By the time the sun becomes a white dwarf, the constellations familiar to us now will not be seen, even if their stars never change in luminosity. Why?

15. Prepare a chart or diagram which exhibits the relative sizes of a typical red giant, the sun, a typical white dwarf, and a neutron star of mass equal to the sun's. You may have to be clever to devise such a diagram.

16. Suppose you had a clock that you could trust to only about 1 s. How could you nevertheless determine the period of the pulsar in Vulpecula to within one part in a million?

17. As observed from the earth, the precise intervals between the pulses of a typical pulsar vary periodically, and in step with the seasons. Can you suggest an explanation? (*Hint:* What is the direction of the pulsar with respect to the direction of the earth's orbital motion?)

18. Suppose five new pulsars are discovered with periods of: (a) 1.32 s, (b) 3.04 s, (c) 0.05 s, (d) 0.97 s, (e) 1.92 s. From which would you most expect to be able to observe pulses of visible light? Explain.

The Bubble nebula, NGC 7635, in Cepheus. *(Lick Observatory)*

CHAPTER 18

Albert Einstein (1879–1955) received the Nobel Prize in 1921 not for his theory of relativity, but for the photoelectric effect. At that time his ideas on relativity were still at the frontier. Einstein believed that such seemingly diverse areas of physics as mechanics and electromagnetic phenomena—and even gravitation—were guided in the same way by underlying principles.

THE GENERAL THEORY OF RELATIVITY—AND BLACK HOLES

In his special theory of relativity (Chapter 7), Einstein showed that different observers in uniform relative motion perceive space and time differently—that is, they disagree with each other on measurements such as length, time, momentum, and energy. The special theory also shows how those measurements by relatively moving observers compare with each other. But the rules of special relativity do not apply to the comparison of measurements by observers that are *accelerated* with respect to each other. Einstein's general theory of relativity, published in 1916, shows how we can extend the special theory to take account of relatively accelerated observers. In doing so, Einstein discloses a new formulation of gravitation, because a gravitational force on a system is indistinguishable from an acceleration of that system—that is, gravitation and acceleration are equivalent.

PRINCIPLE OF EQUIVALENCE

Galileo noted that all bodies, despite their different masses, if dropped together fall to the ground at the same rate. Einstein had the genius to turn that isolated fact[1] into a powerful principle of physics—the principle of equivalence. In doing so, he took the first giant step toward formulating his general theory of relativity.

Because two bodies, falling together side by side approach the ground together, they are obviously *not* accelerated with respect to each other. Thus they are aware of no force acting between them. For example, suppose a brave boy and girl simultaneously jump into a bottomless chasm from opposite sides of its banks. If we ignore air friction, while they fall they accelerate downward together, and feel no external force acting on them. They can throw a ball back and forth between them, aiming always in a straight line, as if there were no gravitation, and the ball, falling along with them, would move directly to its target.

It's very different on the surface of the earth. Everyone knows that a ball, once thrown, falls to the ground. Thus in order to reach its target (the catcher) the ball must be aimed upward somewhat, so that it follows a parabolic arc—falling as it moves forward—until it is caught at the other end.

Because our freely falling boy, girl, and ball are all falling together, we could enclose them in a large box falling with them. Inside that box, no one can be aware of any gravitational force; nothing falls to the ground, or anywhere else, but moves in a straight line in the most simple natural way, obeying Newton's laws. By having our box fall with the boy and girl, we have selected a coordinate system within which there is no force. We have, in fact, removed the force of gravitation by selecting a coordinate system that is accelerating at just the right rate to compensate for

[1] As Julian Schwinger described it: I owe several of the examples in this chapter for describing the general theory of relativity to the unique talent of Professor Schwinger for peeling away the rind and getting to the heart of the idea that describes a fundamental physical concept.

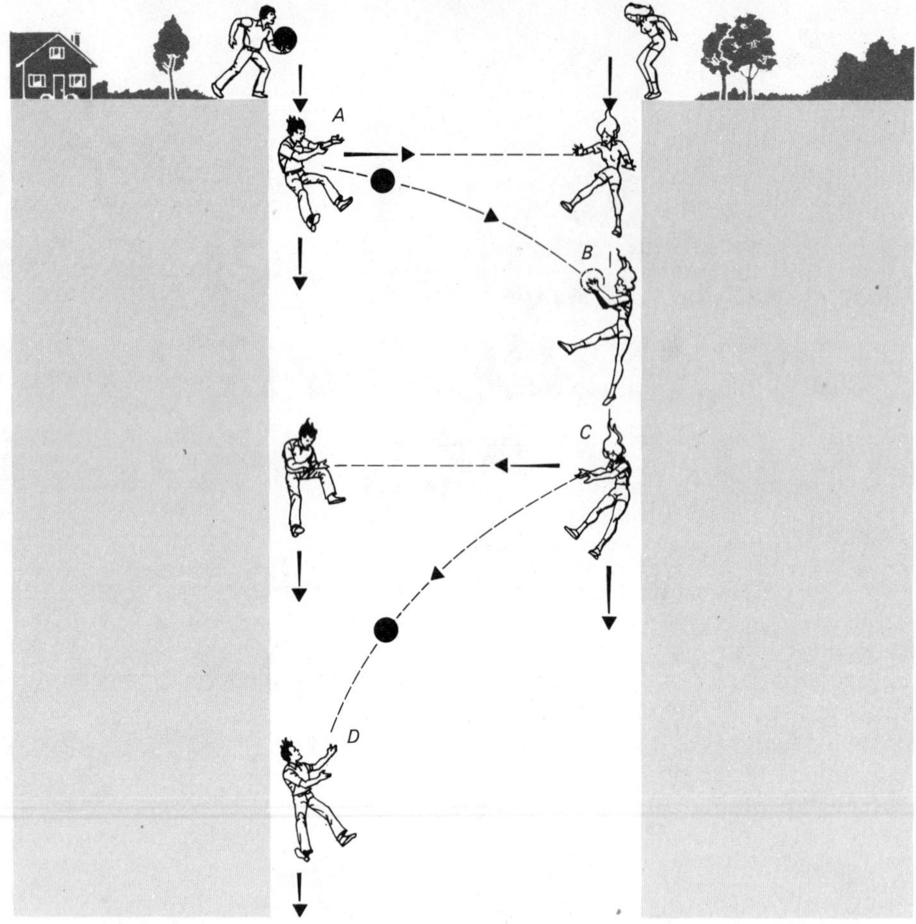

Figure 18.1
The brave couple playing catch as they descend into a bottomless abyss.

gravitation. Here is the principle of equivalence—a force of gravitation is equivalent to an acceleration of the coordinate system of the observer, and such a force can therefore be completely compensated for by an appropriate choice of an accelerated coordinate system. Einstein himself pointed out how a rapidly descending elevator seems to reduce our weight and a rapidly ascending one increases it. In a *freely falling* elevator, with no air friction, we would lose our weight altogether.

This idea is not hypothetical. In 1973 and 1974, astronauts in the Skylab lived for months in just such an environment. The Skylab was, of course, falling freely around the earth, as it continued to do until, unmanned, it finally suffered too much friction with the earth's tenuous upper atmosphere and plunged to a fiery doom in the summer of 1979. But while in free fall the astronauts lived in a seemingly magical world where there were no outside forces. One could give a wrench a shove, and it would move at constant speed across the orbiting laboratory. One could lay a pencil in midair and it would remain there, as if no force acted on it.

Mind, there *was* a force; neither the Skylab nor the astronauts were *really* weightless, for they continually fell around the earth, pulled by its gravity. But since all fell together—lab, astronauts, wrench, and pencil, *locally* all gravitation forces were absent.

This environment can condition one to questionable habits. The wife of one astronaut complained that a few days after her husband returned to earth he was applying shaving lotion to his face, and then laid the bottle "down" in midair—whereupon it promptly fell and shattered on the bathroom floor.

Figure 18.2
In Skylab everything stays put or moves uniformly because there is no apparent gravitation acting inside the laboratory. *(NASA)*

Still the Skylab provides an excellent example of the principle of equivalence—how local effects of gravitation can be removed by a suitable acceleration of the coordinate system. To the astronauts it was as if they were far off in space, remote from all gravitating objects. But what if astronauts *were* in remote space, and were to activate the engines of their ship, producing acceleration. The ship would then push up against their feet, giving the impression of a gravitational tug. If one were to drop a small coin and a hammer, the floor of the ship would obviously move up to meet both objects at the same time; to the astronauts, though, it would seem that the hammer and coin fell to the floor together. To them it would be exactly the same situation as that isolated fact made famous by Galileo—that heavy and light objects fall together. In other words, an acceleration of one's local environment produces exactly the same effect as a gravitational attraction, the two are indistinguishable—again, the principle of equivalence.

Trajectories of Light and Matter

Einstein postulated that the principle of equivalence is a fundamental fact of nature. If so, however, there must be *no* way in which an astronaut, at least by experiments within his local environment, can distinguish between his weightlessness in remote interstellar space and his free fall in a gravitational field about a planet like the earth.

But how about light? If the astronauts shone a beam of light along the length of their ship and if the ship were falling in a free-fall orbit about a planet,

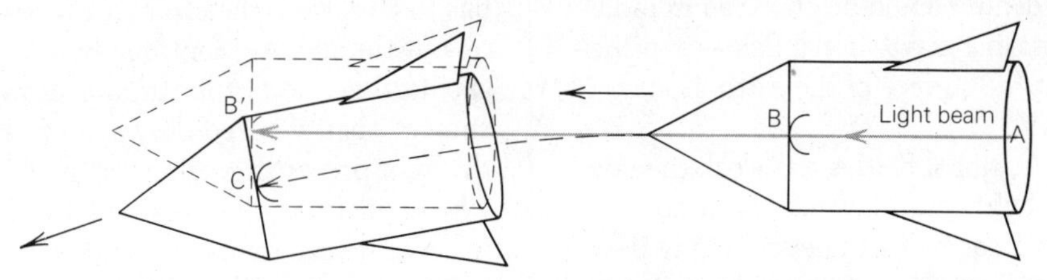

Figure 18.3
If in a spaceship moving to the left (in this figure) in its orbit about a planet, light is beamed from the rear, A, toward the front, B, we might expect the light to strike at B', above the target in the ship, which has fallen out of its straight path in its orbit about the planet. Instead, the light bent by gravity, follows the curved path and strikes at C.

would the ship not then surely fall away from a straight line path, which the beam must follow, causing the light to strike above its target?

Not so, according to Einstein. If the principle of equivalence is correct, there must be no way of knowing whether one is accelerated (any more than he can detect his own absolute motion) and hence the experiment must fail. Thus the light beam *must fall with the ship* if that ship is in orbit about a gravitating body. The idea that light, as well as material bodies, must be affected by gravity led Einstein to the prediction that stars seen by light from them that passes near the sun must appear displaced because of bending of their light by the gravitational field of the sun. This prediction, when formulated precisely, was, as we shall see, eventually confirmed by observation during a solar eclipse.

The Gravitational Redshift

Let us consider another possible experiment in a freely falling laboratory (the Einstein elevator). Suppose we shine a light beam—say, a laser beam of a precise frequency—upward from floor to ceiling. Now the laboratory accelerates downward, gaining speed, so by the time the light beam travels up to the ceiling, that ceiling is moving downward faster than the source on the floor was when the light left it. In other words, the receiver at the ceiling is *approaching* the source (where it was when the light left it). Therefore, wouldn't we expect to find the light at the ceiling blueshifted slightly because of the Doppler effect (Chapter 5)? But this would violate the principle of equivalence, for the blueshift would reveal our downward acceleration and show us we could not be weightless in free space. Therefore, Einstein postulated, there must be a *redshift,* due to the light moving upward against gravity, that exactly compensates the Doppler shift that would otherwise be observed. If so, that gravitational redshift should be observed in radiation climbing upward in a gravitational field—in principle at least—even at the surface of the earth. Is such an effect observed?

The earth's gravitational field is too weak to show the effect on visible light because we know of no source for which the frequency can be so sharply defined that the extremely tiny redshift would be noticeable. Yet, it has been observed, not in visible wavelengths of electromagnetic radiation, but in gamma radiation. In the 1960s, at the Jefferson Physical Laboratory at Harvard University, gamma rays were sent from a source in the basement to a detector at the top of the building. The source of gamma rays was radioactive cobalt; the radiation was confined to a very sharp frequency interval by a technique invented by the Nobel Prize winning physicist Rudolf Mössbauer. If a similar detecting layer of cobalt were placed directly above the emitting layer, the gamma rays would be absorbed by the former. However, the detector was placed at the top of the building, 20 m above the source. The gamma rays, traveling upward against the earth's gravitation, suffered a gravitational redshift, and were not absorbed by the upper cobalt detector. In order to absorb them, the detector had to be moved slowly downward to produce a blueshift to compensate the earth's gravitational redshift. The actual motion of the detecting cobalt needed to make it absorb the gamma rays from the emitting cobalt in the basement was so slow that it would have required a full year to close the 20-m gap between emitter and detector. That speed produced a Doppler shift that agreed with the value needed to compensate for Einstein's predicted redshift to within 1 percent.

According to the principle of equivalence, one should be able to *produce* a gravitational redshift by merely accelerating a spaceship far away from all gravitating bodies. Clearly, such a redshift would be produced: if radiation is beamed in the direction of the spaceship's acceleration, the receiver, where the radiation is absorbed, is moving away from the source, where that same radiation is emitted. Thus, since the source and receiver are separating, there is a redshift (due to the Doppler effect) that is indistinguishable from that produced by a gravitational field that produces the same acceleration. Once more the principle of equivalence is upheld.

Within a freely falling spaceship (like Skylab) or in the Jefferson Laboratory the gravitational field is essentially uniform. Such is not the case, however, for the light we observe leaving a star, because that light has to pass from the strong field near the star's surface on out through the continually weakening one as it gets farther and farther from it. However, Einstein showed that we need only add up the tiny effects as the light passes through each small region within which gravity can be regarded as effectively constant to calculate the total gravitational redshift of light leaving the star. It works out that the wavelengths of light from the sun should be increased by about 2 parts in a million—a redshift too small to be distinguished from other effects.

White dwarf stars, however, being very dense, have a much stronger surface gravity than the sun, and Einstein suggested that the gravitational redshift of the light from white dwarfs might be detectable. It can only be observed, however, for white dwarfs whose radial velocities are known from independent methods so that it can be separated from the Doppler shift due to the stars' motions. Fortunately, several white dwarfs are members of binary star systems, and their radial velocities can be deduced from those of their non-white dwarf companions, for which the gravitational redshift is negligible. The first reliable confirmation of the effect was made in 1954 by UCLA astronomer D. M. Popper, who measured the gravitational redshift of the white dwarf companion of the star 40 Eridani.

The precision of Popper's observations were such as to verify Einstein's predictions to within about 20 percent. Far higher accuracy has been attained recently in the near-earth environment with space-age technology. In the mid-1970s, a hydrogen maser carried by a rocket to an altitude of 10,000 km was used to detect the radiation from a similar maser on the ground. That radiation showed a gravitational redshift due to the earth's field that confirmed the relativity predictions to within a few parts in ten thousand.

Limitations to the Principle of Equivalence

We have seen that the force of gravitation can be compensated in a suitably accelerated coordinate system locally—that is, over dimensions small enough that within them the acceleration produced by gravitation can be regarded as constant. Thus within a freely falling spaceship in orbit about the earth, gravity appears absent, and all bodies behave according to the rules of special relativity; everything either remains at rest or moves uniformly in a straight line—a straight line as defined by the path of a light beam. If there are two spaceships in orbit, but one, say, 100 km above the other, the principle of equivalence applies to each. However, the motions of objects within one ship would *not* appear unaccelerated as seen from the other, for the force of the earth's gravitation varies with distance from the earth, and is appreciably different between the two ships.

Because the ships are in different gravitational fields, the gravitational redshifts within the two of them are different as well. Now all that we have said about frequency of light applies to the rate of all other physical processes as well; the rate of passing of light waves is just one of many ways to measure the passage of time. Time flows differently in different gravitational fields too. A clock in the spaceship of lower orbit (hence in a region of stronger gravity) runs more slowly than one in the other ship. Astronauts in the different ships would therefore disagree on the rate of time passage, as well as on the paths of unaccelerated bodies.

The above considerations suggest how an astronaut could tell that he was in orbit in a gravitational field—even without observing the planet beneath him. Suppose he fires a rocket probe straight ahead. For a while, the probe would continue in essentially the same orbit as the ship and would appear to hang motionless as viewed through a forward porthole. After a time, however, its motion away from the ship would carry the probe into a higher orbit where it would be accelerated downward less strongly, and it would slowly drift upward in the window, and eventually out of view. A differently accelerated reference frame now applies to the probe, just as it does to a separate orbiting spaceship.

Einstein's problem in formulating general relativity was to unify these separate descriptions of motion in different parts of a gravitational field into a connected whole, in order to find how to define a reference frame in which all objects, no matter where they are, are unaccelerated. To succeed, Einstein had to employ two ideas: spacetime and curvature.

SPACETIME

There is nothing mysterious about four-dimensional spacetime. Imagine yourself in the rear seats at an outdoor concert at the Hollywood Bowl. The sound from the orchestra in the shell, hundreds of feet away, takes a goodly fraction of a second to reach you, and the players seem to be behind the beat of the conductor. When a piece is finished, you first hear the applause from people near you, and slightly later from the front of the amphitheater. Because of the finite travel time of sound, all people do not hear the same note of music at the same time; nor do events that appear simultaneous *visually* seem to be audibly simultaneous.

Light also has a finite speed, so we never see an instantaneous snapshot of events around us (as we saw in Chapter 7). The speed of light is so great that within a single room we obtain *effectively* an instantaneous snapshot, but it is certainly not the case astronomically. We see the moon as it was just over a second ago, and the sun as it was about 8 minutes ago. At the same time, we see the stars by light that left them years ago, and the other galaxies as they were millions of years in the past. We do not observe the world about us at an instant in time, but rather we see different things about us as different *events* in spacetime.

As we also saw in Chapter 7, different observers in uniform relative motion do not even agree on the order of events. Two happenings that appear simultaneous to one are not simultaneous to the other. Space and time are inextricably connected. We need to describe the universe not just in terms of three-dimensional space, but in terms of four-dimensional spacetime.

We can easily represent the spatial positions of objects in two dimensions on a flat sheet of paper (for example, the plan of a city). To plot three dimensions on a page, the draftsman uses projections. Architectural drawings of a home generally show three projections: floor plan and two different elevations—say, the house as seen from the east and from the north—to give all necessary information. By the use of perspectives (which rely on our learned experience), we can also give an impression of a three-dimensional view. There is no easy way, though, to draw a four-dimensional perspective to include time.

There is no problem, however, in showing a two-dimensional projection of four-dimensional spacetime. Figure 18.4, for example, shows the progress of a motorist driving to the east across town. How much time has elapsed since he left home is shown on the horizontal axis, and how far he has traveled eastward is shown on the vertical axis. From *a* to *b* he drove at a uniform speed; from *b* to *c* he stopped for a traffic light and made no progress, and from *c* to *d* he drove more slowly because of increased traffic.

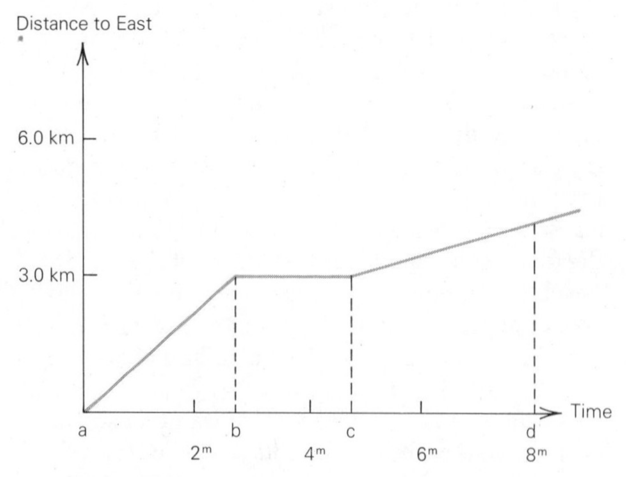

Figure 18.4
The progress of a motorist traveling east across town.

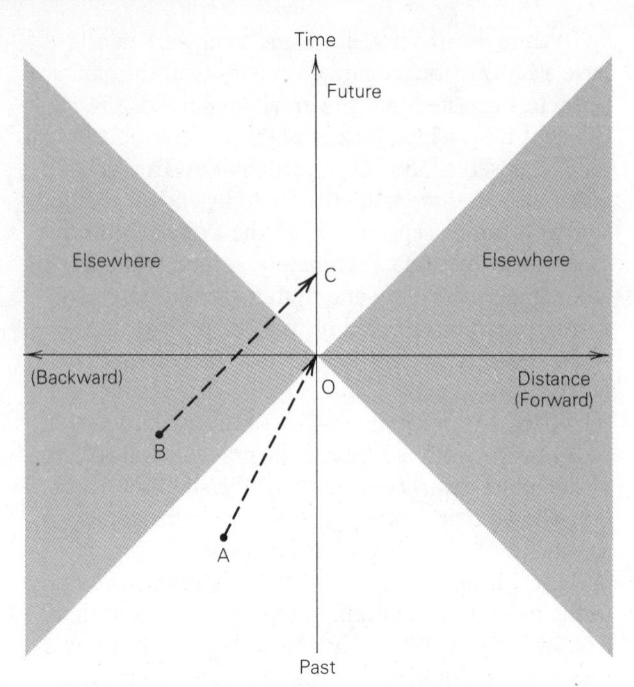

Figure 18.5
A spacetime diagram.

Figure 18.5 shows a rather conventional two-dimensional representation of spacetime. Time increases upward in the figure and one of the three spatial dimensions is shown horizontally. If we measure time in years and distance in light years, light goes one unit of distance in one unit of time, so flows along diagonal lines as shown. "Here and now" is at the origin of the diagram. At this instant we can receive information of a past event along such a line as AO; in this case the messenger was going slower than light, so he covered less distance than light would in the same time. Because nothing can go faster than light, we cannot, right now, know of something happening at point B in spacetime, for the message along BO would have to travel faster than light. We will have to wait until we are at C in the future, before a light or radio beam can get us the word along path BC.

We can also show three dimensions of spacetime

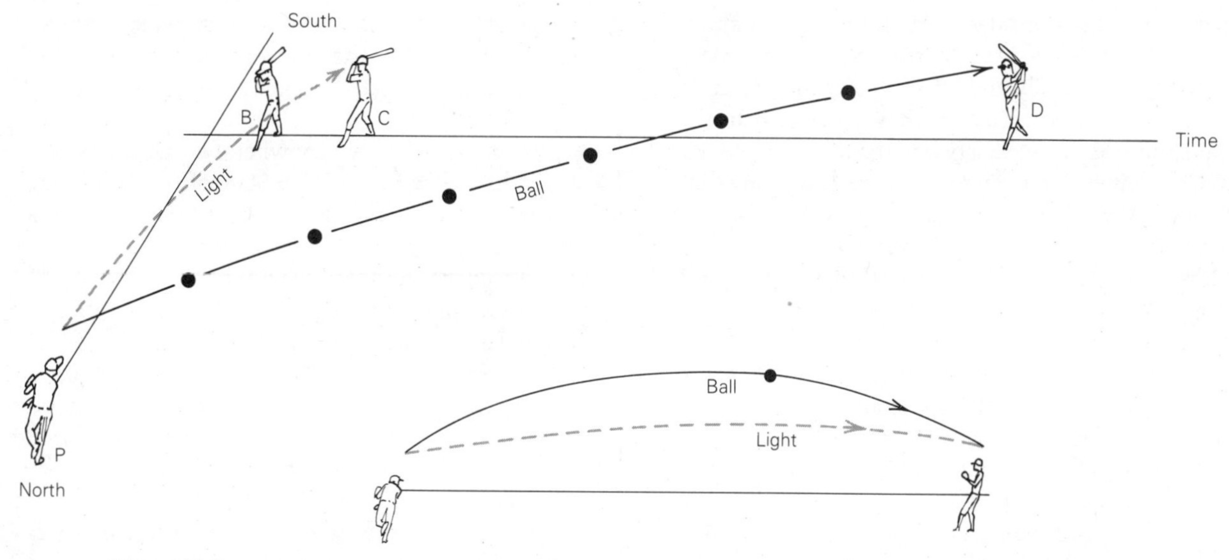

Figure 18.6
The paths of light and a baseball in spacetime. Below are shown the corresponding paths in space alone.

in a perspective drawing, as in Figure 18.6. Here time flows to the right, the height above the ground is upwards in the figure, and one of the two dimensions along the ground is shown obliquely as the north-south line. Suppose a batter is at point B (homeplate) and a baseball pitcher is at P, north of homeplate. The pitcher throws the ball to the batter, but the ball, traveling at a finite speed, flows along path PD in spacetime, arriving at the batter at D. Light, however, travels much faster than the ball, and arrives at the batter at C. In the gravitational field of the earth both the ball and light beam fall to the ground enroute to the plate at the same rate, but the light reaches the plate so quickly that it falls only a slight distance compared to the ball, which takes much longer. Thus in space (lower part of Figure 18.6) the path of the light is much more nearly straight than that of the ball. In spacetime, however, the path of the ball is very long compared to that of the light beam, and it is easy to see how in the uniform gravitational field (as on the baseball diamond) both ball and light fall along paths of the same shape (that is, the same *curvature*) in spacetime.

Now if that ball game were being played on a freely falling platform, that reference frame would be accelerating downward at the same rate the ball and light accelerate, and both would move, in that freely falling system, in straight lines, obeying the laws of special relativity. Those straight line paths of light and the ball would be the kind of straight lines Euclid was talking about in his development of plane geometry.

Actually, on the earth, where things accelerate downward, light falls like everything else. If we used light paths to define straight lines (say, by sighting along a ruler to assure that it is "straight"), technically the laws of Euclidean geometry would not work; if we measured the three sides of a triangle by such straight lines, we would not find its area to be one-half its base times its altitude. In the freely falling system in a uniform gravitational field, however, these difficulties are removed, for there is no force accelerating light in *that* system. In practice, of course, there is no problem on the earth, either, because light travels so fast and falls so slightly no one could measure the difference. But we *could* tell the difference on the surface of a *neutron star*—or even on a white dwarf!

CURVATURE OF SPACETIME

Because of the principle of equivalence, Euclidean geometry applies in a freely falling reference frame in a uniform gravitational field. But freely falling systems at different places on earth fall in different directions, because "down" is always toward the center of the earth. Thus we cannot describe the behavior of objects in different widely separated freely falling systems with spacetime coordinates for which Euclidean geometry holds. Let's consider a familiar analogy. A simple Mercator-type map, with lines of constant latitude running horizontally and lines of constant longitude running vertically, is fine for showing a small area of the earth—say a single city—without noticeable distortion. But such a map cannot show a large area of the curved earth without distortion; everyone knows how distorted and enlarged lands of extreme latitude (those near the poles) appear on the usual flat world maps. We cannot map the earth with plane Euclidean geometry.

Indeed, if we travel far enough in a straight line on the surface of the earth, we end back at our starting point; our path is a *great circle* (the equator is such a great circle). More generally, if we take into account the slight polar flattening of the earth, as well as effects of such irregularities as mountains, our "straight line" path is called a *geodesic*, which means "earth divider."

Einstein showed how to find spacetime coordinates within which all objects move as they would if there were no forces. In a small local region, where a gravitational field is uniform, those coordinates are Euclidean, just as a city plan can be well described with plane geometry. But to describe paths of objects over a large region, where the gravitational field varies, spacetime must be curved, just as we must use curved geometry to describe a large area of the spherical earth.

It is the distribution of matter that determines the nature of a gravitational field, so it is the distribution of matter that determines the geometry of curved spacetime. Some writers on general relativity describe spacetime as being "warped" by matter. Within this curved spacetime, everything moves in the simplest possible way as if no gravitation were there at all. In analogy with earth geometry, the paths of light and material objects in spacetime are called *geodesics*.

People usually say that the geometry of spacetime is determined by matter, but I think this description makes the subject sound unduly mysterious. What meaning can there be to "curved space" if space is the absence of matter? It is not space in itself that is curved; rather it is the system of coordinates that we can conveniently use to describe the motions of objects and light. By selecting non-Euclidean geometry—curved coordinates—we can describe the paths of light and objects as "straight" in the same sense that great circles (or geodesics) are "straight" on the curved surface of the earth.

The mathematics needed to handle the problem, on the other hand, was not available in Euclid's time. The geometry needed to describe curved spacetime was developed after the pioneering work of the great German mathematician, physicist, and astronomer, Karl Friedrich Gauss. Gauss became involved in the invention of new geometry when he was commissioned to survey the German State of Hanover by its king, George IV—also king of Great Britain. The new geometry received its full expression in the hands of Gauss' student, Bernhard Riemann.

Riemann, in applying for a university position at Göttingen, had submitted three possible topics for a lecture he would deliver. Traditionally, the judges selected one of the first two topics offered, however, so Riemann had not bothered to prepare a lecture on the third. That, however, was the very topic that Gauss had been pondering for decades. Consequently, in only a few weeks Riemann wrote out the lecture on that third topic, which was to be his masterpiece: "On the Hypotheses Which Lie at the Foundations of Geometry."

By the end of the nineteenth century the new Riemannian geometry was further facilitated with the invention of tensor calculus. By 1915, Einstein was able to use these new mathematical techniques to derive the *field equations* of general relativity, which describe the curvature of spacetime by matter, and the *geodesic* equations, which describe the unaccelerated paths of objects in spacetime.

TESTS OF GENERAL RELATIVITY

Is general relativity, then, essentially different from Newtonian gravitational theory, or merely a different but equivalent mathematical formulation? Relativity *is* different from Newtonian theory in that the signals that govern gravitational interactions are not instantaneous, but travel with the speed of light, and also, of course, in that matter and energy are equivalent, so that not only matter itself, but also energy contributes to gravitation—that is, to the geometry of spacetime—and energy (light, for example) as well as mass is affected by that geometry. Naturally, where speeds are low compared to that of light, and where the gravitational field is relatively weak—both conditions of which are met throughout most of the solar system—the predictions of general relativity must agree with those of Newton's theory, which has served us so admirably in our technology and in guiding space probes

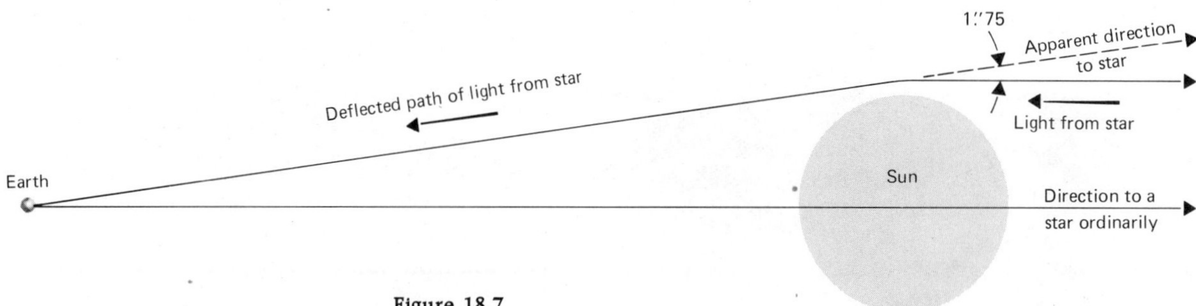

Figure 18.7
Deflection of starlight passing near the sun.

to the other planets. In familiar territory, therefore, the differences between predictions of the two theories are subtle, and consequently very difficult to detect.

Einstein himself proposed three observational tests of general relativity. One is the gravitational redshift, already described, the second is the deflection of starlight that passes close to the sun, and the third is a subtle effect on the motion of the planet Mercury.

Deflection of Starlight

The strength of the gravitational acceleration at the surface of the sun is 28 times its value at the surface of the earth. Einstein calculated from general relativity theory that starlight just grazing the sun's surface should be deflected by an angle of 1."75. Stars cannot be seen or photographed near the sun in bright daylight, but with difficulty they *can* be photographed close to the sun at times of total solar eclipses. Einstein suggested an eclipse observation to test the light deflection in a paper he published during World War I. A single copy of that paper, passed through neutral Holland, reached the British astronomer Arthur S. Eddington. The next suitable eclipse was on May 29, 1919. The British organized two expeditions to observe it, one in West Africa and the other in Brazil. Despite some problems with the weather, those expeditions did obtain successful photographs of stars near the sun. Measures of their positions were then compared to measurements on photographs of the same stars taken at other times of the year when the sun was elsewhere in the sky. The stars seen near the sun were indeed displaced and, to the accuracy of the measurements, the shifts were consistent with the predictions of relativity. It was a triumph that made Einstein a world celebrity overnight.

Eclipse observations to test the relativity effect have continued over the years, but the measures are very difficult to make and the precision of the confirmation is not high. Far higher accuracy has been obtained recently at radio wavelengths. Simultaneous observations of the same source with two radio telescopes far apart can pinpoint the direction of the source very precisely. The United States National Radio Astronomy Observatory at Greenbank, West Virginia, with radio telescopes 35 km apart, observed several remote astronomical radio sources (quasars—see Chapter 20) when the sun was nearly in front of them. The apparent directions of the quasars showed shifts similar to those of stars seen near the sun. The accuracy of these observations is high enough to confirm the Einstein prediction to within 1 percent.

In addition to sources of light or radio waves appearing displaced slightly when seen near the sun, the radiation from them is also delayed slightly in reaching the earth. We have no way of measuring that delay in light from stars or quasars, but we can detect it in the radio pulses broadcast from space probes, because we know where they are and when the signals should arrive at earth. The experiment has been performed with several planetary probes, but most precisely with the Viking landers on Mars (Chapter 10). When Mars is on the far side of the sun, signals from the Vikings must pass through a region of spacetime that is relatively strongly curved by the sun (Figure 18.8) and are observed to be delayed by about 100 microseconds, as if Mars had jumped some 30 km out of its orbit. It is just the delay expected by relativity theory to within 1 part in a thousand.

Advance of Perihelion of Mercury

According to relativity, the energy of motion of a moving body adds to its effective mass, and hence to the force of gravitation on it. Now, Mercury has a fairly eccentric orbit, so that it is only about two-thirds

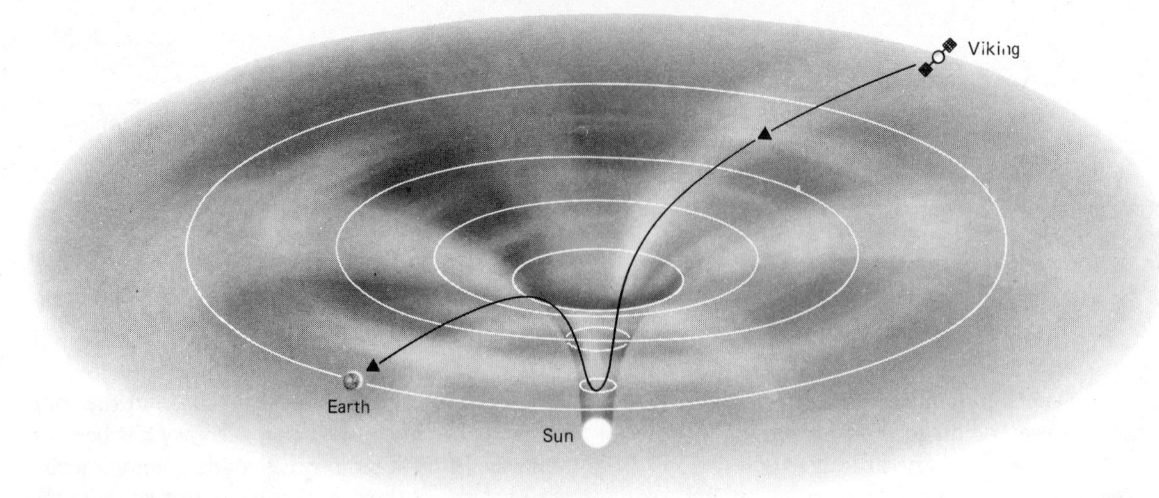

Figure 18.8
Radio signals from Viking are delayed because they have to pass near the sun, where spacetime is curved relatively strongly.

as far from the sun at perihelion as it is at aphelion. As required by Kepler's second law, Mercury moves fastest when nearest the sun, which adds to the gravitation between the two. Consequently, relativity predicts that a very tiny additional push on Mercury, over and above that predicted by Newtonian theory, should occur at each perihelion. The result of this effect is to make the *line of apsides*, which is the long dimension (major axis) of Mercury's orbit, slowly rotate in space, so that each successive perihelion occurs in a slightly

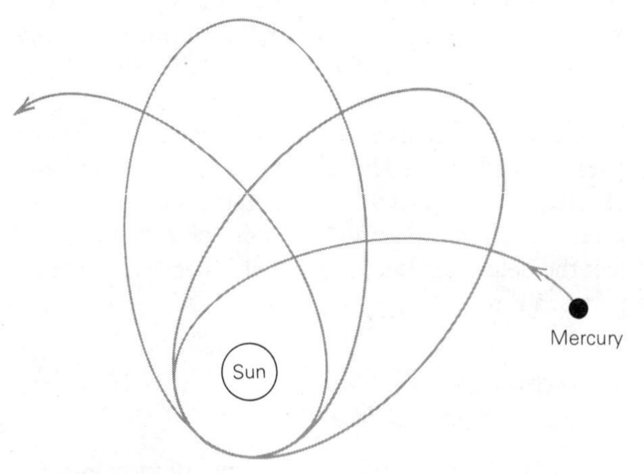

Figure 18.9
Rotation of the line of apsides of the orbit of a planet, such as Mercury, because of various perturbations.

different direction as seen from the sun (Figure 18.9). The prediction of relativity is that the direction of perihelion should change by only 43″ per century; it would thus take about 30,000 years for the line of apsides to make a complete rotation.

The gravitational effects (perturbations) of the other planets on Mercury also produce an advance of its perihelion, and to a far greater extent than the relativistic prediction. According to Newtonian theory, the perihelion of Mercury should advance by 531″ per century. Even in the last century, however, it was observed that the actual advance is 574″ per century. At one time an unknown intramercurial planet was suspected of causing the discrepancy, and the hypothetical planet was even named for the god Vulcan. Vulcan, of course, never materialized, but that 43″ anomaly was entirely explained by relativity. The relativistic advance of perihelion can also be observed in the orbits of several minor planets that come close to the sun.

Additional tests for relativity theory are in the planning or experimental stages—some at the frontier of modern technology. In a satellite experiment expected to fly in the early 1980s, the behavior of a gyroscope will be carefully monitored. Relativity predicts an angular change in orientation of the axis of the gyroscope due to its motion through the earth's gravitational field of only 0″.05 per year, but it is believed that even that small change can be measured accurately enough to check the theory.

GRAVITATIONAL WAVES

Because the geometry of spacetime depends on the distribution of matter, any rearrangement of that matter must result in an alteration of spacetime—that is, it creates a disturbance. Relativity predicts that that disturbance should propagate through space with the speed of light. It is called a *gravitational wave*.

We think we should be able to detect gravitational waves, as we do other kinds of radiation. Whereas electromagnetic radiation can be detected, for example, in the way it sets charges oscillating back and forth, gravitational waves should set material objects vibrating. Detection of gravitational waves, however, is extremely difficult. One reason is that gravitation is an exceedingly weak force (Chapter 1), and its radiation is correspondingly weak. But moreover, the disturbance gravitational waves should produce in an object is far more subtle than simple oscillation; rather than moving back and forth as a unit, the body suffers very slight compressions and lengthenings.

Gravitational waves produced by mass motions on the earth, or even of the earth itself, would be far too weak to detect by technology we can currently imagine. We would need a motion of a very large mass, and at a speed approaching that of light. One possible source could be the sudden collapse of a massive star—perhaps in a supernova explosion; that is, a stellar catastrophe.

The pioneering attempt to detect such gravitational waves from collapsing stars was by Joseph Weber, physicist at the University of Maryland. In the 1960s he suspended large metal cylinders equipped with very delicate sensors. Weber had hoped to detect gravitational waves by the vibrations they set up in the cylinders. To distinguish true gravitational waves from space from purely local disturbances, he placed one cylinder in Maryland and a second one at the Argonne National Laboratory outside Chicago; only signals recorded simultaneously at both stations would be of astronomical origin.

To date, Weber has not detected signals that can unequivocally be attributed to gravitational waves from space. Other laboratories, however, are developing far more sensitive detectors. At Stanford University, for example, an extremely delicate microphone is arranged to pick up the vibrations that a passing gravitational wave sets up in a metal bar. The bar is a practical detector because the two ends would be set vibrating differently, which sets up an oscillation in the bar itself. The gravitational waves passing the earth produced by the collapse of a solar mass star at the center of the Galaxy would displace the ends of the bar by only one ten-millionth the diameter of an atom, but with advanced technology, utilizing principles of superconductivity, the Stanford experiment has a good chance of detecting such gravitational waves if, indeed, they exist.

The Binary Pulsar PSR1913+16

It may be that nature has already provided us with indirect evidence of gravitational radiation. In 1974, R. A. Hulse and J. H. Taylor, of the University of Massachusetts, observing with the 1000-ft radio telescope at Arecibo, Puerto Rico, discovered a remarkable pulsar, now designated PSR1913+16 (the numbers give the coordinates of its location in the sky). The unique thing about PSR1913+16 is that the period of its pulses itself shows cyclic variations over a short time interval of 7^h45^m. These period changes are due to the Doppler effect caused by the pulsar's revolution about another object. When the pulsar is approaching us in its orbit, each successive pulse has less far to travel to reach us on earth, and we receive the pulses slightly closer together than average. Conversely, when the pulsar is moving away from us, we receive the pulses slightly spread out in time. Thus we can analyze the orbital motion of the pulsar just as we do that of a spectroscopic binary star (Chapter 14).

Such an analysis, combined with our knowledge of the expected properties of neutron stars, indicates that the pulsar is in mutual revolution in that 8-hr period about a mute companion of comparable mass that is probably either a white dwarf or another neutron star. Because the size of the orbit is only a little bigger than the diameter of the sun, it is exceedingly unlikely that the companion is a normal star, for then the pulses would be eclipsed by it during part of each revolution (contrary to observation) unless the orbit were almost face on, in which case we should not see appreciable variations in the pulse period.

Now there are two important points about this binary pulsar. First, at its orbital speed of approximately 0.1 percent that of light, it should radiate gravitational waves at a rate great enough to carry appreciable energy away from the system, causing the pulsar and its companion to spiral slowly closer together. Second, pulsars are superb clocks, remaining stable to 0.001 s over several years.

As the pulsar and its companion spiral together, their period of revolution shortens. The shortening is only about one ten-millionth of a second per orbit, but the effect accumulates like a clock that runs a little faster each day. By this writing (late 1979) the system has been under observation for nearly 5 years, and the time or periastron (when the two objects are closest together) has shifted by about a full second of time relative to the time it would now be occurring if the period had remained constant. Pulsars are such good timekeepers that this shift is easily observed.

In fact, the shift in periastron time is just what general relativity predicts it should be due to the emission of gravitational radiation by two stars, each of about 1.5 solar masses, with an orbit like that of the binary pulsar. PSR1913+16 is now regarded as providing strong evidence that gravitational waves do exist—as Einstein predicted.

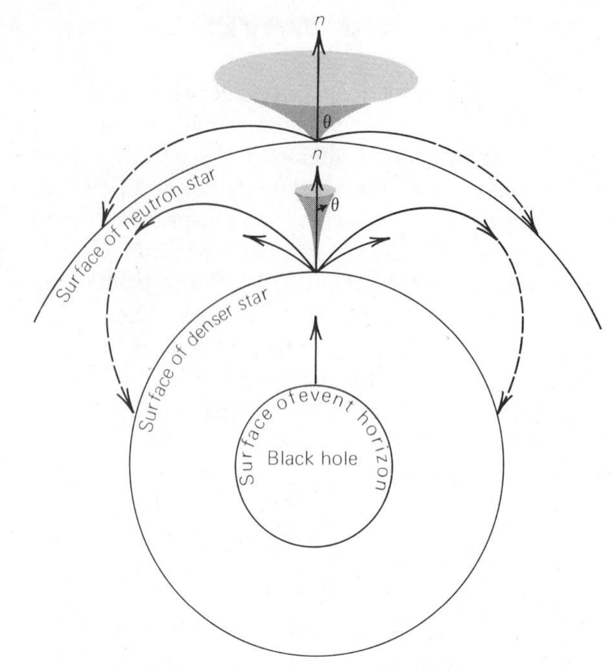

Figure 18.10
Light that leaves the surface of a dense star must flow into a cone of half angle Θ with respect to the normal *n* to the surface. As the star shrinks, the light cone into which radiation must flow to escape is smaller. It shrinks to zero when the star passes through the event horizon.

BLACK HOLES

Until now, the manifestations of general relativity theory we have discussed have been so subtly different from Newtonian theory that one might well wonder if the new ideas about space and time brought about by Einstein are really important enough to concern us. Indeed, many physicists, during the first half century after Einstein introduced his revolutionary theory, regarded the subject as almost academic.

But hold! In the last chapter we encountered pulsars, which provide strong evidence for the existence of neutron stars. Neutron stars have very great density and extremely strong surface gravity. We recall that light grazing the surface of the sun is deflected by about 1″.75. Light grazing the surface of a white dwarf would be deflected by about 1′, and that grazing a typical neutron star by about 30°.

In 1796, the French mathematician Pierre Simon Marquis de Laplace speculated about the properties of an object that had so great a gravitational field that light could not escape at all, but would be bent right around and stay with the object. Laplace's "corps obscurs" were later reconsidered by modern physicists, armed with the new rigor of general relativity theory. John Wheeler, the Princeton physicist who has become intimately associated with general relativity, has dubbed such objects "black holes."

Consider the light radiated from the surface of a neutron star. That which emerges normal (perpendicular) to the surface flows out radially from the star. That emitted at an angle of, say, 30° to the normal (hence, 60° to the surface) leaves the star at an angle somewhat greater than 30° to the normal, because of the gravitational deflection. That light which is emitted 60° or more from the normal, however, is deflected so much that it never escapes the star. Thus, from each point on the surface of the neutron star, that portion of the light which can pass on out into space is emitted into a cone about the normal to the surface of half angle about 60° (Figure 18.10).

Now imagine a more massive star that shrinks to a smaller size and higher density than a neutron star. As the surface gravity increases, the angular cone into which light can be emitted from any point on the star and still escape into space becomes smaller. Eventually, as the star becomes small enough and its gravity strong enough, a point is reached where the cone of light escape shrinks to zero; this happens when the velocity of escape from the star's surface is equal to the velocity of light. As the star contracts still more, light and everything else is trapped inside, unable to escape through that surface where the escape velocity is the speed of light. That surface is called the *event horizon*, and its radius is the *Schwarzschild radius*, named for Karl Schwarzschild, who first described the situation a few years after Einstein introduced general relativity. It is this surface that is the boundary of the black hole. All that is inside is hidden forever from us; as the star shrinks through the black hole it literally disappears from the universe.

The size of the Schwarzschild radius is proportional to the mass of the star. For a star of one solar mass, the black hole is about 3 km in radius; thus the entire black hole, some 6 km in diameter, is about one-third the size of a one-solar-mass neutron star.

The event horizons of larger and smaller black holes—if they exist—have greater and lesser radii, respectively. For example, if a globular cluster of 100,000 stars could collapse to a black hole, it would be 300,000 km in radius, a little less than half the radius of the sun. If the entire Galaxy could collapse to a black hole, it would be only about 3×10^{11} km in radius—about 0.01 pc. On the other hand, for the earth to become a black hole it would have to be compressed to a radius of only 1 cm—about the size of a golf ball. A typical minor planet, if crushed to a small enough size to be a black hole, would have the dimensions of an atomic nucleus!

But should black holes exist? Stars less than about 1.4 solar masses can become white dwarfs. Those of larger mass, we think, can exist as neutron stars, but there is an upper limit to the mass of neutron stars; we think that limit is not over three solar masses. We know that a tiny fraction of all stars have still greater mass. What becomes of them when they exhaust their store of nuclear fuel? Perhaps they eject part of their mass (as a planetary nebula or in a supernova outburst) so that what is left can contract to a white dwarf or a neutron star. But what if they do not? Then we know of no other fate for such massive stars than that they become black holes. Thus we are not certain that any star *must* ever have to become a black hole, but we have good reason to expect that many massive stars, albeit a minority of *all* stars, can end up in that exotic state.

How, then, do we find a black hole, which, of course, we cannot see? We can detect it by its gravitational effects on other stars (as stars collapse into black holes they leave behind their gravitational fields), and this is most easily accomplished in a binary star system.

Candidates for Black Holes

To find a black hole we must: (1) Find a star whose motion (found from the Doppler shift of its spectral lines) shows it to be a member of a binary star system, and to have a companion of mass too high to be a white dwarf of a neutron star. (2) That companion star must not be visible, for a black hole, of course, gives off no light. But being invisible is not enough, for a relatively faint star might be unseen next to the light of a brilliant companion. Therefore, (3) we must have evidence that the unseen star, of mass too high to be a neutron star, is also a collapsed object—one of extremely small size—for then our theory predicts that it must be a black hole—or at least a star on the way to becoming one.

Modern space astronomy has come to the rescue in (3). One way to know we have a small object of high gravity (and possibly a black hole) is if matter falling toward or into it is accelerated to high speed. Near the event horizon of a black hole, matter is moving at near the speed of light. Internal friction can heat it to very high temperatures—up to 100 million degrees or more. Such hot matter emits radiation in the form of X-rays. Modern orbiting X-ray telescopes—especially the Einstein telescope, HEAO 2—can and do reveal such intense sources of X-radiation.

So we want X-ray sources associated with binary stars with invisible companions of high mass. We cannot prove that such a system contains a black hole, but at present we have no other theory for what the invisible massive companion can be if, indeed, the X-rays are coming from gas heated by falling toward it.

We can easily understand the origin of such infalling gas. We have already seen (Chapter 14) how stars in close binary systems can exchange mass. Suppose one star in such a double star system has evolved to a black hole and that the second star has now evolved to a red giant so large that its outer layers pass through that point of no return between the stars and falls to the black hole. The mutual revolution of the giant star and black hole cause the material from the former to flow not directly onto the black hole, but, because of conservation of angular momentum, to spiral around it, collecting in a flat disk of matter called the *accretion disk*. In the inner part of the accretion disk the matter is revolving about the black hole so fast that its internal friction heats it up to the temperature where it emits X-rays. In the course of this friction, some material in the accretion disk is given extra momentum, and escapes from the double star system, and other material loses momentum and falls into the black hole—lost forever to observation from the rest of the universe.

Another way to form an accretion disk in a binary star system is from material ejected from the companion of the black hole as a stellar wind; some of that ejected gas will flow close enough to the black hole to be captured by it into the disk. Such a case, we think, is the binary system containing the first X-ray source discovered in Cygnus—Cygnus X-1. The visible star (Figure 18.11) is a normal B-type star. The spectrographic observations, however, show it to have an unseen companion of mass near 10 times that of the sun. That companion would be a black hole if it were a small, collapsed object. The X-rays from it strongly suggest that it is, for we have no other explanation for the source of those X-rays than gas heated by an infall toward a tiny massive object. Of course we cannot be certain that Cygnus X-1 is a black hole, but many astronomers think that it probably is.

Properties of Black Holes

Much of the modern folklore about black holes is misleading. One idea is that black holes are monsters that go about sucking things up with their gravity. Actually, the gravitational attraction surrounding a black hole at a large distance is the same as that around any other star (or object) of the same mass. Even if another star, or a spaceship, were to pass one or two solar radii from a black hole, Newton's laws would give an excellent account of what would happen to it. It is only very near the surface of a black hole that its gravitation is so strong that Newton's laws break down; for a black hole of the mass of the sun, light would have to come within 3 km of its surface to be trapped. A solar mass black hole, remember, is only 3 km in radius—a very tiny target. Even collisions between ordinary stars, hundreds of thousands of times bigger in diameter, are so rare as to be essentially nonexistent. A star would be far, far safer to us as an

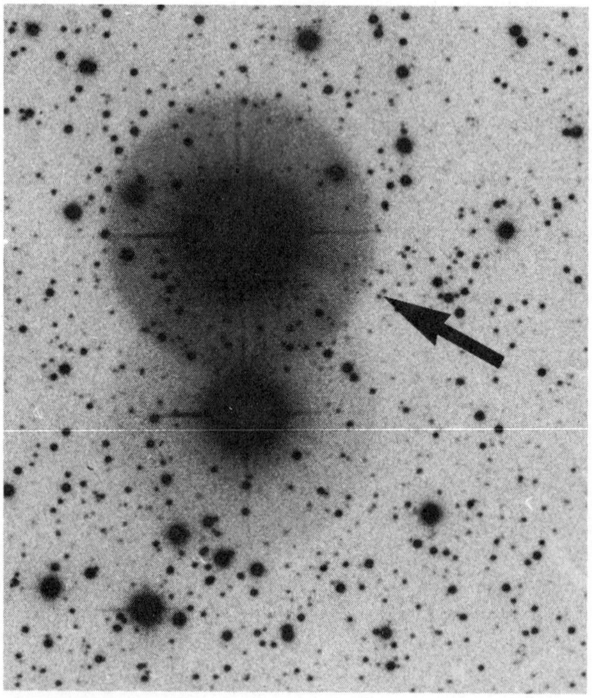

Figure 18.11
In visible light, Cygnus X-1 appears as an ordinary star. *(Hale Observatories)*

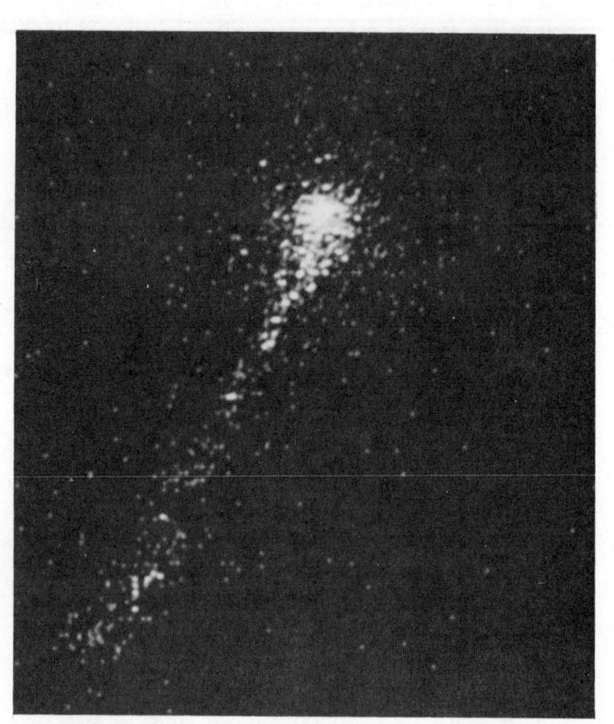

Figure 18.12
Cygnus X-1 in the light of X-rays, as observed by the Einstein X-ray telescope. *(Courtesy of the Harvard–Smithsonian Center for Astrophysics)*

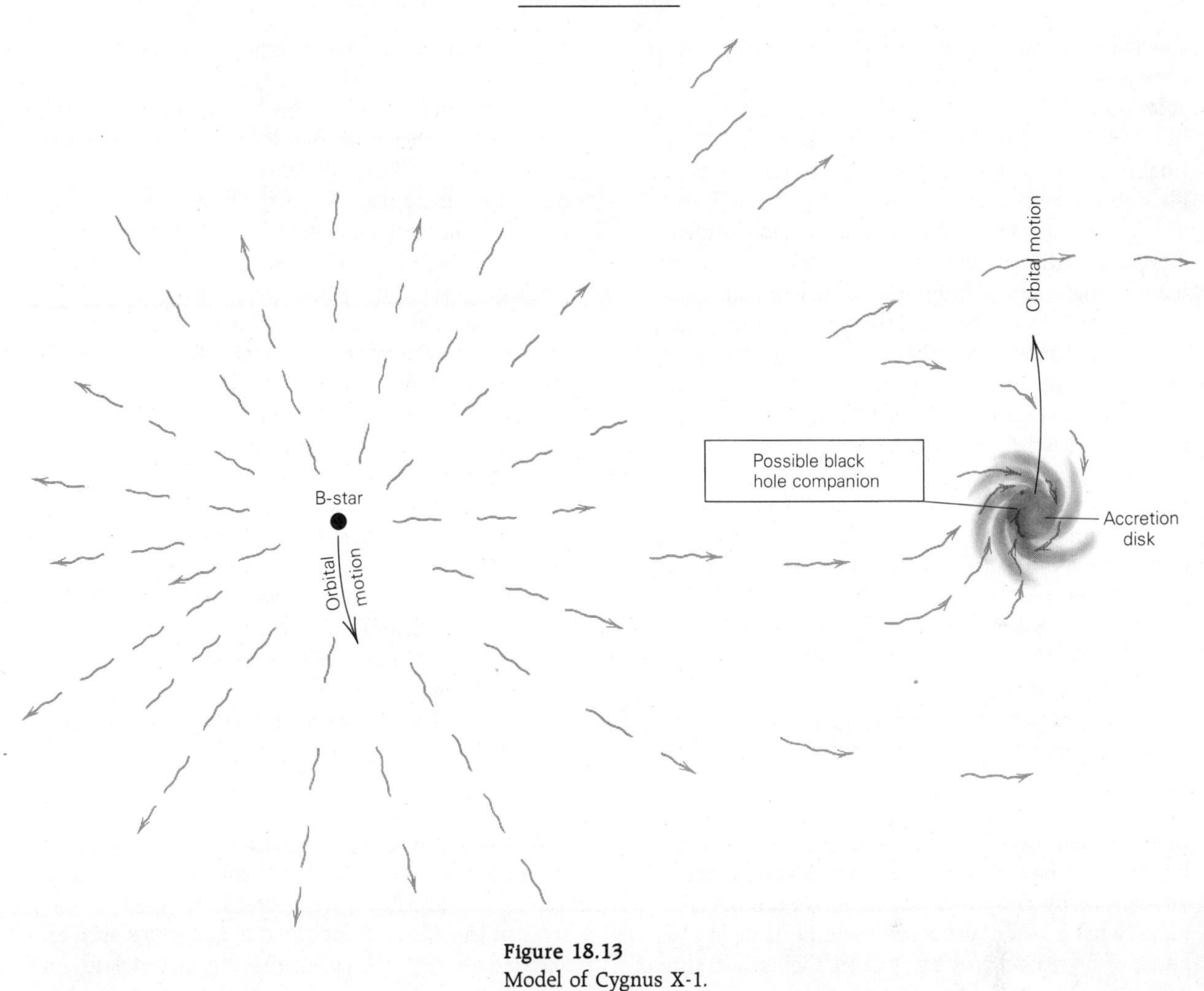

Figure 18.13
Model of Cygnus X-1.

interloping black hole than it would have been in its former stellar dimensions.

A Trip Into a Black Hole

Still, it is interesting to contemplate a trip into a black hole. Suppose that the invisible companion of the star associated with Cygnus X-1 is a black hole of 10 solar masses. What would you see if I bravely fly into it in a spaceship?

At first I dart away from you as though I were approaching any massive star. However, when I near the event horizon of Cygnus X-1—some 30 km in radius, and presumably near the center of the accretion disk—things change. The strong gravitational field around the black hole makes my clocks run more slowly as seen by you. Signals from me reach you at greatly increased wavelengths because of the gravitational redshift. As I approach the event horizon, my time slows to a stop—as seen by you—and my signals are redshifted through radio waves to infinite wavelength; I fade from view as I seem to you to come to a stop, frozen at the event horizon. All matter infalling to a black hole appears to an outside observer to stop and fade at the event horizon, frozen in place, and taking an infinite time to fall through it—including the matter of a star itself that is collapsing into a black hole. For this reason, black holes are sometimes called *frozen stars*.

This, however, is only as you, well outside the black hole, see things. To me, time goes at its normal rate and I crash right on through the event horizon, noticing nothing special as I do so, except for enormously strong tidal forces that rip me apart. At least this is true of 10-solar-mass black holes. If there were a very massive black hole, say thousands of millions of solar masses, its event horizon, although small in astronomical dimensions, would be large enough that its tidal forces are not severe, and an astronaut, in principle, could survive a trip into it.

But in no case is there an escape. Once inside, astronaut, light, and everything else is doomed to remain hidden forever from the universe outside. Moreover, current theory predicts that after entering the black hole, I race irreversibly to the center, to a point of zero volume and infinite density.

Mathematicians call such a point a *singularity*, but many physicists doubt that such singularities of infinite density can exist in the real universe, and suspect that new physics will eventually emerge to help us understand what really happens to matter at extraordinary densities. In particular, mathematical solutions suggesting that black holes can collapse through *wormholes* to emerge as "white holes" in "other universes"—ideas often exploited in the popular literature—now appear to be abstractions without physical basis; recent theoretical studies, in fact, show that even general relativity theory does not predict that black holes emerge elsewhere as white holes.

On the other hand, massive stars collapsing through their own event horizons to become black holes do not have to reach densities beyond those understood by present-day physics. We can never know what goes on inside the event horizon, but we know of no reason why black holes themselves should not exist.

Other Ideas About Black Holes

Black holes need not be limited to stellar masses. There has been considerable consideration of the possibilities of very large amounts of gas collecting together and collapsing into black holes in the centers of globular clusters, galaxies, or even clusters of galaxies.

A mass of gas collapsing into a black hole releases more than 100 times as much energy as can be extracted from that same mass through nuclear fusion. Thus the gravitational collapse of a million solar masses of gas into a black hole at the center of a galaxy could produce a truly prodigious amount of released gravitational potential energy. There is a great deal of speculation, nowadays, about such processes accounting for the energy of quasars and other phenomenal objects. It may well be that through massive black holes general relativity theory will be found to have profound consequences in modern astrophysics.

On the other hand, black holes may exist on the microscopic level as well. We know of no way to produce small-mass black holes today (that is, less than a few solar masses), but the brilliant British theoretical astrophysicist Stephen Hawking suggests that such black holes *could* have been produced in the big bang at the origin of the universe (Chapter 21). If so, they would involve quantum mechanics as well as relativity, and in a most amazing way.

We have seen (Chapter 16) that all fundamental particles have their antiparticles; for example, electrons and positrons, protons and antiprotons, and so on. Whenever a particle and its antiparticle come into contact, they annihilate each other, transforming completely into energy. Similarly, pure energy can be converted into pairs of particles—an electron and a positron, for example. This process is known as *pair production*, and is observed regularly in the nuclear physics laboratory.

Now all this is possible because mass and energy are equivalent. But obviously mass cannot be created from nothing—we need energy to do it. Yet, according to quantum theory, it is possible for matter (or energy) to be created from nothing for an exceedingly brief period of time. This possibility comes about because of the innate uncertainty in nature, at the microscopic level, of the measures of physical quantities such as mass and energy. The principle does not violate conservation laws because any matter that comes into being almost immediately disappears again spontaneously, so on the average mass and energy (combined) are conserved.

But, Hawking points out, what if a positron and electron (say) come into existence momentarily in the vicinity of a black hole. There is a chance that one or the other will fall into the hole and hence not be able to annihilate with its antiparticle, returning the energy it "borrowed" from nature. Its antiparticle, therefore, can escape unscathed. However, many such positrons and electrons so created near black holes and escaping from them do annihilate each other, creating energy. That energy cannot come from nothing; according to Hawking's theory it must come from the black hole itself. Robbing the black hole of energy in this way robs it of mass (for they are equivalent), so the black hole must slowly evaporate through this process of pair production.

As esoteric as this idea may seem, it is generally thought to be possible by theoretical physicists. The process is only important, however, near very tiny black holes. Solar-mass black holes would evaporate in this way at an absolutely negligible rate. In fact, the only black holes that would have had time to so evaporate in the age of the universe would be those of original mass less than about 10^{15} g (like a minor planet). Smaller ones would already be gone; those of about 10^{15} g should be finishing off about now, if they were formed in the big bang in the first place. Because the evaporation rate increases as the mass of the black hole goes down, at the end one would go off explosively, emitting a final burst of gamma radiation.

Nobody knows whether such mini black holes were formed in the early universe, nor, if so, whether the evaporation process Hawking envisions would really occur. If so, we would expect to see bursts of gamma rays from exploding mini black holes from time to time. So far we have not, but the speculation remains an interesting possibility.

We have touched on many ideas in this chapter. Some, near the end, are highly speculative, but the main thrust of general relativity theory now appears to be rather firmly established, even though tests of some of its predictions are still going on. It is probably safe to say that general relativity has come of age and is an important part of modern astronomy.

And general relativity is almost singlehandedly the inspiration of one genius—Albert Einstein. For half a century it was an intellectually stimulating, but still largely academic subject. Today it is in the mainstream, and tomorrow it may well be absolutely fundamental to our understanding of the universe.

SUMMARY REVIEW

Principle of equivalence; removing gravity (locally) in an accelerated coordinate system; weightlessness in Skylab; producing gravity with acceleration; bending of light paths by gravitation; gravitational redshift; Jefferson Laboratory experiment; gravitational redshift in the sun and stars; observations of gravitational redshift in white dwarfs; observation of gravitational redshift of the earth with a rocket in high orbit; local limitations of the principle of equivalence

Spacetime: four-dimensional description of the universe; events in spacetime; graphing spacetime in projection; spacetime diagrams; paths of light and objects in spacetime; curvature; geodesics; curvature of spacetime by matter; new geometry of Gauss and Riemann; field equations; geodesic equations

Tests of general relativity: deflection of starlight; confirmation during solar eclipse observations; observations of quasars; delay of signals from Viking; advance of perihelion of Mercury

Gravitational waves: experiments to detect gravitational waves; the binary pulsar

Black holes: event horizon (Schwarzschild radius); size of event horizon for objects of various mass; predictions of stellar evolution theory; how black holes can be detected; expected X-ray emission; accretion disk; Cygnus X-1; black holes are small targets; an imaginary trip into a black hole; gravitational redshift; frozen stars; singularity; massive black holes; mini black holes; pair production; evaporation of black holes

EXERCISES

1. Consider a bucket nearly full of water. A spring is attached to the middle of the inside of the bottom of the bucket, and at the other end of the spring is a cork. The cork, trying to float to the top of the water, stretches the spring somewhat. Now suppose the bucket, with its water, spring, and cork, is dropped from a high building so that it remains upright as it falls. What happens to the spring and cork? Explain your answer thoroughly.

2. A monkey hanging from a branch of a tree sees a hunter aiming a rifle directly at him. The monkey then sees a flash, telling him that the rifle has been fired. The monkey, reacting quickly, lets go of the branch and drops, so that the bullet will pass harmlessly over his head. Does this act save the monkey's life? Why?

3. Some of the Skylab astronauts exercised by running around the inside wall of their cylindrical vehicle. How could they stay against the wall while running, rather than float aimlessly inside the Skylab? What physical principles are involved?

4. Draw a diagram showing the progress in spacetime of an automobile traveling northward. For the first hour, in city traffic, it goes only 30 km/hr. Then, in the country, for 3 hr it goes 90 km/hr. Finally, because the driver is late, he drives the car for 1 hr at 140 km/hr.

5. Make up a new example of a geodesic in spacetime and show it on a spacetime diagram.

6. The earth moves in its nearly circular orbit of 1 AU radius at a speed of only 1/10,000 that of light. What is the radius of the circle that most nearly matches the path of a beam of light passing the sun at the earth's distance from it?

7. As the binary pulsar loses energy through gravitational radiation, why do its members *speed up* and why does the period get *shorter?*

8. What would be the radius of a black hole with the mass of the planet Jupiter?

9. Why is the time dilation in a gravitational field equivalent to a gravitational redshift? (*Hint:* What is the definition of frequency? How is the frequency of radiation affected by a redshift?)

10. Why would we not expect X-rays from a disk of matter about an ordinary star, or even a white dwarf?

11. If the sun could suddenly collapse to a black hole, how would the period of the earth's revolution about it differ from what it is now?

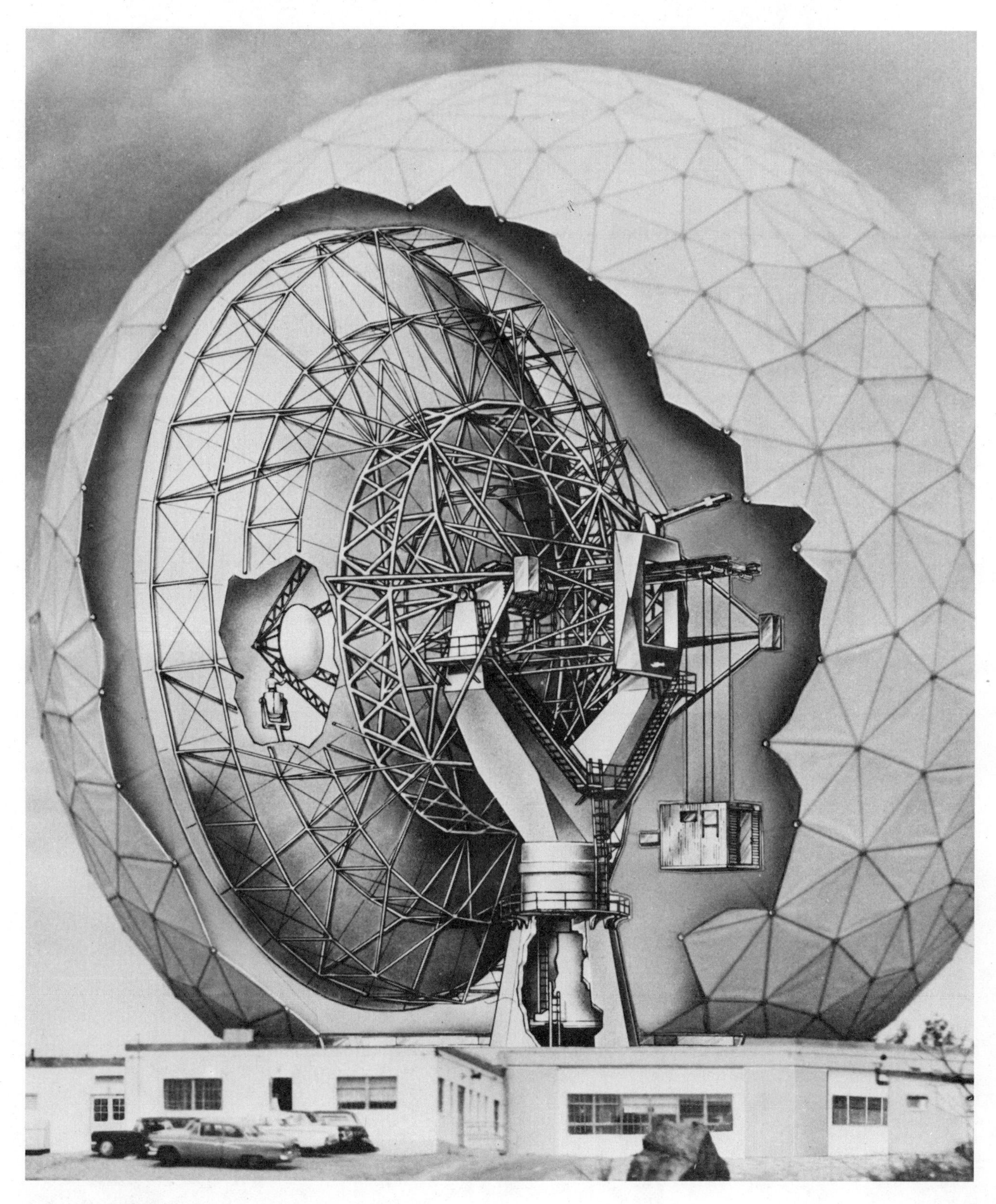

Much of our knowledge of the shifts in the perihelia of planets and other objects in the inner solar system—important in the tests of general relativity—has come from systematic observations with the Haystack Radar Antenna in Westford, Massachusetts. *(Courtesy, Lincoln Laboratories, Massachusetts Institute of Technology)*

CHAPTER 19

Frank Drake (1930–), the American astronomer who heads the National Astronomy and Ionosphere Center, which operates the large radio telescope at Arecibo, Puerto Rico, is a pioneer in the search for extraterrestrial intelligence. His *Project Ozma* (about 1960) was the first organized attempt to detect radio signals from extraterrestrial civilizations.

LIFE IN THE UNIVERSE

It is a legitimate question whether intelligent life, or any kind of life, exists elsewhere than the earth. At the outset we must say that we cannot answer that question with certainty. In this chapter, however, we shall try to explore the possibilities of life and of intelligent beings on other worlds.

LIFE IN THE SOLAR SYSTEM

There has been a great deal of study and a considerable advance in our understanding of the possibilities of life other than on earth in the solar system and beyond. We have learned enough about the other planets to know that we could not exist on any of them (in their natural environments), nor could the more complex forms of animal or plant life that we know on earth. Yet we cannot absolutely rule out the possibility of life forms alien to ours existing even in such seemingly hostile environments as we find even on Jupiter. Although the Viking experiments (Chapter 10) found no life on Mars, laboratory experiments have shown that primitive terrestrial life forms (such as algae) can exist and multiply under the conditions that exist there. On the other hand, there is as yet no hint of any life anywhere beyond the earth. Thus we might begin by looking at the circumstances under which life began here.

Development of Life on the Earth

We have seen (Chapter 9) that the earth, along with the other inner planets, most likely formed from accretion of small solid particles in the solar nebula about 4.6×10^9 years ago. As soon as the solid earth formed, the natural decay of radioactive elements below the surface heated up the crustal rocks, causing the chemical decomposition of some of the minerals present there. In the process, water (H_2O), carbon dioxide (CO^2), and other gases were liberated from chemical compounds. These substances outgas to the surface, especially through volcanism. Water was the most plentiful compound to be released from the earth's crust; once it reached the cooler surface, it condensed to form the oceans where most of it remains today. Next most important is carbon dioxide, with about a tenth the abundance of the water. Much of it dissolved in the oceans and some recombined at the surface temperature of our planet with surface rocks to re-form carbonates, which remain on the ground today. Perhaps only one-fiftieth as abundant as the carbon dioxide was nitrogen gas, but nitrogen is relatively inactive chemically, and now remains in the atmosphere as its major constituent (about 78 percent). Another even more inert gas to be released was argon-40, formed by the radioactive decay of potassium-40; argon today comprises about 1 percent of the atmosphere.

Most experts believe that in the earth's primordial atmosphere methane (CH_4), ammonia (NH_3), and, at least temporarily, hydrogen (H_2) were present, having originated either from outgassing or from chemical activity at the surface. Thus the early atmosphere of the earth may have been made up of nitrogen, carbon dioxide, argon, water, methane, ammonia, and hydrogen. The last three gases would quickly escape or decompose, but their temporary presence, bathed in solar radiation, would create a chemical environment favorable to the formation of more complex molecules. Laboratory experiments have shown that prebiological organic molecules, such as amino acids and sugars will readily form under these conditions, and we can imagine how eventually the far more complex organic substances vital for living organisms (including DNA) can have been built up. Thus we begin to understand how living organisms may have developed in the primordial terrestrial atmosphere or ocean.

Evolution of the Earth's Atmosphere

Of all the planets, only the earth has appreciable free oxygen in its atmosphere. We think that the earth's oxygen came about because of the development of life here. Green plant life, mainly in the oceans, flourished during the first few thousand million years of the earth's existence. This vegetation removed the carbon dioxide from the air by the process of photosynthesis, building itself with the carbon and releasing the oxygen into the atmosphere. When the vegetable organisms die, they decay (or oxidize), removing the oxygen from the air again. However, part of the dead vegetation escaped the decay process, by being preserved in the ground in the form of fossil fuels, where much of it remains today. Thus most of the oxygen is removed from the air by the decay of dead plant matter (and later by decay of animals, by combustion, and by respiration), but a little—somewhere between one part in 10^4 and one part in 10^5—of that produced by photosynthesis remains in the atmosphere, gradually building up the oxygen concentration. It is estimated that oxygen comprised only about 1 percent of the atmosphere 600 million years ago, but since then it has gradually accumulated to about 21 percent. Nitrogen makes up 78 percent, and argon the remaining 1 percent, with water vapor and carbon dioxide present in trace amounts.

If photosynthesis were suddenly to stop, it is estimated that the forces that remove oxygen from the air would take about 10^4 years to deplete our store of this precious gas. That is to say, the turnover time for oxygen in the earth's atmosphere is about ten thousand years. The present oxygen supply required some ten thousand to one hundred thousand times this long to accumulate because of the very slight margin by which production of oxygen by photosynthesis exceeds its removal by decay.

In recent years man has upset this delicate imbalance. It is estimated that now about 15 percent of oxygen removal arises from our consumption of the fossil fuels (coal and oil) dug from the earth. At the present rate it would still take thousands of years to use up our oxygen (the easily available oil and coal will be used up much sooner than this), but the resulting increase in the concentration of carbon dioxide in the atmosphere is measurable. Because this gas is very effective in blanketing the infrared radiation of the earth near the ground, a greater concentration of it could have a profound effect on the climate by increasing the world's average temperature enough to melt the polar caps and flood several countries. (This phenomenon is known as the *greenhouse effect.*)

POSSIBILITY OF INTELLIGENT LIFE IN THE GALAXY

There are probably 10^9 potentially observable galaxies in the universe, and probably many times this number of galaxies too faint to observe. Each may contain thousands of millions of stars, a large fraction of which might have planets. The possibilities of life throughout the universe, therefore, would seem to be enormously greater than in our own Galaxy. Yet because even light requires millions of years to travel the great distances between galaxies, any other societies that we have a chance of discovering are probably in our own Galaxy.

The Number of Galactic Civilizations

Cornell astronomer Frank Drake has pioneered the attempt to estimate the number of potentially communicative civilizations currently extant in the Galaxy. Drake's famous equation identifies seven factors that are required for the estimate. These are the number of stars in the Galaxy, the fraction of those stars with planetary systems, the mean number of planets suitable for life per planetary system, the fraction of those planets suitable for life on which life has actually developed, the fraction of those planets with life on which intelligent organisms have evolved, the fraction of those intelligent species that have developed communicative civilizations, and the mean lifetime of those civilizations in terms of the mean lifetime of the central stars in the planetary systems. The first three factors are essentially astronomical in nature, the next two are biological, and the last two are sociological. We are able to make some educated estimates regarding the astronomical factors, we may be on shaky ground with the biological ones, and we are almost playing numbers games in trying to estimate values for the last two. Yet some interesting estimates can be made, and limits derived.

The mass of the Galaxy (Chapter 15) is probably about 2×10^{11} solar masses. The most common stars are main-sequence M-type stars of mass less than half that of the sun; thus we estimate the total number of stars to be 4×10^{11}; the uncertainty of this figure is a factor of at least 3.

The sun originated from a cloud of gas and dust—the solar nebula—whose rotation caused it to flatten into a disk from which the planets formed. We expect similar formation of planetary systems elsewhere to be commonplace. On the other hand, roughly half of the sun's neighboring stars are members of binary or multiple star systems. It may be that duplicity is an alternative to the formation of a planetary disk, in which case stars in double star systems might not have planets. In any case, it is unlikely that in such systems planets could have stable orbits. Therefore, we shall assume that only half of the stars in the Galaxy have planets.

Not all planetary systems can contain planets suitable for the development of life. The oldest forms of life on earth, such as blue-green algae, are known to have existed at least 3.1×10^9 years ago, and probably even longer back, but 4.5×10^9 years were required before a communicative civilization was realized. On the earth, we had time, because the total lifetime of the sun is about 10^{10} years—its main-sequence life now being about half expended. We would never have made it if the sun were a spectral type B or A star with a lifetime of a few hundred million years or less. On the other hand, the relatively faint main-sequence M stars have luminosities of at most 0.01 that of the sun, and the majority are many times fainter still. For a planet to receive enough energy from such a star to warm it to the point that life has a good chance of developing, that planet would have to be so close to the star that tidal forces produced by the star on it would lock the planet into synchronous rotation (with one side always toward the star). Under these conditions, any atmosphere should migrate from the starlit side around to the dark side, where it would freeze out. We would, therefore, expect planets only around main-sequence stars not too unlike the sun—say, spectral classes from F to K—to be likely to have conditions favorable for life. This requirement eliminates all but about 10 percent of the stars. We shall estimate that there is one habitable planet for every 10 planetary systems.

Many biologists are of the opinion that given the right kind of planet and enough time, the development of life is inevitable. Perhaps it is so, but in the spirit of a devil's advocate, we argue that the certainty of life forming has not been demonstrated as yet. Let us consider the liberal estimate that life, given the right conditions, is certain to develop, but also the devil's advocate guess that it happens only, say, 10 percent of the time.

Similarly, given the emergence of life, there is a widespread view that with enough time and natural selection a highly intelligent species will certainly evolve. Even were it inevitable that an intelligent species evolve on every planet with life, however, how long should it take? On earth, it took 4.5×10^9 years. What if we happened to be quick about it, but that the average intelligent species takes, say, 20×10^9 years? Moreover, of the many parallel lines of evolution on earth, only one (so far) has produced a being with enough intelligence to build a technology. Certainly, one could not rule out that the probability could be as low as 10 percent.

Not all intelligent societies would necessarily develop a technology capable of interstellar communication. We are on the threshold of that capability, and possess a natural curiosity about the rest of the universe. Insects do not appear to have any curiosity at all, however, and it is not certain if this human trait is fundamental to intelligence. Even if a society were curious, it might have good reason for wishing to have nothing to do with any other civilization. Some investigators suppose that half of all intelligent species will form communicative technological societies. As a conservative alternative, let us also assume that only one-tenth of them do.

It is generally agreed that the final factor, the longevity of a civilization, is the most uncertain. Some have speculated that a technology might survive for an average of 10^9 years, so that if the lifetime of a typical star with a life-supporting planet is 10^{10} years, it will support a civilization one-tenth of that time. The only known technology, of course, is our own, and we have only just reached the capability of interstellar communication. Our technology might well end in a few decades. If so, and if we are typical, an average civilization might last only for one hundred-millionth of the life of the planet. A popular compromise estimate is 10^6 years for the longevity of a typical technology. In the spirit of the devil's advocate, we shall suppose that one can equally well defend a longevity of 10^4 years.

We may now take our choice of the estimates of the factors in the equation for the number of communicative civilizations: If we adopt the more or less liberal estimates, we find that there should be 10^6 civilizations extant in the Galaxy today. This is the estimate of astronomers I. S. Shklovsky and Carl Sagan. On the other hand, our slightly more conservative, or devil's advocate estimates lead to only about 20.

If the first estimate is correct, there is an even chance that the nearest civilization is less than 250 LY away. If the second is correct, it may lie tens of thousands of LY away. We cannot, of course, really know the number of civilizations in the Galaxy, nor the distance to the nearest one, but we do see the rationale by which estimates are made, and also see that one can get almost any answer he wants by choosing what could be defended as reasonable estimates of the various factors that enter into the calculation. Indeed, it is entirely possible that we are the *only* technological society in the Galaxy.

Perhaps we are too anthropomorphic in our estimates. We have consistently thought of planets as the only places where life will develop, and that it will necessarily evolve into some kind of beings that have aspirations and interests similar to ours. For an entirely different kind of idea of what intelligent life could conceivably be like, read Fred Hoyle's science fiction novel, *The Black Cloud*.

Interstellar Travel

One means of detecting other Galactic civilizations might be by interstellar travel. We have seen, however, that the nearest neighboring civilization is expected to be at least a few hundred, and probably a thousand, or even tens of thousands of light years away. Because nothing can travel faster than light, a visit to another civilization would involve at least hundreds, and more likely thousands of years.

Now, to be sure, a space traveler's time slows down (with respect to ours) if he travels near enough the speed of light (Chapter 7). To make travel to other possible civilizations feasible in a human lifetime, however, the traveler's time would have to slow down at the very least by a factor of 5 (so that a 400-year round trip could be accomplished in 80 years of the crew's time). This much time dilatation requires a speed of 98 percent that of light, and the energy requirements to reach that speed are absolutely enormous.

Consider, as an example, calculations presented by S. von Hoerner: Suppose we wish to send a moderate payload of 10 tons (three to five automobiles) into interstellar space, and accelerate it to $0.98c$. In this relatively small payload we must provide an environment to provide life support for the crew for several decades. We add another 10 tons for engines and propulsion systems. The total energy required, no matter how it is obtained or how fast it is expended, is about 4×10^{29} ergs—roughly enough energy to supply the entire world's needs (at the present global expenditure rate of energy) for some 200 years. It would probably require the complete annihilation of matter, which we do not know how to accomplish at present. If we wanted the crew to reach its final speed of $0.98c$ at an acceleration equal to the earth's gravity, which would take 2.3 years, it would require the equivalent of 40 million annihilation plants of 15 million watts each, producing energy to be transmitted (with perfect efficiency) by 6×10^9 transmitting stations of 10^5 watts each, and all of this apparatus must be contained within a mass of 10 tons!

These enormous energy requirements apply only if we need to travel close to the speed of light to take advantage of the relativistic time dilation. Interstellar travel is possible if we are willing to take a long time to do it—which requires many generations for the crew. We may wonder about the morality of subjecting the crew's offspring to a life in a spaceship destined for an unknown fate generations in the future. But at least, it is possible.

On the other hand, even now we can send, and have sent, material messages into interstellar space. Both the Pioneer and Voyager spacecraft, for example, after passing Jupiter, have entered orbits on which they will eventually escape the solar system. It is unlikely that they will ever be seen or recovered by another intelligent species, but it is remotely possible. Partly for this reason, each Pioneer carries a plaque bearing line drawings of human beings, and cryptic messages describing the world from which it came, and the Voyagers contain phonograph recordings with messages from and descriptions of earth. It is doubtful that the message on the plaque will ever be decoded or the recording heard, but publicity about them has called attention to the possibility of intelligent life in the Galaxy. More important by far is the message carried by each spacecraft itself; its discovery would convey a great deal of information about the race that launched it, and the state of our technology.

A final possibility for direct communication is a visit to the earth by extraterrestrial visitors. If the nearest civilizations are hundreds of light years away, they cannot have come to see us as a result of learning about us, for even radio waves that we have inadvertently been emitting into space—our radio and television programs—have only been on their way for a few decades. They could have reached only the very nearest stars, and it is highly unlikely that anyone there has received them and dispatched space ships to look us over. If we have been visited, it would have to have been by random selection by interstellar travellers, and it is extraordinarily unlikely that among the millions of stars in the Galaxy for every civilization, we should have been singled out for surveillance.

Yet the popular literature is full of accounts of sightings of UFOs, presumably operated by some intelligence, and even of alleged evidence for highly intelligent beings that have visited the earth and taught people to build such magnificent structures as the pyramids, Easter Island statues, and other marvels. Not only do the latter accounts fail to acknowledge the great amount of work that has been done by competent professional archeologists, but they are highly racist in their implication that earlier civilizations could not have had the talent to create great works of art.

Most scientists are highly skeptical of the extraterrestrial interpretation of reports of lights or erratically accelerating shiny objects in the sky, and of alien beings with unhuman countenances, yet with the human characteristics of two legs, two arms, a head, a mouth, nose, two eyes, and other anthropomorphic features. Hard evidence of objects from space is lacking. Scientists, more than anyone, would delight in finding concrete evidence of alien life—there is so much we have to learn from it! But we still need evidence that can be analyzed by any competent scientist qualified to judge its extraterrestrial origin. Rumors, hearsay, secondhand reports, eyewitness accounts by lay and inexperienced observers, all must be given the benefit of the doubt, but still require positive verification before being taken as final evidence for life in the universe beyond the earth.

Interstellar Communication

On the other hand, if there are other Galactic civilizations, there is a very real possibility that we may be able to communicate with them by radio. We do not necessarily mean two-way communication, for the radio waves would probably require hundreds of years at the very least for their round-trip travel between each question and answer. (Interstellar communication would be between civilizations, not individuals.) On the other hand, if there are communicative civilizations in the Galaxy they may already be trying to communicate, or at least to send one-way messages to other possible civilizations, just to inform them of their existence, and probably to convey much information in addition. With even our present technology we could send such messages ourselves to other stars in the Galaxy. The likelihood of an answer within a human lifetime, however, is almost negligible. Thus, the first step is to try receiving messages.

There is a good chance that we would recognize an intelligent message—for example, a binary-coded broadcast of the number π repeated over and over. The discovery of extraterrestrial intelligence would be one of the most stupendous in the history of mankind. If there were thousands or millions of communicating societies throughout the Galaxy, we could imagine a vast system of intercommunication, whereby many civilizations are sending messages to many other places in the Galaxy, not necessarily with the hope of receiving answers, for the messages themselves may well be received centuries after the sending civilizations had ceased to exist, but to pass on information about life in the Galaxy. It is a romantic and exciting idea, and many of us would like to be in on the network, if, in fact, it exists.

It is my guess that it does not exist. But we will never know if we do not investigate. So if we are willing to expend the energy to find out, despite the small chance of success, the potential rewards might well justify the effort—especially if the equipment developed for the project could be used for other kinds of astronomical research which would be certain to bring fruitful results. Thus, it is worthwhile to ask what is involved in really finding out whether any intelligently coded messages are coming in our direction from other civilizations in the Galaxy.

The problem was explored in a NASA study in 1971 by a group under B. Oliver, called *Project Cyclops*. The committee concluded that a fairly thorough search could be conducted for a cost of a few thousand million dollars. If an appreciable number of communicating civilizations exist, the Cyclops system would have a good chance of detecting one of them. The system would consist of radio telescopes in a vast array about

Figure 19.1
Artist's conception of an aerial view of the antenna array proposed by Project Cyclops in 1971. Each of the many radio telescopes in the system would have an aperture of at least 100 m, and would be designed to receive signals in the 1420 to 1660 megahertz frequency range that might originate from a distance as great as 1000 LY. *(NASA/Ames Research Center)*

5 km in diameter. Even if no civilization were discovered, the system would be enormously important to radio astronomy.

LONGEVITY OF CIVILIZATION ON THE EARTH

Consider the sheet of paper on which the page you are reading is printed; it is roughly 0.1 mm thick. Imagine that you were to cut it into two sheets and stack them on top of one another. Then cut the stack of two sheets and combine the four sheets into a single stack. Next, make a third cut and stack up the eight sheets. If you were to continue the process until 100 cuts had been made, guess how thick the stack of paper would be.

The answer is 10^{10} LY—the distance to a remote quasar (Chaper 20). When we double a number at each step, it is an example of a *geometric progression*. This progression increases slowly at first, but almost suddenly it explosively rises and rapidly approaches infinity as the number of doublings increases. An example of a similar geometric progression is the world

Figure 19.2
An artist's rendering of a closeup of several of the antennas in the proposed Project Cyclops system for detecting radio signals. The building at the right, near the center of the array, is the proposed data collection and processing facility. At present (late 1979) there are no plans to build the system. *(NASA/Ames Research Center)*

population, which presently is doubling about every 35 years (2 percent per year).

In New York City there are a little more than 100 square meters of land area for every person living in that city. Suppose we imagine New York to be more than 100 times as crowded as it is now, so that there is only 1 m^2 per person. Moreover, let us suppose that the same crowding applies over the entire land area of the earth. At the present rate of population growth, the land area of our planet would be filled to this density in about 550 years.

Perhaps one could argue about how crowded the earth can be. An uncontested ultimate limit to the number of people, however, is set by the fact that we are all made of atoms of matter, and matter is conserved; that which comprises our bodies has had to come from the earth itself. Most of the matter in our bodies is in the form of water which, we have seen, outgassed from the earth's crust. More and more of the water needed to build the bodies of people must come from the seas of the world as the population grows. At a doubling time of 35 years, the oceans would be entirely converted to people in about 1200 years (never mind about water to wash with, swim in, drink, and water our fields with, or even for fish). At the same doubling rate, in 1600 years the mass of people would be equal to the mass of the earth, and in 2300 years to that of the solar system. If we could create matter

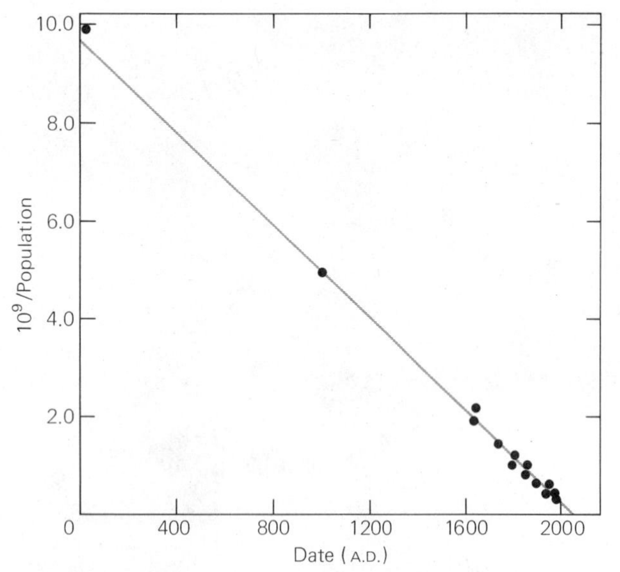

Figure 19.3
The increase of the world's population during the last 2000 years. Against the date is plotted one thousand million (10^9) divided by the population, so that the points are proportional to the reciprocal of the population, or, equivalently, to how much space per person there is on earth. The data are taken from several recent demographic studies. The straight line is a plot of an equation derived by von Förster, Mora, and Amiot, by which the population becomes infinite and the space per person drops to zero late in the year 2026 A.D.

from nothing (to make people), after about 5300 years a great sphere of humanity 150 LY in radius would be expanding at its surface with the speed of light!

The above examples are not meant as predictions but to illustrate how rapidly the numbers increase with a constant doubling time. In fact, during the past thousand years the doubling time for the population of the world has actually been *decreasing* with time. An excellent representation of the world population over the past two millennia is given by a formula graphed in Figure 19.3. A literal interpretation of this formula predicts that the world population will become infinite and the space available to each of us will go to zero in 2026 A.D., on Friday, November 13—a date dubbed "Doomsday."

Of course the world population cannot become infinite; various limitations will slow its present growth rate before 2026. But the problem is that most of the growth occurs so suddenly that the population can become unmanageable without warning; the problem may not be recognized until it is too late to prevent catastrophe.

Nor does expansion into space provide a solution. It has been seriously suggested that we can relieve overpopulation by emigrating to other planets. But even if we wanted to condemn our children to lives in air-tight living enclosures on the moon and Mars, without any hope of playing in the fields or hiking in the woods, we could only extend our time another 35 years, for the entire surface area of the moon and Mars combined is only about the same as that of the land area of the earth. Within about 500 years all possible planets about other stars within 150 LY could be occupied to the extent the earth is now, and to find room for our ever-increasing number of people, we would have to transport them faster than the speed of light to ever more remote planets, which, of course, can not be done.

In short, we are in danger of using up available space (and resources) so suddenly that we will have scarcely any warning to prepare for the onslaught of aggression and suffering that almost inevitably will result. The aggression prompted by population pressure (observed, for example, in experiments with colonies of rats that are allowed to multiply without check) is in itself a threat to survival. Nuclear bombs now stockpiled in the world are equivalent to about 10 tens of TNT per person—enough energy to raise a 10,000-ton apartment house 500 m into the sky to drop on every man, woman and child now living. And the nuclear stockpile is increasing at a faster rate than the population.

In some developed countries, including the United States, the population, while not yet stable, is rising more slowly, but the worldwide growth is unabated. In most of Latin America the population doubles in 20 years. Compared to the United States, Mexico is a poor nation. Yet, to maintain their existing level of poverty, the Mexicans must double their homes, schools, hospitals, roads, factories, and in general, everything in their economy in just 20 years!

Pollution of the atmosphere and exhaustion of our fossil fuels may be altering our climate and potential for food production. The United States, if we give up eating meat, could probably feed its people for another population doubling or so, but this requires maintaining our present efficiency in farming and assumes no climatological degradation of our ability to grow food.

To the problems of overcrowding, the violence it breeds, and energy and food shortages, we should add the gradual increase in the mean temperature of the earth due to the heat produced by the use of energy. This factor alone could alter our climate enough to affect food production in less than a century. Another effect of the present trends of the evolution of our civilization is genetic deterioration (medicine increases the life span, but encourages survival of persons with genetically unfavorable mutations). Even if we could withstand all of these threats to our survival, there remains a possible crisis caused by the boredom and stagnation of a stable society trying to endure without substantial innovation on a completely filled planet for hundreds of centuries.

Possibly these threats to our survival can be circumvented by enlightened action. Or perhaps violent struggles for survival in a chaotic near future will result in a strong, enlightened portion of society surviving to maintain a stable civilization. At present, thoughtful, rational international planning to preserve our planet as an abode for most of our race is missing. It may be overly pessimistic to predict that our technological society is doomed to an early end, but if we continue our present course, it is hard to imagine an alternative. Do other civilizations (if they exist) similarly destroy themselves, or have they learned, as we hope to learn, to preserve their longevity? Perhaps the discovery that another civilization has "made it" will one day be our salvation.

Figure 19.4
An ominous exhibit at the Bishop Museum, Honolulu, Hawaii. *(Bishop Museum)*

SUMMARY REVIEW

Development of life on earth: primordial atmosphere; formation of organic molecules; photosynthesis; evolution of the atmosphere

Intelligent life in the Galaxy: number of civilizations; number of stars; number of stars with planets; number of suitable planets; probability of life forming; probability of intelligent life evolving; fraction of civilizations that are communicative; lifetime of a typical civilization

Interstellar travel: time required; energy required; UFO's

Interstellar communication: how can we communicate?; Project Cyclops

Longevity of civilization on the earth: geometric progressions; world population; other potential hazards

EXERCISES

1. Compare and contrast the conditions in interstellar space and at the surface of a planet as they affect formation of organic molecules.

2. Why might the presence of oxygen in a planetary atmosphere suggest that life exists there? Does lack of oxygen prove the absence of light?

3. Suppose we could carry on two-way radio communication with another civilization. After a question is sent, what is the time we would have to wait before expecting an answer if that civilization is:
(a) on the moon?
(b) on Jupiter (assume Jupiter to be at its nearest to the earth)?
(c) on a planet revolving around Alpha Centauri?
(d) on a planet revolving around Tau Ceti (see Appendix 13 for the distance to the star)?
(e) at the Galactic center?

4. Because of the large number of main-sequence M-type stars, the mean mass of a star could be only 0.25 solar masses. What, then, would be our estimate of the number of stars in the Galaxy?

5. In the case described in Exercise 4, how would our estimate of the number of civilizations be affected?

6. Why could a planet not have a stable orbit about a star in a close binary system?

7. How many Galactic civilizations would you expect if you use our devil's advocate estimates of the various parameters, but assume that an average civilization lasts only 100 years?

8. All considered, what would you judge to be the extreme limits for the number of civilizations in the Galaxy?

9. What wavelengths of electromagnetic radiation might be suitable for interstellar communication, and why? What wavelengths would certainly *not* be suitable? Why?

10. Suggest ways you might code a message to send into space in such a way that another civilization, if it received it, would have a good chance to recognize it as intelligently contrived.

11. Verify the statement in the text that a 0.1-mm sheet of paper cut and stacked 100 times would have a final thickness of 10^{10} LY. (See Appendix 6 for conversion factors.)

12. Take an ordinary sheet of 8½-by-11-inch typewriter or notebook paper, a straight edge, and a pen or pencil. Draw a line (representing a cut) that roughly bisects the paper. Now, perpendicular to your first line, bisect one of the halves. Keep this up, counting the number of times you are able to continue to bisect the smaller and smaller rectangles. Try to imagine continuing to 100 bisections.

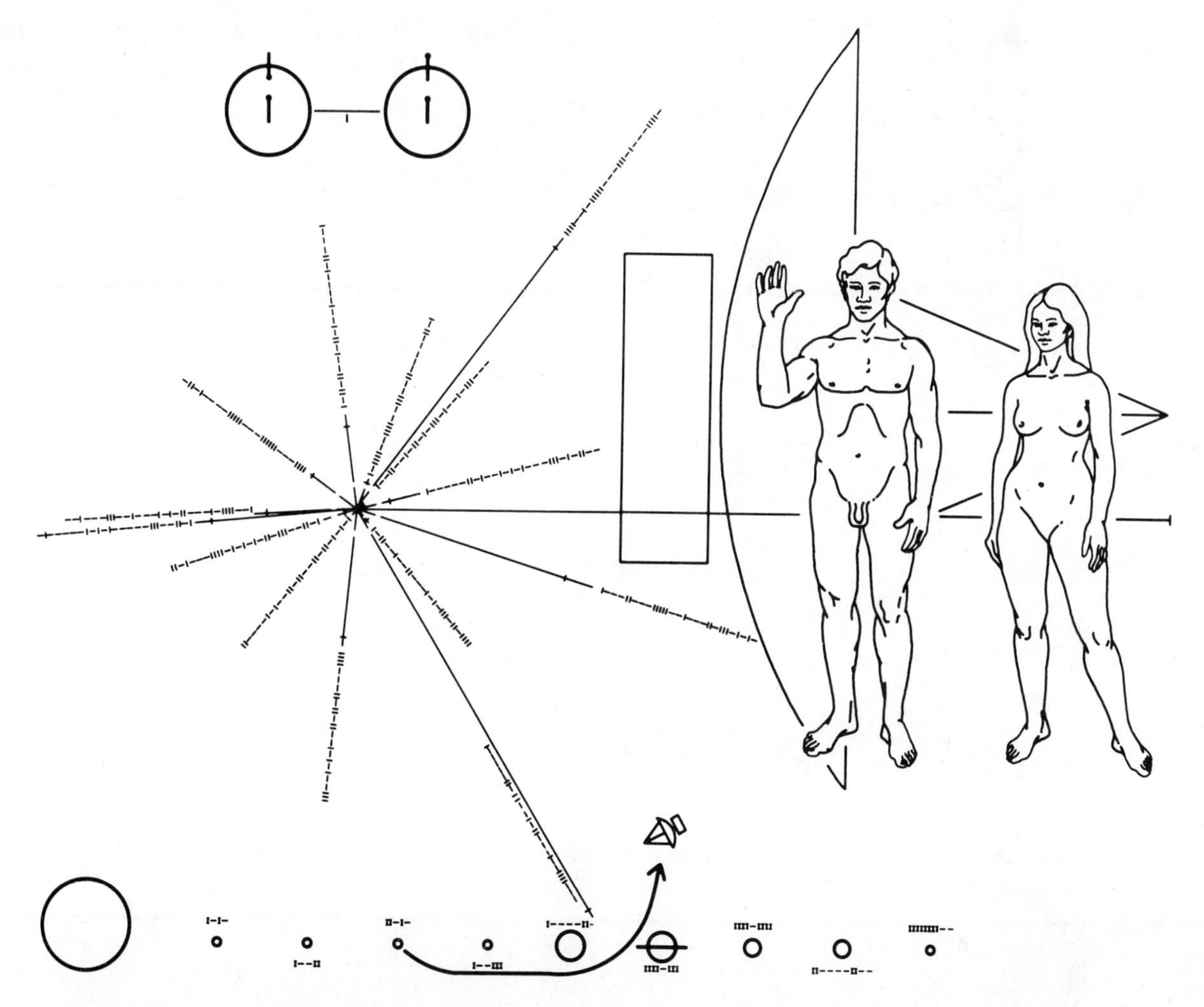

This plaque is borne on each of the Pioneer 10 and 11 spacecraft, both of which will eventually pass out to interstellar space. One purpose of the plaque was to provide information about us to interstellar travellers from another world, on the extremely remote possibility that the Pioneers should ever be encountered. Another purpose was to call the attention of the American public to the possibility of other civilizations. The man, woman, and spacecraft are to scale. Also shown are symbols giving information about our solar system, and about our knowledge of the 21-cm radiation from neutral hydrogen; the directions and periods of several pulsars are also shown to indicate the location of the solar system. The plaque was designed by Frank Drake and Carl Sagan, with the artisitic help of Mrs. Sagan. *(NASA)*

CHAPTER 20

Edwin Powell Hubble (1889–1953) left a career in law to become an astronomer. He showed that our Galaxy is but one of hundreds of millions of "island universes," as had been speculated by the German philosopher Immanuel Kant. Hubble went on to revolutionize twentieth-century cosmology. *(California Institute of Technology)*

THE STRUCTURE OF THE UNIVERSE

The "analogy [of the nebulae] with the system of stars in which we find ourselves . . . is in perfect agreement with the concept that these elliptical objects are just [island] universes—in other words, Milky Ways. . . ."

So wrote Immanuel Kant (1724–1804) in 1755[1] concerning the faint patches of light which telescopes revealed in large numbers. Unlike the true gaseous nebulae that populate the Milky Way (Chapter 15), the nebulous-appearing luminous objects referred to by Kant are found in all directions in the sky *except* where obscuring clouds of interstellar dust intervene. Despite Kant's (and others') speculation that these patches of light are actually systems like our own Milky Way Galaxy, the weight of astronomical opinion rejected the hypothesis, and their true nature remained a subject of controversy until 1924. The realization, only half a century ago, that our Galaxy is not unique and central in the universe ranks with the acceptance of the Copernican system as one of the great advances in cosmological thought.

GALACTIC OR EXTRAGALACTIC?

The discovery and cataloging of nebulae[2] had reached full swing by the close of the eighteenth century. A very significant contribution to our knowledge of these objects was provided by the work of the great German-English astronomer William Herschel (1738–1822) and his only son, John (1792–1871). William surveyed the northern sky by scanning it visually with the world's first large reflecting telescopes, instruments of his own design and manufacture. John took his father's telescopes to the southern hemisphere and extended the survey to the rest of the sky.

Figure 20.1
Herschel's 40-ft telescope. *(Yerkes Observatory)*

Catalogs of Nebulae

One of the earliest catalogs of nebulous-appearing objects was prepared in 1781 by the French astronomer Charles Messier (1730–1817). Messier was a

[1] *Universal Natural History and Theory of the Heavens*.
[2] *Nebula* (plural *nebulae*) literally means "cloud." Faint star clusters, glowing gas clouds, dust clouds reflecting starlight, and galaxies all appear as faint, unresolved luminous patches when viewed visually with telescopes of only moderate size. Since the true natures of these various objects were not known to the early observers, all of them were called "nebulae." Today, we usually reserve the word "nebula" for the true gas or dust clouds (Chapter 15), but some astronomers still refer to galaxies as nebulae or *extragalactic nebulae*.

comet hunter, and as an aid to himself and others in his field he placed on record 103 objects that might be mistaken for comets. Because Messier's list contains some of the most conspicuous star clusters, nebulae, galaxies, and other things in the sky, these objects are often referred to by their numbers in his catalog—for example M31, the galaxy in Andromeda.

In the years from 1786 to 1802, William Herschel presented to the Royal Society three catalogs, containing a total of 2500 nebulae. The *General Catalogue of Nebulae*, published by John Herschel in 1864, contains 5079 objects, of which 4630 had been discovered by him and his father. The *General Catalogue* was revised and enlarged into a list of 7840 nebulae and clusters by J. L. E. Dreyer in 1888. Today most bright galaxies are known by their numbers in Dreyer's *New General Catalogue*–for example, NGC 224 = M31. Two supplements to the *New General Catalogue*, known as the first and second *Index Catalogues* (abbreviated "IC"), were published in 1895 and in 1908.

By 1908 nearly 15,000 nebulae had been cataloged and described. Some had been correctly identified as star clusters and others as gaseous nebulae (such as the Orion nebula). The nature of most of them, however, still remained unexplained. If they were nearby, with distances comparable to those of observable stars, they would have to be luminous clouds, probably of gas, possibly intermixed with stars, within our Galaxy. If, on the other hand, they were very remote, far beyond the foreground stars of the Galaxy, they could be unresolved *systems* of thousands of millions of stars, galaxies in their own right—or as Kant had described them, "island universes." The resolution of the problem required the determination of the distances to at least some of the nebulae.

The Debate Over the Nature of the "Nebulae"

In 1908 construction was completed of the 60-inch (1.5-m) telescope on Mount Wilson, and within a decade the 100-inch (2.5-m) telescope was also in operation. Photographs obtained with these instruments clearly resolved the brightest stars in some of the nearer "nebulae." By 1917 several novae (Chapter 17) had been discovered in the more conspicuous nebulae. If those novae were as luminous as the 26 novae then known to have occurred in our own Galaxy, they, and the nebulae in which they appeared, would have to be at distances of about 1 million LY—far beyond the limits of our galaxy. On the other hand, not all astronomers agreed that real stars had actually been resolved in any of the nebulae. Even the Mount Wilson astronomers who had photographed those stars were not convinced of their true stellar nature and had described them as "nebulous stars."

Two of the major protagonists in the controversy over the nature of the nebulae were Harlow Shapley, of the Mount Wilson Observatory, and H. D. Curtis, of the Lick Observatory. Their opposing views culminated on April 26, 1920, in the famous Shapley-Curtis debate before the National Academy of Sciences. Curtis supported the island-universe theory, and Shapley opposed it. Of course, the controversy was not settled by the debate; according to A. R. Sandage, "Perhaps the fairest statement that can be made is that Shapley used many of the correct arguments but came to the wrong conclusion. Curtis, whose intuition was better in this case, gave rather weak and sometimes incorrect arguments from the facts, but reached the correct conclusion."[3]

The final resolution of the controversy was brought about by the discovery of variable stars in some of the nearer "nebulae" in 1923 and 1924. Edwin Hubble (1889–1953), at Mount Wilson, analyzed the light curves of variables he had discovered in M31, M33, and NGC 6822 and found that they were cepheids (Chapter 13). Although cepheid variables are supergiant stars, the ones studied by Hubble appeared very faint—near magnitude 18. Those stars, therefore, and the systems in which they were found, must be very remote; the "nebulae" had been established as galaxies. Hubble's exciting results were presented in Washington, D.C., to the American Astronomical Society at its thirty-third meeting, which began on December 30, 1924.

DISTANCES AND DISTRIBUTION OF THE GALAXIES

The most direct method of finding the distance to a galaxy is to identify in it an object or objects that are similar to known objects in our own Galaxy and which, presumably, have the same intrinsic luminosities (or absolute magnitudes). By comparing the ap-

[3] A. R. Sandage, *The Hubble Atlas of Galaxies*. Carnegie Institution, Washington, D.C., 1961, p. 3.

Figure 20.2
The Large and Small Magellanic Clouds (top and lower left). *(Sky and Telescope)*

parent magnitude of such an object with its known absolute magnitude, we find its distance, which is, of course, the distance of the galaxy itself, by the method outlined in Chapter 13. This was the procedure followed by Hubble when he derived distances to several nearby galaxies from the cepheid variables in them.

Cepheid variables are our first link to galaxian distances. That is why these relatively rare stars are so important. Unfortunately, however, there are only about 30 galaxies containing cepheids that are near enough so that we can observe those variables, even with the world's largest telescope. The nearest of these (the Clouds of Magellan) are only about 150,000 to 200,000 LY away from the sun, and the most distant are within 20 million LY.

The brightest cepheids have an absolute magnitude of about -6. The most luminous stars in some kinds of galaxies are even brighter, and can be seen, therefore, to a greater distance. The brightest stars found in our own and other galaxies (usually blue supergiant stars) have absolute magnitudes of about -9. Bright novae, at maximum, reach about the same luminosity, and this is also about the absolute magnitude of the most luminous globular clusters. Thus the brightest supergiant stars in a galaxy, occasional nova outbursts that occur in it, and its brightest globular clusters all are distance indicators that can be observed as far away as 80 million LY. Distances to some kinds of galaxies even somewhat further away can be estimated from the angular sizes of H II regions (emission nebulae—Chapter 15) that can be resolved in them. Supernovae (Chapter 17) are temporarily thousands of times as bright as ordinary novae, but we do not yet have reliable independent knowledge of their absolute magnitudes.

The distances that separate galaxies are hundreds of thousands to millions of light-years. There are less than two dozen known galaxies within 2½ million LY, but there are many thousands within 50 million LY; galaxies extend in all directions as far as we can see. The more distant galaxies, of course, appear less conspicuous than the nearer ones. Tiny images of extremely remote galaxies are actually more numerous than images of the foreground stars of our own Galaxy on photographs obtained with large telescopes directed away from the Milky Way. The images of galaxies at the limit of detection can be distinguished from those of faint stars only because the galaxian images are rather fuzzy or elongated and lack the sharp, pointlike appearance of the star images.

In the years following his announcement of the true nature of the "nebulae," Hubble made an extensive study of galaxies and their distribution in space. He realized that it would take thousands of years to photograph the entire sky with the 100-inch telescope and to count all the galaxies that could be observed with that instrument. Instead, he photographed 1283 sample regions of the sky with the 100-inch and 60-inch telescopes, a procedure reminiscent of Herschel's "star gauging" (Chapter 14). The results of Hubble's "galaxy gauging" are shown in Figure 20.3. Each symbol in the figure shows one of the sample regions, and the size and type of the symbol show how many galaxies were visible on the photograph of each region. From the 44,000-odd galaxies Hubble counted in these selected regions, he calculated that nearly 100 million galaxies must exist within the range of the 100-inch telescope. There may be 10^9 galaxies within the reach of the 200-inch (5 m) reflector at Palomar.

Hubble found the largest number of galaxies in the regions near the galactic poles (90° from the great circle running through the Milky Way) and determined that, on the average, fewer and fewer galaxies could be

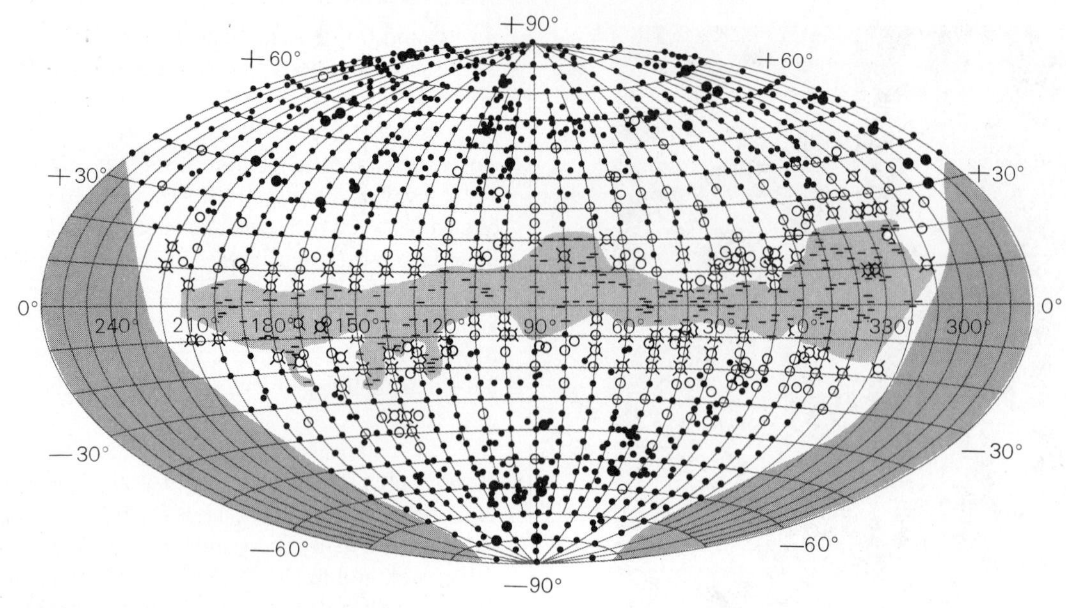

Figure 20.3
Distribution of galaxies. The size of each dot plotted on the map of the sky indicates how many galaxies were counted on a photograph taken of that location with the 2.54-m telescope. Dashes and open circles indicate few or no galaxies, and the irregular region across the center of the map where almost no galaxies are to be seen is the zone of avoidance. This zone coincides with the visible Milky Way, where interstellar dust obscures the exterior galaxies. the empty regions on the plot are those of the sky too far south to survey from Mount Wilson. The numbers indicate galactic coordinants (Appendix 7).

seen in directions successively closer to the Milky Way. This, of course, is an effect of absorption of light by interstellar dust in our Galaxy (Chapter 15). When allowance is made for interstellar absorption, however, Hubble found that the distribution of galaxies, on the large scale, is the same in all directions, and that the mean density of galaxies in space seems to be about the same at all distances. On the small scale, however, galaxies tend to be grouped into *clusters of galaxies*.

All the galaxies in a cluster are at about the same distance—the distance to the cluster itself. We can estimate relative distances to different clusters, therefore, by comparing the apparent magnitudes of their brightest members. If the light received from the brightest galaxies in cluster A, for example, is one-fourth that of the brightest galaxies in cluster B, cluster A is probably twice as distant as B. This method only works, however, for rich clusters, which can be expected to have similar galaxies for their brightest members. Unfortunately, there is no very rich cluster close enough to us so that we can determine its distance by independent means (say, from resolved stars in some of its galaxies). We are not certain, therefore, of the actual luminosities of the brightest galaxies in rich clusters. *Relative* distances to clusters are thus known more reliably than absolute distances.

We have not referred in this section to the well-known correlation between the distances of galaxies and their radial velocities, as indicated by the Doppler shifts of the lines in their spectra. Although distances are often estimated for galaxies from their radial velocities, the method is not really independent, for it depends on the calibration of the velocity-distance relation, which, in turn, must be based on distances measured by other means to selected galaxies with various observed Doppler shifts. The accurate calibration of this relation is one of the major problems of extragalactic astronomy. The velocity-distance relation, and its calibration, is so fundamental to the study of modern *cosmology* that discussion of it is postponed to Chapter 21.

PROPERTIES OF GALAXIES

How Gross Properties Are Determined

The size of a galaxy can be calculated once its distance is known, just as we calculate the diameter of the sun or of a planet. Also, if we know its distance, we can apply the inverse-square law of light and calculate the total luminosity, or the absolute visual magnitude of a galaxy from the amount of light flux we receive from it or from its apparent visual magnitude. Thus a knowledge of galaxian radii and luminosities is provided by a knowledge of galaxian distances.

We determine the masses of galaxies, like those of other astronomical bodies, by measuring their gravitational influences on other objects or on the stars within them. We measure the masses of some nearby galaxies, for example, from their rotations. As an illustration, consider M31, the Andromeda galaxy (Figure 20.4), which is probably quite similar to our own Galaxy. M31 is inclined at an angle of only about 15° to our line of sight, so we see it highly foreshortened. Now the radial velocity of the brilliant nucleus shows that the galaxy as a whole is approaching us at nearly 300 km/s. The regions southwest of the nucleus, however, approach us even faster—at about 500 km/s—so that side of the galaxy is turning toward us and the northeast side away from us. Thus we can measure the rotation of the Andromeda galaxy even though we cannot actually see the stars in it moving during our lifetimes.

It is a straightforward calculation to find the mass that a galaxy must have in order that its constituent parts revolve about its center with their observed speeds. It is analogous to calculating the mass of the sun from the speeds of the planets with Kepler's third law. We find that the mass of M31 is about 3×10^{11} solar masses—perhaps a little greater than the mass of our own Galaxy.

Many galaxies are not highly flattened and rotating rapidly. Nevertheless, the velocities of the stars in such a galaxy depend on its gravitational attraction for them, and hence on its mass. The spectrum of a galaxy is a composite of the spectra of its many stars, whose different motions produce different Doppler shifts. The lines in the composite spectrum, therefore, are broadened, and the amount by which they are broadened indicates the range of speeds with which the stars are moving with respect to the center of mass of the galaxy. Application of mechanical laws, then, enables us to calculate the mass of the galaxy.

Like stars, galaxies are often observed in close pairs. However, we cannot "see" the motion of one galaxy about the other, as we can observe the mutual revolution of the members of a binary star system. Nevertheless, the difference between the radial velocities of the two galaxies in a double system tells us something about its mass. The masses of *clusters* of galaxies can be calculated by a similar technique. The radial velocities of many galaxies in a cluster are first measured. The average of these velocities is that of the center of mass of the cluster, and the differences between the velocities of individual galaxies and this mean value tells us how fast they are moving within the cluster. These speeds depend on the gravitational potential energy of the cluster, so a knowledge of them enables us to estimate the mass of the cluster

Figure 20.4
The Andromeda galaxy, M31, photographed with the 48-inch Schmidt telescope. *(Hale Observatories)*

Spiral Galaxies

Our own Galaxy and M31, which is believed to be much like it, are typical of what are called *spiral* galaxies. Like our Galaxy (Chapter 15), a spiral consists of a nucleus, a disk, a corona, and spiral arms. Interstellar material is usually observed in the arms of spiral galaxies. Bright emission nebulae are present, and absorption of light by dust is also often apparent, especially in those systems turned almost edge-on to our line of sight (Figure 20.5). The spiral arms contain the young

Figure 20.5
NGC 4565, a spiral galaxy in Coma Berenices, seen edge-on. Photographed in red light with the 200-inch telescope. *(Hale Observatories)*

stars, which include luminous supergiants. These bright stars and the emission nebulae make the arms of spirals stand out like the arms of a Fourth-of-July pinwheel. The individual interarm stars are usually not observable at all, save in the nearest galaxies, although their collective light may be appreciable as a uniform glow. Open star clusters can be seen in the arms of nearer spirals, and globular clusters are often visible in their coronas; in M31, for example, more than 200 globular clusters have been identified. Spiral galaxies contain both young and old stars.

Some famous spirals are illustrated in these pages. M51 (Figure 20.6) is seen nearly face-on; NGC 4565 (Figure 20.5) is nearly edge-on. Note the absorbing lane of interstellar dust in NGC 4565—a thin slab in the central plane of the disk—which is silhouetted against the nucleus. M81 (Figure 20.7), like M31 (Figure 20.4), is viewed obliquely.

A large fraction (perhaps half) of spiral galaxies display "bars" running through their nuclei; the spiral arms of such a system usually begin from the ends of the bar, rather than winding out directly from the nucleus. These are called *barred spirals*. A famous example is NGC 1300 (Figure 20.8). The bar in a barred spiral looks like a straight portion of spiral arm and usually contains interstellar matter and young stars. Studies of the rotations of some barred spirals show that their inner parts (out to the ends of the bars) are rotating approximately as solid wheels.

In both normal and barred spirals we observe a gradual transition of morphological types. At one extreme, the nucleus is large and luminous, the arms are small and tightly coiled, and bright emission nebulae and supergiant stars are inconspicuous. At the other extreme are spirals in which the nuclei are small—almost lacking—and the arms are loosely wound, or even wide open. In these latter galaxies, there is a high degree of resolution of the arms into luminous stars, star clusters, and emission nebulae. Our Galaxy and M31 are both intermediate between these two extremes. So far as is known, all spirals and barred spirals rotate in the sense that their arms trail, as does our own Galaxy (Chapter 15).

Figure 20.6
The Sc galaxy NGC 5194 (M51) and its irregular II companion, NGC 5195, photographed with the 200-inch telescope. *(Hale Observatories)*

Figure 20.7
NGC 3031 (M81), spiral galaxy in Ursa Major, photographed with the 200-inch telescope. *(Hale Observatories)*

Figure 20.8
NGC 1300, barred spiral galaxy in Eridanus, photographed with the 200-inch telescope. *(Hale Observatories)*

Elliptical Galaxies

More than two-thirds of the thousand most conspicuous galaxies in the sky are spirals. For this reason it is often said that most galaxies are spirals. Actually, however, the most numerous galaxies in any given volume of space are those of relatively low luminosity, which cannot be seen at large distances, and which, therefore, are not among the brightest-appearing galaxies. Most of these dwarf galaxies fall into the class of *elliptical* galaxies. Moreover, the rich clusters, which contain a good fraction of all galaxies, are composed mostly of ellipticals. Elliptical galaxies are really far more numerous than spirals.

Elliptical galaxies are spherical or ellipsoidal systems; they contain no trace of spiral arms. They resemble the nucleus and corona components of spiral galaxies. Although dust and conspicuous emission nebulae are not easily observed in elliptical galaxies, some do show evidence of sparse interstellar gas in their spectra. In the larger (nearby) ones, many globular clusters can be identified. The elliptical galaxies show various degrees of flattening, ranging from systems that are approximately spherical to those that approach the flatness of spirals. The distribution of light in a typical elliptical galaxy shows that while it has many stars concentrated toward its center, a sparse scattering of stars extends for very great distances and merges imperceptibly into the void of intergalactic space (compare with the corona of our own Galaxy—Chapter 15). For this reason it is nearly impossible to define the total size of an elliptical galaxy. Similarly, it is not obvious how far the corona of a spiral galaxy extends.

The fact that elliptical galaxies are not disk-shaped shows that they are not rotating as rapidly as the spirals. It is widely hypothesized that they are systems that formed from pregalaxian material that had little angular momentum per unit mass—that is, that their original material had low net rotation. Consequently, as such a cloud of primeval material contracted, it did not flatten into a disk, and the density of the material was high enough that it completely (or nearly completely) condensed into stars. In a spiral, on the other hand, a considerable amount of gas (and/or dust) in the flat, rapidly rotating disk was not able to condense into stars at once. This material was presumably formed into spiral arms by the rotation of the galaxy, where it now still slowly condenses into stars at a gradual rate. Elliptical galaxies probably consist entirely of old stars (at least if those galaxies are as old as our Galaxy).

Figure 20.9
NGC 4486 (M87), giant elliptical galaxy in Virgo. Note the many visible globular clusters in the galaxy. *(Kitt Peak National Observatory)*

Figure 20.10
Leo II, a dwarf elliptical galaxy (negative print), photographed with the 200-inch telescope. *(Hale Observatories)*

Elliptical galaxies have a much greater range in size, mass, and luminosity than do the spirals. The rare giant ellipticals (for example, M87—Figure 20.9) are more luminous than any known spiral. Some rich clusters contain one or two supergiant elliptical galaxies of absolute magnitude brighter than -23—more than 10^{11} times the luminosity of the sun, and more than 10 times the luminosity of the Andromeda galaxy. Studies by D. Jenner show that some of these galaxies have masses about 10^{13} times that of the sun. While, as stated above, the diameters of these large galaxies are difficult to define, they certainly extend over at least several hundred thousand light-years.

Elliptical galaxies range all the way from the giants, just described, to dwarfs, which we think are the most common kind of galaxy. An example is the Leo II system, shown in Figure 20.10. There are so few bright stars in this galaxy that even its central regions are transparent. The total number of stars, however (most too faint to show in Figure 20.10), is probably at least several million. Intermediate between the giant and dwarf elliptical galaxies are systems such as M32 and NGC 205, two near companions to M31. They can be seen in the photograph of M31 (Figure 20.4); NGC 205 is the one that is farther from M31.

Irregular Galaxies

About 3 percent of the brightest-appearing galaxies in the northern sky are classed as irregular. They show no trace of circular or rotational symmetry, but have an irregular or chaotic appearance. The best-known examples are the Large and Small Clouds of Magellan (Figure 20.11), our nearest galaxian neighbors. We find many star clusters in these galaxies, as well as variable stars, supergiants, and gaseous nebulae; they contain both old and young stars.

Some irregular galaxies, however, display no resolution into stars or clusters, but are completely amorphous in texture. Their spectra resemble the spectra of type A stars, showing that stars not luminous

TABLE 20.1 Gross Features of Galaxies of Different Types

	Spirals	Ellipticals	Irregulars
Mass (solar masses)	10^9 to 4×10^{11}	10^6 to 10^{13}	10^8 to 3×10^{10}
Diameter (thousands of light-years)	20 to 150	2 to 500 (?)	5 to 30
Luminosity (solar units)	10^8 to 10^{10}	10^6 to 10^{11}	10^7 to 2×10^9
Absolute visual magnitude	-15 to -21	-9 to -23	-13 to -18
Population content of stars	Old and young	Old	Old and young
Composite spectral type	A to K	G to K	A to F
Interstellar matter	Both gas and dust	Almost no dust; little gas	Much gas; much dust in some; less and sometimes no dust in others

enough to be resolved must exist in these galaxies. They generally also show conspicuous dark lanes of absorbing interstellar dust. Examples are M82 (Figure 20.12) and the companion to the spiral galaxy, M51 (Figure 20.6).

Figure 20.11
The Large Cloud of Magellan. *(Hale Observatories)*

Figure 20.12
NGC 3034 (M82), an irregular II galaxy in Ursa Major. *(Hale Observatories)*

Summary

The gross features of the different kinds of galaxies are summarized in Table 20.1. Many of the figures given, especially for mass, luminosity, and diameter, are very rough and are intended only to illustrate orders of magnitude, not precise values.

CLUSTERS OF GALAXIES

Analyses of photographic surveys of the sky have shown that most galaxies belong to groups or clusters.

The richest, but relatively rare, clusters of galaxies have spherical symmetry and show marked central concentration. Most of them probably contain at least

Figure 20.13
Coma cluster of galaxies, photographed with the Palomar Schmidt. *(Hale Observatories)*

1000 members brighter than absolute magnitude −16. A famous example is the Coma cluster (Figure 20.13). Rich clusters consist nearly entirely of elliptical galaxies.

These clusters are often sources of X-rays. Sometimes individual galaxies are found to be X-ray sources, but observations with the Einstein X-ray telescope show that most of the X-rays come from the intracluster space between the galaxies. The nature of the X-ray emission indicates that it is from gas with a temperature of the order 10^8 K.

It is not surprising to find hot gas in rich clusters. Some of it could have been left over from the time of the formation of the cluster, but in any event in the central regions of a rich cluster collisions must commonly occur between galaxies. As far as the widely separated stars are concerned, the galaxies just pass through each other (without stellar collisions), but the interstellar gas must collide and be swept clean of the galaxies. The high relative speeds of the galaxies (typically 1000 km/s) assure that the gas of one colliding galaxy reacts violently enough with that of the other to be heated to the temperature implied by the X-rays. Once there is some gas in the central core of the cluster, it easily sweeps additional gas out of galaxies that pass through it in their orbital motions within the cluster.

Gas sweeping from cluster galaxies explains another phenomenon: the rich clusters often contain some flat galaxies, like spirals, but with no trace of spiral arms. Hubble classed these *SO galaxies*, and considered them to be intermediate between spirals and ellipticals. If spirals in rich clusters are swept clean of gas, as described here, the galaxies have no more source material for star formation, and in a few million years the bright main-sequence stars that make the spiral arms conspicuous will burn out, leaving the galaxies SOs. We are not sure that all SO galaxies are formed in this manner, but their presence in rich clusters can at least be understood in terms of this picture.

Finally, as stars within galaxies age and eject matter into interstellar space, as described in Chapter 17, even an elliptical galaxy gains a certain amount of interstellar gas. That gas is also added to the intracluster gas as those elliptical galaxies pass through it. Evidence for this scenario is provided by the observation of spectral lines of iron at X-ray wavelengths in the intracluster gas in some clusters. Because iron is thought to be produced only by nucleosynthesis in stars, its presence in clusters suggest that it has indeed been swept out of galaxies.

The less rich clusters tend to be more irregular and are more nearly amorphous in appearance, and they possess little or no spherical symmetry or central concentration. These irregular clusters sometimes, however, have several small subcondensations and resemble loose swarms of small clusters. They contain all kinds of galaxies—spirals, ellipticals, and irregulars. These clusters are more numerous than the very rich clusters and range from aggregates of 1000 or so galaxies to small groups of a few dozen members or less. An example of a moderately rich irregular cluster is the Hercules cluster (Figure 20.14), and an example of a typical small one is the *Local Group*, the cluster of galaxies to which our Galaxy belongs.

The Local Group

The Local Group contains 21 known members, spread over a region about 3 million LY in diameter. The two largest members are both spiral galaxies—our own system and M31. Altogether, there are 3 spirals, 4 irregulars, and 14 ellipticals, of which 10 are dwarf elliptical galaxies, including three dwarf companions to

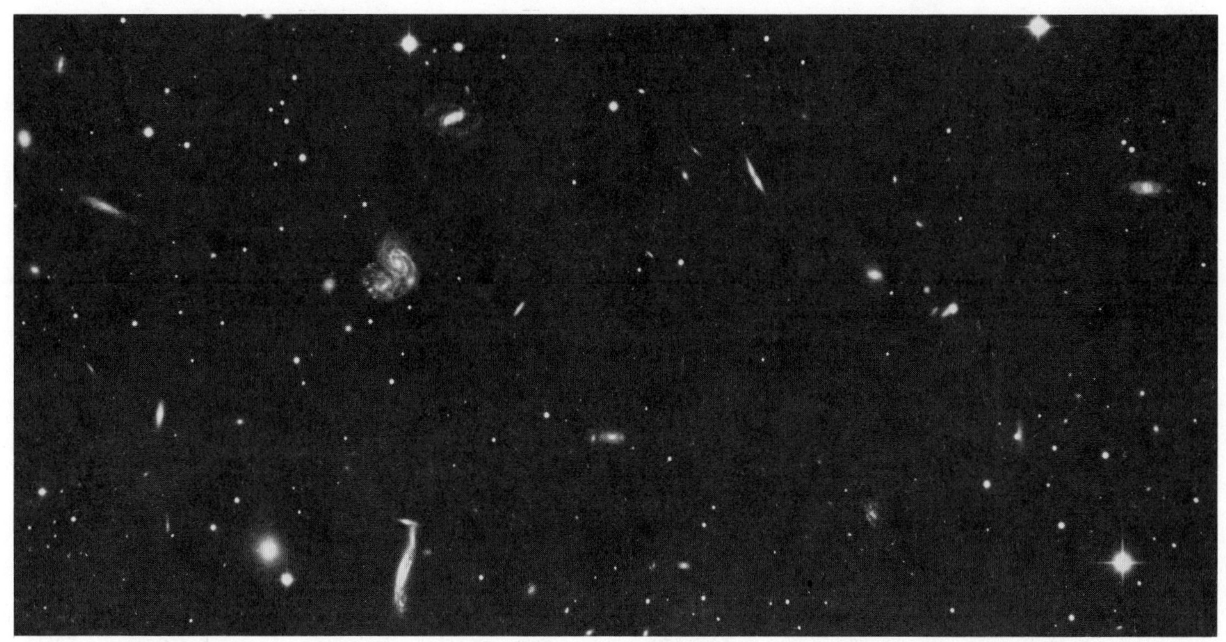

Figure 20.14
Irregular cluster of galaxies in Hercules, photographed with the 200-inch telescope. *(Hale Observatories)*

the Andromeda galaxy, And I, And II, and And III, discovered in 1972 by van den Bergh, and one in Carina, discovered in 1977 on the southern sky survey made with the 1.2-m Schmidt telescope at Siding Spring Observatory, Australia.

Appendix 17 lists the known members of the Local Group that all investigators agree are galaxies. Some of the data are very uncertain and are given in parentheses. Figure 20.15 is a plot of the Local Group; the galaxies have been projected onto an arbitrary plane centered on our Galaxy; then their distances from the center of the plot have been increased so that they are shown at the correct relative distances from us. The distances of And I, And II, and And III are assumed to be the same as that of M31.

The Neighboring Groups and Clusters and the Local Supercluster

Beyond the Local Group, at distances of a few times its diameter, we find other similar small groups of galaxies. Distances to the nearer of these can be determined from cepheid variables that are observed in them. Examples are groups of galaxies centered about each of the bright spirals, M51 and M81 (the galaxies pictured in Figures 20.6 and 20.7, respectively).

The nearest rich cluster of galaxies is the Virgo cluster (so named because it is in the direction of the constellation Virgo). Its distance is probably between 50 and 70 million LY—too remote to observe the cepheid variables in it. The distance to that cluster is estimated from the apparent faintness of the O and B supergiant stars that are barely observable in some of its galaxies, from the apparent brightness of globular clusters in M87, a bright elliptical galaxy in the cluster, and from the angular sizes of the largest H II regions resolved in some of the spiral galaxies. The Virgo cluster appears as a great cloud covering a 10° by 12° region of the sky, within which at least 1000 galaxies can be seen. The brightest of these are giant elliptical galaxies and large spirals; the faintest (observable) are dwarf ellipticals more or less like the dwarf elliptical galaxies in the Local Group.

Altogether, there are several thousand galaxies within about 70 million LY. Many of these galaxies are grouped into the Virgo cluster and a few other somewhat smaller clusters, and most of the rest of them are in relatively small clusters or groups like the Local Group. Beyond 70 million LY, galaxies seem to thin out and are sparse in space until a much larger distance is reached. This great cloud of galaxies, groups, and clusters is an example of a *supercluster,* and is called the *Local Supercluster.* Its existence was sus-

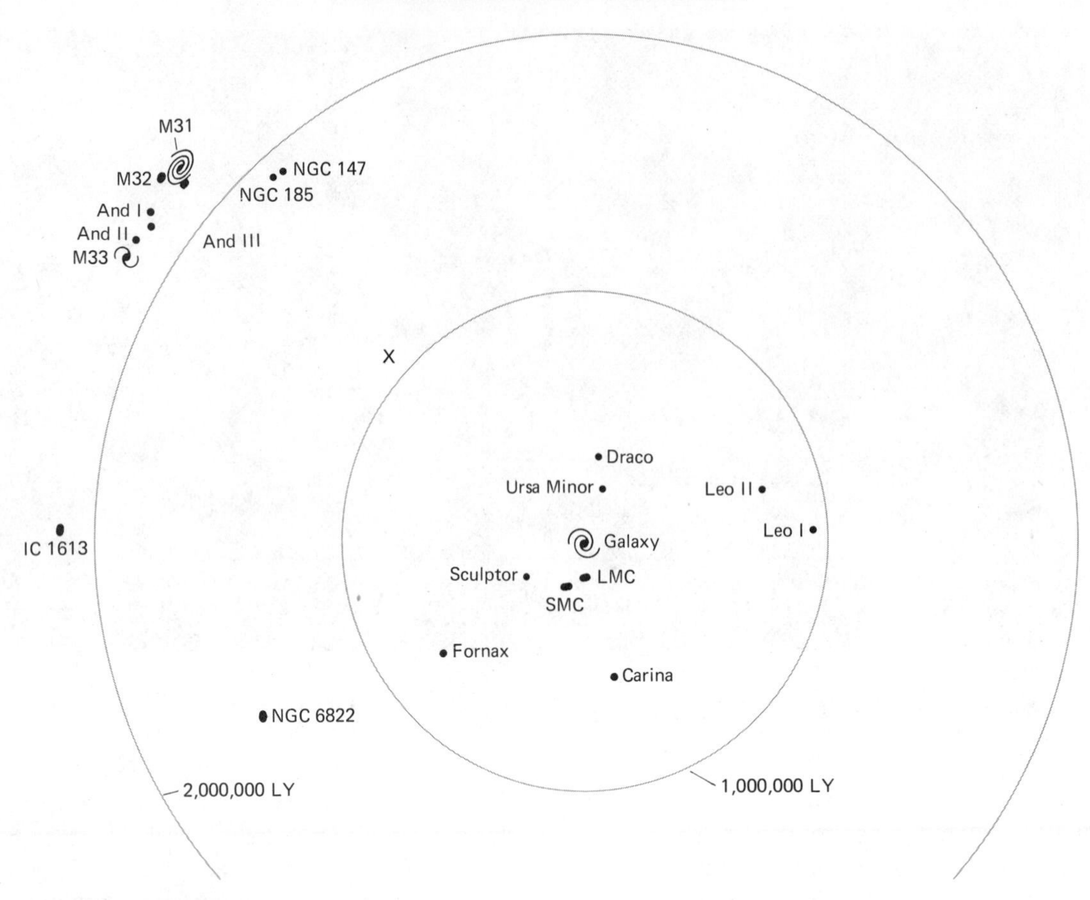

Figure 20.15
Plot of the Local Group. The X shows the approximate location of the center of mass of the group.

pected since the late 1920s, but was first demonstrated by G. de Vaucouleurs, who has studied the system extensively since about 1950. The overall diameter of the Local Supercluster is believed to be between 100 and 300 million LY. Its mass is probably on the order of 10^{15} solar masses.

More Distant Systems of Galaxies

There must be an enormous number of small groups of galaxies that lie beyond the local supercluster, but which are not conspicuous to us. In a typical system like the Local Group, for example, we would see only the one or two brightest galaxies, and would not identify the group as a cluster. Rich clusters of galaxies, on the other hand, stand out conspicuously and are recognized to very great distances.

The nearest rich regular cluster is the Coma cluster (in the constellation Coma Berenices), which lies at a distance of about 400 million LY, and which has a linear diameter of at least 10 million LY. Despite its great distance, we can observe more than 1000 member galaxies. The brightest galaxies in the cluster are two giant ellipticals, whose absolute visual magnitudes are between -23 and -24. The Coma cluster contains more and more member galaxies at magnitudes that are successively fainter. There is every reason to expect there to be dwarf elliptical galaxies present; if so, the total number of galaxies in the cluster might be tens of thousands.

In the 1950s, from the Sky Survey photographs obtained with the 48-inch Schmidt telescope at Palomar, I prepared a fairly complete catalog of the very rich clusters of galaxies that are not hidden by the obscuring clouds of the Milky Way and that lie in the northern three-quarters of the sky that is covered by the Palomar Sky Survey. The most remote clusters in my catalog are at distances of 4×10^9 LY—about 10

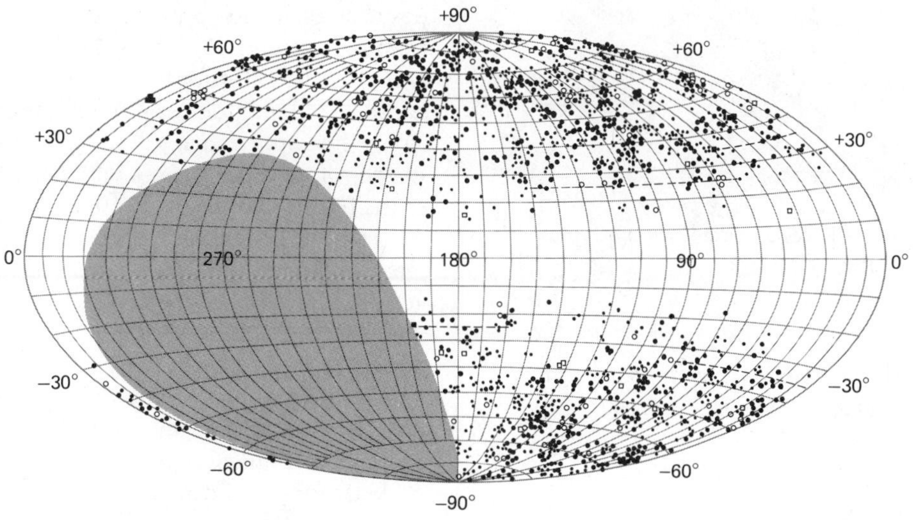

Figure 20.16
Distribution of rich clusters of galaxies. Each symbol on the plot (sky map) represents a cluster of galaxies; the larger the symbol, the closer the cluster. The more distant are probably about 4×10^9 LY away. The empty region along the middle of the plot is the zone of avoidance, as in Figure 20.3. The large, oval, empty region on the left is part of the sky too far south to be surveyed from Palomar Mountain, where the study of rich clusters was carried out by the author. The numbers refer to galactic coordinates.

times the distance of the Coma cluster. In all, 2712 rich clusters are listed. The total volume of space they occupy is about 5×10^{14} times that occupied by our own Galaxy—itself believed to be the entire "universe" just over half a century ago.

The distribution of these rich clusters (Figure 20.16) provides information on the general distribution of matter in the universe. We find that on the large scale the clusters are spread uniformly throughout space in all directions, out to the greatest distance for which our survey of them is complete. The universe, in other words, seems to be homogeneous on the large scale.

On the small scale, however, the clusters are *not* distributed uniformly. Just as we find a tendency for galaxies to be found in clusters, so the clusters themselves are found in superclusters—systems of clusters and groups of galaxies and probably individual galaxies. Like the Local Supercluster, superclusters have diameters of up to 300 million LY; they are separated by comparable distances. Evidence is accumulating that there are very few if any galaxies between superclusters. We think the superclusters are the largest inhomogeneities in the universe. On a much larger scale we find no trace of unevenness; on the whole, the universe appears to be remarkably homogeneous.

RADIO RADIATION FROM GALAXIES

The investigations that followed Jansky's discovery of cosmic radio waves in 1931 (Chapter 6) revealed that continuous radio energy is emitted from the disk and corona of our Galaxy. In addition, since World War II, many thousands of discrete radio sources, each occupying a small region in the sky, have been discovered and cataloged. Most of these discrete sources are known to be extragalactic.

These "radio galaxies" fall into two groups: "normal" and "peculiar" galaxies. The normal radio galaxies are simply ordinary galaxies that, like our Galaxy, emit some of their radiation at radio wavelengths. Radio energy with a continuous spectrum, for example, has been observed to emanate from the disks and coronas of all nearby spirals, and from the nuclei of many of them as well. Their radiation at radio wavelengths amounts, on the average, to a small fraction of their energy output at visible wavelengths. In addition, radiation at the 21-cm line of neutral hydrogen (Chapter 15) has been observed in many galaxies. The Doppler shift of this emission line indicates the same radial velocity for a galaxy as does the shift of a line in its visible spectrum.

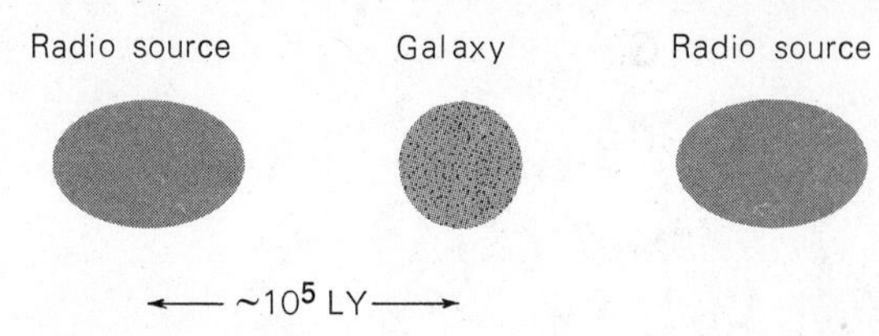

Figure 20.17
Location of radio sources associated with a typical peculiar radio galaxy.

The *peculiar radio galaxies*, usually called simply *radio galaxies*, are galaxies that emit unusually large amounts of radio energy—often more than 10^{44} erg/s (an amount of energy comparable to or greater than their entire optical luminosities). Most radio galaxies resemble ordinary galaxies in all their visual aspects.

The structures of extragalactic radio sources are most remarkable. In some of the galaxies, the radio waves come from single small regions within them, and in a few there are bright central sources surrounded by larger extended regions of radio emission. Usually, however, the radio galaxies are normal-appearing elliptical galaxies with *double* radio sources, with the waves coming from two extended regions on opposite sides of and up to millions of light years away from their centers (Figure 20.17). The radio galaxy NGC 5128 is even stranger in that it is associated with *two pairs* of radio sources—an intense inner pair overlapping the optical image of the galaxy and an asymmetric pair of larger sources with a much greater separation (Figures 20.18 and 20.19).

In some double-source radio galaxies in rich clusters the two radio-emitting regions appear folded together on one side of the galaxy, like a large radio "tail" trailing it. These *head-tail sources* are evidently caused by the motion of the galaxy through intracluster gas; the gas simply drags on the radio-emitting material, making it stream behind as the galaxy plows ahead. The head-tail sources thus provide additional evidence for gas in rich clusters.

The continuous radio emission from extragalactic sources always shows the characteristic increase in intensity with wavelength of synchrotron radiation (Chapter 15). The weak radio energy from normal galaxies can be accounted for (among other ways) by the electrons that would be ejected in collisions between cosmic rays and interstellar matter. These electrons

Figure 20.18
The peculiar elliptical galaxy NGC 5128 in Centaurus, a strong radio source, photographed with the 200-inch telescope. *(Hale Observatories)*

Figure 20.19
Optical and radio images of NGC 5128. The darkest areas within the center circle represent the optical image. Lighter areas, both inside and outside the circle, represent the radio image, the approximate location and extent of regions of radio emission.

Raymond Davis, Jr's, "neutrino telescope." *(Brookhaven National Laboratory)*

The open cluster M16 and associated nebulosity in Serpens. *(Hale Observatories)*

The Pleiades and associated nebulosity in Taurus, photographed with the Palomar Schmidt telescope. *(Hale Observatories)*

The Dumbbell nebula, a planetary nebula in Vulpecula, M27. *(Hale Observatories)*

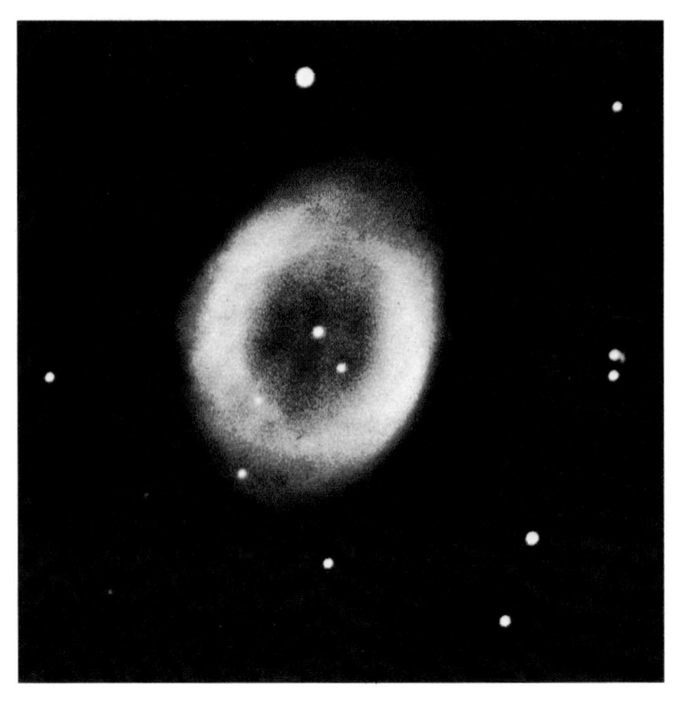

The Ring nebula in Lyra, M57, photographed with the 5-m telescope. *(Hale Observatories)*

The Crab nebula, photographed with the 5-m telescope. *(Hale Observatories)*

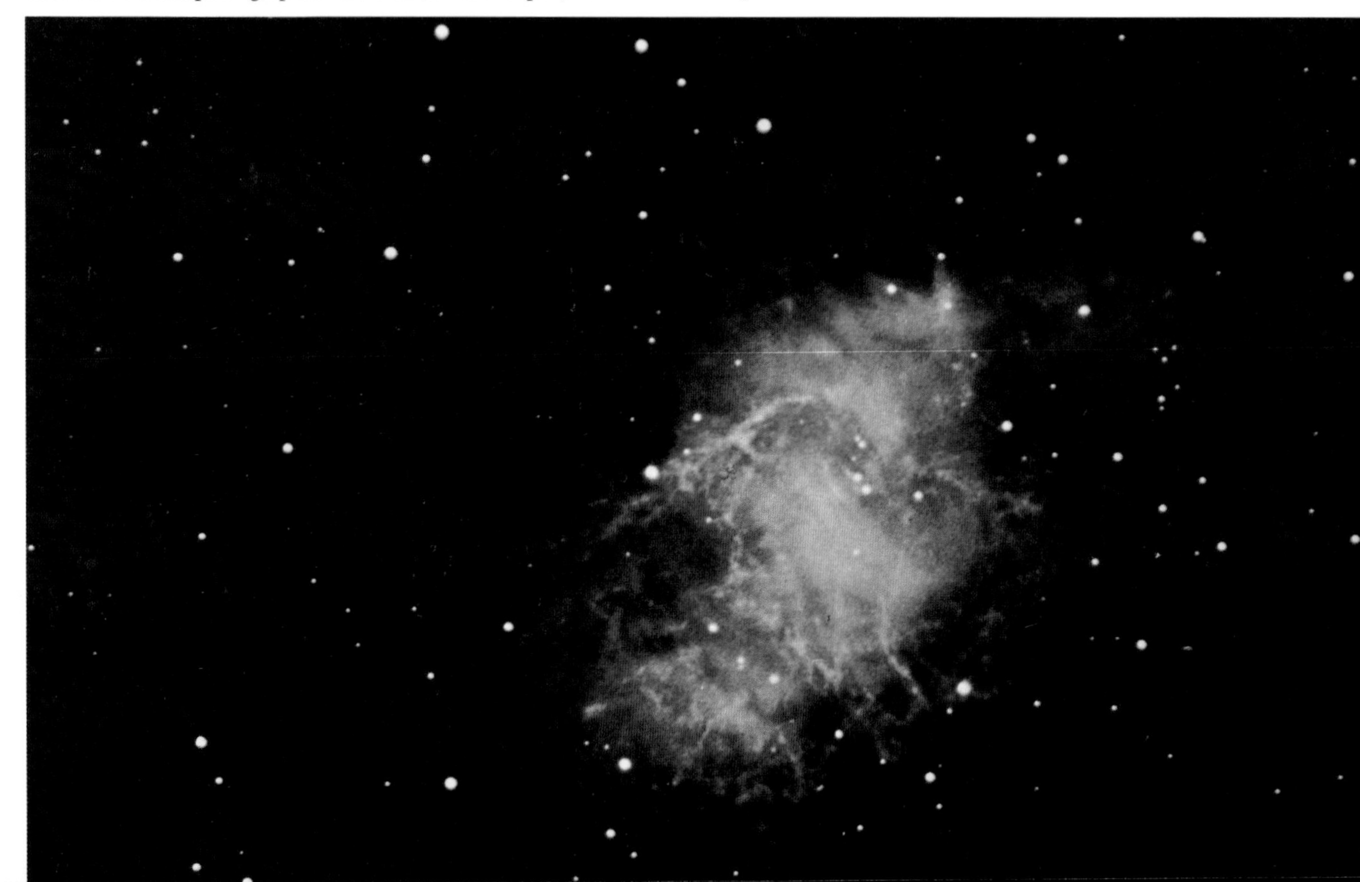

The Large Magellanic Cloud, the nearest external galaxy. *(U.K. Schmit Unit, Royal Observatory Edinburgh)*

M32, a small Local Group elliptical galaxy in Andromeda. *(U.S. Naval Observatory)*

The spiral galaxy M31 in Andromeda, photographed with the Palomar Schmidt telescope. *(Hale Observatories)*

The spiral galaxy M51 in Canes Venatici.
(U.S. Naval Observatories)

The edge-on spiral galaxy NGC 4565.
(U.S. Naval Observatory)

The irregular galaxy M82 in Ursa Major.
(U.S. Naval Observatory)

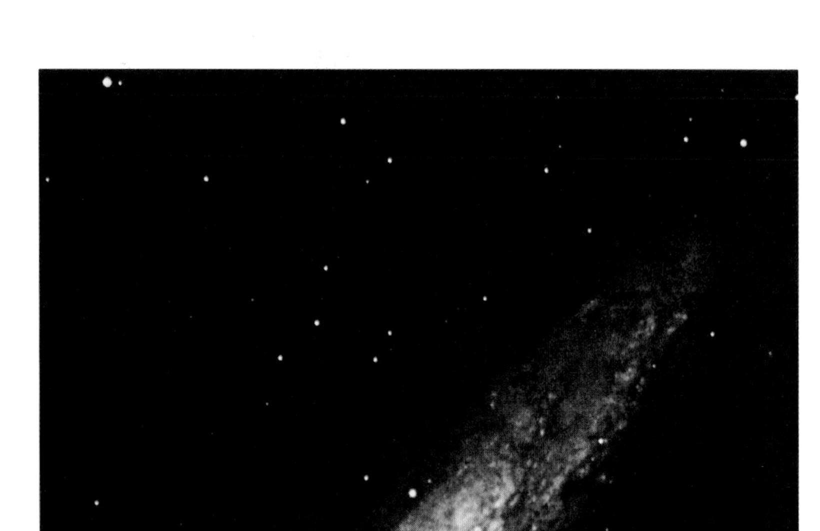

The spiral galaxy NGC 253 in Sculptor.
(Hale Observatories)

would then emit energy as they spiral about the lines of force of intragalactic magnetic fields. It is very difficult, however, to account for the radio emission from some radio galaxies. The amount of energy tied up in the magnetic fields and moving electrons in some radio galaxies is of the order 10^{60} ergs—an appreciable fraction of the total nuclear energy available to a galaxy!

Most astronomers now suspect some kind of violent activity in the nuclear regions of these galaxies is responsible. In fact, emission of radio radiation is a common property of various kinds of unusual galaxies that exhibit obvious evidence of explosive or violent events.

VIOLENT EVENTS IN GALAXIES

Violent extragalactic events can be external ones, as when galaxies collide or interact, or internal. The internal events are, at present, perplexing, and hence very interesting. Yet conventional interpretations of gravitational interactions between galaxies can account for a host of peculiar galaxian characteristics.

Encounters Between Galaxies

In the existing catalogs of peculiar galaxies are many examples of strange-appearing pairs of galaxies interacting with each other. We can now understand many of these in terms of gravitational tidal effects. The effects of tides between pairs of galaxies that chance to pass close to each other (especially likely in clusters) have been studied by astronomers Alar and Juri Toomre.

At first we might expect a tidal interaction between two galaxies to pull matter out of each toward the other. Such bridges of matter may form between the galaxies, but also there are "tails" of material that string out away from each galaxy in a direction opposite to that of the other. Because of the rotation of the galaxies, the tails and bridges can take on unusual shapes, especially when account is taken of the fact that the orbital motions of the galaxies can lie in a plane at any angle to our line of sight. The Toomre brothers have been able to calculate models of interacting galaxies that mimic the appearances of a number of strange-looking pairs actually seen in the sky (Figure 20.20).

Seyfert Galaxies

In 1944 Carl Seyfert described about a dozen spiral galaxies with very unusual spectra. The spectrum of the light from the nucleus of a *Seyfert galaxy* has strong, broad emission lines, which indicate the presence of very hot gas in a small central region. The gas must be rapidly expanding, with speeds of up to thousands of km/s. Some Seyfert galaxies are also strong radio sources, all have strong emission of infrared radiation from their nuclei, and recent observations with the Einstein telescope show them to be strong X-ray sources as well. The visual luminosities of these galaxies are usually about normal for spirals, but when account is taken of the infrared energy they emit, their total luminosities are found to be 100 or so times normal.

Whereas not many Seyfert galaxies are known (several dozen), the Seyfert properties can be recognized easily only in relatively nearby galaxies. It is quite possible that 1 or 2 percent of all spiral galaxies have these active nuclei. Alternately, it is possible that all spiral galaxies (even our own?) have these properties 1 or 2 percent of the time.

Other Galaxies with Active Nuclei

Many galaxies besides Seyferts show evidence of explosive ejection of matter from their nuclei. A possible example is M82 (Figure 20.12), an irregular galaxy with strong radio radiation, and a complex of expanding filaments surrounding its nucleus.

M87 is another interesting galaxy that is a strong radio and X-ray source. Short exposures of it show a luminous jet directed away from its nucleus, and a faint hint of a second radial jet in the opposite direction. Both the nucleus and the brighter jet emit synchrotron radiation, indicating magnetic fields and a source of relativistic electrons.

There are many other galaxies showing evidence of violence in their nuclei. The most spectacular examples are the quasars, which we discuss below. Intermediate between them and such galaxies as M82 and M87 or the Seyfert galaxies is a class known as N-type galaxies. N galaxies have small nuclei that are very bright compared to the main parts of those galaxies; often they appear as stellar images superimposed on faint wispy or nebulous backgrounds. Their bright nuclei indicate that enormous amounts of energy are being emitted from those regions.

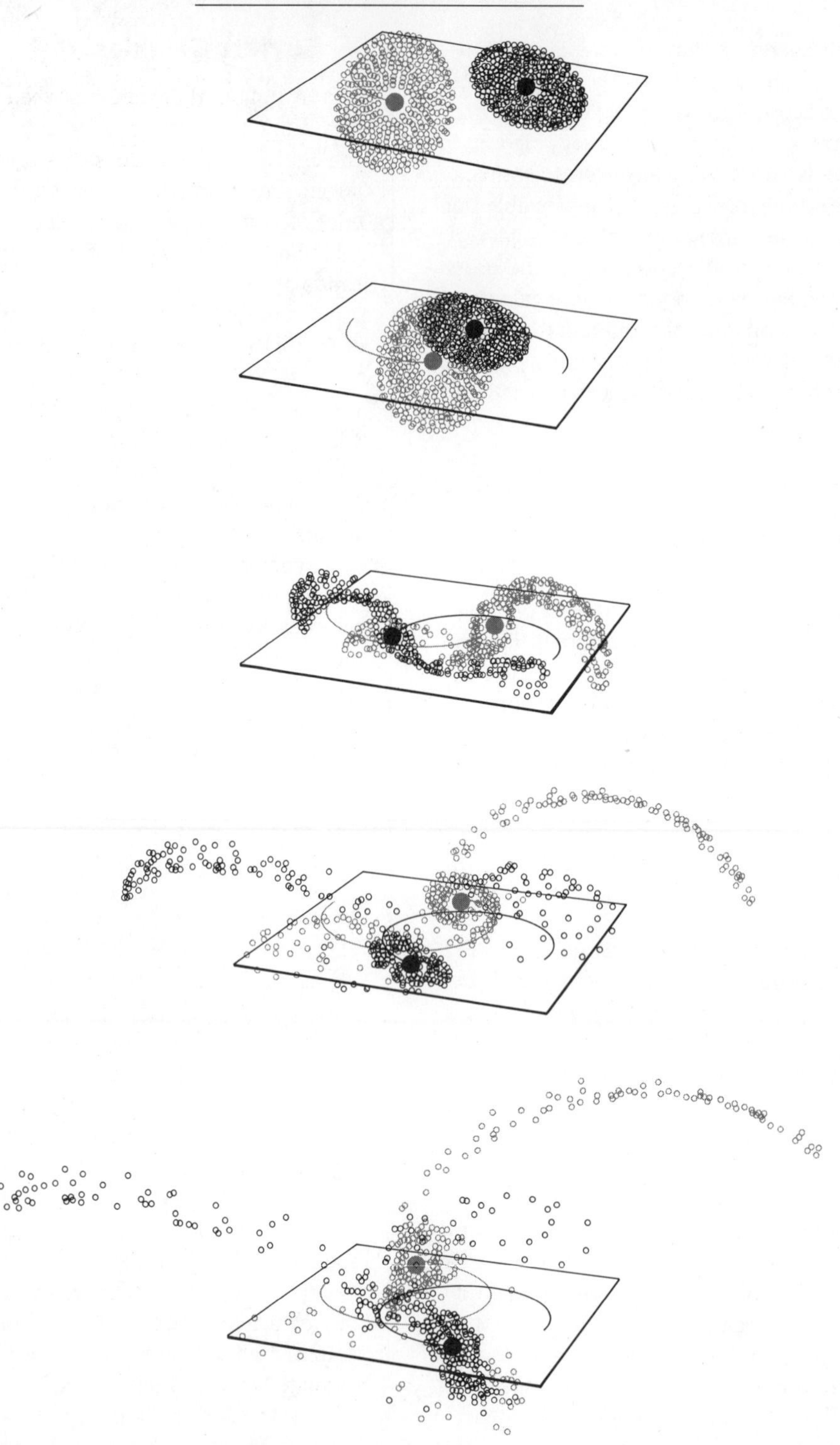

Figure 20.20
A sequence of five frames from a computer-produced motion picture that simulates the tidal distortion of two interacting galaxies. In this computer run, the initial conditions were chosen to see if the strange appearance of the pair of galaxies NGC 4038 and NGC 4039 could be accounted for in terms of tidal effects. Compare the last two frames with the photograph of the galaxies in Figure 20.21. The time interval between successive frames shown here is 200 million years. *(Courtesy Alar and Juri Toomre)*

QUASARS

In 1960 two radio sources were identified with what appeared to be stars. There seemed to be no chance that the identifications were in error, because the precise positions of the radio sources were pinned down by noting the exact instants they were occulted by the moon. By 1963 the number of such "radio stars" had increased to four. They were especially perplexing objects because their optical spectra showed emission lines that at first could not be identified with known chemical elements.

A partial explanation came in 1963 when Maarten Schmidt, at the Hale Observatories (since Director), recognized the emission lines in one of the objects to be the Balmer lines of hydrogen (Chapter 5) shifted far to the red from their normal wavelengths. If the redshift is a Doppler shift, the object must be receding from us at about 15 percent the speed of light! With this hint, the emission lines in the other objects were reexamined to see if they too might be well-known lines with large redshifts. Such proved, indeed, to be the case, but the other objects were found to be receding from us at even greater speeds. Evidently, they could not be neighboring stars; their stellar appearance must be due to the fact that they are very distant. They are called, therefore, *quasi-stellar radio sources*, or simply *quasi-stellar sources* (abbreviated QSS). In modern usage the term has been shortened to *quasar*.

The discovery of these peculiar objects prompted a search for others. By 1974, about 300 had been found. It is estimated that about one-third of all radio

Figure 20.21
The interacting pair of galaxies NGC 4038 and NGC 4039. *(Hale Observatories)*

sources are quasars. By 1979, systematic surveys in sample areas of the sky had shown that over the entire sky there must be about 20,000 quasars brighter than the eighteenth magnitude. The number of still fainter—and presumably more distant—quasars is not known, but there are certainly very many.

Quasars also emit X-rays. Before the Einstein X-ray telescope was launched, a number of discrete sources of X-rays in space had been detected, and a few of these were identified with quasars. But the earlier X-ray satellites (such as Uhuru and HEAO-1) could only detect the brightest X-ray sources. The fainter ones were not resolved, but appeared to blend together to produce a faint diffuse background of X-rays coming from all directions. Some astronomers, in fact, attributed that background of X-radiation to hot gas in space. The Einstein telescope, however, has resolved the background, and found that it is not emission from gas, but from separate discrete sources. About 14 percent are clusters of galaxies. Normal galaxies account for from 1 to 6 percent, and Seyfert galaxies are variously estimated to make up from 5 to 20 percent. The rest, probably more than 70 percent, are now thought to be quasars.

All quasars have spectra that show very large redshifts. The relative shifts of wavelength $\Delta\lambda/\lambda$ range up to 3.53, and for the majority $\Delta\lambda/\lambda$ is greater than 1.0. If we apply the exact formula for the Doppler shift we find that $\Delta\lambda/\lambda = 3.53$ corresponds to a velocity of recession of 91 percent of the speed of light.

We shall see in the next chapter that remote galaxies are receding from us also, and that there is a correlation between their radial velocities and distances. Most investigators regard the redshifts of the quasars as indicative that they are at very great extragalactic distances and that they conform to the same relation between radial velocity and distance that ordinary galaxies do; the vast majority of QSSs, however, have much higher speeds than any known galaxy, and must, therefore, be even more distant than normal-appearing galaxies.

The first known quasars were stellar-appearing objects of extragalactic distance that are also radio sources. Very many quasars, however, radiate too weakly at radio wavelengths to be detected. In 1965 A. R. Sandage, at the Hale Observatories, began a study of faint, very blue objects far from the plane of the Milky Way that had been presumed to be stars in the corona of our Galaxy. Spectra revealed that most such objects are, in fact, stars, but that some have large redshifts, as do the quasars. Thus radio-quiet quasars exist; they are sometimes called *quasi-stellar galaxies (QSGs)*.

It is now known that radio-quiet quasars outnumber the radio quasars by many times, although they are harder to find, because radio emission does not call attention to them. Many investigators believe that the radio-emitting quasars are temporary evolutionary stages of longer-lived radio-quiet ones.

Characteristics of Quasars

Although they differ considerably from each other in luminosity, the QSSs are nevertheless extremely luminous at all wavelengths. In radio energy, many are as bright as the brightest radio galaxies, and in visible light most are far more luminous than the brightest elliptical galaxies—their absolute magnitudes range down to −25 or −26. They are very blue compared to normal stars and galaxies. Most surprising of all is that almost all of them are variable, both in radio emission and visible light. Their variation is irregular, evidently at random, by a few tenths of a magnitude or so, but sometimes flareups of more than a magnitude are observed in an interval of a few weeks. Since quasars are highly luminous, a change in brightness by a magnitude (a factor of 2.5 in light) means an extremely great amount of energy is released rather suddenly. Moreover, because the fluctuations occur in such short times, the part of a quasar responsible for the light (and radio) variations must be smaller than the distance light travels in a month or so; otherwise light emitted at one time from different parts of the object would reach earth at different times (because of the range of distances light would have to travel to reach us), smearing out the variations.

The Significance of the Redshifts

Until now we have assumed that the redshifts of quasars are Doppler shifts, and that their distances are those of hypothetical galaxies of similar redshifts. As we have seen, however, such enormous distances require that these objects have truly astounding luminosities, and, in particular, the variations involve fantastic changes in luminosity over short periods and in relatively small regions. A few investigators, therefore, look to other interpretations for the redshifts.

For example, it has been suggested that quasars are extragalactic, but not at extreme distances. Some evidence has been advanced that they are objects that have been ejected from other nearby galaxies, or even from our own Galaxy. Most astronomers, however, favor the view that quasars do have the enormous distances indicated by their large redshifts.

For one thing, there are other classes of objects that are certainly galaxies but that have some characteristics in common with the quasars. The most important of these are the Seyfert galaxies. None of these has the great luminosity of a typical quasar, but all do, nevertheless, emit large amounts of energy from small regions. Many are radio sources, and some are variable in radio output. Like quasars, they emit X-rays. Finally, they tend to emit especially strongly in the ultraviolet. Many investigators suspect there to be a generic relation between these unusual galaxies and the quasars, in which case the latter are rather extreme examples of peculiar galaxies.

Also, some quasars seem to be associated with clusters of galaxies. In the dozen or so cases where the radial velocities are known for both the clusters and the quasars, they are the same. In these cases, at least, the quasi-stellar objects are evidently members of the clusters, whose redshifts are certainly representative of the general redshift-distance relation. Quasars with distances corresponding to the more common, much larger redshifts, however, would be too far away for us to observe galaxies or clusters of galaxies associated with them.

Figure 20.22
The elliptical galaxy NGC 4486 (M87); this 200-inch telescope photograph was taken with a very short exposure time to show the nuclear "jet."

A still stronger argument is given by Maarten Schmidt. Surveys by him, his former student Richard Green, and others indicate that the numbers of quasars increase extremely rapidly with increasing faintness. Statistically, the fainter objects of a particular class are more distant than brighter ones. In fact, detailed analysis shows that the only possible interpretation is that the density of quasars in space is low near us, but increases rapidly with distance in all directions. Now, if quasars are *not* extremely remote in the expanding universe, but are relatively nearby, we would have to be in a very special place—at the center of a sphere of quasars, with most of them near the surface of the sphere. On the other hand, if quasars have the distances implied by their large redshifts, we are looking far into the past to see them, at a time when they were actually more numerous than now, and all observers everywhere in the universe would see the same general picture. The favored view, therefore, is that quasars are remote phenomena that happened long in the past, but which do not occur much anymore; perhaps they represent an early phase of evolution in galaxies.

Just What Are Quasars?

There is, today, no generally accepted model for quasars, although there are many hypotheses. One theory is that quasars are large masses of gas, probably hundreds to millions of solar masses, in gravitational collapse, the material freely falling together under its mutual gravitation, perhaps into black holes. The energy they radiate comes, of course, from the gravitational potential energy they release as they contract. Another idea is that quasars are galaxies in the process of formation, and still another suggestion is that they are old galaxies in which the stars in their nuclear regions are colliding with each other.

A hypothesis of Phillip Morrison is that quasars are galaxies in which there is a rapidly rotating core of hot gas containing about 10^8 solar masses of material within a radius of about 0.1 LY. This *spinar* model assumes that there is a moderate magnetic field, and that ions and electrons ejected from the rotating gas mass, interacting with the field, give rise to the radiation. As the gas emits particles and loses energy it contracts and spins still faster, so the process continues.

Whatever the correct explanation of quasars, the weight of opinion today is that they, N galaxies, Seyfert galaxies, objects like M87 (with its jet), exploding galaxies like M82, radio galaxies, and even the center of our own galaxy may all be related phenomena connected with active nuclei of galaxies—with quasars at the extreme. Perhaps some or all are temporary phases that can occur in *any* galaxy.

EXTENT OF THE OBSERVABLE UNIVERSE

As far as we can see in all directions we find galaxies and clusters of galaxies. The more intrinsically luminous a galaxy is, the greater is the distance to which it can be observed. At the farthest depth of observable space, we see only the greatest giants among galaxies—that is, the brightest members of individual rich clusters—or perhaps quasars. If we knew the upper limit to the luminosities of galaxies or quasars, we could calculate the distance of those most remote observed objects. At present, we can only guess that they are between 5 and 15×10^9 LY away. In any event, they seem to extend to the greatest range of our optical, radio, and X-ray telescopes, and in much the same way in all directions.

SUMMARY REVIEW

Island universes: galactic or extragalactic?; catalogs of nebulae; debate over the nature of the nebulae; resolution of the controversy

Distance and distribution of galaxies: Cepheid variables; brightest stars; H II regions; Hubble's survey of the distribution of galaxies; distances of clusters of galaxies

Properties of galaxies: size; luminosity; rotation; mass; spiral galaxies; elliptical galaxies (giants and dwarfs); irregular galaxies

Clusters of galaxies: types; X-rays from clusters; gas sweeping; SO galaxies; the Local Group; the Local Supercluster; distribution of rich clusters; superclusters

Radio radiation from galaxies: radiation from normal galaxies; radio galaxies (or peculiar radio galaxies); double source radio galaxies; head-tail sources; stored energy in radio galaxies

Violent events in galaxies: encounters between galaxies; Seyfert galaxies; N galaxies; other galaxies with active nuclei (for example, M82 and M87)

Quasars: discovery; large redshifts; numbers; radiation; radio-quiet quasars; quasars as X-ray sources; the question of the significance of the redshifts; continuity with other types of strange galaxies; distribution in depth; theories of quasars

EXERCISES

1. Why is the term "island universe" a misnomer?

2. Why can we not determine distances to galaxies by geometrical methods, such as those by which we measure the parallaxes of stars?

3. Since globular clusters do not all lie in the plane of our Galaxy, how is it that they can be dimmed at all by interstellar absorption?

4. Make up a chart or diagram of distance indicators for galaxies, and show the distances to which each can be used. Include galaxies at the limit of our greatest telescopes.

5. Starting with the determination of the size of the earth, outline all the steps one has to go through to obtain the distance to a remote cluster of galaxies.

6. Suppose a supernova explosion occurred in a galaxy at a distance of 10^8 pc, and that the supernova reached an absolute magnitude of -19 at its brightest. If we are only now detecting it, how long ago did the event actually occur?

7. Why do we use the *brightest* galaxies in a cluster as indicators of its distance, rather than average galaxies in it? (*Hint:* There are two reasons, one involving the definition of "average," the other involving the distances to typical clusters.)

8. The tenth brightest galaxy in cluster A is at apparent magnitude $+10$, while the tenth brightest galaxy in cluster B is at apparent magnitude $+15$. Which cluster is more distant, and by how many times?

9. How can we determine the inclination of M31 (the Andromeda galaxy) to our line of sight?

10. Where might the gas and dust (if any) in an elliptical galaxy come from?

11. If extragalactic globular clusters exist as "galaxies," as what kind of galaxies would you classify them? How about extragalactic stars?

12. Are there subunits or subcondensations in the Local Group? If so, discuss them.

13. Let us imagine four classes of objects: those that are resolved in photographic images (say, whose angular sizes are more than about 1") and those that are not (smaller than 1"), and for each of these cases, those that have radial velocities less than 1000 km/s, and those that have radial velocities greater than 1000 km/s. Thus we can classify astronomical objects into four groups as follows:

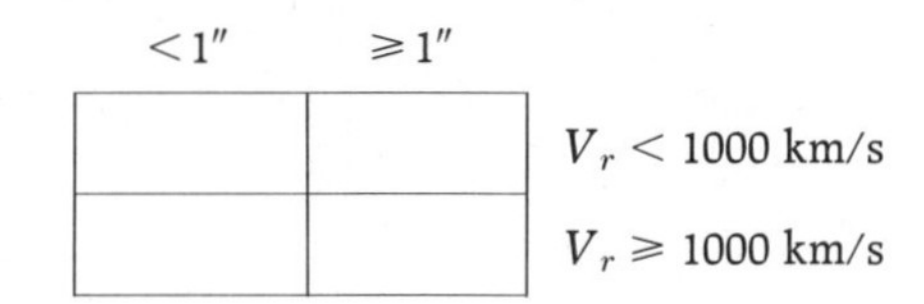

The moon, for example, would go into the upper right-hand box. Now classify each of the following:
(a) the sun
(b) Jupiter
(c) a typical Trojan minor planet
(d) stars in our Galaxy
(e) the Orion nebula
(f) galaxies in the Coma cluster
(g) Seyfert galaxies
(h) quasars
(i) Andromeda galaxy

CHAPTER 21

Milton Humason (1891–1972) drove a mule team up the old dirt trail to Mount Wilson during the construction days of the Observatory. He became interested in the work of astronomers and took a job as janitor, and later night assistant; eventually he became a full astronomer. His collaboration with Hubble in establishing the expansion of the universe won him many honors, including an honorary doctorate. *(California Institute of Technology)*

THE ORIGIN AND EVOLUTION OF THE UNIVERSE

The Reverend Richard Bentley, chaplain to the Bishop of Worcester and contemporary of Newton, was commissioned, for the magnificent fee of £50, to deliver eight sermons proving the existence of God, and opposing the views of "notorious infidels." Bentley thus wrote to Newton to inquire whether he thought intervention from a Supreme Being was not necessary to support the universe from collapse under its own mutual gravitation. Newton had no quarrel with the idea of God, but in the course of this correspondence with Bentley, Newton did express the opinion that if the universe were infinite and uniformly filled with stars the potential of gravitation would be the same everywhere, and the universe could then exist without collapsing.

Einstein's general theory of relativity, however, predicts otherwise. At least in their simplest form, the field equations, when applied to the universe at large, say that gravitation would make a static universe unstable, and it would collapse. But in that case, unless the universe is very young, why has it not collapsed already?

By the 1920s, theoreticians had guessed the answer, and at least one, the Belgian priest, Abbe Georges Lemaître (1894–1966), had pinpointed observations being made in America that should verify the correctness of the answer.

THE EXPANDING UNIVERSE

Those observations were measures of radial velocities of galaxies being made by Vesto M. Slipher at the Lowell Observatory in Arizona. Early in the century, the Observatory director, Percival Lowell, had asked Slipher to observe the spectra of the spiral nebulae. Lowell thought they might be solar systems in the process of formation, and if so perhaps their spectral features would resemble those of planets in our solar system. Thus it was that the Lowell Observatory, established mainly to study planets and search for life, became involved in work of the greatest importance in cosmology.

Slipher began this work in 1912. The first nebula he observed was what we now call the Andromeda galaxy. He found it to be approaching the sun with a speed of about 300 km/s. (Most of this motion is due to the sun's orbital velocity in our Galaxy.) Lowell recognized that this surprising result was important and urged Slipher to observe other "nebulae." By 1925, Slipher had found velocities of more than 40 "nebulae." He did not, of course, at first know the true nature of the objects he investigated, but he found very high velocities for what were later shown to be galaxies. A very few of them showed velocities of approach (negative radial velocities)—these are some of the Local Group galaxies—but Slipher found the vast majority to be moving away from us with speeds up to 1800 km/s.

The Law of Redshifts

In the mid-1920s there seemed to be some evidence that the velocities of the nebulae might be correlated with their distances, but the distances were too poorly known at that time to verify the suspicion. By

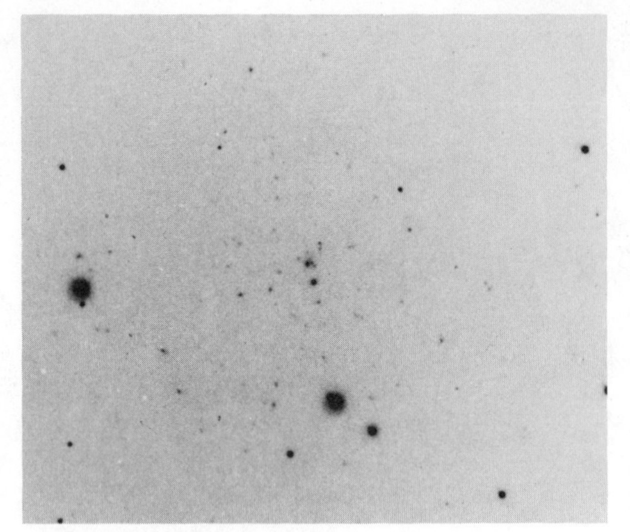

Figure 21.1
A remote cluster of galaxies for which there is a measured redshift (negative print). The cluster is receding from us at 36 percent the speed of light. It was discovered because its brightest member galaxy is a radio source, 3C295. Photographed with an image tube at the 3-m telescope of the Lick Observatory. *(Courtesy H. Ford, UCLA)*

1929, however, Edwin Hubble, at the Mount Wilson Observatory, had determined new estimates for the distances to many of the galaxies whose velocities had been measured by Slipher, and he found that the velocities of recession of those galaxies were, in fact, *proportional* to their distances.

Meanwhile, Milton L. Humason (1891–1972), also at Mount Wilson, was photographing spectra of fainter galaxies and galaxies in clusters with the 100-inch telescope. He and Hubble collaborated, and by 1931 had definitely established that the more distant a galaxy, the greater, in direct proportion, is its speed of recession, as determined by the shift of its spectral lines to the longer (or red) wavelengths. This relation is now known as the *law of redshifts*, or sometimes the *Hubble law*.

As the investigation progressed, more and more remote galaxies of greater and greater speed of recession have been found. The cluster of galaxies shown in Figure 21.1 moves away from us with a speed of 108,000 km/s—36 percent of the speed of light. Since 1975 redshifts of several even more remote clusters have been observed; they have speeds of recession of up to 60 percent the speed of light. The relative distances to clusters of known radial velocity (or redshift) are fairly well established (Chapter 20). To the accuracy of the present observations, these clusters have radial velocities that are proportional to their distances. The constant of proportionality, symbolized H, and called the *Hubble constant*, specifies the rate of recession of galaxies or clusters of various distances. The Hubble constant is now believed to lie in the range 40 to 120 km/s per million parsecs; many astronomers prefer a value of about 50 km/s per million parsecs. In other

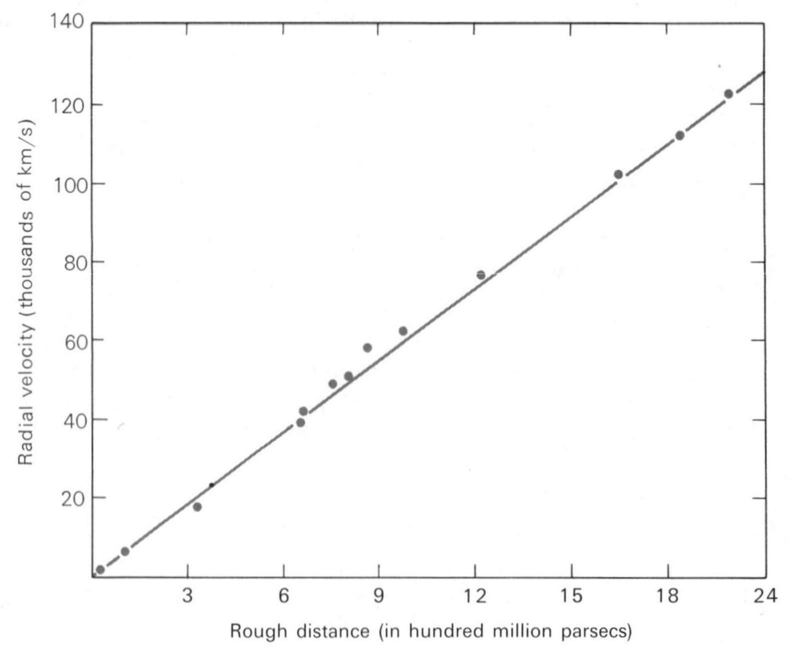

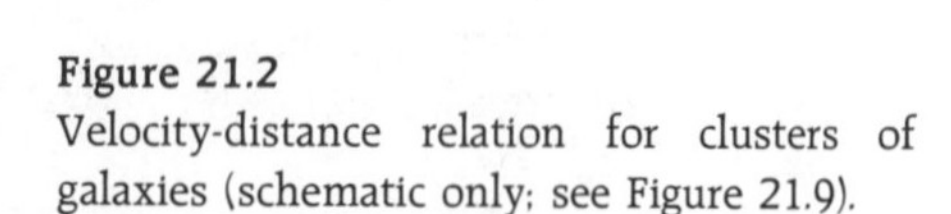

Figure 21.2
Velocity-distance relation for clusters of galaxies (schematic only; see Figure 21.9).

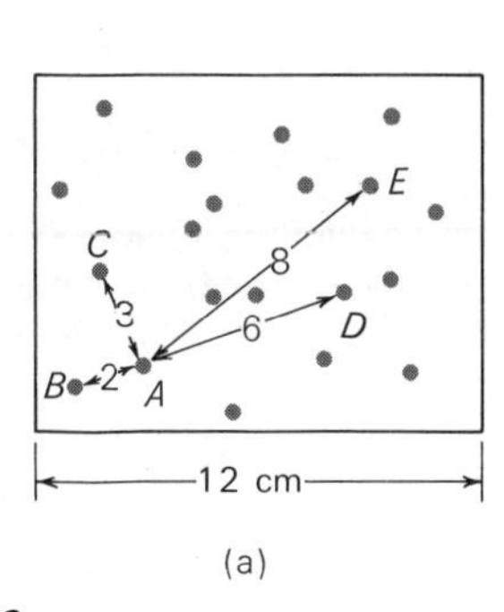

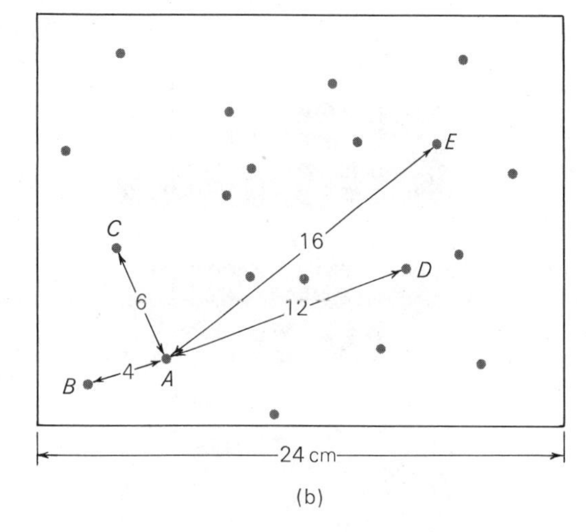

Figure 21.3
Separation of raisins in a hypothetical loaf of raisin bread that is 12 cm in diameter when it is set out to rise *(a)*, and grows to 24 cm in diameter at the end of one hour *(b)*.

words, a cluster moves away from us at a speed of 50 km/s for every million parsecs of its distance. This velocity-distance relation for clusters of galaxies is shown schematically in Figure 21.2.

Interpretation of the Law of Redshifts

The fact that distant galaxies all seem to rush away from us may seem to imply that we must, somehow, be at the "center" of the universe, but such is not the case. A simple and familiar analogy is a balloon whose surface is covered with dots of ink. As the balloon is inflated, the dots all separate, and a tiny insect on any one dot would see all other dots moving away from it; yet none of the dots is at a "center."

For a three-dimensional analogy, consider the loaf of raisin bread shown in Figure 21.3. Before being set out to rise, the loaf is 12 cm in diameter (a). The cook, however, mistakenly put far too much yeast in the dough, and within one hour the loaf has grown in size to a diameter of 24 cm (b). During the expansion of the bread, the raisins in it all separate from each other. Suppose, now, that we are on raisin A (intentionally *not* at the center of the loaf), and that we measure the speeds with which the other raisins recede from us. At the end of the hour, raisin B, originally 2 cm away, has increased its distance to 4 cm; evidently, its speed of recession has been 2 cm/hr. Raisin C has increased its distance from 3 to 6 cm, and so moved at 3 cm/hr, and so on for the other raisins. If we were to plot the data we would find a linear (straight-line) velocity-distance relation for the raisins analogous to that observed for galaxies. The same result (or plot) would be obtained irrespective of which raisin were chosen for "us." We would, in fact, observe an exactly linear velocity-distance relation for the raisins if we could measure their instantaneous velocities and distances, for *any* uniform expansion or contraction of the bread, whether or not it expands (or contracts) at a constant rate.

If the universe were now changing scale—either uniformly expanding or contracting—we should expect to observe a linear relation, or proportionality, between the speeds and distances of galaxies and clusters, just as in the raisin bread analogy. In an expanding universe, galaxies would move away from us at speeds proportional to their distances; in a contracting universe, they would *approach* us at speeds proportional to their distances. The law of redshifts shows that the universe is now expanding; in this expansion the galaxies and clusters are being carried farther and farther from each other.

Note, however, that we observe galaxies, not where they are "now," but at the distances they had when the light by which we see them left those objects. Because of the finite speed of light, those observed distances are millions of years "out of date" for galaxies millions of light-years away. If light traveled with infinite speed, we would observe those objects at much greater distances. Similarly, their radial velocities may have changed from the values indicated by the

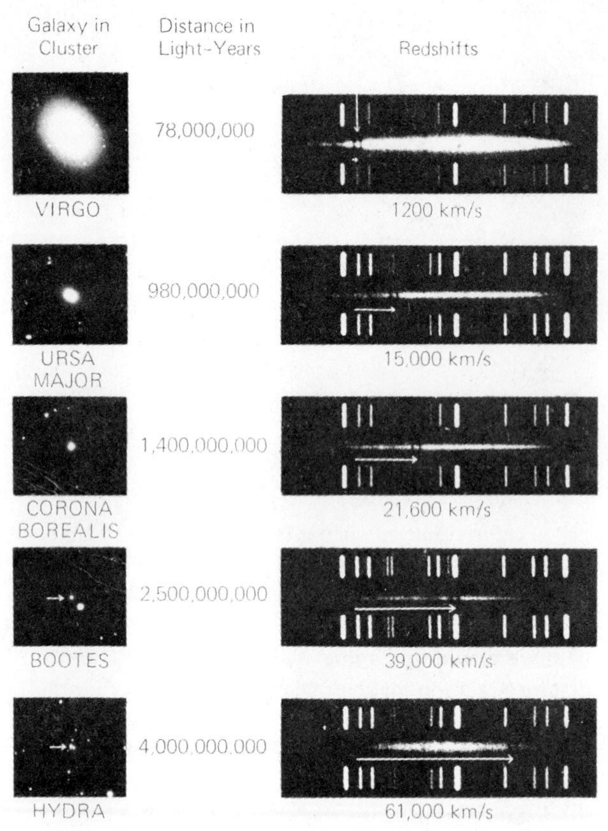

Figure 21.4
(Left) Photographs of individual galaxies in successively more distant clusters; *(right)* the spectra of those galaxies, showing the Doppler shift of two strong absorption lines due to ionized calcium. The distances are estimated from the faintness of galaxies in each cluster, and are provisional. *(Hale Observatories)*

spectra of their observed light. The velocity-distance relation, therefore, should not necessarily be linear for objects at great distances.

The expansion of the universe does not imply that the galaxies and clusters of galaxies themselves are expanding. The raisins in our analogy do not grow in size as the loaf expands. Similarly, their mutual gravitation holds galaxies and clusters together, and they simply separate as the universe expands, just as do the raisins in the bread.

Galaxies in clusters do, of course, have individual motions of their own superimposed on the general expansion. Galaxies in pairs, for example, revolve about each other, and those in clusters move about within the clusters. In fact, a few galaxies in nearby groups and clusters move fast enough within those systems so that they are actually *approaching* us even though the clusters of which they are a part are moving, as units, away.

Theory Predicts an Expanding Universe

Occasionally we hear arguments that the universe may not be expanding—that there could be some unknown cause of the redshifts other than the Doppler effect. But it would be very hard to understand the universe if it were not expanding.

Soon after he introduced the general theory of relativity in 1916, Einstein attempted to apply the new relativistic theory of gravitation to cosmology. He assumed that the universe is homogeneous on the large scale—that is, that matter in it is distributed more or less evenly throughout space. But with this assumption the field equations of general relativity predict that the gravitation of the matter in the universe should cause it to contract. Of course the Hubble law had not yet been discovered and the most natural assumption to make was that the universe is static. Einstein found that the only way he could make the universe static was by modifying the field equations by adding a new term, which provides a cosmic repulsion. The term is called the *cosmological constant;* if it has a positive value it represents a repulsion between distant objects that can balance their mutual gravitation. Einstein assigned just the right value to the cosmological constant to yield a static universe.

In 1922, however, the Russian cosmologist Alexandre Friedmann found nonstatic solutions to the field equations that did not require a cosmological constant, but that applied to a universe that is expanding. In 1927, Georges Lemaître, mentioned above, independently suggested an expanding universe, and also predicted a linear relation between the radial velocities of galaxies and their distances. Lemaître had spent some time in America and was aware of the observations of Slipher, which he thought might confirm the theory.

The Hubble law, therefore, is exactly what was expected from relativity theory. To be sure, the cause of the expansion remains unexplained, but at least at present we can account for the expanding universe in terms of known laws of physics. Einstein is said to have remarked that his modification of his original field equations was "the biggest mistake of my life." Most modern cosmologists tentatively assume that there is no cosmological constant.

COSMOLOGICAL MODELS

Having examined the observational basis for cosmology, we may investigate some of the hypotheses that have been suggested to describe its structure and evolution. Such a hypothesis is called a *cosmological model.*

All cosmological models are based on certain assumptions, without which no hypothesis could be formed. For example, an assumption common to most models, and to all discussed here, is that the redshifts of galaxies are really Doppler shifts and that the universe is really expanding. Another even more basic postulate is the cosmological principle.

The Cosmological Principle

It would be impossible even to begin to formulate a theory of the universe if we did not assume that our observations give us information that applies to the whole universe, not just to our own part of it. In other words, we must assume that the part of the universe that we actually observe is representative of the entire cosmos, and that we are not located in some very unusual place, fundamentally different from the rest of the universe. Stated more generally, we must assume that whereas there will be local variations in the exact details of galaxies and clusters, all observers, everywhere in space, would view the same large-scale picture of the universe. This assumption is known as the *cosmological principle*.

Evolutionary Cosmologies

As the universe expands, it thins out. If no new matter is being created—that is, if the total mass of the universe is constant—and if no previous contraction preceded the present expansion, it follows that all the matter of the universe must have once been close together. This suggests that some original "explosion" may have started the universe expanding. As the matter in it expanded and cooled, it condensed into galaxies and clusters of galaxies. These, like the raisins in the bread, all receded, and are still receding from each other at relative speeds that are proportional to their separations. The idea that such an explosion took place is called the "big bang" theory.

The big bang hypothesis implies that the universe has a finite age—at least since the explosion. Since no galaxy could have a greater age, all galaxies would be aging, and evolving, together. Models based on some version of the big bang theory, therefore, are usually called *evolutionary models* or *evolutionary cosmologies*.

Alternate Cosmological Models

Due in part to the rather limited amount of observational and experimental data at our disposal, there is an almost unlimited range of conceivable cosmological models that could describe the universe to the accuracy of our present knowledge. For example, models with various positive values of the cosmological constant have been worked out. Some models have been suggested that are not even based on the cosmological principle in its strict sense—such as hierarchical models in which clusters are clustered into second-order clusters, which in turn are parts of third-order clusters, and so on to an infinite hierarchy of clustering, all in an expanding universe. Other models have been advanced that involve changing values of physical constants with time, or a gradual increase in the masses of atoms with time.

It is probably healthy to keep in mind all of these possibilities, and others, so that we do not become committed to too narrow a set of conditions on what the universe may be like—a narrowness that could restrict our ability to recognize new ideas that could ultimately lead to new and important advances. But to be fruitful, all models must be susceptible to test (see Chapter 1). It is easiest, especially with the limited available data, to test the simplest models first—and possibly eliminate them. If we find simple models that satisfy the data, we should not reject them until new data are uncovered with which they are not compatible.

The simplest model of all is the *steady state* theory, which received a great deal of interest in the late 1940s and 1950s. The problems of the finite age implied by the evolving cosmologies, and of beginnings and endings, led astronomers Herman Bondi, Thomas Gold, and Fred Hoyle to speculate on the possibility of a cosmological model derived from a single assumption, and extension of the cosmological principle called the *perfect cosmological principle*. It states that the universe is not only the same *everywhere* (except for local small-scale irregularities), but at *all times*. In other words, if we could return to life thousands of millions of years from now, we would find the universe, on the whole, as it is now. As galaxies separate from each other and thin out in the expansion, new matter is continually created in space to keep the overall density of the universe constant. Since nothing changes in time, the theory is called the *steady state*. As we shall see, however, modern observations rule out the steady state theory. We turn, then, to the next simplest models that are seriously considered today—those based on general relativity.

Relativistic Evolving Cosmologies

The relativistic evolving models of the universe are based on the following assumptions: (a) the cosmological principle holds; (b) the mass of the universe is conserved; (c) the correct interpretation of the Hubble law is that the universe is expanding; and (d) general relativity governs the universe. Perhaps we are presumptuous to assume that we already know the ultimate laws of physics, but the field equations of general relativity do allow us to predict, from observational data, how the universe is expanding, thereby changing in scale and density. We shall consider only the simplest relativistic cosmologies, namely those for which the cosmological constant is zero (no repulsion force).

Without repulsion the motions of galaxies after an initial explosion would be slowed down by gravitation in the same way that a rocket fired away from the earth is slowed down by the earth's gravity. If the rocket has an initial speed of about 11 km/s (the "escape velocity"), it will move away from the earth forever. The galaxies, if they were started fast enough in the big bang, could likewise move away from each other forever; if they started slower than some critical escape velocity, however, they would eventually fall back together again.

The cosmological principle dictates that the universe must expand uniformly in all directions. Therefore, the nature of the expansion can be described by a single scale factor. The scale can be represented as the distance between any two corresponding points in the expanding universe; in our raisin bread analogy, it could be the distance between any two raisins. That scale factor is often denoted by the symbol R. Because the expansion changes all distances proportionately, a knowledge of how R varies in time would tell us the history and future of the universe.

The field equations of general relativity allow us to calculate the variation of R with time, $R(t)$, if we know the Hubble constant and the mean density of matter in space. Some sample solutions are shown in Figure 21.5, which is a plot of R against time. In addition to solutions corresponding to the relativistic evolving models, for comparison the solution for the steady-state model is also shown. The zero on the time scale is the present. The units of R are arbitrary, but its changes in time have the correct relative values. Note that at present the scale is increasing (for the universe is expanding).

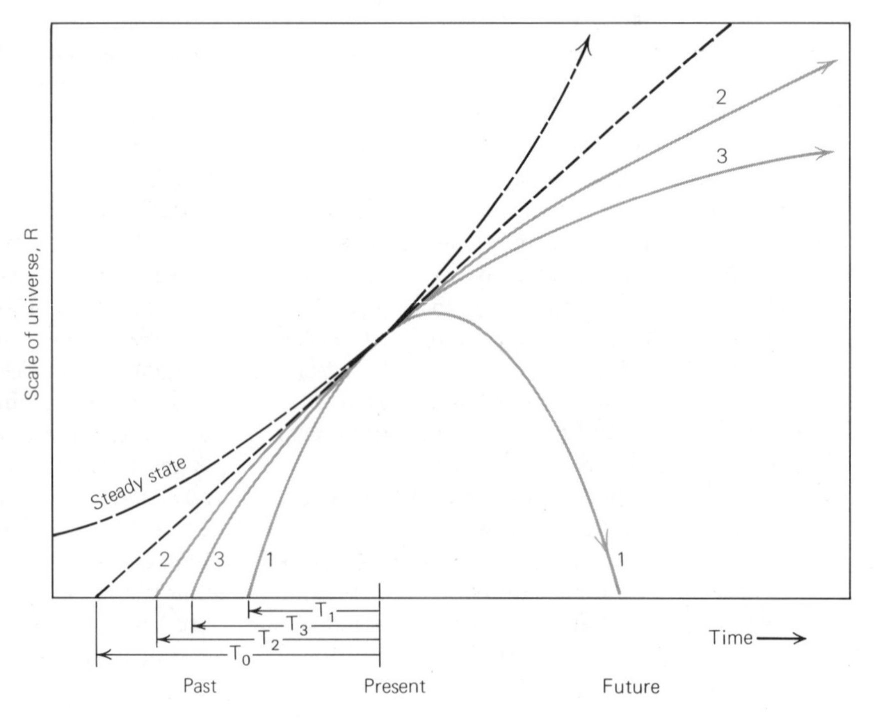

Figure 21.5
A plot of $R(t)$, the scale of the universe, against time for various cosmological models. Curve 1 represents the class of solutions for closed universes, curve 2 represents the class for open universes, and curve 3 is the critical solution for the boundary between open and closed universes. The long-short dashed line is the solution for the steady state model. The evenly dashed line is the case for an empty universe.

TABLE 21.1 Parameters of Models of Evolving Relativistic Cosmological Models

Case (curve in Figure 21.5)	Kind of Universe	Age, T (for $H = 50$ km/s · 10^6 pc) (in units of 10^9 yr)	Sign of Space Curvature, k	Mean Density ($H = 50$ km/s · 10^6 pc) (g/cm^3)
1	closed (oscillating?)	$T < 13.3$	$+1$	$>5 \times 10^{-30}$
2	open	$13.3 < T < 20$	-1	$<5 \times 10^{-30}$
3	flat	$T = 13.3$	0	5×10^{-30}

If the universe were empty of matter, there would be no gravitation, and no slowing of the expansion. If we extrapolate the expansion back to a zero scale (dashed line in Figure 21.5), we would find the expansion of the universe to have begun at a time T_0 in the past; that is, T_0 is the maximum possible age of the universe for the evolutionary models considered here. If $H = 50$ km/s · 10^6 pc, T_0 is about 19.5×10^9 yr.

If there is matter in the universe, its gravitation must slow the expansion. A measure of the effect is given by the mean density of matter in space, which would be the universal density if the atoms of all of the stars and galaxies could be scattered about with complete uniformity. For a high enough mean density the gravitation of matter is enough eventually to stop the expansion, after which the universe will contract. This case is illustrated by the curve labeled 1 in Figure 21.5. If we could imagine a cause for a future big bang, the expansion could be repeated, leading to alternate expansions and contractions (or an *oscillating* universe). In this model, the age T_1 since the last big bang has to be less than $\frac{2}{3}\, T_0$, or, less than 13×10^9 yr (for the same Hubble constant).

If the mean density is too low, on the other hand, the expansion, although ever slowing, will still continue forever (curve 2 in Figure 21.5). The age of the universe is then greater than 13×10^9 yr. The boundary between these two classes of solutions occurs for a critical mean density; in this case the universe barely expands to infinity but only after an infinite time (curve 3). If the Hubble constant, H, is 50 km/s · 10^6 pc, the critical density is about 5×10^{-30} g/cm^3. The age, T_3, for this model is exactly $\frac{2}{3}\, T_0$, or about 13×10^9 yr.

The future of the universe therefore depends (in these models) on how strongly the matter decelerates the expansion.

We saw in Chapter 18 that the presence of matter curves spacetime, and that the shapes of light paths depend on the curvature of spacetime. It is conventional to denote the sign of the curvature by the symbol k, which can take on the values -1, 0, and $+1$. For the critical density that divides open and closed universes, k equals 0, and spacetime is flat, or Euclidean.

For higher densities, k equals $+1$, and spacetime is said to be positively curved. In this case, light travels in closed paths. The geometry of spacetime is such that two light rays leaving one place in parallel directions will eventually converge. Light would be permanently trapped in a region of the universe in this case, and we would, in a sense, be living in a finite universe, but one without boundary or edge.

Alternatively, for a mean density lower than the critical value, light travels on open curves; parallel lines diverge, and space is said to be open and negatively curved, with $k = -1$. The geometries that describe the universe in the cases $k = +1$, 0, and -1, are said to be elliptical, parabolic or Euclidean, and hyperbolic, respectively, corresponding to the kinds of orbits a rocket would have that leaves the earth with a speed less than, equal to, or greater than the escape velocity. The parameters of the various models described here are summarized in Table 21.1

Nature of the Big Bang

Georges Lemaître considered an early model of the big bang itself. He envisioned all the matter of the universe starting in one great bulk he called the *primeval atom*. The primeval atom broke into tremendous numbers of pieces, each of them further fragmenting, and so on, until what was left were the present atoms of the universe, created in a vast nuclear fission. Lemaître thought the left-over radiation from that cosmic fireball was the cosmic rays. In a popular account of his theory he wrote, "The evolution of the world could be compared to a display of fireworks just ended—some few red wisps, ashes and smoke. Standing on a well cooled cinder we see the slow fading of the suns and we try to recall the vanished brilliance of the origin of the worlds."

We know today that the cosmic rays are not from the big bang, but probably from supernovae in our Galaxy. We also know much more about nuclear physics, and that the primeval fission model cannot be correct. Yet Lemaître's vision inspired more modern work, and in some respects was quite prophetic.

In the 1940s the American physicist George Gamow and his associates considered a universe with the opposite kind of beginning—nuclear fusion. Gamow's universe started with fundamental particles that built up the heavy elements by fusion in the big bang. His ideas were close to our modern view, except that the conditions in the primordial universe were not right for atoms to fuse to carbon and beyond, and only hydrogen and helium should have been formed in appreciable abundances. The heavier elements, we think, formed later in stars (Chapter 17).

The modern picture of the big bang is called the *standard model*. At first, when the temperature was over 10^{10} K, the matter consisted only of the fundamental particles, such as protons, neutrons, electrons, positrons, and nuetrinos. After about 100 s, however, the temperature had dropped to 10^9 K, and the particles began to combine to form some heavier nuclei. This nucleosynthesis continued for the next few minutes, during which about 25 percent of the mass of the material formed into helium. Some deuterium also formed (deuterium is an isotope of hydrogen with a nucleus containing one proton and one neutron) but only a small amount—probably less than one part in a thousand. The actual amount of deuterium formed depends critically on the density of the fireball; if it was fairly high, most of the deuterium would have been built up into helium. Scarcely any nuclei heavier than those of helium are expected to have survived. So the composition of the fireball when nuclear building ceased is thought to have been mostly hydrogen, about 25 percent helium, and a trace of deuterium.

For the next million years the fireball was like a stellar interior—hot and opaque, with radiation passing from atom to atom. During this time, the temperature gradually dropped to about 3000 K, and the density to about 1000 atoms/cm^3. At this point the fireball became transparent. The radiation was no longer absorbed and was able to pass freely throughout the universe. After about 1000 million years the matter condensed into galaxies and stars.

Now we must emphasize that the fireball must *not* be thought of as a localized explosion—like an exploding superstar. There were no boundaries and no site of the explosion. It was everywhere. The fireball is still existing, in a sense. It has expanded greatly, but the original matter and radiaion are still present and accounted for. The stuff of our bodies came from material in the fireball. We were and are still in the midst of it; it is all around us.

TESTS OF COSMOLOGICAL MODELS

Tests of cosmological models depend on the differences in the way different models predict that the universe will evolve in time. We cannot, of course, wait long enough to note future changes, but we do (and must) look into the past when we look far into space. Thus we test models by comparing the appearances of the universe at different distances.

Evidence for Evolution in the Universe

One clue that the steady state theory is probably wrong comes from the quasars. If we assume that they obey the same velocity-distance relation that galaxies do (in the professional parlance, that their redshifts are "cosmological") the distribution of quasars in space would have to be uniform to be consistent with a steady state. As we saw in Chapter 20, Schmidt's studies show that the overwhelming majority of quasars—far more than would be expected in a uniform distribution—are at very great distances. In other words, the quasars lie at such distances that nearly all of them represent phenomena occurring long ago in the history of the universe. It is as if the quasars are things that existed early in the evolution of the universe—perhaps things happening to or in galaxies far in the past—and that occur only rarely today. If this analysis is correct, it clearly violates the perfect cosmological principle, and hence the steady state cosmology.

Another test for cosmology is the way the density of matter in space changes in time. Evolutionary cosmologies require that matter gradually thins out in space; remote galaxies, therefore, should appear relatively closer together (as they were in the past) than nearby ones. The steady state theory, however, predicts that the average distance between galaxies never changes and so should be the same for galaxies at all distances. For the very faint, distant galaxies, it is not yet possible to measure distances accurately enough to tell whether the density of the universe is changing in time. On the other hand, many of the sources of radio waves that have been cataloged are probably remote galaxies or clusters of galaxies, and a very great excess in the number of weak cosmic radio sources is incompatible with the steady state.

Radiation From the Primeval Fireball—the Microwave Background

Gamow predicted that as the universe expanded and cooled, the intense radiation in the fireball—radiation that had been continually absorbed and reemitted by the hot gas—would find the universe suddenly transparent, and would subsequently float freely about in space. But to see the source of that radiation, we would have to look far into the past—to the time of the big bang—and hence far off in space where everything is rushing away from us at nearly the speed of light. The light emitted by the hot early universe should therefore now be transformed by the Doppler shift into radio waves.

Now, when that radiation was emitted—just before the universe thinned out enough to become transparent—it was characteristic of that emitted by a blackbody with a temperature of about 3000 K. Today, however, the redshift should change it to resemble that from a cold blackbody just a few degrees above absolute zero.

Gamow's prediction was more or less forgotten because at the time there was no way to observe the radiation. By the mid 1960s, however, Princeton physicist R. H. Dicke realized that microwave radio telescopes could then be built that might detect that dying glow of the big bang. He, P. J. E. Peebles, P. G. Roll, and D. T. Wilkinson confirmed that the theory was correct, and began construction of a suitable microwave receiver on the roof of the Princeton biology building. They were not, however, the first to observe the radiation.

Unknown to them, a few miles away, in Holmdel, New Jersey, Arno Penzias and Robert Wilson, of the Bell Telephone Laboratories, were using the Laboratories' delicate microwave horn antenna to make careful measures of the absolute intensity of radio radiation coming from certain places in the Galaxy. But they were plagued with some unexpected background noise in the system that they could not get rid of. They checked everything, and eliminated the Galaxy as a source, also the sun, the sky, the ground, and even the equipment.

At one point they realized that a couple of pigeons had made their home in the antenna, and nested up near the throat of the horn where it was warmer. Penzias and Wilson could chase the birds away while they observed, but they found that the birds left, as Penzias puts it, a layer of white, sticky dielectric substance coating the inside of the horn. That substance would radiate, producing radio interference. They disassembled the horn and cleaned it, and the unwanted noise did go down somewhat, but it did not go away completely.

Figure 21.6
Robert W. Wilson *(left)* and A. A. Penzias *(right)* with the horn-shaped antenna with which they discovered the microwave background radiation. *(Bell Laboratories)*

Finally, Penzias and Wilson decided that they had to be detecting radiation from space. Penzias mentioned it in a telephone conversation with another radio astronomer, B. Burke, who was aware of his work. Burke got Penzias and Wilson in touch with Dicke, and it was soon realized that the predicted glow from the primeval fireball had been observed. Since then the radiation has been very thoroughly checked throughout the entire radio spectrum, with observations from ground-based radio telescopes, with instruments carried aloft in balloons, and even with a receiver in a U2 reconnaissance airplane. The microwave background radiation closely matches that expected from a blackbody with a temperature of 2.7 K.

Penzias and Wilson received the Nobel Prize for their work in 1978. And perhaps almost equally fitting, just before his death in 1966, Lemaître learned about the discovery of his "vanished brilliance."

A Look Back to the Big Bang

Since this radiation comes from the time when the fireball first became transparent, it is at the farthest point in space to which we can presently observe. If we could see that radiation visually, it would be as if it were coming from an opaque wall, and no radiation from a more distant source could ever reach us—for that source would have to lie farther back into time where it would be behind that opaque wall.

Figure 21.7 is a schematic cross section of four-dimensional spacetime (see Chapter 18). The distance from us, in some arbitrary direction, is shown along the horizontal axis, and time increases along the vertical axis. Light paths are shown as 45° lines, as in Chapter 18. Our world line (geodesic) is upward along the time axis, and we are presently at time t_0. Times t_1, t_2, and t_3 are successively farther back into the past. The world lines of several other hypothetical observers on distant galaxies are shown and are labeled *a, b, c,* and so forth, in order of increasing distance. Note that they all recede from us with time as the universe expands. (Because of the peculiarity of spacetime diagrams of an expanding universe, the diagram is not strictly correct for uniform time and distance measures along the axes, but nevertheless serves to illustrate the situation.)

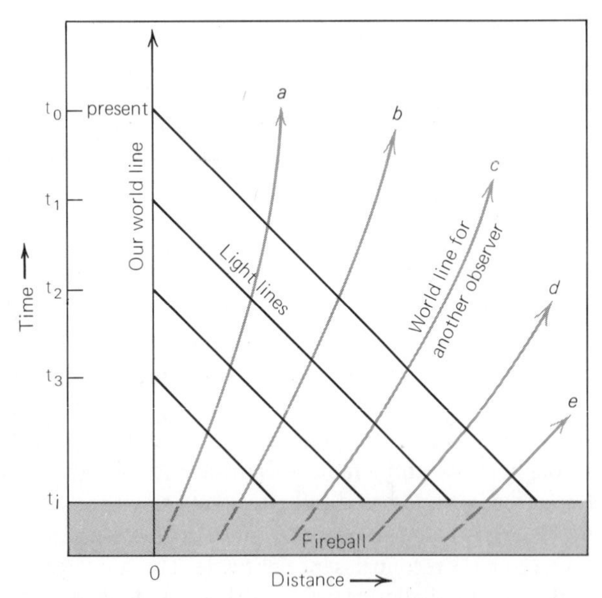

Figure 21.7
Schematic representation of world lines for us and for several other hypothetical observers on distant galaxies in the universe on a two-dimensional cross section of spacetime. The straight diagonal lines are light paths. (See text for discussion.)

When we look out into space, we look back into time along a light geodesic (diagonal lines). Note that the light we see comes from successively more distant objects successively farther into the past at times when they were actually separated less far from us in spatial coordinates. There is a limit, however, to how far back we can look. The fireball is presumed to have become transparent at time t_i, and the shaded region at the bottom of the diagram corresponds to the opaque fireball into which we cannot see. Thus the limiting distance we could observe in a particular direction is at the point where the light path to us intercepts the top of the shaded region at time t_i.

As time goes on, we can see farther and farther away, and more galaxies would come into view (if we had a large enough telescope) for, as the time from the big bang becomes greater we are looking farther into the past to see the fireball, and hence farther away in space. Note that at earlier times such as t_3, t_2, and t_1, we saw only relatively nearer objects, and there were fewer galaxies between us and the threshold provided by the fireball itself. Thus, not only does the universe expand with time, but the part of it accessible to observation becomes greater as well.

On earth, the microwave radiation is very feeble compared with, say, sunlight. But far off in intergalactic space, that radio background is by far and away the most intense radiation around. Moreover, it is extremely *isotropic*—the same in every direction. The observed radiation comes equally from all directions and gives no direction to a "center" of the universe; the universe, its "center," and its origin are all around us. In fact, on the small scale, measurements show that the microwave background is equal in intensity in different directions in the sky to within one part in 10,000.

Motion of the Galaxy

There is, however, a very slight variation in the intensity of the radiation on opposite sides of the sky. This is because of our own motion through space. If you approach a blackbody, its radiation is all Doppler-shifted to shorter wavelengths, and resembles that from a slightly hotter blackbody; if you move away from it, the radiation appears like that from a slightly cooler blackbody. Such an effect has been searched for and observed in the microwave background.

The measurements are very difficult because the signal searched for is very tiny compared to the radiation of the earth's atmosphere. They must, therefore, be made at very high altitude. Princeton physicist D. Wilkinson puts his apparatus in a high-flying balloon, and University of California, Berkeley, physicist George Smoot sends a similar receiver to an altitude of more than 20,000 m in one of NASA's converted U2 planes. Both have found the variation expected due to our motion, and their results are in reasonable agreement. They show that our entire Galaxy is moving with a velocity of about 540 km/s toward the constellation Hydra. It is a very important result, and when it has been thoroughly analyzed it will probably provide new insight about the dynamics of the Local Supercluster.

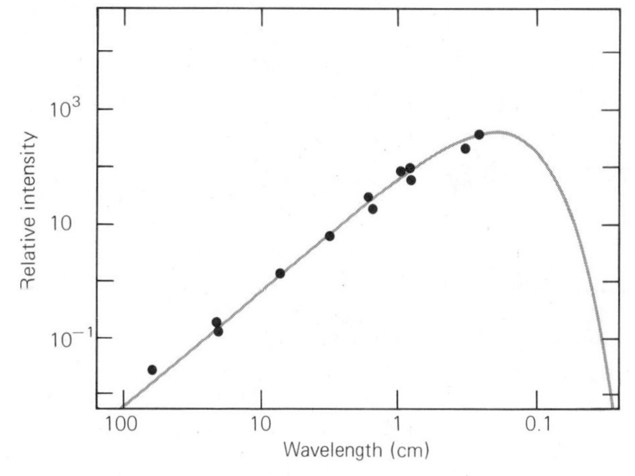

Figure 21.8
Intensity versus wavelength for the microwave background radiation. The points are observed values, and the solid curve is the theoretical one for a blackbody of temperature 2.7 K.

Is the Universe Open or Closed?

There are several observational tests by which we hope to be able to distinguish between the evolving cosmological models. As of this writing, the observations are still not critical enough to reach a definite conclusion, but it is worth describing some examples of these tests.

One is the determination of the mean density of matter in space. We have noted that knowledge of it is sufficient to calculate the way R changes with time from the field equations of general relativity. We can estimate the mean density from the number of galaxies and clusters we observe out to a given distance, and from a knowledge of the masses of these objects. There is considerable uncertainty in the masses of clusters of galaxies, and we do not know how much matter (if any) may exist in intergalactic space. Nevertheless, such estimates indicate a mean density less than 10^{-30} g/cm^3, and probably near 10^{-31} g/cm^3. This is below the critical density, and suggests an open universe, but the estimates are too uncertain to be sure of the conclusion.

We also saw that the production of deuterium in the fireball is very sensitive to the density of the universe within the first few minutes after the big bang. The proportion of deuterium in interstellar space is thought to be a measure of that formed in the fireball, for in stellar interiors it is rather quickly converted to helium. It is very difficult to detect deuterium in space, but careful measures show that the ratio of deuterium to hydrogen is probably in the range 10^{-4} to 10^{-5}. From this crude estimate the density of the fireball at the critical time can be inferred, and from that knowledge it is possible to predict what present-day density would result. The calculation suggests a present density of about 10^{-31} g/cm^3, again pointing to an open universe.

Another less obvious test involves the speeds of receding galaxies at great distances. The evolutionary models we have described predict the expansion to slow down due to the gravitational forces between galaxies. Because of these changes in the rate of expansion of the universe, the radial velocities of remote galaxies should deviate from a relation exactly proportional to their observed distances; that is, a graph of radial velocity versus distance should not necessarily be a straight line for very distant galaxies (see Exercise 6). The exact form of the redshift-distance relation is different as predicted by each cosmological model.

The differences between the observable quantities predicted by various models are small until very large distances are reached—that is, until we look back through an appreciable interval of time. Unfortunately, precise distances of remote objects, as well as their luminosities and other characteristics, are very difficult to determine, and critically accurate observations are required.

One procedure for obtaining relative distances of clusters is from their brightest member galaxies. Figure 21.9 is a plot of the radial velocities (on a logarithmic scale) of nearly a hundred clusters of galaxies against the apparent magnitudes of their brightest members, adapted from a diagram by A. Sandage. The magni-

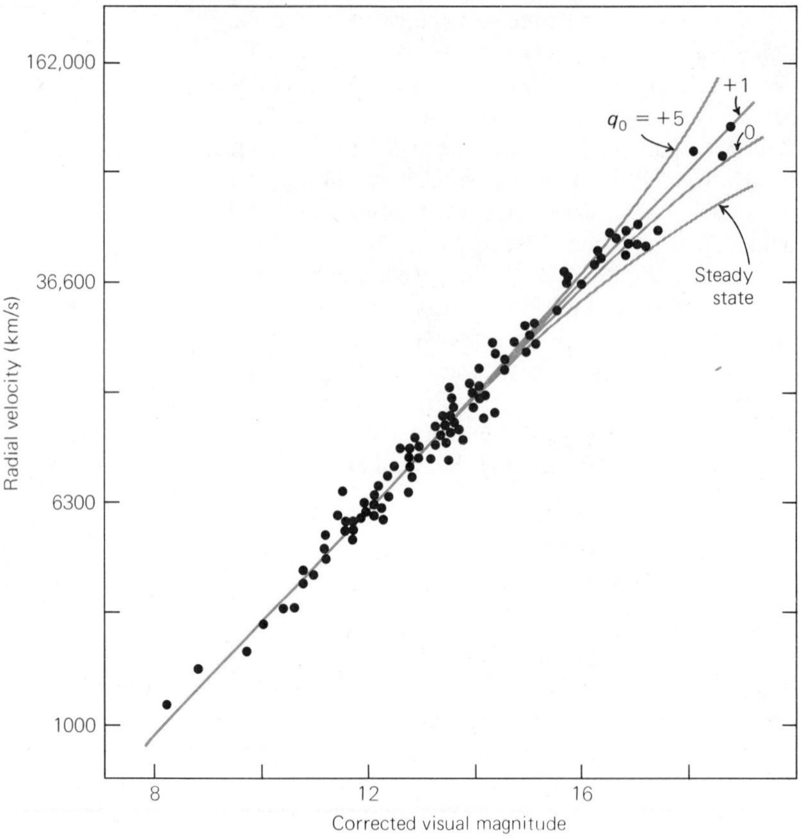

Figure 21.9
Hubble diagram for clusters of galaxies. Ordinates are radial velocities (on a logarithmic scale), and abscissas are the apparent magnitudes of the brightest cluster galaxies, corrected for interstellar absorption. Points are observed values, and the solid curves are the predicted relations for various cosmological models. *(Adapted from a diagram by A. Sandage)*

tudes of the brightest cluster galaxies are indicators of distance (see Chapter 20); they have been corrected for interstellar absorption in our Galaxy and for certain effects of the redshift. The different lines on the plot correspond to predictions of different cosmological models. We can see from the scatter of the points that more precise observations are needed to apply this test. Also we do not yet know how to take account of changes in the luminosities of galaxies due to evolution of the stars in them.

In summary, observational tests of cosmological models appear to rule out the steady state, but are not yet definitive enough to choose between the various evolving models predicted by general relativity. And of course the observations may well be compatible with many other cosmological theories not yet considered.

Conclusion

Present-day evidence suggests that the universe is evolving from a big bang thousands of millions of years ago. We should not, however, take our ideas too seriously. Our thinking in the realm of cosmology may be as rudimentary as that of the Greek philosophers 2500 years ago, who believed that all heavenly motions must occur in perfect circles. It will be very surprising if, after the next century, scientists finally settle on one or another of the specific theories we have described here.

The following words, written by Edwin Hubble in 1936, are still appropriate[1]:

Thus the explorations of space end on a note of uncertainty. And necessarily so. We are, by definition, in the very center of the observable region. We know our immediate neighborhood rather intimately. With increasing distance, our knowledge fades, and fades rapidly. Eventually, we reach the dim boundary—the utmost limits of our telescopes. There, we measure shadows, and we search among ghostly errors of measurement for landmarks that are scarcely more substantial.

The search will continue. Not until the empirical resources are exhausted, need we pass on to the dreamy realms of speculation.

[1] *The Realm of the Nebulae,* Yale University Press, 1936, pp. 201–202. Quoted by permission of the publisher.

SUMMARY REVIEW

The law of the redshifts (Hubble law): early observations by Slipher; Hubble and Humason collaboration; Hubble constant; interpretation as the expansion of the universe; theoretical prediction of the expansion of the universe; cosmological constant

Cosmological models: cosmological principle; evolutionary cosmologies; alternate models; steady state theory; relativistic evolving models; scale of the universe; open universe; closed universe; curvature of spacetime; elliptical, parabolic (flat, Euclidean), and hyperbolic geometries; nature of the big bang; Lemaître's primeval atom; Gamow's fusion model; modern standard model

Tests of cosmological models: evidence for evolution; distribution of quasars and remote radio sources; demise of the steady state; the microwave background; search for the radiation from the big bang; discovery by Penzias and Wilson; confirmation; looking back to the big bang; motion of our Galaxy; is the universe open or closed?; density of matter in space; deuterium in space; form of the Hubble law

EXERCISES

1. A cluster of galaxies is observed to have a radial velocity of 60,000 km/s. Assuming that $H = 50$ km/s per million parsecs, find the distance to the cluster.

2. Plot the velocity-distance relation for the raisins in the bread analogy from the number given in Figure 21.3.

3. Repeat Exercise 2, but use some other raisin than A for a reference, and measure the distances with a ruler. Is your new plot the same as the last one?

4. Why can the redshifts in the spectra of galaxies *not* be explained by the absorption of their light by intergalactic dust?

5. Show that T_0, the maximum possible time since the beginning of the expansion, is equal to $1/H$, that is, the reciprocal of the Hubble constant.

6. Consider the plot of the radial velocities against the distances of remote clusters of galaxies. Are the measured distances and radial velocities that are plotted the *present* values of these quantities for the clusters? In each case, why? If not, try to describe how the diagram might differ if we did plot the present-day values of cluster distances and velocities.

7. Suppose we were to count all galaxies out to a certain distance in space. If the universe were not expanding, the total number counted should be proportional to the *cube* of the limiting distance of our counts. (Why?) Taking account of the finite time required for light to reach us, describe the relation that would be *observed* between the total count and the limiting distance for (a) the steady-state theory, and (b) the evolutionary theory of the universe.

8. Resketch Figure 21.5 for the case of the open universe (curve 2), and indicate on the time axis where the most remote quasars are and also where the 3 K blackbody radiation is coming from (in time).

9. Refer to Figure 21.7. Describe how the background blackbody radiation must have differed at times t_3, t_2, and t_1 from how it appears now.

10. Sketch and discuss a schematic spacetime diagram like that in Figure 21.7, showing the world lines of several galaxies for the case of a closed universe which will contract again sometime in the future.

11. What are the problems in postulating a closed universe if the Hubble constant, H, were 100 km/s $\cdot$ 10^6 pc and the oldest star clusters in our Galaxy were 10^{10} years old?

12. Suppose the universe will expand forever (that is, an open model). Describe what will become of the radiation from the primeval fireball.

CHAPTER 22

John Muir (1838–1914), Scots-American naturalist, was largely responsible for the establishment of Yosemite and other national parks. He wrote, "When we try to pick out anything by itself, we find it hitched to everything else in the universe." *(Sierra Club)*

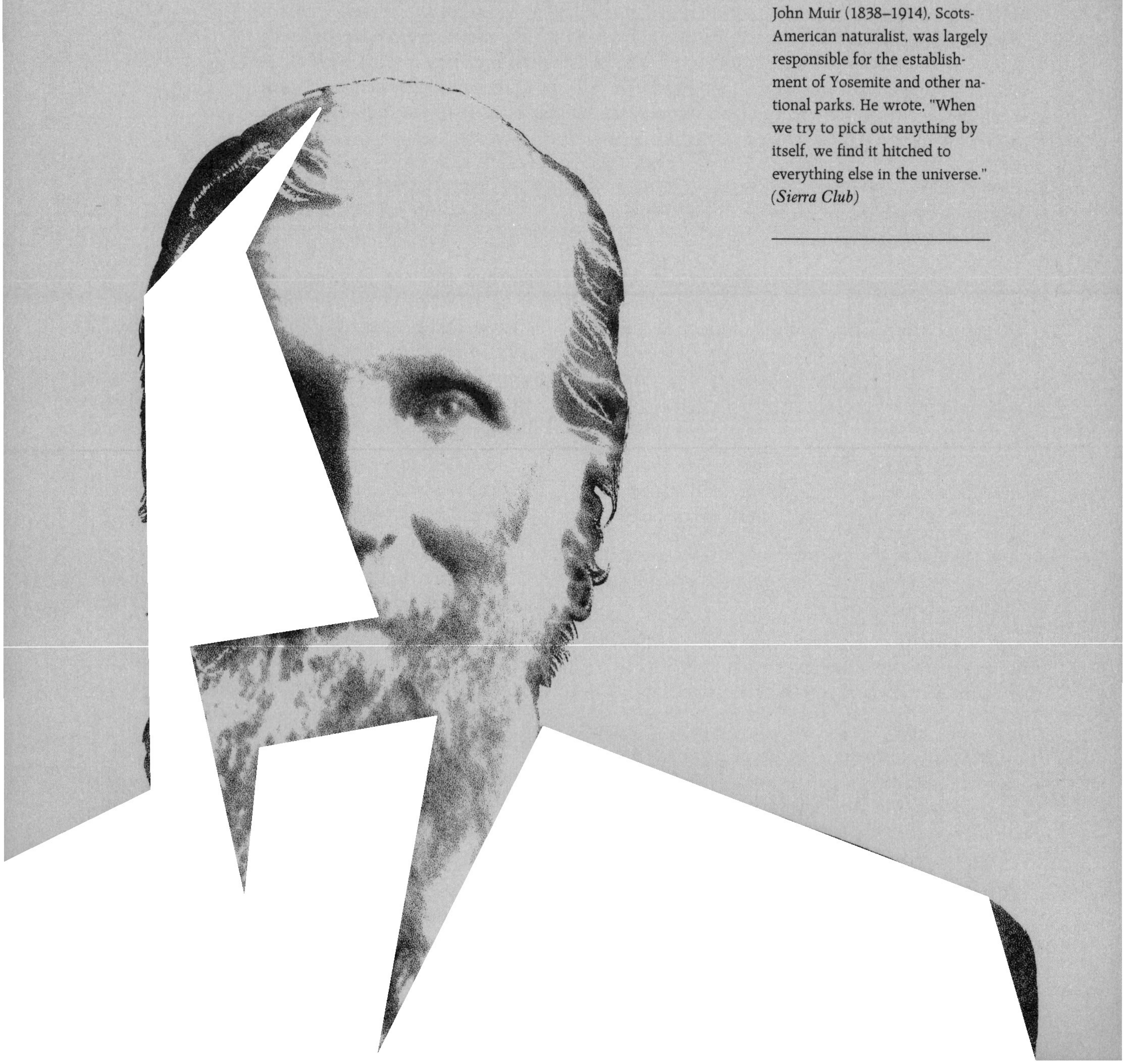

EPILOGUE

The ancients, as we have seen, thought the earth to be in the center of the universe, occupying a very special place in space. They were at least partly right, for the earth is not located typically in the cosmos. Let us imagine that we could view the universe from a really typical place in it—say from within an imaginary space ship with which we could travel at will with the speed of light. (Let us ignore here the limitations placed on such travel by practical considerations, and also the relativistic effects such a speed would have on distance and time.)

Around us, from this typical place, we would find virtually complete blackness—empty space. Perhaps a few faint, indistinct blurs of light might be seen with the naked eye. With a telescope we could make out many more tiny specks of light—all apparently very remote—most, or perhaps all, in small groups or clusters. We pick out one such group for study, and to see it better, race our space ship toward it at the speed of light.

After, say, 2.1 thousand million years, we are close enough to see that group of indistinct specks of light more easily. They are not points, we find, but they present finite angular sizes to our view, and they are separated by distances, on the average, equal to at least 10 times their own sizes. We plunge our ship directly toward the cluster.

In another 6 or 7 million years we are within it. Now we are surrounded by large centers of light—some spheroidal in shape, others flat and resembling great pinwheels in rapid rotation. These centers of light are not brilliant—in fact, they are rather low in surface brightness, and are so far apart that we still feel ourselves in incredibly empty space. Yet we see them clearly—at least with the space ship's telescope. We select one of these patches of light to investigate more closely, and race our ship toward it.

Two million years later we are well inside it, but *still* we find ourselves in empty space. Now, however, there are tiny points of light—thousands of them visible to our eyes, and millions to our telescope—in all directions. In some directions they merge, in projection, into nearly solid faint patches of light, and an irregular pattern of these luminous patches seems to form a complete circle surrounding us. Some of the points of light are in small clusters, and others appear to be emersed in faint nebulous glowing clouds. At random we choose one of those insignificant points of light. Still at the speed of light, we fly toward it.

After, say, 123 years, we are near enough to see that that point is not a point at all, but an intensely hot sphere, pouring blinding amounts of light into space. Our instruments reveal that it is nearly a million miles in diameter and that it has a temperature at its surface of many thousands of degrees. If we approach it any closer, our ship, and we, will surely vaporize. Thus we veer to one side to escape its heat.

Only 8 minutes later we are comfortably far away from that sphere of heat, although it is still blinding to look at. Thus we look around us in other directions, and suddenly find that we are approaching a most remarkable object—a tiny sphere, merely 8000 miles across, mostly blue in color, but with large regions of brown, green, and white. It spins rapidly, turning different portions of its surface into light from the brilliant globe that almost destroyed us. But the heat and light falling on this small world is quite gentle, and makes it a place of warmth and beauty—an abode of life—truly a most atypical place in the universe.

APPENDIX 1

SELECTED BIBLIOGRAPHY

(Technical references are marked with an asterisk.)

General Texts

Abell, G. O., *Exploration of the Universe,* Third Edition. New York: Holt, Rinehart and Winston, 1975.

Hoyle, Fred, *Astronomy and Cosmology, A Modern Course.* San Francisco: W. H. Freeman and Company, 1975.

* Smith, Elske v.P., and Jacobs, K. C., *Introductory Astronomy and Astrophysics.* Philadelphia: Saunders, 1973.

Elementary Texts and Books on Astronomy

Abell, G. O., *Drama of the Universe,* New York: Holt, Rinehart and Winston, 1978.

Alter, D., Cleminshaw, C. H., and Phillips, J., *Pictorial Astronomy.* New York: Crowell, 1974.

Menzel, D. H., *Astronomy.* New York: Random House, 1970.

Verschuur, G. L., *The Invisible Universe.* New York: Springer-Verlag, 1974.

Histories of Astronomy and Astrology

Berendzen, R., Hart, R., and Seeley, D., *Man Discovers the Galaxies.* New York: Neale Watson Academic Publications, 1976.

Galileo, G., *Dialogue on the Great World Systems* (translated by T. Salisbury). Chicago: University of Chicago Press, 1953.

Gallant, Roy A., *Astrology, Sense or Nonsense?* Garden City, New York: Doubleday, 1974.

Gardner, M., *Fads and Fallacies in the Name of Science.* New York: Dover, 1957.

Hoyle, F., *Astronomy.* New York: Doubleday, 1962.

King, H. C., *Exploration of the Universe.* New York: New American Library, 1964.

Koestler, A., *The Sleepwalkers.* New York: Macmillan, 1959.

Newton, I., *Mathematical Principles of Natural Philosophy.* R. T. Crawford, ed. Berkeley, Calif.: University of California Press, 1966.

Pannekoek, A., *A History of Astronomy.* New York: Interscience, 1961.

Shapley, H., and Howarth, H. E., *A Source Book in Astronomy.* New York: McGraw-Hill, 1929.

Standen, Anthony, *Forget Your Sun Sign.* Baton Rouge: Legacy Publishing Co., 1977.

Struve, O., and Zebergs, V., *Astronomy of the Twentieth Century.* New York: Macmillan, 1962.

Celestial Mechanics

Ahrendt, M. H., *The Mathematics of Space Exploration.* New York: Holt, Rinehart and Winston, 1965.

Ryabov, Y., *An Elementary Survey of Celestial Mechanics.* New York: Dover, 1961.

Van de Kamp, P., *Elements of Astromechanics.* San Francisco: Freeman, 1964.

Telescopes and Light

Christianson, W. N., and Hogborn, J. A., *Radio Telescopes.* London: Cambridge University Press, 1969.

Miczaika, G., and Sinton, W., *Tools of the Astronomer.* Cambridge, Mass.: Harvard University Press, 1961.

Minnaert, M., *The Nature of Light and Colour in the Open Air.* New York: Dover, 1954.

Page, T., and Page, L. W., *Telescopes; How to Make Them and Use Them.* New York: Macmillan, 1966.

Steinberg, J. L., and Lequeux, J., *Radio Astronomy.* New York: McGraw-Hill, 1963.

Earth and the Solar System

Hawkins, G. S., *Meteors, Comets, and Meteorites.* New York: McGraw-Hill, 1964.

Menzel, D., *Our Sun.* Cambridge, Mass.: Harvard University Press, 1959.

Watson, F., *Between the Planets.* Cambridge, Mass.: Harvard University Press, 1956.

Whipple, F., *Earth, Moon, and Planets.* Cambridge, Mass.: Harvard University Press, 1968.

Stellar Astronomy and Astrophysics

Aller, L. H., *Atoms, Stars and Nebulae* (rev. ed.). Cambridge, Mass.: Harvard University Press, 1971.

Bok, B. J., and Bok, P. E., *The Milky Way* (4th ed.). Cambridge, Mass.: Harvard University Press, 1973.

Brandt, J. C., *The Sun and Stars.* New York: McGraw-Hill, 1966.

Galaxies, Cosmology, and Relativity

* Couderc, P., *The Expansion of the Universe.* London: Faber and Faber, 1952.

Ferris, Timothy, *The Red Limit.* New York: William Morrow and Company, 1977.

Gardner, Martin, *The Relativity Explosion.* New York: Vintage Books, 1976.

Hodge, P. W., *Galaxies and Cosmology.* New York: McGraw-Hill, 1966.

Hubble, E., *The Realm of the Nebulae.* New Haven, Conn.: Yale University Press, 1936; also New York: Dover, 1958.

Kaufmann, W. J., *The Cosmic Frontiers of General Relativity*. Boston: Little Brown, 1977.

Sandage, A. R., *The Hubble Atlas of Galaxies*. Washington, D. C.: Carnegie Institution, 1961.

* Sciama, D. W., *Modern Cosmology*. London: Cambridge University Press, 1971.

* Sciama, D. W., *The Physical Foundations of General Relativity*. New York: Doubleday, 1969.

Sciama, D. W., *The Unity of the Universe*. New York: Doubleday, 1959.

Shapley, H., *Galaxies* (3rd ed.). Cambridge, Mass.: Harvard University Press, 1972.

Shipman, Harry L., *Black Holes, Quasars, and the Universe*. Boston: Houghton Mifflin, 1976.

Weinberg, Steven, *The First Three Minutes*. New York: Basic Books, 1977.

Life in the Universe

Berendzen, R. (ed.), *Life Beyond Earth and the Mind of Man*. Washington, D. C.: NASA, 1973.

Bracewell, R. N., *The Galactic Club*. Stratford, Calif.: Stanford Alumni Assn., 1974.

Drake, F., *Intelligent Life in Space*. New York: Macmillan, 1962.

Ponnamperuma, C., *The Origins of Life*. New York: Dutton, 1972.

Ponnamperuma, C., and Cameron, A. G. W., *Interstellar Communication: Scientific Perspectives*. Boston: Houghton Mifflin, 1974.

Sagan, C., *The Cosmic Connection*. Garden City, N.Y.: Anchor Press-Doubleday, 1973.

Shklovskii, I. S., and Sagan, C., *Intelligent Life in the Universe*. New York: Dell, 1966.

Star Atlases and Sky Guides

Allen, R. H., *Star Names*. New York: Dover, 1963.

Menzel, D. H., *A Field Guide to the Stars and Planets*. Boston: Houghton Mifflin, 1964.

Norton, W. W., *Sky Atlas*. Cambridge, Mass.: Sky Publishing Company, 1971.

Olcott, W. T., *Olcott's Field Book of the Skies*. New York: Putnam, 1954.

Journals and Periodicals

Astronomy, published monthly by AstroMedia Corp., Milwaukee, Wisconsin.

The Griffith Observer, published monthly by the Griffith Observatory, Los Angeles, California.

Mercury, published bimonthly by the Astronomical Society of the Pacific, San Francisco, California.

Scientific American, published monthly by Scientific American, New York.

Sky and Telescope, published monthly by the Sky Publishing Corporation, Harvard College Observatory, Cambridge, Massachusetts.

Career Information

Students interested in a career in astronomy can obtain an information leaflet from the Executive Officer, American Astronomical Society, 211 FitzRandolph Road, Princeton, New Jersey 08540.

APPENDIX 2

GLOSSARY

aberration (of starlight). Apparent displacement in the direction of a star due to the earth's orbital motion.

absolute magnitude. Apparent magnitude a star would have at a distance of 10 pc.

absolute zero. A temperature of $-273°$ C (or 0 K) where all molecular motion stops.

absorption spectrum. Dark lines superimposed on a continuous spectrum.

accelerate. To change velocity by speeding up, slowing down, or changing direction.

acceleration of gravity. Numerical value of the acceleration produced by the gravitational attraction on an object at the surface of a planet or star.

achromatic. Free of chromatic aberration.

active sun. The sun during times of unusual solar activity—spots, flares, and associated phenomena.

Age of Aquarius. Period (about 2000 years) during which the vernal equinox, moving because of precession, passes through the constellation of Aquarius.

airglow. Fluorescence in the atmosphere.

albedo. The fraction of incident sunlight that a planet or minor planet reflects.

almanac. A book or table listing astronomical events.

alpha particle. The nucleus of a helium atom, consisting of two protons and two neutrons.

altitude. Angular distance above or below the horizon,

measured along a vertical circle, to a celestial object.

amplitude. The range in variability, as in the light from a variable star.

angstrom (Å). A unit of length equal to 10^{-8} cm.

angular diameter. Angle subtended by the diameter of an object.

angular momentum. A measure of the momentum associated with motion about an axis or fixed point.

annular eclipse. An eclipse of the sun in which the moon is too distant to appear to cover the sun completely, so that a ring of sunlight shows around the moon.

anomalistic month. The period of revolution of the moon about the earth with respect to its line of apsides, or to the perigee point.

anomalistic year. The period of revolution of the earth about the sun with respect to its line of apsides, or to the perihelion point.

Antarctic Circle. Parallel of latitude 66½° S; at this latitude the noon altitude of the sun is 0° on the date of the summer solstice.

antimatter. Matter consisting of antiparticles: *antiprotons* (protons with negative rather than positive charge), *positrons* (positively charged electrons), and *antineutrons*.

apastron. The point of closest approach of two stars, as in a binary star orbit.

aperture. The diameter of an opening, or of the primary lens or mirror of a telescope.

aphelion. Point in its orbit where a planet is farthest from the sun.

apogee. Point in its orbit where an earth satellite is farthest from the earth.

Apollo program. The American program (1961–1972) to land men on the moon.

apparent magnitude. A measure of the observed light flux received from a star or other object at the earth.

apparent solar day. The interval between two successive transits of the sun's center across the meridian.

apparent solar time. The hour angle of the sun's center *plus* 12 hours.

apse (or **apsis**; pl. **apsides).** The point in a body's orbit where it is nearest or farthest the object it revolves about.

Arctic Circle. Parallel of latitude 66½° N; at this latitude the noon altitude of the sun is 0° on the date of the winter solstice.

artificial satellite. A man-made object put into a closed orbit about the earth.

ascendant. (astrological term). The point on the zodiac that is on the eastern horizon, just rising at the moment of birth.

ascending node. The point along the orbit of a body where it crosses from the south to the north of some reference plane, usually the plane of the celestial equator or of the ecliptic.

aspect. The situation of the sun, moon, or planets with respect to one another.

association. A loose cluster of stars whose spectral types, motions, or positions in the sky indicate that they have probably had a common origin.

asteroid. A synonym for "minor planet."

asthenosphere. The mantle beneath the lithosphere of the earth.

astigmatism. A defect in an optical system whereby pairs of light rays in different planes do not focus at the same place.

astrology. The pseudoscience that deals with supposed influences on the configurations and locations in the sky of the sun, moon, and planets on human destiny; a primitive religion having its origin in ancient Babylonia.

astrometric binary. A binary star in which one component is not observed, but its presence is deduced from the orbital motion of the visible component.

astrometry. That branch of astronomy that deals with the determination of precise positions and motions of celestial bodies.

astronautics. The science of the laws and methods of space flight.

astronomical unit (AU). Originally meant to be the semimajor axis of the orbit of the earth; now defined as the semimajor axis of the orbit of a hypothetical body with the mass and period that Gauss assumed for the earth. The semimajor axis of the orbit of the earth is 1.000 000 230 AU.

astronomy. The branch of science that treats of the physics and morphology of that part of the universe which lies beyond the earth's atmosphere.

astrophysics. That part of astronomy which deals principally with the physics of stars, stellar systems, and interstellar material. Astrophysics also deals, however, with the structures and atmospheres of the sun and planets.

atmospheric refraction. The bending, or refraction, of light rays from celestial objects by the earth's atmosphere.

atom. The smallest particle of an element that retains the properties which characterize that element.

atomic clock. A time-keeping device regulated by the natural frequency of the emission or absorption of radiation by a particular kind of atom.

atomic mass unit. *Chemical:* one-sixteenth of the mean mass of an oxygen atom. *Physical:* one-twelfth of the mass of an atom of the most common isotope of carbon. The atomic mass unit is approximately the mass of a hydrogen atom, 1.67×10^{-24}g.

atomic number. The number of protons in each atom of a particular element.

atomic time. The time kept by a cesium (atomic) clock, based on the **atomic second**—the time required for 9,192,631,770 cycles of the radiation emitted or absorbed in a particular transition of an atom of cesium-133.

atomic transition. A change in the state of energy of an atom; the atom may gain or lose energy by collision with another particle or by the emission or absorption of a photon.

atomic weight. The mean mass of an atom of a particular element in atomic mass units.

aurora. Light radiated by atoms and ions in the ionosphere, mostly in the polar regions.

autumnal equinox. The intersection of the ecliptic and celestial equator where the sun crosses the equator from north to south.

azimuth. The angle along the celestial horizon, measured eastward from the north point, to the intersection of the horizon with the vertical circle passing through an object.

Baily's beads. Small "beads" of sunlight seen passing through valleys along the limb of the moon in the instant preceding and following totality in a solar eclipse.

ballistic missile. A missile or rocket that is given its entire thrust during a brief period at the beginning of its flight, and that subsequently "coasts" to its target along an orbit.

Balmer lines. Emission or absorption lines in the spectrum of hydrogen that arise from transitions between the second (or first excited) and higher energy states of the hydrogen atoms.

bands (in spectra). Emission or absorption lines, usually in the spectra of chemical compounds or radicals, so numerous and closely spaced that they coalesce into broad emission or absorption bands.

barred spiral galaxy. Spiral galaxy in which the spiral arms begin from the ends of a "bar" running through the nucleus rather than from the nucleus itself.

barycenter. The center of mass of two mutually revolving bodies.

baryons (and antibaryons). The heavy atomic nuclear particles, such as protons and neutrons.

base line. That side of a triangle used in triangulation or surveying whose length is known (or can be measured), and which is included between two angles that are known (or can be measured).

Be star. A spectral type B star with emission lines in its spectrum, which are presumed to arise from material ejected from or surrounding the star.

"big bang" theory. A theory of cosmology in which the expansion of the universe is presumed to have begun with a primeval explosion.

billion. In the United States and France, one thousand million (10^9); in Great Britain and Germany, one million million (10^{12}).

binary star. A double star; two stars revolving about each other.

binding energy. The energy required to completely separate the constituent parts of an atomic nucleus.

biosphere. That part of the earth (its surface, atmosphere, and oceans) where life can exist.

blackbody. A hypothetical perfect radiator, which absorbs and reemits all radiation incident upon it.

black dwarf. A presumed final state of evolution for a star, in which all of its energy sources are exhausted and it no longer emits radiation.

black hole. A hypothetical body whose velocity of escape is equal to or greater than the speed of light; thus no radiation can escape from it.

blink microscope (or **comparator**). A microscope in which the user's view is shifted rapidly back and forth between the corresponding portions of two different photographs of the same region of the sky.

Bode's law. A scheme by which a sequence of numbers can be obtained that give the approximate distances of the planets from the sun in astronomical units.

Bohr atom. A particular model of an atom, invented by Niels Bohr, in which the electrons are described as revolving about the nucleus in circular orbits.

bolide. A very bright fireball or meteor; sometimes defined as a fireball accompanied by sound.

bolometric correction. The difference between the visual (or photovisual) and bolometric magnitudes of a star.

bolometric magnitude. A measure of the flux of radiation from a star or other object received just outside the earth's atmosphere, as it would be detected by a device sensitive to *all* forms of electromagnetic energy.

bremsstrahlung. Radiation from free-free transitions, in which electrons gain or lose energy while being accelerated in the field of an atomic nucleus or ion.

bubble chamber. A chamber in which bubbles form along the electrically charged path of a high-energy charged particle, rendering the track of that particle visible.

burnout. The instant that a rocket stops firing.

calculus. A branch of mathematics that permits computations involving rates of change (*differential* calculus) or of the contribution of an infinite number of infinitesimal quantities (*integral* calculus).

carbon cycle. A series of nuclear reactions involving carbon as a catalyst, by which hydrogen is transformed to helium.

cardinal points. The four principal points of the compass: North, East, South, and West.

Cassegrain focus. An optical arrangement in a reflecting telescope in which light is reflected by a second mirror to a point behind the objective mirror.

cD galaxy. A supergiant elliptical galaxy frequently found in the centers of clusters of galaxies.

celestial equator. A great circle on the celestial sphere 90° from the celestial poles; the circle of intersection of the celestial sphere with the plane of the earth's equator.

celestial mechanics. That branch of astronomy which deals with the motions and gravitational influences of the members of the solar system.

celestial navigation. The art of navigation at sea or in the air from sightings of the sun, moon, planets, and stars.

celestial poles. Points about which the celestial sphere appears to rotate; intersections of the celestial sphere with the earth's polar axis.

celestial sphere. Apparent sphere of the sky; a sphere of large radius centered on the observer. Directions of objects in the sky can be denoted by the positions of those objects on the celestial sphere.

center of gravity. Center of mass.

center of mass. The mean position of the various mass elements of a body or system, weighted according to their distances from that center of mass; that point in an isolated system which moves with constant velocity, according to Newton's first law of motion.

centrifugal force (or **acceleration**). An imaginary force (or acceleration) that is often introduced to account for the illusion that a body moving on a curved path tends to accelerate radially from the center of curvature. The actual force present is the one that diverts the body's motion from a straight line and is directed *toward* the center of curvature. It is, however, legitimate to introduce a fictitious centrifugal force field in a rotating (and hence noninertial) coordinate system.

centripetal force (or **acceleration**). The force required to divert a body from a straight path into a curved path (or the acceleration experienced by the body); it is directed toward the center of curvature.

cepheid variable. A star that belongs to one of two classes (type I and type II) of yellow supergiant pulsating stars.

Ceres. Largest of the minor planets and first to be discovered.

cesium clock. An atomic clock that utilizes a transition in the atom cesium-133.

charm. The name given to a variety of quark; a quark is a hypothetical basic constituent of all nuclear particles.

chromatic aberration. A defect of optical systems whereby light of different colors is focused at different places.

chromosphere. That part of the solar atmosphere which lies immediately above the photospheric layers.

chronograph. A device for recording and measuring the times of events.

chronometer. An accurate clock.

circular velocity. The critical speed with which a revolving body can have a circular orbit.

circumpolar regions. Portions of the celestial sphere near the celestial poles that are either always above or always below the horizon.

cloud chamber. A chamber in which droplets of liquid condense along the electrically charged path of a high-energy charged particle, rendering the track of that particle visible.

Clouds of Magellan. Two neighboring galaxies visible to the naked eye from southern latitudes.

cluster of galaxies. A system of galaxies containing from several to thousands of member galaxies.

cluster variable (RR Lyrae variable). A member of a certain large class of pulsating variable stars, all with periods less than one day. These stars are often present in globular star clusters.

color index. Difference between the magnitudes of a star or other object measured in light of two different spectral regions, for example, photographic *minus* photovisual magnitudes.

color-magnitude diagram. Plot of the magnitudes (apparent or absolute) of the stars in a cluster against their color indices.

coma. A defect in an optical system in which off-axis rays of light striking different parts of the objective do not focus in the same place.

coma (of comet). The diffuse gaseous component of the head of a comet.

comet. A small body of icy and dusty matter, which revolves about the sun. When a comet comes near the sun, some of its material vaporizes, forming a large *coma* of tenuous gas, and often a *tail*.

comparison spectrum. The spectrum of a vaporized element (such as iron) photographed beside the image of a stellar spectrum, and with the same camera, for purposes of comparison of wavelengths.

compound. A substance composed of two or more chemical elements.

conduction. The transfer of energy by the direct passing of energy or electrons from atom to atom.

configuration. Any one of several particular orientations in the sky of the moon or a planet with respect to the sun.

conic section. The curve of intersection between a circular cone and a plane; these curves can be ellipses, circles, parabolas, or hyperbolas.

conjunction. The configuration of a planet when it has the same celestial longitude as the sun, or the configuration when any two celestial bodies have the same celestial longitude or right ascension.

conservation of angular momentum. The law that angular momentum is conserved in the absence of any force not directed toward or away from the point or axis about which the angular momentum is referred—that is, in the absence of a torque.

constellation. A configuration of stars named for a particular object, person, or animal; or the area of the sky assigned to a particular configuration.

contacts (of eclipses). The instants that certain stages of an eclipse begin.

continental drift. A gradual drift of the continents over the surface of the earth due to *plate tectonics*.

continuous spectrum. A spectrum of light comprised of radiation of a continuous range of wavelengths or colors rather than only certain discrete wavelengths.

convection. The transfer of energy by moving currents of a fluid containing that energy.

Coordinated Universal Time. Greenwich Mean Time standardized and regulated by an international agency, the *Bureau International de l'Heure*, on the basis of astronomical observations reported from around the world.

Copernicus satellite. An artificial satellite with scientific instrumentation especially designed for ultraviolet observations in space.

core (of earth). The central part of the earth, believed to be a liquid of high density.

corona. Outer atmosphere of the sun.

corona (or **halo) of Galaxy.** The outer portions of the Galaxy, especially on either side of the plane of the Milky Way.

coronagraph. An instrument for photographing the chromosphere and corona of the sun outside of eclipse.

corpuscular radiation. Charged particles, mostly atomic nuclei and electrons, emitted into space by the sun and possibly other objects.

cosmic rays. Atomic nuclei (mostly protons) that are observed to strike the earth's atmosphere with exceedingly high energies.

cosmogony. The study of the origin of the world or universe.

cosmological constant. A term that arises in the development of the field equations of general relativity, which represents a repulsive force in the universe. The cosmological constant is often assumed to be zero.

cosmological model. A specific model, or theory, of the organization and evolution of the universe.

cosmological principle. The assumption that, on the large scale, the universe at any given time is the same everywhere.

cosmology. The study of the organization and evolution of the universe.

coudé focus. An optical arrangement in a reflecting telescope whereby light is reflected by two or more secondary mirrors down the polar axis of the telescope to a focus at a place separate from the moving parts of the telescope.

Crab nebula. The expanding mass of gas that is the remnant of the supernova of 1054.

crater (lunar). A more or less circular depression in the surface of the moon.

crater (meteoritic). A crater on the earth caused by the collision of a meteoroid with the earth, and a subsequent explosion.

crescent moon. One of the phases of the moon when its elongation is less than 90° from the sun and it appears less than half full.

crust (of earth). The outer layer of the earth.

Cyclops Project. A proposed system of radio antennae for detection of signals from extraterrestrial civilizations.

dark nebula. A cloud of interstellar dust that obscures the light of more distant stars, and appears as an opaque curtain.

daylight saving time. A time one hour more advanced than standard time, usually adopted in spring and summer to take advantage of long evening twilights.

deferent. A stationary circle in the Ptolemaic system along which moves the center of another circle (epicycle), along which moves an object or another epicycle.

degenerate gas. A gas in which the allowable states for the electrons have been filled; it behaves according to different laws from those that apply to "perfect" gases.

density. The ratio of the mass of an object to its volume.

descending node. The point along the orbit of a body where it crosses from the north to the south of some reference plane, usually of the celestial equator or of the ecliptic.

deuterium. A "heavy" form of hydrogen, in which the nucleus of each atom consists of one proton and one neutron.

diamond ring. A flash of sunlight at the instants before and after totality in a solar eclipse while the corona is visible as a complete ring of light around the moon.

differential gravitational force. The difference between the respective gravitational forces exerted on two bodies near each other by a third, more distant body.

differentiation (geological). A separation or segregation of different kinds of material in different layers in the interior of a planet.

diffraction. The spreading out of light in passing the edge of an opaque body.

diffraction grating. A system of closely spaced equidistant slits or reflecting strips which, by diffraction and interference, produce a spectrum.

diffraction pattern. A pattern of bright and dark fringes produced by the interference of light rays, diffracted by different amounts, with each other.

diffuse nebula. A reflection or emission nebula produced by interstellar matter (not a planetary nebula).

disk (of planet or other object). The apparent circular shape that a planet (or the sun, or moon, or a star) displays when seen in the sky or viewed telescopically.

disk of Galaxy. The central disk or "wheel" of our Galaxy, superimposed on the spiral structure.

dispersion. Separation, from white light, of different wavelengths being refracted by different amounts.

diurnal. Daily.

diurnal circle. Apparent path of a star in the sky during a complete day due to the earth's rotation.

diurnal motion. Motion during one day.

diurnal parallax. Apparent change in direction of an object caused by a displacement of the observer due to the earth's rotation.

Doppler shift. Apparent change in wavelength of the radiation from a source due to its relative motion in the line of sight.

draconic month. The period of revolution of the moon about the earth with respect to the nodes of the moon's orbit.

dwarf (star). A main-sequence star (as opposed to a giant or supergiant).

dyne. The metric unit of force; the force required to accelerate a mass of one gram in the amount one centimeter per second per second.

east point. The point on the horizon 90° from the north point (measured clockwise as seen from the zenith).

eccentric. A point, about which an object revolves on a circular orbit, that is not at the center of the circle.

eccentricity (of ellipse). Ratio of the distance between the foci to the major axis.

eclipse. The cutting off of all or part of the light of one body by another passing in front of it.

eclipse path. The track along the earth's surface swept out by the tip of the shadow of the moon (or the extension of its shadow) during a total (or annular) solar eclipse.

eclipse season. A period during the year when an eclipse of the sun or moon is possible.

eclipsing binary star. A binary star in which the plane of revolution of the two stars is nearly edge-on to our line of sight, so that the light of one star is periodically diminished by the other passing in front of it.

ecliptic. The apparent annual path of the sun on the celestial sphere.

ecliptic limit. The maximum angular distance from a node where the moon can be for an eclipse to take place.

Einstein-Rosen bridge. A hypothetical connection in spacetime between two distinct regions of the universe, of which one may be a black hole and one a white hole; a worm hole.

Einstein telescope. The orbiting X-ray telescope (HEAO 2) launched in late 1978.

electric charge. A quantity of electrons or of electrical charge of one sign.

electric current. The flow of electrons.

electric field. The region of space around an electric charge within which an electric force can act on another charged particle.

electromagnetic radiation. Radiation consisting of waves propagated through the building up and breaking down of electric and magnetic fields; these include radio, infrared, light, ultraviolet, X rays, and gamma rays.

electromagnetic spectrum. The whole array or family of electromagnetic waves.

electron. A negatively charged subatomic particle that normally moves about the nucleus of an atom.

electron volt. The kinetic energy acquired by an electron that is accelerated through an electric potential of 1 volt; 1 electron volt is 1.60207×10^{-12} erg.

electroscope. A device for measuring the amount of charge in the air.

element. A substance that cannot be decomposed, by chemical means, into simpler substances.

elements (of orbit). Any of several quantities that describe the size, shape, and orientation of the orbit of a body.

ellipse. A conic section: the curve of intersection of a circular cone and a plane cutting completely through the cone.

elliptical galaxy. A galaxy whose apparent photometric

contours are ellipses and which contains no conspicuous interstellar material.

ellipticity. The ratio (in an ellipse) of the major axis *minus* the minor axis to the major axis.

elongation. The difference between the celestial longitudes of a planet and the sun.

emission line. A discrete bright spectral line.

emission nebula. A gaseous nebula that derives its visible light from the fluorescence of ultraviolet light from a star in or near the nebula.

emission spectrum. A spectrum consisting of emission lines.

energy. The ability to do work.

energy level (in an atom or ion). A particular level, or amount, of energy possessed by an atom or ion above the energy it possesses in its least energetic state.

energy spectrum. A table or plot showing the relative numbers of particles (in cosmic rays or corpuscular radiation) of various energies.

ephemeris. A table that gives the positions of a celestial body at various times, or other astronomical data.

ephemeris time. A kind of time that passes at a strictly uniform rate; used to compute the instants of various astronomical events.

epicycle. A circular orbit of a body in the Ptolemaic system, the center of which revolves about another circle (the deferent).

equant. A stationary point in the Ptolemaic system that is *not* at the center of a circular orbit, and about which a body (or the center of an epicycle) revolves with uniform angular velocity.

equation of state. An equation relating the pressure, temperature, and density of a substance (usually a gas).

equator. A great circle on the earth, 90° from its poles.

equatorial mount. A mounting for a telescope with one axis parallel to the earth's axis, so that a motion of the telescope about that axis compensates for the earth's rotation.

equinox. One of the intersections of the ecliptic and celestial equator.

erg. The metric unit of energy; the work done by a force of one dyne moving through a distance of one centimeter.

eruptive variable. A variable star whose changes in light are erratic or explosive.

Euclidean. Pertaining to Euclidean geometry, or *flat space*.

event. A point in four-dimensional spacetime.

event horizon. The surface through which a collapsing star is hypothesized to pass when its velocity of escape is equal to the speed of light, that is, when the star becomes a black hole.

evolutionary cosmology. A theory of cosmology that assumes that all parts of the universe have a common age and evolve together.

eyepiece. A magnifying lens used to view the image produced by the objective of a telescope.

excitation. The process of imparting to an atom or an ion an amount of energy greater than that it has in its normal or least-energy state.

extinction. Attenuation of light from a celestial body produced by the earth's atmosphere, or by interstellar absorption.

extragalactic. Beyond the Galaxy.

faculus (pl. **faculae**). Bright region near the limb of the sun.

fermions. Subatomic particles, such as electrons, that obey certain laws formulated by Enrico Fermi.

field equations. A set of equations in general relativity that describe the curvature of spacetime in the presence of matter.

filtergram. A photograph of the sun (or part of it) taken through a special narrow-bandpass filter.

fireball. A spectacular meteor.

First Point of Aries. The vernal equinox.

fission. The breakup of a heavy atomic nucleus into two or more lighter ones.

flare. A sudden and temporary outburst of light from an extended region of the solar surface.

flare star. A member of a class of stars that show occasional, sudden, unpredicted increases in light.

flash spectrum. The spectrum of the very limb of the sun observed in the instant before totality in a solar eclipse.

flocculus (pl. **flocculi**). A bright region of the solar surface observed in the monochromatic light of some spectral line; flocculi are now usually called *plages*.

fluxions. Name given by Newton to the calculus.

fluorescence. The absorption of light of one wavelength and reemission of it at another wavelength; especially the conversion of ultraviolet into visible light.

focal length. The distance from a lens or mirror to the point where light converged by it comes to a focus.

focal ratio (speed). Ratio of the focal length of a lens or mirror to its aperture.

focus. Point where the rays of light converged by a mirror or lens meet.

focus of a conic section. Mathematical point associated with a conic section, whose distance to any point on the conic bears a constant ratio to the distance from that point to a straight line known as the *directrix*.

force. That which can change the momentum of a body; numerically, the rate at which the body's momentum changes.

Fraunhofer line. An absorption line in the spectrum of the sun or of a star.

Fraunhofer spectrum. The array of absorption lines in the spectrum of the sun or of a star.

free-free transition. An atomic transition in which the energy associated with an atom or ion and passing electron changes during the encounter, but without capture of the electron by the atom or ion.

frequency. Number of vibrations per unit time; number of waves that cross a given point per unit time (in radiation).

full moon. That phase of the moon when it is at opposition (180° from the sun) and its full daylight hemisphere is visible from the earth.

fusion. The building up of heavier atomic nuclei from lighter ones.

galactic cluster. An "open" cluster of stars located in the spiral arms or disk of the Galaxy.

galactic equator. Intersection of the principal plane of the Milky Way with the celestial sphere.

galactic latitude. Angular distance north or south of the galactic equator to an object, measured along a great circle passing through that object and the galactic poles.

galactic longitude. Angular distance, measured eastward along the galactic equator from the galactic center, to the intersection of the galactic equator with a great circle passing through the galactic poles and an object.

galactic poles. The poles of the galactic equator; the intersections with the celestial sphere of a line through the observer that is perpendicular to the plane of the galactic equator.

galactic rotation. Rotation of the Galaxy.

galaxy. A large assemblage of stars; a typical galaxy contains millions to hundreds of thousands of millions of stars.

Galaxy. The galaxy to which the sun and our neighboring stars belong; the Milky Way is light from remote stars in the Galaxy.

gamma rays. Photons (of electromagnetic radiation) of energy higher than those of X rays; the most energetic form of electromagnetic radiation.

gauss. A unit of magnetic flux density.

gegenschein (counterglow). A very faint, diffuse glow of light opposite the sun in the sky, believed to be caused by sunlight reflected from interplanetary particles.

Geiger counter. A device for counting high-energy charged particles and hence for measuring the intensity of corpuscular radiation.

geodesic. The path of a body in spacetime.

geodesic equations. A set of equations in general relativity by which the paths of objects in spacetime can be calculated.

geomagnetic. Referring to the geometrical center of the earth's magnetic field.

geomagnetic poles. The poles of a hypothetical bar magnet whose magnetic field most nearly matches that of the earth.

giant (star). A luminous star of large radius.

gibbous moon. One of the phases of the moon in which more than half, but not all, of the moon's daylight hemisphere is visible from the earth.

Gliese catalogue. A catalogue of nearby stars compiled by the astronomer W. Gliese.

globular cluster. One of about 120 larger star clusters that form a system of clusters centered on the center of the Galaxy.

globule. A small, dense, dark nebula; believed to be a possible protostar.

granulation. The "rice-grain"-like structure of the solar photosphere.

gravitation. The tendency of matter to attract itself.

gravitational constant, *G*. The constant of proportionality in Newton's law of gravitation; in metric units G has the value 6.668×10^{-8} dyne $\cdot$ cm^2/g^2.

gravitational energy. Energy that can be released by the gravitational collapse, or partial collapse, of a system.

gravitational redshift. The redshift caused by a gravitational field. The slowing of clocks in a gravitational field.

gravitational waves. Oscillations in spacetime, propagated by changes in the distribution of matter.

great circle. Circle on the surface of a sphere that is the curve of intersection of the sphere with a plane passing through its center.

greatest elongation (east or west). The largest separation in celestial longitude (to the east or west) that an inferior planet can have from the sun.

greenhouse effect. The blanketing of infrared radiation near the surface of a planet by, for example, carbon dioxide in its atmosphere.

Greenwich meridian. The meridian of longitude passing through the site of the old Royal Greenwich Observatory, near London; origin of longitude on the earth.

Gregorian calendar. A calendar (now in common use) introduced by Pope Gregory XIII in 1582.

H I region. Region of neutral hydrogen in interstellar space.

H II region. Region of ionized hydrogen in interstellar space.

hadron. A subnuclear particle; one of hundreds now known to exist, of mass from somewhat less to somewhat more than that of the proton.

half-life. The time required for half of the radioactive atoms in a sample to disintegrate.

halo (around sun or moon). A ring of light around the sun or moon caused by refraction by the ice crystals of cirrus clouds.

halo (of galaxy). See corona.

harmonic law. Kepler's third law of planetary motion: the cubes of the semimajor axes of the planetary orbits are in proportion to the squares of the sidereal periods of the planets.

harvest moon. The full moon nearest the time of the autumnal equinox.

head (of comet). The main part of a comet, consisting of its nucleus and coma.

HEAO. High Energy Astronomy Observatory; one of a series of artificial satellites carrying X-ray or gamma-ray detectors. HEAO II is the Einstein telescope.

"heavy" elements. In astronomy, usually those elements of greater atomic number than helium.

Heisenberg uncertainty principle. A principle of quantum mechanics that places a limit on the precision with which the simultaneous position and momentum of a body or particle can be specified.

helio-. Prefix referring to the sun.

heliocentric. Centered on the sun.

helium flash. The nearly explosive ignition of helium in the triple-alpha process in the dense core of a red giant star.

Helmholtz-Kelvin contraction. The gradual gravitational contraction of a cloud or a star, with the release of potential energy.

Hertzsprung-Russell (H-R) diagram. A plot of absolute magnitude against temperature (or spectral class or color index) for a group of stars.

homogeneous. The same everywhere throughout a given volume.

horary astrology. Astrology based on a horoscope drawn up for the place and instant at which a question or idea was first raised. Horary astrology purports to advise on the auspiciousness of an action based on the horoscope of the instant of its conception.

horizon (astronomical). A great circle on the celestial sphere 90° from the zenith.

horizon system. A system of celestial coordinates (altitude and azimuth) based on the astronomical horizon and the north point.

horizontal branch. A sequence of stars on the Hertzsprung-Russell diagram of a typical globular cluster of approximately constant absolute magnitude (near $M_v = 0$).

horoscope (astrological term). A chart showing the positions along the zodiac and in the sky of the sun, moon, and planets at some given instant and place on earth—usually corresponding to the time and place of a person's birth.

hour angle. The angle measured westward along the celestial equator from the local meridian to the hour circle passing through an object.

hour circle. A great circle on the celestial sphere passing through the celestial poles.

house. A division or segment of the sky numbered according to its position with respect to the horizon, and used by astrology in preparing a horoscope.

Hubble constant. Constant of proportionality in the relation between the velocity of remote galaxies and their distances. The Hubble constant is approximately 50 km/s · 10^6 pc.

Hubble law. The law of the redshifts.

hydrostatic equilibrium. A balance between the weights of various layers, as in a star or the earth's atmosphere, and the pressures that support them.

hyperbola. A conic section of eccentricity greater than 1.0; the curve of intersection between a circular cone and a plane that is at too small an angle with the axis of the cone to cut all the way through it, and is not parallel to a line in the face of the cone.

hypothesis. A tentative theory or supposition, advanced to explain certain facts or phenomena, which is subject to further tests and verification.

image. The optical representation of an object produced by light rays from the object being refracted or reflected by a lens or mirror.

image tube. A device in which electrons, emitted from a photocathode surface exposed to light, are focused electronically.

inclination (of an orbit). The angle between the orbital plane of a revolving body and some fundamental plane—usually the plane of the celestial equator or of the ecliptic.

Index Catalogue, IC. The supplement to Dreyer's *New General Catalogue* of star clusters and nebulae.

index of refraction. A measure of the refracting power of a transparent substance; specifically, the ratio of the speed of light in a vacuum to its speed in the substance.

inertia. The property of matter that requires a force to act on it to change its state of motion; momentum is a measure of inertia.

intertial system. A system of coordinates that is not itself accelerated, but which is either stationary or is moving with constant velocity.

inferior conjunction. The configuration of an inferior planet when it has the same longitude as the sun, and is between the sun and earth.

inferior planet. A planet whose distance from the sun is less than the earth's.

infrared radiation. Electromagnetic radiation of wavelength longer than the longest (red) wavelengths that can be perceived by the eye, but shorter than radio wavelengths.

interferometer (stellar). An optical device, making use of the principle of interference of light waves, with which small angles can be measured.

International Date Line. An arbitrary line on the surface of the earth near longitude 180° across which the date changes by one day.

international magnitude system. The system of photographic and photovisual magnitudes, referring to the blue and yellow spectral regions, at one time adopted by international agreement, but now largely superseded by the U, B, V system.

interplanetary medium. The sparse distribution of gas and solid particles in the interplanetary space.

interstellar dust. Microscopic solid grains, believed to be mostly dielectric compounds of hydrogen and other common elements, in interstellar space.

interstellar gas. Sparse gas in interstellar space.

interstellar lines. Absorption lines superimposed on stellar spectra, produced by the interstellar gas.

interstellar matter. Interstellar gas and dust.

ion. An atom that has become electrically charged by the addition or loss of one or more electrons.

ionization. The process by which an atom gains or loses electrons.

ionization potential. The energy required to remove an electron from an atom.

ionosphere. The upper region of the earth's atmosphere in which many of the atoms are ionized.

ion tail (of comet). The relatively straight tail of a comet produced by the interaction of the solar wind with the ions in the comet.

irregular galaxy. A galaxy without rotational symmetry; neither a spiral nor elliptical galaxy.

irregular variable. A variable star whose light variations do not repeat with a regular period.

island universe. Historical synonym for galaxy.

isotope. Any of two or more forms of the same element, whose atoms all have the same number but different masses.

isotropic. The same in all directions.

Jovian planet. Any of the planets Jupiter, Saturn, Uranus, and Neptune.

Julian Calendar. A calendar introduced by Julius Caesar in 45 B.C.

Julian day. The number of the day in a running sequence beginning January 1, 4713 B.C.

Jupiter. The fifth planet from the sun in the solar system.

Kepler's laws. The three laws, discovered by Kepler, that describe the motions of the planets.

kiloparsec (kpc). 1000 parsecs, or about 3260 LY.

kinetic energy. Energy associated with motion; the kinetic energy of a body is one-half the product of its mass and the square of its velocity.

kinetic theory (of gases). The science that treats the motions of the molecules that compose gases.

laser. An acronym for *light amplification by stimulated emission of radiation;* a device for amplifying a light signal at a particular wavelength into a coherent beam.

latitude. A north-south coordinate on the surface of the earth; the angular distance north or south of the equator measured along a meridian passing through a place.

launch window. A range of dates during which a space vehicle can be launched for a specific mission without exceeding the fuel capabilities of that system.

law. A statement of order or relation between phenomena that, under given conditions, is presumed to be invariable.

law of areas. Kepler's second law: the radius vector from the sun to any planet sweeps out equal areas in the planet's orbital plane in equal intervals of time.

law of the redshifts. The relation between the radial velocity and distance of a remote galaxy: the radial velocity is proportional to the distance of the galaxy.

lead sulfide cell. A device used to measure infrared radiation.

leap year. A calendar year with 366 days, occurring every four years, in which one day is added to make the average length of the calendar year as nearly equal as possible to the tropical year.

lepton. A light subatomic particle, such as an electron, neutrino, or muon.

light. Electromagnetic radiation that is visible to the eye.

light curve. A graph that displays the variation in light or magnitude of a variable or eclipsing binary star.

light-year. The distance light travels in a vacuum in one year; 1 LY = 9.46×10^{17} cm, or about 6×10^{12} mi.

limb (of sun or moon). Apparent edge of the sun or moon as seen in the sky.

limb darkening. The phenomenon whereby the sun is less bright near its limb than near the center of its disk.

limiting magnitude. The faintest magnitude that can be observed with a given instrument or under given conditions.

line of apsides. The line connecting the apsides of an orbit (the perifocus and farthest-from-focus points); or the line along the major axis of the orbit.

line of nodes. The line connecting the nodes of an orbit.

line profile. A plot of the intensity of light versus wavelength across a spectral line.

linear diameter. Actual diameter in units of length.

lithosphere. The upper layer of the earth, to a depth of 50 to 100 km, involved in plate tectonics.

Local Group. The cluster of galaxies to which our Galaxy belongs.

local oscillator. The old (classical) idea of an atom absorbing or emitting radiation by setting itself in oscillation or by reducing that oscillation. The local oscillator has been replaced by a different model in the modern quantum theory.

local standard of rest. A coordinate system that shares the average motion of the sun and its neighboring stars about the galactic center.

local supercluster. The cluster of clusters of galaxies, to which the Local Group belongs.

longitude. An east-west coordinate on the earth's surface; the angular distance, measured east or west along the equator from the Greenwich meridian, to the meridian passing through a place.

low-velocity star (or **object**). A star (or object) that has low space velocity; generally an object that shares the sun's high orbital speed about the galactic center.

luminosity. The rate of radiation of electromagnetic energy into space by a star or other object.

luminosity class. A classification of a star according to its luminosity for a given spectral class.

luminous energy. Light.

lunar. Referring to the moon.

lunar eclipse. An eclipse of the moon.

Lyman lines. A series of absorption or emission lines in the spectrum of hydrogen that arise from transitions to and from the lowest energy states of the hydrogen atoms.

Magellanic Clouds. See Clouds of Magellan.

magnetic field. The region of space near a magnetized body within which magnetic forces can be detected.

magnetic pole. One of two points on a magnet (or the earth) at which the greatest density of lines of force emerge. A compass needle aligns itself along the local lines of force on the earth and points more or less toward the magnetic poles of the earth.

magnetometer. A device for measuring magnetic fields.

magnetosphere. The region around the earth or other planet or star occupied by its magnetic field.

magnifying power. The number of times larger (in angular diameter) an object appears through a telescope than with the naked eye.

magnitude. A measure of the amount of light flux received from a star or other luminous object.

main sequence. A sequence of stars on the Hertzsprung-Russell diagram, containing the majority of stars, that runs diagonally from the upper left to the lower right.

major axis (of ellipse). The maximum diameter of an ellipse.

major planet. A Jovian planet.

mantle (of earth). The greatest part of the earth's interior, lying between the crust and the core.

mare. Latin for "sea"; name applied to many of the "sea-like" features on the moon or Mars.

Mariner space probes. A series of space probes launched in the 1960s and early 1970s to explore the planets Mercury, Venus, and Mars.

Mars. Fourth planet from the sun in the solar system.

maser. An acronym for *microwave amplification by stimulated emission radiation;* a device for amplifying a microwave (radio) signal at a particular wavelength into a coherent beam.

mass. A measure of the total amount of material in a body; defined either by the inertial properties of the body or by its gravitational influence on other bodies.

mass defect. The amount by which the mass of an atomic nucleus is less than the sum of the masses of the individual nucleons that compose it.

mass-luminosity relation. An empirical relation between the masses and luminosities of many (principally main-sequence) stars.

Maunder minimum. The interval from 1645 to 1715 when solar activity was very low or absent.

Maxwell's equations. A set of four equations that describe the fields around magnetic and electric charges, and how changes in those fields produce forces and electromagnetic radiation.

mean solar day. Interval between successive meridian passages of the mean sun; average length of the apparent solar day.

mean solar time. Local hour angle of the mean sun *plus* 12 hours.

mean sun. A fictitious body that moves eastward with uniform angular velocity along the celestial equator, completing one circuit of the sky with respect to the vernal equinox in a tropical year.

mechanics. That branch of physics that deals with the behavior of material bodies under the influence of, or in the absence of, forces.

medical astrology. That branch of astrology that deals with supposed connections between planets and zodiacal signs and bodily organs and their diseases.

megaparsec (Mpc). One million (10^6) pc.

Mercury. Nearest planet to the sun in the solar system.

meridian (celestial). The great circle on the celestial sphere that passes through an observer's zenith and the north (or south) celestial pole.

meridian (terrestrial). The great circle on the surface of the earth that passes through a particular place and the north and south poles of the earth.

mesosphere. The layer of the ionosphere immediately above the stratosphere.

Messier catalogue. A catalog of nonstellar objects compiled by Charles Messier in 1787.

meteor. The luminous phenomenon observed when a meteoroid enters the earth's atmosphere and burns up; popularly called a "shooting star."

meteor shower. Many meteors appearing to radiate from a common point in the sky caused by the collision of the earth with a swarm of meteoritic particles.

meteorite. A portion of a meteoroid that survives passage through the atmosphere and strikes the ground.

meteorite fall. The occurrence of a meteorite striking the ground.

meteoroid. A meteoritic particle in space before any encounter with the earth.

micrometeorite. A meteoroid so small that, on entering the atmosphere of the earth, it is slowed quickly enough that it does not burn up or ablate but filters through the air to the ground.

microwave. Short-wave radio wavelengths.

Milky Way. The band of light encircling the sky, which is due to the many stars and diffuse nebulae lying near the plane of the Galaxy.

minor axis (of ellipse). The smallest or least diameter of an ellipse.

minor planet. One of several tens of thousands of small planets, ranging in size from a thousand kilometers to less than one kilometer in diameter.

Mira Ceti-type variable star. Any of a large class of red-giant long-period or irregular pulsating variable stars, of which the star Mira is a prototype.

missile. A projectile, especially a rocket.

model atmosphere (or **photosphere**). The result of a theoretical calculation of the run of temperature, pressure, density, and so on, through the outer layers of the sun or a star.

molecule. A combination of two or more atoms bound together; the smallest particle of a chemical compound or substance that exhibits the chemical properties of that substance.

momentum. A measure of the inertia or state of motion of a body; the momentum of a body is the product of its mass and velocity. In the absence of a force, momentum is conserved.

monochromatic. Of one wavelength or color.

mundane astrology. Astrology applied to nations and kings, rather than to individuals.

muon. A particle that behaves like an electron but is about 200 times as massive.

nadir. The point on the celestial sphere 180° from the zenith.

nanosecond. One thousand-millionth (10^{-9}) second.

natal astrology. Astrology based on the horoscope drawn up for the place and moment of one's birth.

nautical mile. The mean length of one minute of arc on the earth's surface along a meridian.

navigation. The art of finding one's position and course at sea or in the air.

neap tide. The lowest high tides in the month, which occur when the moon is near first or third quarter.

nebula. Cloud of interstellar gas or dust.

nebular hypothesis. The basic idea that the sun and planets formed from the same cloud of gas and dust in interstellar space.

Neptune. Eighth planet from the sun in the solar system.

neutrino. A particle that has no mass or charge but that carries energy away in the course of certain nuclear transformations.

neutron. A subatomic particle with no charge and with mass approximately equal to that of the proton.

neutron star. A star of extremely high density composed entirely of neutrons.

New General Catalogue (NGC). A catalog of star clusters, nebulae, and galaxies compiled by J. L. E. Dreyer in 1888.

new moon. Phase of the moon when its longitude is the same as that of the sun.

Newtonian focus. An optical arrangement in a reflecting telescope where the light is reflected by a flat mirror to a focus at the side of the telescope tube just before it reaches the focus of the objective.

Newton's laws. The laws of mechanics and gravitation formulated by Isaac Newton.

night sky light. The faint illumination of the night sky; the main source is usually fluorescence by atoms high in the atmosphere.

node. The intersection of the orbit of a body with a fundamental plane—usually the plane of the celestial equator or of the ecliptic.

nodical month. The period of revolution of the moon about the earth with respect to the line of nodes of the moon's orbit.

nodical (eclipse) year. Period of revolution of the earth about the sun with respect to the line of nodes of the moon's orbit.

nonthermal radiation. See synchrotron radiation.

north point. That intersection of the celestial meridian and astronomical horizon lying nearest the north celestial sphere.

north polar sequence. A group of stars in the vicinity of the north celestial pole whose magnitudes serve as standards for the international magnitude system.

nova. A star that experiences a sudden outburst of radiant energy, temporarily increasing its luminosity by hundreds to thousands of times.

nuclear. Referring to the nucleus of the atom.

nuclear bulge. Central part of our Galaxy.

nuclear transformation. Transformation of one atomic nucleus into another.

nucleon. Any one of the subatomic particles that compose a nucleus.

nucleosynthesis. The formation and the evolution of the chemical elements by nuclear reactions.

nucleus (of atom). The heavy part of an atom, composed mostly of protons and neutrons, and about which the electrons revolve.

nucleus (of comet). A swarm of solid particles in the head of a comet.

nucleus (of galaxy). Central concentration of stars and gas at the center of a galaxy.

null geodesic. The path of a light ray in four-dimensional spacetime.

objective. The principal image-forming component of a telescope or other optical instrument.

objective prism. A prismatic lens that can be placed in front of a telescope objective to transform each star image into an image of its spectrum.

oblate spheroid. A solid formed by rotating an ellipse about its minor axis.

oblateness. A measure of the "flattening" of an oblate spheroid; numerically, the ratio of the difference between the major and minor diameters (or axes) to the major diameter (or axis).

obliquity of the ecliptic. Angle between the planes of the celestial equator and the ecliptic; about 23½°.

obscuration (interstellar). Absorption of starlight by interstellar dust.

occultation. An eclipse of a star or planet by the moon or a planet.

opacity. Absorbing power; capacity to impede the passage of light.

open cluster. A comparatively loose or "open" cluster of stars, containing from a few dozen to a few thousand members, located in the spiral arms or disk of the Galaxy; galactic cluster.

opposition. Configuration of a planet when its elongation is 180°.

optical binary. Two stars at different distances nearly lined up in projection so that they appear close together, but which are not really dynamically associated.

optics. The branch of physics that deals with light and its properties.

orbit. The path of a body that is in revolution about another body or point.

outgassing. The process by which the gasses of a planetary atmosphere work their way out from the crust of the planet.

Pallas. Second minor planet to be discovered.

Pangaea. Name given to the hypothetical continent from which the present continents of the earth separated.

parabola. A conic section of eccentricity 1.0; the curve of intersection between a circular cone and a plane parallel to a straight line in the surface of the cone.

parabolic speed. See velocity of escape.

paraboloid. A parabola of revolution; a curved surface of parabolic cross section. Especially applied to the surface of the primary mirror in a standard reflecting telescope.

parallactic ellipse. A small ellipse that a comparatively nearby star appears to trace out in the sky, which results from the orbital motion of the earth about the sun.

parallax. An apparent displacement of an object due to a motion of the observer.

parallax (stellar). An apparent displacement of a nearby star that results from the motion of the earth around the sun; numerically, the angle subtended by 1 AU at the distance of the star.

parsec. The distance of an object that would have a stellar parallax of one second of arc; 1 parsec = 3.26 light-years, or 206265 AU.

partial eclipse. An eclipse of the sun or moon in which the eclipsed body does not appear completely obscured.

Pauli exclusion principle. The principle that states that no two subatomic particles of certain types can exist in the same place and time with the same state (or condition).

penumbra. The portion of a shadow from which only part of the light source is occulted by an opaque body.

penumbral eclipse. A lunar eclipse in which the moon passes through the penumbra, but not the umbra, of the earth's shadow.

perfect cosmological principle. The assumption that, on the large scale, the universe appears the same from every place and at all times.

perfect gas. An "ideal" gas that obeys the perfect gas laws.

perfect gas laws. Certain laws that describe the behavior of an ideal gas; Charles' law, Boyle's law, and the equation of state for a perfect gas.

perfect radiator. Blackbody; a body that absorbs and subsequently reemits all radiation incident upon it.

periastron. The place in the orbit of a star in a binary star system where it is closest to its companion star.

perifocus. The place on an elliptical orbit that is closest to the focus occupied by the central force.

perigee. The place in the orbit of an earth satellite where it is closest to the center of the earth.

perihelion. The place in the orbit of an object revolving about the sun where it is closest to the center of the sun.

period. A time interval; for example, the time required for one complete revolution.

period-density relation. Proportionality between the period and the inverse square root of the mean density for a pulsating star.

period-luminosity relation. An empirical relation between the periods and luminosities of cepheid-variable stars.

periodic comet. A comet whose orbit has been determined to have an eccentricity of less than 1.0.

perturbation. The disturbing effect, when small, on the motion of a body as predicted by a simple theory, produced by a third body or other external agent.

phases of the moon. The progression of changes in the moon's appearance during the month that results from the moon's turning different portions of its illuminated hemisphere to our view.

photocell (photoelectric cell). An electron tube in which electrons are dislodged from the cathode when it is exposed to light and are accelerated to the anode, thus producing a current in the tube, whose strength serves as a measure of the light striking the cathode.

photoelectric effect. The emission of an electron by the absorption of a photon by a substance.

photographic magnitude. The magnitude of an object, as measured on the traditional, blue- and violet-sensitive photographic emulsions.

photometry. The measurement of light intensities.

photomultiplier. A photoelectric cell in which the electric current generated is amplified at several stages within the tube.

photon. A discrete unit of electromagnetic energy.

photosphere. The region of the solar (or a stellar) atmosphere from which radiation escapes into space.

photosynthesis. The formation of carbohydrates in the chlorophyll-containing tissues of plants exposed to sunlight. In the process, oxygen is released to the atmosphere.

photovisual magnitude. A magnitude corresponding to the spectral region to which the human eye is most sensitive, but measured by photographic methods with suitable green- and yellow-sensitive emulsions and filters.

pion. A particular kind of *meson*, or subatomic particle of mass intermediate between that of the proton and electron.

Pioneer spacecraft. A series of spacecraft launched to Jupiter and more distant planets and to Venus in the 1970s.

plage. A bright region of the solar surface observed in the monochromatic light of some spectral line; flocculus.

Planck's constant. The constant of proportionality relating the energy of a photon to its frequency.

Planck's radiation law. A formula from which can be calculated the intensity of radiation at various wavelengths emitted by a blackbody.

planet. Any of nine solid bodies revolving about the sun.

planetarium. An optical device for projecting on a screen or domed ceiling the stars and planets and their apparent motions in the sky.

planetary nebula. A shell of gas ejected from, and enlarging about, a certain kind of extremely hot star.

planetoid. Synonym for minor planet.

plasma. A hot ionized gas.

plate tectonics. The motion of segments or plates of the outer layers of the earth over the underlying mantle.

Pluto. Ninth planet from the sun in the solar system.

polar axis. The axis of rotation of the earth; also, an axis in the mounting of a telescope that is parallel to the earth's axis.

polarization. A condition in which the planes of vibration (or the E vectors) of the various rays in a light beam are at least partially aligned.

polarized light. Light in which polarization is present.

Polaroid. Trade name for a transparent substance that produces polarization in light.

Population I and II. Two classes of stars (and systems of stars), classified according to their spectra, chemical compositions, radial velocities, ages, and locations in the Galaxy.

positron. An electron with a positive rather than a negative charge.

postulate. An essential prerequisite to a hypothesis or theory.

potential energy. Stored energy that can be converted into other forms; especially gravitational energy.

precession (of earth). A slow, conical motion of the earth's axis of rotation, caused principally by the gravitational torque of the moon and sun on the earth's equatorial bulge. *Lunisolar precession*, precession caused by the moon and sun only; *planetary precession*, a slow change in the orientation of the plane of the earth's orbit caused by planetary perturbations; *general precession*, the combination of these two effects on the motion of the earth's axis with respect to the stars.

precession of the equinoxes. Slow westward motion of the equinoxes along the ecliptic that results from precession.

primary cosmic rays. The cosmic-ray particles that arrive at the earth from beyond its atmosphere, as opposed to the secondary particles that are produced by collisions between primary cosmic rays and air molecules.

prime focus. The point in a telescope where the objective focuses the light.

prime meridian. The terrestrial meridian passing through the site of the old Royal Greenwich Observatory; longitude 0°.

primeval atom. A single mass whose explosion was postulated by G. Lemaître to have resulted in all the matter now present in the universe.

primeval fireball. The extremely hot opaque gas that is presumed to have comprised the entire mass of the universe at the time of or immediately following the "big bang"; the exploding primeval atom.

Principia. Contraction of *Philosophiae Naturalis Principia Mathematica*, the great book by Newton in which he set forth his laws of motion and gravitation in 1687.

principle of equivalence. A principle of general relativity that states that forces and accelerations are equivalent, and that, in particular, the force of gravitation can be replaced by a suitable acceleration of the coordinate system.

principle of relativity. The assumption basic to special relativity that the laws of physics are the same in all systems in uniform motion with respect to each other.

prism. A wedge-shaped piece of glass that is used to disperse white light into a spectrum.

prolate spheroid. The solid produced by the rotation of an ellipse about its major axis.

prominence. A phenomenon in the solar corona that commonly appears like a flame above the limb of the sun.

proper motion. The angular change in direction of a star per year.

proton. A heavy subatomic particle that carries a positive charge, and one of the two principal constituents of the atomic nucleus.

proton-proton chain. A chain of thermonuclear reactions by which nuclei of hydrogen are built up into nuclei of helium.

protoplanet (or **-star** or **-galaxy**). The original material from which a planet (or a star or galaxy) condensed.

pulsar. A variable radio source of small angular size that emits radio pulses in very regular periods that range from 0.03 to 5 seconds.

pulsating variable. A variable star that pulsates in size and luminosity.

quadrature. A configuration of a planet in which its elongation is 90°.

quantum mechanics. The branch of physics that deals with the structure of atoms and their interactions with each other and with radiation.

quark. A hypothetical fundamental subatomic particle. Quarks of from 1 to 6 different kinds, in various combinations, are presumed to make up all other particles in the atomic nucleus.

quarter moon. Either of the two phases of the moon when its longitude differs by 90° from that of the sun; the moon appears half full at these phases.

quasar. A quasi-stellar source.

quasi-stellar galaxy (QSG). A stellar-appearing object of very large redshift presumed to be extragalactic and highly luminous.

quasi-stellar source (QSS). A stellar-appearing object of very large redshift that is a strong source of radio waves; presumed to be extragalactic and highly luminous.

RR Lyrae variable. One of a class of giant pulsating stars with periods less than one day; a cluster variable.

radar. A technique for observing the reflection of radio waves from a distant object.

radial velocity. The component of relative velocity that lies in the line of sight.

radial velocity curve. A plot of the variation of radial velocity with time for a binary or variable star.

radiant (of meteor shower). The point in the sky from which the meteors belonging to a shower seem to radiate.

radiation. A mode of energy transport whereby energy is transmitted through a vacuum; also the transmitted energy itself, either electromagnetic or corpuscular.

radiation pressure. The transfer of momentum carried by electromagnetic radiation to a body that the radiation impinges upon.

radioactive dating. The science of determining the ages of rocks or other specimens by the amount of radioactive decay of certain radioactive elements contained therein.

radioactivity (radioactive decay). The process by which certain kinds of atomic nuclei naturally decompose with the spontaneous emission of subatomic particles and gamma rays.

radio astronomy. The technique of making astronomical observations in radio wavelengths.

radio galaxy. A galaxy that emits greater amounts of radio radiation than average.

radio telescope. A telescope designed to make observations in radio wavelengths.

ray (lunar). Any of a system of bright enlongated streaks, sometimes associated with a crater on the moon.

Rayleigh scattering. Scattering of light (photons) by molecules of a gas.

recurrent nova. A nova that has been known to erupt more than once.

reddening (interstellar). The reddening of starlight passing through interstellar dust, caused by the dust scattering blue light more effectively than red.

red giant. A large, cool star of high luminosity; a star occupying the upper right portion of the Hertzsprung-Russell diagram.

redshift. A shift to longer wavelengths of the light from remote galaxies, produced by a Doppler shift.

reflecting telescope. A telescope in which the principal optical component (objective) is a concave mirror.

reflection. The return of light rays by an optical surface.

reflection nebula. A relatively dense dust cloud in interstellar space that is illuminated by starlight.

refracting telescope. A telescope in which the principal optical component (objective) is a lens or system of lenses.

refraction. The bending of light rays passing from one transparent medium (or a vacuum) to another.

regression of nodes. A consequence of certain perturbations on the orbit of a revolving body whereby the nodes of the orbit slide westward in the fundamental plane (usually the plane of the ecliptic or of the celestial equator).

relative orbit. The orbit of one of two mutually revolving bodies referred to the other body as origin.

relativistic particle (or **electron**). A particle (or electron) moving at nearly the speed of light.

relativity. A theory formulated by Einstein that describes the relations between measurements of physical phenomena by two different observers who are in relative motion at constant velocity (the *special theory of relativity*), or that describes how a gravitational field can be replaced by a curvature of spacetime (the *general theory of relativity*).

resolution. The degree to which fine details in an image are separated or resolved.

resolving power. A measure of the ability of an optical system to resolve or separate fine details in the image it produces; in astronomy, the angle in the sky that can be resolved by a telescope.

rest mass. The mass of an object or particle as measured when it is at rest in the laboratory.

retrograde motion. An apparent westward motion of a planet on the celestial sphere or with respect to the stars.

retrorockets. Rockets fired from a spacecraft in the direction of its motion to slow it down.

revolution. The motion of one body around another.

right ascension. A coordinate for measuring the east-west positions of celestial bodies; the angle measured eastward along the celestial equator from the vernal equinox to the hour circle passing through a body.

rille (or **rill**). A crevasse or trenchlike depression in the moon's surface.

Roche's limit. The smallest distance from a planet or other body at which purely gravitational forces can hold together a satellite or secondary body of the same mean density as the primary; within this distance the tidal forces of the primary would break up the secondary.

rotation. Turning of a body about an axis running through it.

saros. An 18-year cycle over which solar eclipses of a similar type repeat each other.

satellite. A body that revolves about a larger one; for example, a moon of a planet.

Saturn. The sixth planet from the sun in the solar system.

scale (of telescope). The linear distance in the image corresponding to a particular angular distance in the sky; say, so many centimeters per degree.

Schmidt telescope. A type of reflecting telescope invented by B. Schmidt, in which certain aberrations produced by a spherical concave mirror are compensated for by a thin objective correcting lens.

Schwarzschild radius. See event horizon.

science. The attempt to find order in nature or to find laws that describe natural phenomena.

scientific method. A specific procedure in science: (1) the observation of phenomena or the results of experiments; (2) the formulation of hypotheses that describe these phenomena, and that are consistent with the body of knowledge available; (3) the testing of these hypotheses by noting whether or not they adequately predict and describe new phenomena or the results of new experiments.

Sculptor-type system. A dwarf elliptical galaxy, of which the system in Sculptor is a typical example.

secondary cosmic rays. Secondary particles produced by interactions between primary cosmic rays from space and the atomic nuclei in molecules of the earth's atmosphere.

second-order cluster of galaxies. A cluster of clusters of galaxies; a supercluster.

secular. Not periodic.

seeing. The unsteadiness of the earth's atmosphere, which blurs telescopic images.

seismic waves. Vibrations traveling through the earth's interior that result from earthquakes.

seismograph. An instrument used to record and measure seismic waves.

seismology. The study of earthquakes and the conditions that produce them, and of the internal structure of the earth as deduced from analyses of seismic waves.

seleno-. Prefix meaning to the moon.

semimajor axis. Half the major axis of a conic section.

semiregular variable. A variable star, usually a red giant or supergiant, whose period of pulsation is far from constant.

separation (in a visual binary). The angular separation of the two components of a visual binary star.

Seyfert galaxy. A spiral galaxy whose nucleus shows bright emission lines; one of a class of galaxies first described by C. Seyfert. (They are sometimes radio sources.)

shadow cone. The umbra of the shadow of a spherical body (such as the earth) in sunlight.

shell star. A type of star, usually of spectral-type B to F, surrounded by a gaseous ring or shell.

shower (of cosmic rays). A large "rain" of secondary cosmic-ray particles produced by a very energetic primary particle impinging on the earth's atmosphere.

shower (meteor). Many meteors, all seeming to radiate from a common point in the sky, caused by the encounter by the earth of a swarm of meteoroids moving together through space.

sidereal astrology. Astrology in which the horoscope is based on the positions of the planets with respect to the fixed stars rather than with respect to signs that, as a consequence of precession, slide through the zodiac.

sidereal day. The interval between two successive meridian passages of the vernal equinox.

sidereal month. The period of the moon's revolution about the earth with respect to the stars.

sidereal period. The period of revolution of one body about another with respect to the stars.

sidereal time. The local hour angle of the vernal equinox.

sidereal year. Period of the earth's revolution about the sun with respect to the stars.

sign (of zodiac). Astrological term for any of twelve equal sections along the ecliptic, each of length 30°. Starting at the vernal equinox, and moving eastward, the signs are Aries, Taurus, Gemini, Cancer, Leo, Virgo, Libra, Scorpio, Sagittarius, Capricorn, Aquarius, and Pisces.

simultaneity. The occurrence of two events at the same time. In relativity, absolute simultaneity is seen not to have meaning.

Skylab. An orbiting scientific laboratory occupied by several successive teams of astronauts in the late 1960s and early 1970s.

solar activity. Phenomena of the solar atmosphere: sunspots, plages, and related phenomena.

solar antapex. Direction away from which the sun is moving with respect to the local standard of rest.

solar apex. The direction toward which the sun is moving with respect to the local standard of rest.

solar constant. Mean amount of solar radiation received per unit time, by a unit area, just outside the earth's atmosphere, and perpendicular to the direction of the sun; the numerical value is 1.36×10^6 ergs/cm^2 · s.

solar motion. Motion of the sun, or the velocity of the sun, with respect to the local standard of rest.

solar nebula. The cloud of gas and dust from which the solar system is presumed to have formed.

solar parallax. Angle subtended by the equatorial radius of the earth at a distance of 1 AU.

solar system. The system of the sun and the planets, their satellites, the minor planets, comets, meteoroids, and other objects revolving around the sun.

solar time. A time based on the sun; usually the hour angle of the sun *plus* 12 hours.

solar wind. A radial flow of corpuscular radiation leaving the sun.

solstice. Either of two points on the celestial sphere where the sun reaches its maximum distances north and south of the celestial equator.

south point. Intersection of the celestial meridian and astronomical horizon 180° from the north point.

space motion. The velocity of a star with respect to the sun.

space probe. An unmanned interplanetary rocket carrying scientific instruments to obtain data on other planets or on the interplanetary environment.

space technology. The applied science of the immediate space environment of the earth.

spacetime. A system of one time and three spatial coordinates, with respect to which the time and place of an *event* can be specified; also called *spacetime continuum*.

specific gravity. The ratio of the density of a body or substance to that of water.

spectral class (or **type**). A classification of a star according to the characteristics of its spectrum.

spectral sequence. The sequence of spectral classes of stars arranged in order of decreasing temperatures of stars of those classes.

spectrogram. A photograph of a spectrum.

spectrograph. An instrument for photographing a spectrum; usually attached to a telescope to photograph the spectrum of a star.

spectroheliogram. A photograph of the sun obtained with a spectroheliograph.

spectroheliograph. An instrument for photographing the sun, or part of the sun, in the monochromatic light of a particular spectral line.

spectrophotometry. The measurement of the intensity of light from a star or other source at different wavelengths.

spectroscope. An instrument for directly viewing the spectrum of a light source.

spectroscopic binary star. A binary star in which the components are not resolved optically, but whose binary nature is indicated by periodic variations in radial velocity, indicating orbital motion.

spectroscopic parallax. A parallax (or distance) of a star that is derived by comparing the apparent magnitude of the star with its absolute magnitude as deduced from its spectral characteristics.

spectroscopy. The study of spectra.

spectrum. The array of colors or wavelengths obtained when light from a source is dispersed, as in passing it through a prism or grating.

spectrum analysis. The study and analysis of spectra, especially stellar spectra.

spectrum binary. A binary star whose binary nature is revealed by spectral characteristics that can only result from the composite of the spectra of two different stars.

speed. The rate at which an object moves without regard to its direction of motion; the numerical or absolute value of velocity.

spherical aberration. A defect of optical systems whereby on-axis rays of light striking different parts of the objective do not focus at the same place.

spicule. A narrow jet of rising material in the solar chromosphere.

spiral arms. Arms of interstellar material and young stars that wind out in a plane from the central nucleus of a spiral galaxy.

spiral galaxy. A flattened, rotating galaxy with pinwheel-like arms of interstellar material and young stars winding out from its nucleus.

sporadic meteor. A meteor that does not belong to a shower.

spring tide. The highest tide of the month, produced when the sun and moon have longitudes that differ from each other by nearly 0° or 180°.

Sputnik. Russian for "satellite," or "fellow traveler"; the name given to the first Soviet artificial satellite.

stadium. A Greek unit of length, based on the Olympic Stadium; roughly ⅙ km.

standard time. The local mean solar time of a standard meridian, adopted over a large region to avoid the inconvenience of continuous time changes around the earth.

star. A self-luminous sphere of gas.

star cluster. An assemblage of stars held together by their mutual gravitation.

steady state (theory of cosmology). A theory of cosmology embracing the perfect cosmological principle, and involving the continuous creation of matter.

Stefan's law or **Stefan-Boltzmann Law.** A formula from which the rate at which a blackbody radiates energy can be computed; the total rate of energy emission from a unit area of a blackbody is proportional to the fourth power of its absolute temperature.

stellar evolution. The changes that take place in the sizes, luminosities, structures, and so on, of stars as they age.

stellar model. The result of a theoretical calculation of the run of physical conditions in a stellar interior.

stellar parallax. The angle subtended by 1 AU at the distance of a star; usually measured in seconds of arc.

stimulated emission or stimulated radiation. Photons emitted by excited atoms undergoing downward transitions as a consequence of being stimulated by other photons of the same wavelength.

Stonehenge. An assemblage of upright stones in Salisbury Plain, England, believed to have been constructed by early people for astronomical observations connected with timekeeping and the calendar.

stratosphere. The layer of the earth's atmosphere above the troposphere, where most weather takes place, and below the ionosphere.

strong nuclear force. The force that binds together the parts of the atomic nucleus.

subdwarf. A star of luminosity lower than that of main-sequence stars of the same spectral type.

subgiant. A star of luminosity intermediate between those of main-sequence stars and normal giants of the same spectral type.

subtend. To have or include a given angular size.

summer solstice. The point on the celestial sphere where the sun reaches its greatest distance north of the celestial equator.

sun. The star about which the earth and other planets revolve.

sundial. A device for keeping time by the shadow a marker (gnomon) casts in sunlight.

sunspot. A temporary cool region in the solar photosphere that appears dark by contrast against the surrounding hotter photosphere.

sunspot cycle. The semiregular 11-year period with which the frequency of sunspots varies.

supercluster. See second-order cluster.

supergiant. A star of very high luminosity.

supergranulation. Large-scale convective patterns in the solar photosphere (up to 30,000 km in diameter).

superior conjunction. The configuration of a planet in which it and the sun have the same longitude, with the planet being more distant than the sun.

superior planet. A planet more distant from the sun than the earth.

supernova. A stellar outburst or explosion in which a star suddenly increases its luminosity by from hundreds of thousands to hundreds of millions of times.

surface gravity. The weight of a unit mass at the surface of a body.

surveying. The technique of measuring distances and relative positions of places over the surface of the earth (or elsewhere); generally accomplished by triangulation.

synchrotron radiation. The radiation emitted by charged particles being accelerated in magnetic fields and moving at speeds near that of light.

synodic month. The period of revolution of the moon with respect to the sun, or its cycle of phases.

synodic period. The interval between successive occurrences of the same configuration of a planet; for example, between successive oppositions or successive superior conjunctions.

tachyon. A hypothetical particle that always moves with a speed greater than that of light. (There is no evidence that tachyons exist.)

tail (of comet). Gases and solid particles ejected from the head of a comet and forced away from the sun by radiation pressure or corpuscular radiation.

tangential (transverse) velocity. The component of a star's space velocity that lies in the plane of the sky.

tectonics. See plate tectonics.

tektites. Rounded glassy bodies that are suspected to be of meteoritic origin.

telescope. An optical instrument used to aid the eye in viewing or measuring, or to photograph distant objects.

telluric. Of terrestrial origin.

temperature (absolute). Temperature measured in centigrade (Celsius) degrees from absolute zero.

temperature (Celsius; formerly **centigrade).** Temperature measured on a scale where water freezes at 0° and boils at 100°.

temperature (color). The temperature of a star as estimated from the intensity of the stellar radiation at two or more colors or wavelengths.

temperature (effective). The temperature of a blackbody that would radiate the same total amount of energy that a particular body does.

temperature (excitation). The temperature of a star as estimated from the relative strengths of lines in its spectrum that originate from atoms in different stages of excitation.

temperature (Fahrenheit). Temperature measured on a scale where water freezes at 32° and boils at 212°.

temperature (ionization). The temperature of a star as estimated from the relative strengths of lines in its spectrum that originate from atoms in different stages of ionization.

temperature (Kelvin). Absolute temperature measured in centigrade degrees.

temperature (kinetic). A measure of the speeds or mean energy of the molecules in a substance.

temperature (radiation). The temperature of a blackbody that radiates the same amount of energy in a given spectral region as does a particular body.

tensor. A generalization of the concept of the vector, consisting of an array of numbers or quantities that transform according to specific rules.

terminator. The line of sunrise or sunset on the moon or a planet.

terrestrial planet. Any of the planets Mercury, Venus, Earth, Mars, and sometimes Pluto.

Tetrabiblos. A standard and widely used treatise on astrology by Ptolemy.

theory. A set of hypotheses and laws that have been well demonstrated as applying to a wide range of phenomena associated with a particular subject.

thermal energy. Energy associated with the motions of the molecules in a substance.

thermal equilibrium. A balance between the input and outflow of heat in a system.

thermal radiation. The radiation emitted by any body or gas that is not at absolute zero.

thermodynamics. The branch of physics that deals with heat and heat transfer among bodies.

thermonuclear energy. Energy associated with thermonuclear reactions or that can be released through thermonuclear reactions.

thermonuclear reaction. A nuclear reaction or transformation that results from encounters between nuclear particles that are given high velocities (by heating them).

thermosphere. The region of the earth's atmosphere lying between the mesosphere and the exosphere.

tidal force. A differential gravitational force that tends to deform a body.

tide. Deformation of a body by the differential gravitational force exerted on it by another body; in the earth, the deformation of the ocean surface by the differential gravitational forces exerted by the moon and sun.

ton (American short). 2000 lb.

ton (English long). 2240 lb.

ton (metric). One million grams.

topography. The configuration or relief of the surface of the earth, moon or a planet.

total eclipse. An eclipse of the sun in which the sun's

photosphere is entirely hidden by the moon, or an eclipse of the moon in which it passes completely into the umbra of the earth's shadow.

train (of meteor). A temporarily luminous trail left in the wake of a meteor.

transit. An instrument for timing the exact instant a star or other object crosses the local meridan. Also, the passage of a celestial body across the meridian; or the passage of a small body (say, a planet) across the disk of a larger one (say, the sun).

triangulation. The operation of measuring some of the elements of a triangle so that other ones can be calculated by the methods of trigonometry, thus determining distances to remote places without having to span them directly.

trigonometry. The branch of mathematics that deals with the analytical solutions of triangles.

trillion. In the United States, 1 thousand billions or million millions (10^{12}); in Great Britain, 1 million billions or million-million millions (10^{18}).

triple-alpha process. A series of two nuclear reactions by which three helium nuclei are built up into one carbon nucleus.

Trojan minor planet (or asteroid). One of several minor planets that share Jupiter's orbit around the sun, but located approximately 60° around the orbit from Jupiter.

Tropic of Cancer. Parallel of latitude 23½° N.

Tropic of Capricorn. Parallel of latitude 23½° S.

tropical astrology. The conventional practice of astrology, in which the horoscope is based on the signs that move through the zodiac with precession.

tropical year. Period of revolution of the earth about the sun with respect to the vernal equinox.

troposphere. Lowest level of the earth's atmosphere, where most weather takes place.

tsunami. A series of very fast waves of seismic origin traveling through the ocean; popularly called "tidal waves."

21-cm line. A spectral line of neutral hydrogen at the radio wavelength of 21 centimeters.

Uhuru. An earth satellite equipped for observations at X-ray wavelengths.

ultraviolet radiation. Electromagnetic radiation of wavelengths shorter than the shortest (violet) wavelengths to which the eye is sensitive; radiation of wavelengths in the approximate range 100 to 4000 angstroms.

umbra. The central, completely dark part of a shadow.

uncertainty principle. See Heisenberg uncertainty principle.

universal time. The local mean time of the prime meridian.

universe. The totality of all matter and radiation and the space occupied by same.

Uranus. Seventh planet from the sun in the solar system.

Van Allen layer. Doughnut-shaped region surrounding the earth where many rapidly moving charged particles are trapped in its magnetic field.

variable star. A star that varies in luminosity.

variation of latitude. A slight semiperiodic change in the latitudes of places on the earth that results from a slight shifting of the body of the earth with respect to its axis of rotation.

vector. A quantity that has both magnitude and direction.

velocity. A vector that denotes both the speed and direction a body is moving.

velocity of escape. The speed with which an object must move in order to enter a parabolic orbit about another body (such as the earth), and hence move permanently away from the vicinity of that body.

Venus. The second planet from the sun in the solar system.

vernal equinox. The point on the celestial sphere where the sun crosses the celestial equator passing from south to north.

vertical circle. A great circle in the sky passing through the zenith.

Vikings. A series of spacecrafts that landed laboratories on Mars in 1976. The Viking landers were particularly designed to search for life on Mars.

visual binary star. A binary star in which the two components are telescopically resolved.

volume. A measure of the total space occupied by a body.

Voyagers. A series of spacecrafts that were launched by the United States toward Jupiter and more distant planets in 1977.

Vulcan. A hypothetical planet once believed to exist and have an orbit between that of Mercury and the sun; the existence of Vulcan is now generally discredited.

walled plain. A large lunar crater.

wandering of the poles. A semiperiodic shift of the body of the earth relative to its axis of rotation; responsible for variation of latitude.

watt. A unit of power; 10 million ergs expended per second.

wavelength. The spacing of the crests or troughs in a wave train.

weak nuclear force. The nuclear force involved in radioactive decay. The weak force is characterized by the slow rate of certain nuclear reactions—such as the decay of the neutron, which occurs at the average rate of 11 min.

weight. A measure of the force due to gravitational attraction.

west point. The point on the horizon 270° around the horizon from the north point, measured in a clockwise direction as seen from the zenith.

white dwarf. A star that has exhausted most or all of its nuclear fuel and has collapsed to a very small size; believed to be near its final stage of evolution.

white hole. The hypothetical time reversal of a black hole, in which matter and radiation gush up. White holes are not believed to exist.

Widmanstätten figures. Crystalline structure that can be observed in cut and polished meteorites.

Wien's law. Formula that relates the temperature of a blackbody to the wavelength at which it emits the greatest intensity of radiation.

winter solstice. Point on the celestial sphere where the sun reaches its greatest distance south of the celestial equator.

Wolf-Rayet star. One of a class of very hot stars that eject shells of gas at very high velocities.

worm hole. See **Einstein-Rosen Bridge.**

X rays. Photons of wavelengths intermediate between those of ultraviolet radiation and gamma rays.

X-ray stars. Stars (other than the sun) that emit observable amounts of radiation at X-ray frequencies.

year. The period of revolution of the earth around the sun.

Zeeman effect. A splitting or broadening of spectral lines due to magnetic fields.

zenith. The point on the celestial sphere opposite to the direction of gravity; or the direction opposite to that indicated by a plumb bob.

zenith distance. Arc distance of a point on the celestial sphere from the zenith; 90° minus the altitude of the object.

zero-age main sequence. Main sequence for a system of stars that have completed their contraction from interstellar matter, are now deriving all their energy from nuclear reactions, but whose chemical composition has not yet been altered by nuclear reactions.

zodiac. A belt around the sky centered on the ecliptic.

zodiacal light. A faint illumination along the zodiac, believed to be sunlight reflected and scattered by interplanetary dust.

zone of avoidance. A region near the Milky Way where obscuration by interstellar dust is so heavy that few or no exterior galaxies can be seen.

zone time. The time, kept in a zone 15° wide in longitude, that is the local mean time of the central meridian of that zone. Zone time is used at sea, but over land the boundaries are irregular to conform to political boundaries, and it is called *standard time*.

APPENDIX 3

1. POWERS-OF-10 NOTATION

It is often necessary to deal with very large or very small numbers. For example, the earth is 150,000,000 kilometers from the sun and the mass of the hydrogen atom is 0.000 000 000 000 000 000 000 001 67 g. Instead of writing and carrying so many zeros, the numbers are usually written as figures between 1 and 10 multiplied by the appropriate power of 10. For example, 150,000,000 is $1.5 \times 100{,}000{,}000$, or 1.5×10^8. Similarly, 0.000 000 000 000 000 000 000 001 67 is 1.67/1 000 000 000 000 000 000 000 000 or $1.67/10^{24} = 1.67 \times 10^{-24}$. The rule in reading numbers written in this notation is that the exponent of 10 is the number of places the decimal point is to be moved to the right (if the exponent is positive) or to the left (if the exponent is negative).

2. ANGULAR MEASURE

The most common units of angular measure used in astronomy are the following:

1. arc measure:
   one circle contains 360 degrees = 360°;
   1° contains 60 minutes of arc = 60′;
   1′ contains 60 seconds of arc = 60″.
2. time measure:
   one circle contains 24 hours = 24^h;
   1^h contains 60 minutes of time = 60^m;
   1^m contains 60 minutes of time = 60^s;
3. radian measure:
   one circle contains 2π radians.

A radian is the angle at the center of a circle subtended by a length along the circumference of the circle equal to its radius. Since the circumference of a circle is 2π times its radius, there are 2π radians in a circle.

Relations between these different units of angular measure are given in the following table.

Arc Measure	Time Measure	Radians	Seconds of Arc
57°.2958	$3^h.820$	1.0	206,264.806
15°	1^h	0.2618	54,000
1°	4^m	1.745×10^{-2}	3,600
15′	1^m	4.363×10^{-3}	900
1′	4^s	2.090×10^{-4}	60
15″	1^s	7.27×10^{-5}	15
1″	$0^s.0667$	4.85×10^{-6}	1

APPENDIX 4

ENGLISH AND METRIC UNITS

In the English system of measure the fundamental units of length, mass, and time are the yard, pound, and second, respectively. There are also, of course, larger and smaller units, which include the ton (2000 lb), the mile (1760 yd), the rod (16½ ft), the inch (1/36 yd), the ounce (1/16 lb), and so on. Such units are inconvenient for conversion and arithmetic computation.

In science, therefore, it is more usual to use the metric system, which has been adopted in all countries except the United States. The fundamental units of the metric system are:

length: 1 meter (m)
mass: 1 kilogram (kg)
time: 1 second (s)

A meter was originally intended to be 1 ten-millionth of the distance from the equator to the North Pole along the surface of the earth. It is about 1.1 yd. A kilogram is about 2.2 lb. The second is the same in metric and English units. The most commonly used quantities of length and mass of the metric system are the following:

length

1 km	= 1 kilometer	= 1000 meters	= 0.6214 mile	
1 m	= 1 meter	= 1.094 yards	= 39.37 inches	
1 cm	= 1 centimeter	= 0.01 meter	= 0.3937 inch	
1 mm	= 1 millimeter	= 0.001 meter	= 0.1 cm	= 0.03937 inch
1μ	= 1 micron	= 0.000 001 meter	= 0.0001 cm	= 3.937×10^{-5} inch

also: 1 mile = 1.6093 km
1 inch = 2.5400 cm

mass

1 metric ton	= 10^6 grams	= 1000 kg	= 2.2046×10^3 lb
1 kg	= 1000 grams	= 2.2046 lb	
1 g	= 1 gram	= 0.0022046 lb	= 0.0353 oz
1 mg	= 1 milligram	= 0.001 g	= 2.2046×10^{-6} lb

also: 1 lb = 453.6 g
1 oz = 28.3495 g

APPENDIX 5

TEMPERATURE SCALES

Three temperature scales are in general use:

1. Fahrenheit (F); water freezes at 32° F and boils at 212° F.
2. Celsius or centigrade[1] (C); water freezes at 0° C and boils at 100° C.
3. Kelvin or absolute (K); water freezes at 273 K and boils at 373 K.

All molecular motion ceases at −459°F = −273°C = 0 K. Thus Kelvin temperature is measured from this lowest possible temperature, called ***absolute zero***. It is the temperature scale most often used in astronomy. Kelvins are degrees that have the same value as centigrade or Celsius degrees, since the difference between the freezing and boiling points of water is 100 degrees in each.

On the Fahrenheit scale, water boils at 212 degrees and freezes at 32 degrees; the difference is 180 degrees. Thus to convert Celsius degrees or Kelvins to Fahrenheit it is necessary to multiply by 180/100 = 9/5. To convert from Fahrenheit to Celsius degrees or Kelvins, it is necessary to multiply by 100/180 = 5/9.

Example 1: What is 68° F in Celsius and in Kelvins?

$$68°\text{ F} - 32°\text{ F} = 36°\text{ F above freezing.}$$

$$\frac{5}{9} \times 36° = 20°;$$

thus,

$$68°\text{ F} = 20°\text{ C} = 293\text{ K}.$$

[1] Celsius is now the name used for centigrade temperature; it has a more modern standardization, but differs from the old centigrade scale by less than 0.1°.

Example 2: What is 37° C in Fahrenheit and in Kelvins?

$$37° \text{ C} = 273° + 37° = 310 \text{ K};$$

$$\frac{9}{5} \times 37° = 66.6 \text{ Fahrenheit degrees};$$

thus,

37° C is 66.6° F above freezing

or

$$37° \text{ C} = 32° + 66.6° = 98.6° \text{ F}.$$

APPENDIX 6

SOME USEFUL CONSTANTS

MATHEMATICAL CONSTANTS:

$\pi = 3.1415926536$
1 radian = $57\overset{\circ}{.}2957795$
= $3437\overset{\prime}{.}74677$
= $206264\overset{\prime\prime}{.}806$

Number of square degrees on a sphere = 41 252.96124

PHYSICAL CONSTANTS:

velocity of light	$c = 2.99792458 \times 10^{10}$ cm/s
constant of gravitation	$G = 6.672 \times 10^{-8}$ dyne · cm²/g²
Planck's constant	$h = 6.626 \times 10^{-27}$ erg · s
Boltzmann's constant	$k = 1.381 \times 10^{-16}$ erg/deg
mass of hydrogen atom	$m_H = 1.673 \times 10^{-24}$ g
mass of electron	$m_e = 9.1095 \times 10^{-28}$ g
charge on electron	$\epsilon = 4.803 \times 10^{-10}$ electrostatic units
Stefan-Boltzmann constant	$\sigma = 5.670 \times 10^{-5}$ erg/cm² · deg⁴ · s
constant in Wien's law	$\lambda_{max}T = 0.28979$ cm/deg
Rydberg's constant	$R = 1.09737 \times 10^{5}$ per cm
1 electron volt	eV = 1.6022×10^{-12} erg
1 angstrom	Å = 10^{-8} cm
1 ton TNT	= 4.2×10^{16} erg

ASTRONOMICAL CONSTANTS:

astronomical unit	AU = $1.495978707 \times 10^{13}$ cm
parsec	pc = 206265 AU
	= 3.262 LY
	= 3.086×10^{18} cm

light-year	$\mathrm{LY} = 9.4605 \times 10^{17}$ cm
	$= 6.324 \times 10^{4}$ AU
tropical year	$= 365.242199$ ephemeris days
sidereal year	$= 365.256366$ ephemeris days
	$= 3.155815 \times 10^{7}$ s
mass of earth	$M_\oplus = 5.977 \times 10^{27}$ g
mass of sun	$M_\odot = 1.989 \times 10^{33}$ g
equatorial radius of earth	$R_\oplus = 6378$ km
radius of sun	$R_\odot = 6.960 \times 10^{10}$ cm
luminosity of sun	$L_\odot = 3.83 \times 10^{33}$ erg/s
solar constant	$S = 1.36 \times 10^{6}$ erg/cm$^2 \cdot$ s
obliquity of ecliptic (1900)	$\epsilon = 23°27'8''.26$
direction of galactic center (1950)	$\alpha = 17^h42^m4$
	$\delta = -28°55'$
direction of north galactic pole (1950)	$\alpha = 12^h49^m$
	$\delta = +27°.4$

APPENDIX 7

PRINCIPAL COORDINATE SYSTEMS USED IN ASTRONOMY

There are several astronomical coordinate systems that are in common use. In each of these systems the position of an object in the sky, or on the celestial sphere, is denoted by two angles. These angles are referred to a *reference plane,* which contains the observer, and a *reference direction,* which is a direction from the observer to some arbitrary point lying in the reference plane. The intersection of the reference plane and the celestial sphere is a great circle, which defines the "equator" of the coordinate system. At two points, each 90° from this equator, are the "poles" of the coordinate system. Great circles passing through these poles intersect the equator of the system at right angles.

One of the two angular coordinates of each coordinate system is measured from the equator of the system to the object along the great circle passing through it and the poles. Angles on one side of the equator (or reference plane) are reckoned as positive; those on the opposite side are negative. The other angular coordinate is measured along the equator from the reference direction to the intersection of the equator with the great circle passing through the object and the poles.

The system of terrestrial latitude and longitude provides an excellent analogue. Here the plane of the terrestrial equator is the fundamental plane and the earth's equator is the equator of the system; the North and South terrestrial Poles are the poles of the system. One coordinate, the *latitude* of a place, is reckoned north (positive) or south (negative) of the equator along a meridian passing through the place. The other coordinate, *longitude,* is measured along the equator to the intersection of the equator and the meridian of the place from the intersection of the equator and the Greenwich meridian. The direction (from the center of the earth) to this latter intersection is the reference direction. Terrestrial longitude is either east or west (whichever is less), but the corresponding coordinate in celestial systems is generally reckoned in one direction from 0 to 360° (or, equivalently, from 0 to 24^h).

The following table lists the more important astronomical coordinate systems and defines how each of the angular coordinates is defined.

Astronomical Coordinate Systems

System	Reference Plane	Reference Direction	"Latitude" Coordinate	Range	"Longitude" Coordinate	Range
Horizon	Horizon plane	North point (formerly the south point was used by astronomers)	Altitude, *h*; toward the zenith (+) toward the nadir (−)	±90°	Azimuth, *A*; measured to the east along the horizon from the north point	0 to 360°
Equator	Plane of the celestial equator	Vernal equinox	Declination, δ; toward the north celestial pole (+) toward the south celestial pole (−)	±90°	Right ascension, α or R.A.; measured to the east along the celestial equator from the vernal equinox	0 to 24^h
Ecliptic	Plane of the earth's orbit (ecliptic)	Vernal equinox	Celestial latitude, β; toward the north ecliptic pole (+) toward the south ecliptic pole (−)	±90°	Celestial longitude, λ; measured to the east along the ecliptic from the vernal equinox	0 to 360°
Galactic	Mean plane of the Milky Way	Direction to the galactic center	Galactic latitude, *b*; toward the north galactic pole (+) toward the south galactic pole (−)	±90°	Galactic longitude, *l*; measured along the galactic equator to the east from the galactic center	0 to 360°

APPENDIX 8

SOME NUCLEAR REACTIONS OF IMPORTANCE IN ASTRONOMY

Given here are the series of thermonuclear reactions that are most important in stellar interiors. The subscript to the left of a nuclear symbol is the atomic number; the superscript to the right is the atomic mass number. The symbols for the positive electron (positron) and electron are e^+ and e^-, respectively, for the neutrino is ν, and for a photon (generally of gamma-ray energy) is γ.

1. The Proton-Proton Chains

(Important below 15×10^6 K)

There are three ways the proton-proton chain can be completed. The first (a_1, b_1, c_1) is the most important, but depending on the physical conditions in the stellar interior, some energy is released by one or both of the following alternatives: a_1, b_1, c_2, d_2, e_2, and a_1, b_1, c_2, d_3, e_3, f_3.

(a_1) ${}_1H^1 + {}_1H^1 \rightarrow {}_1H^2 + e^+ + \nu$
(b_1) ${}_1H^2 + {}_1H^1 \rightarrow {}_2He^3 + \gamma$
(c_1) ${}_2He^3 + {}_2He^3 \rightarrow {}_2He^4 + 2{}_1H^1$

or (c_2) ${}_2He^3 + {}_2He^4 \rightarrow {}_4Be^7 + \gamma$
(d_2) ${}_4Be^7 + e^- \rightarrow {}_3Li^7 + \nu$
(e_2) ${}_3Li^7 + {}_1H^1 \rightarrow 2{}_2He^4$

or (d_3) ${}_4Be^7 + {}_1H^1 \rightarrow {}_5B^8 + \gamma$
(e_3) ${}_5B^8 \rightarrow {}_4Be^8 + e^+ + \nu$
(f_3) ${}_4Be^8 \rightarrow 2{}_2He^4$

2. The Carbon-Nitrogen Cycle

(Important above 15×10^6 K)

(a) ${}_6C^{12} + {}_1H^1 \rightarrow {}_7N^{13} + \gamma$
(b) ${}_7N^{13} \rightarrow {}_6C^{13} + e^+ + \nu$
(c) ${}_6C^{13} + {}_1H^1 \rightarrow {}_7N^{14} + \gamma$
(d) ${}_7N^{14} + {}_1H^1 \rightarrow {}_8O^{15} + \gamma$
(e) ${}_8O^{15} \rightarrow {}_7N^{15} + e^+ + \nu$
(f) ${}_7N^{15} + {}_1H^1 \rightarrow {}_6C^{12} + {}_2He^4$

3. The Triple-Alpha Process

(Important above 10^8 K)

(a) ${}_2He^4 + {}_2He^4 \rightarrow {}_4Be^8 + \gamma$
(b) ${}_2He^4 + {}_4Be^8 \rightarrow {}_6C^{12} + \gamma$

APPENDIX 9

ORBITAL DATA FOR THE PLANETS

Planet	Symbol	Semimajor Axis		Sidereal Period		Synodic Period (Days)	Mean Orbital Speed (km/s)	Orbital Eccentricity	Inclination of Orbit to Ecliptic
		Au	10^6 km	Tropical Years	Days				
Mercury	☿	0.3871	57.9	0.24085	87.97	115.88	47.8	0.206	7°004
Venus	♀	0.7233	108.2	0.61521	224.70	583.92	35.0	0.007	3.394
Earth	⊕	1.0000	149.6	1.000039	365.26	—	29.8	0.017	0.0
Mars	♂	1.5237	227.9	1.88089	686.98	779.94	24.2	0.093	1.850
(Ceres)	①	2.7673	414	4.604		466.6	17.9	0.077	10.615
Jupiter	♃	5.2028	778	11.86223		398.88	13.1	0.048	1.305
Saturn	♄	9.5388	1427	29.4577		378.09	9.7	0.056	2.490
Uranus	⛢ or ♅	19.182	2870	84.013		369.66	6.8	0.047	0.773
Neptune	♆	30.058	4497	164.793		367.49	5.4	0.009	1.774
Pluto	♇	39.44	5900	248		366.74	4.7	0.250	17.170

For the mean equator and equinox of 1960.

APPENDIX 10

PHYSICAL DATA FOR THE PLANETS

Planet	Diameter		Mass (Earth = 1)	Mean Density (g/cm³)	Period of Rotation	Inclination of Equator to Orbit	Oblateness	Surface Gravity (Earth = 1)	Albedo	Visual Magnitude at Maximum Light	Velocity of Escape (km/s)
	km	Earth = 1									
Mercury	4,878	0.38	0.055	5.44	58.6 days	~0°	0	0.38	0.06	−1.9	4.3
Venus	12,100	0.95	0.82	5.3	242.9 days	~180°	0	0.91	0.76	−4.4	10.3
Earth	12,756	1.00	1.00	5.52	$23^h56^m04^s$	23°27′	1/298.2	1.00	0.39	—	11.2
Mars	6,800	0.53	0.107	3.9	$24^h37^m23^s$	24°	1/192	0.38	0.15	−2.8	5.1
Jupiter	143,000	11.2	317.9	1.3	9^h50^m to 9^h55^m	3°	1/15	2.64	0.51	−2.5	60
Saturn	121,000	9.49	95.2	0.7	10^h14^m to 10^h38^m	27°	1/9.5	1.13	0.50	−0.4	35
Uranus	47,000	3.69	14.6	1.6	20^h–25^h	98°	1/14	1.07	0.66	+5.6	22
Neptune	45,000	3.50	17.2	2.3	15^h–20^h	29°	1/40	1.41	0.62	+7.9	25
Pluto	3,000	0.24	.002	0.8	6.387 days		?	0.03	?	+14.9	1

APPENDIX 11

SATELLITES OF THE PLANETS

Planet	Satellite	Discovered by	Mean Distance from Planet (km)	Sidereal Period (Days)	Orbital Eccentricity	Diameter of Satellite (km)*	Mass (Planet = 1)†	Approximate Magnitude at Opposition
Earth	Moon	–	384,404	27.322	0.055	3476	0.0123	−12.5
Mars	Phobos	A. Hall (1877)	9,380	0.319	0.021	25	(2.7×10^{-8})	+12
	Deimos	A. Hall (1877)	23,500	1.262	0.003	13	4.8×10^{-9}	13
Jupiter	XIV	Voyager II (1979)	129,000	0.297	0	<40	(10^{-4})	18–19
	V Almalthea	Barnard (1892)	181,300	0.498	0.003	240	(2×10^{-9})	13
	I Io	Galileo (1610)	421,600	1.769	0.000	3640	4.7×10^{-5}	5
	II Europa	Galileo (1610)	670,900	3.551	0.000	3130	2.5×10^{-5}	6
	III Ganymede	Galileo (1610)	1,070,000	7.155	0.002	5280	7.8×10^{-5}	5
	IV Callisto	Galileo (1610)	1,880,000	16.689	0.008	4840	5.6×10^{-5}	6
	VI Himalia	Perrine (1904)	11,470,000	250.57	0.158	(170)	(8×10^{-10})	14
	VII Elara	Perrine (1905)	11,800,000	259.65	0.207	(40)	(4×10^{-11})	18
	X Lysithea	Nicholson (1938)	11,850,000	263.55	0.130	(10)	(1×10^{-12})	19
	XIII Leda	Kowal (1974)	11,110,000	239.2	0.147	(8)	(5×10^{-13})	20
	XII Aranke	Nicholson (1951)	21,200,000	631.1	0.169	(10)	(7×10^{-13})	18
	XI Carme	Nicholson (1938)	22,600,000	692.5	0.207	(15)	(2×10^{-12})	19
	VIII Pasiphae	Melotte (1908)	23,500,000	738.9	0.378	(25)	(8×10^{-12})	17
	IX Sinope	Nicholson (1914)	23,700,000	758	0.275	(15)	(2×10^{-12})	19
Saturn	Janus	A. Dollfus (1966)	157,500	0.749		(350)	(3×10^{-8})	14
	Mimas	W. Herschel (1789)	185,400	0.942	0.020	(500)	6.6×10^{-8}	12
	Enceladus	W. Herschel (1789)	237,900	1.370	0.004	(500)	1.5×10^{-7}	12
	Tethys	Cassini (1684)	294,500	1.888	0.000	(1000)	1.1×10^{6}	11
	Dione	Cassini (1684)	377,200	2.737	0.002	(1000)	2×10^{-6}	11
	Rhea	Cassini (1672)	526,700	4.518	0.001	1600	3×10^{-6}	10
	Titan	Huygens (1655)	1,221,000	15.945	0.029	5800	2.5×10^{-4}	8
	Hyperion	Bond (1848)	1,479,300	21.277	0.104	(400)	2×10^{-7}	14
	Iapetus	Cassini (1671)	3,558,400	79.331	0.028	(1200)	4×10^{-6}	11
	Phoebe	W. Pickering (1898)	12,945,500	550.45	0.163	(300)	5×10^{-8}	14
Uranus	Miranda	Kuiper (1948)	123,000	1.414	0	(200)	1×10^{-6}	17
	Ariel	Lassell (1851)	191,700	2.520	0.003	(600)	1.5×10^{-5}	15
	Umbriel	Lassell (1851)	267,000	4.144	0.004	(400)	6×10^{-6}	15
	Titania	W. Herschel (1787)	438,000	8.706	0.002	(1000)	5×10^{-5}	14
	Oberon	W. Herschel (1787)	585,960	13.463	0.001	(900)	3×10^{-5}	14
Neptune	Triton	Lassell (1846)	353,400	5.877	0.000	6000	3×10^{-3}	14
	Nereid	Kuiper (1949)	5,560,000	359.881	0.749	(500)	(10^{-6})	19
Pluto	Charon	Christy (1978)	20,000	6.387	0	(1500)	(0.1)	16

* A diameter of a satellite given in parentheses is estimated from the amount of sunlight it reflects.
† A mass of a satellite given in parentheses is estimated from its size and an assumed density.

APPENDIX 12

TOTAL SOLAR ECLIPSES FROM 1952 THROUGH 2030

Date	Duration of Totality (min)	Where Visible
1952 Feb. 25	3.0	Africa, Asia
1954 June 30	2.5	North-Central U.S. (Great Lakes), Canada, Scandinavia, U.S.S.R., Central Asia
1955 June 20	7.2	Southeast Asia
1958 Oct. 12	5.2	Pacific, Chile, Argentina
1959 Oct. 2	3.0	Northern and Central Africa
1961 Feb. 15	2.6	Southern Europe
1962 Feb. 5	4.1	Indonesia
1963 July 20	1.7	Japan, Alaska, Canada, Maine
1965 May 30	5.3	Pacific Ocean, Peru
1966 Nov. 12	1.9	South America
1970 March 7	3.3	Mexico, Florida, parts of U.S. Atlantic coastline
1972 July 10	2.7	Alaska, Northern Canada
1973 June 30	7.2	Atlantic Ocean, Africa
1974 June 20	5.3	Indian Ocean, Australia
1976 Oct. 23	4.9	Africa, Indian Ocean, Australia
1977 Oct. 12	2.8	Northern South America
1979 Feb. 26	2.7	Northwest U.S., Canada
1980 Feb. 16	4.3	Central Africa, India
1981 July 31	2.2	Siberia
1983 June 11	5.4	Indonesia
1984 Nov. 22	2.1	Indonesia, South America
1987 March 29	0.3	Central Africa
1988 March 18	4.0	Philippines, Indonesia
1990 July 22	2.6	Finland, Arctic Regions
1991 July 11	7.1	Hawaii, Central America, Brazil
1992 June 30	5.4	South Atlantic
1994 Nov. 3	4.6	South America
1995 Oct. 24	2.4	South Asia
1997 March 9	2.8	Siberia, Arctic
1998 Feb. 26	4.4	Central America
1999 Aug. 11	2.6	Central Europe, Central Asia
2001 June 21	4.9	Southern Africa
2002 Dec. 4	2.1	South Africa, Australia
2003 Nov. 23	2.0	Antarctica
2005 April 8	0.7	South Pacific Ocean
2006 March 29	4.1	Africa, Asia Minor, U.S.S.R.
2008 Aug. 1	2.4	Arctic Ocean, Siberia, China
2009 July 22	6.6	India, China, South Pacific
2010 July 11	5.3	South Pacific Ocean
2012 Nov. 13	4.0	Northern Australia, South Pacific
2013 Nov. 3	1.7	Atlantic Ocean, Central Africa
2015 March 20	4.1	North Atlantic, Arctic Ocean
2016 March 9	4.5	Indonesia, Pacific Ocean
2017 Aug. 21	2.7	Pacific Ocean, U.S.A., Atlantic Ocean
2019 July 2	4.5	South Pacific, South America
2020 Dec. 14	2.2	South Pacific, South America, South Atlantic Ocean
2021 Dec. 4	1.9	Antarctica

Date	Duration of Totality (min)	Where Visible
2023 April 20	1.3	Indian Ocean, Indonesia
2024 April 8	4.5	South Pacific, Mexico, East U.S.A.
2026 Aug. 12	2.3	Arctic, Greenland, North Atlantic, Spain
2027 Aug. 2	6.4	North Africa, Arabia, Indian Ocean
2028 July 22	5.1	Indian Ocean, Australia, New Zealand
2030 Nov. 25	3.7	South Africa, Indian Ocean, Australia

APPENDIX **13**

THE NEAREST STARS

Star	Right Ascension (1950)		Declination (1950)		Distance (pc)	Proper Motion	Radial Velocity (km/s)	Spectra of Components			Visual Magnitudes of Components			Absolute Visual Magnitudes of Components		
								A	*B*	*C*	*A*	*B*	*C*	*A*	*B*	*C*
	h	*m*	°	′		″										
Proxima Centauri	14	26.3	−62	28	1.31	3.86	−16	M5V			+11.05			+15.4		
α Centauri	14	36.2	−60	38	1.35	3.68	−22	G2V	K0V		−0.01	+1.33		+4.4	+5.7	
Barnard's Star	17	55.4	+4	33	1.81	10.34	−108	M5V			+9.54			+13.2		
Wolf 359	10	54.1	+7	19	2.35	4.70	+13	M8V			+13.53			+16.7		
Lalande 21185	10	00.6	+36	18	2.52	4.78	−84	M2V			+7.50			+10.5		
Sirius	6	42.9	−16	39	2.65	1.33	−8	A1V	wd		−1.46	+8.68		+1.4	+11.6	
Luyten 726-8	1	36.4	−18	13	2.72	3.36	+29	M5.5V	M5.5V		+12.45	+12.95		+15.3	+15.8	
Ross 154	18	46.8	−23	53	2.90	0.72	−4	M4.5V			+10.6			+13.3		
Ross 248	23	39.4	+43	55	3.14	1.60	−81	M6V			+12.29			+14.8		
Luyten 789-6	22	35.8	−15	36	3.28	3.26	−60	M7V			+12.18			+14.6		
ϵ Eridani	3	30.6	−9	38	3.31	0.98	+16	K2V			+3.73			+6.1		
Ross 128	11	45.2	+1	06	3.32	1.38	−13	M5V			+11.10			+13.5		
61 Cygni	21	04.7	+38	30	3.38	5.22	−64	K5V	K7V		+5.22	+6.03		+7.6	+8.4	
ϵ Indi	21	59.6	−57	00	3.44	4.33	−40	K5V			+4.68			+7.0		
Procyon	7	36.7	+5	21	3.51	1.25	−3	F5IV-V	wd		+0.37	+10.7		+2.6	+13.0	
BD + 43°44	0	15.5	+43	44	3.55	2.90	+13	M1V	M6V		+8.07	+11.04		+10.3	+13.3	
BD + 59°1915	18	42.2	+59	33	3.55	2.29	+10	M4V	M5V		+8.90	+9.69		+11.2	+11.9	
CD − 36°15693	23	02.6	−36	08	3.58	6.90	+10	M2V			+7.36			+9.6		
τ Ceti	1	41.8	−16	12	3.61	1.92	−16	G8V			+3.50			+5.7		
BD + 5°1668	7	24.7	+5	23	3.70	3.77	+26	M5V			+9.82			+12.0		
CD − 39°14192	21	14.3	−39	04	3.85	3.46	+21	M0V			+6.67			+8.8		
Kapteyn's Star	5	09.7	−45	00	3.91	8.72	+245	M0V			+8.81			+10.8		
Kruger 60	22	26.2	+57	27	3.95	0.86	−26	M3V	M4.5V		+9.85	+11.3		+11.9	+13.3	
Ross 614	6	26.8	−2	46	3.97	1.00	+24	M7V	?		+11.07	+14.8		+13.1	+16.8	
BD − 12°4523	16	27.5	−12	32	4.02	1.18	−13	M5V			+10.12			+12.1		
van Maanen's Star	0	46.5	+5	09	4.18	2.99	+54	wd			+12.37			+14.3		
Wolf 424	12	30.8	+9	18	4.33	1.76	−5	M5.5V	M6V		+13.16	+13.4		+15.0	+15.2	
CD − 37°15492	0	02.5	−37	36	4.44	6.11	+23	M4V			+8.63			+10.4		
BD + 50°1725	10	08.3	+49	42	4.50	1.45	−26	K7V			+6.59			+8.3		
CD − 46°11540	17	24.9	−46	51	4.63	1.06	—	M4V			+9.36			+11.0		
CD − 49°13515	21	30.2	−49	13	4.67	0.81	+8	M1V			+8.67			+10.3		
Luyten 1159-16	1	57.5	+12	50	4.69	2.09	—	M8V			+12.27			+13.9		
CD − 44°11909	17	33.5	−44	17	4.69	1.16	—	M5V			+11.2			+12.8		
BD + 68°946	17	36.7	+68	23	4.69	1.31	−22	M3.5V			+9.15			+10.8		
Ross 780	22	50.6	−14	31	4.78	1.14	+9	M5V			+10.17			+11.8		
Luyten 145-141	11	43.0	−64	34	4.85	2.68	—	wd			+11.44			+13.0		
40 Eridani	4	13.1	−7	44	4.88	4.08	−42	K1V	wd	M4.5V	+4.43	+9.53	+11.17	+6.0	+11.1	+12.7
BD + 20°2465	10	16.9	+20	07	4.90	0.49	+11	M4.5V			+9.43			+11.0		
Lalande 25372	13	43.2	+15	10	4.98	2.30	+15	M4V			+8.50			+10.0		

APPENDIX **14**

THE TWENTY BRIGHTEST STARS

Star	Right Ascension (1980)		Declination (1980)		Distance (pc)*	Proper Motion	Spectra of Components			Visual Magnitudes of Components			Absolute Visual Magnitudes of Components		
							A	*B*	*C*	*A*	*B*	*C*	*A*	*B*	*C*
	h	m	°	′		″									
Sirius	6	44.2	−16	41	2.7	1.33	A1V	wd		−1.46	+8.7		+1.4	+11.6	
Canopus	6	23.5	−52	41	30	0.02	F0Ib-II			−0.72			−3.1		
α Centauri	14	38.5	−60	46	1.3	3.68	G2V	K0V		−0.01	+1.3		+4.4	+5.7	
Arcturus	14	14.8	+19	19	11	2.28	K2IIIp			−0.06			−0.3		
Vega	18	36.2	+38	46	8.0	0.34	A0V			+0.04			+0.5		
Capella	5	15.2	+45	59	14	0.44	GIII	M1V	M5V	+0.05	+10.2	+13.7	−0.7	+9.5	+13
Rigel	5	13.5	−8	13	250	0.00	B8 Ia	B9		+0.14	+6.6		−6.8	−0.4	
Procyon	7	38.3	+5	17	3.5	1.25	F5IV-V	wd		+0.37	+10.7		+2.6	+13.0	
Betelgeuse	5	54.1	+7	24	150	0.03	M2Iab			+0.41v			−5.5		
Achernar	1	37.0	−57	20	20	0.10	B5V			+0.51			−1.0		
β Centauri	14	02.4	−60	17	90	0.04	B1III	?		+0.63	+4		−4.1	−0.8	
Altair	19	49.7	+8	49	5.1	0.66	A7IV-V			+0.77			+2.2		
α Crucis	12	25.5	−62	59	120	0.04	B1IV	B3		+1.39	+1.9		−4.0	−3.5	
Aldebaran	4	34.7	+16	29	16	0.20	K5III	M2V		+0.86	+13		−0.2	+12	
Spica	13	24.2	−11	03	80	0.05	B1V			+0.91v			−3.6		
Antares	16	28.1	−26	23	120	0.03	M1Ib	B4eV		+0.92v	+5.1		−4.5	−0.3	
Pollux	7	44.2	+28	05	12	0.62	K0III			+1.16			+0.8		
Fomalhaut	22	56.5	−29	43	7.0	0.37	A3V	K4V		+1.19	+6.5		+2.0	+7.3	
Deneb	20	40.7	+45	12	430	0.00	A2Ia			+1.26			−6.9		
β Crucis	12	46.6	−59	34	150	0.05	B0.5IV			+1.28v			−4.6		

* Distances of the more remote stars have been estimated from their spectral types and apparent magnitudes, and are only approximate.

Note: Several of the components listed are themselves spectroscopic binaries. A "v" after a magnitude denotes that the star is variable, in which case the magnitude at median light is given. A "p" after a spectral type indicates that the spectrum is peculiar. An "e" after a spectral type indicates that emission lines are present. When the luminosity classification is rather uncertain, a range is given.

APPENDIX 15

PULSATING VARIABLE STARS

Type of Variable	Spectra	Period (Days)	Median Magnitude (Absolute)	Amplitude (Magnitudes)	Description	Example	Number Known in Galaxy*
Cepheids (type I)	F to G supergiants	3 to 50	-1.5 to -5	0.1 to 2	Regular pulsation; period-luminosity relation exists	δ Cep	706
Cepheids (type II)	F to G supergiants	5 to 30	0 to -3.5	0.1 to 2	Regular pulsation; period-luminosity relation exists	W Vir	(About 50; included with type I)
RV Tauri	G to K yellow and red bright giants	30 to 150	-2 to -3	Up to 3	Alternate large and small maxima	RV Tau	104
Long-period (Mira-type)	M red giants	80 to 600	$+2$ to -2	>2.5	Brighten more or less periodically	o Cet (Mira)	4566
Semiregular	M giants and supergiants	30 to 2000	0 to -3	1 to 2	Periodicity not dependable; often interrupted	α Ori	2221
Irregular	All types	Irregular	<0	Up to several magnitudes	No known periodicity; many may be semiregular, but too few data exist to classify them as such	π Gru	1687
RR Lyrae or cluster-type	A to F blue giants	<1	0 to $+1$	<1 to 2	Very regular pulsations	RR Lyr	4433
β Cephei or β Canis Majoris	B blue giants	0.1 to 0.3	-2 to -4	0.1	Maximum light occurs at time of highest compression	β Cep	23
δ Scuti	F subgiants	<1	0 to $+2$	<0.25	Similar to, and possibly related to, RR Lyrae variables	δ Sct	17
Spectrum variables	A main sequence	1 to 25	0 to $+1$	0.1	Anomalously intense lines of Si, Sr, and Cr vary in intensity with same period as light; most have strong variable magnetic fields	α^2C Vn	28

* According to the 1968 edition of the Soviet *General Catalogue of Variable Stars*.

APPENDIX **16**

ERUPTIVE VARIABLE STARS

Type of Variable	Spectra	Duration of Increased Brightness	Normal Absolute Magnitude	Amplitude (Magnitudes)	Description	Example	Number Known in Galaxy*
Novae	O to A hot subdwarfs	Months to years	>0	7 to 16	Rapid rise to maximum; slow decline; ejection of gas shell	GK Per	166
Novalike variables or P Cygni stars	Hot B stars	Erratic	−3 to −6	Several magnitudes	Slow, erratic, and nova-like variations in light; may be unrelated to novae. Gas shell ejected	P Cyg	39
Supernovae	?	Months to years	?	15 or more	Sudden, violent flareup, followed by decline and ejection of gas shell	CM Tau (Crab nebula)	7
R Coronae Borealis	F to K supergiant	10 to several hundred days	−5	1 to 9	Sudden and irregular drops in brightness. Low in hydrogen abundance, but high in carbon abundance	R CrB	32
T Tauri or RW Aurigae	B to M main sequence and subgiants	Rapid and erratic	0 to +8	Up to a few magnitudes	Rapid and irregular light variations. Generally associated with interstellar material. Subtypes from G to M are called T Tauri variables	RW Aur, T Tau	1109
U Geminorum or SS Cygni or "dwarf novae"	A to F hot subdwarfs	Few days to few weeks	>0	2 to 6	Novalike outbursts at mean intervals which range from 20 to 600 days. Those with longer intervals between outbursts tend to have greater amplitudes. Many, if not all, are members of binary-star systems	SS Cyg, U Gem	215
Flare stars	M main sequence	Few minutes	>8	Up to 6	Sudden flareups in light; probably localized flares on surface of star	UV Cet	28
Z Camelopardalis variables	A to F hot subdwarfs	Few days	>0	2 to 5	Similar to U Geminorium, except that variations are sometimes interrupted by constant light for several cycles. Intervals between outbursts normally range from 10 to 40 days.	Z Cam	20

* According to the 1968 edition of the Soviet *General Catalogue of Variable Stars.*

APPENDIX 17

THE LOCAL GROUP OF GALAXIES

Galaxy	Type	Right Ascension (1980)		Declination (1980)		Visual Magnitude (m_v)	Distance		Diameter		Absolute Magnitude (M_v)	Radial Velocity (km/s)	Mass (Solar Masses)
							(kpc)	(1000 LY)	(kpc)	(1000 LY)			
Our Galaxy	Sb	—	—	—		— —	—	— —	30	100	(−21)	—	2×10^{11}
Large Magellanic Cloud	Irr I	5^h	26^m	$-69°$		0.9	48	160	10	30	−17.7	+276	2.5×10^{10}
Small Magellanic Cloud	Irr I	0	51	−73		2.5	56	180	8	25	−16.5	+168	
Ursa Minor system	E4 (dwarf)	15	8.6	+67	11′		70	220	1	3	(−9)		
Sculptor system	E3 (dwarf)	0	58.9	−33	48	8.0	83	270	2.2	7	−11.8		(2 to 4×10^6)
Draco system	E2 (dwarf)	17	19.9	+57	56		100	330	1.4	4.5	(−10)		
Carina system	E3 (dwarf)	6	41.1	−50	57		(170)	(550)	1.5	4.8	(−10)		
Fornax system	E3 (dwarf)	2	38.9	−34	36	8.4	250	800	4.5	15	−13.6	+39	(1.2 to 2×10^7)
Leo II system	E0 (dwarf)	11	12.4	+22	16		230	750	1.6	5.2	−10.0		(1.1×10^6)
Leo I system	E4 (dwarf)	10	7.4	+12	24	12.0	280	900	1.5	5	−10.4		
NGC 6822	Irr I	19	43.8	−14	50	8.9	460	1500	2.7	9	−14.8	−32	
NGC 147	E6	0	32.0	+48	23	9.73	570	1900	3	10	−14.5		
NGC 185	E2	0	37.8	+48	14	9.43	570	1900	2.3	8	−14.8	−305	
NGC 205	E5	0	39.2	+41	35	8.17	680	2200	5	16	−16.5	−239	
NGC 221 (M32)	E3	0	41.6	+40	46	8.16	680	2200	2.4	8	−16.5	−214	
IC 1613	Irr I	1	2.1	+1	51	9.61	680	2200	5	16	−14.7	−238	
Andromeda galaxy (NGC 224; M31)	Sb	0	41.6	+41	10	3.47	680	2200	40	130	−21.2	−266	3×10^{11}
And I	E0 (dwarf)	0	44.6	+37	54	(14)	(680)	(2200)	0.5	1.6	(−11)		
And II	E0 (dwarf)	1	15.2	+33	18	(14)	(680)	(2200)	0.7	2.3	(−11)		
And III	E3 (dwarf)	0	34.2	+36	24	(14)	(680)	(2200)	0.9	0.9	(−11)		
NGC 598 (M33)	Sc	1	32.7	+30	33	5.79	720	2300	17	60	−18.9	−189	8×10^9

APPENDIX 18

THE MESSIER CATALOGUE OF NEBULAE AND STAR CLUSTERS

M	NGC or (IC)	Right Ascension (1980)		Declination (1980)		Apparent Visual Magnitude	Description
		h	m	°	′		
1	1952	5	33.3	+22	01	8.4	"Crab" nebula in Taurus; remains of SN 1054
2	7089	21	32.4	−0	54	6.4	Globular cluster in Aquarius
3	5272	13	41.2	+28	29	6.3	Globular cluster in Canes Venatici
4	6121	16	22.4	−26	28	6.5	Globular cluster in Scorpio
5	5904	15	17.5	+2	10	6.1	Globular cluster in Serpens
6	6405	17	38.8	−32	11	5.5	Open cluster in Scorpio
7	6475	17	52.7	−34	48	3.3	Open cluster in Scorpio
8	6523	18	02.4	−24	23	5.1	"Lagoon" nebula in Sagittarius
9	6333	17	18.1	−18	30	8.0	Globular cluster in Ophiuchus
10	6254	16	56.1	−4	05	6.7	Globular cluster in Ophiuchus
11	6705	18	50.0	−6	18	6.8	Open cluster in Scutum Sobieskii
12	6218	16	46.3	−1	55	6.6	Globular cluster in Ophiuchus
13	6205	16	41.0	+36	30	5.9	Globular cluster in Hercules
14	6402	17	36.6	−3	14	8.0	Globular cluster in Ophiuchus
15	7078	21	28.9	+12	05	6.4	Globular cluster in Pegasus
16	6611	18	17.8	−13	47	6.6	Open cluster with nebulosity in Serpens
17	6618	18	19.6	−16	11	7.5	"Swan" or "Omega" nebula in Sagittarius
18	6613	18	18.7	−17	08	7.2	Open cluster in Sagittarius
19	6273	17	01.4	−26	14	6.9	Globular cluster in Ophiuchus
20	6514	18	01.2	−23	02	8.5	"Trifid" nebula in Sagittarius
21	6531	18	03.4	−22	30	6.5	Open cluster in Sagittarius
22	6656	18	35.2	−23	56	5.6	Globular cluster in Sagittarius
23	6494	17	55.8	−19	00	5.9	Open cluster in Sagittarius
24	6603	18	17.3	−18	26	4.6	Open cluster in Sagittarius
25	(4725)	18	30.5	−19	16	6.2	Open cluster in Sagittarius
26	6694	18	44.1	−9	25	9.3	Open cluster in Scutum Sobieskii
27	6853	19	58.8	+22	40	8.2	"Dumbbell" planetary nebula in Vulpecula
28	6626	18	23.2	−24	52	7.6	Globular cluster in Sagittarius
29	6913	20	23.3	+38	27	8.0	Open cluster in Cygnus
30	7099	21	39.2	−23	16	7.7	Globular cluster in Capricornus

M	NGC or (IC)	Right Ascension (1980)		Decli- nation (1980)		Apparent Visual Magnitude	Description
		h	m	°	′		
31	224	0	41.6	+41	10	3.5	Andromeda galaxy
32	221	0	41.6	+40	46	8.2	Elliptical galaxy; companion to M31
33	598	1	32.7	+30	33	5.8	Spiral galaxy in Triangulum
34	1039	2	40.7	+42	43	5.8	Open cluster in Perseus
35	2168	6	07.5	+24	21	5.6	Open cluster in Gemini
36	1960	5	35.0	+34	05	6.5	Open cluster in Auriga
37	2099	5	51.1	+32	33	6.2	Open cluster in Auriga
38	1912	5	27.3	+35	48	7.0	Open cluster in Auriga
39	7092	21	31.5	+48	21	5.3	Open cluster in Cygnus
40		12	21	+59			Close double star in Ursa Major
41	2287	6	46.2	−20	43	5.0	Loose open cluster in Canis Major
42	1976	5	34.4	−5	24	4	Orion nebula
43	1982	5	34.6	−5	18	9	Northeast portion of Orion nebula
44	2632	8	39	+20	04	3.9	Praesepe; open cluster in Cancer
45		3	46.3	+24	03	1.6	The Pleiades; open cluster in Taurus
46	2437	7	40.9	−14	46	6.6	Open cluster in Puppis
47	2422	7	35.7	−14	26	5	Loose group of stars in Puppis
48	2548	8	12.8	−5	44	6	"Cluster of very small stars"; identifiable
49	4472	12	28.8	+8	06	8.5	Elliptical galaxy in Virgo
50	2323	7	02.0	−8	19	6.3	Loose open cluster in Monoceros
51	5194	13	29.1	+47	18	8.4	"Whirlpool" spiral galaxy in Canes Venatici
52	7654	23	23.3	+61	30	8.2	Loose open cluster in Cassiopeia
53	5024	13	12.0	+18	16	7.8	Globular cluster in Coma Berenices
54	6715	18	53.8	−30	30	7.8	Globular cluster in Sagittarius
55	6809	19	38.7	−30	59	6.2	Globular cluster in Sagittarius
56	6779	19	15.8	+30	08	8.7	Globular cluster in Lyra
57	6720	18	52.8	+33	00	9.0	"Ring nebula; planetary nebula in Lyra
58	4579	12	36.7	+11	55	9.9	Spiral galaxy in Virgo
59	4621	12	41.0	+11	46	10.0	Spiral galaxy in Virgo
60	4649	12	42.6	+11	40	9.0	Elliptical galaxy in Virgo
61	4303	12	20.8	+4	35	9.6	Spiral galaxy in Virgo
62	6266	16	59.9	−30	05	6.6	Globular cluster in Scorpio
63	5055	13	14.8	+42	07	8.9	Spiral galaxy in Canes Venatici
64	4826	12	55.7	+21	39	8.5	Spiral galaxy in Coma Berenices
65	3623	11	17.9	+13	12	9.4	Spiral galaxy in Leo
66	3627	11	19.2	+13	06	9.0	Spiral galaxy in Leo; companion to M65
67	2682	8	50.0	+11	53	6.1	Open cluster in Cancer
68	4590	12	38.4	−26	39	8.2	Globular cluster in Hydra
69	6637	18	30.1	−32	23	8.0	Globular cluster in Sagittarius
70	6681	18	42.0	−32	18	8.1	Globular cluster in Sagittarius

M	NGC or (IC)	Right Ascension (1980)		Decli-nation (1980)		Apparent Visual Magnitude	Description
		h	m	°	′		
71	6838	19	52.8	+18	44	7.6	Globular cluster in Sagitta
72	6981	20	52.3	−12	38	9.3	Globular cluster in Aquarius
73	6994	20	57.8	−12	43	9.1	Open cluster in Aquarius
74	628	1	35.6	+15	41	9.3	Spiral galaxy in Pisces
75	6864	20	04.9	−21	59	8.6	Globular cluster in Sagittarius
76	650	1	41.0	+51	28	11.4	Planetary nebula in Perseus
77	1068	2	41.6	−0	04	8.9	Spiral galaxy in Cetus
78	2068	5	45.7	0	03	8.3	Small emission nebula in Orion
79	1904	5	23.3	−24	32	7.5	Globular cluster in Lepus
80	6093	16	15.8	−22	56	7.5	Globular cluster in Scorpio
81	3031	9	54.2	+69	09	7.0	Spiral galaxy in Ursa Major
82	3034	9	54.4	+69	47	8.4	Irregular galaxy in Ursa Major
83	5236	13	35.4	−29	31	7.6	Spiral galaxy in Hydra
84	4374	12	24.1	+13	00	9.4	Elliptical galaxy in Virgo
85	4382	12	24.3	+18	18	9.3	Elliptical galaxy in Coma Berenices
86	4406	12	25.1	+13	03	9.2	Elliptical galaxy in Virgo
87	4486	12	29.7	+12	30	8.7	Elliptical galaxy in Virgo
88	4501	12	30.9	+14	32	9.5	Spiral galaxy in Coma Berenices
89	4552	12	34.6	+12	40	10.3	Elliptical galaxy in Virgo
90	4569	12	35.8	+13	16	9.6	Spiral galaxy in Virgo
91	omitted						
92	6341	17	16.5	+43	10	6.4	Globular cluster in Hercules
93	2447	7	43.7	−23	49	6.5	Open cluster in Puppis
94	4736	12	50.0	+41	14	8.3	Spiral galaxy in Canes Venatici
95	3351	10	42.9	+11	49	9.8	Barred spiral galaxy in Leo
96	3368	10	45.7	+11	56	9.3	Spiral galaxy in Leo
97	3587	11	13.7	+55	07	11.1	"Owl" nebula; planetary nebula in Ursa Major
98	4192	12	12.7	+15	01	10.2	Spiral galaxy in Coma Berenices
99	4254	12	17.8	+14	32	9.9	Spiral galaxy in Coma Berenices
100	4321	12	21.9	+15	56	9.4	Spiral galaxy in Coma Berenices
101	5457	14	02.5	+54	27	7.9	Spiral galaxy in Ursa Major
102	5866(?)	15	05.9	+55	50	10.5	Spiral galaxy (identification as M102 in doubt)
103	581	1	31.9	+60	35	6.9	Open cluster in Cassiopeia
104*	4594	12	39.0	−11	31	8.3	Spiral galaxy in Virgo
105*	3379	10	46.8	+12	51	9.7	Elliptical galaxy in Leo
106*	4258	12	18.0	+47	25	8.4	Spiral galaxy in Canes Venatici
107*	6171	16	31.4	−13	01	9.2	Globular cluster in Ophiuchus
108*	3556	11	10.5	+55	47	10.5	Spiral galaxy in Ursa Major
109*	3992	11	56.6	+53	29	10.0	Spiral galaxy in Ursa Major
110*	205	0	39.2	+41	35	9.4	Elliptical galaxy (companion to M31)

* Not in Messier's original (1781) list; added later by others.

APPENDIX 19

THE CHEMICAL ELEMENTS

Element	Symbol	Atomic Number	Atomic Weight* (Chemical Scale)	Number of Atoms per 10^{12} Hydrogen Atoms†
Hydrogen	H	1	1.0080	1×10^{12}
Helium	He	2	4.003	8×10^{10}
Lithium	Li	3	6.940	<10
Beryllium	Be	4	9.013	1.4×10^{1}
Boron	B	5	10.82	$<13 \times 10^{2}$
Carbon	C	6	12.011	4.2×10^{8}
Nitrogen	N	7	14.008	8.7×10^{7}
Oxygen	O	8	16.0000	6.9×10^{8}
Fluorine	F	9	19.00	3.6×10^{4}
Neon	Ne	10	20.183	1.3×10^{8}
Sodium	Na	11	22.991	1.9×10^{6}
Magnesium	Mg	12	24.32	3.2×10^{7}
Aluminum	Al	13	26.98	3.3×10^{6}
Silicon	Si	14	28.09	4.5×10^{7}
Phosphorus	P	15	30.975	3.2×10^{5}
Sulfur	S	16	32.066	1.6×10^{7}
Chlorine	Cl	17	35.457	3.2×10^{5}
Argon	Ar(A)	18	39.944	1.0×10^{6}
Potassium	K	19	39.100	1.4×10^{5}
Calcium	Ca	20	40.08	2.2×10^{6}
Scandium	Sc	21	44.96	1.1×10^{3}
Titanium	Ti	22	47.90	1.1×10^{5}
Vanadium	V	23	50.95	1.0×10^{4}
Chromium	Cr	24	52.01	5.1×10^{5}
Manganese	Mn	25	54.94	2.6×10^{5}
Iron	Fe	26	55.85	3.2×10^{7}
Cobalt	Co	27	58.94	3.2×10^{4}
Nickel	Ni	28	58.71	1.9×10^{6}
Copper	Cu	29	63.54	1.1×10^{4}
Zinc	Zn	30	65.38	2.8×10^{4}
Gallium	Ga	31	69.72	6.3×10^{2}
Germanium	Ge	32	72.60	3.2×10^{3}
Arsenic	As	33	74.91	2.5×10^{2}
Selenium	Se	34	78.96	2.5×10^{3}
Bromine	Br	35	79.916	5.0×10^{2}
Krypton	Kr	36	83.80	2.0×10^{3}
Rubidium	Rb	37	85.48	4.0×10^{2}
Strontium	Sr	38	87.63	7.9×10^{2}
Yttrium	Y	39	88.92	1.3×10^{2}
Zirconium	Zr	40	91.22	5.6×10^{2}
Niobium (Columbium)	Nb(Cb)	41	92.91	7.9×10^{1}
Molybdenum	Mo	42	95.95	1.4×10^{2}
Technetium	Tc(Ma)	43	(99)	—
Ruthenium	Ru	44	101.1	68
Rhodium	Rh	45	102.91	32
Palladium	Pd	46	106.4	32
Silver	Ag	47	107.880	4
Cadmium	Cd	48	112.41	71
Indium	In	49	114.82	45

Element	Symbol	Atomic Number	Atomic Weight* (Chemical Scale)	Number of Atoms per 10^{12} Hydrogen Atoms†
Tin	Sn	50	118.70	100
Antimony	Sb	51	121.76	10
Tellurium	Te	52	127.61	2.5×10^2
Iodine	I(J)	53	126.91	40
Xenon	Xe(X)	54	131.30	2.1×10^2
Cesium	Cs	55	132.91	<80
Barium	Ba	56	137.36	1.2×10^2
Lanthanum	La	57	138.92	13
Cerium	Ce	58	140.13	35
Praseodymium	Pr	59	140.92	4
Neodymium	Nd	60	144.27	18
Promethium	Pm	61	(147)	—
Samarium	Sm(Sa)	62	150.35	5
Europium	Eu	63	152.0	5
Gadolinium	Gd	64	157.26	13
Terbium	Tb	65	158.93	2
Dysprosium	Dy(Ds)	66	162.51	11
Holmium	Ho	67	164.94	3
Erbium	Er	68	167.27	7
Thulium	Tm(Tu)	69	168.94	2
Ytterbium	Yb	70	173.04	8
Lutecium	Lu(Cp)	71	174.99	6
Hafnium	Hf	72	178.50	6
Tantalum	Ta	73	180.95	1
Tungsten	W	74	183.86	50
Rhenium	Re	75	186.22	<2
Osmium	Os	76	190.2	5
Iridium	Ir	77	192.2	28
Platinum	Pt	78	195.09	56
Gold	Au	79	197.0	6
Mercury	Hg	80	200.61	$<10^2$
Thallium	Tl	81	204.39	8
Lead	Pb	82	207.21	85
Bismuth	Bi	83	209.00	<80
Polonium	Po	84	(209)	—
Astatine	At	85	(210)	—
Radon	Rn	86	(222)	—
Francium	Fr(Fa)	87	(223)	—
Radium	Ra	88	226.05	—
Actinium	Ac	89	(227)	—
Thorium	Th	90	232.12	2
Protoactinium	Pa	91	(231)	—
Uranium	U(Ur)	92	238.07	<4
Neptunium	Np	93	(237)	—
Plutonium	Pu	94	(244)	—
Americium	Am	95	(243)	—
Curium	Cm	96	(248)	—
Berkelium	Bk	97	(247)	—
Californium	Cf	98	(251)	—
Einsteinium	E	99	(254)	—
Fermium	Fm	100	(253)	—
Mendeleevium	Mv	101	(256)	—
Nobelium	No	102	(253)	—

* Where mean atomic weights have not been well determined, the atomic mass numbers of the most stable isotopes are given in parentheses.
† Provided by L. H. Aller.

APPENDIX **20**

THE CONSTELLATIONS

Constellation (Latin Name)	Genitive Case Ending	English Name or Description	Abbre-via-tion	Approximate Position	
				α	δ
				h	°
Andromeda	Andromedae	Princess of Ethiopia	And	1	+40
Antlia	Antliae	Air pump	Ant	10	−35
Apus	Apodis	Bird of Paradise	Aps	16	−75
Aquarius	Aquarii	Water bearer	Aqr	23	−15
Aquila	Aquilae	Eagle	Aql	20	+5
Ara	Arae	Altar	Ara	17	−55
Aries	Arietis	Ram	Ari	3	+20
Auriga	Aurigae	Charioteer	Aur	6	+40
Boötes	Boötis	Herdsman	Boo	15	+30
Caelum	Caeli	Graving tool	Cae	5	−40
Camelopardus	Camelopardis	Giraffe	Cam	6	+70
Cancer	Cancri	Crab	Cnc	9	+20
Canes Venatici	Canum Venaticorum	Hunting dogs	CVn	13	+40
Canis Major	Canis Majoris	Big dog	CMa	7	−20
Canis Minor	Canis Minoris	Little dog	CMi	8	+5
Capricornus	Capricorni	Sea goat	Cap	21	−20
Carina*	Carinae	Keel of Argonauts' ship	Car	9	−60
Cassiopeia	Cassiopeiae	Queen of Ethiopia	Cas	1	+60
Centaurus	Centauri	Centaur	Cen	13	−50
Cephus	Cephei	King of Ethiopia	Cep	22	+70
Cetus	Ceti	Sea monster (whale)	Cet	2	−10
Chamaeleon	Chamaeleontis	Chameleon	Cha	11	−80
Circinus	Circini	Compasses	Cir	15	−60
Columba	Columbae	Dove	Col	6	−35
Coma Berenices	Comae Berenices	Berenice's hair	Com	13	+20
Corona Australis	Coronae Australis	Southern crown	CrA	19	−40
Corona Borealis	Coronae Borealis	Northern crown	CrB	16	+30
Corvus	Corvi	Crow	Crv	12	−20
Crater	Crateris	Cup	Crt	11	−15
Crux	Crucis	Cross (southern)	Cru	12	−60
Cygnus	Cygni	Swan	Cyg	21	+40
Delphinus	Delphini	Porpoise	Del	21	+10
Dorado	Doradus	Swordfish	Dor	5	−65
Draco	Draconis	Dragon	Dra	17	+65
Equuleus	Equulei	Little horse	Equ	21	+10
Eridanus	Eridani	River	Eri	3	−20
Fornax	Fornacis	Furnace	For	3	−30
Gemini	Geminorum	Twins	Gem	7	+20
Grus	Gruis	Crane	Gru	22	−45
Hercules	Herculis	Hercules, son of Zeus	Her	17	+30
Horologium	Horologii	Clock	Hor	3	−60
Hydra	Hydrae	Sea serpent	Hya	10	−20
Hydrus	Hydri	Water snake	Hyi	2	−75
Indus	Indi	Indian	Ind	21	−55
Lacerta	Lacertae	Lizard	Lac	22	+45

Constellation (Latin Name)	Genitive Case Ending	English Name or Description	Abbre-via-tion	Approximate Position α	δ
				h	°
Leo	Leonis	Lion	Leo	11	+15
Leo Minor	Leonis Minoris	Little lion	LMi	10	+35
Lepus	Leporis	Hare	Lep	6	−20
Libra	Librae	Balance	Lib	15	−15
Lupus	Lupi	Wolf	Lup	15	−45
Lynx	Lyncis	Lynx	Lyn	8	+45
Lyra	Lyrae	Lyre or harp	Lyr	19	+40
Mensa	Mensae	Table Mountain	Men	5	−80
Microscopium	Microscopii	Microscope	Mic	21	−35
Monoceros	Monocerotis	Unicorn	Mon	7	−5
Musca	Muscae	Fly	Mus	12	−70
Norma	Normae	Carpenter's level	Nor	16	−50
Octans	Octantis	Octant	Oct	22	−85
Ophiuchus	Ophiuchi	Holder of serpent	Oph	17	0
Orion	Orionis	Orion, the hunter	Ori	5	+5
Pavo	Pavonis	Peacock	Pav	20	−65
Pegasus	Pegasi	Pegasus, the winged horse	Peg	22	+20
Perseus	Persei	Perseus, hero who saved Andromeda	Per	3	+45
Phoenix	Phoenicis	Phoenix	Phe	1	−50
Pictor	Pictoris	Easel	Pic	6	−55
Pisces	Piscium	Fishes	Psc	1	+15
Piscis Austrinus	Piscis Austrini	Southern fish	PsA	22	−30
Puppis*	Puppis	Stern of the Argonauts' ship	Pup	8	−40
Pyxis* (= Malus)	Pyxidus	Compass on the Argonauts' ship	Pyx	9	−30
Reticulum	Reticuli	Net	Ret	4	−60
Sagitta	Sagittae	Arrow	Sge	20	+10
Sagittarius	Sagittarii	Archer	Sgr	19	−25
Scorpius	Scorpii	Scorpion	Sco	17	−40
Sculptor	Sculptoris	Sculptor's tools	Scl	0	−30
Scutum	Scuti	Shield	Sct	19	−10
Serpens	Serpentis	Serpent	Ser	17	0
Sextans	Sextantis	Sextant	Sex	10	0
Taurus	Tauri	Bull	Tau	4	+15
Telescopium	Telescopii	Telescope	Tel	19	−50
Triangulum	Trianguli	Triangle	Tri	2	+30
Triangulum Australe	Trianguli Australis	Southern triangle	TrA	16	−65
Tucana	Tucanae	Toucan	Tuc	0	−65
Ursa Major	Ursae Majoris	Big bear	UMa	11	+50
Ursa Minor	Ursae Minoris	Little bear	VMi	15	+70
Vela*	Velorum	Sail of the Argonauts' ship	Vel	9	−50
Virgo	Virginis	Virgin	Vir	13	0
Volans	Volantis	Flying fish	Vol	8	−70
Vulpecula	Vulpeculae	Fox	Vul	20	+25

* The four constellations Carina, Puppis, Pyxis, and Vela originally formed the single constellation, Argo Navis.

APPENDIX 21

STAR MAPS

The star maps, one for each month, are printed on six removable sheets at the back of this book. To learn the stars and constellations, the sheet containing the map for the current month should be taken outdoors and compared directly with the sky. The maps were designed for a latitude of about 35° N but are useful anywhere in the continental United States. Each map shows the appearance of the sky at about 9:00 P.M. (Standard Time) near the middle of the month for which it is intended; near the beginning and end of the month, it shows the sky as it appears about 10:00 P.M. and 8:00 P.M., respectively. To use a map, hold the sheet vertically, and turn it so that the direction you are facing is shown at the bottom. The middle of the map corresponds to your zenith (the point overhead), so the stars and constellations in the lower part of the chart (with the printing upright) will match, approximately, the sky in front of you.

These star maps were originally prepared by C. H. Cleminshaw for the *Griffith Observer* (published by the Griffith Observatory, P.O. Box 27787, Los Angeles, California, 90027), and are reproduced here by the very kind permission of the Griffith Observatory, Edwin C. Krupp, Director.

INDEX

Page numbers in boldface refer to principal discussions. Page numbers in parentheses refer to figures.

A

B

C

D

E

F

G

H

I

J

K

L

M

N

O

P

Q

R

S

T

U

V

W

X

Y

Z

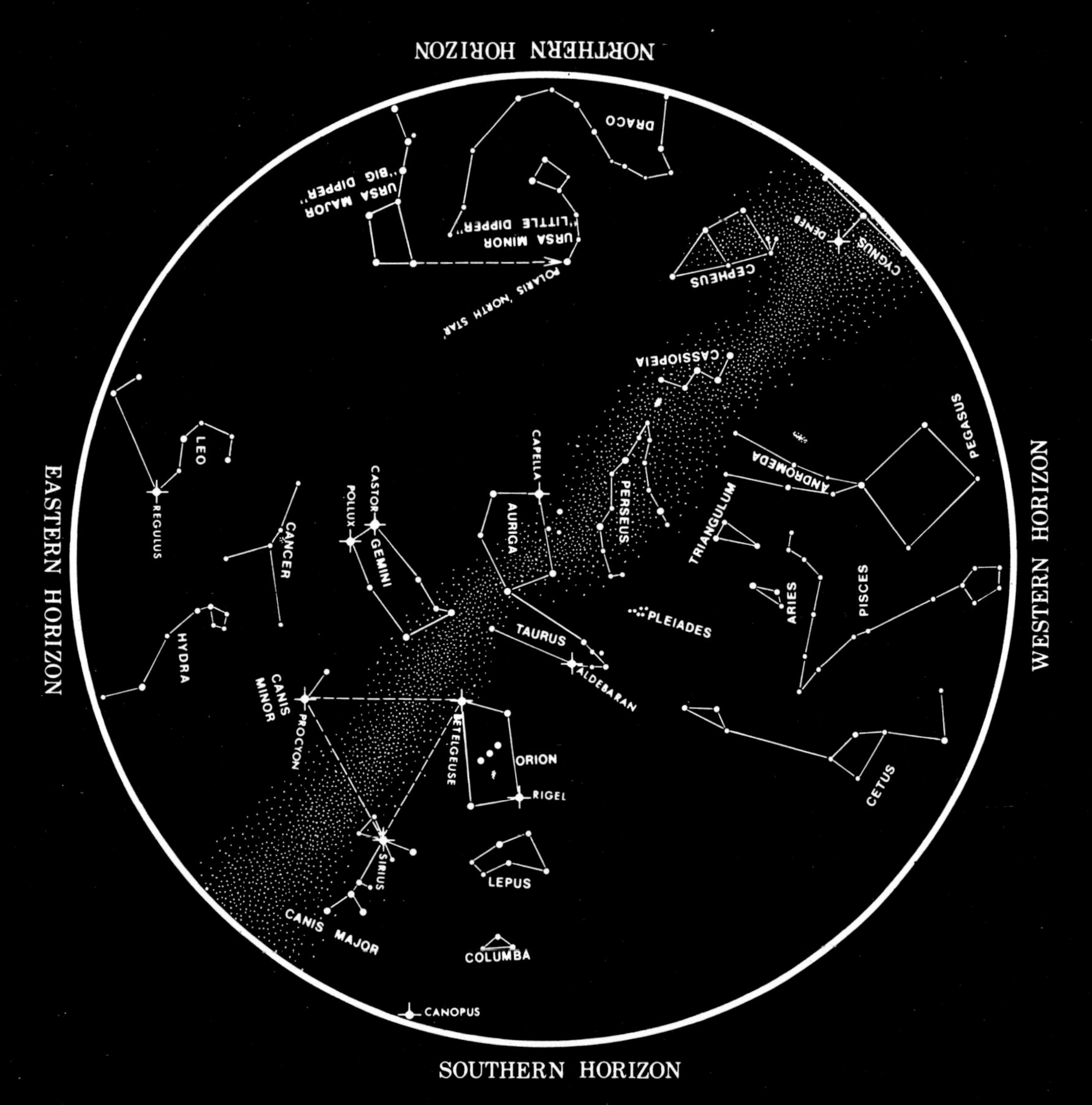

THE NIGHT SKY IN JANUARY

Latitude of chart is 34°N, but it is practical throughout the continental United States.

To use: Hold chart vertically and turn it so the direction you are facing shows at the bottom.

Chart time (Local Standard):
10 p.m. First of month
9 p.m. Middle of month
8 p.m. Last of month

Star Chart from *GRIFFITH OBSERVER*, Griffith Observatory, Los Angeles

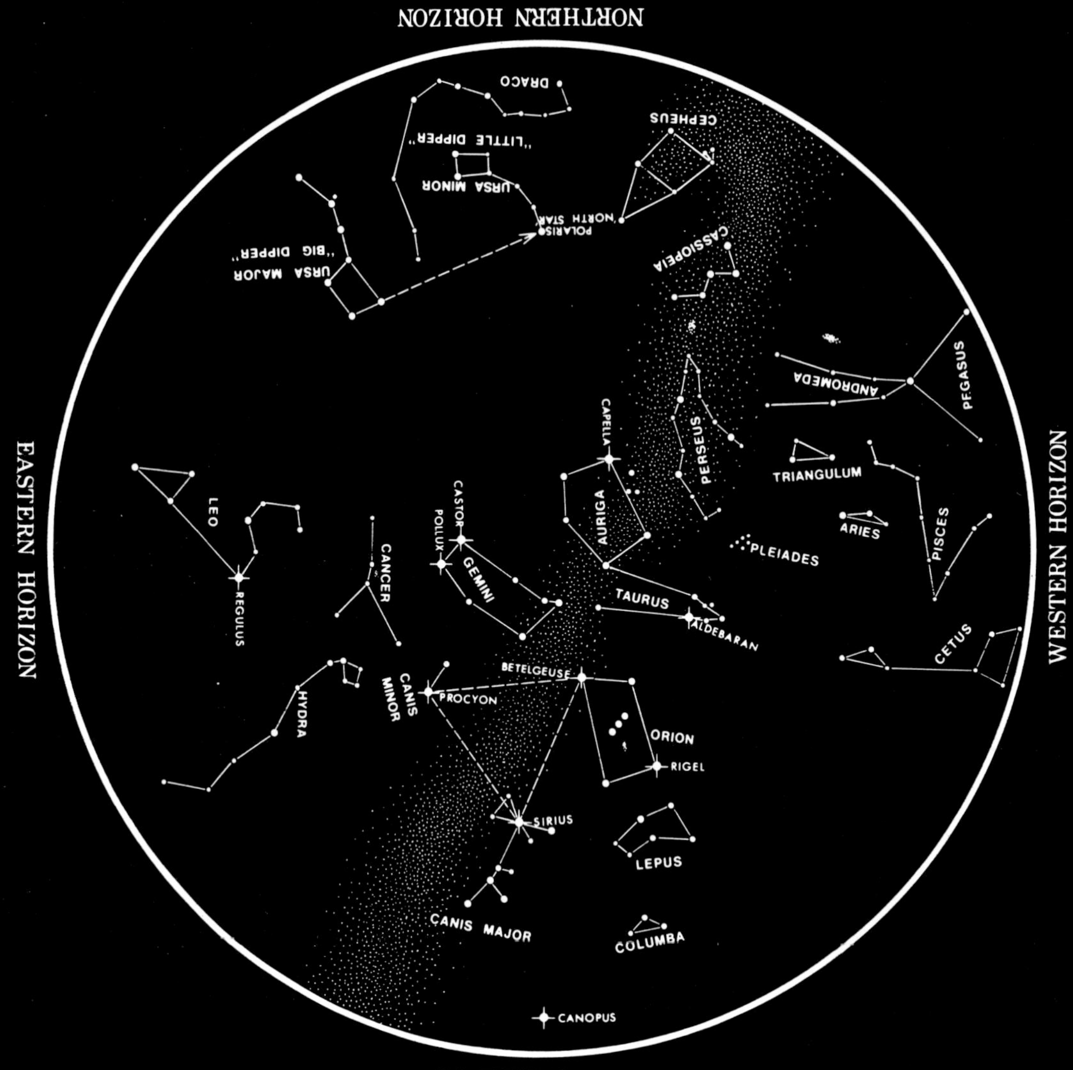

THE NIGHT SKY IN FEBRUARY

Latitude of chart is 34°N, but it is practical throughout the continental United States.

To use: Hold chart vertically and turn it so the direction you are facing shows at the bottom.

Chart time (Local Standard):
10 p.m. First of month
9 p.m. Middle of month
8 p.m. Last of month

Star Chart from *GRIFFITH OBSERVER*, Griffith Observatory, Los Angeles

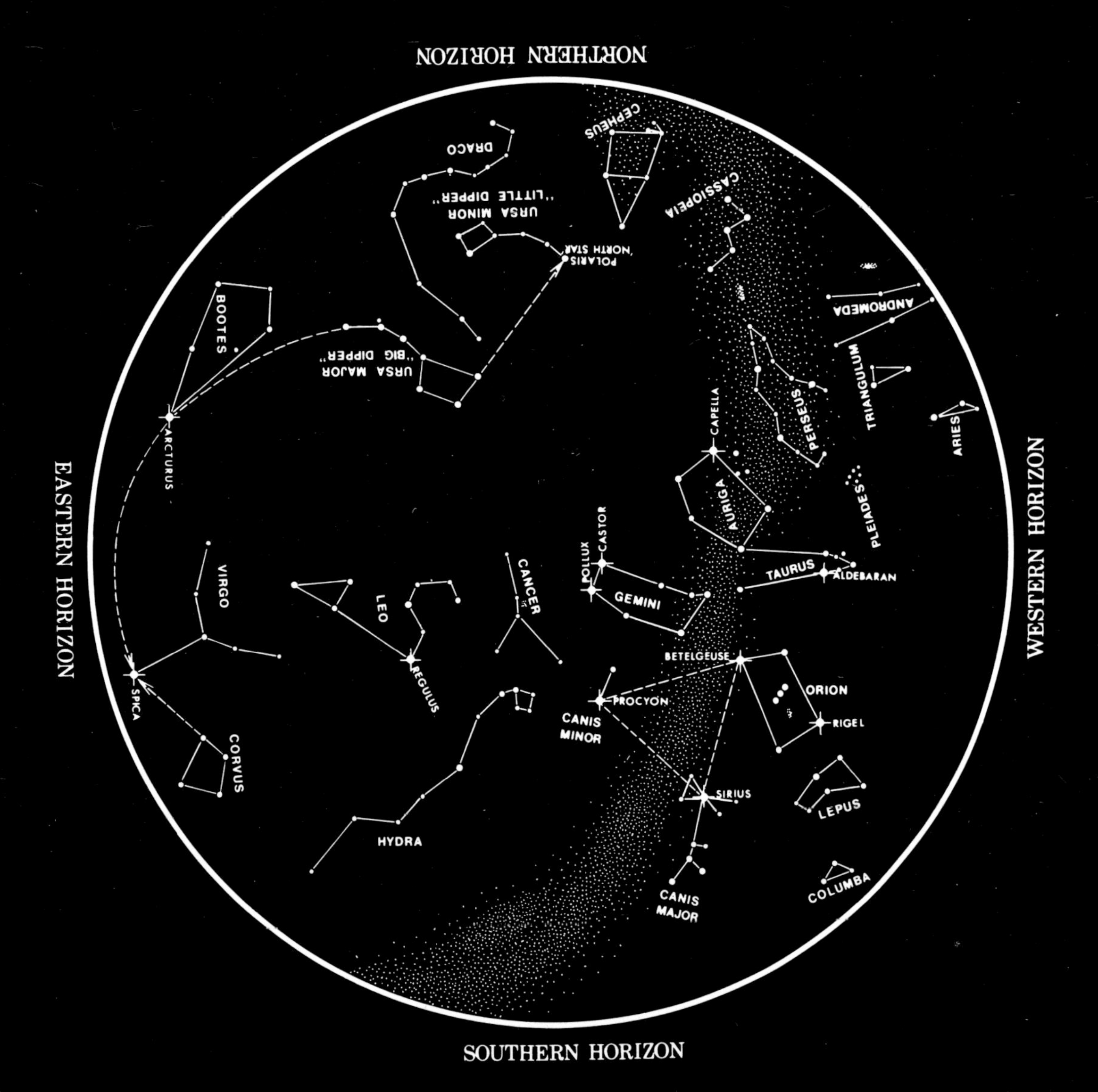

SOUTHERN HORIZON

THE NIGHT SKY IN MARCH

Latitude of chart is 34°N, but it is practical throughout the continental United States.

To use: Hold chart vertically and turn it so the direction you are facing shows at the bottom.

Chart time (Local Standard):
10 p.m. First of month
9 p.m. Middle of month
8 p.m. Last of month

Star Chart from *GRIFFITH OBSERVER*, Griffith Observatory, Los Angeles

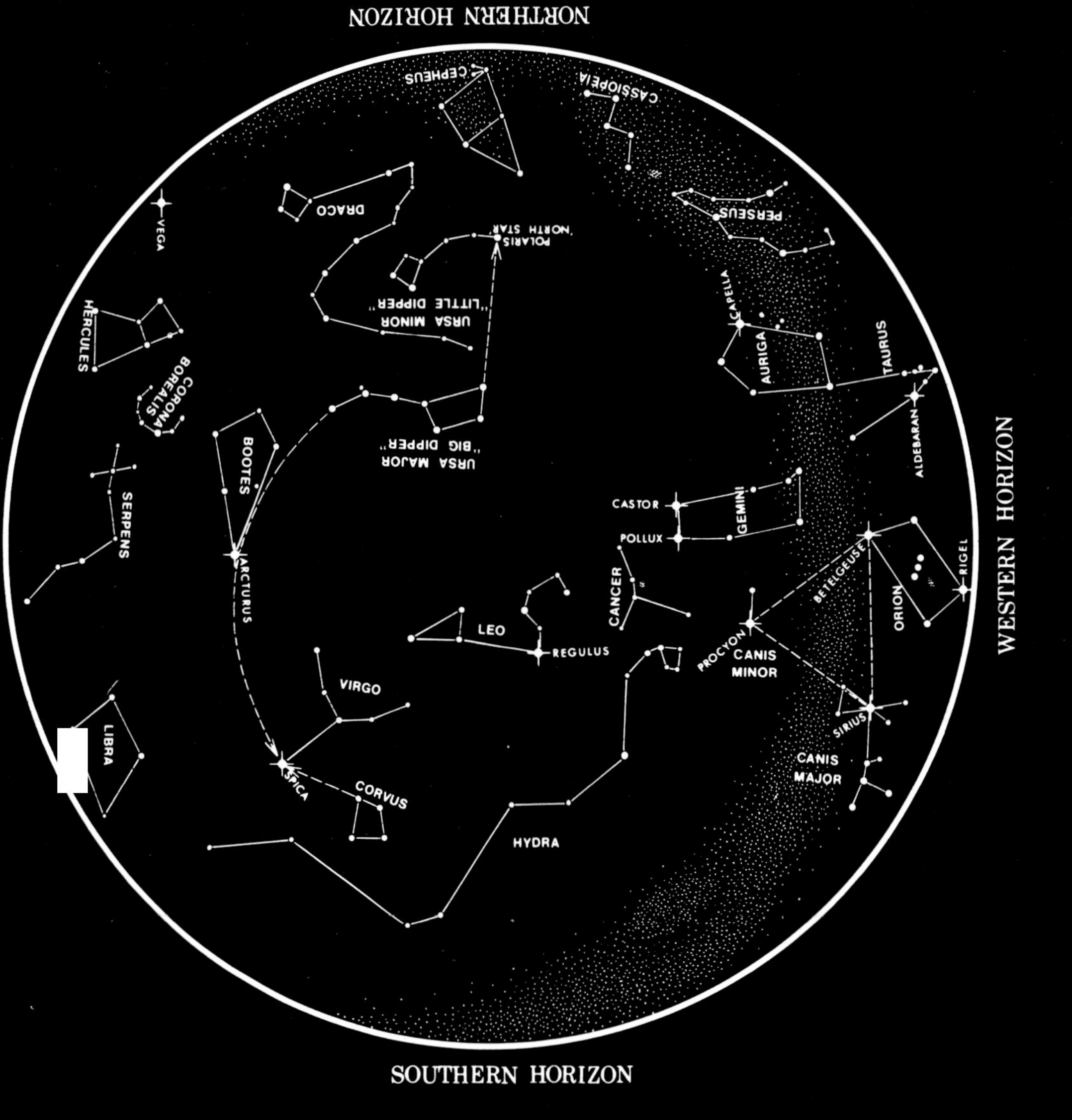

THE NIGHT SKY IN APRIL

Latitude of chart is 34°N, but it is practical throughout the continental United States.

To use: Hold chart vertically and turn it so the direction you are facing shows at the bottom.

Chart time (Local Standard):
10 p.m. First of month
9 p.m. Middle of month
8 p.m. Last of month

SOUTHERN HORIZON

THE NIGHT SKY IN MAY

Latitude of chart is 34°N, but it is practical throughout the continental United States.

To use: Hold chart vertically and turn it so the direction you are facing shows at the bottom.

Chart time (Local Standard):
10 p.m. First of month
9 p.m. Middle of month
8 p.m. Last of month

Star Chart from *GRIFFITH OBSERVER*, Griffith Observatory, Los Angeles

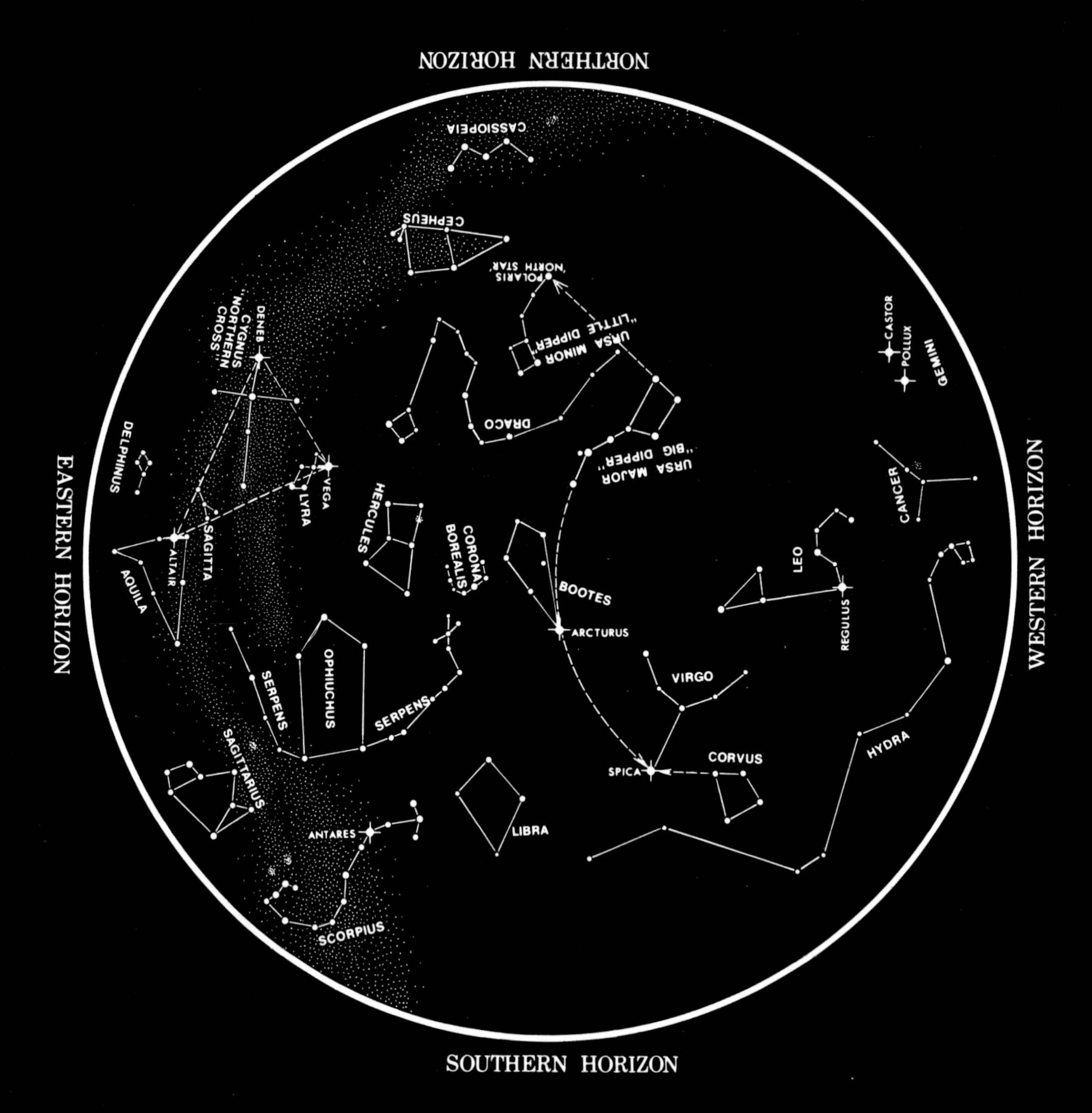

THE NIGHT SKY IN JUNE

Latitude of chart is 34°N, but it is practical throughout the continental United States.

To use: Hold chart vertically and turn it so the direction you are facing shows at the bottom.

Chart time (Local Standard):
10 p.m. First of month
9 p.m. Middle of month
8 p.m. Last of month

Star Chart from *GRIFFITH OBSERVER*, Griffith Observatory, Los Angeles

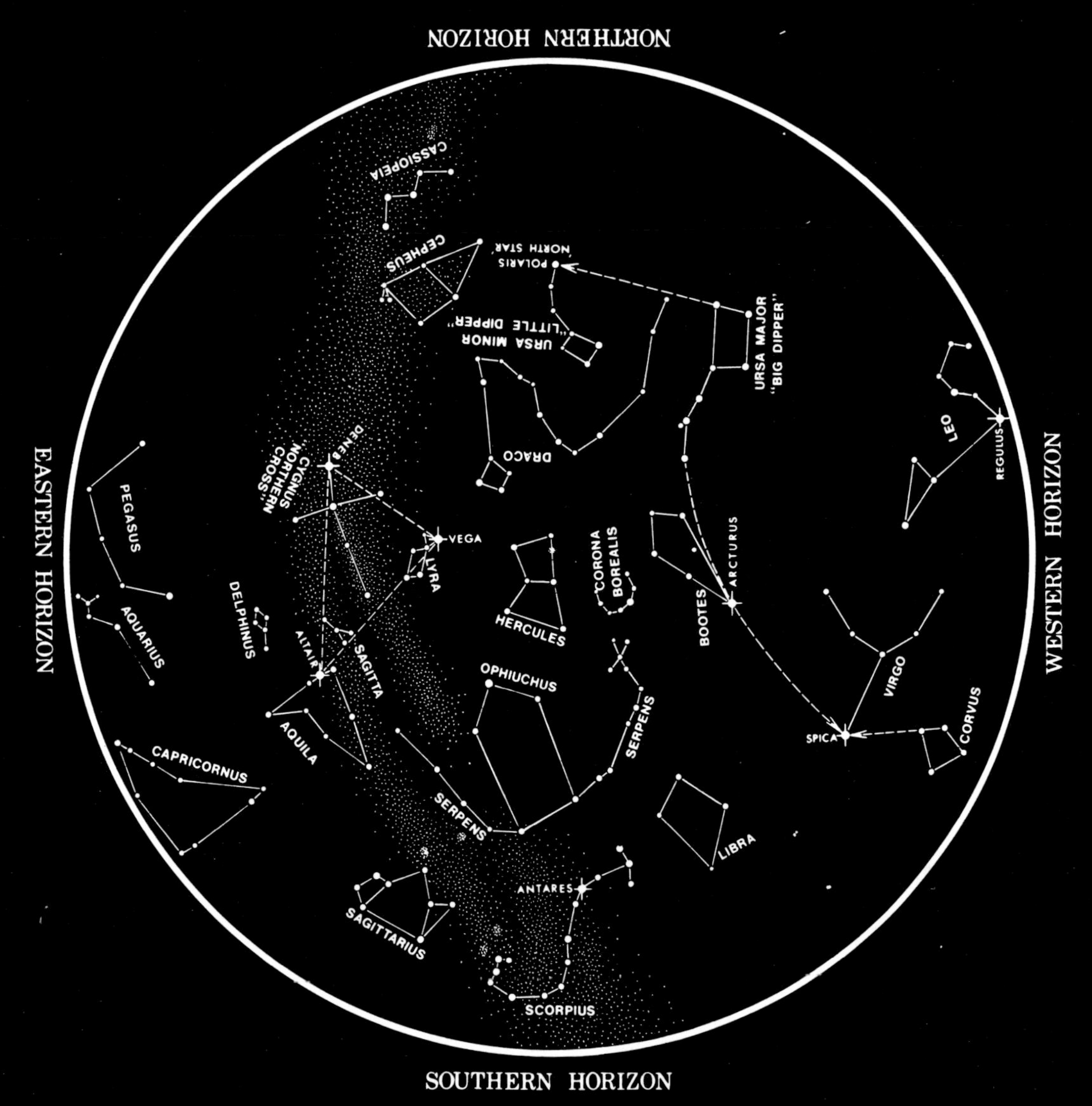

THE NIGHT SKY IN JULY

Latitude of chart is 34°N, but it is practical throughout the continental United States.

To use: Hold chart vertically and turn it so the direction you are facing shows at the bottom.

Chart time (Local Standard):
10 p.m. First of month
9 p.m. Middle of month
8 p.m. Last of month

Star Chart from *GRIFFITH OBSERVER*, Griffith Observatory, Los Angeles

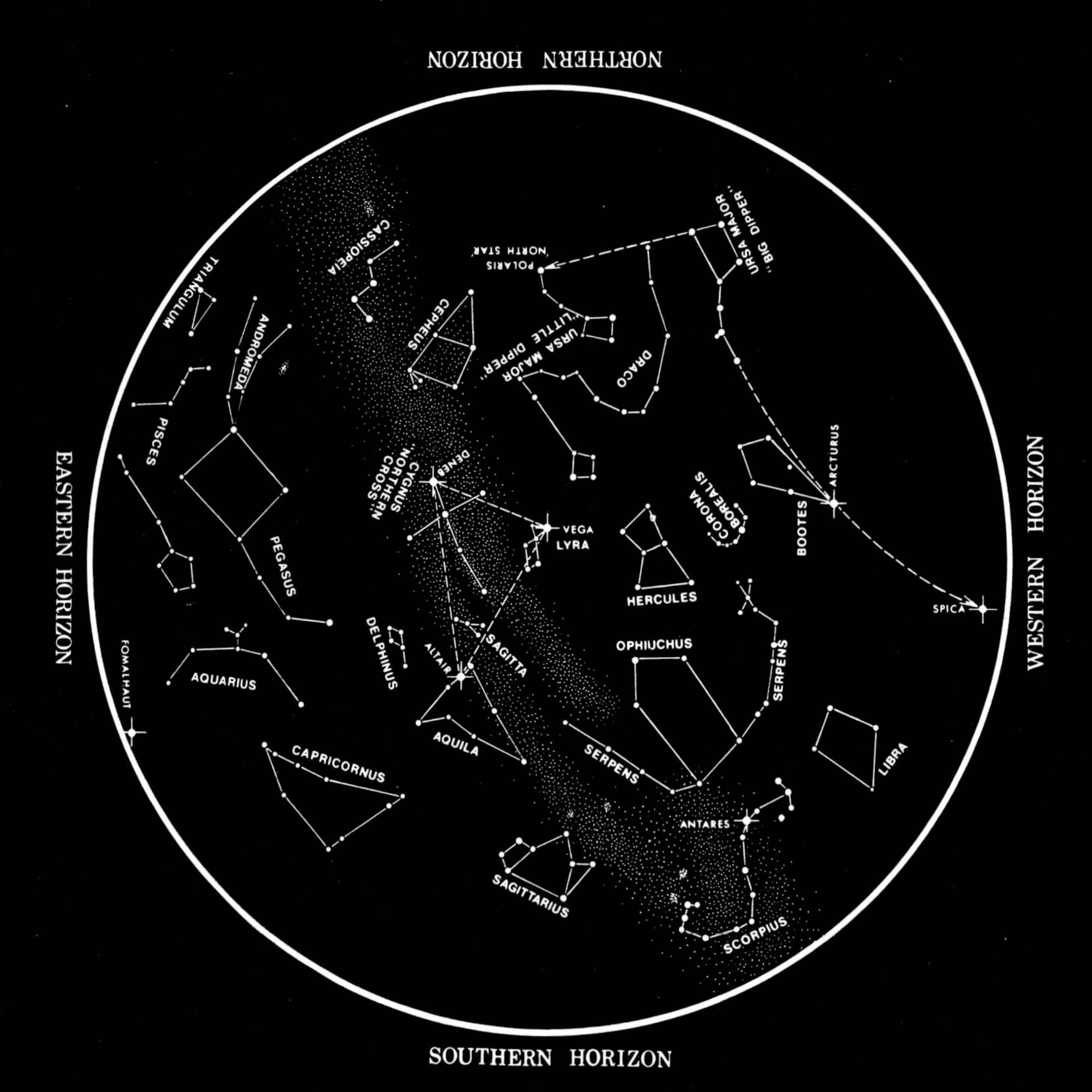

SOUTHERN HORIZON

THE NIGHT SKY IN AUGUST

Latitude of chart is 34°N, but it is practical throughout the continental United States.

To use: Hold chart vertically and turn it so the direction you are facing shows at the bottom.

Chart time (Local Standard):
10 p.m. First of month
9 p.m. Middle of month
8 p.m. Last of month

Star Chart from *GRIFFITH OBSERVER*, Griffith Observatory, Los Angeles

THE NIGHT SKY IN SEPTEMBER

Latitude of chart is 34°N, but it is practical throughout the continental United States.

To use: Hold chart vertically and turn it so the direction you are facing shows at the bottom.

Chart time (Local Standard):
10 p.m. First of month
9 p.m. Middle of month
8 p.m. Last of month

Star Chart from *GRIFFITH OBSERVER*, Griffith Observatory, Los Angeles

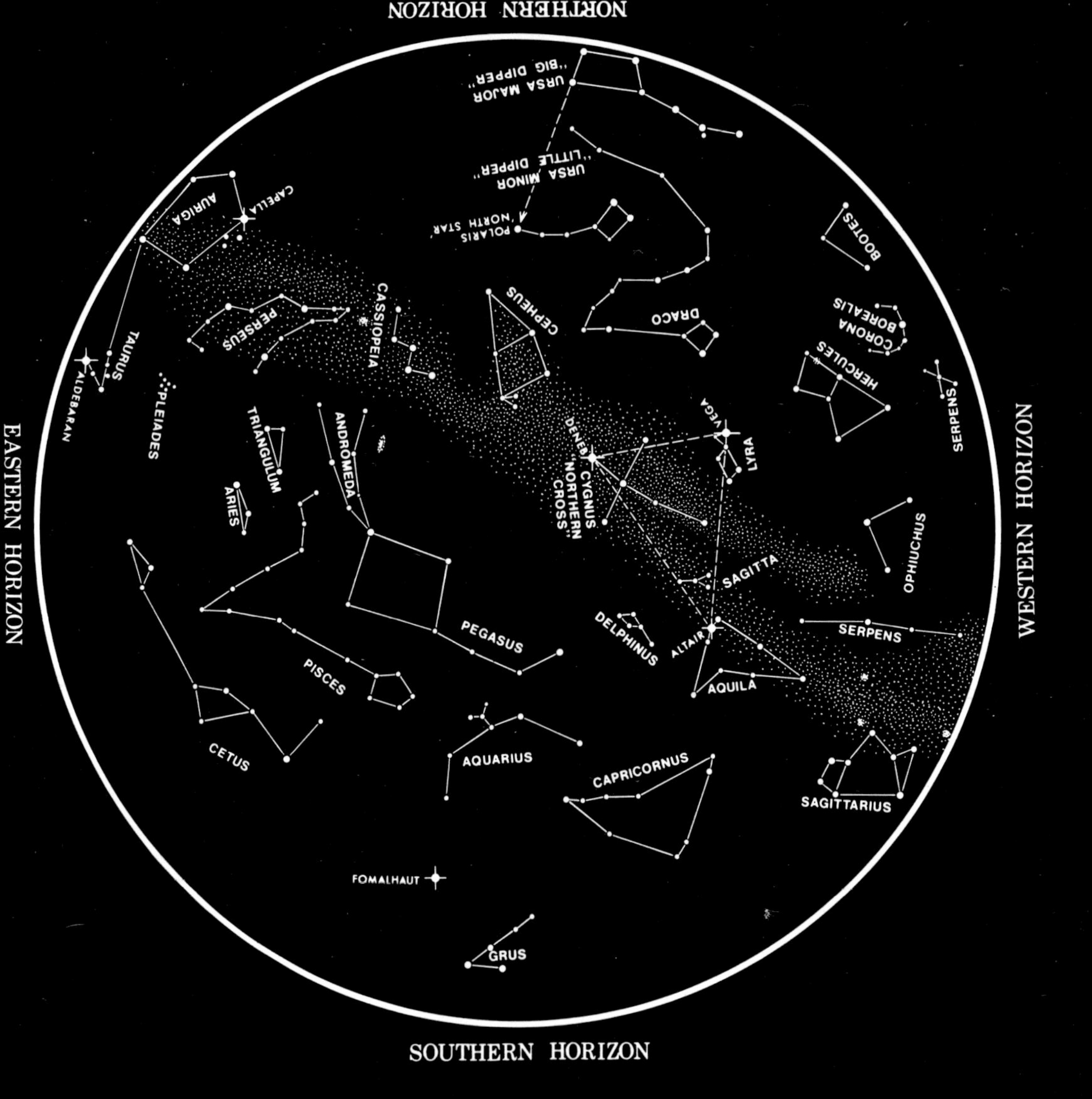

THE NIGHT SKY IN OCTOBER

Latitude of chart is 34°N, but it is practical throughout the continental United States.

To use: Hold chart vertically and turn it so the direction you are facing shows at the bottom.

Chart time (Local Standard):
10 p.m. First of month
9 p.m. Middle of month
8 p.m. Last of month

Star Chart from *GRIFFITH OBSERVER*, Griffith Observatory, Los Angeles

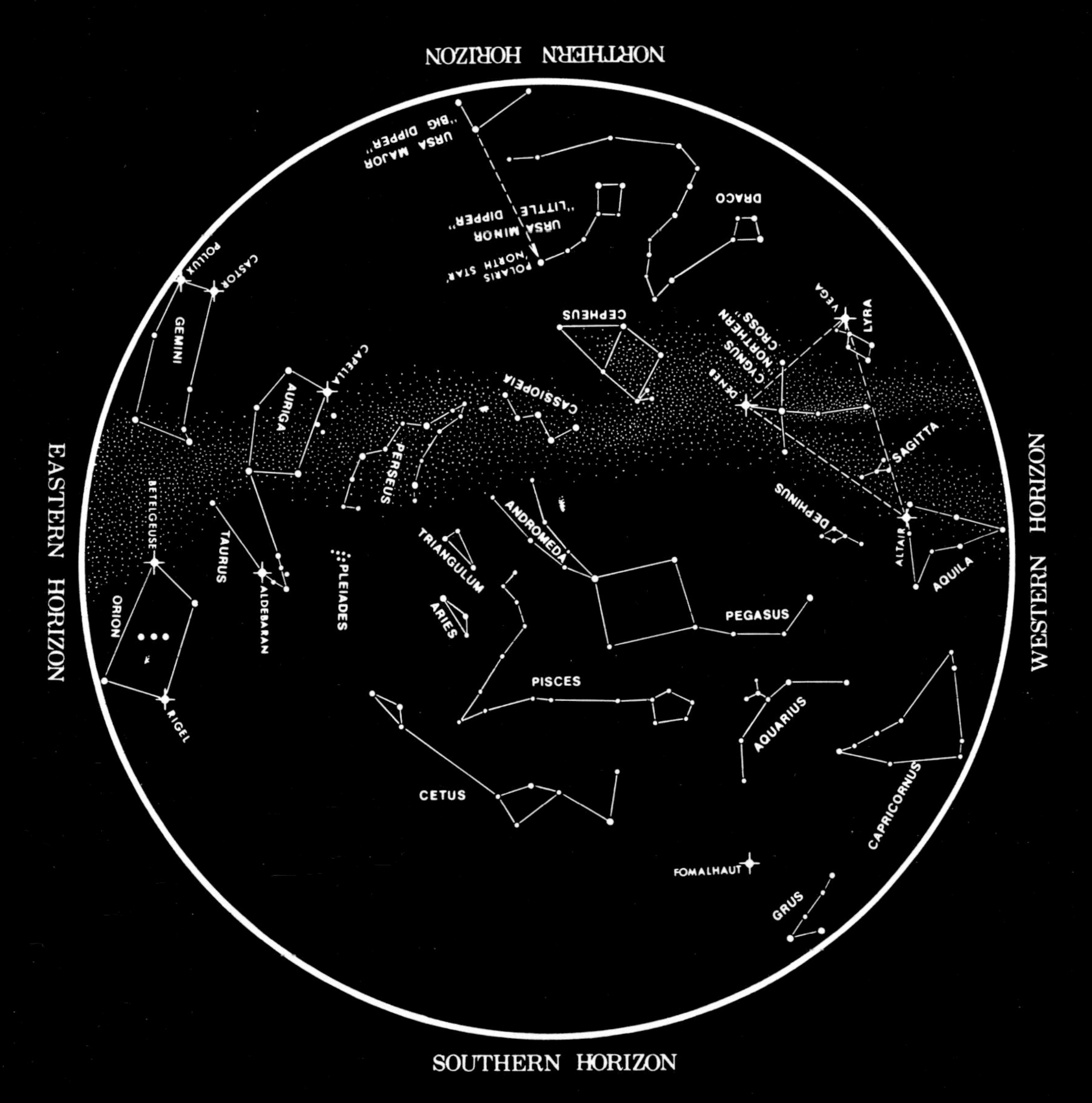

THE NIGHT SKY IN NOVEMBER

Latitude of chart is 34°N, but it is practical throughout the continental United States.

To use: Hold chart vertically and turn it so the direction you are facing shows at the bottom.

Chart time (Local Standard):
10 p.m. First of month
9 p.m. Middle of month
8 p.m. Last of month

Star Chart from *GRIFFITH OBSERVER*, Griffith Observatory, Los Angeles

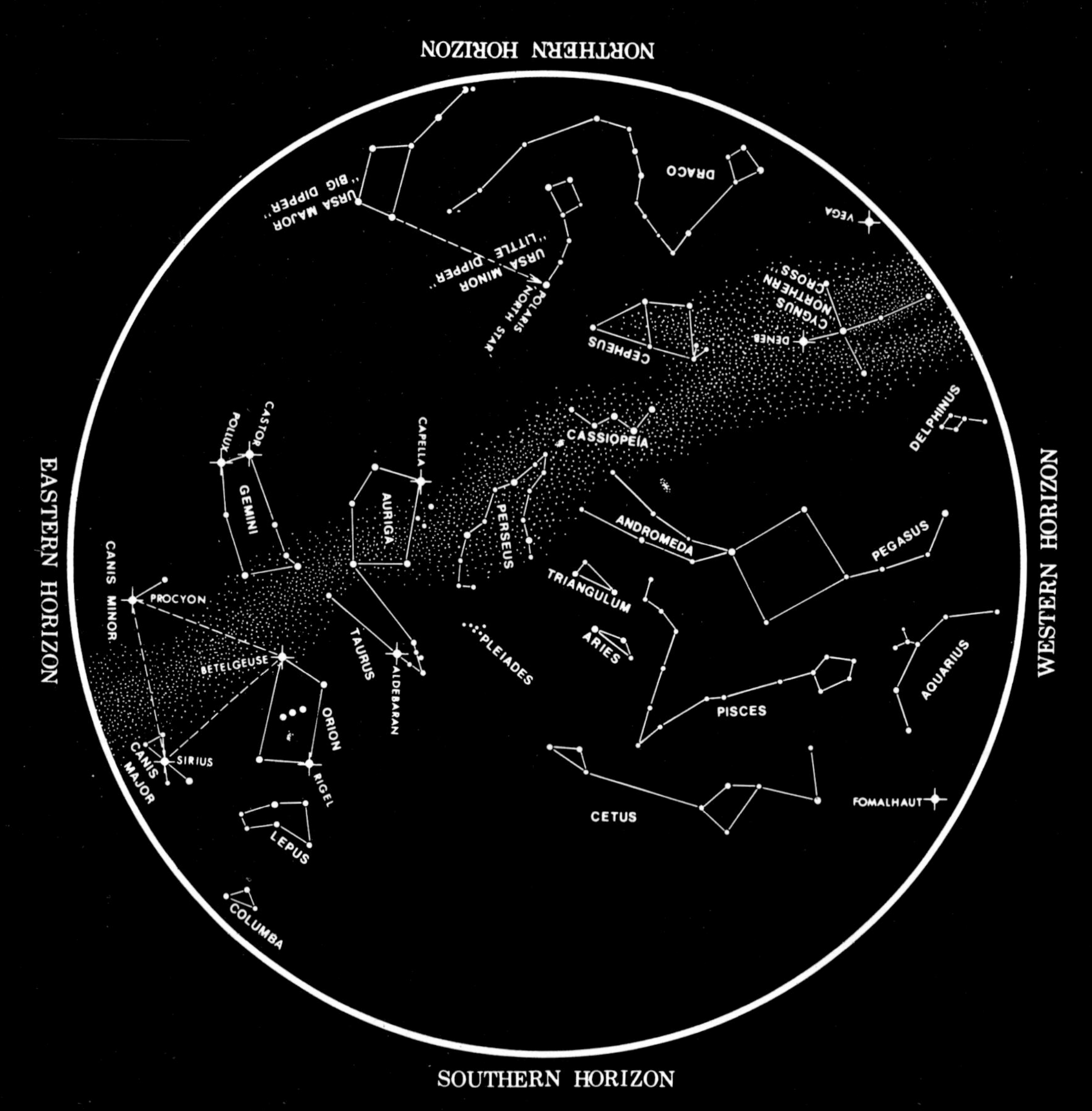

THE NIGHT SKY IN DECEMBER

Latitude of chart is 34°N, but it is practical throughout the continental United States.

To use: Hold chart vertically and turn it so the direction you are facing shows at the bottom.

Chart time (Local Standard):
10 p.m. First of month
9 p.m. Middle of month
8 p.m. Last of month

Star Chart from *GRIFFITH OBSERVER*, Griffith Observatory, Los Angeles